Radar Systems and Radio Aids to Navigation

RADAR SYSTEMS AND RADIO AIDS TO NAVIGATION

A. K. SEN, PhD
and
A. B. BHATTACHARYA, PhD

MERCURY LEARNING AND INFORMATION
Dulles, Virginia
Boston, Massachusetts
New Delhi

Publisher: David Pallai
Mercury Learning and Information
22841 Quicksilver Drive
Dulles, VA 20166
info@merclearning.com
www.merclearning.com
(800) 232-0223

A.K. Sen and A.B. Bhattacharya. *Radar Systems and Radio Aids to Navigation*.
ISBN: 978-1-683921-18-9

Library of Congress Control Number: 2018943747

181920321 This book is printed on acid-free paper in the United States of America

CONTENTS

	Foreword		*xxiii*
	Preface		*xxv*
Chapter 1:	**Navigation**		**1**
	1.1	Introduction	1
	1.2	Navigational Parameters	1
	1.3	Types of Navigational Aids	2
Chapter 2:	**Antennas for Radio Navigational Aids**		**5**
	2.1	Introduction	5
	2.2	Antenna Parameters	5
	2.3	Current Distribution	11
	2.4	Half-Wave Dipole	13
	2.5	Antennas of Length Greater than Half-wave Length	14
	2.6	Parasitic Elements to Increase Directivity	16
	2.7	Folded Dipole	18
	2.8	Parabolic Reflector	19
	2.9	Receiving Antenna	20
	2.10	Microwave Antenna	23
	2.11	Horn Antenna	24
	2.12	Antenna as an Aperture	26
	2.13	One-Dimensional Aperture Distribution	28
	2.14	Circular Aperture	31
	2.15	Parabolic Reflector Antenna	32

	2.16	Lens Antennas	33
	2.17	Pattern Synthesis	36
	2.18	Fourier Integral Synthesis	36
	2.19	Errors in Radiation Patterns	39
	2.20	Cosecant-Squared Antenna Pattern	41
	2.21	Stabilization of Antennas	42
	2.22	Design of Parabolic Reflector Radar Antennas	42
	2.23	Radiation Pattern of a Parabolic Reflector Type Antenna	45
	2.24	Design of Parabolic Reflector Antennas	46
	2.25	Tolerance of Reflectors	49
	2.26	Phase Error Due to Tolerance and Its Effect on Directivity	49
	2.27	Design of the Feeds of Parabolic Reflector Antennas	52
	2.28	Different Types of Feeds	53
	2.29	Dipole Feed	54
	2.30	Waveguide Feeding Method for a Dipole Feed	55
	2.31	Cutler Dual Aperture Feed	55
	2.32	Waveguide Horn Feeds	56
	2.33	Monopulse Feeds	58
Chapter 3:	**Principles of Radar**		**61**
	3.1	Introduction	61
	3.2	Radar Equation	62
	3.3	Radar Frequencies	64
	3.4	Radar Set	65
	3.5	Radar Applications	67
	3.6	Receiver Noise	68
	3.7	Signal to Noise Ratio	70
	3.8	Transmitter Power	76
	3.9	Pulse Repetition Frequency	77
	3.10	Pulse Duration	78
	3.11	Propagation Effects	79

	3.12	Scanning Radars	79
	3.13	Tracking Radar	80
	3.14	Lobe Switching	81
	3.15	Conical Scan	82
	3.16	Monopulse Tracking	85
		3.16.1 Millimeter Wave Tracking Radars	90
	3.17	Accuracy of Radar Measurements in the Presence of Noise	92
		3.17.1 Effect of Pulse Shape on Signal-to-Noise Ratio	93
		3.17.2 Effect of Gating Waveform on Range Accuracy	96
		3.17.3 Accuracy Doppler-Velocity Measurement	99
		3.17.4 Accuracy of Angular Position Measurement	99
		3.17.5 Response of a Matched Filter and an Ambiguity Diagram	101
Chapter 4:	**Radar Targets**		**107**
	4.1	Introduction	107
	4.2	Radar Cross Section	107
		4.2.1 The Rayleigh Region	108
		4.2.2 The Mie Region	109
		4.2.3 The Optical Region	109
		4.2.4 Features in Rayleigh, Mie, and Optical Regions	110
	4.3	Back Scatter Cross Section	110
	4.4	Polarization Scattering Matrix	114
	4.5	Complex Targets	115
	4.6	Cross Section Fluctuations	118
	4.7	Frequency Agility Effects on Target Detection and Tracking	121
	4.8	Radar Cross Section Measurements	124
	4.9	RCS Measurement Systems	124

	4.10	Problems in RCS Measurement	125
	4.11	Sensitivity of RCS Measurement	126
	4.12	Compact Range RCS Measurement	129
	4.13	Instrumentation Radars for RCS Measurement	131
		4.13.1 CW Radar	131
		4.13.2 FM-CW Radar	133
		4.13.3 Gated CW Radar	133
		4.13.4 Pulse Radar	136
Chapter 5:	**Radar Transmitters and Receivers**		**141**
	5.1	Introduction	141
	5.2	The Magnetron Oscillator	142
		5.2.1 Performance Chart and Rieke Diagram	144
		5.2.2 Principle of Operation of a Magnetron	146
	5.3	Klystron Amplifiers	149
		5.3.1 Three-Cavity Klystron	152
		5.3.2 Reflex Klystrons	152
	5.4	Traveling Wave Tube Amplifiers	155
	5.5	Crossed-Field Amplifiers	156
		5.5.1 Amplitron	156
	5.6	Modulators	158
	5.7	Solid State Transmitters	159
	5.8	Noise Figure of a Receiver	160
	5.9	Mixers	162
	5.10	Displays	165
	5.11	Duplexers	166
	5.12	Matched Filter Receiver	169
	5.13	Correlation Detection	172
	5.14	Constant False Alarm Rate (CFAR) Receiver	173
	5.15	Receiver Protector and Sensitivity Time Control	174
Chapter 6:	**Propagation of Radar Waves**		**175**
	6.1	Introduction	175
	6.2	Propagation on Plane Earth	176
		6.2.1 The Effect of Earth's Curvature	179

	6.3	Refraction	180
	6.4	Diffraction	182
	6.5	Anomalous Propagation	185
		6.5.1 Elevated Duct	186
		6.5.2 Evaporation Duct	187
	6.6	Attenuation Due to Atmospheric Gases	187
	6.7	Environmental Noise	188
		6.7.1 Solar Noise	188
		6.7.2 Cosmic Noise	189
		6.7.3 Atmospheric Absorption Noise	190
		6.7.4 Atmospherics and Urban Noise	191
		6.7.5 System Noise Temperature	191
	6.8	Microwave Radiation Hazards	193
Chapter 7:	**Radar Clutter**		**195**
	7.1	Introduction	195
	7.2	Surface Clutter Radar Equations	196
	7.3	Sea Clutter	198
		7.3.1 Theory	199
		7.3.2 Probability Density Function of Sea Clutter	199
		7.3.3 Properties of Sea Clutter	201
	7.4	Land Clutter	203
		7.4.1 Examples	204
		7.4.2 Applications	204
		7.4.3 Frequency Dependence	205
		7.4.4 Properties of Land Clutter	206
	7.5	Effects of Weather on Radar	207
		7.5.1 Radar Equation for Meteorological Echoes	207
		7.5.2 Scattering from Rain	208
		7.5.3 Scattering from Snow	210
		7.5.4 Attenuation by Precipitation	211
		7.5.5 Scattering from Clouds	212
	7.6	Angel Echoes	212

Chapter 8: **MTI Radar** **215**

8.1 Introduction 215

8.2 Operation of MTI Radar 215

8.3 MTI Receiver with Delay Line Canceler 219

8.3.1 Response of the Delay Line Canceler 220

8.3.2 Blind Speeds 220

8.3.3 Double Cancellation 222

8.3.4 Transversal Filter 222

8.4 Multiple or Staggered Pulse Repetition Frequencies 223

8.5 Range-Gated Doppler Filters 224

8.6 Digital Signal Processing 225

8.6.1 Example of an MTI Radar Processor 226

8.7 MTI from a Moving Platform 227

8.8 Limitations to MTI Performance 228

Chapter 9: **Modern Radars** **233**

9.1 Introduction to Pulse Doppler Radar 233

9.2 Block Diagram 234

9.3 Detection of Multiple Targets Moving with Different Velocities 236

9.4 Coherent Integration 238

9.5 Applications 238

9.6 Advantages of Pulse Doppler Radar 239

9.7 Introduction to Frequency Coded Radars 239

9.8 Block Diagram 240

9.9 Discrete Frequency Waveform Coding 241

9.10 Side Lobe Reduction by Weighted Amplitude of the Frequency-Coded Waveform 243

9.11 Matched Filter Realization for Pulse Compression 243

9.12 Waveform Analysis of a Linear-Stepped Frequency Pulse 244

9.13 Applications of Frequency-Coded Radars 245

9.14 Introduction to Phase-Coded Radars 246

	9.15	Phase Coding and Decoding	246
	9.16	Block Diagram of Phase-Coded CW Radar	249
	9.17	Decoders	250
	9.18	Cross-Correlators and Trackers	253
	9.19	Range Trackers	254
	9.20	Comparison of Phase-Code and Linear FM Pulse Compressions	256
	9.21	Introduction to Millimeter Wave Radars	257
	9.22	Propagation of Millimeter Wave Radars	257
	9.23	Military Radars	258
	9.24	Anti-Aircraft Weapons Systems	260
	9.25	Missile Guidance and Seeker Systems	261
	9.26	Beam Rider	262
	9.27	Missile Seeker	262
	9.28	Configurations of Missile Seeker Sensors	265
	9.29	FMCW Sensor	266
	9.30	Radiometric Sensor	266
	9.31	Power Sources for Millimeter Wave Radars	267
	9.32	Jamming and Anti-Jamming Techniques	269
	9.33	Electronic Counter Measures	269
	9.34	Electronic Counter Counter Measure	271
	9.35	Repeater Jamming and ECCM	272
Chapter 10:	**Navigational and Remote Sensing Radars**		**273**
	10.1	Introduction	273
	10.2	Airport Radars	274
		10.2.1 Air Surveillance Radar	274
		10.2.2 Precision Approach Radar	276
	10.3	Meteorological Radar (MET Radar)	277
	10.4	Airborne Radars	278
	10.5	Doppler Navigation	278
	10.6	Doppler Navigation Equipment	285
	10.7	Distance Measuring Equipment	288

10.8	Navy Radar		292
10.9	Remote Sensing Radars		294
10.10	Pattern Synthesis		295
	10.10.1	Discrete Array Synthesis	298
10.11	Phased Array		298
	10.11.1	Directional Pattern	299
	10.11.2	Conformal and Non-Uniform Arrays	303
	10.11.3	Adaptive Arrays	304
	10.11.4	Computer Controlled Phased Array Radar	305
10.12	Remote Sensing of the Earth and Its Atmosphere at Microwaves		306
	10.12.1	Atmospheric Windows for Remote Sensing	306
	10.12.2	Satellite Instruments for Microwave Remote Sensing	307
	10.12.3	Remote Sensing Radars	310
10.13	CW Radar		317
10.14	Imaging Radar		320
	10.14.1	Sources of Backscatter in Imaging Radar	321
	10.14.2	Inverse Synthetic Aperture Radar (ISAR)	322
10.15	Monopulse Radar Imaging		326
	10.15.1	Three-Dimensional Imaging by Monopulse Radar	327
	10.15.2	Stepped-Frequency 3D Monopulse Radar	328
10.16	Multifunction Array Radar (MFAR)		331
	10.16.1	Multifunction Capability of Radar	333
	10.16.2	Design of Fixed Antenna Arrays for Search and Track	333

Chapter 10A: Atmospheric Radars **337**

10A.1 Introduction 337
10A.2 Terminology 338
10A.3 Interpretation of the Product 342
10A.4 Weather Surveillance Radar 343

Chapter 11: Radars for Atmospheric Probing **349**

11.1 Introduction to MST Radars 349
11.2 Early Developments 349
11.3 Doppler Radar Probing with MST Radar 350
11.4 Evolution of MST Radar 350
11.5 MST Radar Systems 351
11.6 Antenna Configuration 353
11.6.1 Echoing Processes 353
11.6.2 Turbulence and Layer Structure Measurements 354
11.6.3 Wind Measurement 355
11.6.4 Forecasting Applications 355
11.7 Introduction to Dual-Polarization Radars 356
11.8 CHILL Radar 356
11.9 Chilbolton Radar 356
11.10 Radar Ceilometers 357
11.11 Dual-Polarization Doppler Radars 357
11.11.1 Cloud and Precipitation Radar Reflectivity at Millimeter Waves 359
11.11.2 Configuration of Dual-Polarization Doppler Radars for Millimeter Waves 363
11.11.3 Estimating Updraft and Particle Size Spectra from Doppler Data 368
11.12 Rainfall Measurement with Meteorological Radar 371
11.12.1 Theoretical Consideration 371
11.12.2 Determination of Radar Parameters 372

	11.12.3	Reflectivity Factor	374
	11.12.4	Signal Averaging	374
	11.12.5	Antenna Gain	375
	11.12.6	Radar Rainfall Adjustment	375
	11.12.7	Factors Affecting Rainfall Measurement	376
	11.12.8	Conclusions	377
11.13	Wind Measurement by Dual Beam Incoherent Radar		378
	11.13.1	The Generalized Dual Beam Fluctuation Spectrum	379
11.14	Turbulence Measurement		384
	11.14.1	Mean and Variance of the Doppler Velocity, and the Spectrum of Turbulent Velocity	385
11.15	Doppler Radar and Its Applications to Meteorology		388
	11.15.1	Doppler Radar Techniques	388
11.16	Examples of Doppler and Fluctuation Spectra		400
11.17	Variance of Sample Mean of Correlated Samples Independence Figure		404
11.18	Measurement of Variance of Doppler Spectrum: R-Meter Qualitative Considerations		408
11.19	Doppler Special Characteristics of Precipitation		412
11.20	Relations of the Doppler Spectral Velocity Parameters to the Various Characteristics of the Particle Size Distribution Relations to D_o		424
11.21	Real-Time Doppler Radar Data Processing and Display		429
11.22	Statistical Properties of Dual-Polarized Radar Signals		430
11.23	Radar Characteristics of Precipitation Echo Due to Receivers		440
11.24	Advanced Meteor Wind Observations Using MST Radars		441

Chapter 11A: Radar Measurement of Rain Intensity, Drop Size, and the Related Attenuation of Radio Waves **443**

11A.1 Introduction 443

11A.2 Radar Measurements of Rainfall 444

11A.3 Theoretical Approach on Rain Rate vis-à-vis Radar Reflectivity 445

11A.3.1 Empirical Rain Rate—Radar Reflectivity Relationships 446

11A.3.2 Drop Size Distribution from Radar Reflectivity 448

11A.4 Attenuation of Radio Waves Due to Rain 449

11A.5 Statistical Model for Rain Rates 453

11A.6 Statistical Model for Rainfall Intensity 454

11A.6.1 Determination of the Parameters R_M, α and β 456

11A.7 R_M and α as calculated from CCIR Rainfall Data 457

11A.8 Return Periods of Precipitation Rate 458

11A.9 Rain Rates in the Tropics Measured by Rain Gauges 461

Chapter 12: Direction Finding **465**

12.1 Introduction 465

12.2 Loop Antenna 465

12.3 Sense Finder 469

12.4 Increasing the Sensitivity of a Direction Finder 471

12.5 Methods of Increasing Loop Voltage 472

12.5.1 Increase of A 472

12.5.2 Increase of N 473

12.5.3 Tuning of the Loop to Increase Loop Voltage 474

12.5.4 Radiation Resistance of the Loop 474

12.5.5 Increasing the Q of the Loop 475

12.5.6 Use of Ferrites to Increase Loop Voltage 475

	12.5.7	Increasing Loop Voltage by Large Fixed Crossed Loops: Bellini Tosi Direction Finder	477
12.6	Errors in Direction Finding		479
	12.6.1	Errors Due to Direction Finding Equipment Antenna Effect	479
	12.6.2	Errors Due to Anomalous Propagation Effects	481
12.7	Automatic Direction Finders		486
	12.7.1	Airborne Automatic Direction Finder Radio Compass	487

Chapter 13: Aircraft Homing System and Instrument Landing System **491**

13.1	Introduction		491
13.2	Switched Cardioid Homing System		491
13.3	Four Course Radio Range		493
13.4	Omnidirectional Ranges		495
	13.4.1	VHF Omnirange (VOR)	495
	13.4.2	Doppler VOR	502
13.5	Tactical Air Navigation (TACAN)		504
13.6	Instrument Landing Aids		505
	13.6.1	Instrument Landing Systems	506
13.7	Ground Controlled Approach		513
	13.7.1	Radar Control	513
	13.7.2	Doppler Direction Finder	513
13.8	Radio Altimeter		515
	13.8.1	Triangular Modulation	515
	13.8.2	FM-CW Altimeter with Sinusoidal Modulation	519
13.9	Microwave Landing System		526
	13.9.1	Introduction	526
	13.9.2	Limitation of ILS	526
	13.9.3	MLS Development	528

	13.9.4	Types of Antennas in MLS	528
	13.9.5	Principle of MLS	530
	13.9.6	System Composition	531
	13.10	Advantages of MLS	533
Chapter 14:	**Hyperbolic Navigation**		**537**
	14.1	Introduction	537
	14.2	LORAN-A	539
	14.3	Measurement of Delay	540
	14.4	Transmitter Power	542
	14.4.1	Accuracy	542
	14.5	LORAN-C	542
	14.6	DECCA	544
	14.6.1	DECCA Chain	544
	14.6.2	Ambiguity of Courses and Lanes	546
	14.6.3	Lane Identification	547
	14.7	DECCA Receiver	548
	14.7.1	Fine Fixing Configuration	549
	14.7.2	Lane Identification Configuration	549
	14.8	OMEGA	550
	14.8.1	Transmission Cycle	551
	14.8.2	Accuracy and Range	552
	14.9	DECTRA	552
	14.10	DERLAC	553
	14.11	Inland Shipping Aids	553
	14.11.1	HIFIX	553
	14.11.2	Accuracy and Range	555
	14.11.3	Lane Identification	555
	14.11.4	Transmitter Power	556
	14.12	SYLEDIS	556
	14.12.1	Principle of SYLEDIS	556
	14.12.2	The Principle of Distance Measurement	557

Chapter 15:	**Satellite Navigation**			**563**
	15.1	Introduction		563
		15.1.1	Doppler Navigation	563
		15.1.2	Global Positioning System	567
		15.1.3	Principles of Operation of GPS Navigation	569
		15.1.4	GPS Segments	571
		15.1.5	Format of GPS Navigation Message	573
		15.1.6	GPS Data Subframe	574
		15.1.7	Sources of Errors in GPS	576
	15.2	Differential Global Positioning System (DGPS)		577
		15.2.1	DGPS Configurations	579
		15.2.2	Applications of GPS and DGPS	579
		15.2.3	GPS Receivers	581
Chapter 16:	**Vessel Traffic Management System**			**583**
	16.1	Introduction		583
		16.1.1	Configuration of VTMS	584
		16.1.2	Operator's Workstation	586
		16.1.3	Radar Tracking System for VTMS	588
	16.2	GPS Transponder System for VTMS		590
Chapter 17:	**Navigation and Modern Measuring Techniques**			**593**
	17.1	Introduction		593
	17.2	Latitude and Longitude of a Place		593
	17.3	Modern Navigational Methods		594
	17.4	Dead Reckoning		596
	17.5	Piloting		596
	17.6	Celestial Navigation		597
	17.7	Marine Chronometer		597
	17.8	Marine Sextant		598
	17.9	Inertial Navigation		599
	17.10	Radio Directions Finder		599
	17.11	Radar Navigation		600
	17.12	Satellite Navigation		601
	17.13	Navigation Application with Mobile Telephony		601

Chapter 18: Millimeter Wave Radars and Radiometers 603

18.1 Introduction 603
18.2 Mm-ware Radars and Radiometers for Defense Applications 604
18.3 Development of a Coherent mm-Wave Receiver for 94 GHz 605
18.4 Tunable mm-Wave Transmitter for 53 GHz 609
18.5 Tunable mm-Wave Receiver 610
18.5.1 IF Main Amplifier and IF Demodulator 611
18.5.2 Antenna and Beam Alignment 612
18.6 Millimeter Wave Video Link at 53 GHz 612
18.7 Generation of FM Signals 613
18.8 FM mm-Wave Receiver 614
18.9 AFT Discriminator (or Detector, Built-In inside the CTV Tuner) 617
18.10 Demodulator of FM Signals 619
18.11 Solar Tracking mm-Wave Radiometers 625
18.11.1 Solar Tracking Radiometer for 53 GHz 626
18.11.2 Dynamic Range of Antenna Temperature and IF Amplifier 626
18.11.3 Dynamic Range of the Detector 627
18.11.4 IF Gain 627
18.11.5 Front-End Converter for a 53 GHz Radiometer 628
18.11.6 Calibration and Drifts in the Radiometer 629
18.12 Solar Tracking Radiometer for 94 GHz 630
18.12.1 Calibration and Drifts 630
18.13 Solar Tracking Radiometer for 140 GHz 630
18.14 Estimation of Solar Brightness Temperature 632
18.14.1 Programmed Solar Tracking 633
18.14.2 Automatic Tracking of the Sun 634

	18.15	Calibration of Millimeter Wave Radiometers	634
		18.15.1 Brightness Temperature and Antenna Noise Temperature	634
		18.15.2 Measurement of Antenna Noise Temperature and Liquid Nitrogen Calibration	635
		18.15.3 Continuous Calibration Points between Liquid Nitrogen and Room Temperature	637
		18.15.4 Liquid and Gaseous Helium Calibration	638
		18.15.5 Calibration of Solar Tracking Radiometers	638
Chapter 19:	**Doppler Radar Development and Remote Sensing Research**		**641**
	19.1	Introduction	641
	19.2	Polarimetric Measurements of Precipitation	642
		19.2.1 Electrical Effects of Storms and Lightning	642
	19.3	Automotive Radar Applications	642
		19.3.1 Measurement	644
	19.4	Weather Radar Correction Procedures	645
		19.4.1 Radar Rainfall Data Preprocessing	646
	19.5	Comparison of Mean Areal Rainfall Estimates	647
	19.6	The Rainfall-Runoff Model	648
	19.7	Radar and Storm Model	650
	19.8	Doppler Weather Radar	651
	19.9	Space Weather Effects on GPS and Ratio Propagation	651
Chapter 20:	**Mathematical Preliminaries Related to Radar Systems and Associated Problems**		**655**
	20.1	Fourier Analysis	655
	20.2	Periodic Function	655
		20.2.1 Fourier Series	655
		20.2.2 Complex Representation	656

	20.2.3 Symmetry Properties	657
	20.2.4 Further Developments	658
20.3	Non-Periodic Functions	660
	20.3.1 Transition to the Fourier Integral	660
	20.3.2 Symmetry Properties	661
	20.3.3 Examples	662
20.4	Singularity Functions	664
20.5	Some Properties of the Fourier Transform	667
20.6	Convolution Theorem	669
20.7	Auto-Correlation Function and Energy Density Spectrum	671
20.8	Applications	673
20.9	Sampling Theorem	674
20.10	Summary	676
20.11	Energy and Power Signals	676
20.12	Random Functions	680
20.13	The Secondary Maximum in the Generalized Fluctuation Spectrum	682
20.14	Dual Beam	683

Appendix A: *Achieving Independence of Successive Samples by Frequency Shifting from Pulse to Pulse* 687

Appendix B: *Review Questions* 703

Appendix C: *Multiple Choice Questions and Answers* 717

References 735

Index 743

FOREWORD

The over-riding consideration in conduct and management of aviation and maritime transport is to ensure safety. With the growth of aviation and particularly with introduction of fast-moving jet-aircraft, this aspect is now of paramount importance. It is, however, not perceived by millions of passengers that a lot of investment, a large number of systems of equipment and dedicated efforts behind the screen are required to enable them reach their destination without any hazard. Thanks to the current progress in technology, it is now possible to monitor and control all flights from take-off to landing provided there is enough money with the provider agencies. The subject of radio navigational aids has gone through a process of rapid development over the past decades. Direction during and hyperbolic radio-navigation aids were in extensive use during those early days. Primary radars were also developed to a stage where reliable service could be given for control and tracking. Development and use of VOR and DME constituted significant milestones in provision of accurate and reliable en route navigation systems. Airborne direction finders working in conjunction with ground based non-directional beacons became an integral part of essential airborne aids. With the advent of the Instrument Landing System, safe landing during adverse weather conditions has been possible. The International Civil Aviation Organization (ICAO) established standards in relation to radio-navigation aids. The different types of facilities have been conveniently classified under categories of *(i)* long distance aids, *(ii)* en-route aids, *(iii)* approach and landing aids, and *(iv)* airport-zone aids. As a result of constant interaction among the designer, sponsoring agencies, users and concerned international organizations, system developments have made rapid strides toward producing a variety of versatile aids such as the Doppler, VOR, MLS, internal navigation equipment, the Omega and Satellite Navigation system, etc. In the field of radar, mono-pulse SSR techniques, data transmission capability and signal processing advances have resulted in highly accurate and versatile systems.

In the present textbook, the authors have covered a number of the above subject starting from the very fundamentals. A separate chapter deals extensively with antennas for navigational aids.

Next, two chapters have been devoted to direction finding principles, homing systems, and ILS. Hyperbolic navigation systems including Omega systems have been described in detail in another chapter. Principles of radar and radar systems have been covered very well in several chapters. Radar target characteristics have been adequately described. The topic of radar clutter will be very useful in understanding of the various responsible for this phenomenon.

The various techniques of radar signal processing have been covered very well. The authors have included a chapter on various radar systems intended for different applications, including remote sensing. On the whole, considering the extent of coverage and excellent treatment of various subjects, this book is expected to meet the long-felt requirements of a suitable university standard textbook for students and a wide cross-section of user communities.

S.C. Majumdar
Director of Communication
National Airport Authority

PREFACE

During the last few years, new application areas of radar systems and radio aids to navigation emerged and led to phenomenal development of these systems in certain directions. Some of these areas include, the Radar Cross Section (RCS) measurement systems based on instrumentation radars, dual polarization, the Doppler radar at millimeter waves for measurements of cloud patterns over an airport, winds, updrafts, and downdrafts in storms and drop-size distribution of rain and hailstones, and special atmospheric probing techniques. Chapters on satellite navigation, vessel traffic management system and mathematical preliminaries are included, as well as atmospheric radars (an introductory chapter), navigation and modern measuring techniques, millimeter wave radars and radiometers, and Doppler radar development, and remote sensing research. These topics are included so the content might be useful to R & D organizations engaged in the development and maintenance of the systems, and also to the user organizations for the maintenance of their operational systems in these areas.

The fundamental objective of the radio navigational aids is to supply the navigator with data about speed of motion, height, position, and other relevant navigational parameters. The techniques of finding these parameters are oft utmost importance to a radio engineer. Moreover, the security of navigation is ensured by introducing certain special radio aids to detect, for example, an impending thunderstorm, which might affect both an aircraft or a ship; or a micro burst or wind shear which prove to be hazardous during the landing phase of an aircraft. The book has been designed so that it can serve the purpose of professionals in the field to a greater extent. Review questions and multiple choice questions and answers have been added in the appendices as an aid to mastering the topics in the text.

CHAPTER 1

NAVIGATION

1.1 INTRODUCTION

The control of the motion of vehicles for guidance through land, sea, or air to the destination is called navigation. The navigation of ships and aircraft, classified as marine navigation and aerial navigation respectively, calls for a variety of navigational aids. For such navigation, the navigator locates his own position in the absence of any identifying signs on the path that are readily available for motion on land. Most of the navigational aids for aerial and marine navigation employ radiowaves that are either picked up by the vehicle or are transmitted from it to get a reply from the destination. The final objective of all these radio navigational aids is to supply the navigator or the control center at the airport or seaport with data about the location, speed of motion, height, and such other navigational parameters of the vehicle. The technique of determining these parameters by radio waves is often described as the radio location technique, although the location of a vehicle is only a first step in the determination of the navigational parameters.

1.2 NAVIGATIONAL PARAMETERS

If we consider a moving vehicle in the air, the significant navigational parameters will be as shown in Figure 1.1. The coordinates of the vehicle P at the instant t relative to the ground control center may be x, y, z in Cartesian coordinates and r, θ, ϕ in the polar coordinates. The polar coordinate is more useful as it directly gives the range, r of the vehicle, elevation angle α or

(90°–θ) and the azimuthal angle or bearing ϕ. Also, the velocity of approach is given by $v_r = -dr/dt$. In the Cartesian system some useful parameters like the height of the vehicle, z, velocity of descent, $v_z = -dz/dt$, and velocity of horizontal travel $v_s = \frac{ds}{dt} = \frac{d\sqrt{x^2+y^2}}{dt}$ can be directly obtained. For marine navigation both z and α are zero.

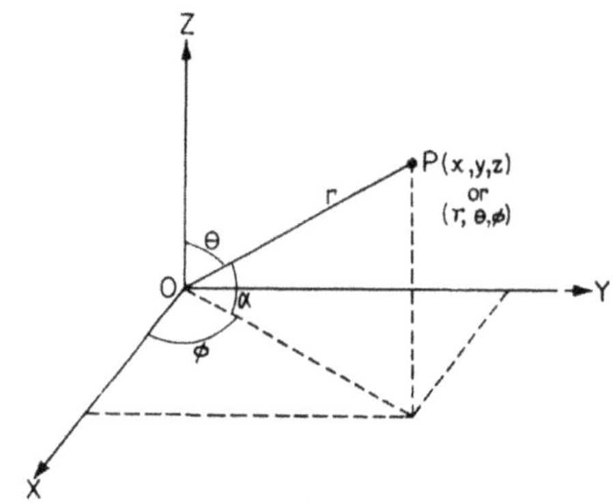

FIGURE 1.1 The target coordinates for significant navigational parameters.

1.3 TYPES OF NAVIGATIONAL AIDS

For most of the radio navigational aids, a radio transmission from one or more control centers is picked up by the vehicle to identify the course of the vehicle. These are called course finding aids. Often a pulsed radio transmission from the control center is scattered back from the vehicle, and the delay in the arrival of the back scattered echo is used to determine the range of the vehicle from the control center. These are called radar aids. These aids are also capable of determining the location of the vehicle in air or in the seas and are in fact very effective aids for the control of the simultaneous landing of a number of aircraft or in the control of ship traffic in the ocean. Besides the course finding and radar aids, a vehicle is often provided with radar aids to measure its distance from the control center. These are called distance measuring equipment (DME). In aerial navigation, the aircraft may also be provided with altitude determining equipment called a radio altimeter. Also, for a blind landing of an aircraft, special radio transmissions from the airport are beamed toward the aircraft, which picks up the signal for its guidance to the runway. These are called instrument landing systems (ILS).

In busy modern airports, special radio monitoring equipment (RME), which can determine the direction of transmission from a number of aircraft simultaneously by employing the doppler effect, helps in air traffic control operations.

Besides the radio navigational aids, the security of navigation is ensured by introducing certain special radio aids to detect, for instance, an impending thunderstorm, which may affect both aircraft and ships, or they may detect an iceberg in the ocean. Specialized radars are used for this purpose. All airports and seaports are equipped with meteorological radars, also called storm warning radars or weather radars, to detect thunderstorms. A marine radar is installed in a ship to detect approaching ships and also to warn the navigators by detecting icebergs.

CHAPTER 2

ANTENNAS FOR RADIO NAVIGATIONAL AIDS

2.1 INTRODUCTION

The operation of a radio navigational aid depends on the generation, propagation, and reception of radio waves. The generation of radio waves from an electrical source of oscillation requires feeding AC electrical power to an antenna or aerial. Reception of the radio waves called for conversion of the radio wave propagated to the receiving location back to the electrical form by a second antenna. The final reception depends largely on the conversion efficiency of the transmitting and receiving antennas as well as their directional patterns. Thus, the antenna characteristics play a major role in determining the performance of radio navigational aids.

A radar antenna acts as a transducer between free space and guided wave propagation. For all practical purposes, the radar antennas fulfill reciprocal but related roles. During transmission the antenna concentrates the radiated energy into a shaped beam, while during reception the antenna collects the energy contained in the echo signal and thereby delivers it to the receiver.

2.2 ANTENNA PARAMETERS

The antenna parameters can be illustrated by considering a short wire of length $l << \lambda$ carrying a uniform alternating current throughout its length, acting as an antenna, as shown in Figure 2.1 (*a*). Such an antenna is called a current element, which can be shown to be equivalent to a Hertzian dipole.

Field Pattern. The distribution of the radiated power in space is usually measured by receiving the radiated field at a fixed distance r for various directions θ and ϕ. The resulting pattern of the electric field is called the field pattern or the radiation pattern of the antenna, as shown in Figures 2.1 (*a*), (*b*), and (*c*). If the rms field measured in a direction is E_r (θ, ϕ), then the power flux in that direction at the distance r is related to the field as

$$F_1 = \frac{P_r(\theta,\phi)}{4\pi r^2} = \frac{E_r^2(\theta,\phi)}{120\pi} \tag{2.1}$$

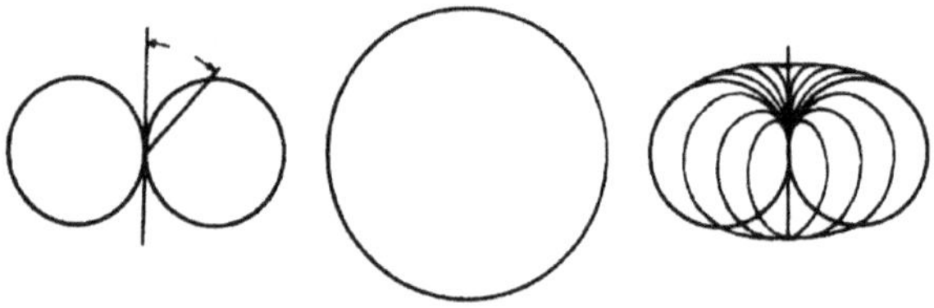

(*a*) Vertical pattern (*b*) Horizontal pattern (*c*) Pattern in three-dimensions
(*a*) Vertical pattern of a short wire ($l << \lambda$) acting as an antenna.
(*b*) Horizontal pattern of a short wire.
(*c*) Three-dimensional pattern of a short wire.

FIGURE 2.1

where 120 π is the resistive characteristic impedance of free space (= 377 Ω). The situation is analogous to the power flow P_L in a circuit of resistance R_0, where $P_L = V_r^2/Ro$, V_r being the voltage. The field E_r of a wave plays the role of the voltage V_r in electric circuits. From equation (2.1), we get

$$E_r(\theta,\phi) = \sqrt{\frac{30\, P_r(\theta,\phi)}{r}} \tag{2.2}$$

Equation (2.2) shows that we can obtain the power pattern from the field pattern.

Beamwidth. For the current element it can be shown that the field pattern or the radiation pattern is given by

$$\begin{aligned} E_r(\theta,\phi) &= E_\theta \\ &= \frac{\omega\, I_0\, ds \sin\theta\, e^{j\omega(t-r/c)}}{4\pi\, \varepsilon r\, c^2} \end{aligned} \tag{2.3}$$

where ω = frequency, I_0 = amplitude of the current, ds = length of the current element, θ = angle between the direction of radiation and the current element, r = distance, c = velocity of propagation, and ε = permitivity of the medium. The pattern is independent of ϕ and is shown in Figure 2.1 (*b*). At an angle $\pm\ \delta$ from the direction of maximum radiation, if the power is halved, then the angular width 2δ is called the beamwidth of the antenna. Equation (2.2) indicated that the field at the half power angles $\pm\ \delta$ will be $1/\sqrt{2}$ times the field in the direction of maximum radiation. Thus, for obtaining the beamwidth, we need not necessarily obtain the power pattern but can have it directly from the field pattern. For this reason, the field pattern, which is directly measured, is more widely employed to specify the directional behavior of an antenna.

Gain. For an antenna the ratio of the power flux in a direction θ, ϕ at a distance r to the power flux at the same distance due to an isotropic antenna fed with same power P_T is defined to be the power gain of the antenna. Accordingly, the power gain $R\ (\theta, \phi)$ is given by

$$G(\theta, \phi) = \frac{P_r(\theta, \phi)/4\pi r^2}{P_T\ /\ 4\pi r^2} = \frac{P_r(\theta, \phi)}{P_T} \tag{2.4}$$

or

$$P_r(\theta, \phi) = P_T G(\theta, \phi)$$

Equation (2.4) shows that the effective radiated power (or ERP) $P_r\ (\theta, \phi)$ is $G\ (\theta, \phi)$ times the power, P_T fed to the antenna. ERP is often termed EIRP (Effective input radiated power), in which I stands for input power to the medium. The effective gain of the antenna is due to redistribution of the radiated power in space. If θ_m, ϕ_m are the directions for which the radiated power is maximum, then the gain $G = G\ (\theta_m, \phi_m)$ is the maximum power gain of the antenna. This is usually referred to simply as the power gain G of the antenna. In practice a part of the transmitter power is lost in the transmission system between the transmitter and the antenna, while another part is lost in the ohmic resistance of the conductor of the antenna. These losses reduce the effective gain of an antenna. For this, the losses are included in the gain of an antenna, and a separate parameter called the directivity (G_D) is introduced, which excludes the losses.

Thus,

$$G = \eta\, G_D \tag{2.5}$$

where η represents the fraction of the power lost in the transmission system and the antenna conductor. The power gain, in fact, arises from the concentration of power in certain directions, which results from the characteristic directional pattern of the antenna. The directivity indicates this power gain arising from the directional pattern only and neglects the ohmic losses of the antenna as well as that in the transmission system and is defined as the ratio of the power density in a direction (θ, ϕ) to that due to an isotropic source fed with the same power. Accordingly, we have

$$G_D(\theta, \phi) = \frac{\text{Power density of antenna in a direction } \theta, \phi}{\text{Power density due to isotropic antenna}} \tag{2.6}$$

If the power in the antenna is concentrated into the solid angle Ω only, the centered around θ_m, ϕ_m, we have from equation (2.4)

$$G_D = \frac{P_T / \Omega}{P_T / 4\pi} \cong \frac{4\pi}{\Omega} \tag{2.7}$$

Combining equations (2.5) and (2.7), we get

$$G = \eta G_D = \eta \frac{4\pi}{\Omega} \tag{2.8}$$

Combining with equation (2.4), we get

$$P_r = P_T G = P_T \, \eta \, G_D \tag{2.9}$$

Effective Aperture. The effective receiving aperture or the effective area is also related to the gain. If A_e denotes the effective area of a lossless antenna, then the gain G is related to the antenna aperture efficiency ρ_a and the physical area A of the antenna by an equation

$$\begin{aligned} G &= \frac{4\pi A_e}{\lambda^2} \\ &= \frac{4\pi \rho_a A}{\lambda^2} \qquad [\text{because } A_e = \rho_a A] \end{aligned} \tag{2.10}$$

where λ is the wavelength.

Aperture Efficiency. The aperture efficiency can be defined as the ratio of the actual antenna directivity to the maximum possible directivity. It is a measure of the effectiveness with which the physical aperture area or the mouth area of the antenna is utilized.

Side Lobe Radiation. For a typical antenna, on either side of the main beam or main lobe, additional but smaller lobes occur. These secondary minor lobes of the antenna pattern are side lobes. In general, low side lobe levels are desired for radar applications, since otherwise a part of the power radiated will be wasted through the side lobes.

Definitions Valid for Any Antenna. The definition of the antenna parameters, which refers to a current element as an antenna, is perfectly general and can be applied to any type of antenna. It may be noted also that all the definitions are based on elementary considerations and are not dependent on the way the radiation pattern or radiation resistance may be deduced by solving the relevant Maxwell field equations.

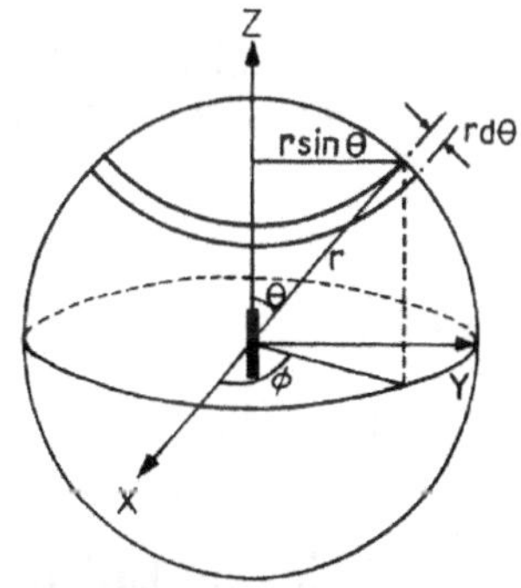

FIGURE 2.2 The geometry of the problem of radiated power flow.

Radiated Power. The total power radiated in all directions can be obtained by integrating $P(\theta, \phi)$ given by equation (2.3) over the limits $\theta = 0$ to π and $\phi = 0$ to 2π and can be shown to be

$$P_r = 40\,\pi^2\, I_0^2 \left(\frac{l}{\lambda}\right)^2 \qquad (2.11)$$

For such integration, the power pattern $P(\theta, \phi)$ need not be known, but it can be deduced from the field pattern $E(\theta, \phi)$ by using the relation of equation (2.2). The geometry of the problem is shown in Figure 2.2. Power flowing through an annular strip of width $rd\theta$ having an area of $2\pi r \sin\theta\, rd\theta$ is integrated over the entire sphere to deduce equation (2.11).

Radiation Resistance. It may be noted that the radiated power is lost in space, and this is drawn from the source of current through the wire. From a circuit theory point of view, if a current of amplitude I_0 flows through a wire of ohmic resistance R_L, the ohmic power loss will be $P_L = \frac{1}{2} I_0^2 R_L$. Likewise, the radiation loss from the wire may be thought of as being due to the power being lost in a fictitious resistance R_r of the wire such that

$$P_r = \frac{1}{2} I_0^2 R_r \tag{2.12}$$

R_r is called the radiation resistance of the antenna. Combining equations (2.11) and (2.12), we get

$$P_r = \frac{1}{2} I_0^2 R_r = 40\,\pi^2 I_0^2 \left(\frac{l}{\lambda}\right)^2$$

from which

$$R_r = 80\,\pi^2 \left(\frac{l}{\lambda}\right)^2 \tag{2.13}$$

Radiating Efficiency. In the presence of ohmic resistance R_L of the wire, the total power consumed from the source is given by

$$\begin{aligned} P_T &= P_L + P_r \\ &= \frac{1}{2} I_0^2 R_L + \frac{1}{2} I_0^2 R_r \end{aligned} \tag{2.14}$$

The radiating efficiency of the antenna

$$= \frac{\text{Power radiated}}{\text{Total power consumed}}$$

$$\eta = \frac{P_r}{P_L + P_r}$$

$$= \frac{\frac{1}{2} I_0^2 R_r}{\frac{1}{2} I_0^2 R_L + \frac{1}{2} I_0^2 R_r} \qquad \text{[from equation (2.14)]}$$

$$= \frac{R_r}{R_L + R_r} = \frac{R_L + R_r - R_L}{R_L + R_r}$$

$$= 1 - \frac{R_L}{R_L + R_r} \tag{2.15}$$

As an example, at 50 Hz,

$$\lambda = 6000 \text{ km, and for an } l \text{ as large as } 60 \text{ km}$$

$$\left(\frac{l}{\lambda}\right)^2 = \left(\frac{1}{100}\right)^2 = 10^{-4}$$

and the ohmic loss R_L is, therefore, likely to exceed greatly the radiation resistance R_L. The antenna will therefore be very inefficient even for a 60 km length. On the other hand, a similar consideration at 50 MHz shows that $\lambda = 6$ m and the antenna a few meters in length will have radiation resistance greater than the ohmic loss. The radiating efficiency will thus be high at higher frequencies because of the term $(l/\lambda)^2$ in equations (2.11) and (2.13).

2.3 CURRENT DISTRIBUTION

The uniform current throughout the length of the current element is rather difficult to realize in practical antenna systems, in which the power is fed at the center of the wire. The generator appears to be open circuited from the viewpoint of DC circuit theory. However, current flows through the wire due to distributed stray capacitance of the wires (shown as dotted lines) when the generator is of an AC type, as shown in Figure 2.3 (*a*). The current will, in fact, be maximum near the generator, as the current through all such capacitance flows through the portion of the conductors near the generator. The current at points removed from the generator decreases due to the number of stray paths for the currents becoming progressively less.

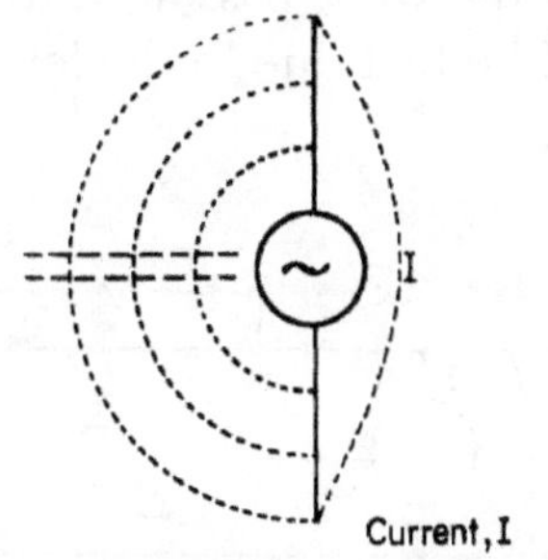

FIGURE 2.3 (*a*) Current flowing through the wire due to distributed stray capacitance.

The current at the open end is, therefore, zero. The distribution of the current along the wire is shown by the curved lines marked *I*. For such a current distribution, the different portions of the wire are not equally effective in radiating power. The total power radiated will in fact be less than that for a current near the generator being maintained throughout the length. Even the radiation pattern will be modified due to the non-uniform current distribution. The pattern is deduced by summing up the radiation fields due to elementary lengths of the wire each carrying a current from the next element following the current distribution along the length. A method of realizing a uniform current through a short wire is indicated in Figure 2.3 (*b*). Radiation from a long and open transmission line connected to a generator may be employed for this purpose. Due to the open termination of the line, there will be standing waves as shown in the figure with amplitude of the current being maximum at the sending end of the line and at S_1, S_2, and S_3 or at S_1', S_2', and S_3', while it is zero or null at N_1, N_2, and N_3, or at N_1', N_2', and N_3'. The instantaneous directions of the current on either side of a null are opposite. Also, the radiation from the transmission line wires tends to cancel out as the current in the opposite wires are oppositely directed everywhere. In fact, at a point equidistant from the wires, the radiation cancels completely. In the plane of the wire, since the radiation from one of the wires will reach earlier relative to that from the other, the cancellation will not be complete. However, as the spacing $s << \lambda$ any such radiation will be small. We note that the current can be considered to be uniform over a short length $l << \lambda$ along the line. Also, the short length of wire to which the generator is connected will also carry a current of uniform amplitude, as $s << \lambda$. In a horizontal plane perpendicular to the wire, passing through the mid point of the wire, *s*, the radiation will be that due to the length *s* only, in which the current is uniform.

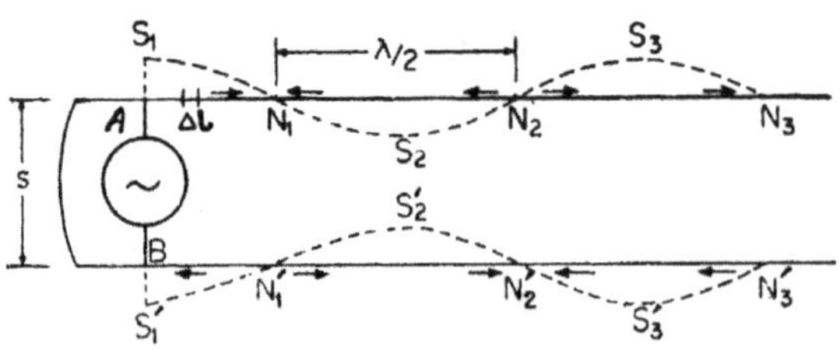

FIGURE 2.3 (*b*) Short wire terminated by an open circuited transmission line.

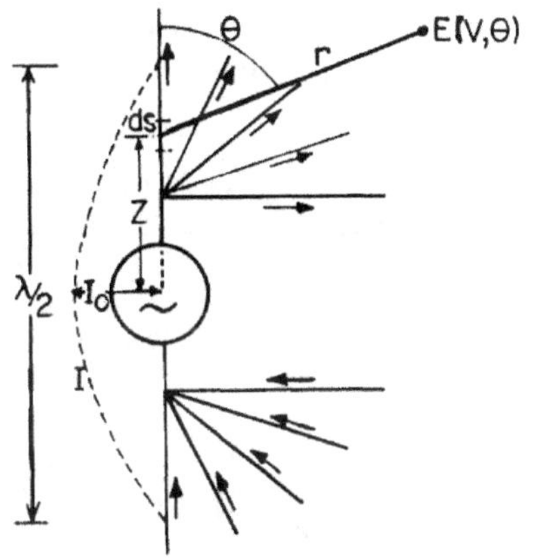

FIGURE 2.3 (*c*) Short wire terminated by open-circuited quarter wave lines degenerated into a half-wave (λ/2) dipole.

2.4 HALF-WAVE DIPOLE

If we cut off the transmission line in Figure 2.3 (*b*) beyond the position of nulls N_1N_1', we get the short wire terminated by open-circuited quarter wave lines as shown in Figure 2.3 (*c*), since $s/2 + AN_1 = \lambda/4$. Now, consider the line to be gradually opened up as shown in the figure to form an antenna of length $\approx 2 \times \lambda/4 = \lambda/2$. Such an antenna is called a half-wave dipole. We note that the instantaneous currents in the upper and lower halves of the wire are now in the same direction, although in the quarter wave opposite. The radiation from the two halves of the wire on either side of the generator, therefore, tends to combine in the half-wave dipole. However, as the total length L is no longer much less than λ, the resultant field has to be calculated by superimposing the radiation from different elementary lengths, assuming appropriate currents and phase paths of the radio waves to the remote point at which the field is considered. The radiation from each element ds can be obtained from equation (2.3). The current distribution is assumed sinusoidal in z and is given by

$$I_z = I_0 \sin z \tag{2.16}$$

as shown in Figure 2.3 (*c*). During the calculation the variations in the amplitude of the field due to the difference in the distance r from different elements is neglected. Only the phase changes of the field with r are considered. The resultant radiation field pattern takes the form:

$$E_\theta = \frac{j\,60\,I_0 e^{-(2\pi/\lambda)r}}{r}\left[\frac{\cos\left(\frac{\pi}{2}\cos\theta\right)}{\sin\theta}\right] \tag{2.17}$$

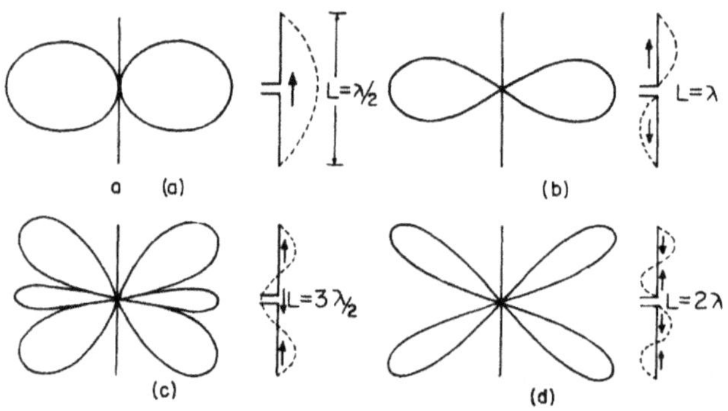

FIGURE 2.4 Radiation pattern and current distribution.

The field is independent of ϕ. The pattern is shown in Figure 2.4 (*a*).

2.5 ANTENNAS OF LENGTH GREATER THAN HALF-WAVE LENGTH

If the length of the antenna, L, is greater than $\lambda/2$, then the instantaneous current in some portion of the wire may be opposite to that of the remaining portions as shown in Figures 2.4 (*b*), (*c*), and (*d*), which show the current distribution and the resulting field patterns for $L = \lambda$, $3\lambda/2$, and 2λ respectively. It may be noted that radiation fields in a direction $\theta = 90°$ cease to be maximum for lengths, L, greater than λ. For $L = 2\lambda$ the radiation field is zero for $\theta = 90°$. Therefore, the length L is usually made equal to $\lambda/2$, hence the popularity of half-wave dipoles.

Feeding of Antennas. For a maximum power transfer to an antenna, its terminal impedance must be matched to the source impedance.

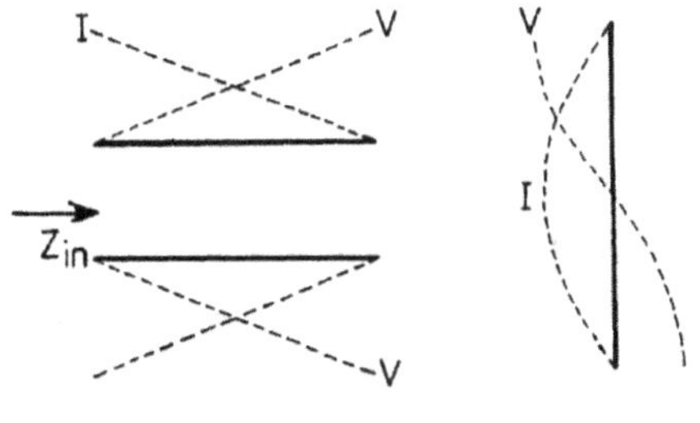

FIGURE 2.5 Current and voltage distribution (*a*) in an open-circuited quarter-wave (λ/4) transmission line; and (*b*) that in a λ/2 dipole.

The terminal impedance, however, depends on the feeding point of the antenna. The dependance can be understood by referring to Figure 2.5, which shows the current and voltage distributions in an open-circuited λ/4 transmission line as well as that in a half-wave dipole formed by opening up the lines. The input impedance of the λ/4 line is given by

$$Z_{in} = \frac{Z_0^2}{Z_R}$$

where Z_0 is the characteristic impedance of the line, and Z_R is the terminating impedance at the far end. At $Z_R = \infty$ for the open-circuited line, $Z_{in} = 0$. The same result could also be derived by dividing the voltage V by the current I at the sending end of the line. At the input terminal $V = 0$ but I is finite; $Z_{in} = O/I = 0$. On opening up the line to form the λ/2 dipole, the impedance at the midpoint would have also been zero from a similar consideration of voltage and current distributions if there were no radiation from the antenna. Due to the radiation power, however, the impedance of the midpoint, instead of being zero, turns out to be 73.1 ohms resistance, as can be shown by a rigorous analysis of the antenna. If the feed point is at an end of the antenna, the voltage and current distributions indicate that the impedance should be infinity. In the presence of radiation loss, the impedance becomes 1100 ohms resistance instead of being infinity. Such an end-fed dipole will be useful in matching it to a source of resistance of 1100 ohms. For a feed point midway between the center and an end, the impedance is between 73.1 and 1100 ohms and also has a reactive part. Usually, an impedance-matching network is connected between the antenna terminal and the source to obtain the desired impedance match. It may be mentioned here that the terminal impedance is increased somewhat by the ohmic loss resistance of the antenna wire, thus making it typically 75 ohms for a half-wave dipole instead of 73.1 ohms, which is only the radiation resistance part.

2.6 PARASITIC ELEMENTS TO INCREASE DIRECTIVITY

A half-wave dipole may be backed by a plane reflector, a plane wire grid, or a single wire as shown in Figures 2.6 (*a*), (*b*). For the plane reflector a radio

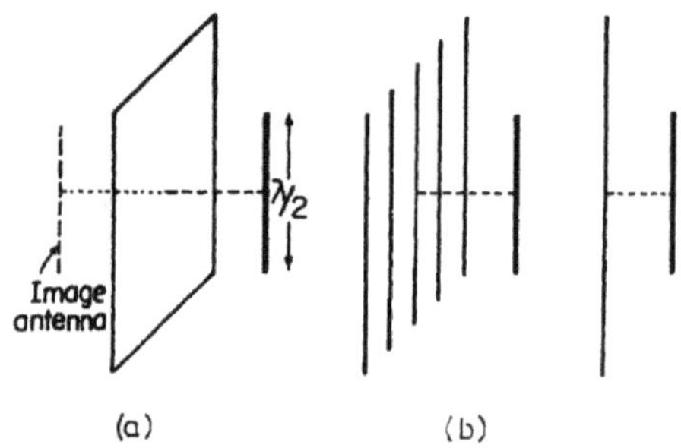

FIGURE 2.6 (*a*) Half-wave dipole backed by a plane reflector. (*b*) A plane wire grid reflector.

wave incident from different angles may be reflected back on to the dipole and can be thought of as due to an image dipole antenna (shown as dotted lines) formed as far behind the reflector as the dipole is in front of it. The instantaneous current in the image antenna is in the same direction as that of the dipole due to a phase change of 180° on the reflection opposite and, therefore, if the distance between the dipole and the reflector is made equal to the radiation from the image antenna, it will add up with that from the dipole in the forward direction indicated by the arrow. In effect, the radiation in the backward direction of the dipole is reflected in the plane sheet to add up with the forward radiation, thus increasing the concentration of energy in a forward direction, and a 3 dB improvement in power gain is, therefore, expected. The plane wire grid shown in Figure 2.6 (*b*) can also increase the gain in a similar way if the spacing between the grids, $s << \lambda$, so that the grid appears to be a plane continuous sheet at the wavelength.

Even a single central wire of the grid, alone, can act as a reflector, if its length is made somewhat larger that λ/2 and is placed at an appropriate distance from the dipole. The radiation of the dipole in the backward direction is in fact picked up by the reflector wire. The length of the wire being > λ/2, the current driven in the wire will lag the induced voltage, the wire behaving inductively. The radiated field from the wire due to this lagging current will be further delayed in reaching the dipole due to the distance it has to travel. A similar delay is introduced for the radiated field of the dipole in reaching the reflector. The total delay due to the distance traveled by the radiation from the dipole to the reflector and back, plus that arising from the phase lag between the induced current and the field causing it, may be adjusted to that required by the ratio wave re-radiated by the reflector to add up to that from the dipole in the diffraction of the backward radiation from the dipole around the reflector wire. Such a back lobe will not be present for a plane sheet reflector or wire grid having dimensions greater than several wavelengths.

The reflector wire is often called a parasitic element as no power is fed to this wire. A parasitic wire having a length shorter than λ/2 can be placed in front of the dipole at an appropriate distance so that its radiated field may add up to that from the dipole in the forward direction. The length being shorter, the parasitic element now behaves capacitively, with its current leading the induced voltage. Such a parasitic element is called a director, which adds further to the directivity of the antenna. The increased forward radiation in the presence of a reflector and director may be further increased by adding a number of directors arranged in the plane containing the antenna and the parasitic elements, each director serving to increase the forward radiation resulting from the preceding elements. An antenna having a reflector and one or more directors, or parasitic elements, is called a Yagi antenna, widely employed in TV reception, as shown in Figure 2.6 (*c*). A directive gain of typical Yagi antennas may be 10 dB for a five-element Yagi having one reflector, three directors, and one fed element, the half-wave dipole. The gain may be increased to 15 dB by increasing the number of directors to twenty-three.

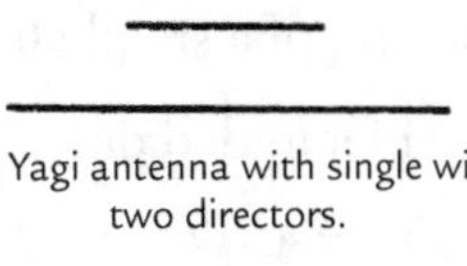

FIGURE 2.6 (*c*) Yagi antenna with single wire reflector and two directors.

2.7 FOLDED DIPOLE

The addition of parasitic elements reduces the terminal impedance of the element of Yagi antennas because of the mutual coupling between the feed and the parasitic elements. Thus, in a five-element Yagi, the terminal impedance of 75 ohms of a half-wave dipole may be reduced to about 20 ohms. To increase the terminal impedance of Yagi to 75 ohms, a folded dipole may be used. The impedance of a folded dipole in the absence of parasitic elements is 300 ohms. The principle of operation of a folded dipole antenna may be explained by referring to Figure 2.7.

The open-circuited transmission line shown in Figure 2.7 (a) is cut off at the antinode of the current at aa', and is then opened up with the far end length

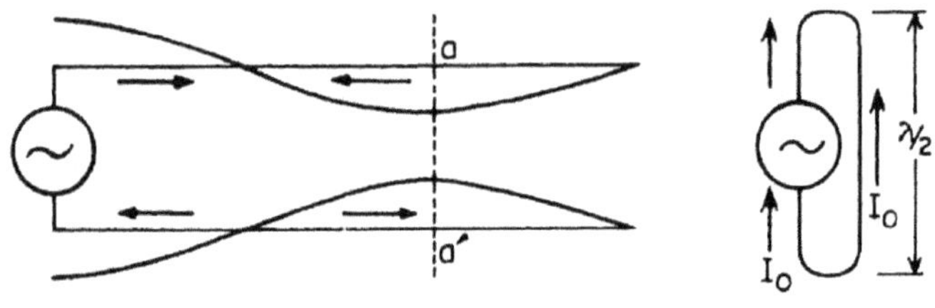

(a) (b)

FIGURE 2.7 (a) Open-circuited transmission line (b) Folded dipole antenna.

λ/4 being folded by 180° to form the folded dipole antenna shown in Figure 2.7 (b). The directions of the current in the fed element and the folded element 2 are the same. The radiated power is, therefore, given by

$$\begin{aligned} P_{rf} &= \frac{1}{2}(2I_0)^2 R_r \\ &= \frac{1}{2} I_0^{\ 2}\, 4R_r \end{aligned} \tag{2.18}$$

where I_0 is the current at the feed point of the fed element. Equation (2.18) indicates that the power radiated is equal to that of a dipole antenna with input current amplitude I_0, but having a radiation resistance R_{rd}, four times that of the impedance R_r, of a simple half-wave dipole.

Thus, the impedance of a folded dipole is given by

$$R_{rd} = 4R_r = 4 \times 73 = 292 \text{ ohms}$$

Taking the ohmic losses into account, the impedance of a folded dipole is about 300 ohms.

2.8 PARABOLIC REFLECTOR

The directive gain of an antenna can be increased greatly by placing it at the focus of a parabolic reflector. A section through the axis of the parabolic reflector as shown in Figure 2.8 indicates the parabolic cross-section of the reflector. In this case, the reflector is often called a parabolic dish. The antenna which serves as the feed antenna of the parabolic dish may be a half-wave

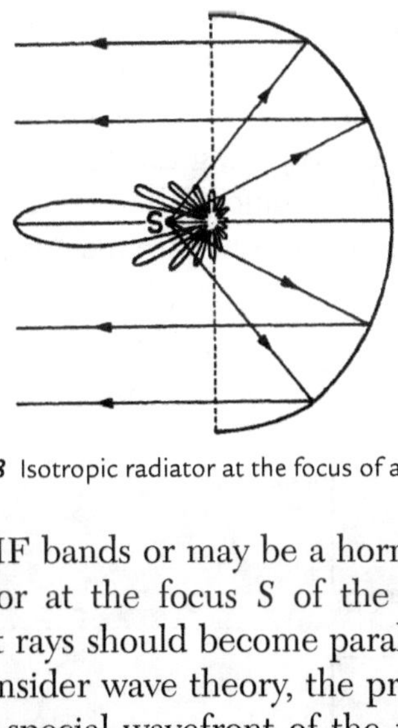

FIGURE 2.8 Isotropic radiator at the focus of a paraboloid.

dipole for VHF and UHF bands or may be a horn antenna at microwaves. For an isotropic radiator at the focus *S* of the paraboloid as shown in Figure 2.8, we note that rays should become parallel after reflection at the dish. However, if we consider wave theory, the presence of the paraboloid effectively converts the special wavefront of the radio wave from the isotropic antenna to a plane wavefront bounded by the mouth area of the dish, which is $\pi D^2/4$, *D* being the diameter of the mouth. The situation is analogous to a circular aperture of diameter *D* illuminated by a plane wavefront. Accordingly, we expect a diffraction pattern of the radio wave as shown in the figure, having a main lobe of width δ given by

$$\delta = 1.22\,\lambda/D \text{ radians} \tag{2.19}$$

$$= 73\,\lambda/D \text{ degress} \tag{2.20}$$

2.9 RECEIVING ANTENNA

A transmitting antenna can also be used in receiving radio waves. In fact, all the antenna parameters remain unaltered when a transmitting antenna serves as a receiving antenna. Such a result arises from the reciprocity relationship between transmitting and receiving antennas, and a separate study of the receiving antenna is therefore unnecessary.

Capture Area. The capture area is the area of the incident wavefront of radio waves through which the power flowing through is captured by a receiving antenna and converted to electrical form.

Relation between Gain and Capture Area

The reciprocity between the transmitting and receiving antennas leads to a relation between the capture area and gain of an antenna. Consider an antenna 1, having a gain G_1 and a capture area A_1 serving as a transmitting antenna, G_2 and A_2 being the corresponding quantities for a receiving antenna 2 placed at distance r from the transmitting antenna as shown in Figure 2.9. The power flux at antenna 2 is given by

$$F_1 = \frac{P_T G_1}{4\pi r^2} \tag{2.21}$$

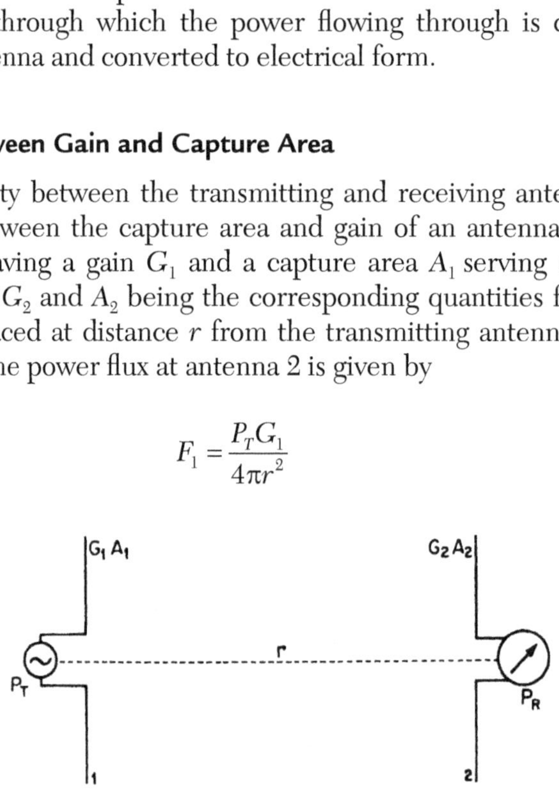

FIGURE 2.9 Transmitting antenna (marked 1) and receiving antenna (marked 2).

The power received in antenna 2 is given by

$$P_{R_1} = F_1 A_2 = \frac{P_T G_1}{4\pi r^{r}} \times A_2$$

Now if the antennas 1 and 2 are interchanged along with P_T and $P_{R,}$ the power flux in antenna 2 placed at the receiving terminal is given by

$$F_2 = \frac{P_T G_2}{4\pi r^2}$$

and the power received

$$P_{R_2} = F_2 A_1$$
$$= \frac{P_T G_2}{4\pi r^2} A_1$$

Due to reciprocity, we expect

$$P_{R_1} = P_{R_2}$$

where $$\frac{P_T G_1}{4\pi r^2} A_2 = \frac{P_T G_2}{4\pi r^2} A_1$$

or $$G_1 A_2 = G_2 A_1$$

or $$G_1 / A_1 = G_2 / A_2 = \text{constant.}$$

It can be shown that the constant is $4\pi/\lambda^2$ and, therefore, we get

$$\frac{G_1}{A_1} = \frac{G_2}{A_2} = \frac{G}{A} = \frac{4\pi}{\lambda^2} \tag{2.22}$$

or $$G = \frac{4\pi A}{\lambda^2}$$

where G and A are the gain and capture area of an antenna.

For a parabolic reflector the mouth area

$$A = \frac{\pi D^2}{4}$$

is nearly equal to the capture area. Putting this in equation (2.22), we get

$$G = \frac{4\pi}{\lambda^2} \frac{\pi D^2}{4} = \frac{\pi^2 D^2}{\lambda^2} \tag{2.23}$$

To be more precise, we should consider that the capture area is slightly less than the mouth area due to the wavefront not being plane near the periphery of the dish. Also, the surface roughness and finite conductivity of the dish tends to reduce the gain. The net effect of these two factors is to reduce the gain by a factor η_A, and equation (2.23) reduces to

$$G = \eta_A \frac{\pi^2 D^2}{\lambda^2} \tag{2.24}$$

Taking the loss η in the transmission systems between the transmitting and receiving antenna inputs as well as the loss due to the finite conductivity of the antenna material and surface irregularity of the dish, we obtain the effective gain as

$$\begin{aligned} G_D &= \eta g = \eta \eta_A \frac{\pi^2 D^2}{\lambda^2} \\ &= \eta_r \frac{\pi^2 D^2}{\lambda^2} \end{aligned} \tag{2.25}$$

η_r, is typically 50–60% for parabolic reflector type antennas.

Bandwidth. The radiation resistance of an antenna determines the antenna's terminal impedance and is, in general, frequency dependent. As a result, an antenna will not remain matched to the source over a wide frequency band. The band over which the mismatch leads to a standing wave ratio of within 2:1 in the feeder system between the transmitter and the antenna is defined to be the bandwidth of the antenna. It may be mentioned here that the standing wave ratio is really the ratio of the power reflected from the load due to mismatch to the power flowing into the load and is given by the reflection coefficient

$$K = \frac{R_a - R_0}{R_a + R_0} = \frac{s-1}{s+1} \tag{2.26}$$

where R_a is the terminal resistance of the antenna, R_0 is the characteristic impedance of the feeder system, and s is the standing wave ratio. It is clear that a higher standing wave ratio will lead to a smaller fraction of the power being delivered to the antenna.

Polarization. The polarization of the radiated electric field from a straight wire antenna is in the plane of the wire and at right angles to the direction of propagation as shown in Figure 2.10 (*a*). The polarization of the electric field is taken as the polarization of the radio wave. Thus, a vertically polarized wave indicates a radio wave in which the electric field is vertical. It may be noted that for a vertical antenna, the polarization is vertical only in the plane perpendicular to the wire or in the azimuthal plane as indicated by E_{90} in Figure 2.10 (*a*). At other angles θ, E_θ is not vertical. The receiving antenna wire must be oriented along the direction of E_θ for maximum pickup, while a vertical orientation of the receiving antenna will introduce a polarization mismatch, causing the pickup to be reduced.

The introduction of a parabolic reflector does not alter the plane of polarization but only increases the effective antenna gain. If two crossed dipoles at right angles to each other as shown in Figure 2.10 (*b*) are fed with currents 90° out of phase with each other, the resultant electric field of the radio wave will have a constant amplitude but will rotate in a plane parallel to the crossed dipoles at the radio frequency of the wave. Such a wave is said to be circularly polarized. If the rotation is clockwise when looking along the direction of propagation of the radio wave, polarization is said to be right circular, while for a counter-clockwise rotation it is called left circular polarization. The receiving antenna for circular polarization is similar in design. The

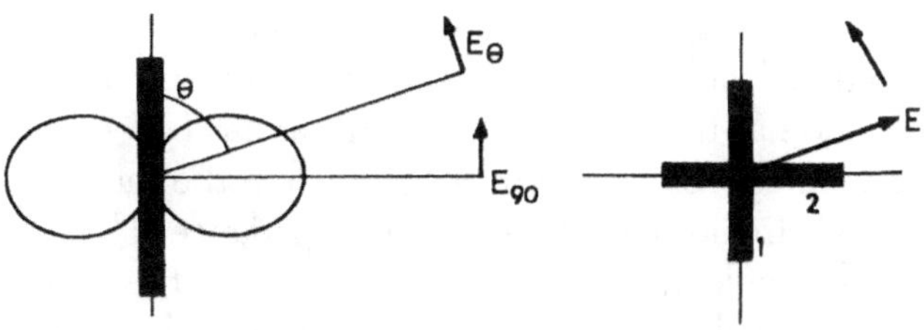

(*a*) Polarization of the radiated electric field due to an electric dipole.

(*b*) Electric field due to two crossed dipoles.

FIGURE 2.10

outputs of the two dipoles are combined after introducing a relative phase difference of ± 90° for receiving the two types of circularly polarized waves, and there must be a polarization match between the transmitting and receiving antenna also for the circularly polarized radio waves. It is easy to see that in the plane of the crossed dipoles, a circularly polarized wave will appear to be plane polarized, while along the perpendicular axis the polarization is circular.

2.10 MICROWAVE ANTENNA

It may be noted that the characteristics of an antenna depend on the ratio of its linear dimensions to the wavelength of operation. At microwaves, λ is of the order of a few centimeters and the antenna becomes very small in dimension. As such, there is considerable scope for increasing the gain, and special antenna structures have been developed for use at microwaves.

2.11 HORN ANTENNA

A microwave antenna widely employed in radars is shown in Figure 2.11 (*a*). This is called a horn antenna. Microwave power is fed to the rectangular pipe called a waveguide, which is coupled to the horn at the forward part. The microwave propagates through the waveguide by successive reflection in the walls with negligible attenuation and reaches the input opening of the horn. The ratio of the electric field E to the magnetic field H of the electromagnetic field has the dimension of ohms and is called the wave impedance. It plays a similar role in electromagnetism as the ratio of voltage to current in electrical circuits, called the circuit impedance. The impedance of the wave inside the guide may be higher and lower than the wave impedance expected in free space, depending on where the wave is launched into the waveguide. In either case, the horn provides an impedance match between the wave impedance of the radiated microwave from the horn and the wave impedance inside the waveguide. In fact, the impedance varies continuously as the wave travels along the horn from its input to the mouth and brings about the impedance match. Under such a matched condition, the maximum amount of microwave power will be transferred from the guide to the free space. The situation is analogous to the maximum power transfer in electrical circuits obtainable under an impedance matched condition between the source and the load. The feeding of microwaves inside the guide calls for an electric or magnetic probe inside the waveguide passed through a suitable location on the guide wall. The probe is called an exciter, the dimension and location of which is critically dependent on the nature of the desired wave field inside the guide as well as on the desired terminal impedance of the exciter. The source of microwave electrical power is also matched to this terminal impedance of the exciter to ensure maximum power transfer from the source to the exciter. In this way, the microwave source effectively transfers the maximum electromagnetic power to the free space due to the proper impedance match established through the excite, the waveguide, and the horn, where

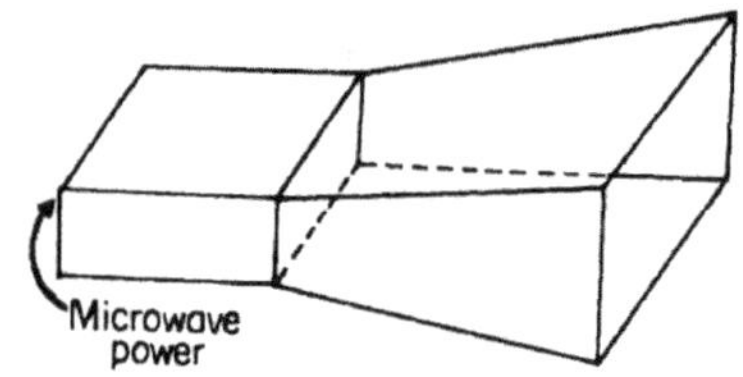

FIGURE 2.11 (*a*) A microwave pyramidal antenna.

the exciter serves in the duel capacity of converting electrical power of the source to electromagnetic power in the guide and in transforming the circuit impedance toward the source end to the wave impedance at the guide end. The length of the guide is kept just sufficient enough to connect the source to the horn, to minimize microwave power loss during transmission through the waveguide.

The gain of the horn antenna may be estimated from the mouth area of the horn being illuminated by a plane wavefront, which in reality not perfectly plane, but may be spherical or cylindrical. However, for usual flare angles the error due to the departure from the plane wave assumption may be neglected. The diffraction pattern expected from such an illuminated aperture is the radiation pattern of the horn. The gain of a horn may be typically 15–30 dB. In radar applications a horn antenna is used as a feed with its mouth placed at the focus of a parabolic dish with its mouth facing the dish.

To obtain an idea of the dimensions of the waveguide and the horn at different wavelengths, we may consider the evolution of the microwave components as shown in Figures 2.11 (*b*) and (*c*). In a transmission line shown in Figure 2.11 (*b*), the spacing *s* is made much smaller than the wavelength, λ, so that any direct radiation from the two wires may cancel each other at a remote point, because of the oppositely directed instantaneous currents in the two wires. At microwaves, the spacing becomes small so as to make it difficult to keep it constant over a considerable length. In such a situation, a number of quarter wavelength (λ/4) sections of a similar transmission line, each of which is shorted at the far end, may be connected to the main line to add to the mechanical rigidity of the line. The shorted sections will not affect the main line electrically, as the input impedance of a λ/4 line shorted at the far end is infinity.

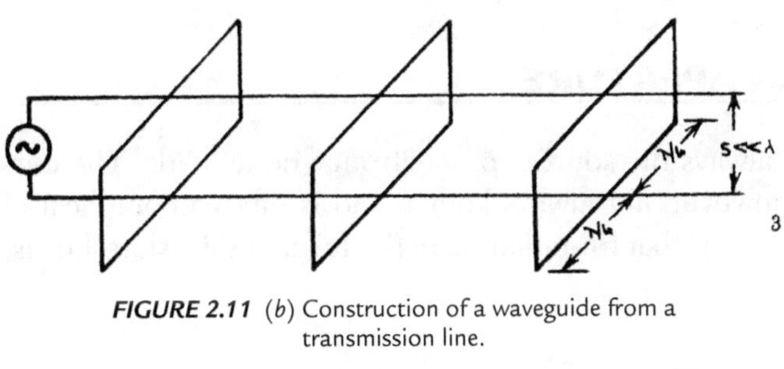

FIGURE 2.11 (*b*) Construction of a waveguide from a transmission line.

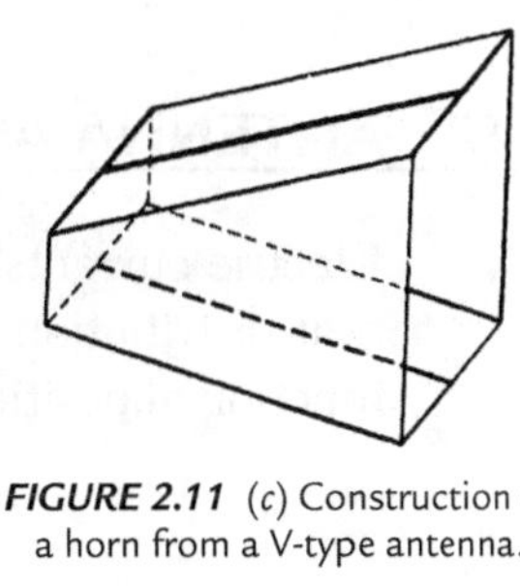

FIGURE 2.11 (*c*) Construction of a horn from a V-type antenna.

If the number of such shorted sections is increased to infinity, we shall eventually obtain a rectangular pipe, thus evolving the waveguide. A completely enclosed space bounded by the four conducting walls of the waveguide ensures a run radiation loss while at the same time attenuation is also minimized, since conducting walls, instead of wires, are now available. A rigorous analysis of the fields inside a waveguide, however, shows that the energy within the guide is propagated as a wave by refraction in the walls rather than as an electrical signal, and that a waveguide exhibits a low frequency cutoff below which the transmission through the waveguide will be nil. The waveguide cutoff phenomenon prevents the use of waveguides of convenient physical dimensions to be used at frequencies below about 1000 MHz.

The dimensions of the waveguide thus turns out to be $\lambda/2 \times s \times l$, where $\lambda/2 \times s$ is the cross-sectional dimension and l is the length of the waveguide.

A horn may be considered similarly to be evolved from a tapered transmission line. However, a rigorous analysis shows that, here, the taper may be introduced in both the E and H planes as shown in Figure 2.11 (*a*) to obtain a nearly circular cross-section of the main antenna beam. This is called a pyramidal horn. The horn may also be shaped conically as shown in Figure 2.11 (*d*), and it is called a conical horn. A conical horn is fed from a waveguide of a circular cross-section. A conical horn may be excited in the same way as a pyramidal horn to generate a plane polarized wave. It is, however, specially suited to generating a circularly polarized wave when the plane polarized wave in a rectangular waveguide connected to the source is first converted to a circularly polarized wave before feeding it into the horn. The conversion of plane to circular polarization is achieved by a circular polarizer, which splits a plane polarized wave into two equal-amplitude orthogonal components and introduces a 90° phase shift between them to produce circularly polarized radio waves.

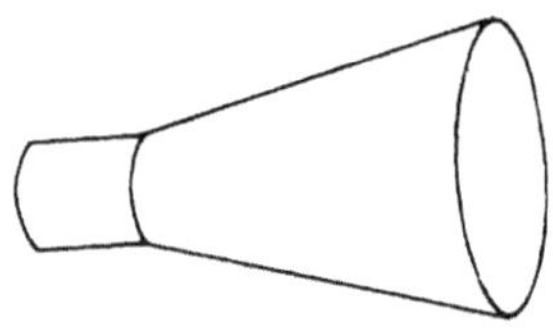

FIGURE 2.11 (*d*) Conical horn.

2.12 ANTENNA AS AN APERTURE

Electric currents are always the sources of electromagnetic fields. The current distribution is, however, not always known and is often a complicated function of position, such as that for radiators in the form of slits, slots, horns, reflectors, and lenses. For such cases, the theoretical estimate of the antenna parameters are more conveniently made by an aperture theory. It is known from the Huygens principle that each point on a wavefront acts as a new source of spherical waves. Subsequently, around 1900 A. E. H. Love and H. M. McDonald extended the Huygens principle for application in electromagnetic theory. Their idea was developed further in the mid-thirties by S. A. Schelkunoff, who called it the equivalence theorem. This theory shows how to replace an actual source by a set of equivalent sources spread over a specified closed surface. In field calculation these equivalent sources are treated in the same way as any primary source distribution.

Accordingly, the field intensity $E(\phi)$ due to the radiation from an antenna treated as an aperture is a function of the amplitude and phase of the current distribution of such equivalent sources of current across the aperture. By adding the contributions vectorially from these current elements constituting the aperture gives the field intensity in terms of an integral.

The region in the neighborhood of the aperture extending several antenna diameters is the near field. The near field is followed by the Fresnel region, where rays from the radiating aperture to the target are not parallel. The farthest region from the aperture is the far field or the Frounhofer region, where the radiating source and the target are at a sufficiently large distance from each other. Radar antennas are operated in this region.

A plot of the electric field intensity $| E(\theta, \phi) |$ is known as the field intensity pattern of the antenna. When the square of the field intensity $| E(\theta, \phi) |^2$ is plotted, one may get the power radiation pattern $P(\theta, \phi)$. The internal for electric field intensity in the Fraunhofer region is expressed by a Fourier transform. Let in the rectangular aperture of Figure 2.12, the width in the z dimension is a and the angle in the yz plane when measured from the y-$axis$ is ϕ. Assuming $a >> \lambda$, the far-field electric field intensity is

$$E(\phi) = \int_{-a/2}^{a/2} A(z) \exp\left[j\, 2\pi \frac{z}{\lambda} \sin\theta \right] dz \tag{2.27}$$

where $A(z)$ is the current at distance z. $A(z)$, the aperture distribution or the illumination, can be written as a complex quantity including both amplitude and phase. We may thus write,

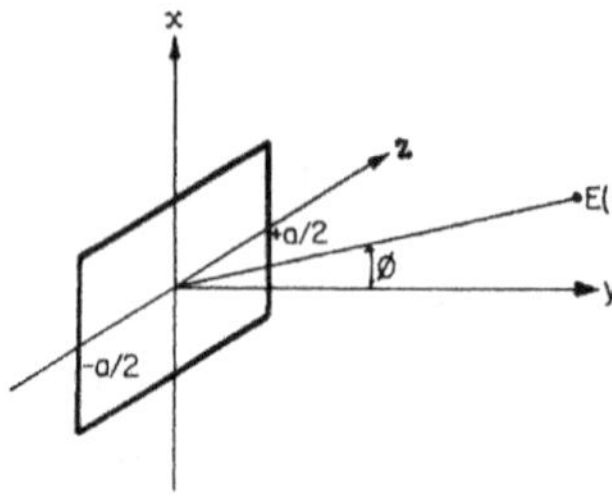

FIGURE 2.12 Rectangular aperture.

$$A(z) = |A(z)| \exp j\,\psi(z) \tag{2.28}$$

where $|A(z)|$ is the amplitude and $\psi(z)$ is the phase distribution.

Equation (2.27) is similar to the inverse Fourier transform. Hence the theory of Fourier transforms can be applied to the calculation of the field intensity patterns if the aperture distribution is known. The Fourier transform of a function $f(t)$ can be defined as

$$F(t) = \int_{-\infty}^{\infty} f(t) \exp(-j\,2\,\pi f t)\,dt$$

while the inverse Fourier transform is

$$f(t) = \int_{-\infty}^{\infty} F(f) \exp(j\,2\pi\,ft)\,df \tag{2.29}$$

Since the aperture distribution is zero beyond $z = \pm\, a/2$, hence the limits of equation (2.27) can be extended over the infinite interval from $-\infty$ to $+\infty$.

The Fourier transform allows the aperture distribution $A(z)$ to be found for a given field intensity pattern, since

$$A(z) = \frac{1}{\pi} \int_{-\infty}^{\infty} E(\phi) \exp\left[-j\,2\pi \frac{z}{\pi} \sin\phi\right] d\,(\sin\phi) \tag{2.30}$$

2.13 ONE-DIMENSIONAL APERTURE DISTRIBUTION

If it is assumed that the aperture extends in one dimension only, then this will represent the distribution across a line source or the distribution in a plane of a rectangular aperture. If the phase distribution across the aperture is constant, then the antenna pattern is

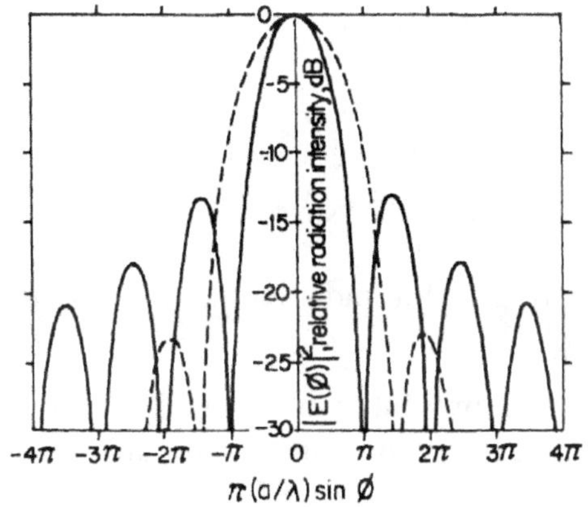

FIGURE 2.13 Radiation pattern due to a uniformly illuminated rectangular aperture (solid curve); and that due to a co-sinusoidal distribution of the illumination (dashed curve).

$$E(\phi) = A_0 \int_{-a/2}^{a/2} \exp j\left[2\pi \frac{z}{\pi} \sin\theta\right] dz$$

$$[A_0 = \text{constant value of the aperture distribution}]$$

$$= A_0 \frac{\sin[\pi(a/\lambda)\sin\phi]}{(\pi/\lambda)\sin\phi} = A_0 a \frac{\sin[\pi(a/\lambda)\sin\phi]}{\pi(a/\lambda)\sin\phi}$$

If we normalize to make $E(0)=1$, we get

$$A_0 = 1/a$$

$$\therefore \quad E(\phi) = \frac{\sin[\pi(a/\lambda)\sin\phi]}{\pi(a/\lambda)\sin\phi} \tag{2.31}$$

It is seen from equation (2.31) that the rhs is of the form $(\sin x)/x$. This is shown by the solid curve of Figure 2.13. The figure shows that the intensity of the first side lobe is 13.2 dB below the peak value. The voltage pattern of the previous equation is positive over the entire main lobe but changes its sign. The radiation pattern for the cosine aperture distribution is given by

$$A(z) = \cos\frac{\pi z}{a}(z) < \frac{a}{2} \tag{2.32}$$

The normalized radiation pattern can be written as

$$E(\phi) = \frac{\pi}{4}\left[\frac{\sin(\psi + \pi/2)}{(\psi + \pi/2)} + \frac{\sin(\psi - \pi/2)}{(\psi - \pi/2)}\right] \tag{2.33}$$

where $\psi = \pi\,(a\lambda)\sin\phi$

Some properties of the radiation pattern due to aperture distributions are listed in Table 2.1.

The information presented in Table 2.1 reveals that the gain of the uniform distribution is greater than the gain for any other distribution.

TABLE 2.1 Characteristics of Radiation Pattern Due to Various Aperture Distributions.

Type of distribution $\lvert z \rvert < 1$	Relative gain	Half-power beam width (deg.)	Intensity of first side lobe (dB below maximum intensity)
Uniform; $A(z) = 1$ Cosine; $A(z) = \cos^n(\pi z/2)$	1	$51\ \lambda/a$	13.2
$n = 0$	1	$51\ \lambda/a$	13.2
$n = 1$	0.810	$69\ \lambda/a$	23
$n = 2$	0.607	$83\ \lambda/a$	32
$n = 3$	0.575	$95\ \lambda/a$	40
$n = 4$	0.515	$111\ \lambda/a$	48
Parabolic:			
$A(z) = 1 - (1 - \Delta)z^2$			
$= 1.0$	1	$51\ \lambda/a$	13.2
$= 0.8$	0.994	$53\ \lambda/a$	15.8
$= 0.5$	0.970	$56\ \lambda/a$	17.1
$= 0$	0.833	$66\ \lambda/a$	20.6
Triangular; $A(z) = 1 - \lvert z \rvert$	0.75	$73\ \lambda/a$	26.4
Circular; $A(z) = \sqrt{1 - z^2}$	0.865	$58.5\ \lambda/a$	17.6
Cosine squared plus pedestal:			
$0.33 + 0.66\cos^2(\pi z/2)$	0.88	$63\ \lambda/a$	25.7
$0.08 + 0.92\cos^2(\pi z/2)$	0.74	$76.5\ \lambda/a$	42.8

λ = Wavelength $\quad$ a = aperture width

2.14 CIRCULAR APERTURE

Let us consider here the antenna pattern due to two-dimensional distribution across a circular aperture. The polar coordinates (r, θ) can be used for describing the aperture distribution $A(r, \theta)$, where r represents the radial distance from the center of the circular aperture, while θ is the angle measured in the plane of the aperture. The field intensity at a distance R is written as

$$E(r) = \int_0^{2\pi} d\theta \int_0^{r_0} A(r,\theta)\, \exp\left[j\frac{2\pi R}{\lambda} \right] r dr \tag{2.34}$$

Again, for a circular aperture with uniform distribution, the field intensity is

$$E(\phi) = \int_0^{2\pi} d\theta \int_0^{r_0} A(r,\theta) \exp\left[j\frac{2\pi r}{\lambda} \sin\phi \cos\theta \right] r dr$$

$$= \pi\, r_0^{\,2}\, 2J_1(\xi)/r \tag{2.35}$$

where $\quad \xi = 2\pi\,(r_0/\lambda)\sin\phi$

and $\quad J_1(\xi) =$ first-order Bessel function.

A plot of the normalized radiation pattern is shown in Figure 2.14 (*a*). It is seen from the figure that the first side lobe is 17.5 dB below the main lobe and the beam width is 58.5/*D*.

The radiation pattern for a typical paraboloid reflector antenna illustrating the main beam and side lobe radiation is shown in Figure 2.14 (*b*).

Aperture Blocking. If an obstacle is in front of an antenna, it can alter the aperture illumination and the radiated pattern. This is what is meant by blocking or shadowing. An example is the blocking caused by the feed and its supports in reflector type antennas. Aperture blocking degrades the performance of the antenna by raising the side lobes, lowering the gain, and filling in the mulls.

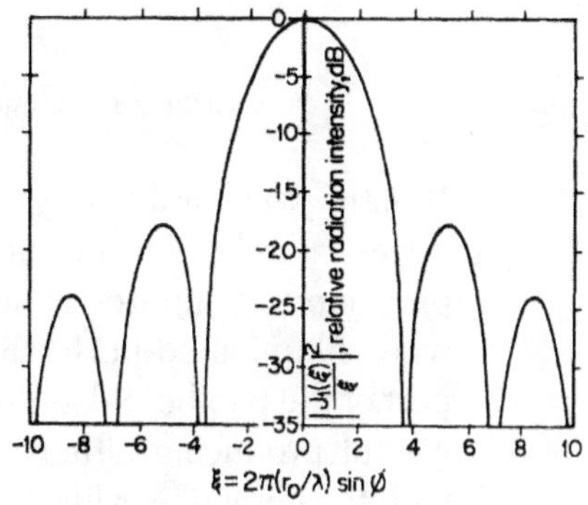

FIGURE 2.14 (*a*) Radiation pattern due to a circular aperture illuminated uniformly.

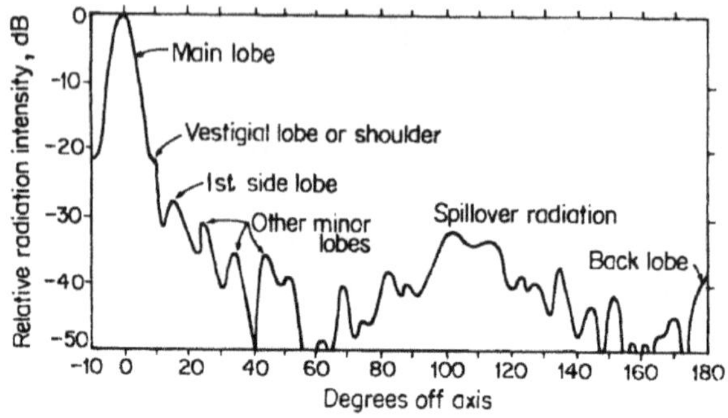

FIGURE 2.14 (*b*) The main beam and side lobe radiation patterns of a practical paraboloid reflector antenna.

2.15 PARABOLIC REFLECTOR ANTENNA

It is a widely used microwave antenna. The parabolic reflector antenna is shown in Figure 2.15. The parabola is illuminated with a source of energy called a feed. The feed is placed at the focus of the parabola, and it is directed toward the surface of the reflector.

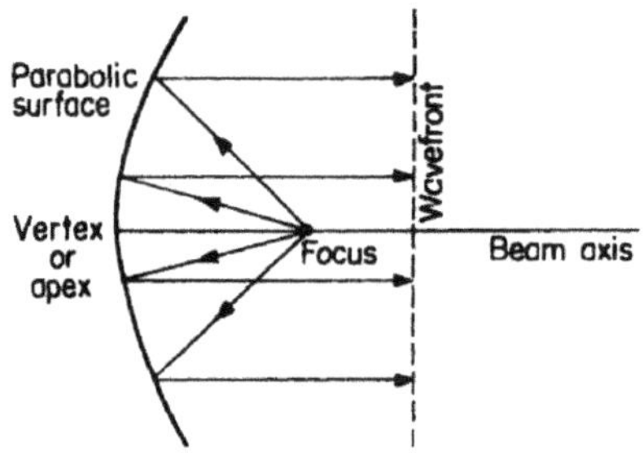

FIGURE 2.15 Geometry of a ray path in a paraboloid reflector antenna.

Rotating the parabolic curve shown in the previous figure about its axis produces a parabola of revolution termed a circular parabola or paraboloid. The paraboloid generates a nearly symmetrical pencil-beam antenna pattern when illuminated perfectly by a point source at the focus. By using only a part of the paraboloid, one can obtain an asymmetrical beam shape. Another way of producing either symmetrical or asymmetrical antenna patterns is with the parabolic cylinder, which can be generated by moving the parabolic content parallel to itself.

There are variations of parabolic reflectors such as pillboxes, cheeses, hog horns, and so forth.

The parabola is suitable for microwave antennas because of the following reasons:

1. any ray from the focus point can be reflected in a direction parallel to the axis of the parabola, and
2. the distance traversed by a ray from the focus to the parabola and by reflection to a plane perpendicular to the axis of the parabola is independent of its path.

2.16 LENS ANTENNAS

There are three types of microwave lenses applicable to radar. These are:

1. dielectric lenses,
2. metal-plate lenses, and
3. lenses with a non-uniform index of refraction.

Dielectric Lens. The homogeneous dielectric lens antenna is shown in Figure 2.16. It is similar to the conventional optical lens, At the focus of the lens, a point produces a plane wave on the opposite side of the lens. The index of refraction of a dielectric lens can be defined as

$$\frac{\text{Speed of light in free space}}{\text{Speed of light in the dielectric medium}}$$

Its value is equal to the square root of the dielectric constant. Materials like polyethylene, teflon, and so forth are suitable for small microwave lenses. As the velocity of propagation in the air is greater than that in the dielectric medium, a converging lens is thicker at its middle than at the outer edges.

A limitation of the lens is its thick size and heavy weight, both of which can be reduced by stepping or zoning the lens as shown in Figure 2.17. Due to zoning a 360° change of phase at the aperture has no effect on the aperture phase distribution.

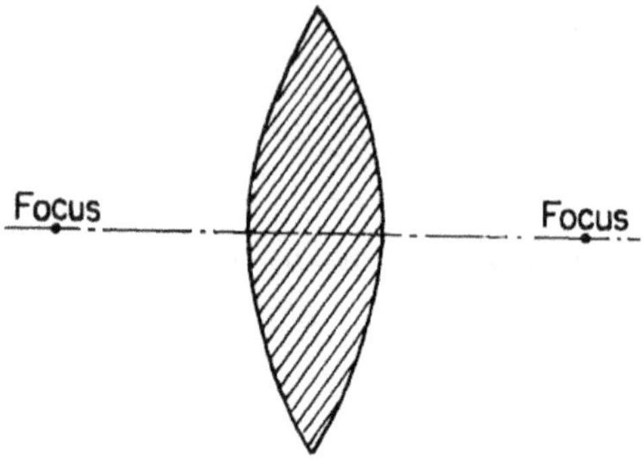

FIGURE 2.16 Converging dielectric lens antenna.

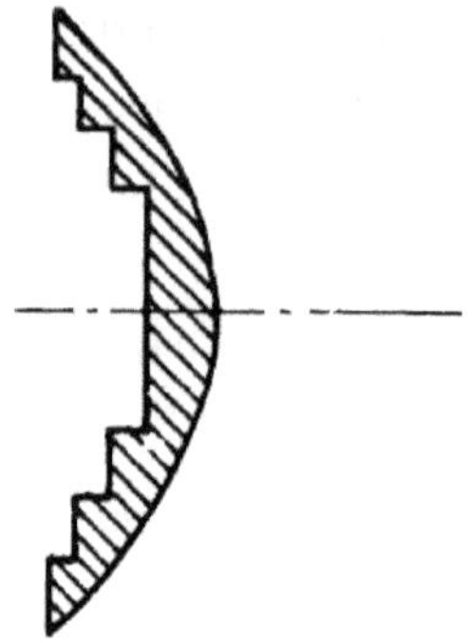

FIGURE 2.17 Zoned dielectric lens antenna.

Lens antennas can alternately be constructed by artificial dielectrics instead of using ordinary dielectric materials. The artificial dielectric consists of dielectric particles of macroscopic size or discrete metallic particles. The particles may be strips, disks, rods, or spheres imbedded in a material of low dielectric constant like polystyrene foam. If the particles are metallic spheres of radius a and spacing s between centers, the dielectric constant becomes

$$\varepsilon = 1 + \frac{4\pi a^3}{s^3} \tag{2.36}$$

Lenses constructed by using artificial dielectrics are usually less weight than solid dielectrics, and so the artificial dielectrics are preferred for a larger antenna.

Metal-Plate Lens. An artificial dielectric is constructed with parallel plate waveguides. This is shown in Figure 2.18. In a parallel-plate waveguide, the phase velocity is greater than that in the free space. So the index of refraction becomes less than unity.

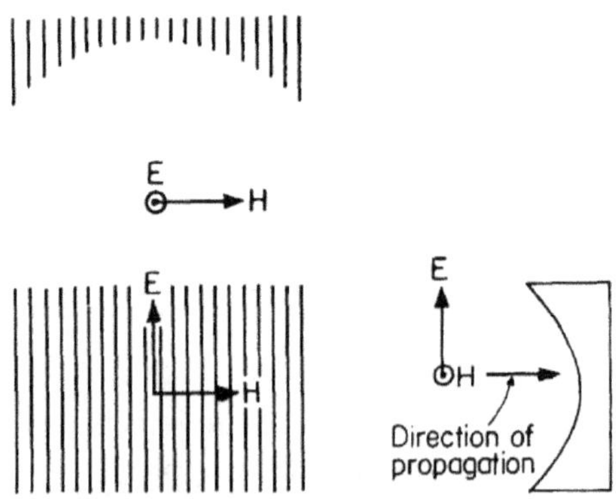

FIGURE 2.18 Parallel plate E-plane metal lens; plan, elevation, and side views.

The metal-plate lens shown in the figure in an E-plane less than the electric field vector is parallel to the plates. Generally the surface contour of a metal plate lens is not parabolic as in the case of the reflector. The spacing s between the plates must lie between $\lambda/2$ and λ. The index of refraction for the metal plate lens can be written as

$$\eta = \left[1 - \left(\frac{\lambda}{2s}\right)^2\right]^{1/2} \tag{2.37}$$

where λ = wavelength in air.

The previous equation shows that η is always less than unity. At the upper limit, when $s = \eta$, the index of refraction becomes 0.866. The value of η will be less when the spacing is closer and then the lens will be thinner. Equation (2.37) shows that for a value of $s = \lambda/2$, the index of refraction is zero and so the waveguide is beyond the cutoff. For practical purposes a compromise value of between 0.5 and 0.6 is frequently selected, corresponding to plate spacings of 0.557 λ and 0.625 λ respectively.

Luneburg Lens. This is a type of lens where the index of refraction can be varied within the lens. When microwave frequencies are taken into consideration, it is possible to control η (since $\eta = \sqrt{\varepsilon}$), and such lenses with a non-uniform are practical.

In radar the most important variable η lenses is that due to Luneburg which is spherically symmetrical. It has the property that a plane wave incident on the sphere is focused to a point at the diametrically opposite side on the surface. In a similar way, a transmitting source on the surface of the sphere can be converted to a plane wave on passing through the lens. This is shown in Figure 2.19. This type of lens can also generate a number of fixed beams.

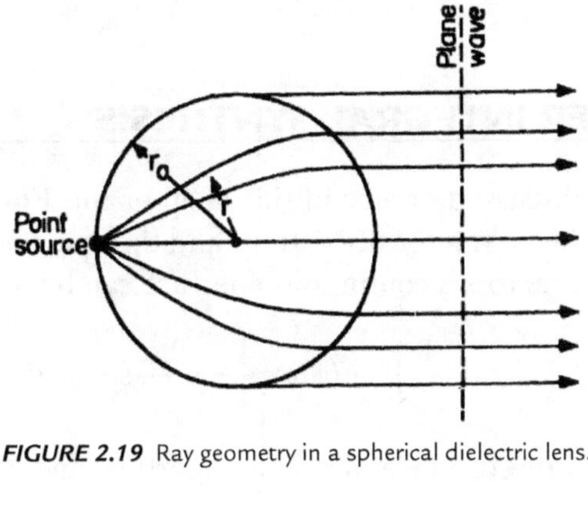

FIGURE 2.19 Ray geometry in a spherical dielectric lens.

The index of refraction changes with the radial distance in a Luneburg lens of radius r_0 by the equation

$$\eta = \left[\left(2 - \frac{r}{r_0}\right)^2\right]^{1/2} \tag{2.38}$$

The index of refraction becomes maximum at the center and decreases on the periphery. Three-dimensional practical Luneburg lenses are constructed by using a large number of spherical shells, each of a constant index of refraction.

2.17 PATTERN SYNTHESIS

In antenna design the term pattern synthesis means a proper distribution of current across a finite-width aperture. The methods of pattern synthesis are classified into two categories, depending on whether the aperture is continuous or an array.

The following discussion is applicable either to linear one-dimensional apertures or to rectangular apertures where the distribution is separable, that is

$$A\,(x, z) = A(x)\,A(z) \tag{2.39}$$

The synthesis techniques when applied to array antennas assume a uniform spaced isotropic element, which is generally taken to be a half-wave length. But if the element is not isotropic but has a pattern $E_c(\theta)$ and the desired overall pattern is $E_d(\theta)$, then the pattern to be found by synthesis is $E_d(\theta)/E_c(\theta)$.

2.18 FOURIER INTEGRAL SYNTHESIS

We have discussed earlier in this chapter the Fourier integral relationship between the field intensity pattern and the aperture distribution. The distribution $A(z)$ across a continuous aperture can be written as

$$A(z) = \frac{1}{\lambda}\int_{-\infty}^{\infty} E(\phi)\exp\left[-j\frac{2\pi z}{\lambda}\sin\phi\right] d(\sin\phi) \tag{2.40}$$

where z is the distance along the aperture and $E(\phi)$ is the field intensity pattern.

If the aperture distribution is considered over the finite aperture dimension d, then the resulting antenna pattern would be

$$E_a(\phi) = \int_{-d/2}^{d/2} A(z) \exp\left[j\, 2\pi \frac{z}{\lambda} \sin\phi \right] dz \tag{2.41}$$

Substituting equation (2.40) in equation (2.41) and changing the variable of integration from ϕ to ξ for avoiding confusion, the antenna pattern becomes

$$E_a(\phi) = \frac{1}{\lambda} \int_{-d/2}^{d/2} \int_{-\infty}^{\infty} E(\xi) \exp\left[j \frac{2\pi z}{\lambda} (\sin\phi - \sin\xi) \right] d(\sin\xi)\, dz \tag{2.42}$$

Interchanging the order of integration, the antenna pattern becomes

$$E_a(\phi) = \frac{d}{\lambda} \int_{-\infty}^{\infty} E(\xi) \frac{\sin\left[\pi\left(\frac{d}{\lambda}\right)(\sin\phi - \sin\xi) \right]}{\pi\left(\frac{d}{\lambda}\right)(\sin\phi - \sin\xi)} d(\sin\xi) \tag{2.43}$$

where $E_a(\phi)$ represents the Fourier integral pattern approximating the desired pattern $E(\phi)$ when $A(z)$ is restricted to a finite aperture of dimension d.

It has been established by Ruze that the approximation to the antenna pattern as derived on the basis of the Fourier integral has the property that the mean square deviation between the desired and the approximate pattern is minimum. The Fourier series is used to synthesize the pattern of a discrete array. However, the method is restricted to arrays with element spacing in the vicinity of a half wavelength.

Woodward-Levinson Method. The antenna-synthesis technique which was based on sampled values was introduced by Levinson and was later developed by Woodward.

Figure 2.20 (*a*) reveals the pattern $E(\phi)$ by a series of values spaced λ/d radians apart. The sampled values $E_s(n\lambda/d)$, where n is an integer which determines the antenna pattern, are shown in Figure 2.20 (*b*). The antenna pattern $E_a(\lambda)$ can be constructed from the sample values with a pattern of the form $(\sin\psi)/\psi$, where $\psi = \pi\,(d/\lambda)\sin\phi$. The function $(\sin\psi)/\psi$ is called the composing function. The antenna pattern is now given by

$$E_a(\phi) = \sum_{n=-\infty}^{\infty} E_s\left(\frac{n\lambda}{d}\right) \frac{\sin\left[\pi\left(\frac{d}{\lambda}\right)\left(\sin\phi - \frac{n\lambda}{d}\right)\right]}{\pi\left(\frac{d}{\lambda}\right)\left(\sin\phi - \frac{n\lambda}{d}\right)} \tag{2.44}$$

That means the antenna pattern is reconstructed from a sum of (sin ψ) ψ composing functions spaced λ/d radians apart, as shown in Figure 2.20 (*c*).

By substituting the antenna pattern of equation (2.44) into equation (2.40), one may add in the aperture distribution which becomes,

$$A_s(z) = \frac{1}{d} \sum_{n=-\infty}^{\infty} E_s\left(\frac{n\lambda}{d}\right) \exp\left[-j\frac{2\pi nz}{d}\right] \tag{2.45}$$

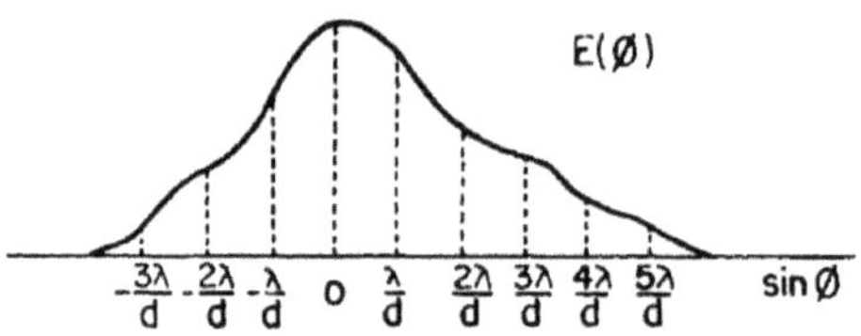

FIGURE 2.20 (*a*) Radiation pattern with sampled values spaced by λ/d radians, where d = aperture.

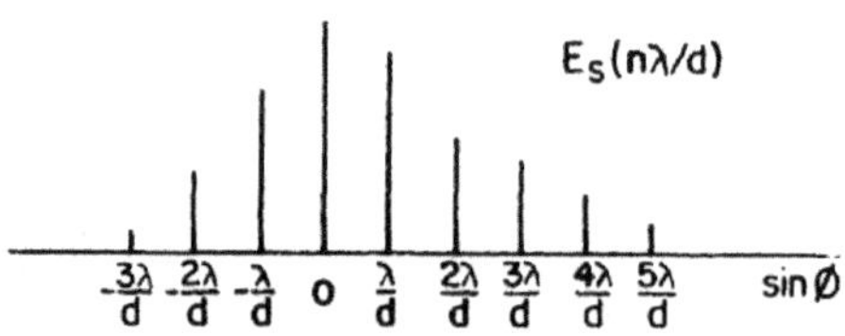

FIGURE 2.20 (*b*) Sampled values of an electric field which specify the radiation pattern of Figure 2.20 (*a*).

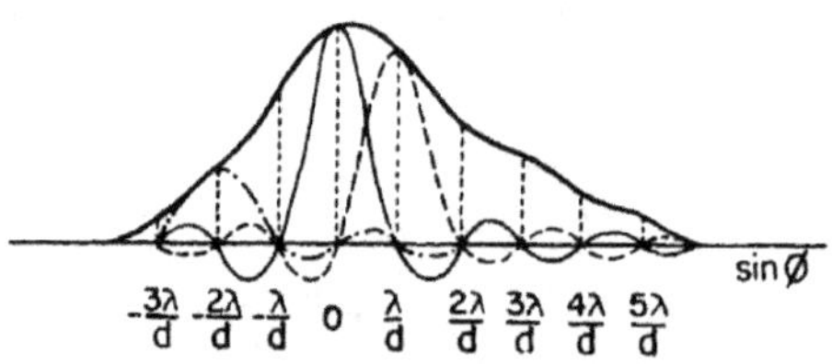

FIGURE 2.20 (*c*) Reconstructed pattern using the (sin ψ/ψ) composing function to approximate the radiation pattern of Figure 2.20 (*a*).

Thus, the aperture distribution has uniform amplitude and is proportional to the sampled values. The difference between Fourier Integral Synthesis and the W-L method is that the former gives a radiation pattern while the latter gives antenna patterns.

2.19 ERRORS IN RADIATION PATTERNS

Errors in the aperture illumination are either systematic or random. Examples of systematic errors include:

1. aperture blocking in reflector antennas owing to the feed and its supports;
2. diffraction at the steps in zoned-lens antenna;
3. mutual coupling between the elements of an array; and
4. periodicities in the construction of an antenna.

Similarly, examples of random errors include:

1. RF measurement errors incurred to adjust an array;
2. random distortion of the antenna surface;
3. wall-spacing errors in metal plate lenses;
4. errors in the manufacture of the antenna.

Although the random errors are relatively small, their effects on the side lobe radiation are considerably large. In any particular design the systematic errors are usually the same from antenna to antenna.

Systematic Errors. The effect of systematic errors on the antenna radiation pattern is found by modifying the aperture distribution for taking into account the known errors. For example, a linear phase error across the antenna aperture causes the beam position to tilt in angle.

Random Errors in Reflectors. The effect of random errors in the antenna radiation pattern has been shown by Ruze. He showed that the gain of a circular aperture with an arbitrary phase error is given by

$$G = G_0(1 - \delta^{-2}) \tag{2.46}$$

where G_0 = gain of the antenna in the absence of errors

δ = phase errors in radians.

Using an antenna model where the reflector is distorted by a large number of random Gaussian-shaped bumps, Ruze showed that the radiar pattern can be written as

$$G(\theta,\phi) = G_0(\theta,\phi)^{e-\bar{\delta}^2} + (2\pi c/\lambda)^2 e^{-\bar{\delta}^2} \sum_{n=1}^{\infty} \frac{[\bar{\delta}^2]^n}{n!n} e^{-(\pi cu/\lambda)^{2/n}} \tag{2.47}$$

where $G_0(\theta, \phi)$ = no error radiation pattern

$= \rho_0 (\pi D/\lambda)^2$ (Its axial value)

D = antenna diameter = aperture efficiency

C = correlation interval of the error

$u = \sin\theta$.

In equation (2.47) the mean square phase error $\bar{\delta}^2$ is assumed to be Gaussian. The first term in the equation represents no error radiation pattern while the second term represents the disturbing pattern. For small phase errors when the first term of the series ($n = 1$) is considered, equation (2.47) becomes

$$G(\theta,\phi) = G_0(\theta,\phi)e^{-\bar{\delta}^2} + \left(\frac{2\pi C}{\lambda}\right)\bar{\delta}^2\, e^{-(\pi cu/\lambda)^2} \tag{2.48}$$

From equation (2.47), the reduction in gain on the axis can be expressed as

$$\frac{G}{G_0} = e^{-\bar{\delta}^2} + \frac{1}{\rho_a}\left(\frac{2D}{D}\right)^2 e^{-\bar{\delta}^2} \sum_{n=1}^{\infty} \frac{[\bar{\delta}^2]^n}{n!n} \tag{2.49}$$

and when the second term of equation (2.49) is neglected, the gain becomes

$$G = G_0 e^{-\bar{\delta}^2} + \rho_a \left(\frac{\pi D}{\lambda}\right)^2 e^{-(4\pi\varepsilon/\lambda)^2} \tag{2.50}$$

where ε = effective reflector tolerance measured in the same units as λ.

For a particular reflector size D, the gain increases as the square of the frequency until the exponential term becomes significant. Differentiation of equation (2.50) reveals that the maximum gain corresponds to a wavelength

$$\lambda_m = 4\pi\varepsilon \tag{2.51}$$

The gain at this wavelength will be 4.3 dB below what it would be in the absence of errors. Then the maximum gain is given by

$$G_{max} = \frac{\rho_a}{43}\left(\frac{D}{\varepsilon}\right)^2 \tag{2.52}$$

Thus, the antenna gain is limited by the mechanical tolerance to which the surface can be constructed.

The construction tolerance of an antenna is in practice described by the peak error rather than the rms error. By careful experiment it has been found that the ratio of the peak to rms error is about 3:1.

2.20 COSECANT-SQUARED ANTENNA PATTERN

The design of a cosecant-squared antenna pattern is a good application of synthesis techniques. The cosecant-squared pattern is approximated with a reflector antenna by using more than one feed or by shaping the surface. In Figure 2.21 a common method of producing the cosecant-squared pattern is shown. In the figure, the upper half of the reflector is a parabola. It reflects

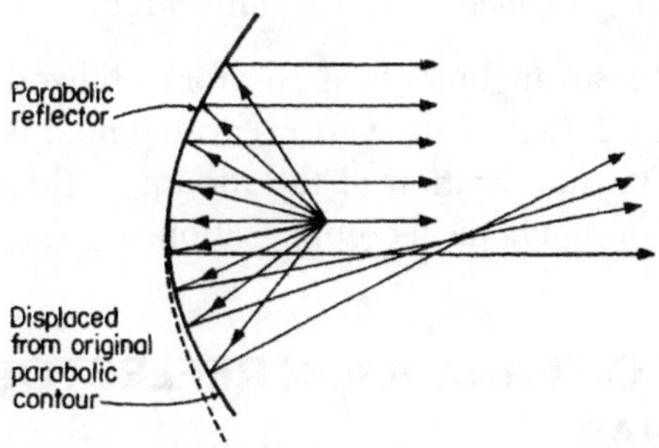

FIGURE 2.21 Generation of cosecant-squared antenna by a displacement of the reflector surface from its parabolic form.

energy from the feed in a direction parallel to the axis. However, the lower half is distorted from the parabolic contour so as to direct a part of energy in the upward direction.

2.21 STABILIZATION OF ANTENNAS

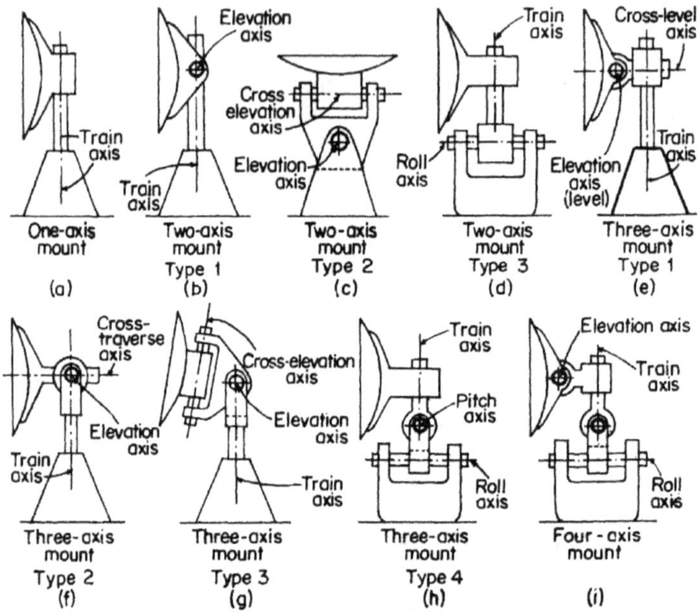

FIGURE 2.22 Various types of stabilized antenna mounts.

Stabilization is the use of a servomechanism for controlling the angular position of an antenna. The requirements for stabilization depend partly on the nature of the application. It differs whether the radar is meant for tracking, height finding, surface search, or air search.

Nine possible arrangements of axes for stabilized antenna mounts are shown in Figure 2.22. Out of the various arrangements, the one-axis mount is the simplest. The stabilization of the mount in the figure is assumed to be by some form of mechanical compensation.

2.22 DESIGN OF PARABOLIC REFLECTOR RADAR ANTENNAS

As most widely used radar antennas are basically parabolic reflector antennas, which have a high efficiency, low noise temperature, and a wide bandwidth dependent mainly on the bandwidth of the antenna feed, a more detailed information about the design of parabolic antennas is included in this section.

Typical structures of parabolic reflector antennas for radar are shown in Figure 2.23.

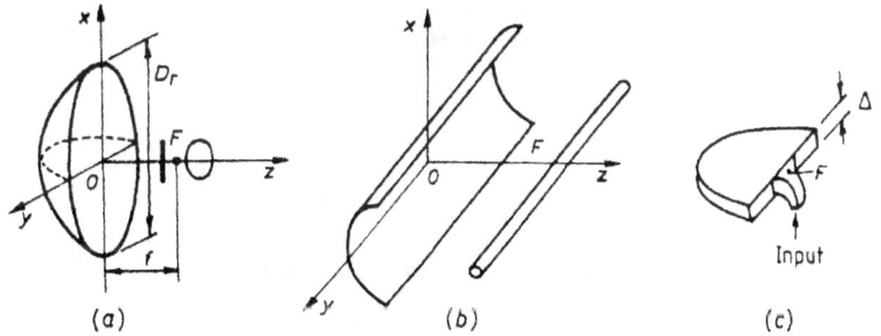

FIGURE 2.23 Parabolic reflector antennas for radar: (*a*) paraboloid of revolution ('dish'), (*b*) parabolic cylinder, (*c*) pillbox or cheese.

The reflector in Figure 2.23 (*a*) is a paraboloid of revolution and it is the most widely used type of reflector antenna for radar, satellite communication, and microwave LOS links. Figure 2.23 (*b*) shows a radar antenna with a parabolic cylindrical reflector to produce a fan beam with its plane parallel to the axis of the cylinder, which is horizontal. Such an antenna is used in radars which require only angular resolution in elevation. This antenna may also be oriented with its axis vertical to have a radar antenna with a high azimuth resolution. Alternatively, one can use the antenna shown in Figure 2.23 (*c*) to have a good azimuth resolution and a very wide beam in elevation, and such an antenna is referred to as a "pillbox" or "cheese" antenna, which is very useful for mechanical scanning of the radar in azimuth only for aircraft flying at a wide range of elevation angles, such as that used in airport landing areas.

The geometric properties of a parabolic reflector surface shown in Figure 2.23 (*a*) convert the rays reflected, emerging from the feed at the focal point into a beam of rays parallel to its axis. The path length from the focus to the parabolic reflector back to the circular aperture line passing through the edges of the paraboloid is constant for any angle θ with respect to the axis of the paraboloid as shown in Figure 2.24 (*a*). In this way a plane wavefront is formed over the aperture area from the spherical wavefront radiated from the feed, within the angle $2\theta_{max}$ shown in the figure in the cross-section, along the axis of the paraboloid.

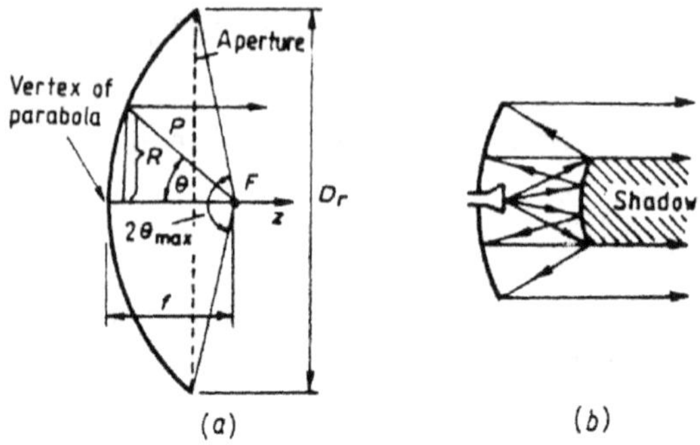

FIGURE 2.24 Contour of reflector antennas: (*a*) parabolic reflector antenna, (*b*) Cassegrain antenna.

In rectangular coordinates the equation of the paraboloid of revolution is

$$x^2 + y^2 = 4fz$$

while in spherical coordinates, with the origin of the coordinates located at the focus, the equation of the paraboloid of revolution is

$$\rho(\theta) = 2f/(1 + \cos\theta).$$

There are, in fact, two types of parabolic reflector antennas: (*a*) a long-focal-length type in which the focal point is located outside the antenna, $f > \frac{D_r}{4}$ and the aperture angle $2\theta_{max} < \pi$, and (*b*) a short-focal-length type, in which the focal point lies between the reflector and its aperture plane, such that $f < D_r/4$ and $2\theta_{max} > \pi$.

The long-focal-length type parabolic reflector antenna is more widely employed, as it is more efficient in utilizing the surface area of the reflector, while the short-focal-length type is preferred where the reduction of noise and interference picked up from the backside of the antenna, used as a receiving antenna, is essential. The pickup from the backside may also be reduced by using a hyperbolic subreflector within the focal point and a horn brought near the surface of the reflector at its center, facing the subreflector. Such an antenna is called a Cassegrain antenna, which is widely used in satellite communication where the noise pickup from the backside is a crucial problem that has to be reduced. Figure 2.23 (*b*) shows the structure of a Cassegrain antenna. For long-range radars where the return signal from the maximum range may be marginal, a Cassegrain antenna will be useful in reducing the noise pickup from the backside during scanning of a long-distance overhead target.

2.23 RADIATION PATTERN OF A PARABOLIC REFLECTOR TYPE ANTENNA

The radiation pattern of a parabolic reflector type antenna is calculated by the method of equivalent surface electric and magnetic currents, with the radiation integral expressed in one of the following forms:

1. Integration carried out over the surface area of the flat aperture and over the rear or shadow side of the paraboloid.
2. Integration carried out over the curvilinear surface which extends in on the illuminated and shadow sides of the paraboloid.

The curvilinear surface current on the illuminated side of the parabolic reflector is shown in Figure 2.25 (*a*) for an electric dipole feed and in Figure 2.25 (*b*) for a magnetic dipole feed.

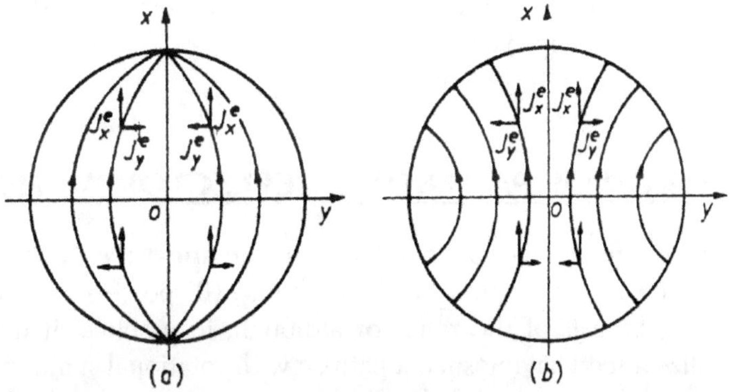

FIGURE 2.25 Surface current distribution in the aperture of paraboloidal reflectors: (*a*) with electric dipole feed and (*b*) with magnetic dipole feed.

It may be mentioned here that the electric surface current on the rear or shadow side of the parabolic reflector is relatively small compared to that on the side illuminated by the feed of the parabolic reflector antenna. In practice, the shadow side current distribution is, therefore, neglected.

For derivation of the radiation pattern from the current distribution over the plane aperture of the antenna, the equivalent electric and magnetic currents are assumed to be in phase with one another and their distribution is assumed to follow the ratio $F(\theta, \phi)/\rho(\theta)$, where $F(\theta, \phi)$is the field strength radiation pattern of the feed and $\rho(\theta)$ is the distance from the focus to a point on the surface of the paraboloid. This approach is called the **aperture field method**.

Alternatively, one can use the **induced current method** with the current distribution in the curvilinear surface of the reflector, as indicated in Figure 2.25 (*a*) and Figure 2.25 (*b*). The electric surface current is approximately equal to twice the tangential component of the magnetic field intensity due to waves emerging from the feed. However, in this induced current approach, the amplitude and phase distribution of the induced current are a more complex function of coordinates than that in the aperture field approach. In spite of this complexity, the induced current approach is useful where the derivation of the antenna pattern is required to be more precise. The complexity of the induced current distribution is taken care of by using a computer to derive the radiation pattern.

It may be mentioned here that neither of the two approaches for derivation of antenna patterns as previously indicated is precise, as the surface current distribution in each of the approaches is estimated approximately. Still, the aperture field method is less accurate because of the geometrical-optics approach involved in this case, while the pattern should have been derived from a wave approach.

2.24 DESIGN OF PARABOLIC REFLECTOR ANTENNAS

The design of a parabolic reflector by the **aperture field** method presumes that the radiation pattern of the feed, $F(\theta)$ possesses rotational symmetry about the axis of the reflector although, in practice, it may be difficult to realize a feed having such a pattern with rotational symmetry. Assuming the rotational symmetry of the feed pattern, the amplitude distribution in the aperture can be found by the equations

$$I(\theta) = F(\theta)/\rho(\theta) = F(\theta)\,(1+\cos\theta)/2f$$
$$\sin\,\theta = F/\rho(\theta) = R(1+\cos\theta)/2f$$

where R = radial coordinate in the aperture plane

θ = angle between the axis of the paraboloid and the line from the focus to the current point on the paraboloid reflector surface.

The amplitude distribution so estimated may be used to calculate the beamwidth of the antenna pattern and the side lobe level. The amplitude distribution also determines the aperture efficiency K_f. One can also determine the feed efficiency K_f, which is the fraction of the radiated power from the feed reaching the reflector, as given by

$$K_f = \int_0^{\theta_{max}} F^2(\theta)\sin\theta\, d\theta \Big/ \int_0^{\pi} F^2(\theta)\sin\theta\, d\theta < 1$$

K_f accounts for the spillover, or the loss of the feed power as it spills over the reflector edge. The efficiency parameters of the antenna for aperture and feed derived as K_{ae}, a and K_{ae}, f, respectively are dependent on the relative focal length of f/Dr, Dr being the aperture diameter, as shown in Figure 2.26 as dotted lines. Figure 2.26 also shows the resultant overall efficiency of the parabolic reflector antenna in the solid line, by combining both the aperture and feed efficiencies, indicating an optimal K_{ae} overall efficiency for a relative focal length $[f/D_r]_{opt}$.

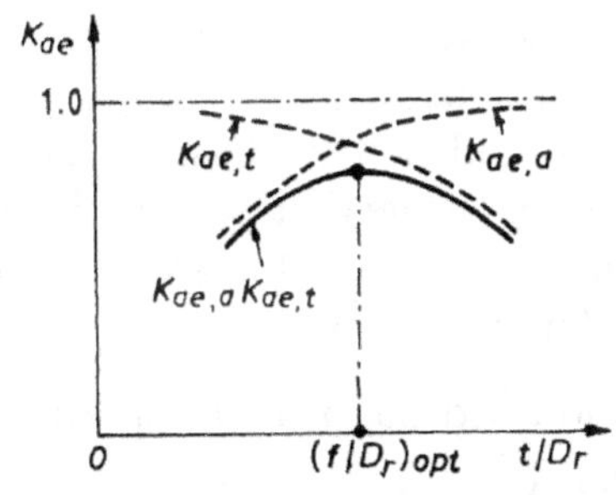

FIGURE 2.26 Area efficiency as a function of focal length.

The dotted curves for K_{ae}, a in the figure indicate how the aperture efficiency increases with the relative focal length of f/D_r, while the dotted curve for k_{ae}, f shows how the feed efficiency decreases with relative focal length f/D_r. At low values of the relative focal length, the feed efficiency tends to approach unity or 100%, but then the radiation pattern of the feed will be narrow compared to the aperture angle of the reflector, and therefore the reflector aperture is not fully utilized due to a very non-uniform amplitude distribution with the radio illumination strongly tapering down the edges. On the other hand, if the relative focal length is large, the amplitude distribution on the reflector is made more uniform, thereby increasing the aperture efficiency, but then the reflector intercepts only a part of the total power radiated by the feed while the rest, going outside the reflector, is wasted, increasing the scatter fraction of the antenna. Calculation, in fact, shows that if the feed is in the form of a half-wave dipole fitted with a subreflector as shown in Figure 2.23 (*a*), the optimal relative focal length is $(f/D_r) = 0.38$. The illumination level at the aperture edge with this optimal relative focal length is 0.33 or 10 dB down relative to the center of the aperture. The overall aperture efficiency with this optimal relative focal length is 0.83, and the resulting beamwidth of the parabolic reflector antenna is given by

$$\Delta\theta = 70\lambda/D_r$$

with the level of the first side lobe, which is largest, being about 22 to 24 dB down relative to the main antenna beam.

The approximate design method for the parabolic reflector antenna previously indicated is used for the sketch design of antennas with small size reflectors of diameter 2–3 m for microwaves. The small loss in aperture efficiency for such an approximate design may be compensated by an increase in aperture size. However, for the design of parabolic reflector antennas with larger reflector diameters, as the cost of the reflector increases rapidly with the diameter, a more accurate design procedure is required to truly obtain a maximal aperture efficiency. Thus, for large reflector antennas the entire reflector antenna and its feed must be optimized by more exact calculations, which would take into account the greatest possible number of relevant factors as indicated in the following:

1. The amplitude distribution in the aperture and aperture efficiency K_{ae}, a.
2. Feed efficiency, K_{ae}, f.
3. Aperture blockage area by the feed with its supporting structure.
4. Spillover of electric currents in the shadow side of the reflector producing side lobes on the backside of the reflector, as can be accounted for by diffraction factor K_{diff}, derived from diffraction theory.
5. Cross polarization, accounted for by a factor K_{cp}.
6. The errors, systematic and random, in the distribution of equivalent electric currents in the aperture.
7. Effect of radio-transparent protective enclosures of the antenna such as a 'radome,' which introduces additional phase errors in the current distribution in an equivalent aperture.
8. Surface tolerance of the reflector introducing phase errors.

As this design parameter is a critical one in fabricating the parabolic reflector and any subreflector, the next section includes more detailed design equations related to surface tolerance and its effect.

2.25 TOLERANCE OF REFLECTORS

The surface of a parabolic reflector should ideally have a parabolic contour indicated in the cross-sectional plane passing through the axis of the paraboloid as shown in Figure 2.27. However, in practice, the surface cannot be made perfectly smooth but will have a random deviation, Δ, from the parabolic contour, an example of which is also shown in the figure at an angle θ from the focal point, relative to the axis of the parabola. Such deviations, in fact, occur all over the surface of the paraboloid randomly in magnitude and position on the paraboloid. Thus, even if the average contour may be made parabolic, the true contour shows random deviations represented by Δ, varying in magnitude, sign, and position, with a mean deviation being zero. Then the maximum deviation from the mean design value can be shown not to exceed

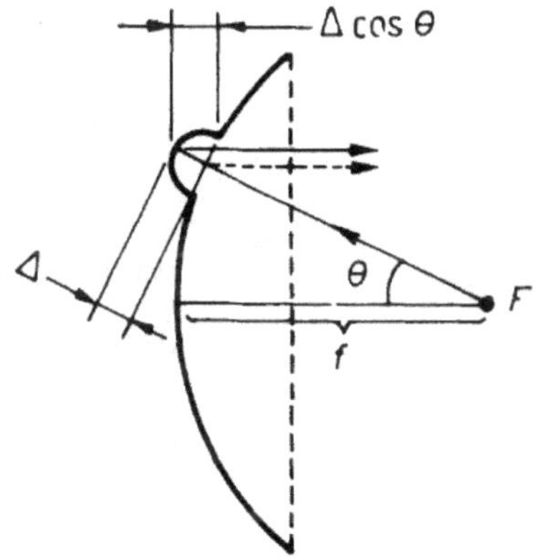

FIGURE 2.27 Surface contour of a parabolic reflector.

$$\Delta_{max} = 2.6\bar{\Delta}$$

with a probability of 99%. Δ_{max} is usually referred to as the tolerance on the surface contour of a parabolic reflector.

2.26 PHASE ERROR DUE TO TOLERANCE AND ITS EFFECT ON DIRECTIVITY

The random variation of the reflector contour from a parabolic shape, expressed as tolerance Δ, introduces phase error, which in turn degrades the directivity of the antenna beam pattern. The path length from the feed phase center to the reflector and then back from the reflector to the aperture varies in a random manner, producing an rms value of the random phase error in the aperture as given by

$$\phi \approx 2\beta\bar{\Delta} = 2\frac{2\pi}{\lambda}\bar{\Delta} = 4\pi\bar{\Delta}/\lambda$$

As the correlation radius for discontinuities on the surface of the reflector is usually several times the wavelength, the directivity D of the parabolic reflector antenna, considering the random phase errors, may be given by

$$\begin{aligned} D &= (4\pi S/\lambda^2) g \exp(-\overline{\phi}^2) \\ &= (\pi D_r/\lambda)^2 g \exp[-(4\pi\overline{\Delta}/\lambda)^2] \end{aligned} \tag{2.53}$$

where D_r = reflector diameter

g = resultant area efficiency factor subject to the effect of all other factors that impair the directivity of the antenna, except the random errors.

The $\exp(-\overline{\phi}^2)$ may be called Kg. The equation (1) for directivity indicates that the phase error degrades the directivity. In fact, the directivity of an ideal reflector producing no phase error will be given by

$$D = (\pi D_r/\lambda)^2 g \tag{2.54}$$

which shows that the directivity of an ideal parabolic reflector in the direction of maximum radiation increases inversely as the wavelength squared or directly as the frequency squared. In a realistic reflector with rms contour error $\overline{\Delta}$ and reflector diameter D, a decrease in wavelength will gradually slow down the rise in directivity until the directivity starts decreasing abruptly with further decrease of wavelength. The maximum directivity D_{max} may be derived from equation (2.53) to be given by

$$D_{max} = 0.023\, g\, (D_r/\overline{\Delta})^2 \tag{2.55}$$

occurring at a wavelength

$$\lambda_{\min} = 4\pi\,\overline{\Delta}$$

and at $$K_0 = -4.34 \text{ dB}.$$

At shorter wavelengths Kg decreases rapidly and decreases the directivity, indicating how random errors in the parabolic surface contour of the reflector limit the directivity of a parabolic reflector antenna and set the limit for the lowest wavelength or the highest frequency that may be used with a reflector with the specified rms contour error. Regarding the lowest frequency or the highest wavelength λ_{max} that can be used with the parabolic reflector antenna, the physical optics concepts in the calculation of the reflector radiation pattern by the induced current method indicate that λ_{max} will be given by

$$\lambda_{max} = D_r/3 \tag{2.56}$$

During the manufacture of parabolic reflector antennas by various methods, the statistics data indicate that the ratio of rms error in the surface contour of the parabolic reflector to the reflector diameter, λ/D_r, tends to be constant; the value of the constant for the most common methods of reflector manufacture is given by

$$\bar{\Delta}/D_r \approx 0.4\times10^{-3}$$

indicating that for a reflector of diameter $D_r = 1$ m, the tolerance on the surface contour may be $\Delta_{max} \approx 1$ mm.

To obtain higher precision of about an order of magnitude better with

$$\bar{\Delta}/D_r \leq 0.4\times10^{-4} \tag{2.57}$$

a finer reflector technology has to be invoked with "surface fitting."

For a very precise parabolic reflector antenna manufactured with a smaller size reflector like that needed at millimeter waves, fabrication by spinning a molten plastic material until it cools and solidifies may be used, with the plastic parabolic surface being subsequently metallized. Such a metallized parabolic plastic reflector of diameter 8.5 m may have an accuracy as high as that given by

$$\bar{\Delta}/D_r = 2.4\times10^{-8}$$

which is better in precision by about four orders of magnitude compared to that with the finer reflector technology indicated previously in equation (2.57).

The variations in directivity of parabolic reflector antennas with frequency of operation, for different values of rms surface contour error of the reflectors is shown in Figure 2.28, for g = 1.

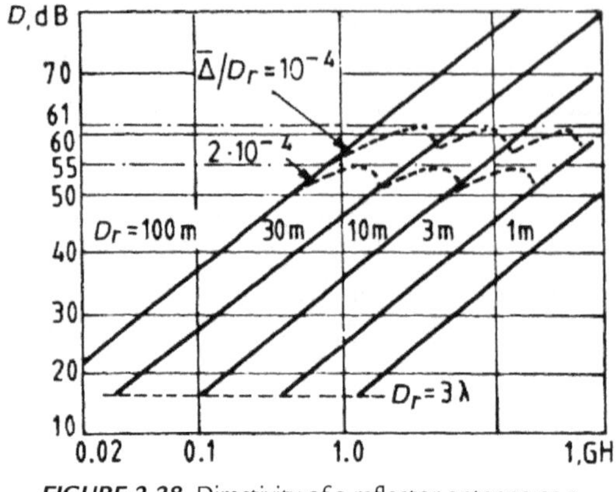

FIGURE 2.28 Directivity of a reflector antenna as a function of frequency.

It may be mentioned here that the reduction of directivity due to random phase error does not affect the shape of the main beam of the radiation pattern, which remains nearly undistorted, but increases the sidelobe radiation.

All the effects of random contour errors of a parabolic reflector on the radiation pattern and directivity of the antenna may also occur due to any random surface contour error of the hyperbolic subreflector of Cassegrain feed, and similar concepts may in fact be used to calculate the effects of surface contour errors of the hyperbolic subreflectors of cassegrain feeds, for which expressions may be somewhat different. Usually, however, the precision of subreflectors is made very high, which is also not difficult to achieve as the diameter of the hyperbolic subreflector is much smaller than the parabolic reflector. As a result, the surface contour error of the subreflector may be neglected in practice.

2.27 DESIGN OF THE FEEDS OF PARABOLIC REFLECTOR ANTENNAS

The performance of a parabolic reflector antenna is critically dependent on the performance of the feed in addition to the tolerance of the reflector. As the feed is at the focal point of the paraboloid, any minor degradation of the feed performance may affect the performance of the parabolic reflector antenna.

In fact, a feed should be designed satisfying the following crucial requirements:

1. The radiation pattern of the feed should produce the desired amplitude distribution over the aperture and produce minimal radiation beyond the aperture.
2. The phase center of the feed should be located at the focal point of the parabolic reflector.
3. The cross-sectional area of the feed perpendicular to the axis of the antenna should be small enough to ensure minimal aperture blocking.
4. The power handling capability of the feed should be high enough to avoid any risk of breakdown with the full power fed to the feed during transmission.
5. The bandwidth of the feed should be wide enough to match with that of the radio system during transmission or reception.
6. The feed must be rugged enough to survive in adverse weather conditions; for high it may be placed inside a low loss dielectric enclosure.

2.28 DIFFERENT TYPES OF FEEDS

For parabolic reflector antennas, different types of feeds may be used such as (*a*) dipole feeds, (*b*) waveguide horn feeds, (*c*) off-axis feeds, (*d*) monopulse feeds (*e*) Cassegrainian feeds, and (*f*) Gregorian feeds and quasi-paraboloidal feeds.

Of these feeds, the dipole feed may be useful for VHF/UHF parabolic antennas, while waveguide horn feeds are most widely used for microwave/millimeter wave parabolic antennas. Aperture blocking due to the feeds can be avoided by using an off-axis feed, which is also called an off-set feed. For radar scanning to locate and track a target, a monopulse feed is universally used. In space communication and radio astronomy, where the antenna is usually oriented with its beam toward the sky, thermal noise picked up from the backside of the parabolic reflector by diffraction often creates a problem by increasing the antenna noise temperature. This problem may be minimized by using a Cassegrainian feed, which reduces the noise and interference pickups from the backside of the parabolic reflector. For high aperture efficiency, instead of a Cassegrain feed, a Gregorian feed would be preferred, while for the highest aperture efficiency, a quasi-paraboloidal feed would be the best one. Such feeds producing high aperture efficiencies and low backside noise pickup are useful in high sensitivity and long-range radars used in defence applications and in radar astronomy in addition to being widely employed in radio astronomy.

The structure of the various types of feeds for parabolic reflector antennas are covered in the next sections.

2.29 DIPOLE FEED

A dipole feed for a parabolic reflector antenna usually consists of a driven half-wave dipole antenna backed by a subreflector in the form of a disc or a parasitically excited reflector, as shown in Figures 2.29 (*a*) and (*b*).

The design of the waxial feed indicated in Figure 2.29 (*a*) shows a quarter wave skirt balun, while Figure 2.29 (*b*) shows the design using a slee balancing stub, which provides a better balancing effect with, however, a lower power-handling capacity. In both the designs of Figures 2.29 (*a*) and (*b*), the phase center is located between the subreflector and the reflector, the optimum aperture of the reflector being 140°–160°.

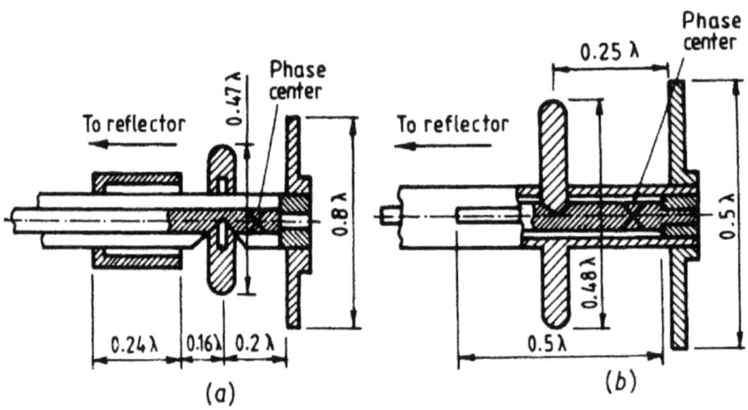

FIGURE 2.29 Coaxial-fed dipole feeds: (*a*) $A = 0.47\lambda_0$, $B = 0.19\lambda_0$, $C = 0.80\lambda_3$, $D = 0.24\lambda_0$, $E = 0.16\lambda_0$, (*b*) $A = 0.48\lambda_0$, $B = 0.25\lambda_0$, $C = D = 0.5\lambda_0$, X = phase center.

For a dual polarization feed with two simultaneous orthogonal polarization facilities, instead of a single dipole, a crossed turnstile dipole pair may be used with each dipole being energized by a separate coaxial line, the two lines being laid in a common enclosure.

2.30 WAVEGUIDE FEEDING METHOD FOR A DIPOLE FEED

The feeding of a dipole feed by a rectangular waveguide may be made by a structure shown in Figure 2.30. The TE_{10} rectangular waveguide cross section is $0.15 \times 0.7\ \lambda$, while the dipole length is 0.7λ, installed in a PTFE Poly Tetra Fluoro Ethylene TEFLON sleeve placed inside the waveguide. A shorted coaxial line with the waveguide as its inner conductor is used in place of a subreflector. The shorted coaxial line operating in the TE_{11} mode will have radiation characteristics close to a Huygens source with a cardioid radiation pattern with a beamwidth of about 130° between 10 dB points, the backward radiation being 30 dB down. Such a feed provides area efficiency of the parabolic antenna, about 0.65, operating over a bandwidth around 12% with a VSWR (voltage standing wave ratio) less than 1.5. This type of feed can be used also for circular polarization with the waveguide made circular in cross section and load with a dielectric and the dipole replaced by a turnstile structure.

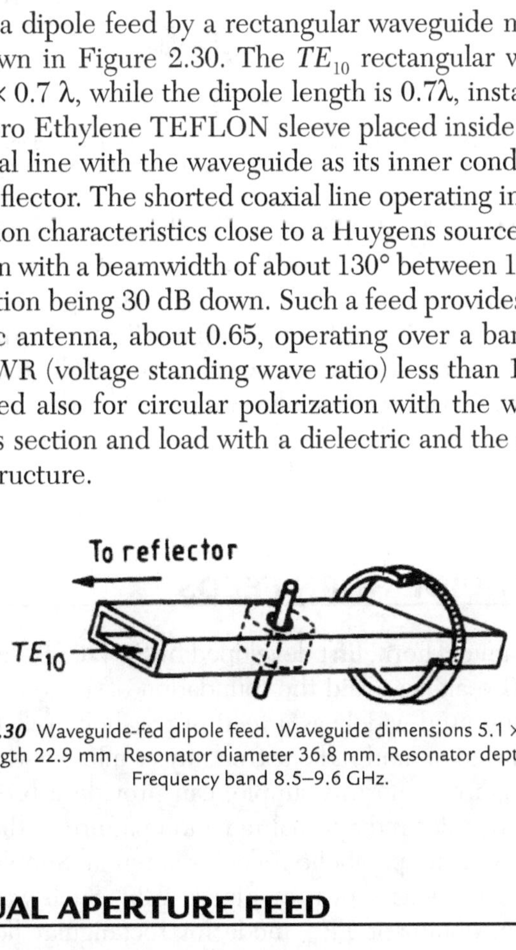

FIGURE 2.30 Waveguide-fed dipole feed. Waveguide dimensions 5.1 × 22.9 mm. Dipole length 22.9 mm. Resonator diameter 36.8 mm. Resonator depth 8.9 mm. Frequency band 8.5–9.6 GHz.

2.31 CUTLER DUAL APERTURE FEED

The structure of the cutler feed is shown in Figure 2.31. A cylindrical resonant cavity with two curved slots cut with the curvature and relative positions at the end of the cavity nearest to the reflector, as shown in the figure, is used to obtain the same radiation pattern width around 130° in the *E*

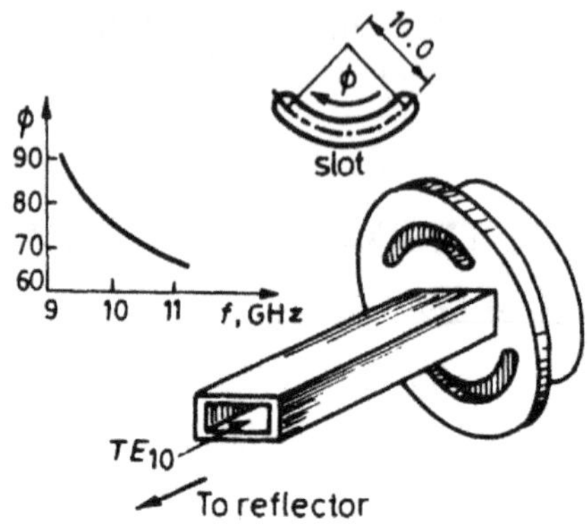

FIGURE 2.31 Dual-aperture (Cutler) feed. Waveguide dimensions 5.1 × 22.5 mm. Resonator length 22.2 mm. Resonator depth 4.9 mm. Flange diameter 44.5 mm.

and H planes, by properly making the diameter of the circular shield and the dimensions of the curved slots. For matching with the feed, the guide is tapered in the narrow dimension with a tuning screw provided in the wall of the resonant cavity farthest from the reflector. With this type of cutler dual slot feed, the aperture efficiency of the parabolic reflector antenna would be around 0.57 with some spillover, however with bandwidth limited to about 3%. The cutler dual aperture feed is convenient for use at low levels of radiation in the microwave bands.

2.32 WAVEGUIDE HORN FEEDS

The pyramidal horn, first developed by Sir J. C. Bose in Calcutta, India more than 100 years ago, laid the foundation of a wide variety of horn antennas now being used widely as a feed of parabolic reflector antennas for radar, communication, and radio astronomy at microwaves and millimeter waves. The design of a horn is simple, can provide a broad bandwidth, and can also create a desired optimal radiation pattern of the horn feed to have low side lobes of the parabolic reflector antenna. Simple waveguide horn feeds have structures as shown in Figure 2.32. Such waveguide horn feeds are excited in dominant TE_{10} mode for rectangular horns and TE_{11} mode in circular horns. A limitation of such simple horn feeds is that they do not provide axial symmetry for the radiation pattern, but have different E and H plane patterns and appreciable cross-polarization of the radiated field to receive cross-polarized returns from a radar target. In communication systems the cross-polarized components of a simple horn feed tend to mix up the independent signal channels in the E and H planes during reception by an antenna.

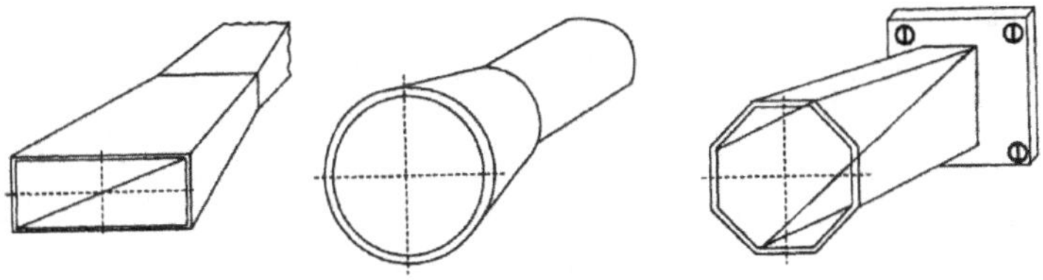

FIGURE 2.32 Waveguide horn feeds.

The cross-polarization problems can be minimized by using a circular corrugated horn feed as shown in Figure 2.33 (a), with its symmetric radiation pattern shown in Figure 2.33 (b). Such feeds are now most widely used in parabolic reflector antennas.

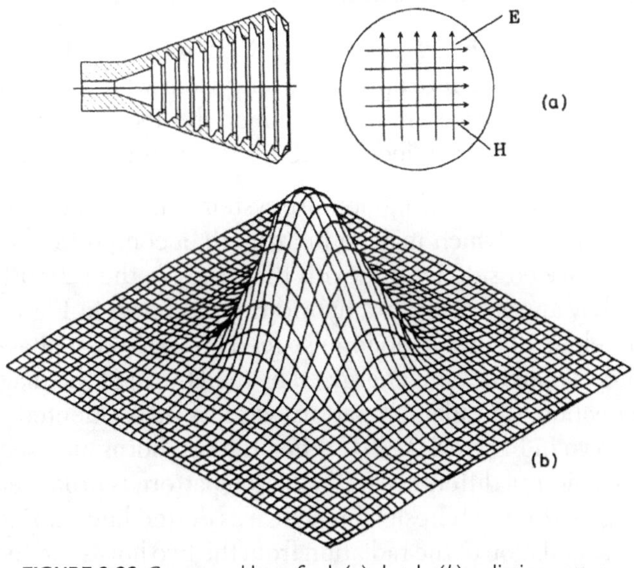

FIGURE 2.33 Corrugated horn feed: (*a*) sketch, (*b*) radiation pattern.

The depth of corrugation and the corrugation spacing affect only slightly the shape of the radiation pattern if the depth of corrugation is not close to half wavelength. The boundary conditions at the interface between the corrugated surface structure and the inner space of the corrugated horn are nearly the same, allowing the electric and magnetic lines of force to be bent slightly over the cross section. This produces the axial symmetry of the radiation pattern with negligible cross-polarization. Calculations and experimental results show that, with the operating frequency changing by as much as 50%, the radiation pattern width of a corrugated horn feed will change by only 3% to 7%, depending on the aperture angle and horn length. The design parameters of a corrugated horn feed can be optimized further by placing it in the aperture of a contoured dielectric lens to modify suitably the illumination of the parabolic reflector by the horn lens to reduce the side-lobe level. The use of such a lens may raise the aperture efficiency of the parabolic reflector antenna to nearly 0.9, besides protecting the feed from weather by sealing it with a dielectric lens.

2.33 MONOPULSE FEEDS

Monopulse feeds are covered in Section 3.16 in Chapter 3, in describing the structure of a monopulse tracking radar where such feeds are employed. However, the structure of the monopulse feed needs further input for the development of such feeds, and this is covered in this section.

In a one-dimensional monopulse system, intended for low radar targets only in one plane, which may be in azimuth, a composite feed consisting of two horns, placed side by side with their axes in the azimuthal plane, are energized by a magic *T* or by a hybrid ring as shown in Figure 2.34 (*a*). The two horn elements of the monopulse feed are excited in phase and in equal magnitude, when they are energized by the H-arm of the magic *T* to produce the summation (Σ) radiation pattern with its peak oriented along the boresight as shown in Figure 2.34 (*b*). When the two horns are excited by the *E*-arm of the magic *T*, a difference (Δ) radiation pattern is produced with the radiation being zero at the boresight as shown as dotted lines in Figure 2.34 (*b*), due to the cancellation of the radiation from the two horns excited in opposite phase.

The target is illuminated only the sum (Σ) pattern while the echo signals are received parallelly by two receivers, one of which collects the signal due sum beam while the other collects the difference beam return signal to locate the target.

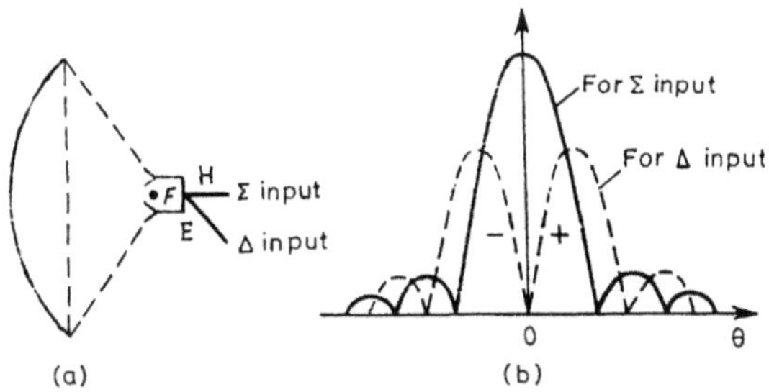

FIGURE 2.34

In practice, monopulse leads used in radars are two-dimensional, requiring four horn feeds, with additional horn feeds placed with their axial planes orthogonal to the first pair of horn feeds, as shown in Figure 3.13 under Section 3.16 in Chapter 3.

CHAPTER 3

PRINCIPLES OF RADAR

3.1 INTRODUCTION

Radar is an electronic system for radio location or the detection of objects by radio waves. The location is achieved by measuring the distance or range of an object from the radar as well as its angular positions, the azimuth, and the elevation. Often, the range, height, and the azimuth are measured by radar for the radio location. The term RADAR is an acronym for Radio Detection And Ranging first used during World War II, when it was developed primarily to detect enemy aircraft which were targets of attack. It may be mentioned here that the basic principle of radar was invented much earlier by Appleton and Barnett as early as 1926 to detect and measure the height of the ionized layers of the earth's upper atmosphere from the ground. This was further improved by Briet and Tuve to develop an ionosonde, which transmits pulses of radio waves upward, and measures the delay of the radio echo reflected from the ionized layer. Radar operates exactly in the same manner with an object or target replacing the ionized layer. The target, in general, acts as a scatter of radio waves and only in certain special cases behaves as a reflector. If the time taken by the pulse of a radio wave to travel to the target and return is given by T_R, the range R is given by

$$R = \frac{cT_R}{2} \tag{3.1}$$

where c is the velocity of the radio wave. Since the electromagnetic energy of the radio waves propagate with the speed of light, c in the above equation is 3×10^8 m/sec. The factor 2 in the denominator is due to the way radio waves are propagated. If the range R is in kilometers (km) or in nautical miles (nmi) and the time T_R is in microseconds, then equation (3.1) becomes,

$$\begin{aligned} R\ (\text{km}) &= 0.15\ T_R\ (S) \\ R\ (\text{nmi}) &= 0.081\ T_R\ (S) \end{aligned} \tag{3.2}$$

3.2 RADAR EQUATION

The radar equation, also called the range equation, relates the range from which a radar can detect a target to the characteristics of the radar system. Figure 3.1 shows a radar system radiating a pulse of radio waves toward the target.

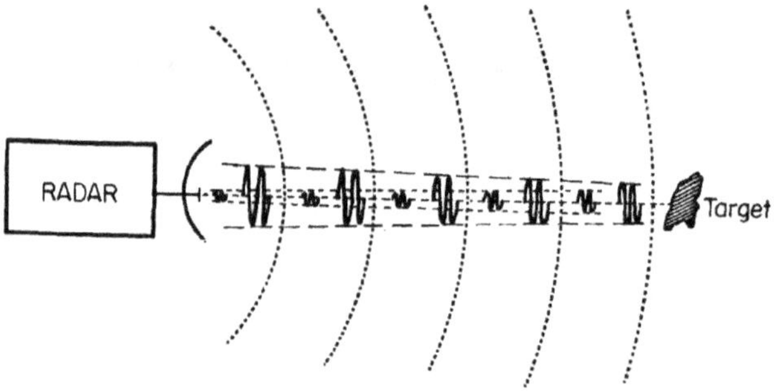

FIGURE 3.1 Pulse radar ranging.

The radar waves propagate with a spherical wavefront and become weak due to spreading on reaching the target. If we imagine a spherical surface of radius r passing through the target, then for an isotropic antenna, the power P_t fed to it will be radiated isotropically over an imaginary spherical surface to produce a power flux F_i at the target given by

$$F_t = \frac{P_t}{4\pi r^2} \tag{3.3}$$

where $4\pi r^2$ is the surface area of the sphere.

Usually, however, the antenna of the radar system is directive, concentrating radiation in the direction of the target into a sharp beam. The power radiated toward the target is, therefore, increased by a factor G relative to what is expected with a non-directive antenna like an isotropic radiator fed with the same power P_t, where G is the gain of the antenna. The effective radiated power, P_r is then given by

$$P_r = P_t \times G$$

The power flux at the target accordingly becomes

$$F_i = \frac{P_t G}{4\pi r^2} \tag{3.4}$$

A part of the incident power is intercepted by the target, which reradiates it in different directions with an amount depending on the property of the target. The part of the power returned by the target back along the direction of incidence is a measure of the echoing power of the target. This is defined as the equivalent area of the incident wavefront on the target, the power flowing through which when fed to an isotropic radiator will produce the same echo power as is actually returned by the target. The echoing power is also called the radar cross section of the target. If σ is the radar cross section, the power flux at the radar antenna due to the echo will be given by

$$F_r = \frac{P_t G}{\pi r^2} \times \frac{\sigma}{4\pi r^2} \tag{3.5}$$

If the effective collecting area or the capture area of the antenna is A_e, then the echo power received is given by

$$\begin{aligned} P_r - F_r A_e &= \frac{P_t G}{4\pi r^2} \times \frac{\sigma}{4\pi r^2} \times A_e \\ &= \frac{P_t G \sigma A_e}{(4\pi)^2 r^4} \end{aligned} \tag{3.6}$$

The previous equation indicates that P_r decreases with increasing range and may eventually become too weak to be detectable by the radar. If S_{min} denotes this minimum detectable received echo power, the maximum range r_{max} of the radar beyond which target will not be detectable is given by

$$r_{max} = \left[\frac{P_t G \sigma A}{(4\pi)^2 S_{min}} \right]^{1/4} \tag{3.7}$$

Equation (3.7) is the fundamental form of the radar equation, from which it is seen that the important antenna parameters are the transmitting gain G and receiving effective area A, which are related as

$$G = \frac{4\pi A}{\lambda^2} \tag{3.8}$$

Combining equations (3.8) and (3.7), we get two alternative forms of range equations as

$$r_{max} = \left[\frac{P_t A^2 \sigma}{4\pi \lambda^2 S_{min}} \right]^{1/4} \tag{3.9}$$

and

$$r_{max} = \left[\frac{P_t G^2 \lambda^2 \sigma}{(4\pi)^3 S_{min}} \right]^{1/4} \tag{3.10}$$

Equations (3.7), (3.9), and (3.10) are the three different forms of the radar equation. The appropriate form to be considered will depend on whether the effective area or the gain is constant with the wavelength. Usually A is constant and equation (3.8) indicates that the range $r_{max} \propto 1/\sqrt{\lambda}$ if P_t, σ and S_{min} are kept constant. This means that at shorter wavelengths, the range will be larger. This justifies the use of microwaves having wavelengths as short as a few centimeters in radar applications. In the previous equations some important factors which affect range are not explicitly included and, therefore, these simplified versions of the radar equations are unable to describe adquately the performance of practical radars.

3.3 RADAR FREQUENCIES

In general, radars are operated at frequencies in the range 220 MHz to 35 GHz. This is, of course, not necessarily the limit, since radars are operated outside this range. For example, ground wave HF radars might be operated at frequencies as low as 2 MHz, sky wave HF over-the-horizon (OTH) radars operate at 5 to 20 MHz, while at the other end of the spectrum millimeter radars are operated at 94 and 140 GHz and laser radars, or LIDAR, operate at much higher frequencies in the optical region. In Table 3.1, the place of usual radar frequencies in the electromagnetic spectrum is presented. Letter codes such as *L*, *S*, *C*, and so on are employed to designate radar frequency bands. These codes were originally used to guard military secrecy. Now these are used for some convenient short nomenclature. Table 3.1 shows the standard radar-frequency letter band nomenclature.

TABLE 3.1

Band	Frequency Range	Radar Bands
HF	3—30 MHz	
VHF	30—300 MHz	138—144 MHz
		216—225 MHz
UHF	300—1000 MHz	420—450 MHz
		890—942 MHz
L	1000—2000 MHz	1215—1400 MHz
S	2000—4000 MHz	2300—2500 MHz

Band	Frequency Range	Radar Bands
		2700—3700 MHz
C	4000—8000 MHz	5250—5925 MHz
X	8000—12000 MHz	8500—10680 MHz
K_u	12.0—18.00 GHz	13.4—14.0 GHz
K	18—26.5 GHz	24.05—24.25 GHz
K_a		33.4—36.0 GHz
mm	40—300 GHz	Q, U, V, E, W, F, D, and G bands

*Q (33—50 GHz), U(40—60 GHz), V(50—75 GHz), E(60—90 GHz), W(75—110 GHz), F(90—140 GHz), D(110—170 GHz), G(140—220 GHz).

Manufacturers of millimeter wave products include K and K_a bands in millimeter waves, while those manufacturing microwave products include 1—40 GHz in microwaves.

3.4 RADAR SET

The block diagram of a typical pulse radar set is shown in Figure 3.2. Its different parts are described as follows:

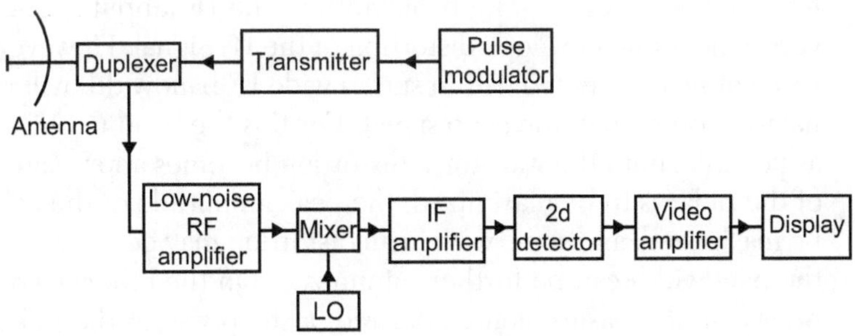

FIGURE 3.2 Pulse radar set.

Transmitter and Modulator. The transmitter may be an oscillator like a magnetron which is turned on and off (pulsed) by the modulator for generating a repetitive train of pulses.

Antenna. The electrical waveform generated by the transmitter travels through a transmission line or a waveguide to the antenna from which it is radiated into space as pulses of radio waves. For both transmitting and receiving operations, a single antenna is normally used for which a special transmit/receive switch or duplexer is introduced.

Duplexer. It protects the receiver from damage caused by the high power of the transmitter and also serves to channel the returned echo signals to the receiver when the transmitter output is off.

Receiver. The receiver is of the superheterodyne type. To reduce the noise contribution of the mixer, its first stage called the front end might be a low noise RF amplifier like a parametric amplifier or a low noise transistor amplifier followed by a balanced mixer. Recently special low noise field effect transistors made from gallium arsenide are being used as GaAs FET (field-effect transistor) preamplifiers. In many radar applications the front end preamplifiers may be excluded. This is in fact useful in a noisy environment where a reduction of receiver noise will not have any noticeable effect due to the rather high external noise. As an input stage the mixer has a greater dynamic range, less vulnerability to external electromagnetic interference, and less susceptibility to overload when compared to a front end RF preamplifier. The mixer and the local oscillator convert the RF signal to an intermediate frequency or IF.

IF Amplifier. The IF amplifier amplifies the mixer output at the intermediate frequency without producing appreciable distortion in the pulse waveform of the IF signal. The bandwidth of the IF amplifier should ideally be very wide for no envelope distortion of the IF Signal. However, the inherent receiver nose accepted within such a wide IF bandwidth will reduce the signal to noise ratio of the echo signal. For this the bandwidth is made as small as possible, until the waveform distortion becomes appreciable. If the range of the radar is to be maximized, the peak amplitude of the echo signal must be maximized, and some waveform distortion may be tolerated. Accordingly, the bandwidth can be further minimized. On the other hand, if high range precision of measurement is desired, a steep rise of the pulse waveform is useful, for which a larger IF bandwidth will be essential, although it will inevitably reduce the signal to noise ratio.

The center frequency of a typical IF amplifier is in the range 30—60 MHz, while the bandwidth is of the order of 1 MHz.

Detector. The envelope of the IF signal is extracted by an envelope detector. A diode detector may be used for the purpose. However, since the shape of the pulse envelope is important, to reduce distortion in the pulse waveform, the detector load is frequency compensated by inductances using the shunt peaking and series peaking techniques as employed in video detectors of TV systems.

Video Amplifier. In this stage the detected echo signal is amplified to a level where it can be properly displayed on a cathode ray tube screen. Here also the waveform distortion is minimized by video compensation techniques.

Displays. The simplest form of display is called the A-scope, where the echo amplitude produces a deflection of the electron beam in the CRO (cathode-ray oscilloscope) in the y-direction. The A-scope is a deflection modulated display and it is more suited to tracking radar applications.

Another type of display is the B-scope display, which utilizes the rectangular coordinates to display range and elevation angle. The echo in this display modulates the intensity of the electron beam, causing an intensity modulation of the spot on the screen.

The most widely used form of display is perhaps the plan position indicator, abbreviated as PPI. This effectively maps, in polar coordinates, the plan position or the location of the target in azimuth and range. The PPI display is also intensity modulated, in which the echo modulates the electron beam like that in a B-scope. In PPI the electron beam rotates in synchronism with that of the axis of the main beam of the antenna pattern. In most radars, the PPI display is supplemented by an alternative display which is similar to the B-scope, but the elevation angle is replaced by the height of the target. This is called the range height display, abbreviated as RH display.

3.5 RADAR APPLICATIONS

The chief uses of radar are for military purposes, although they have been increasingly important in civil applications for marine and aerial navigation. The major areas of radar applications are discussed as follows:

Military. The major role of radar for military application has been for surveillance, navigation, and also for the control and guidance of weapons.

Air Traffic Control (ATC). Radar is used throughout the world for the safety of aircraft and controlling air traffic. It has been used with the ground control approach (GCA) systems for guiding aircraft to a safe landing during adverse weather conditions.

Aircraft Navigation. Weather radar used on aircraft to outline the precipitation regions to the pilot is a classical form. Weather radars are also employed in airports to warn the pilot of thunderstorms. Besides these, radar is also utilized for terrain avoidance and terrain following during aircraft navigation. Sometimes, ground-mapping radars of moderately high resolution are used.

Marine Navigation. Marine radars are used to enhance the safety of ship travel by warning about collision with other ships and boys. Shore-based radars are used for surveillance of harbors.

Remote Sensing. Remote sensing is the sensing of geophysical objects or the environment from a remote location. Thus, radar was used to probe the moon and the planets. The ionospheric sounder, an important adjunct for short wave communications, is basically a remote sensing radar. Remote sensing radars borne in satellites are now used for remote sensing of earth resources, water resources, and environmental pollution.

3.6 RECEIVER NOISE

Noise is the unwanted energy which interferes with the ability of the receiver to detect an echo or any desired signal and, therefore, limits the useful sensitivity of the receiver. It may originate within the receiver or may enter via the receiving antenna with the desired signal. A part of the receiver noise is generated by the active devices in the receiver. The other part originating from the random thermal motion of the conduction electrons in the ohmic portions of the circuit is called the thermal agitation noise or the Johnson Noise. This nose is random in nature and is directly proportional to the resistance (R), the temperature (T) of the circuit, and the receiver bandwidth (B). The mean square noise voltage across a resistance is given by

$$e_n^2 = 4kTBR \tag{3.11}$$

where k is the Boltzman's constant.

The maximum noise power available from the resistance is given by

$$P_n = \frac{e_n^{\ 2}}{4R} = kTB \tag{3.12}$$

P_n is independent of the resistance R.

The noise in devices of the receiver as well as that received by the antenna is expressed in terms of equivalent temperatures T_R and T_A respectively at which a resistance is to be raised to obtain the same available noise power at the receiver input. The spectrum of the noise power is flat, extending ideally over an infinite bandwidth. However, the effective RF bandwidth of the receiver as determined by the bandwidth B of the IF amplifier limits the spectral components. The noise power at the output of the IF amplifier depends on the shape of the IF response characteristics, $H(f)$, and we define an equivalent noise bandwidth B_n as given by

$$B_n = \frac{\int_{-\infty}^{\infty} |H(f)|^2 \, df}{|H(f_0)|^2} \tag{3.13}$$

where f_0 is the frequency of maximum response. B_n is really an ideal rectangular pass band, through which the noise power passed through is the same as that of the actual IF amplifier with non-ideal response. Accordingly,

$$B_n |H(f_0)|^2 = \int_{-\infty}^{\infty} |H(f)|^2 \, df$$

from which equation (3.13) follows.

The noisiness of the receiver is assessed by its noise figure F_n, which can be defined as

$$\begin{aligned} F_n &= \frac{\text{Noise power at the output of practical receiver}}{\text{Noise output of ideal receiver at standard temperature}} \\ &= \frac{N_0}{kT_0 B_n G_a} \end{aligned} \tag{3.14}$$

where G_a is the available gain. The value of T_0 is taken to be 290°K. The noise N_0 is measured at the linear portion of the receiver preceding the detector. The available gain G_a is the ratio of the output signal S_0 to the input signal S_i when both the input and output are matched. The noise figure F_n in equation (3.14) can alternately be written as,

$$F_n = \frac{S_i / N_i}{S_a / N_a} \tag{3.15}$$

Rearranging equation (3.15) we can write the input signal as,

$$S_i = \frac{kT_0 B_n F_n S_0}{N_0} \tag{3.16}$$

If the minimum detectable signal is denoted by S_{min}, then,

$$S_{min} = kT_0 B_n F_n \left(\frac{S_0}{N_0} \right)_{min} \tag{3.17}$$

Using equation (3.17), the maximum radar range R_{max} can be written as

$$R^4_{max} = \frac{P_t G_a A_e}{(4\pi)^2 \, kT_0 B_n F_n \left(\frac{S_0}{N_0} \right)_{min}} \tag{3.18}$$

3.7 SIGNAL TO NOISE RATIO

The statistical noise theory can be conveniently applied to get the signal to noise ratio at the output of the IF amplifier. Let us consider an IF amplifier with bandwidth B_{IF} followed by a second detector and a video amplifier with bandwidth B_v as shown in Figure 3.3. In order to pass all the video modulation, B must be greater than $B_{IF}/2$.

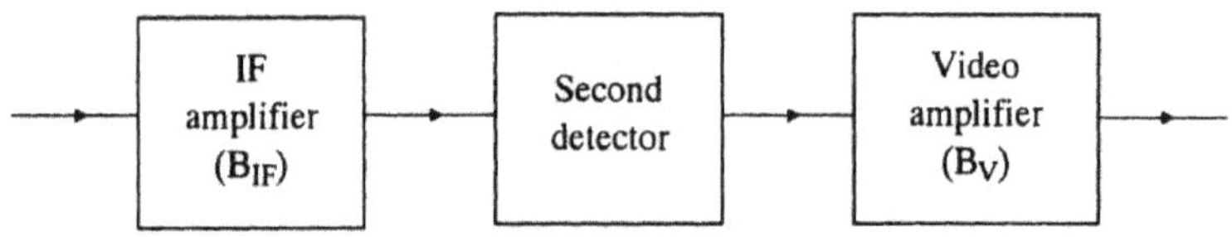

FIGURE 3.3 Second detector and video amplifier of a pulse radar set.

Assuming the noise entering the IF filter to be Gaussian, we can write the probability-density function as

$$p(v) = \frac{1}{\sqrt{2\pi \Psi_0}} \exp \frac{-v^2}{2\Psi_0} \tag{3.19}$$

where $p(v)\,dv$ is the probability of finding the noise voltage v between v and $(v + dv)$, and Ψ_0 is variance or the mean square value of the noise voltage. If Gaussian noise is made to pass through a narrow band IF filter, then the envelope of the noise voltage can be written as

$$p(r)=\frac{R}{\Psi_0}\exp\left(-\frac{R^2}{2\Psi_0}\right) \tag{3.20}$$

where R represents the amplitude of the envelope of the filter output. The previous equation is a form of the Rayleigh probability density function. The probability that the envelope of the noise voltage will be within V_1 and V_2 is given by

$$\text{Probability} \qquad (v_1 < R < v_2) = \int_{v_1}^{v_2} \frac{R}{\Psi_0}\exp\left(-\frac{R^2}{2\Psi_0}\right) \tag{3.21}$$

Again, the probability that the noise voltage envelope will exceed the threshold voltage V_T is

$$\begin{aligned}\text{Probability} \qquad (V_T < R < \infty) &= \int_{v_T}^{\infty} \frac{R}{\Psi_0}\exp\left(-\frac{R^2}{2\Psi_0}\right) dR \\ &= \exp\left(-\frac{V_T^2}{2\Psi_0}\right) = P_{fa}\end{aligned} \tag{3.22}$$

When the noise will cross the threshold, the probability of a false alarm as denoted by P_{fa} in equation (3.22) is obtained. The average time interval between crossings of the threshold by noise alone is known as the *false alarm time*. Denoting it by T_{fa} we can write,

$$T_{fa} = \underset{N\to\infty}{Lt}\ \frac{1}{N}\sum_{k=1}^{N} T_k \tag{3.23}$$

Here T_k is the time required between crossings of the threshold V_T by the noise envelope when the slope is positive. The false alarm probability can alternately be defined as,

$$P_{fa} = \frac{\text{Duration of time the envelope is actually above the threshold}}{\text{Total time it could have been above the threshold}}$$

$$= \frac{\sum_{k=1}^{N} t_k}{\sum_{k=1}^{N} T_k}$$

$$= \frac{<t_k>_{av}}{<T_k>_{av}} = \frac{1}{T_{fa}B} \tag{3.24}$$

The terms t_k and T_k are defined in Figure 3.4. Equating equations (3.22) and (3.24), we have

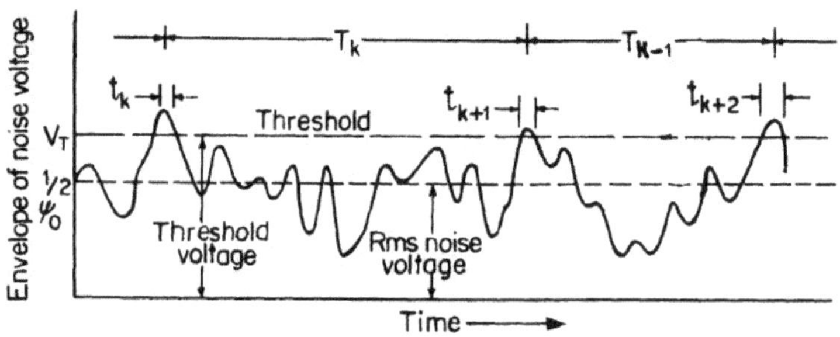

FIGURE 3.4 False alarms due to noise in detector output.

$$T_{fa} = \frac{1}{B_{IF}} \exp \frac{V_T^2}{2\Psi_0} \tag{3.25}$$

The false alarm probability of a practical radar is very small. For a 1 MHz bandwidth there are 10^6 noise pulses per second. So the false alarm probability of pulse is 10^6 if the false alarm time of more than 1 second is obtained.

Next, let us consider a sine-wave signal of amplitude A to be present along with the noise at the input to the IF filter. The signal frequency is the same as IF midband frequency f_{IF}. The probability density function of the output of the envelope detector can be written as,

$$p_s(R) = \frac{R}{\Psi_0} \exp\left(\frac{-R^2 + A^2}{2\Psi_0}\right) I_0\left(\frac{RA}{\Psi_0}\right) \tag{3.26}$$

where $I_0(Z)$ is the modified Bessel function of zero order and argument Z. When Z is large, the asymptotic expansion for $I_0(Z)$ is

$$I_0(Z) \approx \frac{e^2}{\sqrt{2\pi Z}}\left[1 + \frac{1}{8Z} + \ldots\ldots\right]$$

When the signal is absent, $A = 0$. Then equation (3.26) reduces to equation (3.20). Equation (3.26) is called the Rice probability distribution function.

The probability that envelope R will exceed a predetermined threshold V_T, is the probability of detection p_d, which is given by

$$P_d = \int_{v_T}^{\infty} p_s(R)\, dR = \int_{v_T}^{\infty} \frac{R}{\Psi_0} e^{-(R^2 + A^2)/2\Psi_0} I_0\left(\frac{RA}{\Psi_0}\right) dR \tag{3.27}$$

Assuming RA/Ψ_0 and $A >> |R - A|$ and neglecting terms in A^{-3} and beyond, a series approximation of the integral yields

$$P_d = \frac{1}{2}\left(1 - erf\frac{V_T - A}{\sqrt{2\Psi_0}}\right) + \frac{\exp[-(V_T - A)^2 / 2\Psi_0}{2\sqrt{2\pi}(A/\sqrt{\Psi_0})} \times \left[1 - \frac{V_T - A}{4A} + \frac{1 + (V_T - A)^2 / \Psi_0}{8A^2 / \Psi_0}\right] \tag{3.28}$$

where the error function is defined as

$$erf\ z = \frac{2}{\sqrt{\pi}} \int_0^z e^{-u^2} du$$

The signal-to-noise voltage ratio, $A/\Psi^{1/2}$ in equation (3.28) may be expressed in terms of signal to noise power ratio, S/N as follows:

$$\begin{aligned} A/\Psi^{1/2} &= \frac{\sqrt{2A_{rms}}}{\text{rms noise voltage}} = \sqrt{\frac{2(A_{rms})^2}{(\text{rms noise voltage})^2}} \\ &= \left(\frac{25}{N}\right)^{1/2} \end{aligned} \tag{3.29}$$

Also, from equation (3.22) we have

$$V_T^2 / 2\Psi_0 = (1 / P_{fa}) \tag{3.30}$$

using equations (3.29) and (3.30) in equation (3.28), the resulting relationship between P_d, S/N, and P_{fa} may be plotted as a family of curves shown in Figure 3.5.

The relationship of P_d with S/N and P_{fa} shown in Figure 3.5 is true for a single pulse echo receiver. In practice many pulses are returned from the target on each radar scan, and these can be used to improve the detection probability.

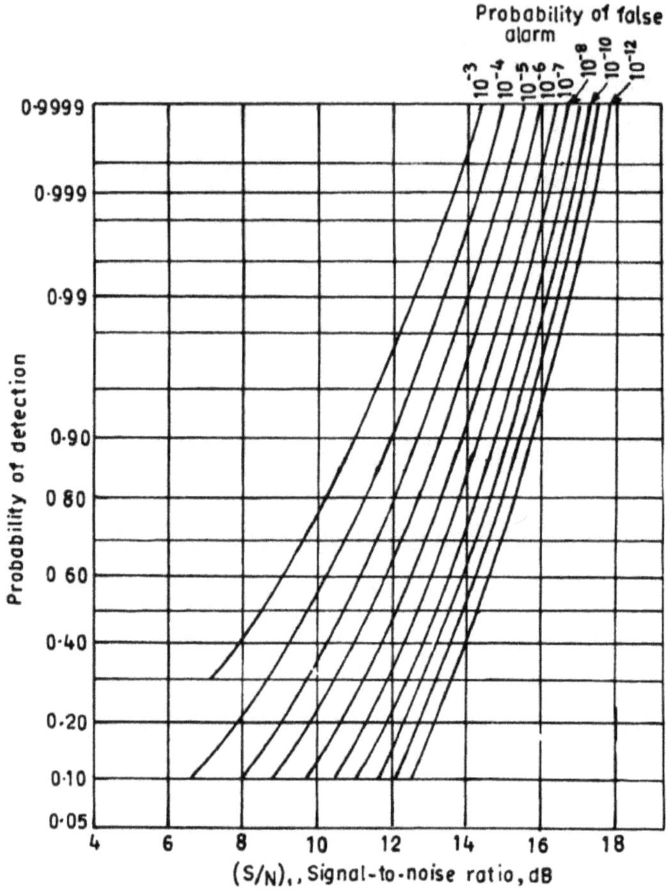

FIGURE 3.5 Probability of detection p_d for a sine wave signal in noise as a function of the signal to noise power ratio (S/N) and the probability of false alarm, p_{fa}.

If θ_S is the antenna scan rate in deg/s, θ_B is the beamwidth in degree, and f_p is the pulse repetition frequency in Hz, the number of pulses returned from the target per second is given by

$$n_B = \frac{f_p}{\theta_s} \times \theta_B$$

because f_p, pulses are contained in θ_S deg which is scanned in 1 sec.

If the scan rate in RPM is ω_m then we have

$$\theta_s = \frac{\omega_m \times 360}{60} = 6\omega_m .$$

from which
$$n_B = \frac{\theta_B f_p}{6\omega_m} \tag{3.31}$$

Typical values of the parameters of a search radar are $\omega_m = 5$ RPM, $f_p = 300$ Hz, and $\theta_b = 1.5°$. Using these in equation (3.31) we have $n_{sB} = 15$ hits from a point target during each scan. All these n_{sB} pulses may be summed up to improve the detection of the target, and the process is called integration. The integration may be made before detection at the IF level or may be done after detection. In predetection integration the phases of individual echo pulses are preserved, and if they are superimposed at the same position on the cathode ray tube display, they will add up coherently while the noise with add up incoherently. Under ideal conditions predetection of n pulses will improve the signal to noise power ratio n times, and the process is called coherent integration. The persistence of the CRT phosphor helps the integration of the n pulse waveforms. After detection the phase information of the pulses are lost and, therefore, post detection integration will be less efficient than predetection or coherent integration. Post detection integration may be called noncoherent integration. Marcum defined the efficiency of a post detection integrator relative to that of an ideal predetection integrator as given by

$$E_i(n) = \frac{(S/N)_1}{n(S/N)_n} \tag{3.32}$$

where the suffixes 1 and n indicate cases with single pulse and n pulse respectively. The integration improvement factor with post detection integration is given by

$$\frac{(S/N)_1}{(S/N)_n} = n\, E_i(n) \tag{3.33}$$

Substituting $(S/N)_n$ for (S_0/N_0) min in equation (3.18), the radar equation with n pulses integrated is given by

$$r_{max}^4 = \frac{P_t\, G\, A_e \sigma}{(4\pi)^2 k\, T_0 B_n F_n (S/N)_n} \tag{3.34}$$

Using equation (3.33) this becomes

$$r_{max}^4 = \frac{P_t\, G\, A_e \sigma\, n\, E_i(n)}{(4\pi)^2 k T_0 B_n F_n\, (S/N)_1} \tag{3.35}$$

Using the relation $P_{av} = \dfrac{P_t \tau}{T_p} = P_t \tau\, f_p$

the range equation can also be expressed as

$$r_{max}^4 = \frac{P_{av}\; G\, A_e \sigma\, n E_i(n)}{(4\pi)^2\, k\, T_0 B_n F_n (B_n \tau)\, (S/N)_1\, f_p} \tag{3.36}$$

The product $B_n\, \tau$ is generally of the order of unity.

3.8 TRANSMITTER POWER

The power P_t in a radar equation is the peak power. If A_e is the amplitude of the microwave signal during pulse transmission, then the instantaneous peak power is αA_e^2. The rms value of the peak power, P_t, is half as large.

The average power is also important in radar, which is the average transmitter power over the pulse-repetition period. The average power P_{av} is related to the peak power by an equation,

$$P_{av} = \frac{P_t \tau}{T_p} \tag{3.37}$$

where τ is the width of the rectangular pulse and $T_p (= 1/f_p)$ is the pulse repetition period. Equation (3.27) can be rewritten as

$$\frac{P_{av}}{P_t} = \frac{\tau}{T_p} \tag{3.38}$$

$$= \tau\, f_p \tag{3.39}$$

This ratio is called the duty cycle of the radar. A pulse radar might typically have a duty cycle of 0.01, while a CW radar has a duty cycle of unity.

In terms of the average power, the radar equation can be written as

$$r_{max}^4 = \frac{P_{av}\; G\, A_e \sigma\, n\, E_i(n)}{(4\pi)^2\, k\, T_0 F_n (B_n \tau)\, (S_0 / N_0)_{\min} \cdot f_p} \tag{3.40}$$

where n is the number of echoes or hits integrated and $E_i(n)$ is the integration efficiency.

Here the bandwidth and the pulse width are grouped together. The product of the two widths is generally of the order of unity.

3.9 PULSE REPETITION FREQUENCY

It is determined mainly by the maximum range at which the targets are desired. If the pulse repetition frequency (PRF) is made high enough, the likelihood of obtaining target echoes is increased. If the echo signal is received after an interval more that the pulse repetition period, such signals are called multiple time around echoes. Let us consider the three targets labeled A, B, and

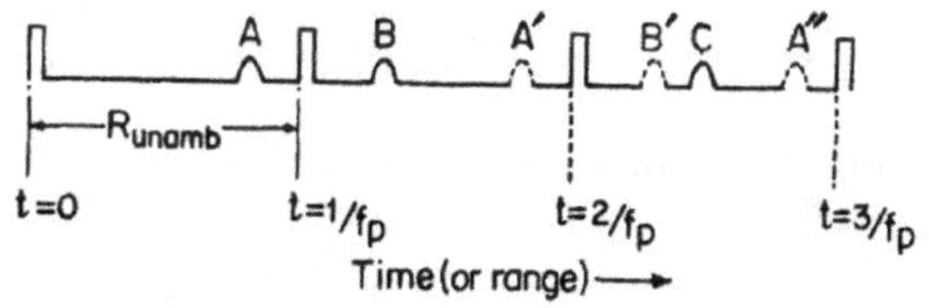

(*a*) **A-scope display showing multiple time around echoes from targets A, B and C at ranges. $R < R_{unamb}$, $R_{unamb} < R < 2\,R_{unamb}$ and $2\,R_{unamb} < R < 3\,r_{unamb}$.**

(*b*) (*c*)

FIGURE 3.6 (*b*) Three targets *A*, *B* and *C* appearing on A-scope
(*c*) Multiple time around echoes in A-Scope with changing pulse repetition frequency.

c as in Figure 3.6 (a). In the figure the target A is located within the maximum unambiguous range R_{unamb} of the radar, target B is at a distance greater than R_{unamb} but less than $2R_{unamb}$, while the target c is greater than $2R_{unamb}$ but less than $3R_{unamb}$. In Figure 3.6 (b) the appearance of three targets on an A-scope is sketched. The multiple time around echoes on the A-scope are not possible to be distinguished from proper target echoes within the maximum unambiguous range. One of the methods of distinguishing multiple time around echoes from unambiguous echoes is to operate with a varying pulse repetition frequency. Echoes from multiple time around targets is spread over a finite range as shown in Figure 3.6 (*c*). If T is the interpulse period, then the maximum unambiguous range is given by $r_{max} = cT/2$.

Ambiguities can be resolved indirectly by observing the variation of the echo signal with time. Echoes from the longer range of the ambiguous echo will exhibit larger fading.

3.10 PULSE DURATION

The duration of a pulsed transmission in a radar determines the accuracy of range determination. Two close targets will be resolved if the separation in range Δr is given by $\Delta r \geq c\tau/2$. Thus, the minimum pulse duration is given by

$$\tau_{min} = \frac{2\Delta r}{c}$$

A short pulse duration also helps in reducing the minimum range of the radar, given by

$$r_{min} = \frac{c\,\tau}{2}$$

Therefore, a target closer than r_{min} will produce an echo which will overlap with the transmitted pulse as shown in Figure 3.6.

A short pulse duration will, however, call for an increased bandwidth requirement of the receiver to pass the large spectral bandwidth of the short pulse. The bandwidth is, in fact, related to τ as $B_n = 1/\tau$. A larger bandwidth in turn increases the noise power accepted by the receiver, increasing the minimum detectable signal power. This effectively reduces the range.

Also a short pulse will fail to excite the phosphorescent screen adequately, which is made sluggish to retain its illumination for a period of a few pulse intervals in order to attain continuity of the display. Thus, for a shorter pulse the transmission must be stronger for the echo to be detectable or perceivable on the screen. Otherwise, the range will be reduced.

From the previous discussion, it is clear that a short pulse ultimately tends to reduce the range, although it helps in increasing the range accuracy and in reducing the minimum range, which are often considered very important, particularly for airborne radars and in guided missiles. In such cases, laser radar or LIDAR (Light Detection And Ranging) is employed to achieve an extremely short pulse duration coupled with high peak power. Millimeter, wave, and infrared radars are also used for this purpose.

3.11 PROPAGATION EFFECTS

In order to analyze the radar performance, it is better to assume that the radar and target are both located in free space, though there are only a few radar applications which strictly follow the free space conditions. For most of the cases the earth's surface and the medium involved have a significant effect. When the propagation is in the non-free space conditions, they are of three different categories. These are:

1. refraction of the radar wave by the earth's atmosphere,
2. lobe structure caused by the interference between the direct wave and the reflected wave, and
3. attenuation of the radar wave when it propagates through the earth's atmosphere.

For most of the applications of radar at microwave frequencies, the attenuation is not sufficient for affecting the radar performance. During adverse weather, however, the attenuation and reflection of radar signals in rain is often a limiting factor. The presence of the curved earth's surface not only limits the line of sight range, but it can also cause major modification by breaking the antenna elevation pattern into many lobes. The propagation effects for most of the cases must be properly taken into consideration.

3.12 SCANNING RADARS

A radar may be designed to scan its antenna beam through the target repetitively and display the positions of the moving target in successive scans on a long persistence oscilloscope screen to obtain the track of the target in plan. For a plan position indicator (PPI) mode of display, which essentially displays the plan position of the target with the radar at the center of the display, the track of the target is clearly discernible in the azimuthal plane on the screen. The PPI display can be looked upon as a polar plot of the target in range and azimuthal position as shown in Figure 3.7. The concentric circles in the display are the range markers.

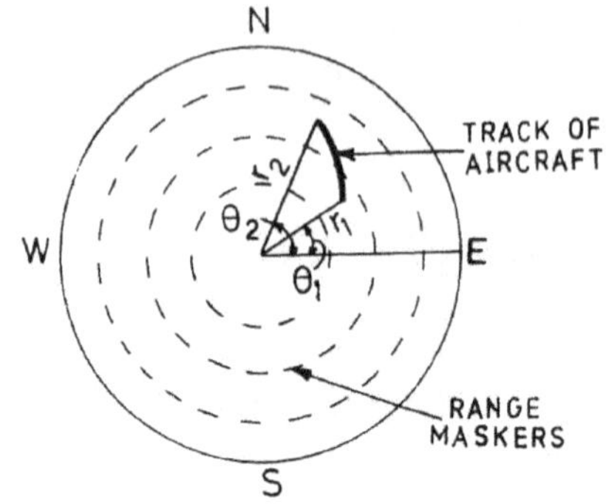

FIGURE 3.7 PPI display in air surveillance radar.

To generate such a display, each transmitted pulse is used to trigger a radial sweep in the oscilloscope screen, on which is displayed the position of the target as a bright spot produced by intensity modulation of the cathode ray beam of the oscilloscope. Azimuthal position of the trace during a sweep is changed continuously at a rate synchronous with that of the azimuthal position of the antenna beam, by simultaneous circular defecting signals applied to the pair of defecting electrodes. If the *X* and *Y* deflecting plates are supplied with an AC signal driving the antenna motor but with a 90° phase difference, the circular deflection of the spot on the screen will be produced, with the spot circling at the rotation frequency of the scanning antenna beam.

The tracking of the target by scanning will, however, be a slow process and may be suitable only for monitoring the position and guiding an aircraft to the airport by radar control support from the airport.

3.13 TRACKING RADAR

For tracking of fast moving targets like supersonic aircraft and also for anti-aircraft weapons systems, a continuous and fast tracking of the target is necessary. Increasing the radar operating frequency to the millimeter wave bands reduces the size of the antenna, which allows much faster scanning rates. Continuous tracking is essential where the target is to be fired by a missile as in an anti-aircraft gun, or when a missile is to be destroyed by tracking it with radar to fire on the missile track in the air by radar control from the ground. The information about range, elevation, and azimuth angles, and the Doppler frequency shift, are utilized by a tracking radar to measure the target coordinates, determine the target path in space, and predict future positions of the target. This allows the missile to be fired at an orientation so that the missile may hit the target accurately, after moving as a projectile through the air. For greater precision a mobile tracking radar may be mounted on the missile itself as an airborne missile seeker to find the target, track it, and home in on it accurately. Missile seeker radars are discussed further in Chapter 9 on modern radars.

Various types of tracking radars have been developed. Basically, tracking may mean angular tracking of the target, range tracking, or doppler frequency tracking for moving targets. The technologies are different for each type of tracking. However, the role of an automatic feedback control system to track the appropriate target parameter *viz.* angular position, range, or doppler shift due to line of sight velocity of the target is common to all the systems. Usually tracking radar means an angular tracking radar or angular tracker. Two types of such angular trackers are (a) continuous trackers and (b) track-while-scan (TWS) radar. In continuous trackers, the antenna beam orientation is controlled by a servomechanism activated by an error signal generated by one of the techniques called (a) lobe switching, (b) monopulse, or (c) conical scan. Gunfire controls make use of such a continuous tracking radar. In the track-while-scan radar, target coordinates are measured each time the antenna beam scans through the target, and the target track is then reconstructed by joining the successive positions of the target displayed on a long persistence oscilloscope screen or by computer controlled interpolation.

3.14 LOBE SWITCHING

In the lobe switching technique, the antenna beam is switched alternately between two positions such that there is an overlap region of the beams as shown in Figure 3.8.

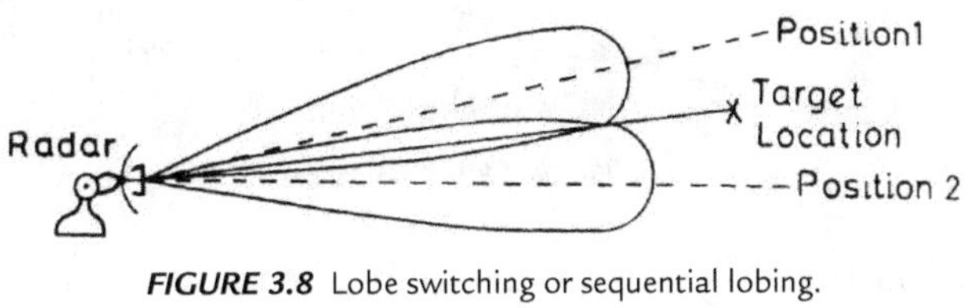

FIGURE 3.8 Lobe switching or sequential lobing.

Lobe switching is often termed sequential lobing to distinguish it from the technique of simultaneous lobing or monopulse. In lobe switching the difference in amplitude between the radar returns obtained in the two positions would be zero on the equisignal line. Any deviation from this line will make the amplitudes unequal. Top obtain an error signal capable of actuating the antenna drive motor, the sign of the error must change for target positions on either side of the equisignal line.

This can be accomplished by (a) detecting the amplitudes for two positions; (b) separating into two channels 1 and 2 corresponding to positions 1 and 2 respectively, by a switch working synchronously with the lobe switching; (c) converting to DC signals representing the peak amplitude of the echo by sample and hold or other integrating techniques for the two channels; and (d) subtracting the DC output of the channel from that of the other.

For tracking in three dimensions two more lobes 3 and 4 must be provided for the antenna to track in a plane perpendicular to that for the first two lobes 1 and 2.

Lobe switching is employed in early airborne interception radars and in ground based antiaircraft fire control radars, but it is no longer used in modern radar tracking applications.

3.15 CONICAL SCAN

Instead of switching the antenna lobe between two positions, as in lobe switching, the beam position can be steered continuously, making the beam axis trace out a cone in space in the target region as shown in Figure 3.9.

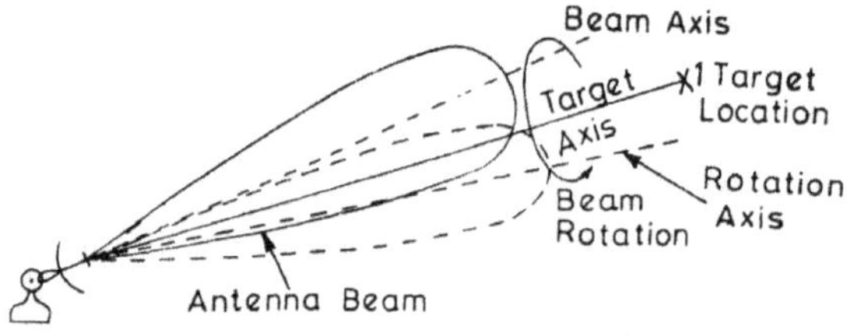

FIGURE 3.9 Conical scan tracking beam.

In a non-tracking radar, the antenna beam would be along the axis of the parabolic dish with the antenna beam along the same axis as shown as dotted lines in the figure. For a conical scan, the beam axis is changed by an angle called the squint angle, and rotated about the axis of the dish, which is also the axis of rotation of the beam. If the target happens to coincide with the axis of rotation, then the conical scan motion of the beam axis will not produce any change in the echo intensity, which will, however, be smaller compared to what is expected for the target position on the beam axis. For any deviation of the target position from the axis of rotation, the echo intensity will undergo a sinusoidal variation at the rotation frequency, with a phase depending on the target position. This is illustrated in Figure 3.10.

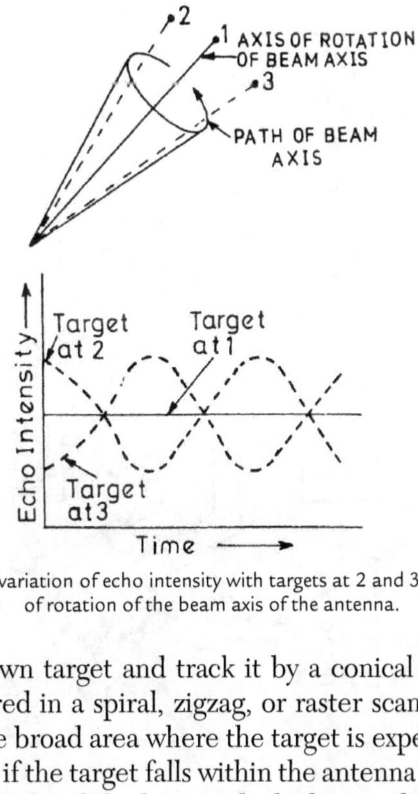

FIGURE 3.10 Sinusoidal variation of echo intensity with targets at 2 and 3, which are not on the axis of rotation of the beam axis of the antenna.

To locate an unknown target and track it by a conical scan tracking radar, the antenna is steered in a spiral, zigzag, or raster scan mode to search for the target within the broad area where the target is expected. During such a scanning operation, if the target falls within the antenna beam, the searching scan motion is stopped and the beam is locked on to the target by using the conical scan motion of the antenna. Such a scanning system is often referred to as the search-and-lock system, which is employed in missile seekers and antiaircraft weapons systems.

A schematic diagram of a conical scan radar is shown in Figure 3.11.

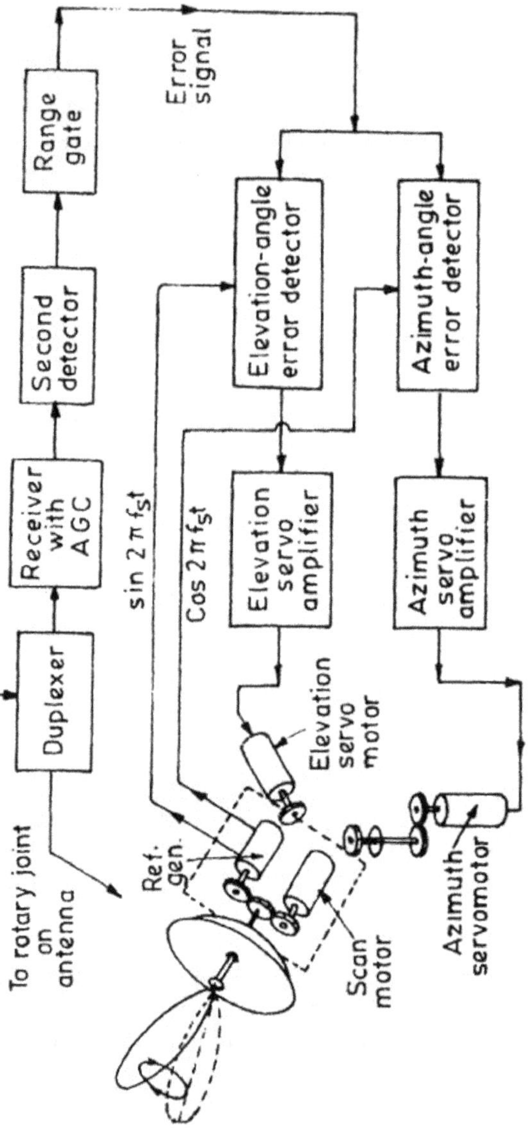

FIGURE 3.11 Conical scan tracking radar.

A cassegranian feed with a rotating effentric subreflector is used to generate the conical scan antenna beam. The rotation is made by gearing to a scan motor as shown in the figure. A generator driven by a gear coupled to the rotating gear of the subreflector generates two sinusoidal reference signals at the rotation frequency f_s 90° out of phase with each other: sin $2\pi f_s t$ and cos $2\pi f_s t$. These are applied to a pair of phase-sensitive detectors for the incoming echo intensity received through the range gate of the radar receiver. One of the detectors produces a DC voltage proportional to the elevation angular error while the other is proportional to the azimuthal angular error. These error signals are amplised by servo amplifiers to drive the azimuth and elevation drive motors of the antenna in a direction to correct the respective errors automatically by a servo feedback control system and direct the axis of rotation of the beam toward the target with a very small residual error. A minimum residual error is inevitable in any feedback control system. The positions of the rotation axis of the antenna beam in elevation and azimuth are displayed by making use of standard angle transducers like synchros, potentiometers, or analog-to-digital angle encoders. The position information is often used to predict future target position and fire it from the radar site, as in fire control radars. To extract the modulation of intensity over a train of echo pulses, a sample and hold circuit is used. This is also referred to as a box car generator.

3.16 MONOPULSE TRACKING

The monopulse tracking of a radar antenna is basically a simultaneous beam switching technique in which a pair of closely spaced horn feeds of the antenna produce a pair of off axis beams that are switched simultaneously but alternately in phase and out of phase to generate antenna beams as shown in Figures 3.12 (*a*) and (*b*) in polar and Cartesian coordinates.

A phase sensitive detector in the radar receiver senses the phase of the echo signal and produces outputs varying with the angular deviation of the target from the axis of the dish, as shown in Figure 3.12 (*c*). The S-curve produced as shown in Figure 3.12 (*c*) is employed to control the antenna orientation in the plane passing through the axis of the horn feeds using a closed loop servo feedback control system.

For tracking of the radar antenna in the orthogonal plane, a second pair of spaced horn feeds are employed, similar to the first pair, but oriented with their axes passing through the orthogonal plane.

A schematic diagram of a monopulse tracking radar capable of tracking in attitude and azimuth is shown in Figure 3.13.

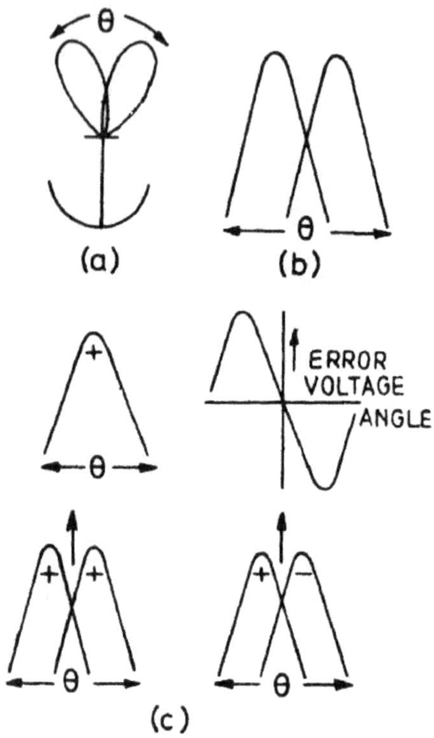

FIGURE 3.12 Monopulse antenna patterns and error signal.

The four horn feeds marked 1, 2, 3, and 4 form an altitude-azimuth monopulse feed. Horns 2 and 3 are connected to the *E*-arms of a hybrid junction I. Likewise, the horns 1 and 4 are connected to the *E*-arms of a second hybrid junction II. The sum and difference outputs from each of the hybrid junctions I and II are further processed by a pair of hybrid junctions III and IV to obtain the (1) Sum channel (1 + 2 + 3 + 4), (2) Elevation difference channel (2 – 3) + (1 – 4), and (3) Azimuth difference channel (2 + 3) – (1 + 4) as shown in Figure 3.12. The three-channel outputs of the monopulse feed system are processed by three identical radar receivers. The sum channel signal is detected by an amplitude detector. The elevation and azimuth difference channels are detected by phase sensitive detectors whose outputs indicate the elevation and azimuth angle errors that are used to drive, respectively, the elevation and azimuth servo motor drives for the antenna, with a polarity required to correct the angle errors through a closed loop servo feedback control system in the same way as in a conical scan tracking radar.

An alternative to using the three separate radar receivers for the three channels is to use a single radar receiver up to the IF stage in a time shared scheme. The receiver will then be switched to receive sequentially the signals in the three channels, and is separated at the IF outputs by a synchronous sequential switch for feeding into amplitude detectors, and the two phase-sensitive detectors for elevation and azimuth difference signals.

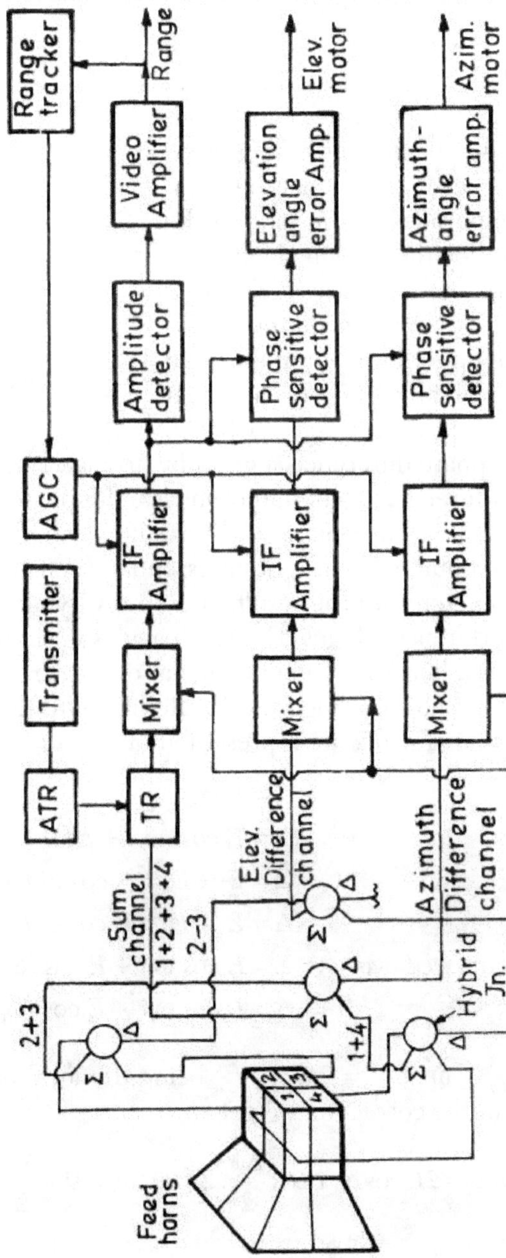

FIGURE 3.13 Monopulse tracking radar.

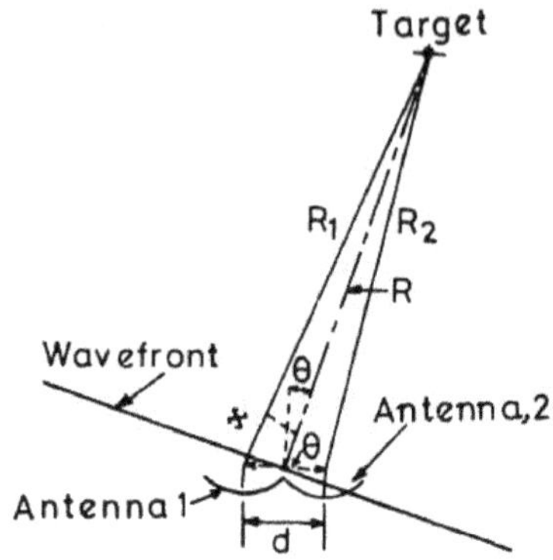

FIGURE 3.14 Phase-comparison monopulse radar.

In monopulse tracking, the amplitudes of the signals received in two or more antenna feeds are compared. Instead of the amplitudes, the phases of the signals received by spaced antennas having beam axes parallel to each other may be compared. The amplitudes of the target echo signals received by the antenna beams are almost the same, but the phases differ, as shown in Figure 3.14, in which the ranges r_1 and r_2 of the target from antennas 1 and 2 are nearly equal if the range of the target is large compared to the spacing between the feeds. The phase difference is given by $\Delta\phi = 2\pi/\lambda.\ d \sin\theta$ where $d \sin\theta$ is the range difference. As the target moves, the phase difference changes to produce a significant change in the resultant amplitude of the sum of the two antenna outputs. The system essentially constitutes a two-element radar interferometer, with the resultant output varying with target position, producing interference fringes. For small angles θ, $\sin\theta \approx \theta$ and the phase difference $\Delta\phi$ is a linear function of angular error. A receiving arrangement to measure $\Delta\phi$ is shown in Figure 3.15. Referring to the figure, if the echo signals in antennas 1 and 2 are represented by

$$E_1 = E \cos \omega_c t$$

and $$E_2 = E \cos(\omega_c t + \Delta\phi)$$

Then we have $$\Sigma = E_1 + E_2 = E \cos \omega_c t + E \cos(\omega_c t + \Delta\phi)$$

$$\Delta = E_1 - E_2 = E \cos \omega_c t - E \cos(\omega_c t + \Delta\phi)$$

At the IF outputs we have $$\Sigma = E \cos \omega_i t + E \cos(\omega_i t + \Delta\phi)$$

and $$\Delta = E \cos \omega_i t - E \cos(\omega_i t + \Delta\phi)$$

where $\omega_i = \omega_{Lo} - \omega_c$, $\omega_{Lo} = 2\pi. f_{Lo}$, f_{Lo} being the local oscillator frequency. At the output of the correlator, we get

$$\Sigma \times \Delta = 2E \cos \omega_i\, t \cos\frac{\Delta\phi}{2}.\, 2E \sin \omega_i\, t. \sin\frac{\Delta\phi}{2}$$

$$= 2E \sin\frac{\Delta\phi}{2} \cos\frac{\Delta\phi}{2}.\, 2E \sin \omega_i\, t. \cos \omega_i\, t$$

$$= E^2 \sin\Delta\phi \sin 2\,\omega_i\, t$$

At the output of the amplitude detector, we have $E_d = E^2 \sin \Delta\phi \approx E^2 \Delta\phi$ for $\Delta\phi << 1$. Such a receiver configuration for obtaining $\Delta\phi$ cancels the components of IF noise present at correlator inputs originating from the two independent receiver channels. The voltage E_d represents $\Delta\phi$ with its proper sign and is amplified by a servo amplifier to drive the antenna servo motors of the two antennas simultaneously, to form the servo feedback loop.

The sum channel IF output can be diverted also to the amplitude detector of a receiver to obtain the echo intensity, and the transmitter can be connected by a Duplexer to the sum point of the hybrid of junction. For both elevation and azimuth tracking, a second unit fed from a second pair of spaced antennas must be used.

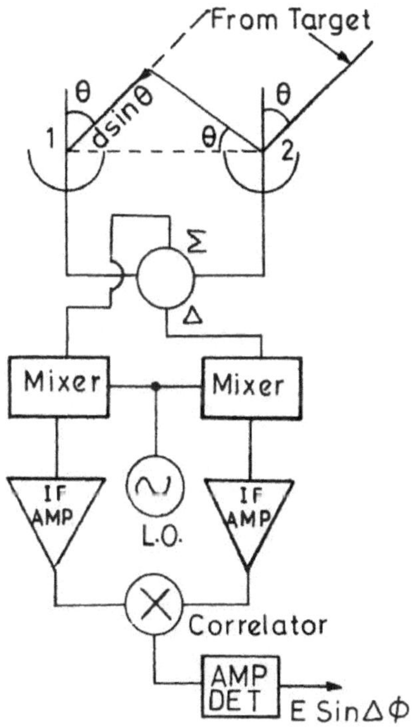

FIGURE 3.15 Radio interferometers.

The precision of tracking in such a radio interferometric tracking radar is much higher than that possible with lobe switching, conical scan, or monopulse radar. In radio interferometry the primary antenna beam is split into a number of finer beams of which the largest central beam is utilized for the tracking. The width of this beam can be reduced by increasing the spacing between the antennas up to a certain limit, beyond which the gain of the central beam will be almost as large as that of the two adjacent beams on either side of it. This is illustrated in the echo signal in Figure 3.16. Figure 3.16 (*a*) shows that the left hand beam (*L*) and right hand beam (*R*) are each smaller than the central beam, while with increased spacing the pattern becomes as shown in Figure 3.16 (*b*), which shows that *L* and *R* beams receive echo signals comparable to the central beam, and the latter is, therefore, not distinguishable.

Reducing the beamwidth of each antenna will help in improving the distinguishability with the larger spacing as shown in Figure 3.16 (*c*). The behavior is identical to that experienced in a double slit diffraction pattern in optics, where the primary beamwidth is determined by the width of individual slits,

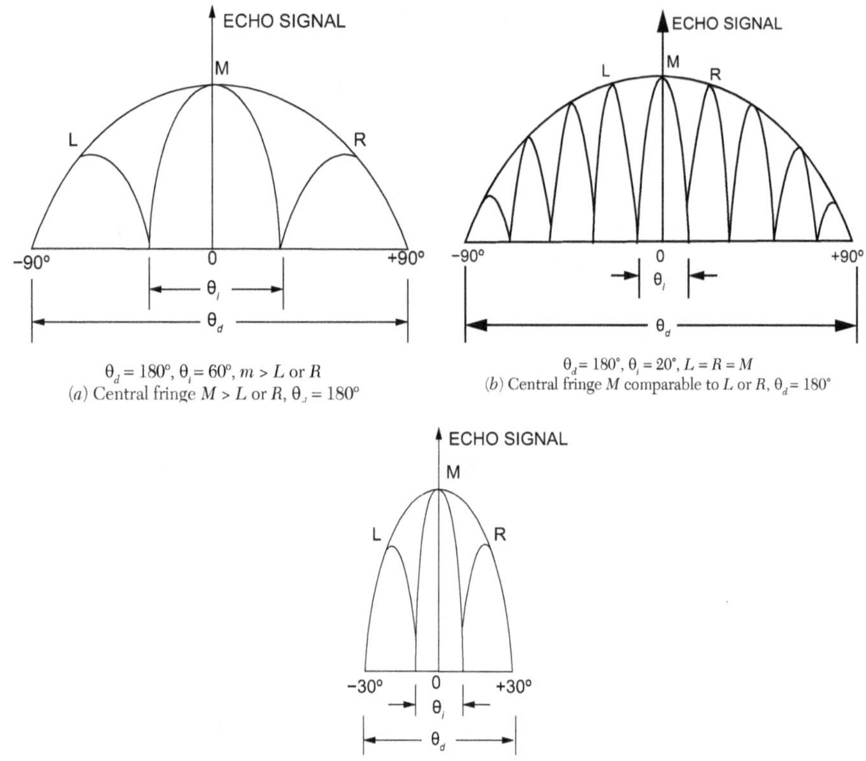

$\theta_d = 180°$, $\theta_i = 60°$, $m > L$ or R
(*a*) Central fringe $M > L$ or R, $\theta_d = 180°$

$\theta_d = 180°$, $\theta_i = 20°$, $L = R = M$
(*b*) Central fringe M comparable to L or R, $\theta_d = 180°$

$\theta_d = 60°$, $\theta_i = 20°$, $M > L$ or R
(*c*) Central fringe $M > L$ or R with θ_d reduced to 60°.

FIGURE 3.16

while the number of fringes and therefore the width of a fringe is determined by the spacing between the centers of the slits.

Radio interferometric tracking is widely employed for tracking of space probes in deep space missions where the range is rather large and where other tracking methods would fail to attain the required precision.

3.16.1 Millimeter Wave Tracking Radars

At millimeter waves, a tracking radar has got a greater angular resolution for an antenna of given diameter and reduced multipath effect, which allows a lower angle tracking capability compared to that at microwaves. Atmospheric attenuation is, however, higher at millimeter waves. Nevertheless, with the available power levels the millimeter wave tactical radars still perform well under inclement weather and battlefield conditions.

A number of millimeter tracking radars have been successfully designed and developed. These include: (a) SEA TRACKS system, which combines a low elevation surveillance/fire control radar system with a 35 GHz monopulse tracking radar, (b) STARTLE radar, which is a monopulse tracking radar at 94 GHz, integrated with the main gun control system on a M60A3 tank (further details included in Table 9.3 in Chapter 9), and (c) TRAKX dual frequency range instrumentation radar, which can track simultaneously at 9 and 35 GHz bands, using computer control.

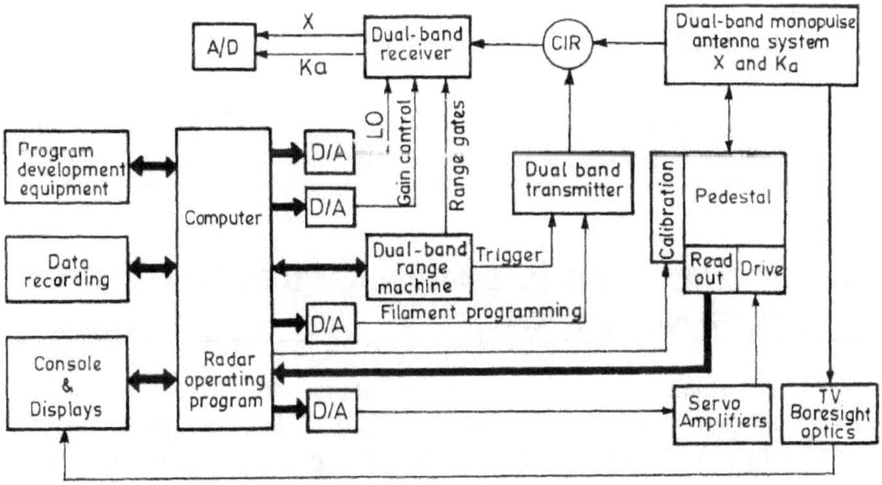

FIGURE 3.17 Block diagram of the TRAKX dual band monopulse tracking radar system with computer control.

A schematic diagram of the TRAKX dual band tracking radar system is shown in Figure 3.17.

Three channel receivers are employed in each band of the dual band monopulse tracking radar, and computer control allows independent tracking in range, azimuth, and elevation. A TV camera is employed for optical boresighting of the target, made simultaneously with the tracking of the radar. This often helps the recognition of the target under clear weather conditions. The range machine provides gating and trigger pulses for the receiver and transmitter respectively under the computer control. The computer gets in the echo pulses from each of the three receivers for each of the bands 9 and 35 GHz and processes it to obtain the amplitude, range, elevation, and azimuth, which are taken out at appropriate output ports of the computer.

The characteristics of the TRAKX dual band monopulse tracking radar are shown in Table 3.2.

TABLE 3.2 TRAKX Radar Characteristics.

Frequencies	9 and 35 GHz
Power	300 kW (beak at *X* band)
	130 kW (peak at K_a band)
Pulse width	0.25 μ sec
Pulse repetition frequency	160 to 1600 Hz in steps of 160 Hz
Beamwidth	1° (*X* band), 0.25° (K_a band)
Tracking	Three channel monopulse
Tracking range	140 km (*X* band), 52 km (K_a band) for
	1-m^2 target on a normal day.

3.17 ACCURACY OF RADAR MEASUREMENTS IN THE PRESENCE OF NOISE

The accuracy of radar measurements is limited by the presence of noise which, in fact, reduces the accuracy of time delay measurement, which determines the range, and that of doppler velocity measurement, which determines range rate, and also the accuracy of amplitude measurements, which determines radar cross section, or RCS. Of these the accuracy of range is, perhaps, the most important aspect in locating distance of the target.

The range accuracy in a pulsed radar in the presence of noise can be estimated by referring to Figure 3.18, which shows a realistic pulse shape with finite rise and fall times. The range *R* is derived from the measurement of the round trip delay time of the pulse, T_0, from the transmitter to the target and back, through the relation $T_0 = \frac{2R}{c}$, *c* being the velocity of the propagation of a radar pulse, assumed to be equal to the velocity of light, 3×10^8 m/s. Thus, the accuracy of time delay measurement is indicative of the accuracy of range measurement, the two errors being proportional to each other as given by $\Delta T_0 = \frac{2\Delta R}{c}$.

The dotted pulse in the figure indicates the pulse with the noise added to it. The shape of the realistic pulse is not rectangular and has finite rise and fall times, as an ideal rectangular pulse with zero rise and fall times can neither be generated by a practical systems nor can it be utilized in a receiver with finite bandwidth. The effect of noise is to modify the shape of the pulse and to shift the time of threshold crossing to produce an error ΔT_0 in time delay measurement as shown in Figure 3.18. For a large signal-to-noise ratio, the shape of the pulse with noise is the same as that of the pulse without noise, and the maximum slope occurring near the threshold-crossing point for the uncorrupted pulse of amplitude A, as shown in Figure 3.18 at the output of the video filter, is given by

$$M = \frac{A}{t_r} \tag{3.41}$$

where t_r is the rise time of the pulse. For the pulse corrupted with noise, the corresponding slope is $n\,(t)/\Delta T_0$, where $n\,(t)$ is the noise voltage around the threshold crossing and $\Delta\ T_0$ is the error in time delay measurement. As the two slopes are equal we have

$$\frac{A}{t_r} = \frac{n\,(t)}{\Delta T_0} \tag{3.42}$$

or

$$\Delta T_0 = \frac{n\,(t)}{A / t_r} \tag{3.43}$$

or

$$\sqrt{(\Delta T_0)^2} = \delta T_0 = \frac{T_r}{(A^2 / n^2)^2} \tag{3.44}$$

$$= \frac{t_r}{(2S / N)^{1/2}} \tag{3.45}$$

$A^2 / \sqrt{x^2}$ in equation (3.44) is the video signal to noise ratio which is twice the IF signal to noise ratio, 2*S*/*N* equation (3.45). From equation (3.44) we find that for a constant pulse amplitude A, the rms time delay error is proportional to the rise time t_r and is independent of the pulse width. Therefore, a decrease in rise time t_r or an increase of signal to noise ratio *S*/*N* will improve the accuracy of time delay measurement.

3.17.1 Effect of Pulse Shape on Signal-to-Noise Ratio

In deriving the expressions for range resolution and minimum range in terms of pulse duration, as presented in Section 3.10, the shape of the pulse is assumed to be rectangular, which ideally requires an infinite bandwidth of the receiver to maintain the rectangular shape of the pulse in the receiver.

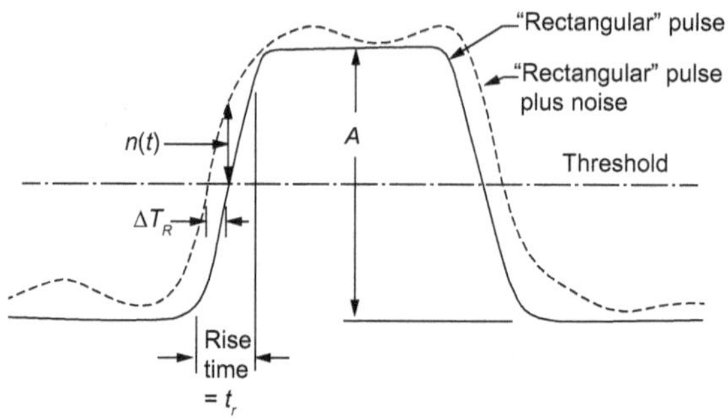

FIGURE 3.18 Measurement of time delay using the leading (or trailing) edge of the pulse. The solid curve represents the echo pulse uncorrupted by noise. The dashed curve represents the effort of noise.

In practice, however, the finite bandwidth of the receiver required to obtain a good signal to noise ratio inevitably distorts the pulse shape, lengthening the duration of the pulse by an amount proportional to the reciprocal of the bandwidth. A finite and optimal bandwidth of the receiver, required for a large range capability of the radar, will attenuate the spectral sidebands of the pulse outside the bandwidth of the receiver, resulting in the lengthening of the pulse in the time domain. This sideband cutting also effectively reduces the energy of the echo pulse as well as the receiver noise passing through the finite bandwidth. For an optimal bandwidth the S/N ratio for the echo pulse is largest. With such an optimal bandwidth filter there would be a wastage of transmitter power due to sideband cutting if an ideal rectangular pulse, having the optimal bandwidth and whose sidebands are not ultimately allowed to pass through the receiver, is generated. Considerable saving in transmitter power may therefore be possible by generating a bandwidth limited pulse, which will not be degraded further by the finite optimal bandwidth of the receiver. Thus, under the constraints of bandwidth limitation in the receiver, it would be worthwhile to study the effect of pulse shapes in reducing the wastage of transmitter power due to sideband cutting by the receiver bandpass characteristics. Even the shape of the receiver passband may be modified to study the loss in S/N ratio for various shapes of the passband compared with a matched filter, which is supposed to produce the optimal S/N ratio.

It can be shown by using the Schwartz inequality that the frequency response function which maximizes the signal-to-mean-noise power ratio of a matched filter is given by

$$H(f) = G_a S^*(f)\, e^{-j2\pi f t_1}$$

when the input noise is stationary and 'white,' and the ratio R_f of the peak signal power to noise power is given by

$$R_f = \frac{|s_0(t)|^2_{max}}{N}$$

irrespective of the shape of the input signal waveform,

where $s(f) = \int_{-\alpha}^{\alpha} s(t)e^{-j2\pi ft}\,dt, S^* =$ complex conjugate of $S(f)$, $t_1 =$ fixed value of time at which signal is observed to the maximum, $G_a =$ constant equal to maximum filter gain, $s_0(t)_{max} =$ maximum value of output signal voltage, and $N =$ mean noise power at receiver input.

The output of a matched filter may also be shown to be proportional to the input signal cross-correlated with a replica of the transmitted signal, except for the time delay t_1. For two signals $y(\lambda)$ and $s(\lambda)$, each of finite duration, the cross-correlation function is, in fact, given by

$$R(t) = \int_{-\alpha}^{\alpha} \gamma(\lambda)\, s(\lambda - t)\, d\lambda \tag{3.46}$$

while the output of a filter with impulse response $h(t)$ with an input $Y_{in}(t) = s(t) + n(t)$ is given by

$$Y_0(t) = \int_{-\alpha}^{\alpha} Y_{in}(\lambda)\, h(t - \lambda)\, d\lambda \tag{3.47}$$

With a matched filter $h(\lambda) = s(t_1 - \lambda)$, the equation becomes

$$Y_0(t) = \int_{-\alpha}^{\alpha} Y_{in}(\lambda)\, s(t_1 - t + \lambda)\, d\lambda = R(t - t_1) \tag{3.48}$$

Thus, the matched filter produces cross-correlation between the received signal with noise and a replica of the transmitted signal.

The efficiency of nonmatched bandpass filters of various shapes of the passband compared with matched filters may be estimated theoretically, and the results obtained are summarized in Table 3.3.

TABLE 3.3 Efficiency of Nonmatched Filters Compared with Matched Filters.

Pulse shape	Shape of passband of filter	Optimum B	Loss in S/N ratio compared with matched filter, dB
Rectangular	Rectangular	1.37	0.85
Rectangular	Gaussian	0.72	0.49
Gaussian	Rectangular	0.72	0.49
Gaussian	Gaussian	0.44	0.00
Rectangular	Single tuned, one stage	0.40	0.88
Rectangular	5 single tuned cascaded stages	0.672	0.50

The efficiency of a filter is measured as the ratio of peak signal-to-noise ratio with the filter to the peak signal to-noise ratio with the matched filter. An interesting result revealed from the table is that for a Gaussian pulse shape, the matched filter also has a Gaussian passband, resulting in no loss in SNR in dB.

3.17.2 Effect of Gating Waveform on Range Accuracy

A radar receiver usually employs a gate having a controllable width and position to allow the receiver to pass on the echo pulse only when it arrives at the receiver gate. For this purpose the gate width is made of the order of the pulse width. Any receiver noise passing through the gate is cut off except during the period when the gate is open. The shape of the gating waveform has an effect on the accuracy of ranging. A block diagram of a receiver with a gating facility is shown in Figure 3.19. The input signal $y(t)$ is gated

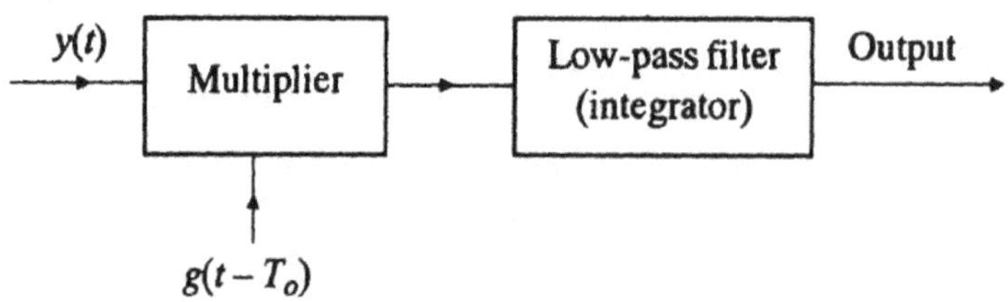

FIGURE 3.19 Receiver for measuring range (time delay) using a gating signal $g(t - T_o)$ where T_R is the estimate of the true time delay.

through the multiplier by feeding it with a gating waveform g $(t - T)$, where $T = \frac{2R}{C}$ is the estimate of the true delay time T_0. The purpose of gating is to aid in extracting the true estimate of the delay time T_0. The echo signal is given by $y(t) = s(t - T_0) + n(t)$, where $s(t - T_0)$ is the signal in the absence of noise. The low pass filter output due to the signal only is given by

$$s_0(T - T_0) = \int g(t - T)\, s(t - T_0) \tag{3.49}$$

while the contribution due to noise $n(t)$ is

$$n_0(T) = \int g(t - T)\, n(t)\, dt \tag{3.50}$$

Assuming $T - T_0 = \Delta T$, the form of the optimum gating signal, $s_0(\Delta T)$, should be an odd function, with $s_0(\Delta T) = 0$ for $\Delta T = 0$ if we want to realise a range error detector with the gating pulse for application to an automatic control of gate position, the gate being set to a position for minimum range error. The variation of receiver output as a function T is shown in Figure 3.20, from which the slope of the receiver output without noise is M. The slope in the presence of noise $n(t)$ is also nearly M near the zero crossing point.

The ratio of the rms voltage $\sqrt{n_0^2}$ to the slope M of the output $s_0(\Delta T)$ for $\Delta T = \theta$ may be taken as a measure of rms error in time measurement and hence that of range.

Thus, $$\delta T = \sqrt{(\Delta T)^2} = \frac{\sqrt{n_0^2}}{M}$$

where $$M = \left[\frac{\delta s_0(\Delta T)}{\delta \Delta T}\right] \Delta T = \theta$$

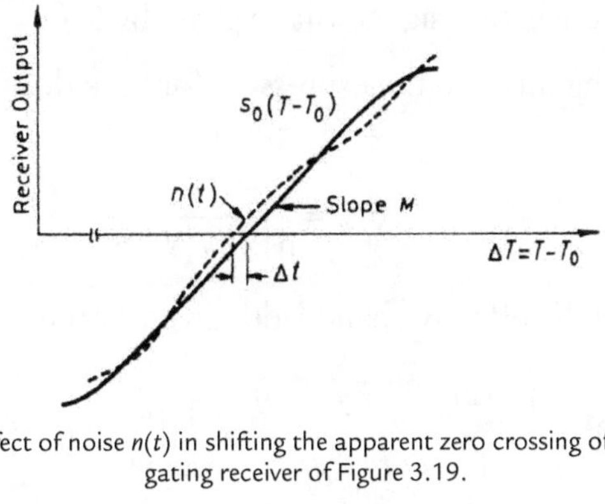

FIGURE 3.20 Effect of noise $n(t)$ in shifting the apparent zero crossing of the output of the gating receiver of Figure 3.19.

The error is illustrated in Figure 3.20, which shows the effect of noise $n(t)$ shifting the zero crossing of the output $s_0\,(T - T_0)$ of the gating receiver.

It can be shown by using the calculus of variation that the optimum gating function will have a Fourier transform given by

$$S_g(f) = \frac{j2\pi f\, S(f)}{[N_i(f)]^2} \tag{3.51}$$

where $S(f)$ and $N_i(f)$ are the Fourier transforms of $s(t - T_0)$ and $n(t)$ respectively.

Equation (3.51) indicates that the optimum gating waveform may be obtained by time differentiation of the received waveform when the noise spectrum is time invariant. If the spectrum of the received pulse is $S(f)$, then by the convolution theorem the spectrum of the gated output pulse would be given by

$$S_0\,(f) = S_g(f)\; S(f)$$

or

$$S_0(f) = \frac{j\;2\pi f\; S(f)\, S(f)}{[N_i(f)]^2}$$

$$= j2\pi\, f\; S(f) \frac{S^*(f)}{|N_i(f)|^2} \tag{3.52}$$

$$= \frac{j2\pi\, f\,|\,S(f)\,|^2}{|\,N_i(f)\,|^2} \tag{3.53}$$

The last factor in equation (3.52) is the frequency response function of a matched filter for the input signal with spectrum $S(f)$. Thus, if such a matched filter is inserted at the input of the multiplier as shown dotted, then the optimum gating signal will be given by $S(f) = j\;2\pi f$.

Substituting this into the expression for time delay error, it can be shown that

$$\delta T_R = \frac{1}{\beta\,(2E\,/\,N_0)^{1/2}} \tag{3.54}$$

where β is the effective bandwidth given by the equation

$$\beta^2 = \frac{\int_{-\infty}^{\infty} (2\pi f)^2\,|\,S(f)\,|^2\, df}{\int_0^{\infty} |\,S(f)\,|^2\, df} = \frac{1}{E}\int_{-\alpha}^{\alpha} (2\pi f)^2\,|\,s(f)\,|^2\, df$$

3.17.3 Accuracy Doppler-Velocity Measurement

The dopper velocity is derived from the doppler shift of a carrier due to a moving target or a moving radar system, the two parameters being proportional to each other, as indicated equation (8.9) in Section 8.3.2 on blind speeds in the chapter on MTI Radar in this book. Manasse (1955) showed using inverse probability that the minimum rms error in frequency measurement is given by

$$\delta f = \frac{1}{\alpha\,(2E/N_0)^{1/2}} \tag{3.55}$$

where

$$\alpha^2 = \frac{\int_{-\infty}^{\infty} (2\pi t)^2\, s^2(t)\, dt}{\int_{-\infty}^{\infty} s^2(t)\, dt} \tag{3.56}$$

The value of α^2 for an ideal rectangular pulse of width τ is $\pi^2\tau^2/3$ from which the rms frequency error for the rectangular pulse is given by

$$\delta f = \frac{\sqrt{3}}{\pi^{\tau}(2E/N_0)^{1/2}} \tag{3.57}$$

This indicates that a longer duration of the pulse will lead to a higher accuracy of frequency measurement, making it smaller, which in turn will improve the accuracy of doppler velocity measurement. The improvement would be similar for a Gaussian-shaped pulse, for which the expression for frequency error is given by

$$\delta f = \frac{1.18}{\pi^{\tau}(2E/N_0)^{1/2}} \tag{3.58}$$

indicating that only the numerator on the right-hand side is reduced to 1.18 for the Gaussian pulse instead of $\sqrt{3}$, shown in the expression for a rectangular pulse.

3.17.4 Accuracy of Angular Position Measurement

As the Fourier transform describes the relationship between the radiation pattern and aperture distribution of an antenna, the estimation of angular error becomes analogous to that of time delay and frequency error, each of which depends on a Fourier transform, as shown in Sections 3.17.2 and 3.17.3.

The angular error in one coordinate plane xz for an antenna with a rectangular receiving aperture, as shown in Figure 3.21, can be estimated by

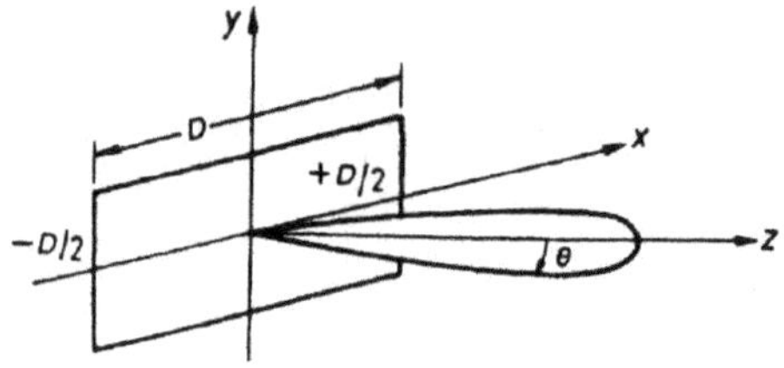

FIGURE 3.21 Rectangular receiving aperture of width D and amplitude distribution $A(x)$, giving rise to radiated pattern $G_v(\theta)$.

assuming an amplitude distribution function, $A(x)$, across the aperture. The voltage gain, as a function of the angle θ in the xz plane, is proportional to

$$G_i(\theta) = \int_{-D/2}^{D/2} A(x) \exp\left(j2\pi \frac{x}{\lambda} \sin\theta \right) dx \tag{3.59}$$

For a sharp beam with beamwidth θ, we have $\sin\theta \simeq \theta$.

Substituting this in equation 3.59, we have

$$Gv^{(\theta)} = \int_{-D/2}^{D/2} A(x) \exp \frac{e^{j2}\ \pi^{x\,\theta}}{\lambda} dn \tag{3.60}$$

This is analogous to the inverse Fourier transform relating frequency spectrum $S(f)$ with the time waveform $s(t)$, through the relation

$$s(t) = \int_{-\infty}^{\infty} S(f)\ e^{j2}\ \pi^{ft}\ df \tag{3.61}$$

The signal voltage from a fixed point source will be proportional to $G_v(\theta)$, which will be equal to $G_v(w_t)$ if the antenna scans in an x^2 plane at a uniform rate w_s. Comparing equations 3.60 and 3.61, if t and f in equation 3.61 are identified as $\frac{\theta}{\lambda}$ and x respectively in equation 3.60, then the angular rms error can be obtained by analogy to equation 3.54 for the time-delay accuracy and is given by

$$\delta\left(\frac{\theta}{\lambda}\right) = \frac{1}{\gamma(2e/N_0)^{1/2}} \tag{3.62}$$

where γ is the effective aperture width given by

$$\gamma^2 = \frac{\int_{-\infty}^{\infty} (2\pi x)^2 \,|\, A\,(n)\,|^2 \; dn}{\int_{-\infty}^{\infty} |\, A\,(n)\,|^2 \; dn} \tag{3.63}$$

From equations 3.62 and 3.63 it can be shown that the angular accuracy of the antenna is given by

$$\delta\theta = \frac{\sqrt{3}\lambda}{\pi D\,(2E\,/\,N_0)^{1/2}} \tag{3.64}$$

3.17.5 Response of a Matched Filter and an Ambiguity Diagram

The output of a matched filter is equal to the cross correlation between the received and transmitted signals. With a high signal to noise ratio the matched filter output is given by

$$Y_0(t) = \int_{-\infty}^{\infty} s_r(t)\, s^*(t - T_0)\, dt \tag{3.65}$$

where $s_r(t)$ is the received signal, $s(t)$ is the transmitter signal, $s^*(t)$ is the complex conjugate of $s(t)$, and to is the estisuated delay time. In complex form the transmitted signal may be expressed as $u(t)e^{j2}\,\pi^{fot}$, where $u(t)$ is the complex modulation function and f_0 is the carrier frequency. Assuming the received echo signal to be of the same form as the transmitted signal with time delay T_0 and the doppler frequency shift fd, the form of the received signal is given by

$$s_r(t) = n(t - T_0)\, e^{j2}\; \pi^{(f_0 + ft)(t - T_0)}. \tag{3.66}$$

In this expression the round trip loss of the signal from the transmitter to the target and back has been ignored, and in estimating the waveform of the matched filter output, the path loss will have n role except when the path loss is too large to affect the waveform by noise level. The matched filter output with received signal represented by equation 3.66 is given by

$$srm^{(t)} = \int_{-\infty}^{\infty} u(t - T_0)\, e^{j2\pi\;(f_0 + fd)(t - T_0)}\; [n\,(t - T_0)\, e^{j2\pi\, f_0\,(t - T_0^{\,*})}\;]\, dt$$

$$= \int_{-\infty}^{\infty} u(t - T_0)\, n^*(t - T_o)\, e^{j2\pi\,(f_0 + fd)(t - T_0)}\; e^{-j2\pi\, f(t - T_o)}\; dt$$

To get an idea of the form of the matched filter response, we may set $T_o = \theta$ and $f_o = \theta$, so that $T_o - T_o$ becomes $-T_o = T_R$, when the output of the matched filter is given by

$$x(T_R, f_d) = \int_{-\alpha}^{\alpha} n(t)\, n^*(t + T_R)\, e^{j2\pi fdt}\, dt \tag{3.67}$$

The ambiguity function is then $[\, x(T_R, fd)]^2$ which can be used to assess the properties of the transmitted waveform and its effect on target resolution, measurement accuracy, and response in clutter. A plot of the ambiguity function is called the ambiguity diagram.

WORKED OUT PROBLEMS

1. (*i*) The receiver of a radar has a noise figure 6 dB. If the IF bandwidth B of the receiver is 3 MHz, calculate the minimum detectable power.

Noise figure $F = 6 \text{ dB} = 10 \log_{10} r$

where $$r = \frac{(S/N)\,i}{(S/N_0)} = F \text{ in ratio}$$

$$F = \text{Antilog}\,\frac{6}{10} = 4 \text{ in ratio}$$

Minimum detectable noise power $P_{min} = kT_r B$

where the receiver noise temperature $T_r = (F-1)\,T$

$$\therefore \quad P_{min} = k(F-1)\,T\,B$$
$$= 1.38 \times 10^{-23}\,(4-1)\,(273+17)\,3 \times 10^6$$

where $T = 17°\text{C} = (273 + 17)\text{ K} = 290\text{ K}$

Now $1.38 \times 10^{-23} \times 290 \simeq 4 \times 10^{-21}$

$$\therefore \quad P_{min} = 4 \times 10^{-21} \times (4-1) \times 3 \times 10^6$$
$$= 36 \times 10^{-15}\text{ W.}$$

(*ii*) If the radar is designed for operation at 10 GHz with an antenna of diameter 2m, calculate the peak pulse power required to have a maximum range of 1000 km with a target of cross-sectional area 20 m^2.

We have $$R^4 = \frac{P_t A^2 \sigma}{4\pi \lambda^2 P_{min}}$$

or $$(1000 \times 10^3)^4 = \frac{P_t \left[\frac{\pi D^2}{4}\right]^2 \sigma}{4\pi \lambda^2 P_{min}} = \frac{P_t \left[\frac{\pi 2^2}{4}\right]^2 \times 20}{4\pi\lambda^2 P_{min}}$$

$$= \frac{P_t \pi^2 \times 20}{4\pi (.03)^2\, 36 \times 10^{-15}}$$

$$\simeq \frac{P_t \times 10 \times 20}{4\pi \times 9 \times 10^{-4} \times 36 \times 10^{-15}} \text{ assuming } \pi^2 \simeq 10$$

$$= \frac{P_t \times 2 \times 10^2 \times 10^{15} \times 10^4}{4\pi \times 9 \times 36}$$

$$= \frac{P_t \times 2 \times 10^{21}}{(36)^2\, \pi} = \frac{P_t \times 2 \times 10^{21}}{4072}$$

or $$P_t = \frac{10^{24} \times 4072}{2 \times 10^{21}} = 2036 \times 10^3 \text{ W} \simeq 2 \times 10^6 \text{ W} = 2 \text{ M W}.$$

(iii) With the 3 MHz bandwidth of the radar receiver, calculate the highest range resolution realizable with the radar.

Pulse duration $$\tau \geq \frac{1}{B}$$

$$\simeq \frac{1}{3 \times 10^6}$$

$$\simeq 0.33\ \mu \text{ sec.}$$

For highest range resolution $\tau = 0.33\ \mu$ sec.

Then the range resolution $$\Delta r = \frac{c\,\tau}{2}$$

$$= \frac{3 \times 10^8 \times 0.33 \times 10^{-6}}{2} \text{ m}$$

$$\simeq 0.5 \times 10^2 \text{ m}$$

$$= 50 \text{ m.}$$

(*iv*) For a maximum unambiguous range of 1000 km, calculate the PRF required for the radar.

If the interpulse period is T then the maximum unambiguous range

$$r_{u_{max}} = \frac{cT}{2}$$

or

$$1000 \times 10^3 = \frac{3 \times 10^8\, T}{2}$$

$$T = \frac{2 \times 10^6}{3 \times 10^8} = .6667 \times 10^{-2} \text{ sec}$$

$$= 6.667 \times 10^{-3} \text{ sec}$$

$$= 6.667\ \mu\ \text{sec}.$$

The PRF for a maximum unambiguous range of 1000 km $= \dfrac{1}{T}$

$\therefore$

$$\text{PRF} = \frac{1}{6.667 \times 10^{-6}} = \frac{100 \times 10^4}{6.667}$$

$$\cong 15 \times 10^4 = 150 \text{ kHz}.$$

(*v*) With the interpulse period required for attaining the maximum unambiguous range of 1000 km, calculate the duty cycle of the pulse. Duty cycle

$$= \frac{\tau}{T} = \frac{0.33}{6.667} = \frac{1}{20} = .05.$$

(*vi*) With a 10 GHz radar if the antenna diameter is 2m, calculate the beam width of the antenna

$$\theta = 70\frac{\lambda}{D} \text{degrees} = 70\frac{.03}{2} = \frac{7 \times .3}{2} = \frac{2.1}{2} \simeq 1°.$$

2. A radar receiver has a low noise RF amplifier (LNA) of noise figure 6 dB and gain 20 dB followed by a mixer of conversion loss 6 dB. If the noise figure of the IF preamplifier is 3 dB with a gain of 26 dB and that of the IF main amplifier is 10 dB, calculate the noise figure of the receiver and also its noise temperature. Indicate contributions of the mixer and the IF preamplifier and IF main amplifier toward the receiver noise figure relative to that of LNA. The receiver noise figure

$$F = F_1 + \frac{F_2 - 1}{G_1} + \frac{F_3 - 1}{G_1\, G_2} + \frac{F_4 - 1}{G_2\, G_3} \tag{1}$$

Here $F_1 = 6$ dB

$$F_1 \text{ in ratio} = \text{Antilog } \frac{6}{10} = 4$$

F_2 = Noise figure of mixer = conversion loss of mixer

$= 6$ dB $= 4$ in ratio

$G_1 = 20$ dB $= 100$

$F_3 = 3$ dB $= 2$ (ratio), $G_3 = 26$ dB $= 400$ (ratio for IF preamplifier)

$$G_2 = -6 \text{ dB} = \frac{1}{4} \text{ (ratio)}$$

$F_4 = 10$ dB $= 10$ (ratio), for IF main amplifier

$$\therefore \quad F = 4 + \frac{4-1}{100} + \frac{2-1}{100 \times \frac{1}{4}} + \frac{10-1}{100 \times \frac{1}{4} \times 400}$$

$$= 4 + .03 + .04 + .0009 \tag{2}$$

$$= 4.0709$$

From equation (1) it is clear that the contribution of

the mixer is $\frac{.03}{4.0709} \simeq .0075 = 0.75\%$

while that of the IF preamplifier is $\frac{.04}{4.0709} \simeq .01 = 1\%$

The contribution of the IF main amplifier $= \frac{.0009}{4.0709} \simeq 2.25 \times 10^{-2}\%$

The noise temperature of the radar receiver is

$$T_R = (F-1)\,T \simeq (4.0709 - 1)\,290$$

$$= (3.0709)\,290$$

$$= 890.561 \text{ K}$$

$$\simeq 891 \text{ K.}$$

CHAPTER 4

RADAR TARGETS

4.1 INTRODUCTION

The range equation of the radar gives the maximum range at which a target may be detected with a given probability. It includes the radar cross section (RCS) of the target, which is nothing but a measure of the proportion of the incident energy scattered back to the radar. Of course, the returned energy varies with transmitted wavelength, target geometry, orientation, and reflectivity.

4.2 RADAR CROSS SECTION

The radar cross section of a target σ may be defined as

$$\sigma = \frac{\text{Power reflected toward source/unit solid angle}}{\text{Incident power flux/ } 4\pi}$$

Utilizing the previous definition, let us consider the RCS of a perfectly conducting isotropic scatterer. The product of the incident power flux P_I and its geometric projected area A_I gives the power intercepted by the radiator. For the isotropic scatterer, we may thus write

$$\sigma_i = \frac{P_I A_I / 4\pi}{p_I / 4\pi} = A_I \tag{4.1}$$

The previous equation shows that the RCS of an isotropic reflector is equal to its geometric projected area.

The scattering properties of a sphere illustrate the nature of the RCS wavelength dependence. It has been found that for the RCS there are three distinct regions of behavior for a sphere of radius r.

1. *The Rayleigh region.* In this region $2\pi r/\lambda < 1$ and the RCS is proportional to λ^{-4}.
2. *The Mie or resonance region.* In this region $2\pi r/\lambda \cong 1$ and the RCS oscillates between the ever diminishing values for decreasing wavelengths.
3. *Optical region.* In this region λ is very small compared to the dimension of the sphere, and the fluctuations of the Mie region converge to a constant optical cross section πr^2 for all shorter wavelengths.

The radar cross section of a sphere as a function of its circumference measured in wavelengths is shown in Figure 4.1.

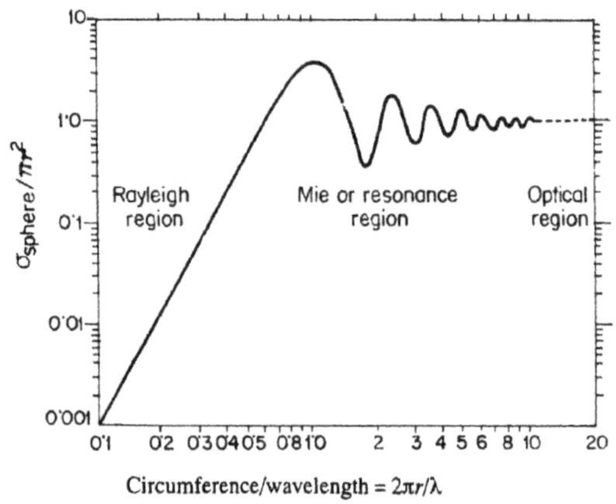

FIGURE 4.1 Radar cross section of a sphere of radius r at a wavelength λ.

4.2.1 The Rayleigh Region

Though the radar targets do not fall, in general, within the Rayleigh region, others scatter like the precipitation particles do. Siegel established that the critical parameter in the determination of RCS is the volume. He showed that for most bodies of revolution the RCS along the axis of symmetry is

$$\sigma = \frac{4}{\pi} K^4 V^2 F^2 \tag{4.2}$$

Where $K = \frac{2\pi}{\lambda}$

V = volume of the scatterers in m^3

F = dimensionless factor which depends on the gross shape of the body.

The previous approach is less accurate if the scatterer becomes more flat since a flat surface has zero volume but a non-zero cross section.

In the Rayleigh region, the RCS of any object is sensitive to polarization and aspect angle if one major dimension is very small in comparison to wavelength.

4.2.2 The Mie Region

In between the Rayleigh region and the optical region there is a region where the geometry of the body is a critical factor and none of the previous two methods can be conveniently applied to the problem of RCS estimation. A few techniques for RCS estimation are available in this region, but they are too complicated and no simple generalizations are possible.

4.2.3 The Optical Region

In this region most of the radar targets occur. It is so named because the ray techniques of geometric optics are applied to the problem of RCS estimation. From a consideration of power reduction, the RCS is found to be

$$\sigma \approx \pi R_1 R_2 \tag{4.3}$$

where R_1 and R_2 are the principal radii of curvature at the surface normal. The RCS behavior with wavelength is monotonic in the optical region, and the behavior may be classified for many objects in terms of principal radii of curvature.

The RCS of large complex reflectors can be approximated by breaking the body into individual reflectors, assuming that the parts do not interact. The total RCS in that case can be expressed as the vector sum of the individual cross sections,

$$\sigma = \left| \sum_k \sqrt{\sigma_k} \exp\left(\frac{j4\pi d_k}{\lambda} \right) \right|^2 \tag{4.4}$$

where σ_k = RCS of the kth scatterer

d_k = distance between the kth scatterer and the receiver.

Optical approximation may be used in the calculation of specular returns of their side lobes.

4.2.4 Features in Rayleigh, Mie, and Optical Regions

From Figure 4.1 we have for the Rayleigh region $\frac{\text{circumference}}{\text{wavelength}} = \frac{2\pi a}{\lambda} = \frac{1}{3}$, $\delta = 0.1\,\pi\,r^2 = \frac{1}{10} \times$ projection of the sphere. From (1), (2), and (3) we have that for $\frac{2\pi r}{\lambda} = 0.7$, $\delta = \pi r^2$ = projection of the sphere, and for $\frac{2\pi r}{\lambda} = 0.7, \delta = \pi r^2 = \frac{1}{1000} \times$ projection of sphere.

In the Mie or resonance region, the RCS is maximum for the circumference wavelength $\frac{2\pi r}{\lambda} = 1$ for which $\sigma = 4\pi r^2 = 4 \times$ projection of the sphere.

The RCS of a metallic sphere of radius r in the optical region for $\frac{\text{circumference}}{\text{wavelength}} = \frac{2\pi r}{\lambda} = 10$ is given by $\delta = \pi r^2$ = projection of the sphere, which is in fact, valid for $\frac{2\pi r}{\lambda} \geq 10$.

4.3 BACK SCATTER CROSS SECTION

In the case of a sphere, the RCS is not dependent on aspect, it is viewed. But the cross section of other objects will depend on the direction as viewed by the radar. As a radar receives the echo scattered back along the direction of incidence on the target, the RCS is also called the echoing area or back scatter cross section.

A plot of the back scatter cross section of the long thin rod as a function of aspect is shown in Figure 4.2. The length of the rod is 39 λ while its diameter is 4/λ, and it is made of silver. If instead of silver the rod is made of steel, the first minimum would be about 5 dB below that shown. The radar cross section of the thin rod or similar objects is small when viewed end on (i.e., $\theta = 0°$) as the physical area involved is small. But at near end on, the incident waves couple onto the scatterer which travels down the length of the object and hence reflects from the discontinuity at the far end, which gives rise to a traveling wave component.

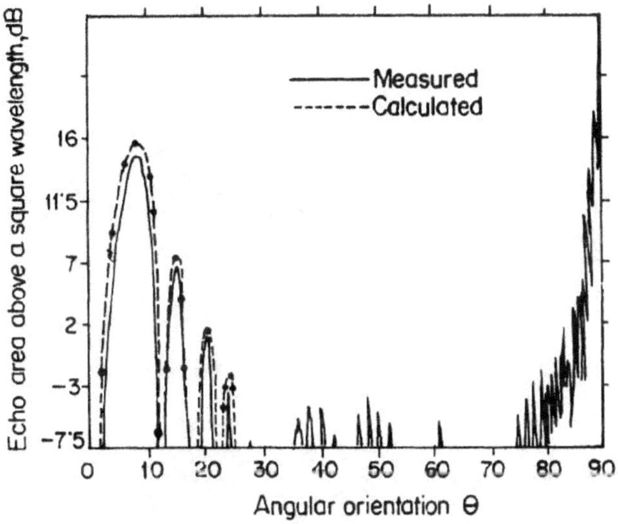

FIGURE 4.2 Radar cross section of a long thin rod.

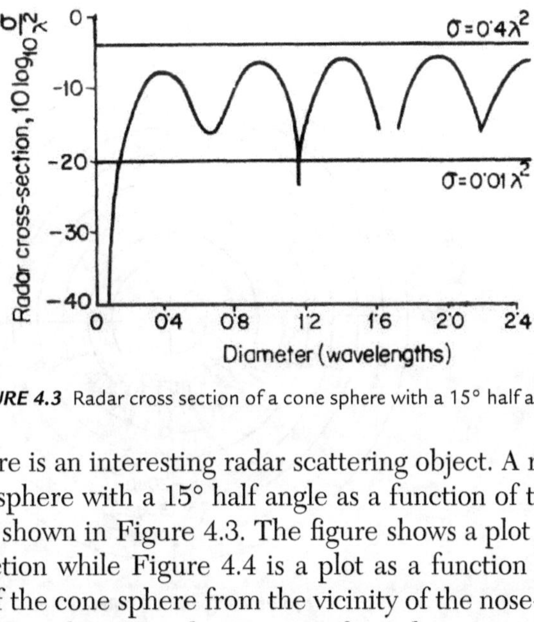

FIGURE 4.3 Radar cross section of a cone sphere with a 15° half angle.

The cone sphere is an interesting radar scattering object. A radar cross section of a cone sphere with a 15° half angle as a function of the diameter in wavelengths is shown in Figure 4.3. The figure shows a plot of the nose-on radar cross section while Figure 4.4 is a plot as a function of aspect. The cross section of the cone sphere from the vicinity of the nose-on direction is very low. Scattering from any object occurs from discontinuities, and hence the back scattering of the cone sphere is from the tip and from the join between the cone and the sphere.

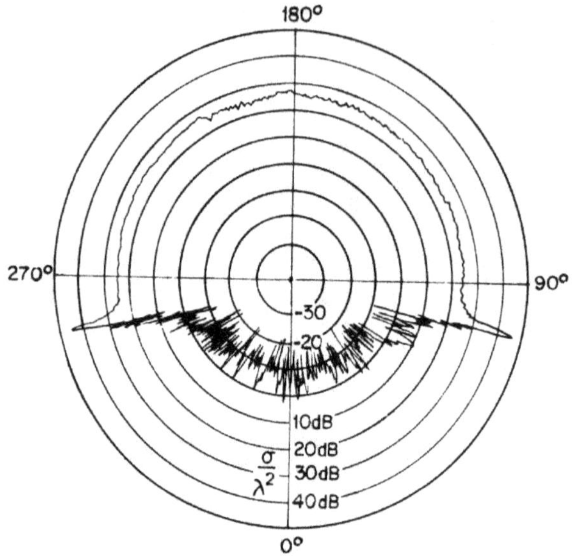

(*a*) Horizontal (perpendicular) polarization

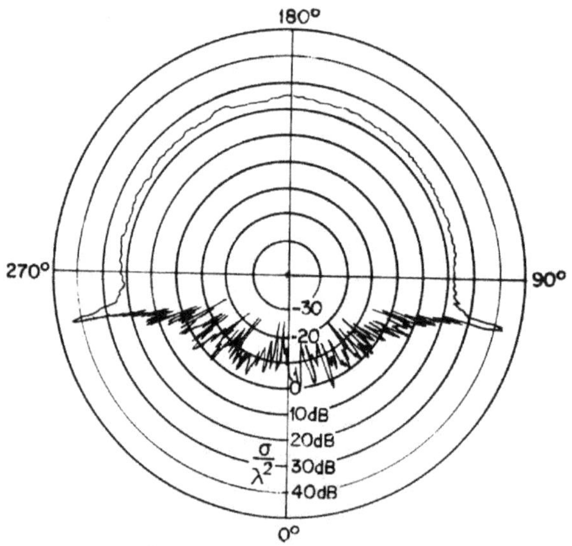

(*b*) Vertical (parallel) polarization.

FIGURE 4.4 Experimentally measured radar cross section of a cone sphere of a 12.5° half angle and a 10.14 λ radius base.

The nose-on radar cross section is low and it decreases with the square of the wavelength. Over a relatively large angular region, the cross section is small. If the cone sphere is viewed at near perpendicular incidence to the cone surface, that is, for $\theta = 90° - \alpha$, where α is the cone half angle, a large specular return is obtained. The maximum value of the nose-on cross section of the cone sphere is about $0.4\ \lambda^2$ and the minimum value is $0.01\ \lambda^2$ for a wide range of half angles for frequencies above the Rayleigh region. If a typical value of $0.1\ \lambda^2$ is taken, then the cross section at the S-band ($\lambda = 0.1$ m) is 10^{-3} m^2 and at the X-band ($\lambda = 3$ cm) the cross section is 10^{-4} m^2, so the cone sphere, in theory, can have a very low back scatter energy. To realize the very low value of the radar cross section for a cone sphere, the tip of the cone must be sharp and the surface must be smooth. Also the join between the cone and the sphere must have a continuous first derivative, and there must be no holes or windows on the surface. In Figure 4.5, a comparison of the nose-on cross section of several cone-shaped objects is shown.

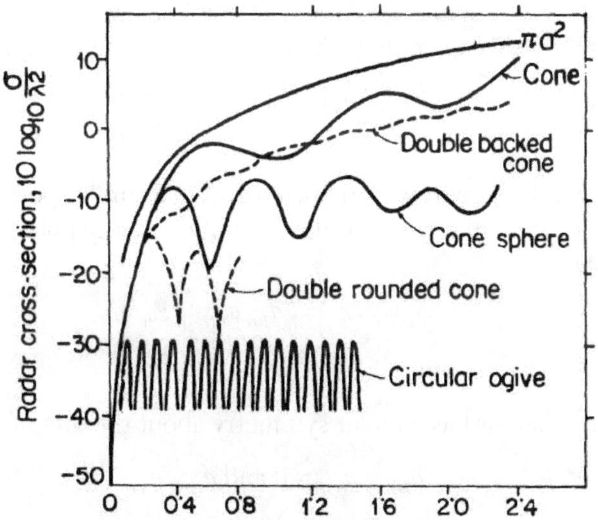

FIGURE 4.5

Shaping the target with the cone sphere is a good method for reducing the cross section of the radar. Materials like carbon fiber composites further reduce the cross section of the targets.

4.4 POLARIZATION SCATTERING MATRIX

Let us consider that a linearly polarized plane wave is incident on a target. The incident wave field at the target is resolved into components of electric field E_H^T and E_V^T along the H and V axes respectively. Generally the component of scattered electric field E due to the incident H component only will have both H and V components.

The H component of the scattered field can then be expressed as

$$E_H^S = a_{HH} E_H^T \tag{4.5}$$

Similarly, we can write the V-component as

$$E_V^S = a_{HV} E_H^T \tag{4.6}$$

The first subscript in a_{HV} is for the transmitted component while the second is for receive. The vector sum of E_H^S and E_V^S can provide the total scattered energy. If both the components of the incident field are found to be present, then the scattered field can be stated in terms of the scattering matrix. We thus have

$$\begin{pmatrix} E_H^S \\ E_V^S \end{pmatrix} = \begin{pmatrix} a_{HH} & a_{HV} \\ a_{VH} & a_{VV} \end{pmatrix} \begin{pmatrix} E_H^T \\ E_V^T \end{pmatrix}$$

For circular or ellipical polarization also, a similar polarization matrix exists. If we express in terms of right and left circularly polarized waves E_R and E_L, we have

$$\begin{pmatrix} E_R^S \\ E_L^S \end{pmatrix} = \begin{pmatrix} a_{RR} & a_{RL} \\ a_{LR} & a_{LL} \end{pmatrix} \begin{pmatrix} E_R^T \\ E_L^T \end{pmatrix}$$

If the reflector has circular symmetry about the line of sight axis, then

$$a_{HV} = a_{VH} = 0 \text{ and } a_{HH}, a_{VV} \neq 0$$

$$a_{RR} = a_{LL} = 0 \text{ and } a_{RL}, a_{LR} \neq 0.$$

Also, for most of the cases, regardless of symmetry

$$a_{HV} = a_{VH}$$

$$a_{RL} = a_{LR}.$$

An important consequence of the previous properties is that a radar where the same circularly polarized antenna is used for both transmission and reception will receive practically no power from spherical targets like raindrops. What actually happens is that a circularly polarized wave incident on a spherical scatterer is reflected as a circularly polarized wave with the opposite sense of rotation and is, therefore, rejected by the antenna which transmitted it. The reversal of the sense of rotation of polarization occurs due to the phase changes of 180° in each of the cross polarized components constituting the circularly polarized incident wave. This property rejection of echo from spherical targets in circularly polarized antennas is used to ascertain the ellipticity of raindrops and at the same time to reject rain clutter on both military and air traffic control radars.

Many experiments have been made to study the reduction of the radar cross section (RCS) of raindrop echoes using same sense of circular polarization (a_{RR} or a_{LL}). The reduction for most of the experiments has been 15 to 30 dB. For surface radars and low elevation angle beams (0–10°) the reduction is small. Some results are shown in Table 4.1 for different frequency bands and elevation angles.

It may be mentioned here that recent measurements of the ratio of horizontal to vertical reflectivity by dual polarization radar developed in the U.K. indicated that the raindrops are in reality not spherical but are mostly oblate spheroids. Accordingly, the clutter rejection mentioned previously cannot be complete.

4.5 COMPLEX TARGETS

Radar cross sections of complex targets like ships, aircraft, cities, and terrain are rather complicated. With the help of digital computers, the target cross sections may be computed or they can be measured experimentally with full-scale targets. However, the measurement is more convenient on scale models at the proper scaled frequency. A complex target comprises a large number of independent objects which scatter energy in various directions. If the separation between the individual scattering objects is greater compared to the wavelengths, then the phases of the individual signals at the radar receiver will change, resulting in a "scintillating echo."

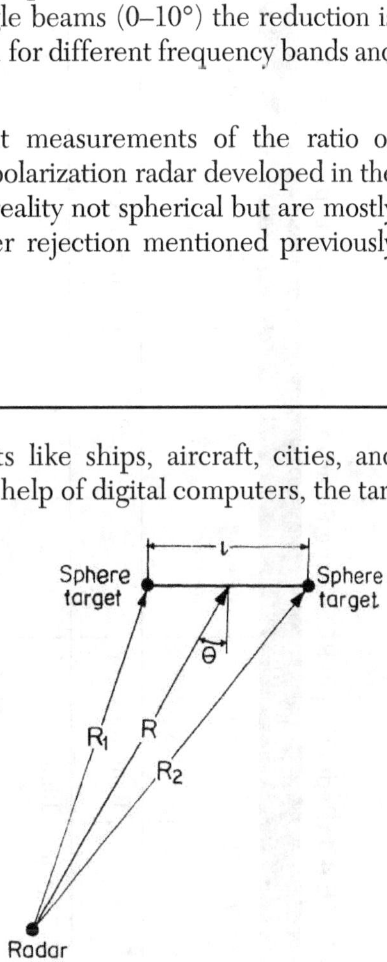

FIGURE 4.6 Geometry of a two scatterer complex target.

TABLE 4.1 Reduction in RCS Using Polarization Properties (Experimental Data).

Precipitation				Targets			
Frequency band	*Type*	*Elevation degree*	σ_{RL}/σ_{RR} *in dB*	*Type*	*Aspect degree*	$\frac{\sigma_{HH} \text{ or } \sigma_{VV}}{\sigma_{RR} \text{ or } \sigma_{LL}}$ *in dB*	$\frac{\sigma_{HH} \text{ or } \sigma_{VV}}{\sigma_{HV} \text{ or } \sigma_{LL}}$ *in dB*
L, 24 cm	Rain wet snow		22-30	Piston Aircraft land	-3 el	6-8	7-8
C, 5.6 cm	Rain	0-10	17	Land Vehicles	-3 el		6-10
				Trees Ships	-3 el		5-16
				Ships	-3 el		0-6
					-3 el		– (2-8)
					-6 el		3-5
X, 3.2 cm	Thunderstorm	high	15	Aircraft	Nose and tail	Av 2.5	2.5-12
	Rain						
	Bright band		26-28	Jet aircraft	Nose	4.5-6	10
	Fire snow		13-20	Ships	0 el.	6	
			26	Ships	–(3-6) el		7-15
K_a 1.2 cm	Rain	high	26				
K_a 0.86 cm	Rain	30	17-18	Jet and piston	Tail	3-5	7-11
	Bright band	30	5-11	Aircraft	Nose	2-4	8-10
	Dry snow	30	12-15	Trees	-3 el		4
				Vehicles	-3 el		8-14

Let us consider the scattering due to a complex target of two equal spheres separated by a distance l, as shown in Figure 4.6. It is assumed that,

$$l < \frac{c\tau}{2}$$

where c is the velocity of propagation and τ is the pulse duration. It is further assumed that l is small compared to R, the distance from radar to target, and also, $R_1 \approx R_2 \approx R$. Let the cross sections of the two targets designed as σ_0 be equal. Then the composite cross section σ_r of the two scatterers is

$$\frac{\sigma_r}{\sigma_0} = 2\left[1 + \cos\left(\frac{4\pi l}{\lambda}\sin\theta\right)\right] \tag{4.7}$$

The value of σ_r/σ_0 may vary from zero to a maximum of four times the cross section of an individual scatterer. Polar plots for the two-scatterer complex target are shown in Figure 4.7 for

(a) $l = \lambda$ *(b)* $l = 2\lambda$ and *(c)* $l = 4\lambda$.

The radar cross section of an actual target is far more complicated than the simple two-scatter target discussed previously since practical targets are composed of many individual scatterers with different properties. The most realistic way for getting the radar cross section of an aircraft is to measure the actual target in flight. Radars at L, S, C, and X bands illuminate the aircraft target in flight. It is seen that a small change in viewing aspect or frequency results in large fluctuations in cross section. Values of cross sections for different targets at microwave frequencies are shown in Table 4.2. In the table one value is given, though there may be large variations.

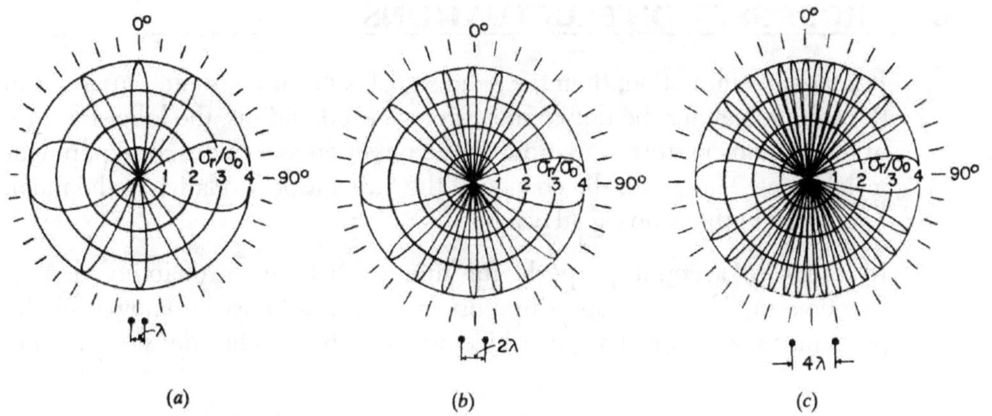

FIGURE 4.7 Polar plots of σ_r/σ_0 for the two-scatterer complex target.

TABLE 4.2 Radar Cross Sections at Microwave Frequencies

	m^2
Conventional, unmanned winged missile	0.5
Small, single engine aircraft	1
Small fighter or 4 passenger jet	2
Large fighter	6
Medium bomber or medium jet airline	20
Large bomber or large jet airliner	40
Jumbo jet	100
Small open boat	0.02
Small pleasure boat	2
Cabin cruiser	10
Pickup truck	200
Automobile	100
Bicycle	2
Man	1
Bird	0.01
Insect	10.5

The radar cross section of an automobile at *X* band is greater than that of an aircraft or boat, in general. The cross section may vary from 10 to 200 m^2 at *X* band, and it increases with increasing frequency.

4.6 CROSS SECTION FLUCTUATIONS

If a target is in motion, then the echo signal is never constant. Variations in the echo signal may be due to meteorological conditions, the lobe structure of the antenna pattern, variations in the target cross sections, or equipment instabilities. Thus, with the change of the target aspect relative to the radar, a variation in the echo signal occurs.

In order to account properly for fluctuations, the probability density function and the correlation properties with time must be known for the particular target and type of trajectory. The probability density function provides the probability of finding any particular value of a target cross section between σ and $\sigma + d\sigma$. The auto-correlation function, on the other hand, describes the degree of correlation of the cross section with time or number of pulses.

Swerling calculated the detection probabilities for four different fluctuation models or cross sections. These are as follows:

Model 1. The echo pulses received for a target have constant amplitude but are independent from scan-to-scan. Echo fluctuation of this type is referred to as scan-to-scan fluctuation. The probability density function can be expressed as

$$P(\sigma) = \frac{1}{\sigma_{av}} e^{-\sigma/\sigma_{av}} \quad \sigma \geq 0 \tag{4.8}$$

Where σ_{av} = average cross section over all target fluctuations.

Model 2. The probability density function for the target cross section is given by equation (4.8), but in this case the fluctuations are more rapid and are independent from pulse-to-pulse (instead of from scan-to-scan).

Model 3. The fluctuation in this case is independent from scan-to-scan as in model 1, but the probability density function is written as

$$P(\sigma) = \frac{4\sigma}{\sigma_{av}^2} e^{-2\sigma/\sigma_{av}} \tag{4.9}$$

Model 4. The fluctuation is pulse-to-pulse following equation (4.9).

In all the previous models, the value of the cross section to be put in the radar equation is the average cross section. The signal-to-noise ratio required for getting a specified probability of detection without exceeding a specified false-alarm probability may be calculated for each model of target behavior. For the purpose of comparison, the non-fluctuating cross section may be called model 5.

A comparison of the five previous models of probability density function for a false-alarm number $nf = 10^8$ for $n = 10$ pulses integrated is shown in Figure 4.8. If the detection probability is high, then all four models in which the target cross section is not constant need a greater signal-to-noise ratio than the constant cross section of model 5.

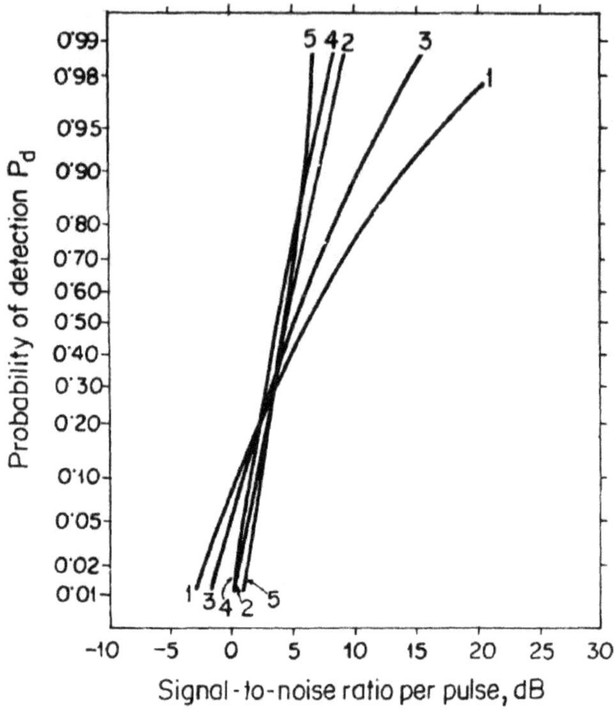

FIGURE 4.8 Comparison of detection probabilities for five different models of target fluctuations for $n = 10$ pulses integrated and false alarm number $n_f = 10^8$.

The probability density function expressed earlier in equations (4.8) and (4.9) are the special cases of the chi-square distribution of degree $2m$. The probability density function is

$$P(\sigma) = \frac{m}{(m-1)!\sigma_{av}} \left(\frac{m\sigma}{\sigma_{av}}\right)^{m-1} \exp\left(-\frac{m\sigma}{\sigma_{av}}\right), \sigma > 0 \tag{4.10}$$

It is also known as the gamma distribution. In statistics $2m$ is the number of degrees of freedom and is an integer. The chi-square distribution is a mathematical model which can be used for representing the statistics of the fluctuating radar cross section.

The chi-square distribution with $m = 1$ (Swerling models 1 and 2) is the Rayleigh distribution which results from a large number of independent scatterers. Again, the chi-square distribution with $m = 2$ (Swerling models 3 and 4) is indicative of scattering from one large scatterer and a collection of small independent scatters. It is, however, the Rice distribution that follows from such a model. The Rice probability density function can be expressed as

$$P(\sigma) = \frac{1+s}{\sigma_{av}} \exp\left[-s - \frac{\sigma}{\sigma_{av}}(1+s)\right] I_0\left[2\sqrt{\frac{\sigma}{\sigma_{av}} s(1+s)}\right] \sigma > 0 \tag{4.11}$$

where s = ratio of the radar cross section of the single dominant scatterer to the total cross section of the small scatterers.

$I_0\left[2\sqrt{\frac{\sigma}{\sigma_{av}} s(1+s)}\right]$ = modified Bessel function of zero order.

The log-normal distribution is also considered for the target echo fluctuation, which is expressed as

$$p(\sigma) = \frac{1}{\sqrt{2\pi}\, s_d \sigma} \exp\left\{-\frac{1}{2{s_d}^2}\left[\text{In}\left(\frac{\sigma}{\sigma_m}\right)\right]^2\right\}, \sigma > 0 \tag{4.12}$$

where s_d = standard deviation of $\text{In}\left(\frac{\sigma}{\sigma_m}\right)$

σ_m = median of σ.

The ratio of mean to median σ is exp. $\left(s_d^2 / 2\right)$.

It is certainly difficult to be precise about any statistical model that can be applied to any particular target.

4.7 FREQUENCY AGILITY EFFECTS ON TARGET DETECTION AND TRACKING

The capability of a radar to change its frequency from pulse-to-pulse is called its agility. During a single look at the target, a frequency agility of diversity radar, in fact, transmits several pulses on various carrier frequencies. After detecting the received signals, those are linearly integrated or combined with some logic technique. In the communication field, the diversity techniques have been undoubtedly accepted as a valuable tool to reduce the fading losses. But in radar technology the benefits are not so obvious.

In detection and tracking radars frequency agility can:

1. reduce range and angle tracking errors owing to the finite target extent and multipath effects,
2. reduce the probability of the aspect angle of a target which gives a null in the radar cross section,
3. allow some improvement in the target signal-to-clutter fluctuations for incoherent pulse radar, and
4. deny the possibility of concentrating all of the jamming power at a single carrier frequency.

The radar cross section versus frequency near a null for a 54-in. satellite is shown in Figure 4.9. The deep null at 85.7° aspect occurs only over a frequency span of about 40 MHz. If the reflecting points are uniformly distributed along a line target of length L, then the correlation coefficient ρ of the square-low echoes is

$$\rho(\Delta f) = \frac{\sin^2[2\pi \Delta f (L \sin\theta)/c]}{[2\pi \,\Delta f (L \sin\theta)/c]^2} \tag{4.13}$$

where Δf = carrier frequency shift

$L \sin\theta$ = projection of target length on radial dimension

c = velocity of light.

The previous equation is based on the assumption that the pulse length $c/2$ is greater than $L \sin\theta$. The critical frequency shift can now be written as

$$\Delta f_c = \frac{c}{2L\sin\theta} = \frac{c}{2L_0} \tag{4.14}$$

In equation (4.14) the term $c/2L_0$ is the inverse of the radial tangent extent in radar time, and L_0 is the effective radial extent of the scatterers.

Considering few approximations, the spectrum of the angle error signals for a tangent of linearly distributed scatters may be written as

$$W_n(f) = \frac{\bar{U}^2}{2f_m}\left(2 - \frac{3f^2}{f_m} + \frac{3f^2}{f_m^{\,2}} - \frac{f^3}{f_m^{\,3}}\right) \text{ for } f \le 2f_m$$
$$= 0, \text{ elsewhere}$$

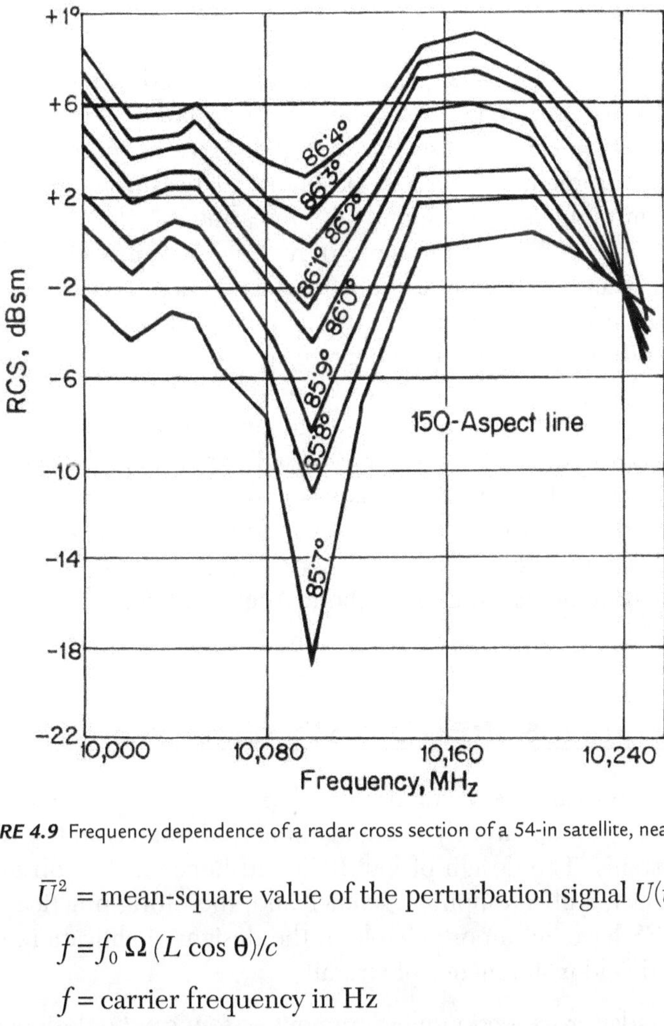

FIGURE 4.9 Frequency dependence of a radar cross section of a 54-in satellite, near null.

where, $\overline{U}^2$ = mean-square value of the perturbation signal $U(t)$

$f = f_0\, \Omega\, (L \cos \theta)/c$

f = carrier frequency in Hz

Ω = rotation rate of the line target about its center

$L \cos \theta$ = projection of the target perpendicular to line of sight.

The variance of $U(t)$ is found to be independent of Ω while the spectral width is directly proportional to Ω. For aircraft targets where the length is comparable to the wingspan, the same amount of frequency shifting per pulse ($\sim c/2L_0$) improves the detectability and reduces angle and range glint considerably.

4.8 RADAR CROSS SECTION MEASUREMENTS

Scattering of radar signals from an object produces a radar echo from which, knowing the range of the object, information about the radar reflectivity of the object can also be extracted. Current interest in radar reflectivity measurement is centered around: (*a*) measurements of the back scatter from rain, snow, land, and sea appearing as "clutter," (*b*) basic radar cross section (RCS) measurements of real and scale models of objects like aircraft, tanks and other military vehicles, missile nose cones, and models of objects having various shapes and sizes and made of various materials, (*c*) diagnostic high-resolution measurements in range and cross range to support target modeling and Radar Cross Section Reduction (RCSR) efforts, and (*d*) polarization measurements to support target identification and clutter suppression studies. Many types of experimental techniques are used to undertake these various classes of RCS measurements, including (*a*) tower based and airborne systems, (*b*) indoor experiments in a compact range, (*c*) outdoor experiments in an open range, (*d*) Real Aperture Radar (RAR) measurements, and (*e*) Synthetic Aperture Radar (SAR) measurements.

4.9 RCS MEASUREMENT SYSTEMS

The pressing need for the development of low-RCS military aircraft and spacecraft apparently initiated a renewed interest in the RCS measurement systems. The design of low RCS and hence a "low observable" vehicle is, in fact, still an empirical science and, therefore, practical measurements of RCS have an important role in the design, qualification, quality assurance, and field maintenance of aircraft.

A radar cross section measurement system can be designed for use at VHF, UHF, microwave, or millimeter wave and is based on an instrumentation radar at the frequency band of interest. The distinction between an instrumentation radar and a conventional radar is that the target range is usually known when an RCS measurement system is being used, and its task is to measure the RCS level of a target at known range, indicated by the amplitude and sometimes the phase of the radar return or echo signal. In contrast to this, the task of a conventional radar is to detect the presence of a target of unknown RCS level and measure its range.

4.10 PROBLEMS IN RCS MEASUREMENT

Radar cross section measurements are inherently difficult for the following reasons:

1. The large path loss which the transmitted radar signal has to suffer, causing the echo signal to decrease inversely as the fourth power of the target range as given by

$$P_r = P_t \frac{G^2\lambda^2}{(4\pi)^3 R^4}\sigma \tag{4.15}$$

where P_r = Received power, λ = Wavelength of the radar signal
P_t = Transmitted power, R = Range
G = Antenna Gain, σ = Radar Cross Section of target.

2. The range of the target has to be made sufficiently large to realize the classical far-field test range for which the phase error of illumination across the test aperture, covering the target as well as that of the scattered radiation across the receiving antenna, may less than one eighth of a wavelength. This condition is met when the range

$$R \geq \frac{2E^2}{\lambda} \tag{4.16}$$

where E = extent or width of the test aperture and
λ = wavelength of the radar test signal.

3. To keep the amplitude of illumination across the target uniform, the gain of the antenna cannot be increased beyond a limit where the beamwidth of the radiation pattern would become too narrow to produce a tapered illumination of the target. To keep the amplitude tapered across the test aperture as well as across the radar antenna for the scattered return signal to less than 1 dB, the gain of the antenna G must satisfy the relation

$$G = < (R/E)^2 \tag{4.17}$$

Thus, the combined effect of path loss or range attenuation, the far field range requirement, and the limitation on antenna gain result in a low power level for the received radar echo. It is thus difficult to make far-field RCS measurements requiring high receiver sensitivities and high transmit powers receive an echo signal which may overcome the receiver noise. The problem is aggravated for large targets of low radar cross sections. A large target calls for (*a*) increased far-field range or (*b*) a decreased antenna gain to attain a broader beamwidth required to cover the target or a combination of (*a*) and (*b*). Either of these factors reduces the level of echo signal scattered by the target in addition to the reduction caused by the low radar cross section of the target.

4.11 SENSITIVITY OF RCS MEASUREMENT

The RCS measurement sensitivity is defined to be the minimum radar cross section $\sigma_{\min}$, of a target which produces a detectable echo signal using the range equation (1). The sensitivity is given by

$$\sigma_{min} = \frac{(4\pi)^3 R^4}{G^2 \lambda^3}\left(\frac{S}{P_t}\right)_{min} \tag{4.18}$$

where the minimum detectable echo power P_r of equation (1) is assumed to be equal to the receiver sensitivity S.

The RCS measurement sensitivity σ_{min} is modified in practice due to (*a*) added losses L, from cables and other system losses, and (*b*) any gain G_p, due to ground plane effect. Including these in equation (4.18), it is modified to

$$\sigma_{min} = \frac{(4\pi)^3 R^4 L}{G^2 \lambda^3 G_p}\left(\frac{S}{P_t}\right)_{min} \tag{4.19}$$

Equation (4.19) for RCS measurement sensitivity indicates that the ratio of receiver sensitivity to transmitter power, $\frac{S}{P_t}$, is a key parameter and should be minimized to reduce σ_{min}, thereby obtaining a better RCS measurement sensitivity. This ratio $\frac{S}{P_t}$ can therefore be used as a figure of merit for the radar. The equation also shows that an increase of antenna gain would improve the RCS measurement sensitivity by lowering σ_{min}. However, the increase of antenna gain may bring excessive amplitude taper of the target illumination due to a consequent reduction of the antenna beamwidth.

It may be noted that the dependence of σ_{min} on wavelength cannot be properly assessed from equation (4.19), because the antenna gain which is constrained by equation (4.17) is a function of λ. The effect of antenna gain on σ_{min} and its beamwidth on the amplitude taper of target illumination in relation to the wavelength λ can be incorporated in equation (4.19) as follows:

The beamwidth of an antenna is given by

$$\theta = \alpha\frac{\lambda}{D} \tag{4.20}$$

where θ = Antenna beamwidth

α = constant dependent on amplitude taper and antenna configuration. $\alpha \approx 0.73$ for 1 dB one way amplitude taper for a circularly symmetric aperture antenna

D = Diameter of antenna aperture

The beamwidth of the antenna must be greater than the angle subtended by the target for obtaining a nearly uniform target illumination. Accordingly, we have

$$\theta \geq \frac{E}{R} \tag{4.21}$$

Combining equations (4.20) and (4.21) we have

$$\theta = \alpha\frac{\lambda}{D} \leq \frac{E}{R}$$

[taking the equality sign in equation (4.21)]

or

$$D \leq \frac{\alpha\lambda R}{E} \tag{4.22}$$

Now, gain of the antenna is given by

$$G = \frac{4\pi A}{\lambda^2}\eta_A \tag{4.23}$$

where A is the area of the antenna and η_A is the efficiency.

Substituting $A = \dfrac{\pi D^2}{4}$ in equation (4.22)

we have
$$G = \frac{4\pi}{\lambda^2} \cdot \frac{\pi D^2}{4} \cdot \eta_A$$

$$= \frac{\pi^2 D^2}{\lambda^2} \eta_A \tag{4.24}$$

The maximum permissible gain of the antenna will be obtained by combining equations (4.22) and (4.24), and we then have

$$G \leq \frac{\pi^2 \alpha^2 \lambda^2 R^2}{\lambda^2 E^2} \eta_A$$

or
$$G_{max} = \frac{\pi^2 \alpha^2 R^2 \eta_A}{E^2} \tag{4.25}$$

Substituting the optimum value of G equal to G_{max} given by equation (4.25), in equation (4.19) we get an alternate expression for RCS measurement sensitivity given by

$$\sigma_{min} = \frac{64 E^4 L}{\pi \alpha^4 \eta_A{}^4 \lambda^2 G_p} \left(\frac{S}{P_t} \right)_{min} \tag{4.26}$$

This shows that under optimal conditions of antenna gain, the RCS measurement sensitivity is independent of the range R but is a function of the target extent E. Also, the sensitivity depends on wavelength λ, improving at longer wavelengths with $\sigma_{min} \propto 1/\lambda^2$. The use of a longer wavelength, however, creates other disadvantages which are not apparent in equation (4.26). In fact, the RCS measurement systems of shorter wavelengths in the millimeter wave region have certain unique advantages over those at VHF, UHF, and microwave bands:

1. The equipment is smaller in size and weight than microwave equipment.
2. Wide bandwidths of the order of a few GHz are possible at millimeter waves, permitting high-range resolution and doppler processing. The range resolved RCS data of a target obtained sequentially for different orientations of the target may be combined to construct a three-dimensional RCS model of the target.
3. A high angular resolution with a high gain of the antenna is feasible at millimeter waves, and these qualities can be exploited to realize an imaging radar to resolve the finer details of a target. High-resolution

scanning in two dimensions coupled with the high-range resolution along the third dimension results in a three-dimensional precision radar image of the target obtained at the fast scanning rate realizable at millimeter waves.

4. Target size can be scaled down in the ratio of wavelengths given by $r = \frac{\lambda_{mmwave}}{\lambda_{microwave}}$, to facilitate RCS measurements within a small laboratory space. The scaled down RCS measurements are, however, subject to the limitations that the target properties at millimeter waves may not truly reflect those expected at microwaves with a full scale model. Even so, scale model measurements are now being used extensively to study the effect of varying the shape and size of the target in relation to wavelength. Attempts are also being made to deduce the full-scale RCS at microwaves from measured scale model RCS at millimeter waves by taking into account the frequency dependence of the target reflectivity.

4.12 COMPACT RANGE RCS MEASUREMENT

For RCS measurement in an open range, the range must be sufficiently large to satisfy the far-field condition, given by $R \geq 2D^2/\lambda$, for which the wavefront incident on the target may be assumed to be near a plane. An alternative method of creating the far-field conditions at a compact range with the target at a short distance from the transmitter makes use of a parabolic reflector to convert the spherical wavefronts near the transmitter to a plane wavefront as shown in Figure 4.10. Within the test zone of volume $\Delta V = \Delta x \, \Delta y \, \Delta r$ the beam may be considered parallel and, therefore, the wavefront would be plane. Such a compact range RCS measurement system is being widely employed for precision modeling of RCS within the bounds of a small laboratory.

In fact a small room with floor space of only 12 ft × 8 ft is required for compact range RCS measurement in the microwave and millimeter wave bands down to 1 GHz. Commercial versions of the equipment are, in fact, available for operation in the frequency range 1—95 GHz.

The compact range configuration produces a test zone limited by the range and cross range in an orthogonal direction by limiting the dimension of the reflector. Standard models of the equipment manufactured by Scientific Atlanta, Inc. provide a quiet zone of length 4 ft and diameter 4 ft for RCS measurement over 2—18 GHz requiring a floor space of only 12 ft × 8 ft.

Improved reflectors for compact range are now capable of providing quiet zones which are 6 ft long and 4 ft in diameter. Also larger reflectors operating up to 94 GHz have been developed to provide a quiet zone 12 ft long and 8 ft in diameter. The length of a compact range would appear to be unlimited from the figure which shows a parallel. Beam, formed by the reflector, illuminating the target. However, really we are to consider a wave theory instead of the ray-concept, when the beam would be expected to diverge beyond a certain range, as shown by the dotted lines, thus limiting the longitudinal extent of the test zone.

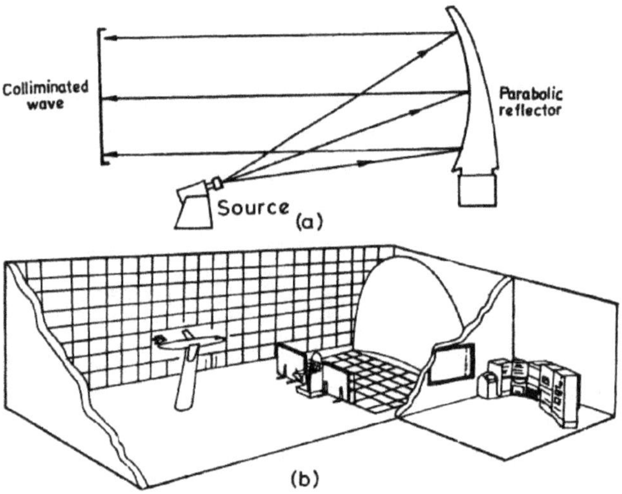

(*a*) Illustration of a compact range concept.
(*b*) Typical compact range configuration (drawing Courtesy of Scientific Atlanta.)

FIGURE 4.10.

With the introduction of the compact range RCS measurement system, we now have the ability to test the RCS of targets in a small indoor chamber of length about 12 ft, which normally would require outdoor ranges of hundreds or even thousands of feet long. Also, the indoor measurements allow operation in a precisely controlled environment of temperature, pressure, and humidity to minimize the effect of atmospheric variations on the measurement in compact range. Measurements may, however, suffer due to residual background radar return, or clutter produced by the anechoic chamber, tending to limit the measurement sensitivity, and special care must, therefore, be taken in the design of the anechoic chamber to minimize the clutter.

4.13 INSTRUMENTATION RADARS FOR RCS MEASUREMENT

RCS measurement is made with an instrumentation radar which is specially designed to measure the RCS level of a target at a known range. In contrast to this the design of conventional radar is aimed at detecting the presence of a target and to measure its range without depending on the knowledge of its RCS level. Some assumption of the RCS level of probable targets is however, necessary to specify the range capability of a conventional radar. The advent of low RCS aircraft called STEALTH, in recent years, has apparently imposed a new challenge to the design of conventional operational radars. Instrumentation radars capable of measuring the RCS of STEALTH models have been designed.

TYPE OF INSTRUMENTATION RADARS

The instrumentation radars can be classified into the following basic types: (*a*) CW Radar, (*b*) FMCW Radar, (*c*) Gated CW Radar, and (*d*) Pulse Radar. In order to select a particular type which would be best suited for a particular service, the following key performance criteria are considered:

(*i*) Frequency coverage, (*ii*) Range coverage, (*iii*) Measurement capability, (*iv*) Measurement sensitivity, (*v*) Measurement, (*vi*) Dynamic range, (*vii*) Clutter isolation, (*viii*) Target resolution, (*ix*) Measurement time, and (*x*) Ease of operation.

In describing the various types of instrumentation, radars of each of these criteria will be discussed.

4.13.1 CW Radar

The simplest type of instrumentation radar is a CW Radar, a block diagram of which is shown in Figure 4.11.

The phase of the received carrier will be shifted with respect to that transmitted by the total phase path for the round trip propagation of the CW signal from the target. If r is the range of the target, then the phase path is given by

$$\phi = \frac{2\pi}{\lambda} 2r$$

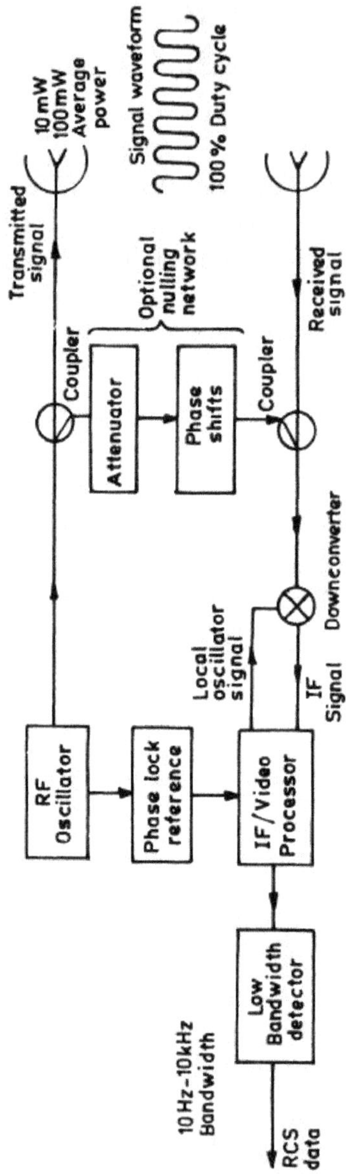

FIGURE 4.11 In CW radar—the simplest category of instrumentation radars—the transmitted signal is a constant tone modulation in neither frequency nor amplitude.

However, if ϕ exceeds 2π, which will usually be the case for the ranges involved in practice, it will not be possible to determine the range involved unless the transmitted pulse is coded in phase or amplitude in some way to resolve the ambiguity. A CW radar will however be useful in the measurement of RCS, by assuming the known range of the target. Even then the radar has the limitation that it cannot distinguish the target return from undesired clutter due to return from objects around the target. A CW radar can detect doppler shift due to a moving target, but this has no significance in RCS measurement where the target is fixed in position.

4.13.2 FM-CW Radar

In FM-CW Radar, the RF signal is frequency modulated, usually by a linear sweep with a triangular modulating waveform or often with a sinusoidal modulation like that in a radio altimeter (included in Art. 13.8). A block diagram of the FM-CW instrumented radar for RCS measurement is shown in Figure 4.12.

The additional information content in the radar signal due to the frequency modulation allows the target range to be determined by the radar as in the radio altimeter. For multiple targets at different ranges, it is possible to receive the return from each target by appropriate signal processing. This capability is useful in RCS measurement against clutter originating at distances other than that of the test target.

4.13.3 Gated CW Radar

In the gated CW radar the transmitter is pulse modulated by using a short modulating pulse at a fixed repetition rate. The receiver is gated to receive the echo pulse from only the desired range where the target is located, thus eliminating clutter echoes from objects at other ranges. Both the position and duration of the gate is controllable. A block diagram of a gated CW instrumentation radar for RCS measurement is shown in Figure 4.13.

However, increasing range resolution in this way also reduces at the some time the target illumination energy.

High-resolution RCS data for various target orientations can be obtained by software to generate a high-resolution three-dimensional model of the RCS. Usually, the pulse width is selected to be long enough so that the received energy represents and equal weighting of all target points along the range. The receiver bandwidth in the gated CW receiver is narrow, even though the spectral bandwidth of the short radar pulse, chosen to illuminate the entire target or a part of it, is lange. The receiver response will, therefore, be lengthened considerably in the time domain to encompass a number of successive pulses. The response will in fact be an average of many return pulses. This will, however, be without any adverse effect on the RCS measurement and, as such, an average represents the RCS, particularly for stationary targets. The return will in fact be equivalent to that due to a CW transmission with the same average power requiring a small bandwidth as in a CW Radar, while at the same time allowing a high-range resolution and clutter rejection, attained by adjusting the position and duration of the gating pulse. The only limitation of this type of instrumentation radar is that the data rate would be

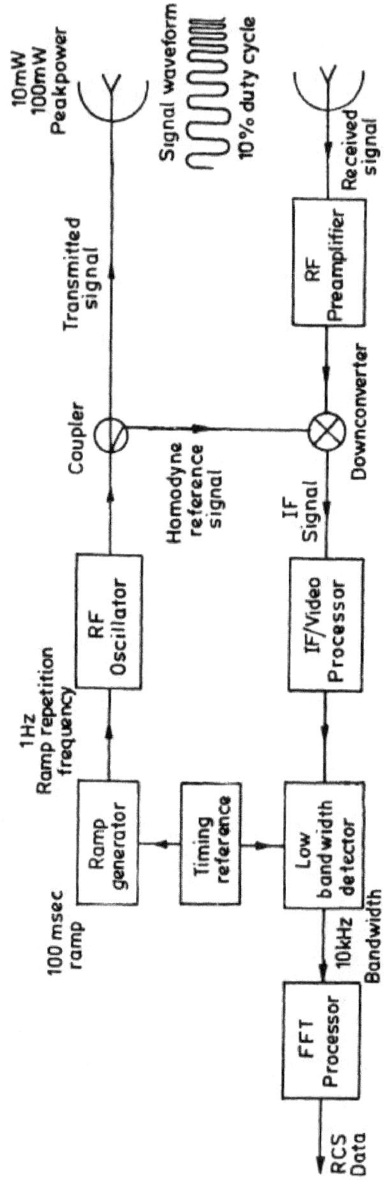

FIGURE 4.12 In this coherent FM-CW radar, a sample of the swept frequency transmitted RF is mixed with the received RF signal to produce an audio-frequency IF.

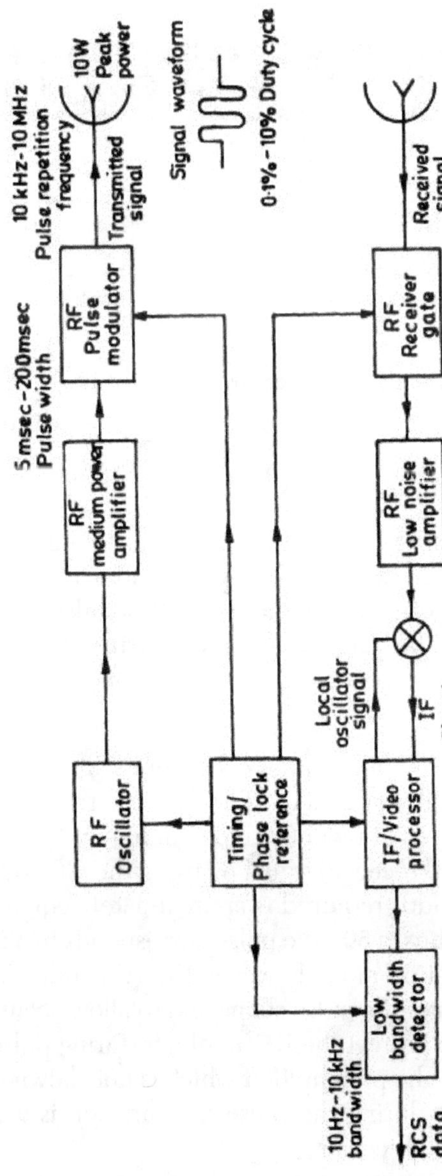

FIGURE 4.13 In gated CW radars, the RF signal is amplitude modulated. A moderately short pulse of RF energy is transmitted to the target at a fixed repetition rate, so the pulse width and FRF become operational parameters of the radar.

lower because of smaller receiver bandwidth, and as such RCS measurement time will be longer compared to what would be possible with receivers allowing the full spectrum of the echo pulse to be received. Scientific Atlanta manufactured such a gated instrumentation radar for RCS measurement up to 94 GHz (Model 2087 RCS measurement system). The specifications of the system are given in Table 4.3.

TABLE 4.3

Characterization of Gated CW RCS Measurement System Model 2087 CW RCS
Transmitter Rise time ≤ 4 nsec
Transmitter fall time ≤ 4 nsec
Transmitter pulse width/Receiver Range gate width 10 to 99 nsec in 1 nsec steps, independent, adjustable
Receiver Range gate delay 30 to 299 nsec in 1 nsec steps
Transmitter PRF 3.25 MHz
Signal source stability crystal controlled to 1×10^7
Dynamic Range 80 dB with software and hardware gating

4.13.4 Pulse Radar

A pulse instrumentation radar is similar to a gated CW radar except that the receiver bandwidth in the pulse radar is wide enough to accept the significant spectral components of the pulse, allowing accurate detection of each radar pulse without distorting the pulse shape. A block diagram of a pulse instrumentation radar for RCS measurement is shown in Figure 4.14.

The radar employs a common coherent oscillator called COHO and a common stable oscillator called STALO for both the transmitter and receiver channels as in a MTI radar (included in Figure 8.3). This allows the coherent detection of the received signal by using the COHO as the reference signal. The IF bandwidth required is approximately equal to the reciprocal of the pulse width. Thus, a 50 nsec pulse corresponds to a 20 MHz bandwidth. For an instrumentation radar, however, the IF bandwidth is made several times larger to preserve the pulse shape and to allow accurate RCS measurements to be made. To protect the RF amplifier during pulse transmission, a limiter is used prior to the preamplifier which is not shown in Figure 4.14. Blanking of the receiver during the pulse transmission is also employed to prevent saturation of the receiver.

A coherent instrument radar for RCS measurement is now available commercially. The characteristics of such a coherent RCS measurement radar system manufactured by Hughes aircraft company (MMS-300) are shown in Table 4.4.

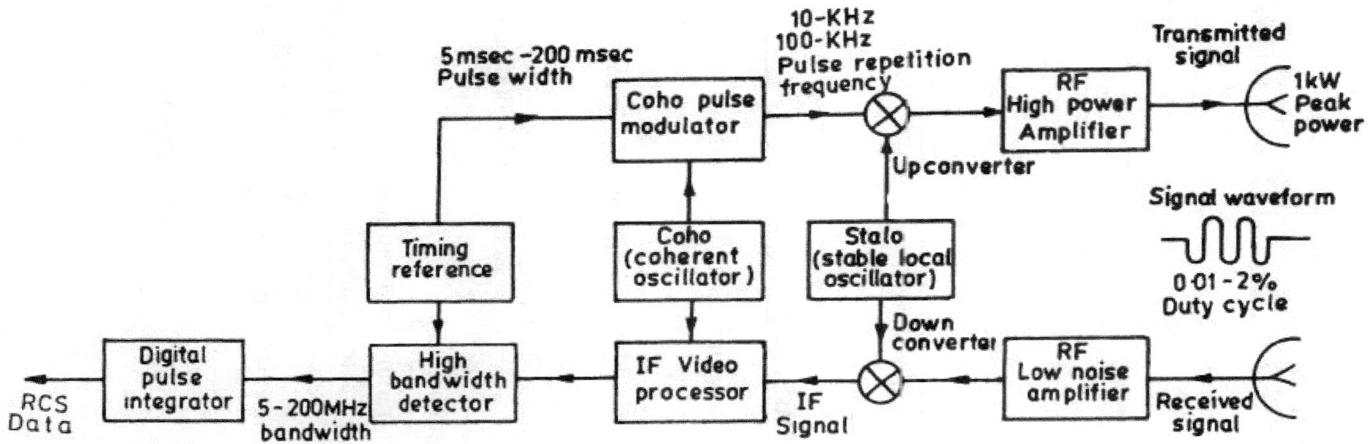

FIGURE 4.14 In coherent pulse radars, the power amplifier transmits a short burst of modulated RF energy provided by the pulse modulator.

TABLE 4.4

Characteristics of a Coherent Instrumentation Radar for RCS Measurement (MMS-300)	
Frequency bands	0.1—100 GHz
IF bandwidth	2—400 MHz
Transmitter power	0.4—1.0 W peak
Frequency drift	< 1 part in 10^9/day
PRF	100 Hz to 1 MHz (100 KHz for 95 GHz)
Pube width	5 nsec to > 1 nsec (50 nsec for 95 GHz)
Rise/Fall times	1 nsec
Duty cycle	100 dB
Dynamic range with pulse integration	90 dB

WORKED OUT PROBLEMS

1. (*i*) Three metallic spheres of diameters, *tdf* ≃ 1 mm, 10 mm, and 20 cm, are used as calibrators for RCS measurement with radar systems. If the frequency of the radar is 9.549 GHz, calculate in which of the regions-Rayleigh, Mie, and optical regions—the spheres may act as a calibrator.

$$\lambda = \frac{c}{f} = \frac{3\times10^{10}}{9.549\times10^{9}}\text{cm} = 3.1416902 \text{ cm}$$

$$\simeq \pi \text{ cm}$$

For $d = 1$ mm, radius $r = 0.5$ mm.

For $d = 0.5$ mm, $\frac{2\pi r}{\lambda} = \frac{2\pi\times.05}{\pi} = 0.1$ which is in the Rayleigh Region.

For $d = 10$ mm, $r = 5$ mm, and $\frac{2\pi r}{\lambda} = \frac{2\pi\times0.5}{\pi} = 1$ which is in the Mie region.

For $d = 20$ cm, $r = 10$ cm and $\frac{2\pi r}{\lambda} = \frac{2\pi\times10}{\pi} = 20$ which is in the optical region.

(*ii*) Calculate the RCS values for $r = 0.5$ mm and $r = 10$ cm.

For $r = 0.5$ mm, as $\frac{2\pi r}{\lambda} = 0.1$ is in the Rayleigh region for which $\frac{\sigma}{\pi r^2} = .001$

We have
$$\begin{aligned}\sigma &= .001 \times \pi \times (0.5)^2 \text{ mm}^2\\ &= .001 \times .25 \times \pi \text{ mm}^2\\ &= .00025\,\pi \text{ mm}^2\\ &= 7.85 \times 10^{-4} \text{ mm}^2\end{aligned}$$

For $r = 10$ cm, as $\frac{2\pi r}{\lambda} = 20$ is in the optical region

$$\frac{\sigma}{\pi r^2} = 1 \text{ or } \sigma = \pi \times (10)^2 \text{ cm}^2 = 100\,\pi \text{ cm}^2 = 314.16 \text{ cm}^2.$$

CHAPTER 5

RADAR TRANSMITTERS AND RECEIVERS

5.1 INTRODUCTION

There are various requirements and system constraints that enter into the selection and design of a transmitter. The first radar developed prior to World War II used the grid controlled vacuum tube for operation at VHF. The magnetron oscillator, which triggered the development of microwave radar, has been a widely used radar transmitter, particularly for mobile systems. The klystron amplifiers offered higher power at microwaves than available from magnetron. As an amplifier the klystron permitted the use of sophisticated waveforms. The klystron was followed by the traveling wave tube, which had the advantage of its wider bandwidth. There is in fact no universal transmitter suitable for all radar applications, but it must be of adequate power for getting the desired radar range, and it should also satisfy the requirements imposed by the system applications.

The function of the radar receiver is to detect wanted echo signals by separating the unwanted signal. Receivers amplify the desired signals to a level where target information can be displayed to an operator or be used in an automatic data processor. A receiver should be designed so as to generate as little internal noise as possible. It must also have sufficient gain, phase, and amplitude stability, tuning range, ruggedness, and dynamic range. In short, we can say that radar transmitter or receiver design and implementation are not easy tasks for all practical purposes.

5.2 THE MAGNETRON OSCILLATOR

The magnetron power oscillator is widely used in radar. The classical magnetron is of convenient size and weight, low cost, and high efficiency. The basic structure of the classical magnetron is shown in Figure 5.1. In the figure, the anode (1) is a large block of copper in which are cut holes (2) and slots (3) which serve the purpose of the resonant circuits. The holes correspond to the inductance L and slots to the capacitance C. The cathode (4) is a cylinder of an oxide-coated thermionic emitter. It is rugged to withstand the heating and disintegration due to the back bombardment of the electrons and positive ions. The electrons interact with the crossed DC electric and magnetic fields in the interaction space (5) where they are bunched and give up their energy to the RF field. The RF power is extracted by means of a coupling loop (6) in one of the cavities. Straps (7) are metal rings connected to alternate segments of the anode block for improving the stability and efficiency of the tubes.

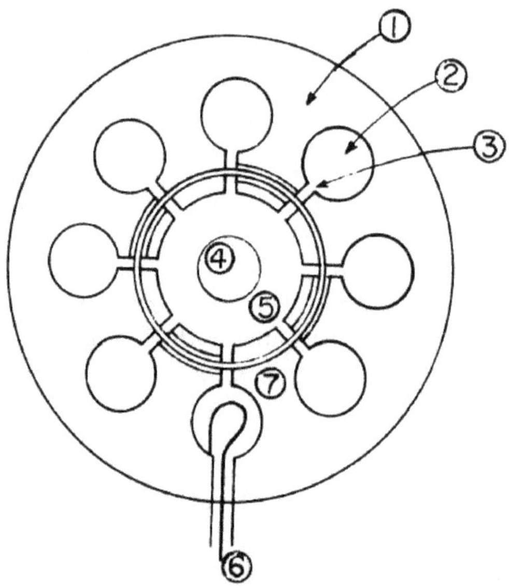

FIGURE 5.1 Cross section of a classical cavity magnetron.

Sometimes, instead of hole and slot resonators, vanes or slots may be used. Figure 5.2 shows two such magnetron resonators:

(*a*) is the vane type, while

(*b*) is the rising sun type with alternated large and small slot lengths.

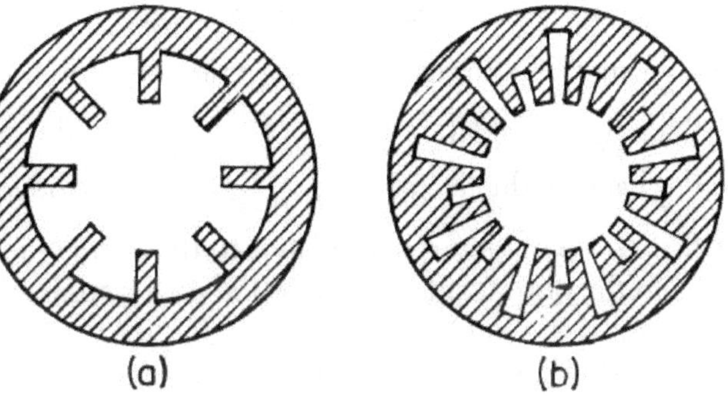

FIGURE 5.2 Magnetron resonator cavity structures.

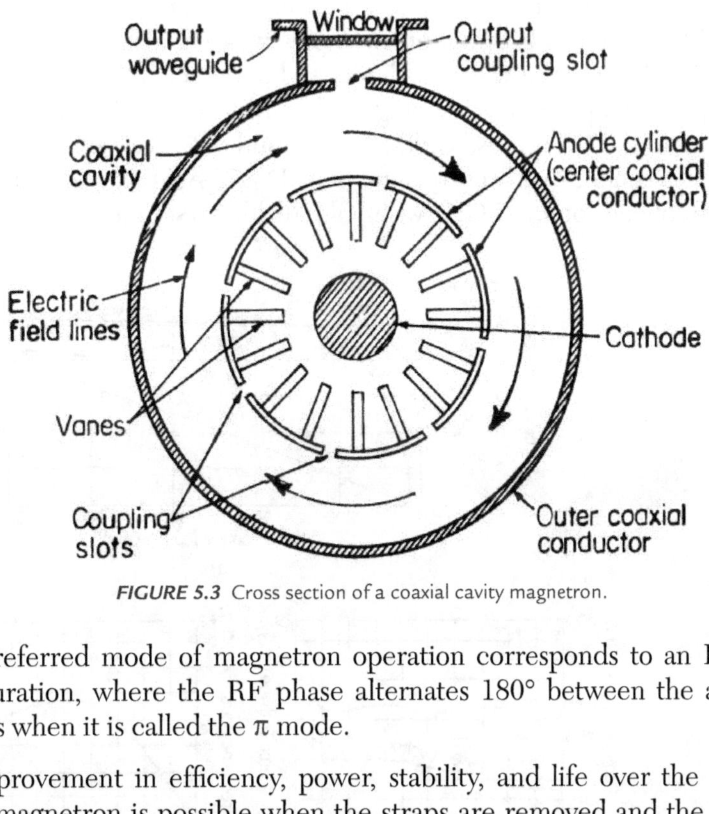

FIGURE 5.3 Cross section of a coaxial cavity magnetron.

The preferred mode of magnetron operation corresponds to an RF field configuration, where the RF phase alternates 180° between the adjacent cavities when it is called the π mode.

An improvement in efficiency, power, stability, and life over the conventional magnetron is possible when the straps are removed and the π mode is controlled by coupling alternate resonators as shown in Figure 5.3. This is known as a coaxial magnetron. Here the output power is coupled from the stabilizing coaxial cavity operating in the TE_{011} mode with the electric field lines closed on themselves and concentric with the circular cavity.

5.2.1 Performance Chart and Rieke Diagram

The following four parameters determine the operation of the magnetron:

(*i*) the magnetic field (*ii*) the anode current

(*iii*) the load conductance and (*iv*) the load susceptance.

The first two parameters are related to the input side and the last two are related to the output side. A plot of the observed magnetron quantities as a function of the input circuit parameters for fixed load is known as the performance chart, while a plot of the observed quantities as a function of the load conductance and susceptance for a fixed magnetic field and anode current is known as the load diagram or the Rieke diagram.

(*a*) Variation of power output, efficiency, and peak voltage of the magnetron SFD-341 as a function of the input voltage with an operating frequency of 5.65 GHz, pulse width, 2.15 sec, and duty cycle 0.0009; (*b*) variation of peak power output and peak voltage of the SFD-341 with frequency for an operating current 23.9 A, pulse width, 1.8 sec, and duty cycle 0.0009; and (*c*) variation of frequency with current for a SFD-377 AX-band coaxial magnetron at a frequency of 9.737 GHz with pulse width 1.0 sec and duty cycle 0.001.

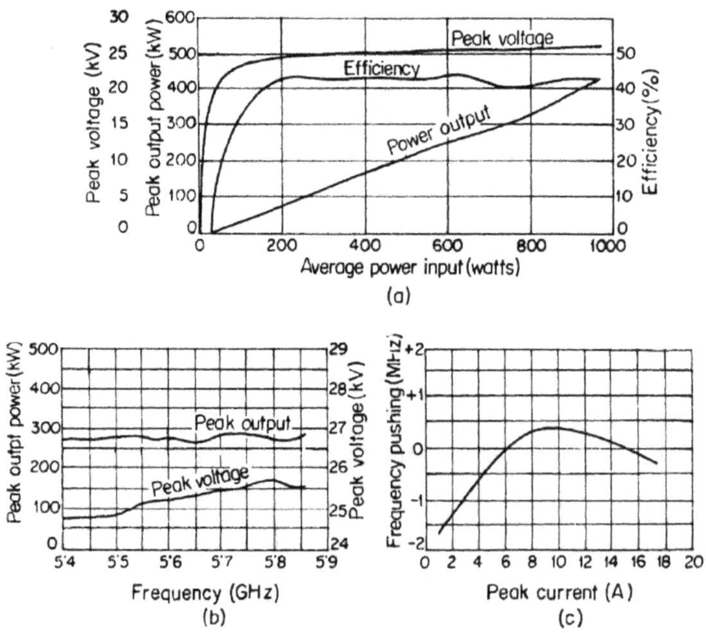

FIGURE 5.4 Performance characteristics of the coaxial magnetron.

The performance characteristics of the coaxial magnetron are shown in Figure 5.4 (*a*). The power output, efficiency, and the peak anode voltage are plotted as a function of the magnetron input power for a fixed frequency and with the magnetron waveguide-load matched. The peak voltage in the figure is found to vary only slightly with a change in input power, while the power output varies nearly linearly. The variation of power output and the voltage with the tuning frequency is plotted in Figure 5.4 (*b*) while the variation of the oscillator frequency with the anode current, the so-called pushing figure, is shown in Figure 5.4 (*c*).

The effect of load on magnetron characteristics is shown by the Rieke diagram in Figure 5.5. Its coordinates are the load conductance and susceptance. It is obtained by varying the voltage-standing wave ratio (VSWR) of the RF load with the frequency, keeping the peak current constant. The region of highest power on the Rieke diagram is called the sink, and the low power region is called the antisink region.

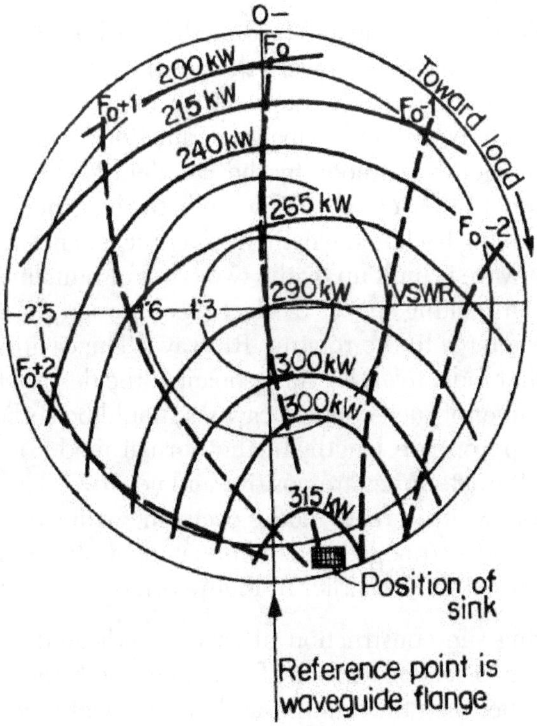

FIGURE 5.5 Rieke diagram for the coaxial magnetron SED-341 with duty cycle 0.0009, Peak current 24 A, and pulse width 2.25 sec at an operating frequency of 5.56 GHz.

5.2.2 Principle of Operation of a Magnetron

The magnetron is basically a crossed-field oscillator with DC electric and magnetic fields at right angles to each other. It was developed well before the other tube types such as klystrons and traveling-wave tubes, which are based on electron bunching by the velocity modulation technique. The magnetrons are specially suitable for pulsed operation and capable of generating peak powers of the order of megawatts. The cross section of a practical form of vane type of magnetron indicated earlier in Figure 5.2 (*a*) is shown in greater detail in Figure 5.6. It consists of (1) a coaxial cathode, (2) a multicavity anode block, (3) a permanent magnet to provide *a* DC magnetic field along the axis of the cathode, (4) a radial DC electric field at right angles to the magnetic field, and (5) a means of coupling the RF power to an external load. The vanes of the structure act as delay lines reduce the value of the phase velocity that the electromagnetic waves has, to below its free space value. This facilitates interaction between the rotating electron cloud and the rotating forward traveling RF wave. For oscillation a feedback mechanism is provided by two types of electron stream bunching processes. First, electrons lagging and leading the ideal position tend to accelerate and decelerate respectively toward the position where a maximum conversion of DC to RF energy occurs—a process leading to the bunching of electrons at an optimum position by velocity modulation. Second, some of the electrons emitted from the cathode by the DC field may be in proper phase to give up energy to the rotating RF wave with the other electrons, which absorb energy falling back to the cathode—a process known as crossfield bunching. The RF wave is built up by the two types of bunching of electrons which, in fact, form rotating spokes of electrons between the cathode and anode and transfer energy to the rotating RF wave. The cavities which are coupled to each other inductively are all resonant at the desired frequency of operation. The number of such coupled cavities should be sufficient to provide a delay line of appropriate length. In the normal mode of operation called the π mode, alternate vanes are positive and negative. The possibility of improper operation, with alternate modes occurring with two consecutive vanes being positive and two negative, is eliminated by strapping alternate vanes by a conductor, as shown earlier in Figure 5.1.

The rising sun construction of cavities indicated earlier in Figure 5.2 (*b*) leads to two sets of resonant frequencies with the tube operating in the π mode in between the π mode, and the adjacent modes of operation are thus increased compared to that in an unstrapped system.

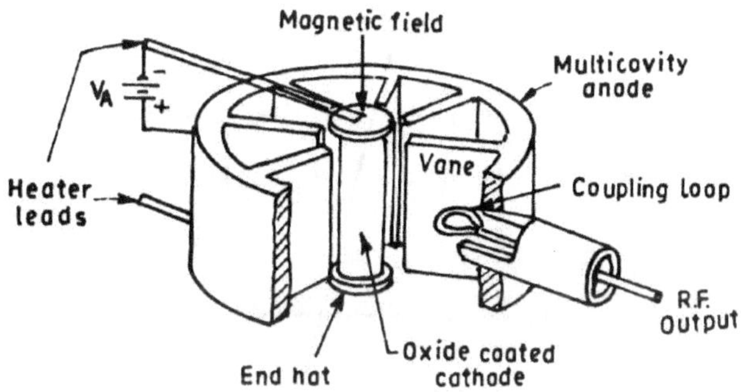

FIGURE 5.6 Detailed configuration of a magnetron oscillator.

The number of cavities, N, may be 8-12 or more with the larger number being chosen for lower voltage operations. The resonant frequency of isolated cavities is 2-3 times that of the final π mode operating frequency, where the DC magnetic field is provided by a permanent magnet producing a field, B_0, of the order of 2000—3000 Gauss. The cyclotron frequency, ω_c is related to B_0 as $\omega_c = \frac{eB_0}{m}$ from which the radius r_b of the top of the Brillouin cloud near the cathode can be obtained using the relation

$$r_b = \frac{r_c}{\sqrt{1 - \frac{\omega_\pi}{\omega_c}\frac{4}{N}}} \tag{5.1}$$

where r_c is the cathode radius.

The Brillouin cloud top velocity v_b is estimated from the relation

$$v_b = \frac{2\omega_\pi r_b}{N} \tag{5.2}$$

from which the equivalent voltage v_B can be determined by using the relation

$$v_B = \left(\frac{v_b}{5.93 \times 10^7}\right) \text{volts} \tag{5.3}$$

Usually $V_B << V_A$, the anode voltage and the anode radius, r_a, is calculated from

$$\text{In}\frac{r_a}{r_b} = \frac{V_A - V_B}{\frac{w_c^2}{4(e/m)} \cdot \frac{r_b^4 - r_c^4}{r_b^2}} \tag{5.4}$$

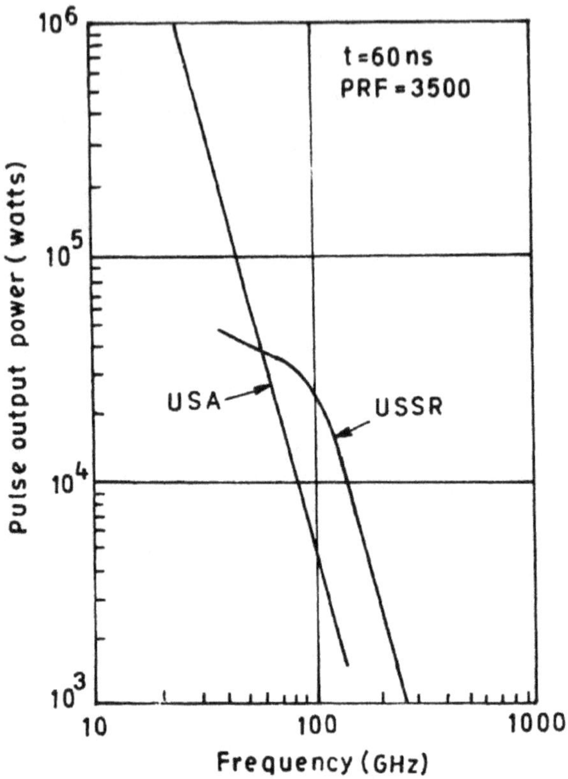

FIGURE 5.7 Power output of millimeter wave magnetrons versus frequency.

Thus, the basic design parameters of the magnetron may be determined from equations (5.1) to (5.4).

The depth of the cavity may be chosen to be $\lambda_0/4$ where λ is the operating wavelength, and the pitch per cavity, $\frac{2\pi r_a}{N}$, is obtained from r_a estimated from equation (5.4).

It may be mentioned here that after the design of a magnetron is made for a particular frequency of operation, it can be scaled to obtain the design parameters at higher frequencies up to the millimeter wave region. The only limitation is that with increasing frequencies the dimensions, which become progressively small, may reach a stage where fabrication is difficult and/or an element becomes so thin that power dissipation problems crop up and the power output decreases with increasing frequencies as shown in Figure 5.7.

5.3 KLYSTRON AMPLIFIERS

The limitation of vacuum tube amplifiers like triodes, tetrodes, or pentodes, arising from the fuite transit time of the electrons between the cathode and grid, was overcome with the development of the klystron by Varian and Hausen in 1938, based on the concept of electron bunching in a field free space.

The principle of operation of a Klystron amplifier can be explained by referring to the two-cavity configuration shown in Figure 5.8. In the klystron amplifier electrons emitted by an electron gun/cathode are accelerated by the electric field between the anode and the cathode and passed through the first cavity, serving as a buncher, into a field free drift space where the electrons are bunched and then passed through the output resonator to a collector. The bunching process is illustrated in Figure 5.9. The RF input signal is coupled to the buncher, of which gap distance d is kept small enough so that the electron transit time $\frac{d}{v_o} << $ time period $\frac{1}{f}$ of the RF signal, f being

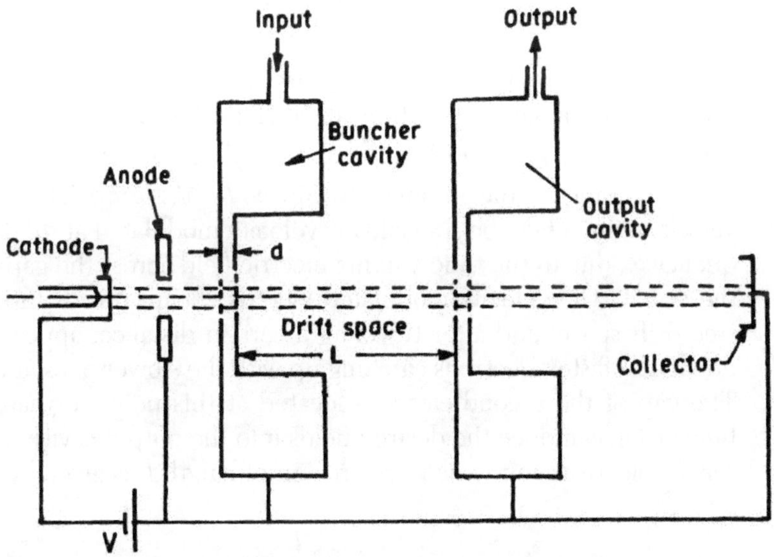

FIGURE 5.8 Two-cavity klystron configuration.

the frequency of operation to which the cavity is tuned. The velocity of the electrons coming out of the buncher is given by

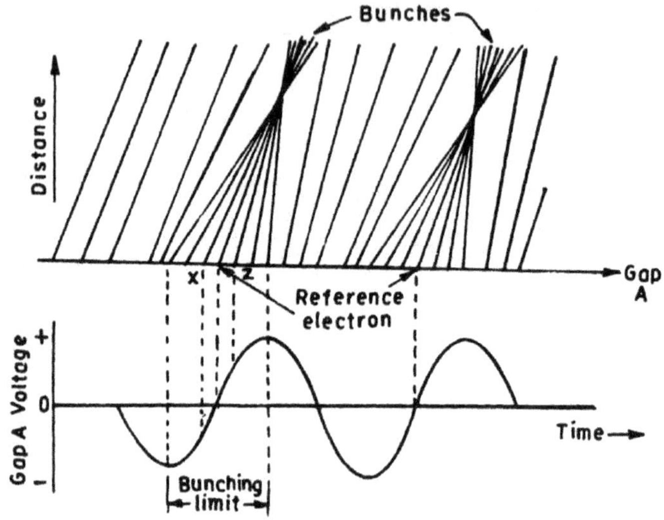

FIGURE 5.9 Bunching in a klystron amplifier.

$$v = v_o + e/m\, E_o \frac{d}{V_0} \beta \sin \omega t_o \tag{5.5}$$

where the electric field across the buncher gap, d, is $E = -E_o \sin \omega t$ due to the RF input signal, t_o is the time at which the electron is at the center of the gap, $v_o = \sqrt{2e\,V/m}$, and V is the anode voltage relative to the cathode and is the coupling coefficient given by Sin $(\pi fd/v_o)/(\pi fd/v_o)$. The electron beam velocity will, in fact, be sinusoidally velocity modulated at the RF signal frequency ω, due to the time-varying electric field across the gap produced by the signal. The velocity-modulated electron beam travels though the field free drift space, and after traveling a certain distance, appears in bunches due to the faster electrons catching up with the slower ones at that distance. The gap of the second cavity is located at this point so that the electron bunches may induce the desired fields into the output cavity. The drift time for the electrons in the field free region of length L is given by

$$t_e - t_o = \frac{L}{v_o(1 + (2v_g / V)\beta \sin \omega t_o)} \tag{5.6}$$

where v_g (E.d) is the peak RF voltage to cross the buncher gap. Assuming $v_g/V << 1$, the exit current or beam current drawn by the collector $I_e(t_o)$ can be shown to be

$$I_e(t_o) = I_o \frac{dt_o}{dt_e} = \frac{I_o}{1-(\omega L/v_o)(2v_g/V)\beta \cos \omega t_o} \tag{5.7}$$

$$= \frac{I_o}{1 - X \cos \omega t_o} \tag{5.8}$$

where I_o is the input current at the beginning of the drift space and $X = (\omega L/v_o)\,(2v_g/V)\,\beta$ is called the bunching parameter, as it determines the intensity and location of the electron bunches. The waveforms of the beam current for various values of the bunching parameter in the range $X = 0$ to $X = 1.84$ are shown in Figure 5.10, from which it is evident that for bunching parameters exceeding unity, the beam current contains a very high harmonic content. Such a waveform may be useful if the klystron is to be operated as a harmonic generator with the output cavity tuned to the particular harmonic of interest. For operation as an amplifier, however, a value of $X = 1$ would be preferred to avoid harmonic content. Also, the output cavity is to be tuned to fundamental frequency f, which is equal to that of the input cavity.

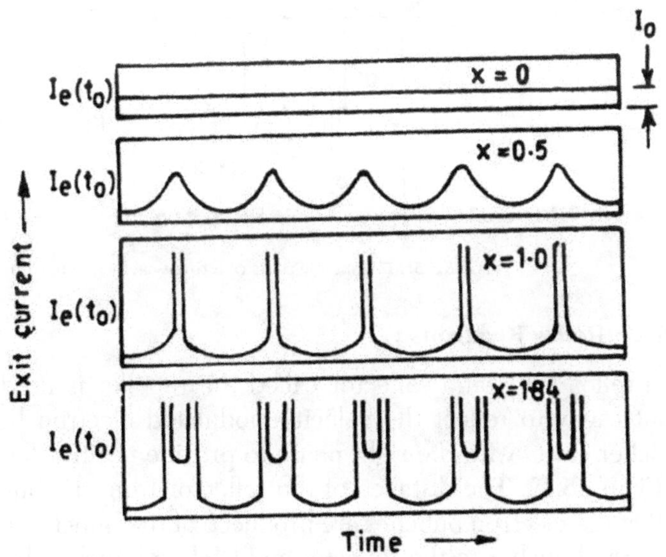

FIGURE 5.10 Beam current waveforms.

It may be mentioned here that the two-cavity klystron amplifier can be operated as an oscillator by providing a positive feedback from the output to the input cavity through a coaxial cable.

5.3.1 Three-Cavity Klystron

The power gain of a two-cavity klystron is limited to about 10 dB, and the efficiency is less than 30%. To improve these performance characteristics, the drift space can also be supported by a cavity, as shown in Figure 5.11, to form a three-cavity klystron amplifier. The electrons emitted by the cathode are focused into a narrow cylindrical beam by the electron gun, consisting of the cathode, the control electrode, and the anode, similar to that employed in the two-cavity klystron. Such multicavity klystrons have been developed to obtain a power gain > 35 dB and power levels > 50 MW peak at a high efficiency in the microwave band. However, most of the klystrons in use are of the two-cavity type for use as an amplifier, and that of the reflex-tape, which operates only as an oscillator.

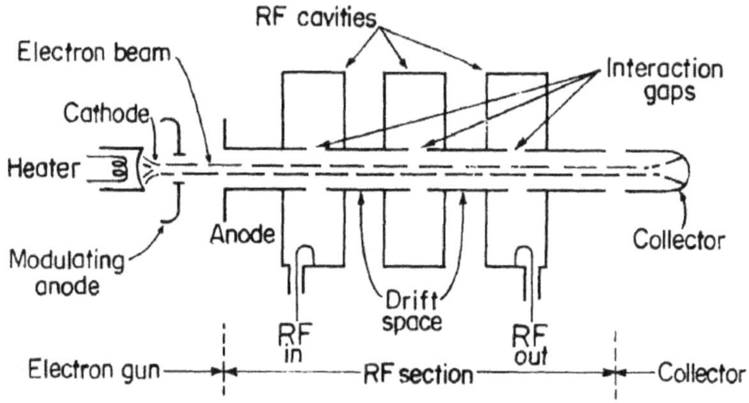

FIGURE 5.11 Principle parts of a three-cavity magnetron.

5.3.2 Reflex Klystrons

In a reflex klystron a reflector called the repeller is used in place of the input cavity to reflect the velocity-modulated electron beam back to the buncher cavity with the right phase to provide positive feedback, as shown in Figure 5.12. The distance of the reflector from the buncher is adjusted so that the electron bunches are produced at the buncher by two-way travel from the buncher to the reflector and back to buncher, by a process called reflection bunching. This is illustrated in Figure 5.13.

The electrons reflected back and arriving at the buncher cavity gap in bunches sustain the oscillation. Power is coupled out of the buncher cavity by loop coaxial line. At millimeter wave the output coupling by slots are more popular.

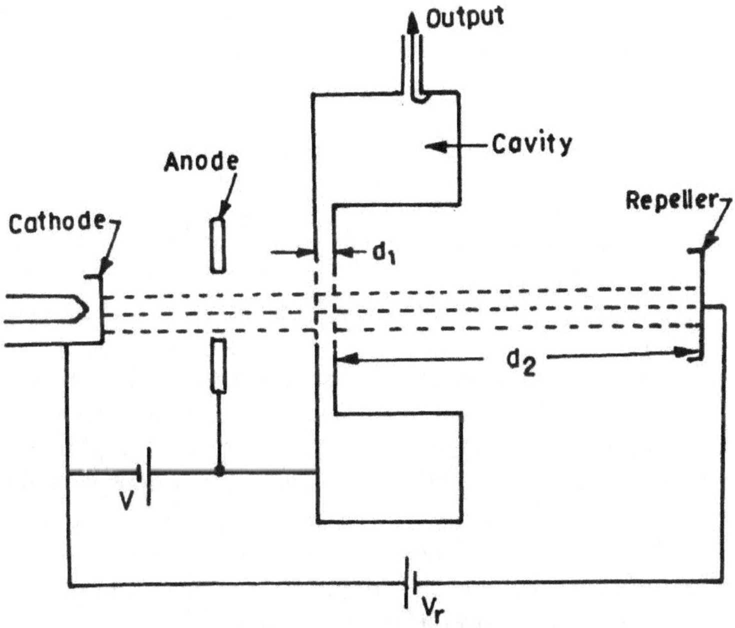

FIGURE 5.12 Schematic of a reflex klystron.

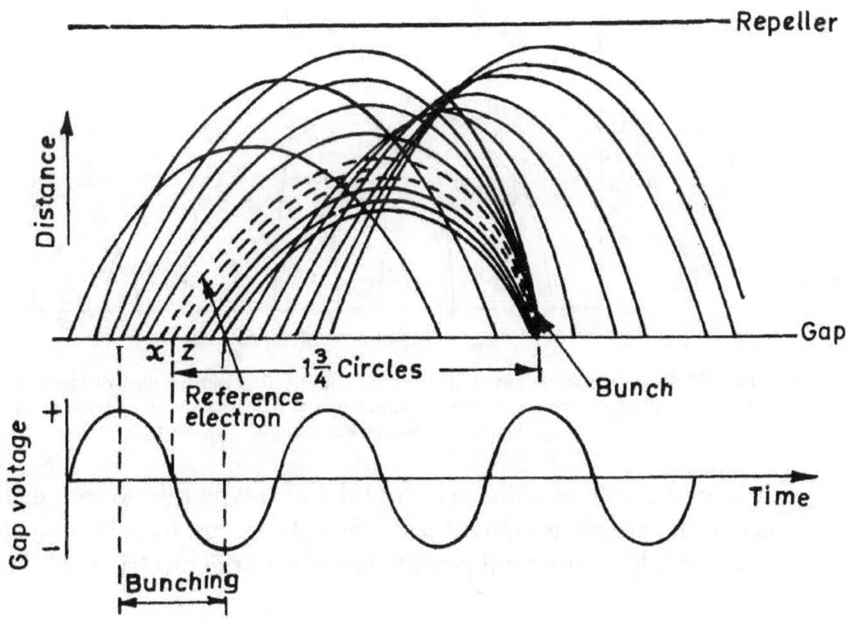

FIGURE 5.13 Bunching for a reflex klystron in the $1\frac{3}{4}$ cycle mode.

It can be shown that the average transit time τ_r in the gap-repeller space for optimum power transfer to the gap fields is given by

$$\tau_r = \frac{2mv_o d_2}{e(V+V_r)} = \frac{N+\frac{3}{4}}{f} \tag{5.9}$$

where d_2, V, and V_r are buncher-repeller distance, buncher voltage, and repeller voltage respectively, as shown in Figure 5.11, N takes integer values including zero, and f is the buncher cavity resonance frequency. The variation of the repeller voltage will thus allow an electronic tuning over a small range. The range is in fact small due to the high θ of the cavity, which results in a typical variation of power output with repeller voltage, as shown in Figure 5.14. Tuning over a larger range is provided by a mechanical tuning arrangement in addition to the electronic tuning. The different peaks of power output shown in Figure 5.14 correspond to different modes of reflection bunching. One of these modes shows a maximum output, and mechanical tuning would allow this frequency of peak output power to be shifted in a desired direction over a sizable range.

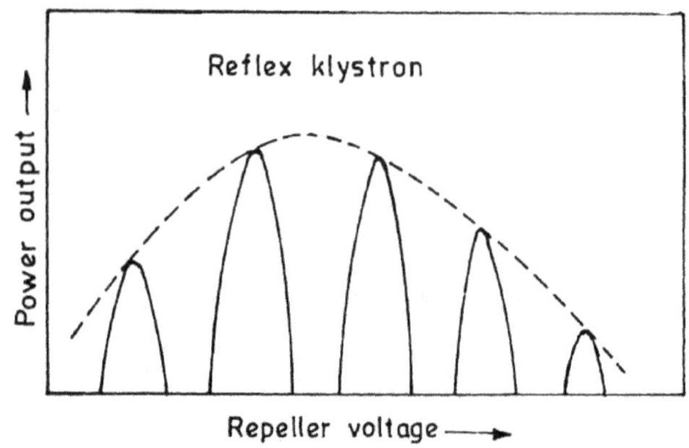

FIGURE 5.14 Variation of output power of a Reflex Klystron with repeller voltage showing different modes of reflection bunching producing subsidiary sharp peaks of output power. Mechanical tuning would shift the peak of the dotted envelope.

It is an example of a linear beam tube or 0-type tube whose characteristic is that the electrons emitted from the cathode are formed into a cylindrical beam which receives full potential energy of the electric field. Transit time effects limiting the high frequency amplification of conventional triodes or pentodes are effectively utilized in klystrons.

The principle parts of the klystron are shown in Figure 5.11. As shown, the cathode emits a stream of electrons which is focused into a narrow cylindrical beam by the electron gun consiting of a cathode, a modulating anode or control grid, and the anode. The emission density of the electron from the cathode is less than that required for the electron beam needing a large area cathode surface to obtain a high density electron. The modulating anode or other beam control electrode is usually included as part of the electron gun structure for pulsing the electron beam on and off. The RF cavities which correspond to the LC resonant circuits are at anode potential. The electrons are removed by the collector electrode after the beam has given up its RF energy to the output cavity.

The speeded up electrons during the peak of one cycle catch up with those slowed down during the previous cycle, and thus the electrons of the velocity-modulated beam become density-modulated or 'bunched' after moving through the drift space. After the bunched electron beam delivers its RF power to the output cavity, the electrons are removed by the collector electrode. Power is extracted from the output cavity and delivered to the load by a coupling loop or a waveguide.

Klystrons are focused generally with lightweight permanent magnets. In a high-power klystron, normally two to five percent of the beam power is intercepted by the interaction structure of the body of the tube.

5.4 TRAVELING WAVE TUBE AMPLIFIERS

A traveling wave tube (TWT) is shown in Figure 5.15. It is another example of a linear beam or 0-type tube. Electrons emitted from the cathode of the TWT are focused into a beam and pass through the RF interaction circuit. This is known as the slow wave structure or periodic delay line. Just as with a klystron, an axial magnetic field is provided to maintain the electron-beam focus. After delivering the DC energy to the RF field, electrons are removed by the collector electrode. A helix is depicted as the slow wave structure. A modification of the helix known as a ring bar circuit has been used in TWTs in order to get higher power and efficiencies.

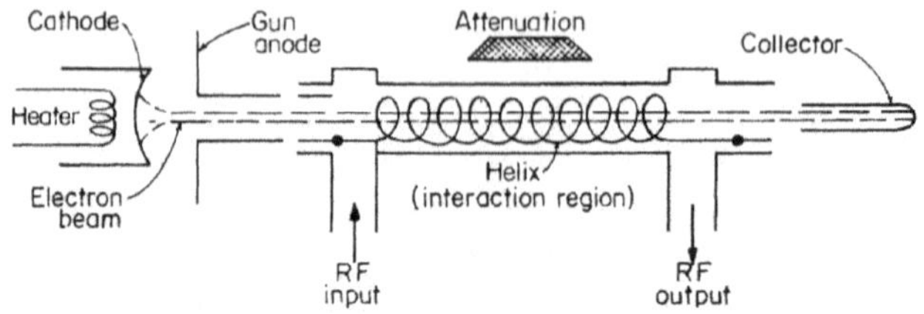

FIGURE 5.15 Principal parts of a traveling wave tube.

For high power TWTs the popular form of slow-wave structure is the coupled cavity circuit. An example of a TWT designed for radar applications is the S-band variant VA-125 A, which is capable of 3 MW of peak power over a 300 MHz bandwidth.

5.5 CROSSED-FIELD AMPLIFIERS

There are different types of crossed-field amplifier (CFA). In a typical form it resembles the magnetron oscillator, with an exception that there are two external couplings. Electrons, after emitting from the cathode under the action of the crossed electric and magnetic fields, form into rotating electron bunches or spokes which drift along a slow wave circuit in phase with the RF signal. This allows energy to be transferred to the RF wave to provide amplification. Instead of the collector electrode, in this case the spent electrons terminate on the slow wave anode structure.

The conversion efficiency of a CFA can be defined as,

$$\text{Efficiency} = \frac{\text{RF power output} - \text{RF drive power}}{\text{DC power input}}$$

The insertion loss of a CFA made inactive by omitting the DC supply, is small (< 0.5 dB). Hence the RF drive will appear at the output of the tube with a low attenuation under the inactive condition.

Characteristics of a CFA are illustrated by variant SFD-257 which operates over the frequency range 5.4 to 5.7 GHz, peak power 1 mW, duty cycle 0.001, and an efficiency of more than 50%.

5.5.1 Amplitron

An amplitron is a cross-field amplifier which is somewhat similar in design to a magnetron, except that the RF circuit is non-reentrant, with input and output ports as shown in Figure 5.16. A microwave or millimeter wave signal applied to the input port is amplified and coupled out at the output port. The straps connect anode vanes alternately. Typical bandwidths of the order of 10% are obtainable at microwaves with an amplitron for an amplification of 10 dB and an efficiency as high as 85%, with somewhat lower values at millimeter waves. The maximum output power of a CFA like an amplitron decreases with increasing frequency as shown in Figure 5.17, which also includes a comparison with the performances of the klystron, TWT, and triode amplifiers.

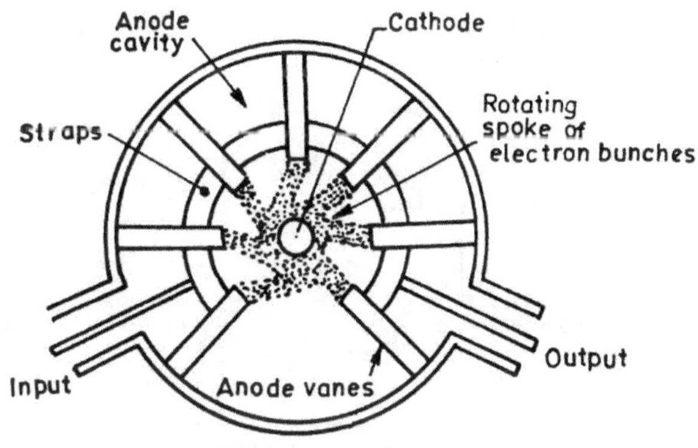

FIGURE 5.16 Amplitron amplifier schematic.

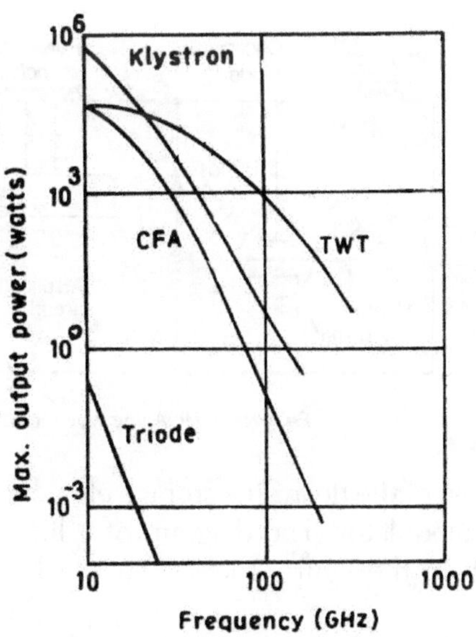

FIGURE 5.17 Maximum power output of tube sources as a function of frequency.

5.6 MODULATORS

The function of a modulator is to turn the transmitting tube on and off for generating the desired waveform. If the transmitted waveform is a pulse, then the modulator is called a pulser. For example, the magnetron modulator must be designed to handle full pulse power.

The basic elements of a pulse modulator are shown in Figure 5.18. Energy from an external source is accumulated in the energy storage element. The rate at which energy can be delivered to the storage element is limited by the charging element. At the proper time the switch is closed so that the stored energy is rapidly discharged through the load to form the pulse.

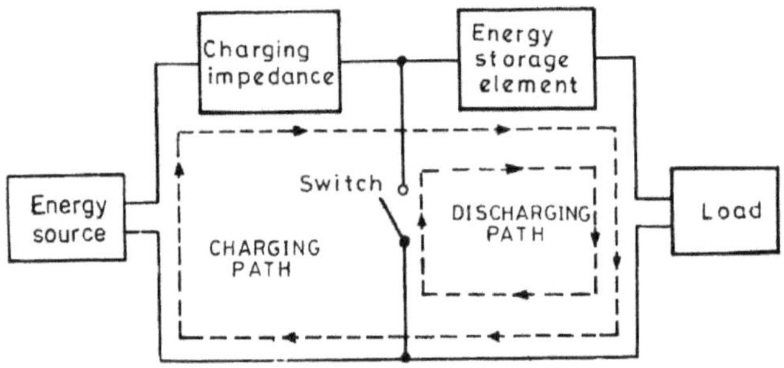

FIGURE 5.18 Basic elements of a radar pulse modulator.

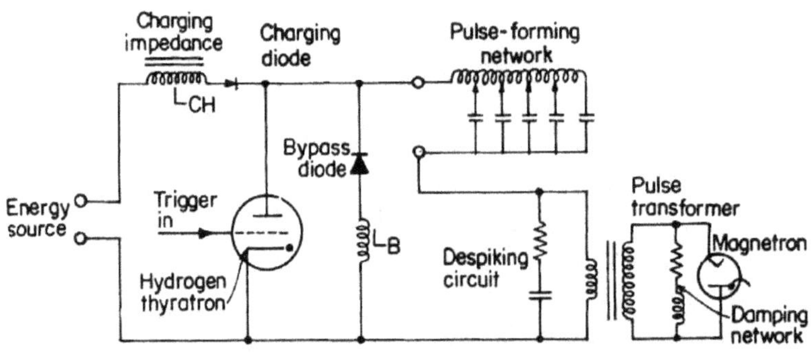

FIGURE 5.19 A line type modulator.

A combination of the delay line storage element and gas tube switch is called a line type modulator. The diagram of a line type modulator is shown in Figure 5.19. In the figure, the switch is a hydrogen thyratron, but it can also be a mercury ignitron, silicon controlled rectifier, or saturable reactor. A gas tube like a thyratron or ignitron can handle high power and presents a low impedance when conducting. The bypass diode and the inductance L_B connected in parallel with the thyratron are used to dissipate any charge remaining in the capacitor due to the mismatch of the tube. The mismatch of the pulse forming network causes a spike to appear at the leading edge of the pulse. The despiking circuit minimizes this effect while the damping network reduces the trailing edge of the pulse and thus prevents post-pulse oscillations.

5.7 SOLID STATE TRANSMITTERS

There are two categories of solid state devices which can be considered as potential sources of microwave power for radar applications. These are transistor amplifiers and single-post microwave diodes that can be operated as oscillator or as negative resistance amplifiers.

At the L-band the CW power can be obtained from a single microwave transistor of several tens of watts. In order to increase the power output, the transistors can be operated in parallel. Theoretically, the power output of a microwave transistor varies inversely as the square of the frequency or 6 dB per octave. An X-band GaAs FET amplifier producing 4 watts of output power was recently available.

The Gunn and LSA bulk-effect diodes and the Trapatt and Impatt avalanche diodes are operated as oscillators or as negative resistance amplifiers. At the higher microwave frequencies, they are considered suitable. LSA diodes have somewhat greater peak power, but their average power and efficiency are low.

There are several methods for employing solid state devices as radar transmitters, for example,

1. direct replacement of a vacuum tube,
2. multiple modules in a mechanically scanned array with the power combined in space, and
3. multiple modules in an electronically steered phased-array radar with the power combined in space.

5.8 NOISE FIGURE OF A RECEIVER

The noise figure of a receiver can be described as a measure of the noise produced by a practical receiver as compared with the noise of an ideal receiver. The noise figure F_n of a linear network can be written as

$$F_n = \frac{S_{in} / N_{in}}{S_{out} / N_{out}} \tag{5.5}$$

where S_{out} and S_{in} are respectively the available output and input signal power, while the ratio of the available output to input power is the available gain G; N_{in} is the available input noise power whose value is kT_0B_n, while N_{out} is the available output noise power. Hence, equation (5.5) can be rewritten as

$$F_n = \frac{N_{out}}{kt_0B_nG} \tag{5.6}$$

where k = Boltzmann's constant = 1.38×10^{-23} J/deg

T_0 = Standard temperature of 290 K

B_n = Noise bandwidth

The previous equation allows two different but equivalent interpretations of the noise figure. It may be interpreted as the degradation of the signal to noise ratio caused by the receiver, or it may be considered as the ratio of the actual available output noise power to the noise power which would be available if the network amplified the thermal noise. The noise figure can alternately be expressed as

$$F_n = \frac{kT_0B_nG + \Delta N}{kT_0B_nG} = 1 + \frac{\Delta N}{kT_0B_nG} \tag{5.7}$$

where ΔN is introduced by the network itself and hence it is the additional noise. Commonly the noise figure is expressed in decibels, that is, 10 log F_n. Sometimes instead of the term noise figure, the term factor is also used when F_n is expressed as a ratio.

The Noise Figure of Networks in Cascade. Let us consider two networks in cascade, each of same noise bandwidth B_n but of different noise figures F_1 and F_2 and available gains G_1 and G_2. This is shown in Figure 5.20.

FIGURE 5.20 Two cascaded networks having different gain and noise figures.

To find the overall noise figure F_0 of the two circuits in cascade, we may write the output noise N_0 of two circuits as

$$\begin{aligned} N_0 &= F_0 G_1 G_2 k T_0 B_n \\ &= \text{Noise from network 1 at output of network } 2 + \text{Noise } N_2 \text{ introduced by the network 2} \\ &= kT_0 B_n F_1 G_1 G_2 + \Delta N_2 \\ &= kT_0 B_n F_1 G_1 G_2 + (F_2 - 1) kT_0 B_n G_2 \end{aligned}$$

or

$$F_0 = F_1 + \frac{F_2 - 1}{G_1} \tag{5.8}$$

If the gain of the first network is large, then one may neglect the contribution of the second network. For the design of multistage receivers, this concept is important. If N number of networks are in cascade, then the noise figure can be shown in the previous manner to be

$$F_0 = F_1 + \frac{F_2 - 1}{G_1} + \frac{F_3 - 1}{G_1 G_2} + \ldots\ldots + \frac{F_n - 1}{G_1 G_2 \ldots\ldots G_N - 1} \tag{5.9}$$

Noise Temperature. Noise introduced by a network can be expressed as an effective noise temperature, T_e. This is a fictional temperature at the input of the network which accounts for the noise N at the output. Thus, we have

$$\Delta N = kT_e B_n G$$

and from equation (5.7)

$$F_n = 1 + \frac{T_e}{T_0}$$

or,

$$T_e = (F_n - 1) T_0 \tag{5.10}$$

We may now define the system noise temperature T_s, which is the effective noise temperature of the receiver system including the effects of antenna temperature T_a. If T_e represents the receiver effective noise temperature, then

$$T_s = T_a + T_e = T_0 F_s \tag{5.11}$$

where F_s = system noise figure

The effective noise temperature of a receiver consisting of a number of networks in cascade is

$$T_e = T_1 + \frac{T_2}{G_1} + \frac{T_3}{G_1 G_2} + \ldots \tag{5.12}$$

where T_1 and G_1 are the effective noise temperature and gain of the first network, T_2 and G_2 are those for the second network, and so on.

Noise Figure Measurement. The receiver noise figure is measured with a broad band noise source of known intensity. The noise figure F_n can be shown to be

$$F_n = \frac{T_2/T_0 - 1}{Y - 1} \tag{5.13}$$

where $$Y = N_2/N_1.$$

The noise figure is found by measuring

N_1 = the noise power output N_1 of the receiver when an impedance at $T_0 = 290°K$ is connected to the receiver input and

N_2 = the noise power output N_2 when a matched noise generator at T_2 is connected to the receiver input.

Radar sensitivity may be visually displayed by measuring the receiver noise to display the normal range rings on PPI, where the radar can detect targets reliably. For the measurement of the receiver noise figure, generally a signal generator is inserted by a directional coupler ahead of the duplexer and other RF components.

5.9 MIXERS

Instead of a low-noise RF amplifier, the first stage of many radar superheterodyne receivers is simply the mixer. The function of a mixer is to convert RF energy to IF energy with minimum loss. An integral part of the mixer is the local oscillator.

The *conversion loss*, L_c of a mixer can be defined as

$$L_c = \frac{\text{Available RF power}}{\text{Available IF power}} \tag{5.14}$$

Thus, it is a measure of the efficiency of the mixer in converting RF signal power into IF. The conversion loss of microwave crystals in single-ended mixers varies from 5 to 6.5 dB. If the signal and image frequencies are both terminated in matched loads, then the crystal mixer is called "broad band."

The noise temperature ratio of a crystal mixer can be defined as

$$t_r = \frac{\text{Actual available IF noise power}}{\text{Available noise power from an equivalent resistance}}$$

$$= \frac{F_c k T_0 B_n G_c}{k T_0 B_n}$$

$$= F_c G_c = \frac{F_c}{L_c} \quad (5.15)$$

where F_c = crystal mixer noise figure

L_c = conversion loss = $\frac{1}{G_c}$

From 100 kHz down to a small fraction of hertz, t_r varies approximately inversely with frequency. This is called flicker noise or 1/*f* noise. The noise temperature ratio above 500 kHz approaches a constant value. At 30 MHz, which is a typical radar IF, it might range from 1.3 to 2.0.

From equation (5.15) $$F_c = t_r L_c \quad (5.16)$$

The overall noise figure thus depends not only on the mixer stage but also on the noise figure of the IF stage and the mixer conversion loss. It is obtained from the expression for the noise figure $t_r L_c$ and gain $1/L_c$. The receiver noise figure with a mixer front end is then given by

$$F_0 = F_1 + \frac{F_2 - 1}{G_1}$$

$$= L_c \left(t_r + F_{IF} - 1\right) \quad (5.17)$$

The receiver noise figure for low noise temperature-ratio diodes is nearly equal to the conversion loss times the IF noise figure.

Balanced Mixers. Due to the nonlinear action of the mixer, noise that accompanies the local oscillator (L_0) signal can appear at the IF frequency. If the receiver sensitivity is to be maximized, then the L_0 noise must be removed. A method of eliminating this L_0 noise without any disadvantage of a narrow bandwidth filter is the balanced mixer, which uses a hybrid junction, a magic *T*, or an equivalent. Figure 5.21 illustrates the principle of the balanced mixer based on the magic *T*. At one of the diodes, the sum of the RF and L_0 signals appears, and at the other diode, the difference of the two is obtained. In the figure, the balanced diodes are shown reversed so that IF outputs can be added while the local oscillator noise at the two diode mixers is in phase and will be canceled at the output.

In order to get a reactive termination without narrow-bandwidth components, one may use the image recovery mixer. It is also called an image-enhanced mixer or product return mixer. This is shown in Figure 5.21. The RF hybrid junction produces a 90° phase difference between the L_0 inputs to the two mixers on the left of the circuit. On the right the IF hybrid junction imparts another 90° phase differential so that the images cancel, but the IF signals add in phase from the two mixers. The image-recovery mixer is attractive as a receiver front end due to its high dynamic range, less susceptibility to burnout, and low intermodulation products.

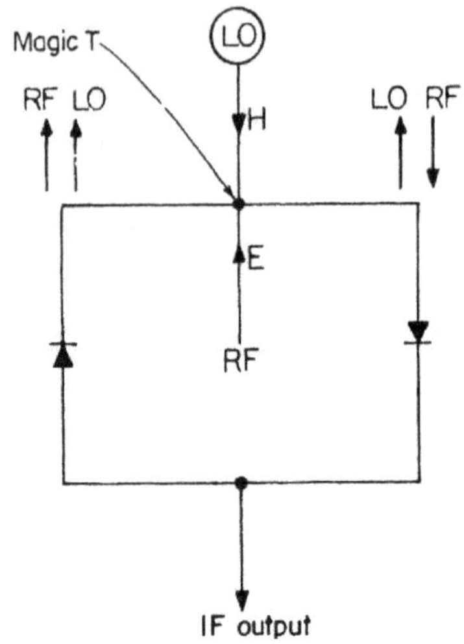

FIGURE 5.21 Double balanced mixer based on the magic *T*.

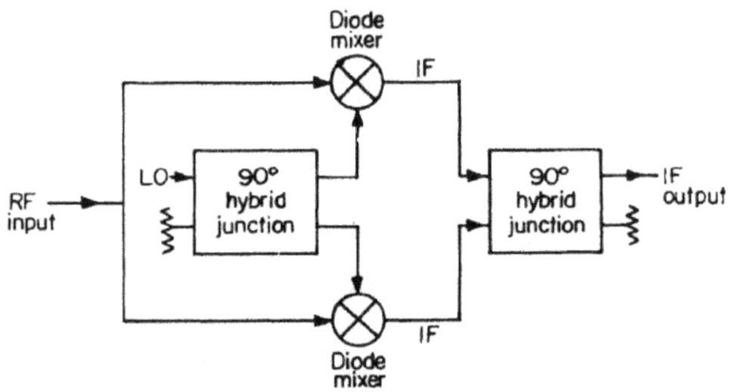

FIGURE 5.22 Image recovery mixer.

Noise Figure Due to RF Losses. The RF losses might be due to the receiver transmission line, duplexer, pre-selector filter, rotary joint, monitoring devices, or loss in the radome. The noise figure in the presence of RF losses may be written as

$$F_L = \frac{N_{out}}{kT_0B_nG} \tag{5.18}$$

where $$= N_{out} = kT_0B_n$$

and $$G = \frac{1}{L_{RF}}$$

Hence, we get $F_L = L_{RF}$ (5.19)

The noise figure of a receiver preceded by RF losses is

$$F_0 = F_1 + \frac{F_2 - 1}{G_1}$$

Putting $F_1 = L_{RF}$, $G_1 = 1/L_{RF}$, and $F_2 = F_r$ (receiver noise figure), we get

$$F_0 = L_{RF} + (F_r - 1)\,L_{RF} = F\,L_{RF} \quad (5.20)$$

5.10 DISPLAYS

The cathode ray tube (CRT) is almost universally used as the radar display. There are two fundamental CRT displays:

1. the deflection modulated CRT, and
2. the intensity modulated CRT.

The focusing and deflection of the electron beam can be made electrostatically, magnetically, or by a combination of the two.

The color emitted by the screen is given in Table 5.1. If an extremely long persistence is desired in a CRT display, one may use a storage tube. In fact, an image placed on a storage tube will remain indefinitely until erased. Recently, digital storage facility has also been available in CRT.

Information from a CRT display depends on several factors like the brightness of the display, density and character of the background noise, pulse repetition rate, signal clipping, scan rate of the antenna beam, decay time of the phosphor, size viewing distance, dark adoption, ambient illumination, display size, and operator fatigue. Nowadays there has been a growing interest in applying solid state technology for a radar display, replacing the vacuum tube CRT.

5.11 DUPLEXERS

A duplexer is a device which permits a single antenna to serve both the transmitter and the receiver. On transmission it protects the receiver from damage, and on reception it channels the echo signal to the receiver.

TABLE 5.1 Radar CRT Phosphor Characteristics.

Phosphor	Fluorescent Color	Phosphorescent Color	Persistence
P1	Yellowish green	Yellowish green	Medium
P7	Blue	Yellowish green	Blue, medium short; Yellow long
P12	Orange	Orange	Long
P13	Reddish orange	Reddish orange	Medium
P14	Purplish blue	Yellowish green	Blue, medium short; Yellowish orange, medium.
P17	Blue	Yellow	Blue, short; Yellow long
P19	Orange	Orange	Long
P21	Reddish orange	Reddish orange	Medium
P25	Orange	Orange	Medium
P26	Orange	Orange	Very long
P28	Yellowish green	Yellowish green	Long
P32	Purplish blue	Yellowish green	Long
P33	Orange	Orange	Very long
P34	Bluish green	Yellowish green	Very long
P38	Bluish green	Yellowish green	Very long
P39	Yellowish green	Yellowish green	Long

For high power applications, duplexers employ a form of gas discharge device. Sometimes solid state devices are also used. The transmitter peak power in a typical duplexer might be a megawatt or more, while the maximum safe power that can be tolerated at the receiver might be less than a watt.

There are two basic methods which allow the use of a common antenna for both transmitting and receiving. The first one is applied to the branch-type duplexer and the balanced duplexer, while the other uses a ferrite circulator for separating the transmitter and receiver and a receiver protector consisting of a gas TR-tube and a diode limiter.

Branch-Type Duplexer. It is one of the earliest duplexers. The principle of the duplexer can be explained by considering Figure 5.23. In the figure, TR is the transmit receive switch and ATR is the anti-transmit receive switch, both of which are gas discharge tubes. When the transmitter is turned on, the TR and ATR tubes ionize so that they break down or fire. In the fired condition, TR acts as a short circuit so that the transmitter power is prevented from entering the receiver. As shown, TR is kept a quarter wavelength from the main transmission line so that it appears as a short circuit at the receiver but as an open circuit at the transmission line. Since ATR is also displaced a quarter wavelength from the main transmission line, during the fired condition the short circuit it produces appears as an open circuit on the transmission line and hence has no effect on the transmission. During reception the transmitter is off, and so neither TR nor ATR is fired. The open circuit at ATR places a short circuit of the line, which in turn places an open circuit presented to the junction of the line with the line from TR. The open circuit at TR keeps the receiver matched to the antennas to receive the echo signal when the transmitter is disconnected due to the open circuit at ATR.

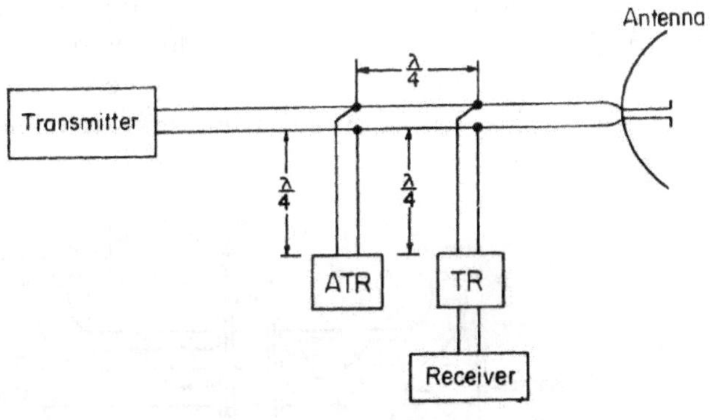

FIGURE 5.23 Branch type duplexer.

The branch type duplexer has limited bandwidth and power handling capacity. In low-cost radars this type of duplexer is generally used.

Balanced Duplexers. They are based on the short-slot hybrid function as shown in Figure 5.24. This short-slot hybrid may be considered as a broad band directional coupler. During transmission by the first short-slot hybrid, junction power is divided equally into each waveguide [Figure 5.24 (a)]. On reception the TR tubes are unfired so that the echo signals pass through the duplexer and into the receiver [Figure 5.24 (b)].

This duplexer has wide bandwidth. Its power handling capability is also greater.

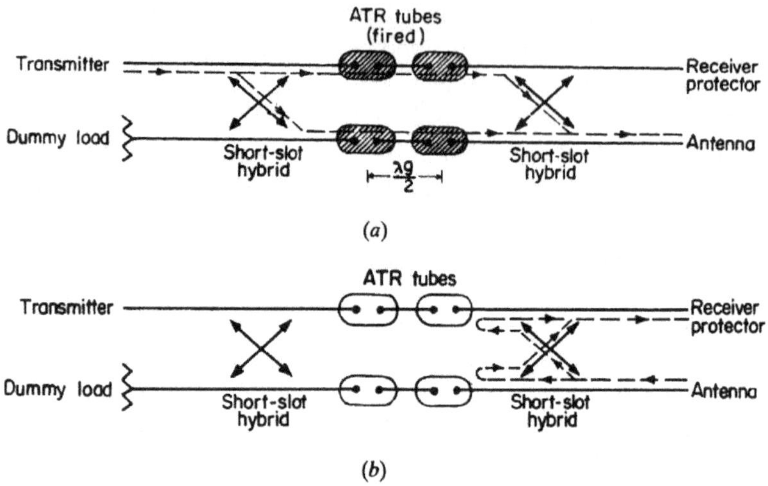

FIGURE 5.24 Balanced duplexer using dual TR-tubes (*a*) Transmit condition, (*b*) Receive condition.

An alternative form of balanced duplexer uses two hybrid junctions and four ATR tubes. During transmission the ATR tubes located in a mount between two short-slot hybrids ionize and allow high power to pass to the antenna [Figure 5.25 (*a*)]. On the other hand, during reception the ATR tubes present a high impedance and, as a result, the echo-signal power is reflected to the receiver. This type of duplexer has less bandwidth but higher power handling capability.

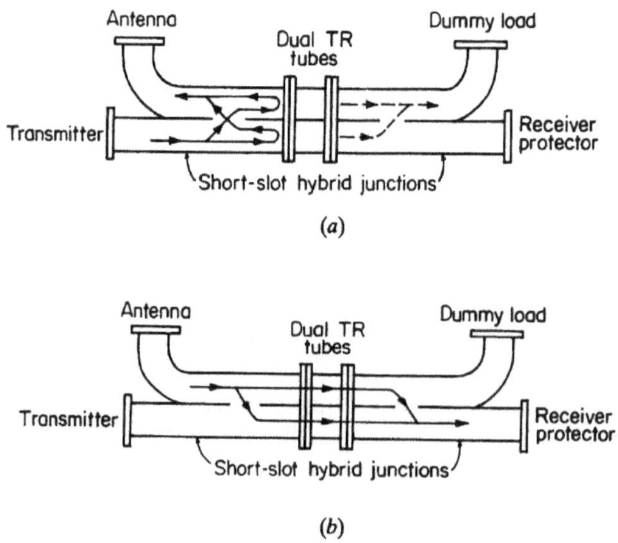

FIGURE 5.25 Balanced duplexer using ATR tubes (*a*) Transmit condition, (*b*) Receive condition.

5.12 MATCHED FILTER RECEIVER

If the frequency response function of a network maximizes the output peak signal-to-mean noise power ratio, then it is called a matched filter. This criterion is utilized for the design of almost all radar receivers.

Let $H(f)$ denote the frequency response function. For a received waveform $s(t)$ with a given ratio of signal energy E to noise energy N_0, we have

$$H(f) = G_a S^*(f)\exp(-j\,2\pi f t_1) \tag{5.21}$$

where G_a = maximum filter gain (a constant)

$S^*(f)$ = complex conjugate of $S^*(f)$

$S(f) = \int_{-\infty}^{\infty} s(t)\exp.(-j\,2\pi f t)\,dt,$

voltage spectrum of input signal

t_1 = fixed time at which signal attains maximum.

Ignoring the constant G_a, equation (5.21) can be written as

$$|H(f)|\ \exp[-j\phi_m(f)] = |\,S(f)\,|\exp\{j[\phi_s(f) - 2\pi f t_1]\} \tag{5.22}$$

In the previous equation, the frequency spectrum of the received signal is written as an amplitude spectrum $|\,S(f)\,|$ and phase spectrum exp $[-j\phi_s(f)]$. Similarly, the matched filter frequency response function is written in terms of its amplitude spectrum as $|\,H(f)\,|$ and phase spectrum exp $[-j\phi_m(f)]$.

From equation (5.22), we have

$$|\,H(f)\,| = |\,S(f)\,|$$

and

$$\phi_m(f) = S(f) + 2 - \phi_s(\pi) f t_1 \tag{5.23}$$

The matched filter can also be specified by its impulse $h(t)$. We can write

$$h(t) = \int_{-\infty}^{\infty} H(f)\exp(j\,2\pi f t)\,df \tag{5.24}$$

This is the inverse Fourier transform of the frequency-response function.

Substituting equation (5.24) in (5.21), we get

$$h(t) = G_a \int_{-\infty}^{\infty} S^*(f) \exp[j 2\pi f (t_1 - t)] \, df$$

$$= G_a \int_{-\infty}^{\infty} S(f) \exp[j 2\pi f (t_1 - t)] \, dt \left[\text{because } S^*(f) = S(-f)\right]$$

$$= G_s S(t_1 - t) \tag{5.25}$$

Thus, we must always have $t < t_1$.

Derivation of the Matched-Filter Characteristic. Using the Schwartz inequality one can derive the matched-filter frequency response function. We have attempted here to establish that the frequency-response function of the linear, time-invariant filter, which maximizes the output peak-signal-to-mean-noise ratio, is

$$H(f) = G_a S^*(f) \exp(-j 2\pi f t_1) \tag{5.26}$$

We, in fact, wish to maximize the ratio

$$R_f = \frac{|s_0(t)| max^2}{N}$$

where, $|s_0(t)|_{max}$ = maximum value of output signal voltage
N = mean noise power at receiver output.

The output voltage of a filter is,

$$|s_0(t)| = \left| \int_{-\infty}^{\infty} S(f) H(f) \exp(j 2\pi f t) \, dF \right| \tag{5.27}$$

where $S(f)$ is the Fourier transform of the received signal.

The mean output noise power is

$$N = \frac{N_0}{2} \int_{-\infty}^{\infty} |H(f)|^2 \, df \tag{5.28}$$

where N_0 = input noise power per unit bandwidth. The factor 1/2 appears before the integral, because the limits extend from $-\infty$ to $+\infty$.

Substituting equations (5.27) and (5.28) in equation (5.26), and assuming that the maximum value $|s_0(t)|^2$ occurs at $t = t_1$, we get

$$R_f = \frac{1 \int_{-\infty}^{\infty} S(f) H(f) \exp|(j 2\pi f t_1) \, df|^2}{\left(\frac{N_0}{2}\right) \int_{-\infty}^{\infty} |H(f)|^2 \, df} \tag{5.29}$$

If P and Q are two complex functions, then Schwartz's inequality states,

$$\int P^* P\,dx \int Q^* Q\,dx \geq \left| \int p^* Q\,dx \right|^2 \tag{5.30}$$

The previous equality sign applies when $P = kQ$, where k is a constant. Let us put,

$$P^* = S(f)\exp(j\,2\pi f t_1)$$

and $$Q = H(f)$$

Also recalling that, $\int P^* P\,dx = \int |P|^2\,dx$

We get $$R_f \leq \frac{\int_{-\infty}^{\infty} |H(f)|^2\,df \int_{-\infty}^{\infty} |S(f)|^2\,df}{\left(\frac{N_0}{2}\right)\int_{-\infty}^{\infty} |H(f)|^2\,df}$$

$$\leq \frac{\int_{-\infty}^{\infty} |S(f)|^2\,df}{(N_a/2)} \tag{5.31}$$

Now from Parseval's theorem,

$$\int_{-\infty}^{\infty} |S(f)|^2\,df = \int_{-\infty}^{\infty} s^2(t)\,dt = \text{signal energy} = E$$

Hence we have, $R_f \leq \dfrac{2E}{N_0}$ (5.32)

The frequency-response function which maximizes the peak signal-to-noise ratio R_f is obtained by noting that the equality sign in equation (5.30) is applicable when $P = kQ$ or,

$$H(f) = G_a\,S^*(f)\exp(-j\,2\pi f t_1)$$

where the constant $k = 1/G_a$.

Matched Filter and the Correlation Function. The output of a matched filter is not the replica of the input signal. It may be shown that the output is proportional to the input signal, cross-correlated with a replica of the transmitted signal, except for the time delay t_1. If $y(\lambda)$ and $s(\lambda)$ are two signals, each of finite duration, then the cross-correlation function $R(t)$ can be defined as,

$$R(t) = \int_{-\infty}^{\infty} y(\lambda)\, s(\lambda - t)\, d\lambda \tag{5.33}$$

The output $y_0(t)$ of a filter with the impulse response $h(t)$ for the input $y_{in}(t) = s(t) + n(t)$ is expressed as,

$$y_0(t) = \int_{-\infty}^{\infty} y_{in}(\lambda)\, h(t - \lambda)\, d\lambda \tag{5.34}$$

For a matched filter,

$$h(\lambda) = s(t_1 - \lambda)$$

then, equation (5.34) becomes

$$y_0(t) = \int_{-\infty}^{\infty} y_{in}(\lambda)\, s(t_1 - t + \lambda)\, d\lambda$$

$$= R(t - t_1) \tag{5.35}$$

Thus, a cross-correlation is formed by the matched filter.

5.13 CORRELATION DETECTION

The matched filter receiver can be replaced by a cross-correlation receiver which performs the same mathematical operation. The block diagram of a cross-correlation receiver is shown in Figure 5.26. The input signal $y(t)$ is multiplied by the transmitted signal $s(t - T_r)$ and the product is passed through a low-pass filter for performing the integration. Targets at other ranges or time delays can be found by varying T_r.

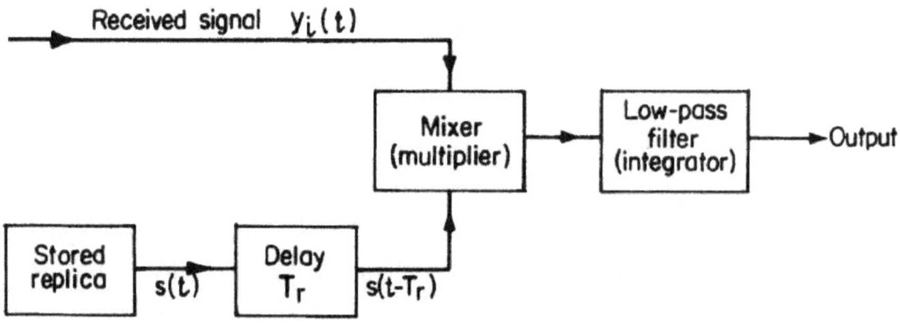

FIGURE 5.26 Cross-correlation receiver.

5.14 CONSTANT FALSE ALARM RATE (CFAR) RECEIVER

This receiver is obtained by observing the noise or clutter background in the vicinity of the target and adjusting the threshold according to the measured background. The cell-averaging CFAR is illustrated in Figure 5.27. It does not sum all the range cells but sums the cells ahead of the test cell separately from the sum of the cells following the test cell. The number of taps used in a cell averaging CFAR may vary from 16 to 20. It may be thought of as using the outputs of the sampled cells to estimate the unknown amplitude of the background noise or clutter.

When the target echo is large, energy can spill over into the adjacent range resolution cells. This affects the measurement of the average background, and for that reason the range cells surrounding the test cell are often omitted.

CFAR is used widely to prevent clutter and noise interference from saturating the display of an ordinary radar. Moreover, it is needed in ADT or track-while-scan systems. However, the CFAR introduces an additional loss in comparison to optimum detection. Also it maintains the false-alarm rate constant at the expense of the probability of detection, and hence it causes targets to be missed.

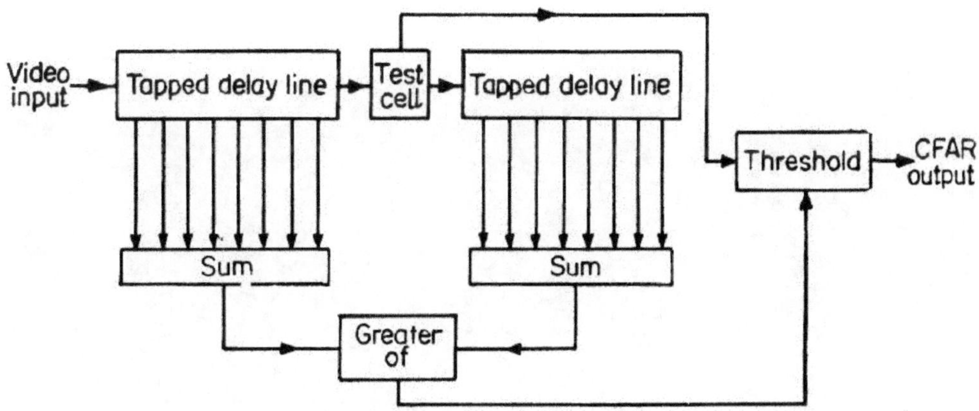

FIGURE 5.27 Cell averaging CFAR.

5.15 RECEIVER PROTECTOR AND SENSITIVITY TIME CONTROL

The duplexers based on TR and ATR devices have recovery times varying from a fraction of a microsecond to several microseconds. The attack time, which depends on the ionizing time of the gas or gases filling the devices, is of the order of 10 nsec. With fast rise times of high power RF pulses, a receiver protector is necessary in addition to the duplexer to limit the receiver input within a nanosecond, and fast-acting PIN diodes are useful for the purpose.

The PIN diode protector may also be used to serve an additional purpose of achieving a sensitivity time control (STC), which is a time dependent RF attenuator, by inserting a variable bias derived from the sweep voltage into the limiter diode, so as to increase the gain with increasing range of the echo. This will tend to make the echo level independent of ranges, a desirable feature in radar display. A PIN diode STC allows the STC to be applied at RF, which is more desirable than an STC introduced at the IF level. The RF stages may be overloaded by echoes from close ranges if STC is applied only to the IF stages.

CHAPTER 6

PROPAGATION OF RADAR WAVES

6.1 INTRODUCTION

During their propagation, radar electromagnetic waves are affected by the earth's surface and its atmosphere. They are, in fact, affected by the reflection or back scatter of energy from the earth's surface and the bending of the ray paths due to refraction in the atmosphere. They are also affected by scattering from snow, rain, birds, and various other clutter-producing objects. The signal-to-noise ratio is influenced by the propagation conditions as well as by the external noise environment in which the radar is located. In the presence of noise, the propagation introduces errors in the radar measurements and also modifies the free-space performance of the radar. It is better to classify between two different regions when considering the radar propagation. The first is the interference of the optical region, which is within the line of sight of the radar, while the other is the diffraction region, which lies beyond the radio horizon or beyond the line of sight range of the radar. It is important to realize, for all practical purposes, how propagation phenomena in the presence of the noise environment can influence the radar performance. In fact, this is one of the major factors in determining how successfully a radar performs in a particular application.

6.2 PROPAGATION ON PLANE EARTH

Let us consider the earth as a flat reflecting surface with the radar antenna situated at a height h_a. We assume that the target is at a distance r from the radar and it is at a height h_r. The energy radiated from the antenna arrives at the target via the direct path and the reflected path, as illustrated in Figure 6.1. In this analysis we assume that the lengths of the direct and the reflected paths are nearly equal, so that the amplitudes of the two signals are almost the same. The reflection coefficient of the surface τ may be written as

$$\tau = \rho e^{j\psi} \tag{6.1}$$

where the real part ρ describes the change in amplitude, while the argument gives the phase shift on reflection.

In the figure, A is the position of the radar antenna while its image is mirrored by the ground at A'. The difference between the reflected path AMB *or* $A'MB$ and the direct path AB is,

$$\Delta = 2h_a \sin\alpha,$$

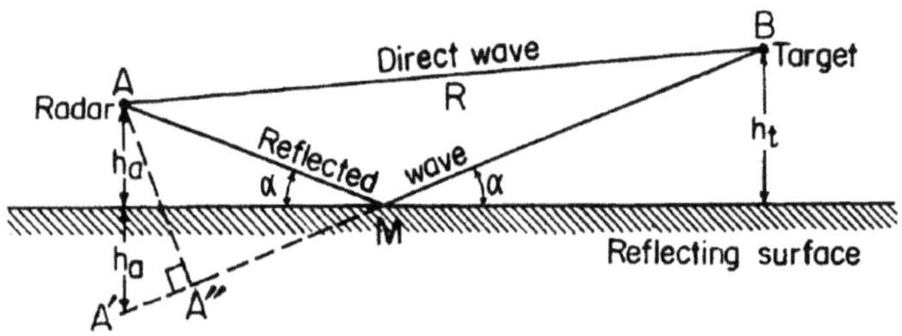

FIGURE 6.1 Radio wave propagation over a plane surface.

when $$R >> h_a \alpha \text{ is small,} \tag{6.2}$$

$$\sin\alpha = \tan\alpha = \frac{h_a + h_t}{r}$$

and then $$\Delta = \frac{2h_a(h_a + h_t)}{r} \tag{6.3}$$

Again, if $$h_1 >> h_a$$

then, $$\Delta = \frac{2h_a h_t}{r} \tag{6.4}$$

The phase difference Ψ_d corresponding to the path difference of the previous equation is,

$$\Psi_d = \frac{2\pi}{\lambda}\frac{2h_a h_t}{r} \text{ radians} \tag{6.5}$$

To get the total phase difference Ψ we must add a phase shift Ψ_r due to the reflection of the wave at M which is assumed to be 180° or radians. Thus, we have

$$\Psi = \Psi_d + \Psi_r = \frac{2\pi}{\lambda}\frac{2h_a h_t}{r} + \pi \tag{6.6}$$

The result of the two signals of unity amplitude and Ψ phase difference is

$$[2\,(1 + \cos\Psi)]^{1/2}$$

Now, the field strength (volts/meter) at the target in the presence of surface reflection is given by the ratio

$$\eta = \frac{\text{field strength at the target in presence of surface reflection}}{\text{field strength at target in free space}}$$

Therefore, the ratio of the power incident on the target at B to that incident on the target if located in free space is

$$\begin{aligned} \eta^2 &= 2\left(1 - \cos\frac{4\pi h_a h_t}{\lambda r}\right) [\text{because } \cos(\Psi_d + \pi) = -\cos\Psi_d] \\ &= 4\sin^2\frac{2\pi h_a h_t}{r} \end{aligned} \tag{6.7}$$

Due to reciprocity, the path from target to radar is the same as that from the radar to the target. Thus, the power ratio at the radar is

$$\eta^4 = 16\sin^4\left(\frac{2\pi h_a h_t}{\lambda r}\right) \tag{6.8}$$

The radar equation (6.6) describing the received echo power must be multiplied by the propagation factor η_4 of equation (6.8). As the magnitude of the sine function varies from 0 to 1, the factor changes from 0 to 16. The received signal strength also changes from 0 to 16. Therefore, the fourth power relation between range and echo signal causes a variation of radar range from 0 to 2 times the range of the same radar in free space.

The power ratio becomes maximum when the argument of the sine term in equation (6.8) is equal to $\frac{\pi}{2}, \frac{3\pi}{2}, \ldots, (2n+1)\frac{\pi}{2}$ where $n = 0, 1, 2, \ldots$. The maxima occur when

$$\frac{4h_a h_t}{\lambda r} = 2n+1$$

Similarly, the minima or the nulls occur when

$$\frac{2h_a h_t}{\lambda r} = n$$

The presence of a plane reflecting surface thus causes the continuous elevation coverage to break up into a lobed structure. This is shown in Figure 6.2. The angle of the lowest lobe is nearly equal to $\lambda/4h_a$. Therefore, for a low angle coverage the radar antenna height should be high and the wavelength of the radiated energy small.

Including the propagation factor η, the simple form of the radar equation can be written as

$$\begin{aligned} P_r &= \frac{P_t\, G^2 \lambda^2 \sigma}{(4\pi)^3\, r^4} 16 \sin^4 \frac{2\pi h_a h_t}{\lambda r} \\ &\cong \frac{4\pi P_t\, G^2 \sigma (h_b\, h_t)^4}{\lambda^2\, r^8} \text{ (for small angles)} \end{aligned} \tag{6.9}$$

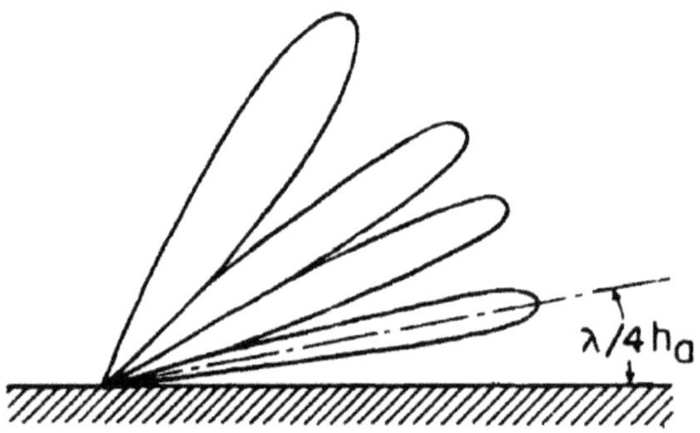

FIGURE 6.2 Vertical radiation pattern in the presence of a plane reflecting surface.

Equation (6.9) indicates that the received signal power for targets at low angles varies as the eighth power of the range. This has been experimentally verified for ship targets at short distance.

The previous analysis is based on many simplifying assumptions, and therefore care should be taken in applying the previous result in any realistic situation. Theoretically the nulls in the lobe structure are at zero field strength, but in practice the nulls are filled in and the lobe maxima is reduced due to non-perfect reflecting surfaces.

For a perfectly conducting surface, such as the sea, the reflection coefficient is $\rho_0 \exp[-2k^2 \bar{h}^2 \sin^2 \psi]$.

where ρ_0 = complex reflection coefficient for a smooth surface

$k = \frac{2\pi}{\lambda}$ (λ = wavelengths)

$\bar{h}^2$ = mean square surface height

Ψ = grazing angle.

This expression confirms well with the experimental data.

The reflection of electromagnetic waves from the sea can be separated into a coherent and an incoherent component. The coherent component has a reflection coefficient whose amplitude and phase depend on the given geometry and sea state, and the fluctuating or the incoherent component of a surface scattered signal is characterized by random phase and amplitude fluctuations.

6.2.1 The Effect of Earth's Curvature

In order to predict radar coverage, the curvature of the earth cannot be neglected, particularly for coverages at low elevation angles near the horizon.

We know that there are two regions of interest in the radar propagation, namely the interference region and the diffraction region. The interference region is located within the line of sight of the radar. In this case the direct and reflected waves interfere, producing a lobed radiation pattern. This is similar to that for the plane earth, except that the lobing is not so pronounced. The other region lies beyond the interference region below the radar line of sight and is known as the shadow or the diffraction region. Radar signals in the diffraction region are rapidly attenuated. Only a few microwave radars can penetrate the diffraction region to a great extent owing to the severe losses.

By analytical means one can predict the effect of the round earth on radar coverage. Figure 6.3 shows a particular case, the computed vertical plane radar coverage patterns for horizontal polarization.

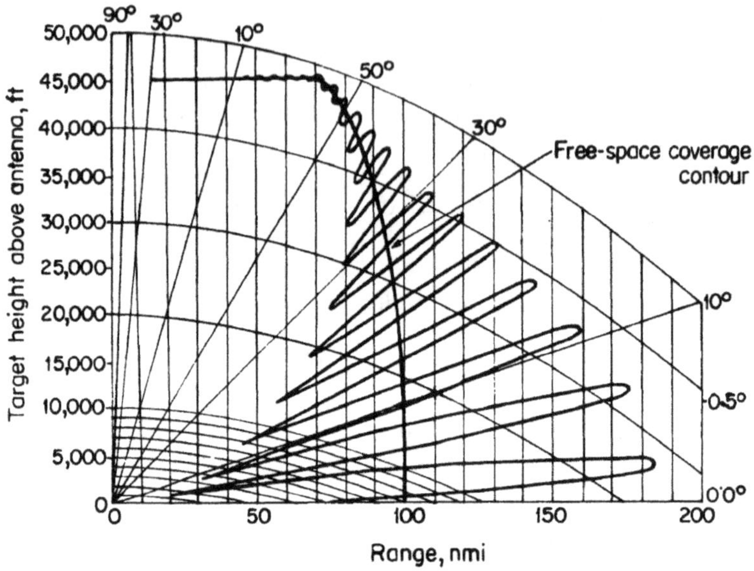

FIGURE 6.3 Vertical plane coverage diagram for horizontal polarization with an operating frequency 1300 MHz, antenna height 50 ft, antenna beam width 12°, and beam maximum pointing on the horizon.

6.3 REFRACTION

We know that in free space radar waves travel in a straight line. But electromagnetic waves propagating within the earth's atmosphere are generally bent or refracted. Due to refraction the distance to the horizon is extended, thus increasing the radar coverage as shown in Figure 6.4 (*a*). Moreover, an angular error is caused by refraction as illustrated in Figure 6.4 (*b*).

Refraction or the bending of the radar waves in the atmosphere is due to the variation of the velocity of propagation with altitude. A term *index of refraction* is therefore introduced, which may be defined as the velocity of propagation in free space to that in the medium. At microwave frequencies, the index of refraction n for air which contains water vapor is expressed as

$$(n-1)\,10^6 = N = \frac{77.6p}{T} + \frac{3.73\times 10^5 e}{T^2} \tag{6.10}$$

where, p = barometric pressure in m bar

e = partial pressure of water vapor in m bar

T = absolute temperature in °K.

In equation (6.10) the parameter $(n - 1)\ 10^6 = N$ is the scaled-up index of refraction and is termed as refractivity. The index of refraction generally decreases with increasing altitude. A typical value of this near the surface of the earth is 1.0003, the corresponding refractivity being 300. The decrease of the velocity of propagation increases with altitude. This causes the radio waves to bend downward, resulting in an increase in the effective radar range.

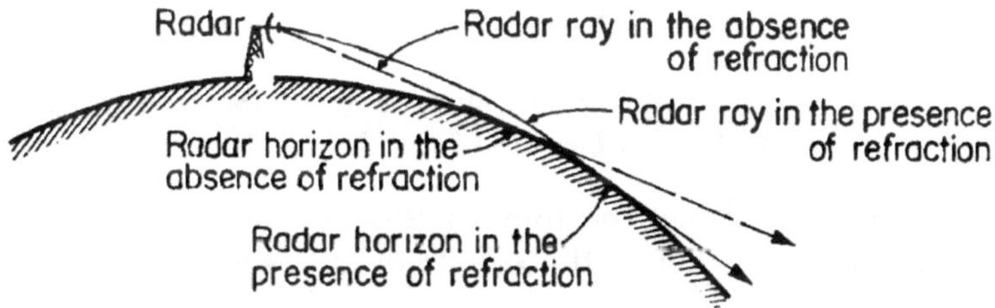

FIGURE 6.4 (a) Extension of the ratio horizon due to reflection of radar waves by the atmosphere.

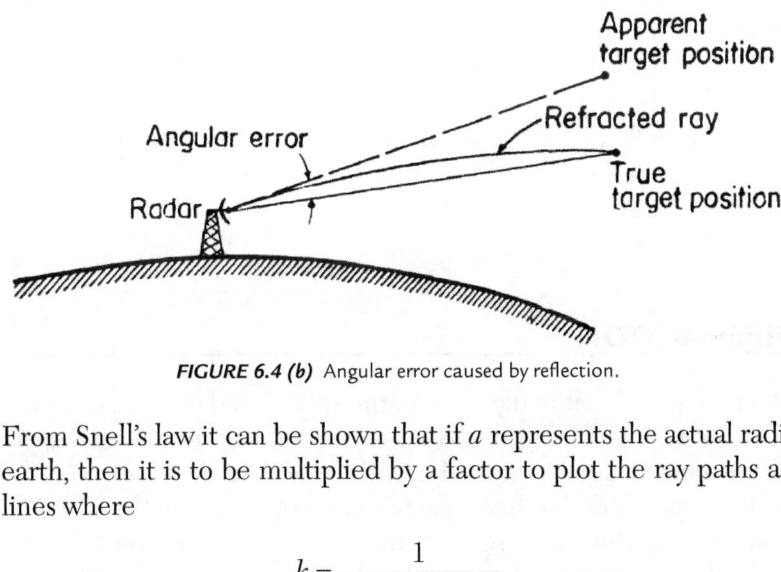

FIGURE 6.4 (b) Angular error caused by reflection.

From Snell's law it can be shown that if a represents the actual radius of the earth, then it is to be multiplied by a factor to plot the ray paths as straight lines where

$$k = \frac{1}{1 + a\,(dn / dh)} \tag{6.11}$$

where dn/dh = rate of change of refractive index with height. The vertical gradient of the refractive index dn/dh is negative, in general. If this gradient is assumed constant with height and having a value of 39×10^{-6}/meter, then the value of k becomes 4/3.

The distance d to the horizon for a radar of height h can be shown as

$$d = \sqrt{2k_a h} \tag{6.12}$$

where k_a is the effective radius of the earth.

In a more appropriate refractivity model, the refractivity varies exponentially with height by the equation,

$$N = N_S \exp\left[-c_e(h - h_s)\right] \tag{6.13}$$

where N_s is the refractivity at the surface of the earth, h the altitude of the target, h_s the altitude of radar, and $C_e (= l_a\, N_s/N_1)$ is a constant.

In has been found that the previous exponential model provides a more accurate determination of the effects of atmospheric refraction than the linear model.

It must be pointed out that the use of the correct atmospheric model is important in a height-finder radar, particularly when the target is at long distances. Refractions are, in fact, responsible for the bending of the radar rays, causing an apparent elevation angle different from the true one. Therefore, some corrections are required to get a better estimate of elevation angles, range, or height. Refraction is troublesome when the angle of elevation is low, specially at or near the horizon. It is usually neglected if the angle is greater than 3° to 5°.

6.4 DIFFRACTION

We have noted that in the earth's atmosphere, radar waves can propagate by refraction beyond the geometrical horizon. An alternative mechanism which allows radar range to extend beyond the geometrical horizon is diffraction, which depends on the frequency or more precisely on the size of the object compared to the wavelength. If the frequency is lowered, the wave is diffracted more. The mechanism of diffraction is, in fact, highly important for worldwide communications at very low frequencies (VLF).

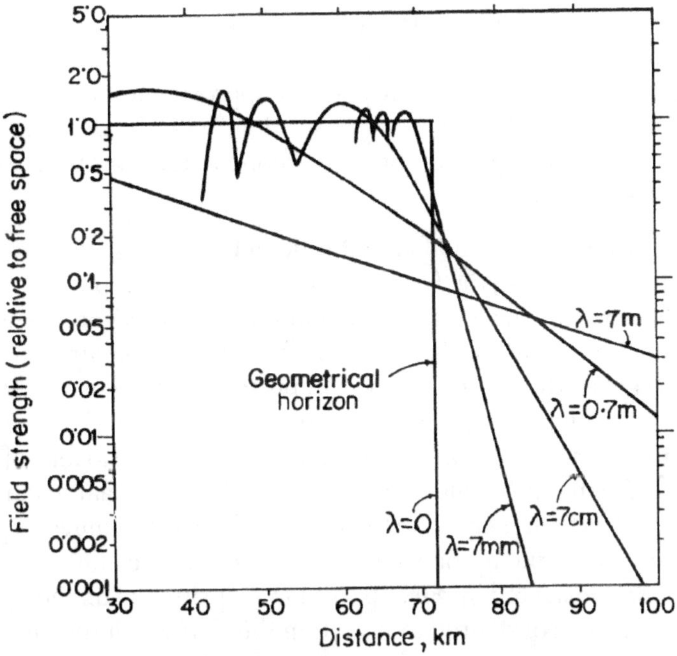

FIGURE 6.5 Theoretical estimate of field strength as a function of distance from the transmitting antenna for vertical polarization, equal antenna heights of 100 m, $k = 1$, ground conductivity = 10^{-2} mho/m, and dielectric constant = 4.

A plot of the electric field strength at the target as a function of the distance from the transmitting antenna is in Figure 6.5. In this case, it is assumed that both the radar antenna and the target are at a fixed height. The distance between the radar and the target along the line of sight is expressed as

$$d_0 = \sqrt{2kah_1} + \sqrt{2kah_2} \tag{6.14}$$

where a is the radius of the earth while h_1, h_2 are the heights of the radar antenna and target respectively.

The point of tangency of the line of sight with the earth gives the geometrical or optical horizon. Between the radar and the geometrical horizon, the field strength at optical frequencies ($\lambda \approx 0$) is the same as in free space. Beyond the horizon, the field does not penetrate. Hence for optical frequencies, the geometrical horizon represents an approximate boundary between the regions of propagation and no propagation. With the decrease of frequency, as shown in the figure, more and more energy propagates beyond the geometrical horizon.

In the diffraction region, if low altitude radar coverage is desired beyond the geometrical horizon, the frequency must be as low as possible. The decrease in radar coverage owing to attenuation of electromagnetic energy is shown in Figure 6.6. This we have drawn for a radar operating at frequency of 500 MHz. In the figure, curve 1 represents the locus of the geometrical line of sight for $k = 4/3$. Curve 2 is the constant radar signal contour in the diffraction region, assuming that the radar is capable of a free-space range of 220 nautical miles. Curves 3, 4, and 5 are similar to curve 2, but they are for radars with ranges of 110, 55, and 27.5 nautical miles respectively. The approximate boundary between the interference region and the diffraction region is represented by curve 6. If a target is situated on the right of curve 6, then it may be considered to be within the diffraction region.

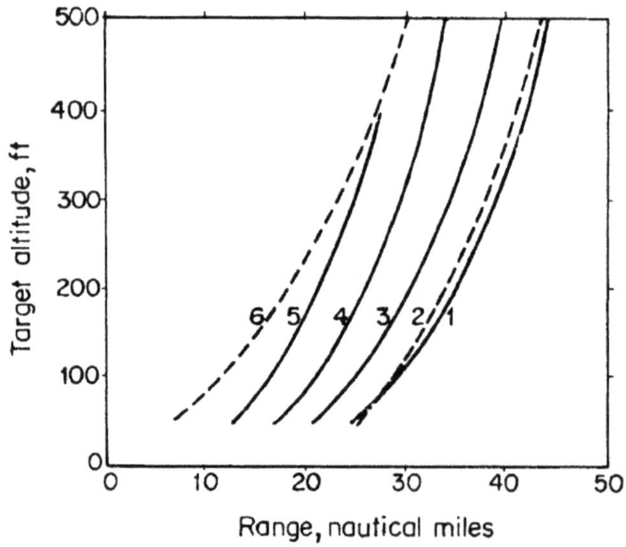

FIGURE 6.6 Contours of radar coverage for an antenna height of 200 ft above the curved earth.

6.5 ANOMALOUS PROPAGATION

The anomalous or the abnormal propagation of electromagnetic waves is termed as super-refraction, trapping, or ducting. According to the definition for normal atmospheric refraction, the gradient dn/dh varies from 0 to -0.787×10^{-7} or when the effective earth's radius varies from $k = 1$ to $k = 2$. Super-refraction corresponds to the gradient dn/dh in the range -0.787×10^{-7}/m to -1.57×10^{-7}/m or $k = 2k = a$. Trapping or ducting occurs when dn/dh -1.57×10^{-7}/m. The radar ranges are greatly extended with ducted propagation, as shown in Figure 6.7. If with increasing height the atmospheric index of refraction were increased, the rays would curve upward and the radar range would be reduced in comparison to the normal condition. This rarely occurs, and it is known as sub-refraction. By anomalous or non-standard propagation we mean any of the previous propagation conditions other than normal. Normally it is used to describe the cases where the radar range is extended far beyond normal.

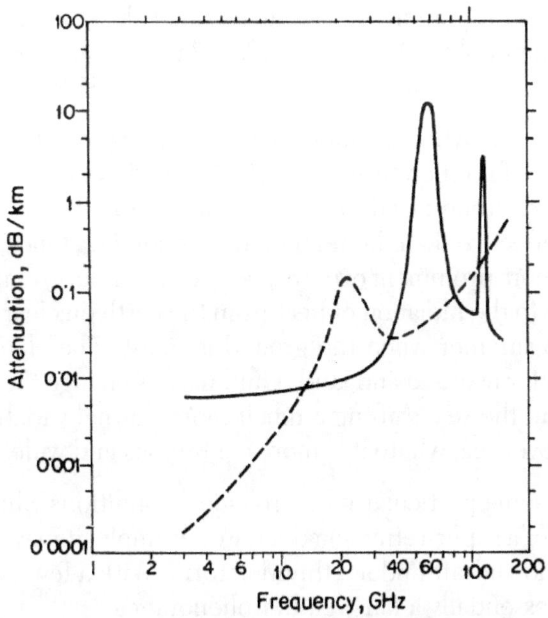

FIGURE 6.7 Attenuation of electromagnetic energy by atmospheric gases in an atmosphere at 76 cm pressure. Dashed curve shows absorption due to water vapor for 1% (7.5 g water/m^3) water vapor density.

If a super-refracting duct lies close to the ground, then it is called a surface duct or ground based duct. When the duct lies above the surface, it is called an elevated duct. Over water the surface duct is often called the evaporation duct due to the water vapor evaporated from the surface. Apparently surface ducts are more common than elevated ducts. In order to propagate energy within the duct, the angle that is to be made by the radar ray with the duct should be less than 1°, in general, so that power may be coupled. An extension of the radar range within the duct causes a reduction of the coverage in other directions.

Generally, atmospheric ducts are of the order of 10 or 20 meters in height. An approximate model of propagation in atmospheric ducts giving the maximum wavelength λ_{max} that can propagate in a surface duct of depth d is written as

$$\lambda_{max} = 2.5\left(-\frac{\Delta n}{\Delta h}\right)^{1/2} \tag{6.15}$$

where λ_{max}, Δh and d are in the same units. As for example, at X band ($\lambda_{max} = 3$ cm). Equation (6.15) suggests that the duct must be at least 10 m thick, at S band ($\lambda = 10$ cm) it must be 22 m, and at UHF ($\lambda = 70$ cm) it must be 80 m thick when we assume $-\Delta n/\Delta h$ to be $1.57 \times 10-^{7}$/m.

A duct is formed when the index of refraction rapidly decreases with altitude. An increase of temperature with height, called the temperature of inversion, must be very pronounced to produce ducts. However, water vapor gradients are more effective than the temperature gradient. Hence, super-refraction is generally more prominent over seas, especially in warm climates. Ducting over land is due to the radiation of heat from the earth on clear nights, particularly during the summer when the ground is moist. The character of ducting is likely to differ over sea and land. Land masses change the temperature more quickly than the sea, causing a much more diurnal variation of ducting over land than over sea, where it is more continuous and widespread.

There are some particular meteorological conditions which may lead to the formation of a super-refracting duct. For example, it may be produced by the diverging downdraft under a thunderstorm. With a few exceptions, however, ducting is essentially a fair-weather phenomena.

6.5.1 Elevated Duct

An elevated duct is found in a tradewind region between the mid-ocean, high pressure cells and the equatorial doldrums. The tradewind temperature inversion is seen over the eastern part of the tropical oceans, where there is a slow-sinking of high altitude air. Two tradewind areas where investigations have been made lie between Brazil and the Ascension Islands and between Southern California and Hawaii. For getting the maximum advantage of propagation in an elevated duct, the radar and the target should be fixed at an altitude near that of the duct.

6.5.2 Evaporation Duct

It is mainly due to the water vapor or humidity evaporated from the sea just above its surface. This duct exists over the sea almost all the time. It has been found that the height of the duct is within the range of 6 to 30 meters depending on the typographical condition, season of the year, time of the day, and the wind speed. The height and strength of the duct is found to vary with wind speed. If the wind speed is from 8 to 15 knots, then it produces a low, moderately strong duct, while winds from 20 to 30 knots produces a higher but weaker duct. A passing thundersquall or rain shower does not wipe out the duct. The thicker the duct, the lower the frequency allowed through it. The lower microwave frequencies below the L-band are not strongly affected by the evaporation duct.

The height of the evaporation duct is calculated by measuring the surface water temperature at some convenient height, the air temperature, wind speed, and relative humidity. From these one can conclude about the propagation conditions beyond the normal horizon. In the presence of the duct, the relative phases between the direct and reflected rays are different, and focusing can alter the relative amplitude of the two components.

Although both the elevated and the evaporation ducts result in extended radar ranges, their consequences are often bad. It is often difficult to predict the presence of extended ranges. During ducting conditions, the extended ranges may bring in ground clutter from longer ranges.

6.6 ATTENUATION DUE TO ATMOSPHERIC GASES

When no precipitation is present, the attenuation of radar energy in a clear atmosphere is produced by the oxygen and water vapor molecules. When propagated over a distance r and back, the reduction in radar signal power

$$\alpha = \exp(-2\alpha r)$$

where α is the one-way attenuation coefficient.

If the attenuations, α, are plotted in dB/km against the frequency in GHz, we may obtain curves as shown in Figure 6.7. In the figure, attenuation of electromagnetic energy due to atmospheric gases in an atmosphere of 76 cm of pressure is shown. The solid curve represents the absorption due to oxygen while the dashed curve is for absorption due to water vapor in an atmosphere containing 1% water vapor molecules.

The figure clearly indicates that the resonance peaks for oxygen molecules occur around 60 GHz and 119 GHz, while the peaks for the water vapor occur around 22.25 GHz and 183 GHz. The attenuation becomes increasingly important at frequencies above 10 GHz. The attenuation of the atmospheric gases decreases with increasing altitude. So the attenuation experienced by a radar depends on the target and on the range. The attenuation is greatest with a ground based radar when the antenna points to the horizon, while it is least when it points to the zenith.

6.7 ENVIRONMENTAL NOISE

The external noise level at microwave frequencies is relatively low, and the sensitivity of radar receivers is determined by the internal noise. Characteristics of the sources of ambient or external noise are discussed as follows.

6.7.1 Solar Noise

The sun is a strong emitter of electromagnetic radiation whose intensity changes with time. The blackbody radiation at a temperature of about 6000°K accounts for the minimum level of solar noise observed.

Unlike most other noise mechanisms, the solar noise power increases with frequency. The solar noise storms which occur during sunspots and flares can increase the solar noise level. Sometimes the solar noise can be of significant magnitude to affect the sensitivity of low noise radar receivers. Discrete ratio sources known as the ratio stars are too weak at radar frequencies with the usual antennas, but they have been used in conjunction with sensitive receivers in determining pointing and focusing corrections of large antennas.

6.7.2 Cosmic Noise

Electromagnetic radiation in the form of a continuous background of noise, which arrives from extra terrestrial sources such as our own galaxy, extra-galactic sources, and the radio stars, is called cosmic noise. With increases of frequency cosmic noise decreases, and at frequencies above *L*-band it can generally be neglected. However, it can be a serious limitation when the radar is operated at VHF or lower bands. The variation of cosmic noise with frequency is shown by a dotted line in Figure 6.8.

It has been found that magnitude of the cosmic noise depends on the portion of the celestial sphere in which the antenna points. When looking to our own galaxy, it becomes maximum, while observing along the center about which the galaxy revolves, it reduces to a minimum at the pole. In Figure 6.8 a plot of the maximum and minimum brightness temperatures of the sky as seen by an ideal single polarization antenna on earth is made. Under ideal conditions, a highly directive antenna viewing a distributed source of radiation would receive a noise power of value kT_BB, where k is the Boltzmann's constant, T_B = the brightness temperature of the source, and B is the bandwidth of the receiver. The antenna temperature, for a practical antenna, is defined as the integral of the brightness temperature over all angles, weighted by the antenna pattern.

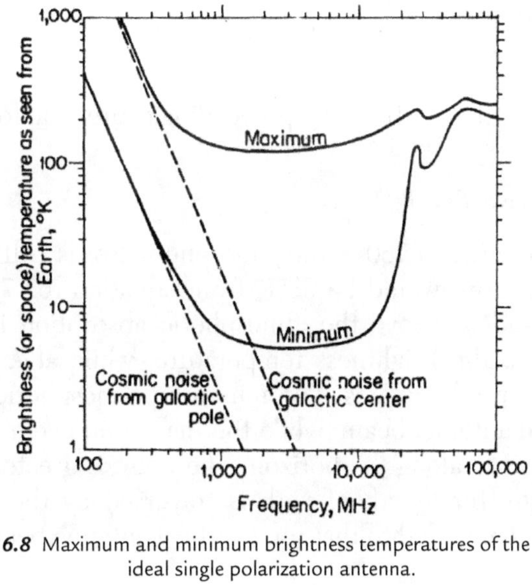

FIGURE 6.8 Maximum and minimum brightness temperatures of the sky for an ideal single polarization antenna.

6.7.3 Atmospheric Absorption Noise

Let us consider an absorbing atmosphere at ambient temperature surrounded by an imaginary blackbody at some temperature. In passing through the atmosphere, the energy is attenuated by a loss L. Now, from the imaginary blackbody, the noise power available

$$N_a = kT_aB_n \text{ (where } B_n = \text{bandwidth)}$$

The noise power after passing through the atmosphere

$$N_aL = \frac{kT_aB_n}{L}$$

The amount of power absorbed by the atmosphere

$$N_a - N_aL = \Delta N = kT_aB_n\left(1 - \frac{1}{L}\right)$$

where ΔN is the noise power radiated by the atmosphere itself.

From the definition of the effective noise temperature we may write, $\Delta N = kT_eB_nG$ (T_e = effective noise temperature)

$$= \frac{kT_eB_n}{L}\left(\text{where } \frac{1}{L} = G, \text{the gain}\right)$$

$$= kT_aB_n\left(1 - \frac{1}{L}\right) \text{from the preceeding expression for } \Delta N \qquad 6.16$$

Hence, we have $T_e = T_a(L-1)$ (6.17)

If at a temperature of 260°K the atmospheric loss is 1 dB, then the effective noise temperature would be 68°K from equation (6.17). It has been seen that at X-band or above the atmospheric absorption is the predominant contributor to the brightness temperature, while at L-band or lower the cosmic noise predominates. The minimum atmospheric absorption occurs for a vertical antenna beam, while the maximum occurs when the antenna beam is directed along the horizon. The noise is greater along the horizon because a greater length of path is traversed by the beam through the atmosphere, compared to that for a vertical antenna beam.

6.7.4 Atmospherics and Urban Noise

A lighting discharge radiates a significant RF noise power. It has been found that at any moment there are an average of 1800 thunderstorms in progress in different parts of the globe causing 100 lightning flashes in every second. All these lightning strikes give rise to the noise spectrum, which is large at broadcast and shortwave radio frequencies. Noise arising due to the electromagnetic radiation from lightning strikes is known as atmospheric noise. The spectrum due to the atmospheric noise rapidly falls, with frequency showing a peak at about 10 kHz. It has a little consequence at frequencies above 50 kHz. Thus, one can conclude that the atmospheric noise is seldom an important factor in the radar design except for radars in the lower region.

Urban noise is the another noise source predominant at the lower radar frequencies. This is also known as manmade noise. Noise due to electric razors, automobile ignitions, power tools, and fluorescent lights are the examples belonging to this group. It has little importance at UHF and higher frequencies.

6.7.5 System Noise Temperature

Some noise sources which are to be considered when computing the system effective noise temperature are illustrated in Figure 6.9.

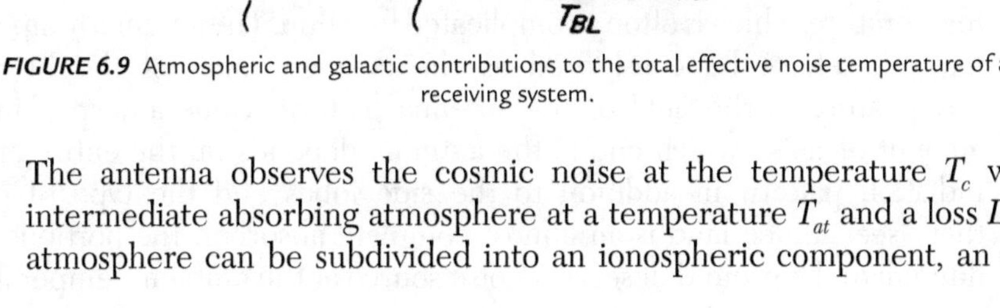

FIGURE 6.9 Atmospheric and galactic contributions to the total effective noise temperature of a receiving system.

The antenna observes the cosmic noise at the temperature T_c with an intermediate absorbing atmosphere at a temperature T_{at} and a loss L_{at}. The atmosphere can be subdivided into an ionospheric component, an oxygen component, and a water vapor component. The sum of the temperatures due to cosmic noise and the atmospheric noise $[T_c + T_{at}(L_{at} - 1)]$ is known as the ideal antenna temperature. It is also called the space temperature or the brightness temperature. As indicated in the figure, the RF losses L_{rf} include the antenna, radome, and duplex losses and also the transmission line loss. In the figure, T_{re} represents the effective noise temperature of the receiver. Now, assuming that the noise contributions enter the receiver through the main beam only, the effect of the side lobe can be ignored, and then the total system effective noise temperature T_e can expressed as

$$T_e = T_c + (L_{at} - 1)T_{at} + (L_{rt} - 1)T_{rf}L_{at} + T_{re}L_fL_{rf}L_{at} \tag{6.18}$$

The previous equation can be applied to an ideal antenna with no side lobes. If the main beam views the 'cool' sky, but the side lobes view the 'hot' earth, then the total noise power due to the side lobes becomes greater than the noise power in the main beam. Usually the contributions to the total effective system noise temperature can be classified as:

1. the effective noise temperature of the receiver itself
2. the effective space noise temperature, and
3. the effective noise temperature contributions due to RF lossy components.

The total antenna temperature is obtained by integrating the temperature as viewed by the antenna, weighted by the antenna gain over the entire sphere. Mathematically, this can be expressed as

$$T_a = \frac{\int T_B(\theta,\phi)\, G(\theta,\phi)\, d\Omega}{\int G(\theta,\phi)\, d\Omega} \tag{6.19}$$

where the solid angle, $d\Omega = \sin\theta\, d\theta\, d\phi$ and $T_B(\theta, \phi)$ is the brightness temperature, which is often a complicated function. The antenna temperature may, therefore, be considered as the average value of the brightness temperature in the field of the antenna pattern. Generally speaking, the amount of noise which enters the antenna depends on the entire antenna radiation pattern in addition to the side lobes and the type of objects they "see." Since land is an almost complete absorber, the portions which illuminate the ground observe a noise source at the ambient temperature.

6.8 MICROWAVE RADIATION HAZARDS

It has been noted that high power microwave energy can produce spectacular effects. If the microwave energy is of sufficient intensity, it becomes a health hazard, producing biological damage in human beings. On living tissue heating is the chief effect of excessive microwave radiation. Many instances have been noted of the formation of cataracts in animals' eyes by exposure to microwave radiation. The viscous material of the eyeball is affected by heat due to exposure to microwave radiation.

If the frequency is below 400 or above 3000 MHz, then the body absorbs less than half the incident energy. However, lower frequencies pass through while the higher frequencies are reflected at the skin's surface. The percentage of radiation absorbed in the range 1000 to 3000 MHz can approach up to 100 percent, depending on the subcutaneous layers of fat and thickness of the skin. If the whole body is immersed in microwave radiation, a sensation of warmth serves as a warning before the damage of the localized parts of the body. But if only a part of the body is exposed, the sensation of warmth may or may not be felt, depending on the frequency. With the increase of the radar power, the biological damage increases. To avoid any serious harm, therefore, proper safety precautions must be taken. In the United States armed services, the average power density for the maximum safe continuous exposure level is 10 m W/cm^2. This value is based on the experimental results by considering the thermal effects. It must, however, be pointed out that the microwave radiation is non-ionizing and is widely different from the ionizing radiations like X-rays.

CHAPTER 7

RADAR CLUTTER

7.1 INTRODUCTION

Any unwanted radar echo is called clutter. Such echoes can "clutter" the radar output and thus make it difficult to detect the desired targets. Examples of clutter include the reflections from land, sea, rain, birds, insects, and chaff. Clutter can also be due to clean-air turbulance and other atmospheric effects as well as due to ionized media like the aurora and meteor trails. Unwanted echoes might also be obtained from "point" or fixed targets such as poles, towers, and similar objects.

Echoes from land or sea are called surface clutter, while those from rain or other atmospheric phenomena are called volume clutter. For surface clutter the cross-section per unit area σ° can be defined as

$$\sigma^\circ = \frac{\sigma_c}{A_c} \tag{7.1}$$

where σ_c is the radar cross-section from the area A_c. For volume clutter the cross-section per unit volume or the reflectivity η can be written as

$$\eta = \frac{\sigma_c}{V_c} \tag{7.2}$$

where σ_c is the radar cross-section from the volume V_c.

It should be mentioned, however, that the radar echoes from rain, birds, and so forth are not always undesired. For example, reflections due to storm clouds are wanted by the radar observer to investigate the rainfall rate over a large area. This information is sent to aircrafts and ships to help the navigation.

7.2 SURFACE CLUTTER RADAR EQUATIONS

In Figure 7.1, as shown, let the radar be illuminating the surface at a grazing angle α and let the width of the illuminated area, A_c, be determined by the azimuthal beamwidth θ_B. The power C received from the clutter is given by

$$C = \frac{P_t G A_e \sigma_c}{(4\pi)^2 r_4} \tag{7.3}$$

where P_t is the transmitter power, G is the antenna gain, A_c is the antenna effective aperture, σ_c is the clutter cross-section, and r is the range.

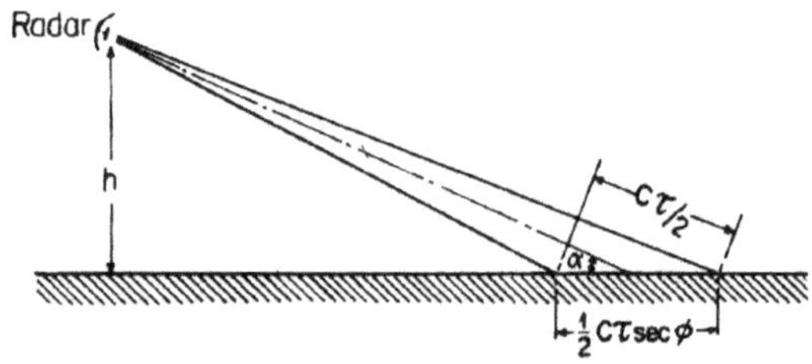

(*a*) **Surface intercepted by radar pulse.**

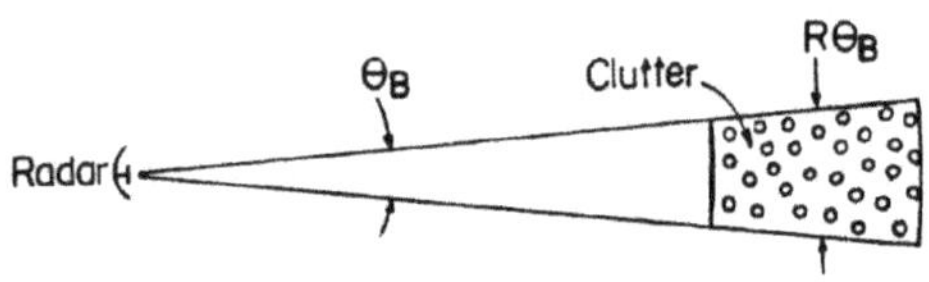

(*b*) **Clutter patch of individual scatterers.**

FIGURE 7.1 Geometry of radar clutter.

The value of σ_c in equation (7.3) can be expressed as

$$\begin{aligned}\sigma_c &= \sigma^\circ A_e \\ &= \sigma^\circ r\, \theta_B\, (c\, \tau/2) \sec \alpha\end{aligned} \tag{7.4}$$

where c is the velocity of propagation and π is the radar pulse width.

Substituting equation (7.4) in equation (7.3), we get

$$C = \frac{P_t G A_e \sigma^\circ \theta_B (C\tau/2) \sec\alpha}{(4\pi)^2 r^3} \tag{7.5}$$

Equation (7.5) indicates that the echo from surface clutter is inversely proportional to the cube of the range.

For a target of cross-section σ_t the signal power returned is given by

$$S = \frac{P_t G A_e \sigma_t}{(4\pi)^2 r^4} \tag{7.6}$$

The signal to clutter ratio *S/C* can now be easily obtained by taking the ratio of expressions in equations (7.6) and (7.5). Thus,

$$\frac{S}{C} = \frac{\sigma_t}{\sigma^\circ r \theta_B (C\tau/2)\sec\alpha} \tag{7.7}$$

If the maximum range of the radar occurs when the signal-to-clutter ratio becomes minimum, equation (7.7) can be rewritten as

$$r_{max} = \frac{\sigma_t}{\left(\frac{S}{C}\right)_{min} \sigma^\circ \theta_B \left(\frac{C\tau}{2}\right) \sec\alpha} \tag{7.8}$$

Equation (7.7) is a different form of the radar equation where the range appears as the first power rather than as the fourth power in the usual valid form (for receiver noise limited operation).

Next, let us consider the case when the radar finds surface clutter at perpendicular incidence, that is, at a grazing angle of 90°. The clutter area will be determined by θ_B and α_B in two principal planes. The area A_c in equation (7.1) can be written as

$$A_c = \left(\frac{\pi}{4}\right) r\,\theta_B\, r\,\alpha_B \sin\alpha \tag{7.9}$$

where α is the grazing angle. The factor ($\pi/4$) is for the elliptical shape of the area. Again, it can be shown that

$$G = \frac{\pi^2}{\theta_B \alpha_B} \tag{7.10}$$

Now by equations (7.1) and (7.3), we get

$$C = \frac{P_t G A_e \sigma^\circ A_c}{(4\pi)^2 r^4} \qquad [\text{because } \sigma_c = \sigma^\circ A_c]$$

$$= \frac{P_t A_e \sigma^\circ A_c}{(4\pi)^2 r^4}$$

$$= \frac{P_t A_e \sigma^\circ}{(4\pi)^2 r^4} \cdot \frac{\pi^2}{\theta_B \alpha_B} \left(\frac{\pi}{4}\right) r\,\theta_B\, r\,\alpha_B \sin\alpha$$

[Putting the values of G and A_c]

$$= \frac{\pi P_t A_e \sigma^\circ \sin\alpha}{64 r^2} \tag{7.11}$$

Equation (7.11) shows that the clutter power C varies inversely as the square of the range. This equation is applicable to the signal received from the ground by radar altimeter for clutter unlimited operation.

Instead of the grazing angle for the geometry of surface clutter, sometimes the incidence angle and the depression angle are taken into consideration (Figure 7.2). The grazing angle is defined with respect to the tangent to the surface. On other hand, the incidence angle is defined relative to the normal to the surface while the depression angle is defined with respect to the local horizontal at the radar. If the earth's surface is considered flat, then obviously the grazing angle and the depression angle become identical.

Land clutter is mainly dependent on the nature of the terrain and the local conditions. Forests, mountains, urban areas, cultivated lands, and deserts can all produce different types of radar echoes in the form of clutter.

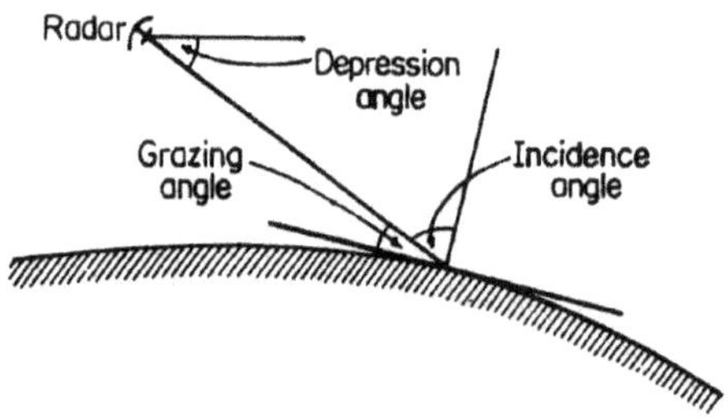

FIGURE 7.2 Geometry of radar and surface clutter showing various angles.

7.3 SEA CLUTTER

The echo from a sea surface depends on the wave height, wind speed, length of time, and distance. It also depends on frequency, polarization, grazing angle, and also the size of the area under observation.

7.3.1 Theory

Various models proposed earlier to explain the mechanism of sea clutter, but none of those were adequate in explaining experimental observations.

The *Composite Surface Model* proposed and established later on has been more successful. In this model the sea is considered to be composed of a number of waves of different wavelengths and amplitudes.

The composite surface model is based on the application of classical scattering theory. Here the radar wavelength λ_r is related to water wavelength λ_w by the equation

$$\lambda_r = 2\lambda_w \cos \alpha \ (\alpha = \text{grazing angle}) \tag{7.12}$$

Equation (7.12) is known as the Bragg back scattering resonance condition due to its similarity to the X-ray scattering as found by Bragg. Hence, the radar is only responsive to those water waves which satisfy equation (7.12). At a zero degree grazing angle, an X-band radar (λ = 3 cm) would back scatter from waves of 1.5 cm in length. Such short water waves are called *capillary waves*. On the contrary, at a zero degree grazing angle, UHF radar (λ = 70 cm) responds to water waves of 35 cm length. These are called *gravity waves*, as the earth's gravitational force controls the dimensions of such large waves. The effect of large waves is responsible for tilt the scattering surface of the smaller water waves. This is shown in Figure 7.3. The tilting effect is particularly important at a low grazing angle.

This theory accounts for the better experimental results obtained with vertical polarization than with horizontal polarization.

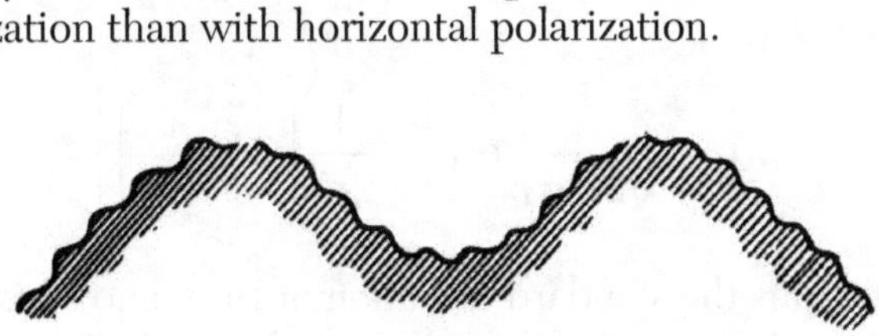

FIGURE 7.3 Composite rough-surface model of the sea.

7.3.2 Probability Density Function of Sea Clutter

The amplitude of sea clutter is described in terms of probability density function. The statistical fluctuations of the envelope v of such echo at the output of the envelope detector is expressed by the Rayleigh probability density function as

$$p(v) = \frac{2v}{\sigma^2} \exp.\left(-\frac{v^2}{\sigma^2}\right) \tag{7.13}$$

where σ = standard deviation of the envelope v. Clutter which conforms to this model is known as *Rayleigh clutter*. If the area illuminated by the radar has a dimension comparable to or smaller than the water wavelength, then the probability function deviates from Rayleigh.

An example of sea clutter statistics is revealed in Figure 7.4. The figure shows that the actual clutter has a greater probability of a large value of sea clutter than does Rayleigh clutter. That means the actual distribution has higher tails than the Rayleigh.

With a radar of high resolution and at the higher sea states, the log normal probability density has been proposed for modelling the clutter echo. For a normal clutter cross-section, the probability density function is

$$p(\sigma_c) = \frac{1}{\sqrt{2\pi}\sigma\sigma_c} \exp\left[-\frac{1}{2\sigma^2}\left(\text{In}\frac{\sigma_c}{\sigma_m}\right)^2\right] \tag{7.14}$$

where σ_c is the clutter cross-section, σ is the standard deviation of In σ and σ_m is the median value of σ_c.

When equation (7.14) represents the clutter input then the probability density function of the voltage amplitude at the output of the envelope detector is,

$$p(v) = \frac{1}{\sqrt{2\pi}\,\sigma v} \exp\left[-\frac{1}{2\sigma^2}\left(2\,\text{In}\frac{v}{v_m}\right)\right] \tag{7.15}$$

where σ remains the standard deviation of In σ_c and v_m is the median value of v. In Figure 7.4, we have shown a fit of the log-normal probability density function for σ = 6 dB to actual data for sea state 2 to 3. For the purpose the median value of the theoretical distribution is equated to the median of the actual data.

A few other analytical probability density functions have been suggested to model the real statistical variations received from clutter. One of these is the contaminated normal probability density function, which can be fitted to the data of Figure 7.4 for sea states.

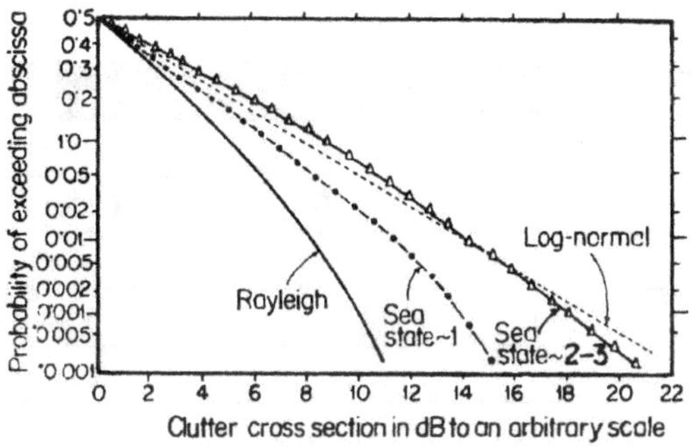

FIGURE 7.4 Statistics of vertical polarization X-band sea clutter for two sea states.

In the Weibull clutter model, on the other hand, the amplitude probability density function of the voltage v out of the envelope detector can be expressed as

$$p(v) = \alpha \text{ In } 2\left(\frac{v}{v_m}\right)^{(\alpha-1)} \exp\left[-\text{In } 2\left(\frac{v}{v_m}\right)^{\alpha}\right] \tag{7.16}$$

where β = a parameter that relates to the skewness of the distribution (Weibull parameter).

v_m = median value of the distribution.

An idea of the statistics of the clutter is necessary to properly design a constant false alarm rate (CFAR) receiver.

7.3.3 Properties of Sea Clutter

The relative uniformity of sea over the oceans makes it simpler to deal with sea clutter than with land clutter. But it is rather difficult to collect information at sea under controlled and reproducible conditions. In particular, the clutter due to large waves can be observed only with radars in the UHF bands, which are rarely employed in ships. Ground-based HF radars are now used in coastal regions to study the sea state. Some important information regarding the radar echoes from sea are discussed as follows:

1. *Average value of* $\sigma°$. A plot of the mean cross-section per unit area $\sigma°$ as a function of the grazing angle for various frequencies and polarizations is made in Figure 7.5. This is a composite average of sea clutter data and is derived from a body of data extending from 10 to 20 knots wind speed. The sea state is a measure of the wave height as given in Table 7.1.

TABLE 7.1

Sea state	Wave height (*m*)	Descriptive term
0	0	Calm, glassy
1	0—0.1	Calm, rippled
2	0.1—0.5	Smooth, wavelengths
3	0.6—1.2	Slight
4	1.2—2.4	Moderate
5	2.4—4.0	Rough
6	4.0—6.0	Very rough
7	6.0—9.0	High
8	9.0—14	Very high
9	Over 14	Phenomenal

Sea clutter is sometimes described in terms of the wind speed but is not a complete measure in itself.

2. *Variation with grazing angle.* According to the grazing angle, three identified scattering regions are there. At an almost vertical incidence ($\alpha = 90°$), the radar echo is relatively large. This is known as the quasi-specular region. When the grazing angle is of the order of several degrees, $\sigma°$ rapidly decreases with angle. This is known as the interference region, since the direct wave and the scattered wave from the sea surface form destructive interference. The angular region between the quasi-specular region and the interference region is called the diffuse or plateau region.

3. *Variation with frequency and polarization.* For vertical polarization, except at a low grazing angle, $\sigma°$ is independent of frequency (Figure 7.5). For horizontal polarization the clutter echo decreases with frequency, which is most pronounced at lower grazing angles.

4. *Variation with wind.* The value of $\sigma°$ is found to increase with an increase of the wind speed, except at the grazing angle near normal incidence. When the wind speed is less than 5 knots, the back scatter is quite low. With the increase of wind speed from about 5 to 20 knots, the back scatter increases rapidly. However, at higher wind speeds the increase is less rapid. The wind effectively controls the amplitude of the waves.

The energy back scatter from the sea also depends on the wind direction relative to the direction of the radar antenna beam.

5. *Effect of pulse width.* Theoretically, the sea back scatter per unit area is independent of the illuminated area. But in practice, the area illuminated by the radar affects the nature of the back scatter, particularly for short-pulse radars. For a low resolution radar, the sea echo is not spatially uniform, but is composed of occasional individual target-like echoes of short duration.

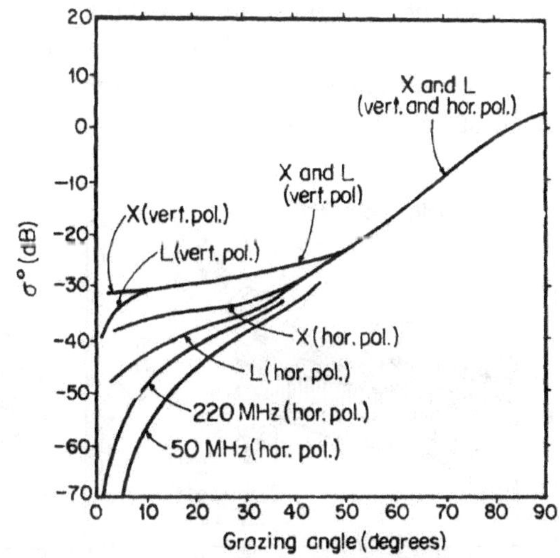

FIGURE 7.5 Composite of $\sigma°$ data for average conditions with wind speeds between 10 and 20 knots.

7.4 LAND CLUTTER

Land clutter is generally more complicated than sea clutter. The back scatter from land is considerably greater than that from sea and so, over the usual grazing angles, it is more difficult to detect targets over the land than over the sea.

It is also very difficult to classify and quantify land clutter. This is due to the fact that the radar echo from land depends on the particular type of terrain as described by its dielectric properties and the roughness. Cultivated fields, mountains, cities, roads, deserts, forests, and so on all have different scattering properties. In fact, the radar cross section of a farmer's field varies before and after ploughing and also before and after harvesting.

7.4.1 Examples

Usually the data for land clutter is reported in terms of the cross section per unit area, $\sigma°$, similar to sea clutter. An example of clutter $\sigma°$ for broad classes of terrain is shown in Figure 7.6, which is applicable to X-band clutter. The boundaries of the different regions are wide enough to indicate the wide variation of the data.

There are also enough experimental data of $\sigma°$ of crops. It must be pointed out, however, that the radar back scatter depends not only on the particular type of crop but also on the moisture content of the soil, the time of day, and even on the state of its growth.

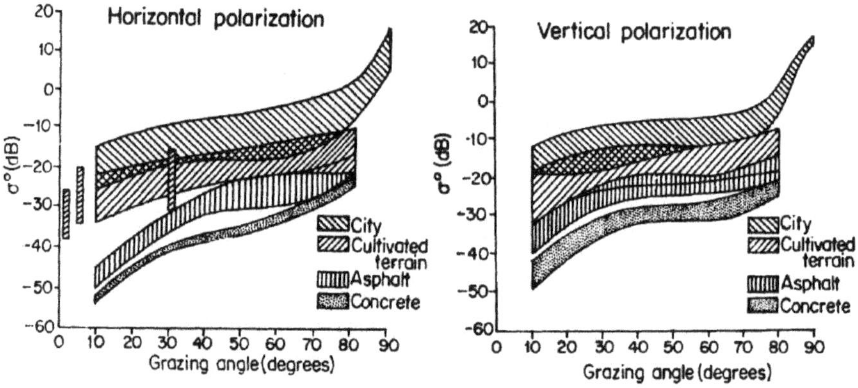

(*a*) Horizontal polarization (*b*) Vertical polarization

FIGURE 7.6 Boundaries of measured radar returned at the X-band from various land clutter.

7.4.2 Applications

Information about radar back scatter due to land is necessary for various applications. Some important applications are:

1. The detection of surface targets over land.
2. The detection of aircraft over land.
3. The detection of terrain features.
4. Altimeters for measuring the height of aircraft or spacecraft.
5. Mapping or imaging radars that utilize high resolution antenna systems.
6. Remote sensing with imaging radars and scatterometers for obtaining specific information about surface conditions.

7.4.3 Frequency Dependence

The frequency dependence of land clutter can be best understood by considering Figure 7.7 on next page. In the figure frequency dependence of land clutter have been plotted for a particular grazing angle. In these measurements clutter due to urban areas are independent of frequency. Also rural terrain shows no frequency dependence from L to X-band. In practice, hill, desert, and cultivated land exhibit a linear dependence with frequency.

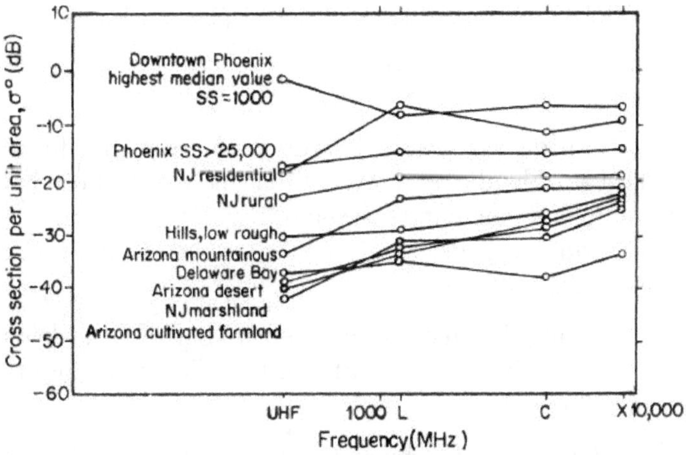

(*a*) **Vertical polarization.**

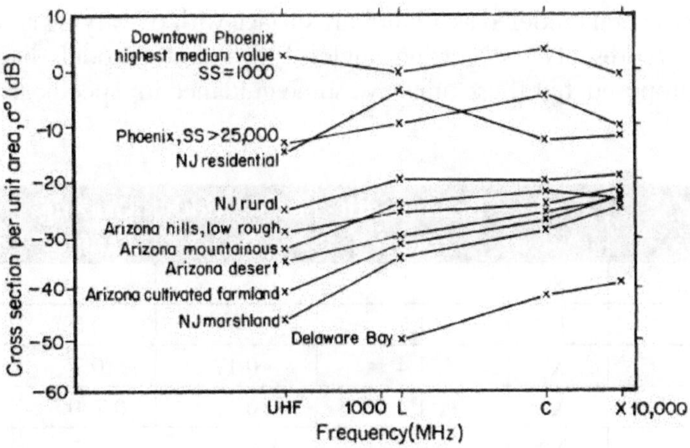

(*b*) **Horizontal polarization.**

FIGURE 7.7 Median values of σ° as a function of frequency, with grazing angle = 8°.

7.4.4 Properties of Land Clutter

Some important information regarding radar echoes from land are discussed as follows:

1. *Effect of resolution.* In order to describe the radar scattering from clutter, the use of $\sigma°$ implies that the clutter is uniform, but in reality it is not so. If the radar resolution is sufficiently great for resolving the areas of lower clutter from the areas of greater clutter, targets within the clear areas can be easily detected and tracked. The ability to resolve the strong clutter regions into discrete areas, between which targets can be detected, is termed as *inter clutter visibility.* It is very difficult to establish a quantitive measure of this effect.

2. *Probability density function.* Most of the statistics for sea clutter are applicable to land clutter, except that the physical mechanisms are different. It has been found that the statistical variation of $\sigma°$ from homogeneous terrain (e.g., deserts) can be described by the Gaussian probability density function (pdf).

   The Weibull pdf, which is intermediate between Rayleigh and log normal, is applicable to land clutter. A list of Weibull clutter parameters is given in Table 7.2.

3. *Land Clutter Models.* The wide variety and complexity of terrain due to the weather, time of day, season of the year, and so on makes it difficult in reality to establish a satisfactory theoretical model. In fact, land clutter has been modeled as a Lambert surface with $\sigma°$ varying as $\sin^2 \alpha$, where β represents the grazing angle. A few other models have also been proposed, but these only give some guidance for specific land clutter.

TABLE 7.2

Terrain/ sea state	Freq.	Beam width (deg.)	Pulse width (μs)	Grazing angle	Weibull Parameter
Rocky mountains	S	1.5	2	—	0.52
Wooded hills	L	1.7	3	0.5	0.63
Forest	X	1.4	0.17	0.7	0.51–0.53
Cultivated land	X	1.4	0.17	0.7–0.5	0.61–2.0
Sea state 1	X	0.5	0.02	4.7	1.45
Sea state 3	K_u	5	0.1	1–30	1.16–1.178

7.5 EFFECTS OF WEATHER ON RADAR

Radar return from rain, snow, or hail has high importance. Tornado warnings and hurricane tracking are good examples of applications where radar has proved its worth in saving property and life.

7.5.1 Radar Equation for Meteorological Echoes

The radar equation in a simplified form can be written as

$$P_t = \frac{P_t G^2 \lambda^2 \sigma}{(4\pi)^3 r^4} \tag{7.17}$$

where the symbols have the usual significance. To extend the previous equation for meteorological targets, let us assume that rain, hail, snow, or other hydrometeers are a large number of independent scatterers of the cross section located within the radar resolution cell. If $\Sigma\sigma_i$ represents the average total back scatter cross section of the particles per unit volume, then the radar cross section σ can be written as

$$\sigma = V_m \, \Sigma\sigma_i \tag{7.18}$$

where V_m = volume of the radar resolution cell. If the volume V_m is occupied by a radar beam of horizontal beamwidth θ_B, vertical beamwidth ϕ_B, and a pulse of duration τ then we can write

$$V_m \approx \frac{\pi}{4}(r\phi_B)(c\tau/2) \tag{7.19}$$

where c is the velocity of propagation. The factor ($\pi/4$) has been included for the elliptical shape of the beam area. Assuming a Gaussian-shaped antenna pattern, the volume V_m in equation (7.19) is reduced by a factor of 2 In 2. Then for a Gaussian beam shape equation (7.17) becomes

$$\begin{aligned} \overline{P}_r &= \frac{P_t G^2 \lambda^2 \theta_B \phi_B c\tau}{1024(l_n{}^2)\pi^2 r^2} \sum_i \sigma_i \\ &= \frac{P_t G \lambda^2 c\tau}{1024(l_n{}^2) r^2} \sum_i \sigma_i \qquad \left[\text{putting } G = \frac{\pi^2}{\theta_B \phi_B}\right] \end{aligned} \tag{7.20}$$

The bar over P_t represents the average received power.

If the radar wavelength is large in comparison to the circumference of a scattering particle of diameter D, then the radar cross section is expressed as

$$\sigma_i = \frac{\pi^5 D^6}{\lambda^4} |K|^2 \tag{7.21}$$

where $|K|^2 = \frac{\varepsilon - 1}{\varepsilon + 2}$ (ε= dielectric constant of the scattering particles). In equation (7.21) the value of $|K|^2$ for water changes with wavelength and temperature. At a wavelength of 10 cm and for a temperature of 10°C, the value of $|K|^2$ is about 0.93. For ice its value at all temperatures is 0.197.

Putting the value of σ_i, from equation (7.21) in equation (7.20), we get

$$\bar{P}_r = \frac{\pi^5 P_t G c\tau}{1024(l_n^{\,2}) r^2 \lambda^2} (K)^2 \sum_i D^6 \tag{7.22}$$

The particle diameter thus appears as the sixth power, and so in any distribution of precipitation particles, the small number of large drops will contribute most to the echo power.

In traversing the range r and back, the two-way attenuation of the radar signal is exp. $(-2\beta r)$, where β denotes the one-way attenuation coefficient. If β is not a constant over the path, then the total attenuation is expressed as the integrated value over the two-way path.

7.5.2 Scattering from Rain

Equation (7.22) is used for measuring with the radar the sum of the sixth power of the raindrop diameters in unit volume. This sum is called the radar reflectivity factor, and is denoted by Z, that is,

$$Z = \sum_i D^6 \tag{7.23}$$

If R represents the rainfall rate, then Z is related to R by the equation

$$Z = a\, R^b \tag{7.24}$$

where a and b are two constants. One of the widely accepted forms of equation (7.24) is

$$Z = 200\, R^{1.6} \tag{7.25a}$$

The previous equation is applicable to stratiform rain. For thunderstorm rain,

$$Z = 486\ R^{1.37} \tag{7.25b}$$

and for orographic rain,

$$Z = 31\ R^{1.71} \tag{7.25c}$$

We thus find that the choice of Z-R relationship should be made on the basis of the particular type of rain. Putting equation (7.25*a*) in equation (7.22) with $|K|^2 = 0.93$, we get

$$\bar{P}_r = \frac{2.4 P_t G \tau R^{1.6}}{r^2 \lambda^2} \times 10^{-8} \tag{7.26}$$

where R is in mm/hr, r and λ in meters, P_t in watts, and τ in seconds. This equation indicates how one can utilize the radar output for measuring the rainfall.

The back scatter cross section per unit volume versus rainfall rate for different wavelengths is plotted in Figure 7.8.

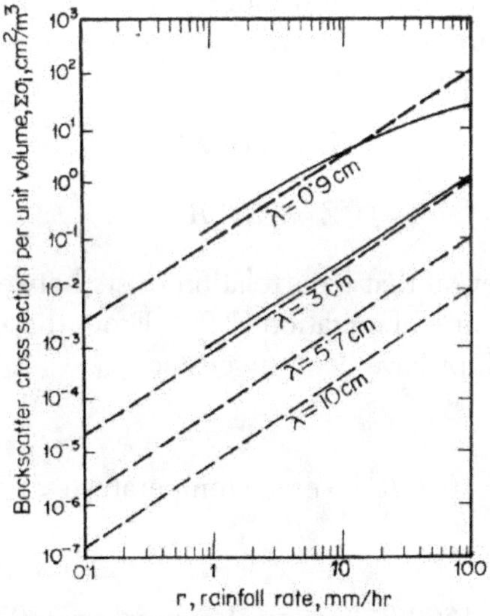

FIGURE 7.8 Exact (solid curves) and approximate (dashed curves) back scattering cross section per unit volume of rain at a temperature 18°C.

It must be noted here that the reflectivity factor z of equation (7.23) is based on the assumption of Rayleigh scattering. If the scattering is not of the Rayleigh type, then instead of Z we may introduce a term Z_c, called the equivalent radar reflectivity factor, where

$$Z_c = \frac{\lambda^4 \eta}{\pi^5 |K|^2} \tag{7.27}$$

Here η is the back scatter cross section per unit volume or the actual radar reflectivity, and $|K|^2$ is taken as 0.93.

7.5.3 Scattering from Snow

Dry snow particles are either single or aggregated ice crystals. If R represents the snowfall rate, then the relationship between Z and R can be written as given by equation (7.24),

$$Z = c\, R^d \tag{7.28}$$

where c and d are two new constants.

For snow, there are only a few more measurements than for rain, and from these few measurements different values have been proposed for c and d. Two important expressions are:

$$Z = 2000\, R^2$$

and

$$Z = 1780\, R^{2.21} \tag{7.29}$$

Investigations reveal that a correlation exists between surface temperature and the coefficient c of equation (7.28). From these measurements the following relationships have been suggested:

For dry snow:

$$Z = 1050\, R^2 \text{ [average temperature} < 0°\text{C]} \tag{7.30}$$

For wet snow:

$$Z = 1600\, R^2 \text{ [average temperature} > 0°\text{C]} \tag{7.31}$$

In general, the radar is less affected by snow and ice than by rain. This is because the factor $|K|^2$ is less for ice than for rain, and snowfall rates are generally less than that for rainfall rates.

7.5.4 Attenuation by Precipitation

If the particles are small in size compared to the wavelength so that the Rayleigh scattering is applicable, then for that frequency range the attenuation due to absorption can be expressed as

$$\text{Attenuation}\left(\frac{\text{dB}}{\text{km}}\right) = 0.434\left\{\frac{\pi^2}{\lambda}(\Sigma D^3)\right\}\text{Img}(-K) \tag{7.32}$$

where D is the particle diameter in cm, λ is the wavelength in cm, and the factor K depends on the dielectric constant of the particle. In the previous equation the summation is over 1 m^3 and Img $(-K)$ represents the imaginary part of $(-K)$. At 10°C IMG $(-K)$ for water is 0.00688 for 10 cm wavelength (S band) and 0.0247 for 3.2 cm wavelength (X band).

Equation (7.32) gives a quite good result for rain attenuation at the S-band or longer wavelengths. However, the rain attenuation is generally small and unimportant at longer wavelengths, and so the previous relation is of limited use. In Figure 7.9 the rainfall attenuation is plotted against the wavelength. The computation of rain attenuation is based on the exact formulation of spheres as developed by Mie.

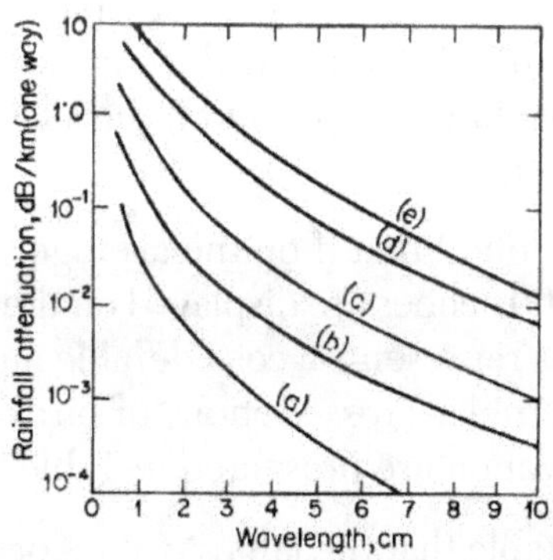

FIGURE 7.9 One-way attenuation (dB/km) of rain at a temperature of 18°C. (*a*) Drizzle - 0.25 mm/hr; (*b*) light rain - 1 mm/hr; (*c*) moderate rain - 4 mm/hr; (*d*) heavy rain - 16 mm/hr; (*e*) excessive rain - 40 mm/hr.

In the atmosphere, the attenuation due to ice particles is smaller than that due to rain. According to Gunn and East, the attenuation due to snow is,

Attenuation (at 0°C in dB/km)

$$= \frac{0.00349 R^{1.6}}{\lambda^4} + \frac{0.0024 R}{\lambda} \tag{7.33}$$

where R is the snowfall rate in mm/hr and λ is the wavelength in cm.

7.5.5 Scattering from Clouds

Cloud droplets are generally of a diameter less than 100 μm, and so for the prediction of cloud echoes, Rayleigh scattering can be suitably applied at radar frequencies. We know that in Rayleigh scattering the back scatter is directly proportional to the sixth power of the diameter. Furthermore, the diameter of cloud droplets is approximately one-hundredth of the diameter of raindrops. Hence the echoes due to fair-weather clouds have no importance.

At millimeter wavelengths weak echoes may be obtained from a deep intense fog, but at a 3 cm wavelength and higher, such echoes are insignificant.

7.6 ANGEL ECHOES

Angel echoes are obtained due to different reasons. But they are broadly classified into two different types:

1. *Dot angels.* These are point targets due to birds and insects.
2. *Distributed angels.* These have substantial horizontal or vertical extents. Such angel echoes are formed due to inhomogeneities of the atmospheric refractive index. Birds and insects can also sometimes appear as distributed angels.

Birds. It has been found that if on the average only one bird is present per square mile, over 300 echoes are displayed on the PPI within a 10 mile radius from the radar. This represents a considerable amount of clutter. Some typical examples of the radar cross sections of birds taken at three frequencies with vertical polarization are presented in Table 7.3.

It is seen from the table that the largest values occur at the *S* band. The back scatter from birds fluctuates over large values. The mean-to-median ratio of the cross section, which is a measure of the fluctuation in the cross section, is independent of the magnitude of the radar cross section but is a function of the size of the bird relative to the radar wavelength.

The term bird activity modulation (BAM) is applied to the clear wave forms obtained from birds. The BAM pattern of a bird in flight gives considerably stable spectral components.

TABLE 7.3
Radar Cross Section of Birds

Bird	Frequency Band	Mean Radar cross section (cm^2)	Median Radar cross section (cm^2)
Sparrow	X	1.6	0.8
	S	14	11
	UHF	0.02	0.02
Pigeon	X	15	6.4
	S	80	32
	UHF	11	8.0
Grackle	X	16	6.9
	S	25	12
	UHF	0.57	0.45

Insects. These can be easily detected by radar and can clutter the display. Therefore, insects can reduce the capability of a radar for detecting the desired signal. Appreciable echoes are found when the body length of an insect is greater than a third of the radar wavelength. Insect echoes are generally obtained at a station of lower altitudes, near dawn and twilight. Most of the insects are incapable of flight at a temperature above 90°F or below 40°F.

The adverse effects of both birds and insects are reduced by sensitivity time control (STC). The study of such objects with a radar is called radar ornithology.

Clear-air turbulence. The echoes from clear-air turbulence (CAT) are too weak and are viewed only by high-power radars. The radar cross section (m^2) per cubic volume due to a turbulent medium is expressed as

$$\eta = 0.38\, C_n^{\,2} \lambda^{-1/3} \tag{7.34}$$

where the structure constant $C_n^{\,2}$ is a measure of the intensity of the refractive index fluctations and λ represents the radar wavelength. The CAT is a potential danger to aviation.

CHAPTER 8

MTI RADAR

8.1 INTRODUCTION

A pulse radar which utilizes the Doppler frequency shift for discriminating moving targets from fixed ones, appearing as clutter, is known as moving target indication radar, abbreviated as MTI Radar. It usually operates with ambiguous Doppler measurement but with unambiguous range measurement. The opposite is generally the case for a pulse Doppler radar, which also discriminates moving targets from clutter by Doppler frequency shift measurements. The design of an MTI Radar is much more challenging than a simple pulse radar or a CW radar. Its basic concepts were introduced during World War II. MTI is a necessity in high-quality air-surveillance radars, which may operate in the presence of clutter.

8.2 OPERATION OF MTI RADAR

A CW radar, as shown in Figure 8.1 (*a*), essentially consists of a transmitter, a receiver, an indicator, and the antennas. It may be converted into a pulse radar by providing a power amplifier and a modulator (Figure 8.1 [*b*]). The basic difference between the two is that a small portion of the CW oscillator power that generates the transmitted pulses is diverted to the receiver in order to take the place of the LO. This CW signal acts as the coherent reference signal required for detecting the Doppler shift. By the term coherent we mean that the phase of the transmitted signal is preserved in the reference signal which is, in fact, the distinguishing feature of the coherent MTI radar.

If A_1 denotes the amplitude and f_1 the carrier frequency, then the CW oscillator voltage can be represented by $A_1 \sin 2\pi f_1 t$, where f_1 is the carrier frequency and t is the time.

So the reference signal,

$$V_{ref} = A_2 \sin 2\pi f_2 t \tag{8.1}$$

and the Doppler-shifted echo signal

$$V_{echo} = A_3 \sin[2\pi(f_1 \pm f_d)t - \frac{4\pi f_t r_o}{c} \tag{8.2}$$

where A_2 = amplitude of the reference signal

A_3 = amplitude of the signal received at a range r_o

f_d = Doppler frequency shift

and c = velocity of propagation

because the round-trip path place lag = $2\pi f_1 \cdot \frac{2r_o}{c}$

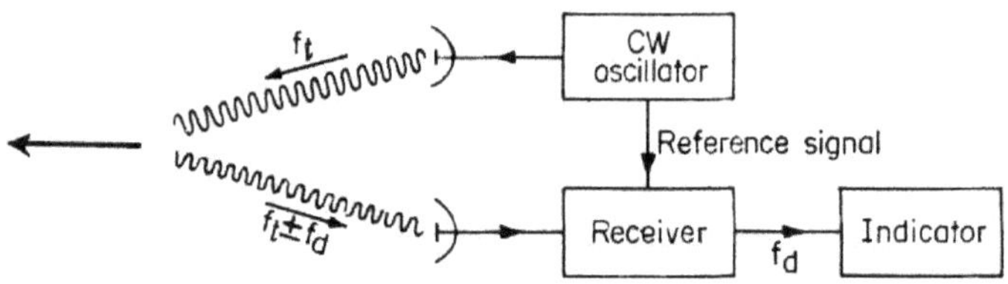

(*a*) CW radar.

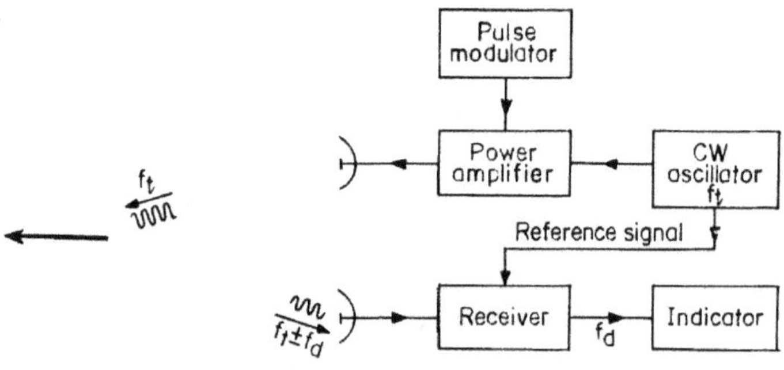

(*b*) Pulse radar using Doppler information.

FIGURE 8.1

In the mixer stage of the receiver, the reference signal and the target echo signal are heterodyned. The low frequency component, which is the difference frequency signal from the mixer, is of interest and is given by

$$V_{diff} = A_4 \sin\left(2\pi f_d t - \frac{4\pi_0 f_t r_o}{c}\right) \tag{8.3}$$

where A_4 is the amplitude. The difference frequency is equal to the Doppler frequency f_d.

Equations (8.1) and (8.3) all represent sine-wave carriers on which the pulse modulation is imposed. Accordingly, A_2, A_3, and A_4, in reality, represent the pulse waveforms.

One may distinguish the moving targets from the stationary targets by examining the video output on an A-scope (amplitude vs. range), in which a single sweep might appear as shown in Figure 8.2 (*a*). Figure 8.2 (*a*) to (*e*) exhibit clearly the successive sweeps of an MTI radar A-scope display representing the echo amplitude as a function of time, while Figure 8.2 (*f*) reveals the super position of many sweeps. The arrows in Figure 8.2 indicate the positions of moving targets.

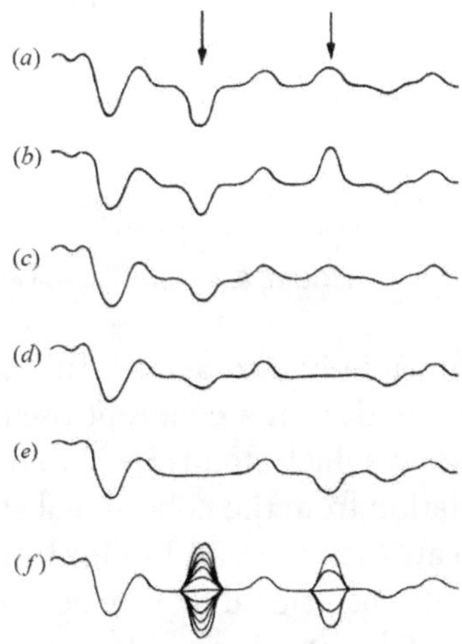

FIGURE 8.2 Successive sweeps of an MTI radar A-scope display (*a* to *e*); (*f*) Superposition of many sweeps.

The block diagram of an MTI radar with a power amplifier as the transmitter is shown in Figure 8.3 (*a*) while the same with a power oscillator as the

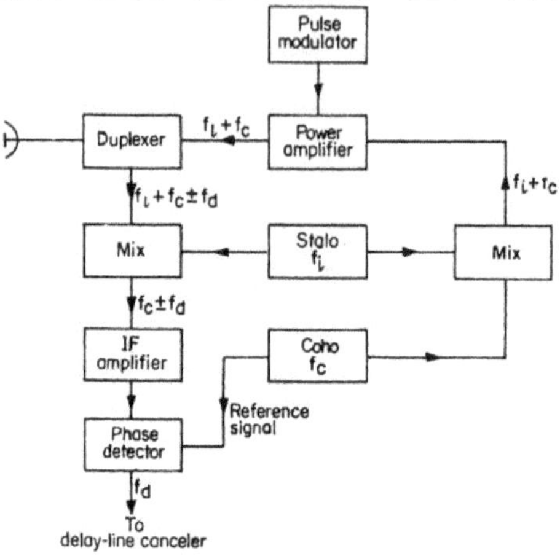

(*a*) MTI radar with power amplifier.

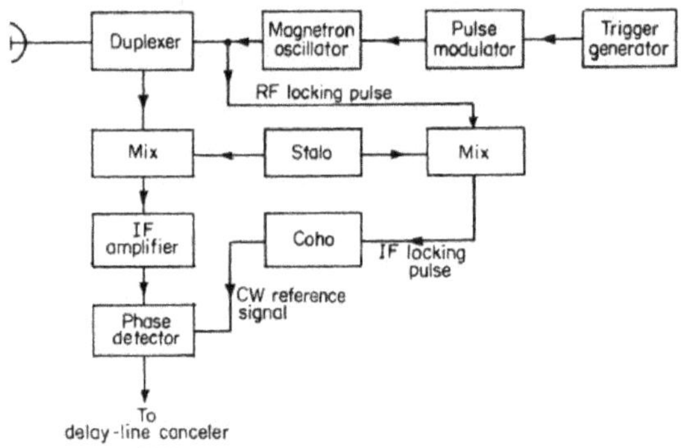

(*b*) MTI radar with power oscillator transmitter.

FIGURE 8.3 Block diagram of MTI radars.

transmitter is shown in Figure 8.3 (*b*). In Figure 8.3 (*a*) the transmitted signal, at f_c, is generated from a coherent oscillator, abbreviated as coho. The function of the stalo, which stands for stable oscillator, at f_b, is to provide the frequency translation from the coho signal at f_c to the transmitted frequency $f_c + f_b$. The stalo also serves as an LO for the receiver section to generate an intermediate frequency signal at f_c which, in turn, is detected by a phase detector with the coho signal serving as a reference signal. The output of the phase detector is proportional to the phase difference between the two input signals. In Figure 8.3 (*b*), on the other hand, a part of the transmitted signal is mixed with the stalo output for producing an IF beat signal whose phase is directly related to the phase of the transmitter. The IF pulse is applied to the coho so that the phase of the coho CW oscillation is locked in step with the phase of the IF reference pulse. This type of MTI radar has wide applications. Since both the amplitude and phase of the IF output is useful in the delay line canceler, such a radar is also called a vector processing MTI.

The necessity of the coho and stalo arises from the fact that the Doppler shift in the received echo signal due to target motion is likely to be a very small percentage of the transmitted carrier frequency. Hence to detect such a small Doppler shift, we need extremely stable transmitter and receiver frequencies. The introduction of the coho and stalo minimizes this stringent requirement of the frequencies by ensuring the cancellation of the drifts Δf_c and Δf_1 in the coho and stalo respectively in the MTI radar.

8.3 MTI RECEIVER WITH DELAY LINE CANCELER

A method frequently employed for extracting Doppler information suitable for display on the PPI scope is with a delay line canceler. This is shown in Figure 8.4. It acts as a filter to eliminate the DC component of fixed targets and to pass the AC component of moving targets. The video portion of the MTI receiver can be divided into two channels, one is normal video and in the other there is a time delay equal to one pulse repetition period. The outputs of the two channels are subtracted. The output of the subtractor is a bipolar video as at the input. It is converted to unipolar video by means of a full-wave rectifier.

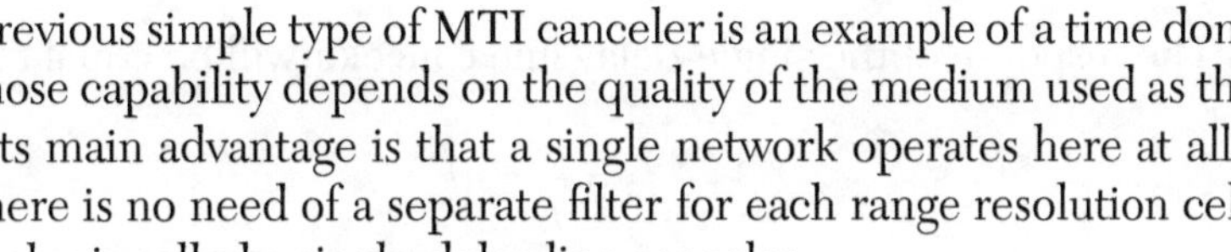

FIGURE 8.4 MTI receiver with delay line canceler.

The previous simple type of MTI canceler is an example of a time domain filter whose capability depends on the quality of the medium used as the delay line. Its main advantage is that a single network operates here at all ranges and there is no need of a separate filter for each range resolution cell. Such a canceler is called a single-delay line canceler.

8.3.1 Response of the Delay Line Canceler

The video signal received from a particular target at a range r_o can be expressed as

$$V_1 = k \sin (2\pi f_d t - \phi_o) \tag{8.4}$$

where k is the amplitude of the video signal and ϕ_o is the phase shift.

The signal from the previous transmission which is delayed by a time, T, is given by

$$V_2 = k \sin[2\pi f_d (t - T) - \phi_o] \tag{8.5}$$

The output from the subtractor

$$\begin{aligned} V &= V_1 - V_2 \\ &= 2k \sin \pi f_d T \cos\left[2\pi f_d\left(t - \frac{T}{2}\right) - \phi_o\right] \end{aligned} \tag{8.6}$$

Equation (8.6) indicates that the output from the single-delay line canceler is a cosine wave at a Doppler frequency f_d with an amplitude $2k \sin \pi f_d T$. The amplitude of the canceler video output is thus a function of the Doppler frequency shift and the pulse repetition interval. The ratio of the amplitude of the output from the delay line canceller, $2k \sin \pi f_d T$, to the amplitude of the normal radar video, k, is shown in Figure 8.5.

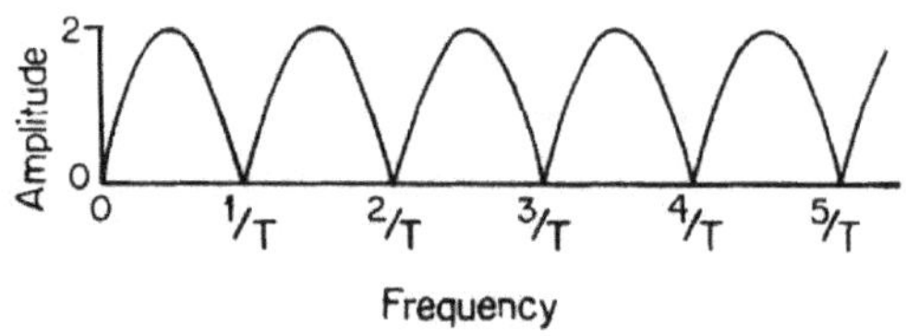

FIGURE 8.5 Frequency response of a single-delay line canceller; T = delay time.

8.3.2 Blind Speeds

The response of the single-delay line canceler will be equal to zero, when

$$f_d = \frac{n}{T} = nf_p \tag{8.7}$$

where $n = 0, 1, 2, \ldots$ and f_p is the pulse repetition frequency. The delay line canceler eliminates the DC component due to clutter in addition to any moving target whose Doppler frequency happens to be the same as the prf. The relative target velocities resulting in zero MTI response are called blind speeds. The nth blind speed V_n can be expressed as

$$v_n = \frac{n\lambda}{2T} = \frac{n\lambda f_p}{2} \tag{8.8}$$

where $n = 1, 2, 3, \ldots$

If λ is in meters, f_p in Hz, and the relative velocity in knots, the blind speeds are written as

$$v_n = \frac{n\lambda f_p}{1.02} = n\lambda f_p \tag{8.9}$$

The occurrence of blind speeds is one of the limitations of MTI radar which does not occur with CW radar. Blind speeds are not easy to avoid. In Figure 8.6, the first blind speed v_1 is plotted as a function of maximum unambiguous range, with radar frequency as the parameter. If the first blind speed were 600 knots, then at a frequency of 300 MHz, the maximum unambiguous range would be 130 nautical miles, at 3000 MHz the range would be 13 nautical miles, and at 10,000 MHz this range would be 4 nautical miles.

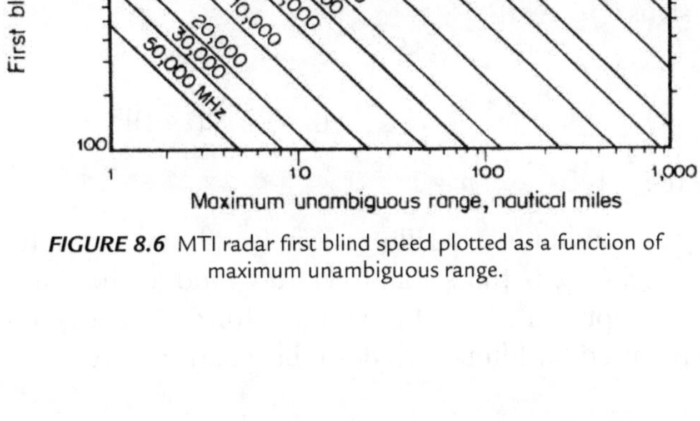

FIGURE 8.6 MTI radar first blind speed plotted as a function of maximum unambiguous range.

8.3.3 Double Cancellation

The diagram of a double-delay line canceler or briefly a double canceler is shown in Figure 8.7 (*a*), while a three-pulse canceler is shown in Figure 8.7 (*b*).

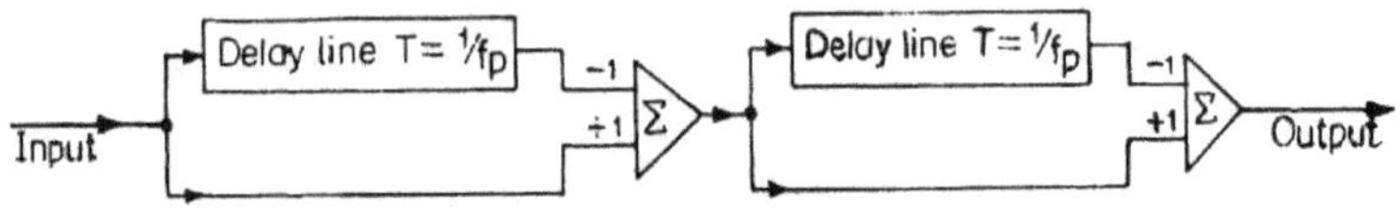

FIGURE 8.7 (*a*) Double-delay line canceler.

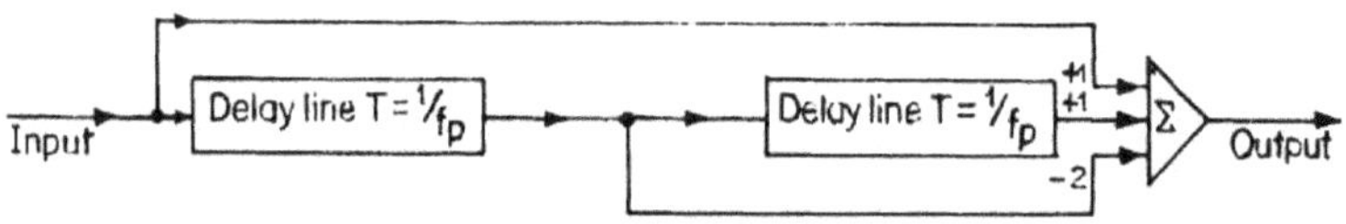

FIGURE 8.7 (*b*) Three-pulse canceler.

The output of the two single-delay cancelers in cascade is just the square of that from a single canceler, and so the frequency response is $4 \sin^2 \pi f_d T$. The response of the single-delay line canceler and the double line canceler, when compared carefully, provides an additional cancellation of clutter which is offered by the double canceler.

8.3.4 Transversal Filter

The general form of the transversal filter for MTI signal processing is shown in Figure 8.8. The three-pulse canceler as shown is an example of the transversal filter. It is also known as a non-recursive filter, a feed forward filter, a finite memory filter, or a delay line filter. With n delay lines the weights of a transversal filter that gives a response $\sin^n \pi f_d T$ are the coefficients of the expansion of $(1-x)^n$ which represent the binomial coefficients with alternating signs.

$$w_i = (-1)^{i-1} \frac{n!}{(n-i+1)!(i-1)!} \tag{8.10}$$

where $i = 1, 2, \ldots\ldots n + 1$

The previous transversal filter maximizes the average ratio $I_c = (S/C)_{out}/(S/C)_{in}$, where $(S/C)_{out}$ is the signal-to-clutter ratio at the output of the filter, and $(S/C)_{in}$, represents that at the input. However, a large number of delay lines are required for filters with desirable characteristics.

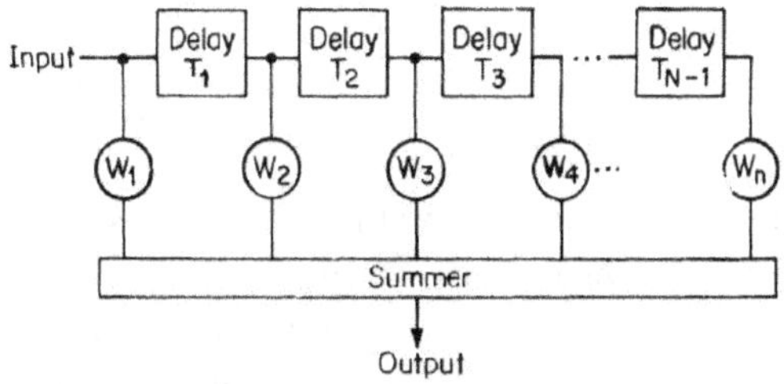

FIGURE 8.8 Transversal filter for MTI signal processing.

8.4 MULTIPLE OR STAGGERED PULSE REPETITION FREQUENCIES

In the design of MTI Doppler filters, the use of more than one pulse repetition frequency reduces not only the effect of blind speeds but also allows a sharper low-frequency cut off in the frequency response. The blind speeds of two radars operating at the same carrier frequency will differ if their pulse repetition frequencies are different. A single radar will also have blind speeds in succession if the prf is staggered, with the successive prfs not being the same.

An example of the average response of an MTI radar which is operating with two pulse repetition frequencies on a time shared basis is shown in Figure 8.9.

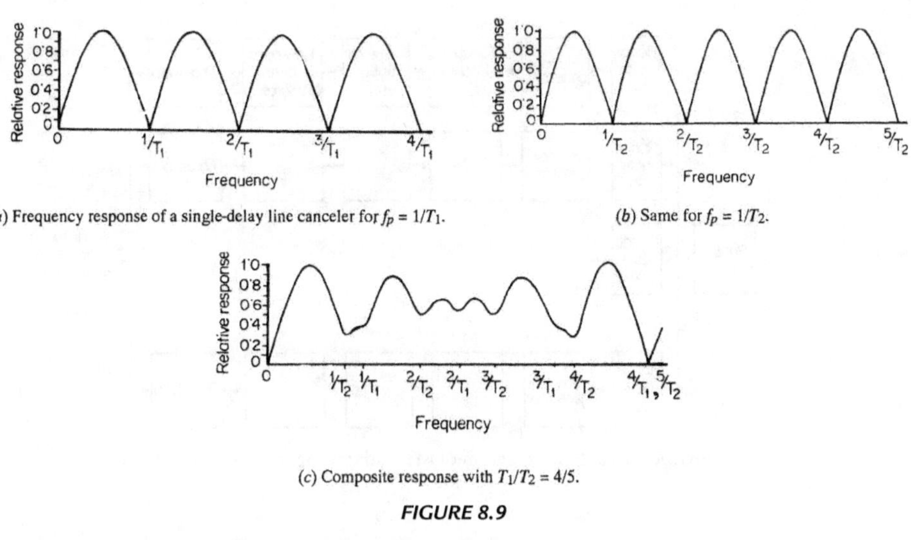

(*a*) Frequency response of a single-delay line canceler for $f_p = 1/T_1$.

(*b*) Same for $f_p = 1/T_2$.

(*c*) Composite response with $T_1/T_2 = 4/5$.

FIGURE 8.9

Here we have shown in Figure 8.9 (a), the frequency response of a single-delay line canceler for $f_d = 1/T_1$; in Figure 8.9 (b), the same for $f_p = 1/T_2$; and in Figure 8.9 (c), composite response with $T_1/T_2 = 4/5$.

A great disadvantage of the staggered prf is its inability to cancel second-time-around clutter echoes, because it does not appear at the same range from pulse to pulse and so produces uncanceled residue. Second-time-around clutter echoes may be removed by using a constant prf, which is employed only over those angular sectors where second-time-around clutter is expected, or by changing the prf every scan period, or by changing the prf each time the antenna scans half a beamwidth.

8.5 RANGE-GATED DOPPLER FILTERS

In order to separate moving targets from stationary clutter, the delay line canceler has been widely used in MTI radar. The loss of range information and collapsing loss can be eliminated by quantizing the time into small intervals. This process is known as range gating, where the width of each gate depends on the range accuracy desired. After quantizing the radar return into range intervals, the output from each gate is applied to a narrow band filter. In Figure 8.10, a block diagram of the video of an MTI radar using multiple range gates followed by clutter rejection filters is shown. Here the output of the phase detector is sampled sequentially by the range gates.

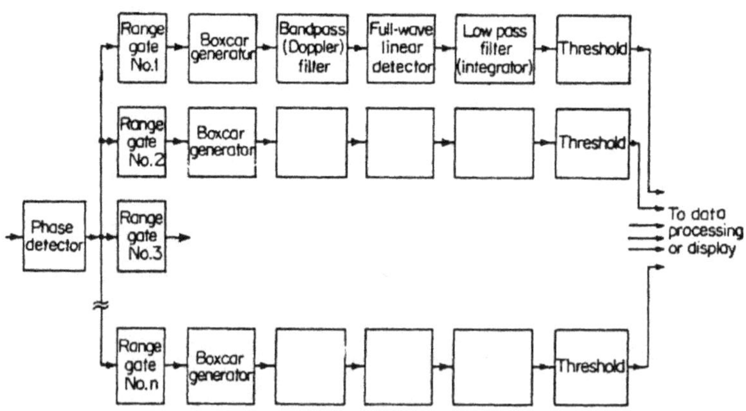

FIGURE 8.10 Block diagram of MTI radar using range gates and filters.

Each range gate acts as a switch or a gate which opens and closes at the proper time. The output of the range gate is stretched in a circuit known as a boxcar generator. Its function is to aid in the filtering and detection process enhancing the fundamental of the modulation frequency and eliminating the harmonics of the pulse repetition frequency. The clutter rejection filter is nothing but a bandpass filter whose bandwidth depends on the extent of the expected clutter spectrum.

The frequency-response characteristic of an MTI radar using range gates and filters is shown in Figure 8.11. The shape of the rejection band is mainly determined by the shape of the bandpass filter.

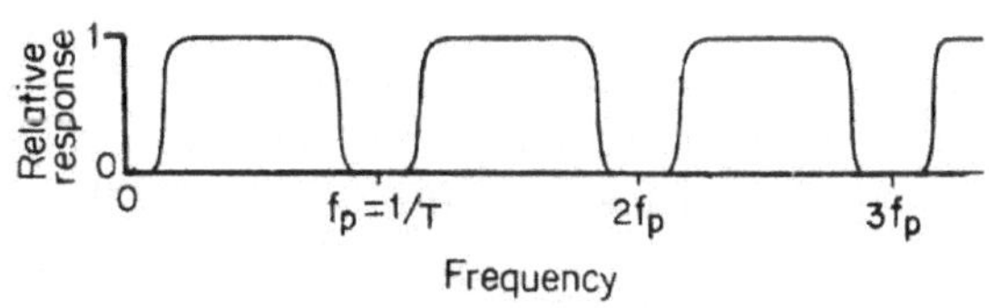

FIGURE 8.11 Frequency response characteristic of an MTI using range gates and filters.

It must be pointed out that the MTI radar using range gates and filters is more complex than an MTI with a single-delay line canceler. A better MTI performance is achieved from a better match between clutter filter characteristic and clutter spectrum.

8.6 DIGITAL SIGNAL PROCESSING

An introduction of digital processing to MTI radar allowed a significant increase in the options open to the designer of signal processing. The major advantages of a digital MTI processor are due to its use of digital delay lines. A block diagram of a digital MTI processor is shown in Figure 8.12.

In the figure, as shown, from the output of the IF amplifier the signal is split into two channels, one for the in-phase channel denoted I and the other for the quadrature channel denoted Q. The outputs of the two detectors are 90° out of phase. The quadrature channel eliminates the effects of blind phases. Thus, the convenience of the digital processing signal permits the quadrature channel to be included without significant burden. Following the phase detector, the bipolar video signal is sampled within each range resolution cell.

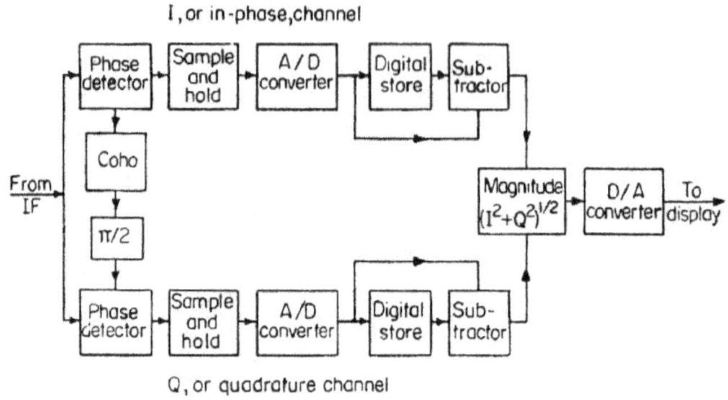

FIGURE 8.12 A digital MTI signal processor.

These voltage samples are converted to a series of digital words by an A/D converter. In a digital memory the digital words are stored for one pulse repetition period and are then subtracted from the digital words of the next sweep. By taking the square root of $(I^2 + Q^2)$, the digital outputs of the I and Q channels are combined. This can also be done alternatively, which is considered adequate for most cases, by taking $|I| + |Q|$. The combined output is next converted to an analog signal by the *D/A* converter. The unipolar video output is then ready to be displayed.

8.6.1 Example of an MTI Radar Processor

The moving target detector (MTD) is an MTI radar processor, a block diagram of which is shown in Figure 8.13. In the figure, the input on the left is from the output of the I and Q analog to-digital converter. The zero-velocity clutter is eliminated by the three-pulse canceler and the eight-pulse Doppler filter bank thereby generates eight overlapping filters covering the Doppler interval. The three-pulse canceler ahead of the filter bank eliminates stationary clutter and hence reduces the necessary dynamic range of the Doppler filter bank.

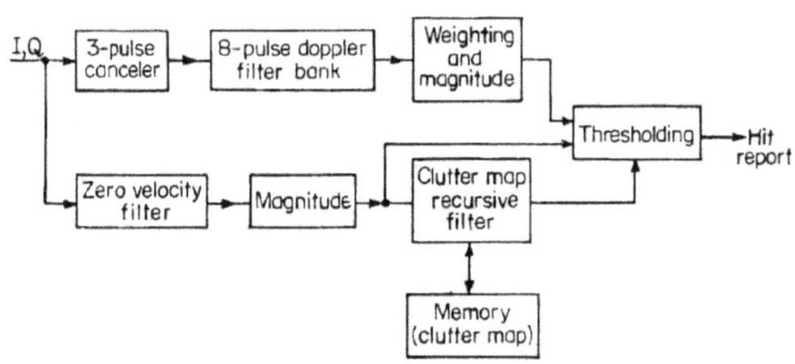

FIGURE 8.13 A moving target detector (MTD) signal processor.

8.7 MTI FROM A MOVING PLATFORM

When the radar is mounted on a ship or an aircraft and it is in motion, the detection of a moving target in presence of clutter becomes more difficult than when it is stationary. The Doppler frequency shift of the clutter varies with the direction of the antenna in azimuth and the elevation angle to the clutter. The design of an MTI is more difficult with an airborne radar in comparison to a shipborne radar due to the higher speeds and the greater range of elevation angles. In addition to the shift of the center frequency of the clutter, its spectrum is also widened. The spectral width can be obtained approximately by considering the differential of the Doppler frequency,

$$f_d = 2\,(v/\lambda)\cos\theta$$

or

$$\Delta f_d = \frac{2v}{\lambda}\sin\theta\,\Delta\theta \tag{8.11}$$

where v = platform speed

λ = wavelength

and θ = azimuth angle.

If the beamwidth is taken as $\Delta\theta$, then Δf_d represents a measure of the width of the Doppler frequency spectrum. The effect of platform velocity may be considered as having two components; one is in the direction the antenna is pointing, and the other is normal to the direction where the antenna is pointing.

An MTI radar when used on a moving platform is called AMTI, where the letter "A" stands for airborne.

Compensation for Clutter Doppler Shift. It the clutter Doppler shift is not DC, the null of the frequency response of the MTI processor shifts accordingly. When the center of the clutter Doppler frequency is shifted by an amount f_e, the effect on the improvement factor is shown in Figure 8.14 for a three-pulse delay line canceler. In the figure, f_c denotes the mean frequency of the clutter spectrum, σ_c is the standard deviation of the clutter spectrum, and f_p is the pulse repetition frequency. Two methods are applied generally for providing the Doppler frequency compensation. In one method the frequency of the coherent oscillator is changed to compensate for the shift in the clutter Doppler frequency. The other implementation is to insert a phase shifter in one branch of the delay line canceler and hence adjust its phase for shifting the null of the frequency response.

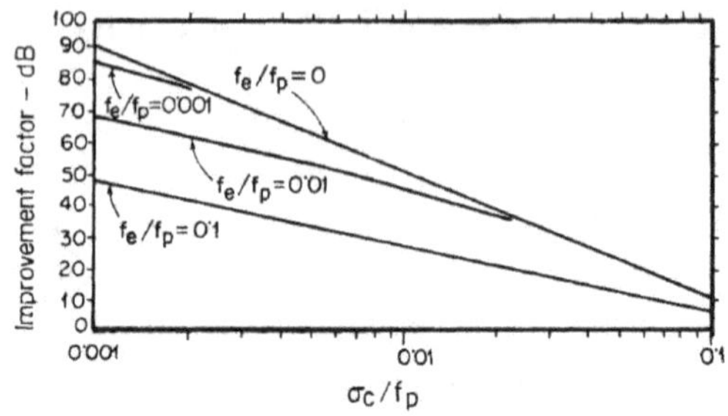

FIGURE 8.14 Effect of a non-zero clutter Doppler frequency on the improvement factor of a three-pulse canceler.

Compensation for Clutter Doppler Spread. Equation (8.11) shows that the spread in the clutter spectrum is a function of the angle θ between the velocity vector of the moving platform and the antenna beam pointing direction. Furthermore, it depends on the wavelength. The standard deviation of the clutter spectrum due to platform motion can be expressed as

$$\sigma_{pm} = 1.6\frac{v_x}{a} \tag{8.12}$$

where v_x = horizontal component of the velocity perpendicular to the antenna pointing direction,

and a = effective horizontal aperture width.

8.8 LIMITATIONS TO MTI PERFORMANCE

An improvement in the signal-to-clutter ratio of an MTI is affected by several factors other than the design of the Doppler signal processor. Equipment instabilities, internal fluctuation of clutter, scanning modulation, and limiting in the receiver can all detract from the performance of an MTI radar.

1. *Equipment instabilities.* Pulse-to-pulse changes in the amplitude, frequency, or phase of the transmitter signal lower the improvement factor of an MTI radar. If the echo from stationary clutter on the first pulse is $A \cos \omega t$ and from the second pulse is $A \cos (\omega t + \Delta\phi)$, then the difference between the two

$$= A \cos \omega t - A \cos (\omega t + \Delta\phi)$$

$$(\Delta\phi = \text{change in oscillator phase})$$

$$= 2A \sin\left(\frac{\Delta\phi}{2}\right) \sin\left(\omega t + \frac{\Delta\phi}{2}\right)$$

For small phase errors, the amplitude of the resultant difference

$$2A\sin\left(\frac{\Delta\phi}{2}\right) \cong A\,\Delta\phi$$

So, the limitation on the improvement factor due to oscillator instability is

$$I = \frac{1}{(\Delta\phi)^2} \tag{8.13}$$

This would apply to the coho locking or to the phase change which is introduced by a power amplifier.

2. *Internal fluctuation of clutter.* There are many types of clutter which are not absolutely stationary, like that due to buildings, water towers, hills, mountains, and so forth. Echoes from rain, sea, trees, vegetation, and so on fluctuate with time, and these limit the performance of an MTI radar. Most of the fluctuation clutter targets can be represented by a model consisting of many independent scatterers situated within the resolution cell of the radar.

   Experimentally measured power spectra of clutter signals may be approximately written as

$$W(f) = |g(f)|^2 = |g_o|^2 \exp\left[-a\left(\frac{f}{f_o}\right)^2\right] \tag{8.14}$$

   where $W(f)$ = clutter power spectrum as a function of frequency
   $g(f)$ = Fourier transform of the input waveform
   f_o = radar carrier frequency
   a = parameter which depends on the clutter.

   The expression for the improvement factor for an N-pulse canceler with $N_1 = N - 1$ delay lines can be written as

$$I_{NC} = \frac{2N_i}{N_l!}\left(\frac{f_s}{2\pi\sigma_c}\right)^{2N_l} \tag{8.15}$$

3. *Antenna scanning modulation.* As the antenna scans by a target, it observes the target for a finite time t_o

where
$$t_o = \frac{n_B}{f_p} = \frac{\theta_B}{\theta_s} \tag{8.16}$$

Here n_B = number of hits received

f_p = pulse repetition frequency

θ_B = antenna beamwidth

and θ_S = antenna scanning rate

The received pulse train of duration t_o has a frequency spectrum whose width is in proportion to $1/t_o$. So even for perfectly stationary clutter, there will be a finite width of the clutter spectrum owing to the finite time on the target. When the clutter spectrum is too wide, it affects the improvement factor. This limitation is also called scanning modulation or scanning fluctuations.

The limitations to the improvement factor caused by antenna scanning are

$$I_{1s} = \frac{n_B^2}{1.388} \quad \text{(single canceler)}$$

$$I_{1s} = \frac{n_B^4}{3.852} \quad \text{(double canceler)} \tag{8.17}$$

4. *Limiting in MTI radar.* Before the MTI processor, a limiter is generally employed in the IF amplifier for preventing the residue from large clutter echoes. An ideal MTI radar should reduce the clutter to a level comparable to receiver noise. If the limit level relative to noise is set higher than the improvement factor, clutter residue obscures part of the display, while if it is set too low there may be a "black hole" effect on the display.

The limiter provides a constant false alarm rate (CFAR), and thus it serves a very essential part in obtaining usable MTI performance. The use of the limiter eliminates the amplitude information of the IF output holding it constant to the limiting level and, therefore, such an MTI radar may be called a phase processing MTI, since only the phase information is retained after limiting.

5. *Envelope processing in MTI radar.* If the envelope of the IF output is detected by an envelope detector, the phase information of the output is of no use, and such an MTI is, therefore, also called an amplitude processing MTI, or incoherent MTI. An advantage of this system is that the LO need not be as stable as in other systems which employ synchronous or coherent detection. However, the disadvantage is that the signal to noise ratio with the envelope detector is less and, therefore, clutter must be present in a relatively larger amount in the envelope processing MTI for the detection of a moving target. The clutter serves the same function as the reference signal in the coherent MTI that employs a coherent oscillator, the coho.

WORKED OUT PROBLEMS

1. (*i*) An MTI Radar is operating at 10 GHz with a PRF of 1000 Hz. Calculate the lowest three blind speeds.

Blind speed $v_n = \frac{n\lambda f_p}{2}$

Here $f = 10 \text{ GHz} = 10^{10} \text{ Hz}$

$$\lambda = \frac{c}{f} = \frac{3\times10^{10}}{10^{10}} = 3\,\text{cm} = 0.03\,\text{m}$$

$$1 = \frac{1\times0.03\times1000}{2}\ \text{m/s}$$

$$= \frac{30}{2} = 15\ \text{m/s}$$

$$= \frac{15\times60\times60}{1000}\ \text{km/hr}$$

$$= 54\ \text{km/hr}$$

Blind speed $v_2 = 2v = 2 \times 54 = 108$ km/hr

and Blind speed $v_3 = 3v_1 = 3 \times 54 = 162$ km/hr.

(*ii*) The MTI Radar is used by traffic control police to measure the speed of cars. If the Doppler frequency shift measured from the moving car is 1.6 kHz, calculate the speed of the car.

The speed of the car $v_r = \dfrac{f \times \lambda}{2}$

$$= \frac{1600 \times .03}{2} \text{ m/s}, \quad \because \lambda = 3 \text{ cm for } f = 10 \text{ GHz}$$

$$= 800 \times .03 \text{ m/s}$$

$$= 24 \text{ m/s}$$

$$= \frac{24 \times 60 \times 60}{1000} \text{ km/hr}$$

$$= 86.4 \text{ km/hr.}$$

(*iii*) Indicate how the approaching and receding cars producing the same Doppler frequency shift may be distinguished.

The approaching car will produce an increase of carrier frequency due to the Doppler effect while the receding car will produce a decrease in carrier frequency. In the receiver, the carrier is converted to IF, which is then fed to a frequency discriminator, producing + *ve* or – *ve* DC voltages proportional to the Doppler shift of frequency, with the + *ve* shift produced by the approaching vehicle and the – *ve* shift produced by the receding vehicle, if the discriminator is designed to have its DC output increase with a positive shift of the carrier frequency, shifted on the higher side.

CHAPTER 9

MODERN RADARS

The demand for a higher range and range resolution, the requirement of measuring the velocity of the target, and the need for security for electromagnetic interference and interception apparently initiated the development of a wide range of radar systems in recent years. These modern radar systems can be classified into four types:

(*i*) Pulse Doppler radar
(*ii*) Frequency coded or chirp radar
(*iii*) Phase coded radar and
(*iv*) Millimeter wave radar.

The following section covers the essential features and the principles of operation of these modern radars.

9.1 INTRODUCTION TO PULSE DOPPLER RADAR

A pulse radar determines the range of a target by measuring the round trip delay of a transmitted RF pulsed signal returned by the target received as an echo signal. The envelope of the echo is compared with that of the transmitted RF pulse to determine the round trip delay and, therefore, any Doppler shift in the frequency of the carrier due to a target motion will not be recognized by the envelope detector used to detect the echo pulse. However, information about the target motion is contained in the received echo signal in the form of a Doppler shift of the carrier, similar to that occurring in a CW radar. To extract this information, the conventional pulse radar needs to be modified by introducing a coherent or phase stable (hence frequency stable as well) oscillator in the transmitter and the receiver chains, so that by the time the echo is returned, the LO derived from the transmitter may not have undergone a sizable drift of frequency or phase. The amount of Doppler shift is usually small compared to the carrier frequency under the usual conditions of target motion. Such a radar which combines the range discrimination capability of a pulse radar with the frequency discrimination capability of a CW radar is called a pulse Doppler radar, which is similar to MTI radar except that the unambiguity of range is mitigated by a bank of Doppler filters to detect the Doppler shift.

9.2 BLOCK DIAGRAM

The block diagram of a pulse Doppler radar is shown in Figure 9.1. A very stable local oscillator called the STALO (**STA**ble **L**ocal **O**scillator), which is 'phase stable' or 'coherent,' is a crucial subsystem which allows the phase of the transmitted wave to be maintained by a locking mixer, the RF output of which locks a coherent oscillator called COHO (**COH**erent **O**scillator) which, in turn, serves as a reference phase for the detector at the intermediate frequency, as shown in the figure. The phase detector measures the difference in phase between the two RF signals during a pulsed echo. Due to target motion the phase path of the echo changes from pulse to pulse, and hence the phase difference also changes by the same amount. The rate of change of phase $d\phi/dt = \omega_d = 2\pi f_d$, where f_d is the Doppler shift of the carrier. In reality, samples of the phase difference, Q_1, Q_2, Q_3 are measured during successive samples and, therefore, at least two successive samples are necessary to recognize any pulse-to-pulse change in phase.

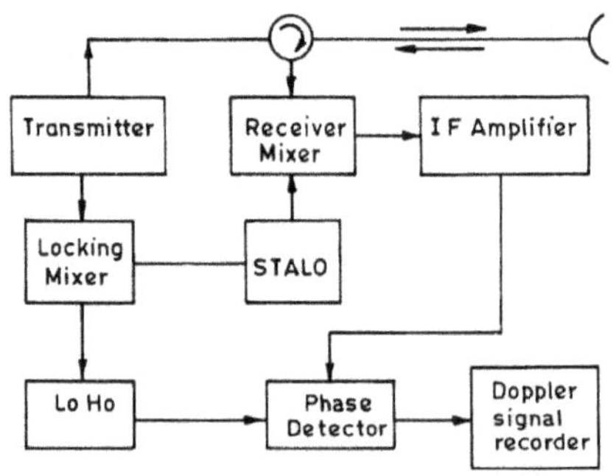

FIGURE 9.1 Block diagram of a pulse Doppler radar.

For a Doppler shift of frequency f_d the phase difference would be changing sinusoidally at f and to measure it at least $2f_d$ samples would be required (Nyquist Sampling Theorem). The PRF of the pulses transmitted, which is effectively equal to the sampling rate, must therefore be $\geq 2f_d$, and we have PRF $\geq 2f_d$. Hence, the Doppler frequency shift is given by

$$f_d \leq \frac{\text{PRF}}{2} = \frac{1}{2T} \tag{9.1}$$

where T is the interpulse period, and the highest Doppler shift of frequency is given by

$$f_{d\,max} = \frac{\text{PRF}}{2} \tag{9.2}$$

If the electromagnetic wave emitted by the antenna has a phase ϕ_o, the phase of the received echo will be

$$\phi = \phi_o + \frac{2\pi}{\lambda} \times 2r$$

where r is the range of the target. The rate of change of phase is equal to the Doppler shift of frequency $\omega_d = 2\pi f_d$. Hence,

$$\omega_d = 2\pi\, f_d = \frac{d\phi}{dt} = \frac{4\pi}{\lambda}\frac{dr}{dt} = \frac{4\pi}{\lambda} v_r \tag{9.3}$$

or

$$f_d = \frac{2v_r}{\lambda} \tag{9.4}$$

where v_r is the radial velocity of the target. Equations (9.3) and (9.4) are the same as those deduced in connection with the MTI radar. Combining (9.4) and (9.1), we have

$$\frac{2v_r}{\lambda} \leq \frac{\text{PRF}}{2} \tag{9.5}$$

or

$$v_{r\,max} = (\text{PRF})\frac{\lambda}{4} \tag{9.6}$$

If a 3.25 cm Doppler radar has a PRF = 4000/sec, the largest Doppler shift of frequency and velocity it can measure are, from equations (9.2) and (9.5), 2000 Hz and 32.5 m/sec respectively. The maximum velocity is proportional to the product of PRF and wavelength as indicated in equation (9.6). Hence, to measure higher velocities a longer wavelength and a higher PRF must be employed.

Again, as the maximum unambiguous range of a pulse radar is given by

$$r_{max} = \frac{1}{2}\left(\frac{c}{\text{PRF}}\right) \tag{9.7}$$

where c is the velocity of the radiowave, the maximum Doppler shift of frequency is related to the maximum unambiguous range by the relation

$$f_{max} = \frac{1}{4}\frac{c}{r_{max}} \tag{9.8}$$

obtained by combining equations (9.2) and (9.7). Also, the maximum radial velocity of target that can be measured by combining equations (9.6) and (9.7) is given by

$$v_{max} = \frac{\lambda}{8}\frac{G}{r_{max}} \tag{9.9}$$

For a pulse Doppler radar operating at 10 cm and PRF = 8,000/sec, v_{max} = 200 m/sec from equation (9.6), while the maximum unambiguous range would be 18.7 km from equation (9.9).

9.3 DETECTION OF MULTIPLE TARGETS MOVING WITH DIFFERENT VELOCITIES

When a number of targets move with different velocities, the resultant echo is a superposition of a corresponding number of pulse trains, each with a characteristic Doppler shift representing the velocity of the target involved. In such a situation a range gate may be used to select only those pulse trains coinciding in time with the echoes from a particular target at particular range. The position of the range gate may, in fact, be locked to the echo from a particular target by a delay lock loop network. A narrow band tunable filter following the range gate may be adjusted to select a single spectral line corresponding to the Doppler shift of the particular target chosen. All those pulse trains which pass the range gate but do not have the same Doppler shift as that of the target chosen will then be attenuated. In practice, a tunable bandpass filter is replaced by bank of narrow band filters having frequencies differing by small steps. The desired filter is then selected by electronic switching. Likewise, the range gate position is also selected discretely by electronic switch. A block diagram of a pulsed Doppler radar with a discretely selectable range gate and bandpass filter is shown in Figure 9.2. The received signals are converted to a convenient first IF, and then converted again to a second IF, which is convenient for the highly selective final Doppler bandpass filtering after range gating. The repetitive pulse waveform of the transmitted pulses result in a spectrum containing a number of significant spectral line speeds at intervals of $1/T = \text{PRF}$ on either side of the carrier frequency, as shown in Figure 9.3. In the presence of a Doppler shift caused by target motion, each spectral line of the received echo will be shifted by the same amount.

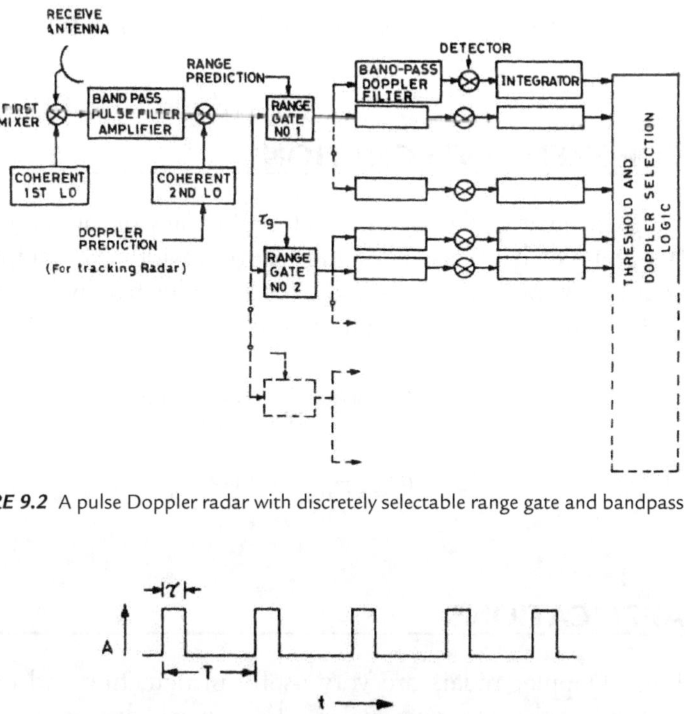

FIGURE 9.2 A pulse Doppler radar with discretely selectable range gate and bandpass filter.

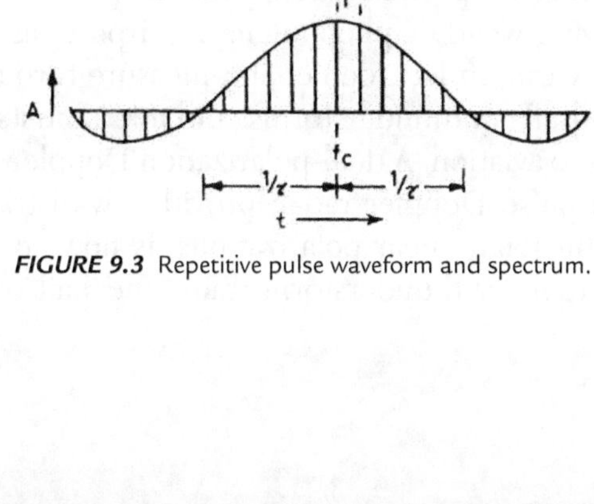

FIGURE 9.3 Repetitive pulse waveform and spectrum.

The Doppler filter which is appropriate to this shift passes a single spectral line of the echo signal from that particular target, provided it is also range gated properly. Selection of the particular Doppler filter is made by Doppler selection logic. It may also be possible to adjust the frequency of the second coherent oscillator to cause a Doppler-shifted spectral line from any desired target to pass through a particular Doppler filter. Likewise, an adjustment of the input to a range gate may be made to allow the echo from the target at a desired range to pass through a particular range gate.

A refinement of the Doppler filter may be made by introducing a comb filter having a multiple resonances coincident with the spectral lines of a Doppler-shifted echo signal, thus extracting the energy in all the spectral lines.

9.4 COHERENT INTEGRATION

If the receiver is to coherently integrate many of the target echo pulses, N_c, the Doppler bandpass filter must have a width $b \leq$ reciprocal of the total width of N_c pulses. Accordingly, the Doppler bandwidth $b << 1/N_cT$. For a 5 kHz PRF and for pulse coherent integration *for* $N_c = 50$,

$$N_cT = 50 \times \frac{1}{5000} = \frac{1}{100} \text{ sec.}$$

Hence,

$$b \leq \frac{1}{N_cT} = 100 \text{ Hz.}$$

9.5 APPLICATIONS

Pulse Doppler radars are very useful in detecting and estimating aircraft motion, separating and locking to a particular aircraft approaching or receding in a group. In addition, pulse Doppler radars are widely employed for observing weather phenomena in airports as a meterological warning radar. They can, in fact, detect and measure turbulence in the air, updrafts and downdrafts in thunderstorms, and microbursts, all of which prove to be hazardous to aviation. A dual-polarization Doppler radar, which is a modified form of a pulse Doppler radar, provided with transmission and reception in two orthogonal linear polarizations, is now considered one of the most powerful tools in thunderstorm, rain, and hail observations. The ratio of the reflectivities for the horizontal and vertical polarizations may be used to distinguish ice and rain regions in a thunderstorm cell. Such radars operating in the S-band are in use in the United Kingdom and the United States. A millimeter wave dual-polarization pulse Doppler radar at 35 GHz can be used to observe the structure and motion of clouds over airports up to a height of about 10 km, and is often termed a Ceilometer.

9.6 ADVANTAGES OF PULSE DOPPLER RADAR

A pulse Doppler radar has the following advantages over pulse, CW, and MTI radar:

1. The ability to reject unwanted echoes either by range gating or by Doppler selection.
2. The ability to measure range and velocity unambiguously over predetermined limits, even in the presence of multiple targets.
3. Coherent integration of echo pulses by the Doppler filter increases the signal to noise ratio, and echoes from a greater range are detectable.

9.7 INTRODUCTION TO FREQUENCY CODED RADARS

The carrier frequency of a pulse radar may be modulated or coded to improve the range capability of the radar. A linear frequency-modulated pulse signal in which the carrier frequency increases linearly for the duration of a pulse may be processed by a matched filter to obtain a narrow pulse, compressed in the time domain, with a corresponding increase in the peak amplitude of the pulsed signal. The process is referred to as the pulse compression technique, which is widely employed to compress the echo pulse in a radar system with a linear FM pulsed transmission. Such a radar is also referred to a "pulsed FM" or "chirp" (linear FM) radar. In the radar the energy contained in the long pulsed electromagnetic wave is in fact compressed in the time domain by the pulse compression technique to obtain the resolution of a short pulse, although the transmitted energy is spread over a longer time, requiring only a rather small peak transmitter power to obtain a reasonable range capability. The entire signal energy of a pulse of radiation is effectively "dumped" into a small time duration after the compression of the pulse by a matched filter.

9.8 BLOCK DIAGRAM

The block diagram of a linear FM pulse compression radar is shown in Figure 9.4. The waveform and frequency versus time plots at various stages of the radar are shown in Figure 9.5. A linear FM pulsed signal $f_2(t)$ is derived from IF pulse $f_1(t)$ by passing it through a dispersive delay network. The waveforms of $f_1(t)$ and $f_2(t)$ are given by

$$f_1(t) = \frac{\sin(\mu\,\tau\,\omega t/2)}{(\mu\,\tau\,\omega t/2)} \cos\omega_{IF}\, t \tag{9.10}$$

$$f_2(t) = \cos\left[\omega_{IF}\, t + \int \mu\, t dt\right] \text{for} -\frac{\tau_\omega}{2} \le t \le \frac{\tau_\omega}{2}$$

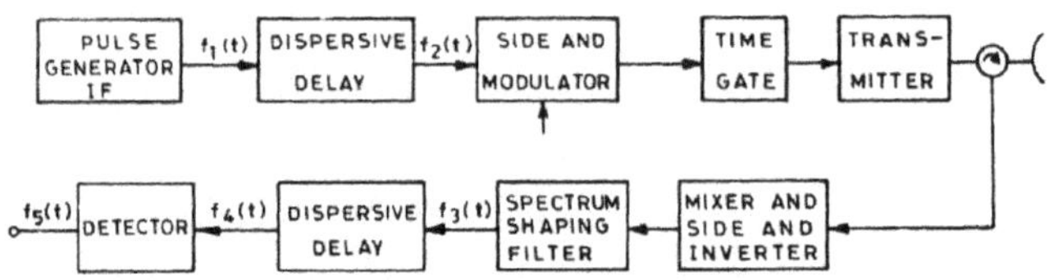

FIGURE 9.4 Block diagram of a pulsed FM radar.

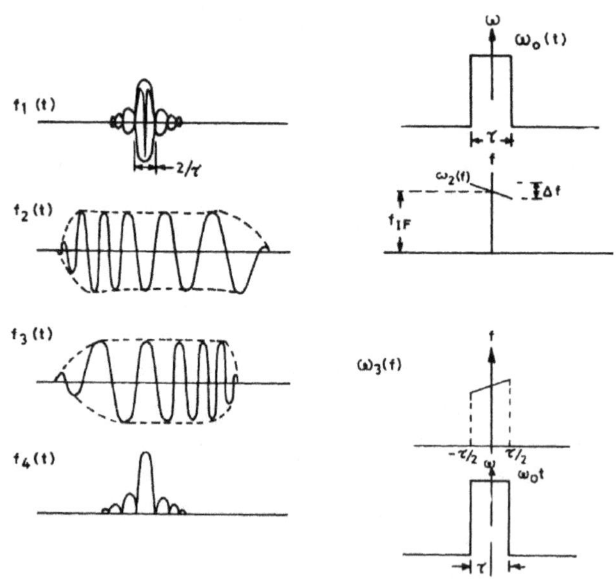

FIGURE 9.5 Waveform and frequency versus time plots at various stages of the radar.

$$= \cos\left[\omega_{IF}\, t + \frac{\mu t^2}{2}\right] \text{for} -\frac{\tau_\omega}{2} \le t \le \frac{\tau_\omega}{2} \quad (9.11)$$

The signal $f_2(t)$ will then be "converted up" to the desired carrier frequency ω_o by a sideband modulator, which is a balanced mixer, and then passed through a time gate of duration τ_ω for onward transmission to the final power amplifier stage of the transmitter. The received signal is "down converted" by a mixer-cum-sideband inverter, in which the LO frequency is kept higher than the carrier frequency ω_0 by ω_{IF}. This effectively produces a spectrum of $f_3(t)$, with the higher frequencies producing a lower value of ω_{IF}. This inversion allows a dispersive delay network identical to that in the transmission to reproduce the transmitted IF pulse waveform $f_1(t)$. The spectrum-shaping filter in the receive chain is employed to reduce the time side lobes in $f_4(t)$ and $f_5(t)$ to a tolerable level at the cost of a slight reduction in signal-to-noise ratio and an increase in compressed pulses width.

9.9 DISCRETE FREQUENCY WAVEFORM CODING

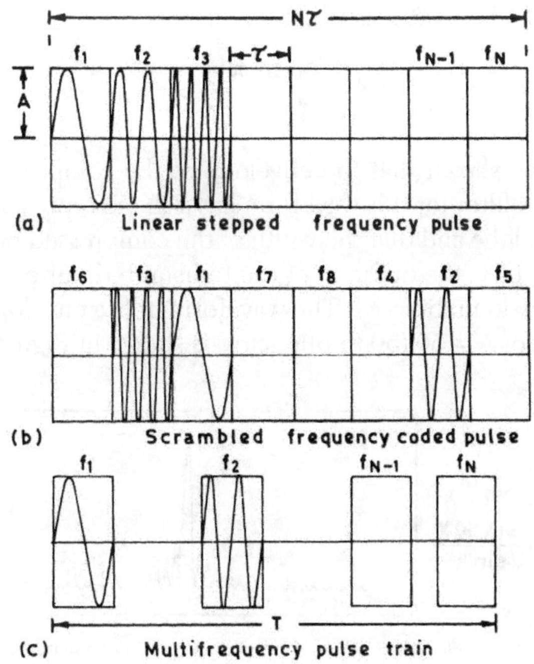

FIGURE 9.6 Envelope waveform for discrete frequency coding.

For a pulse waveform containing N rectangular time segments of equal amplitudes and length τ but having a different frequency f_n, the envelope of the waveform will be as shown in Figure 9.6. Assuming the time of origin as the midpoint of the first segment, the waveform of the nth segment is given by

$$v_n(t) = A_e^{j2\pi(f_n t + \phi_n)} \text{ for } (n - 3/2)\,\tau < t < \left(n - \frac{1}{2}\right)\tau$$

where A = segment amplitude, f_n = frequency of the nth segment, ϕ_n = phase of the nth segment, and $f_n - f_{n-1} = 1/\tau$. The Fourier transform of the nth segment can be written

$$V_n(f) = A\int_{(n-1/3)\tau}^{(n-1/2)\tau} e^{j2\pi(f_n + f_n t + \phi_n)}\, e^{-j2\pi ft}\, dt \tag{9.12}$$

$$= A\,\tau \frac{\sin\pi[(f_0 - t)\,\tau + n]}{\pi[(f_0 - f)\,\tau + n]}\, e^{j[2\pi\,(f_0 - f + n/\tau)(n-1)\,\tau]} \tag{9.13}$$

after integration of (9.12), replacing f_n by $f_o + nf$ for a frequency spacing of Δf and assuming $\phi_n = n\phi_o = 0$. The spectrum of the entire frequency-coded pulse waveform is given by

$$V(f) = A\,\tau \sum_{n=1}^{N} \frac{\sin\pi(f\tau - n)}{\pi(f\tau - n)}\, e^{j[2\pi(f\tau - n)(1-n)]} \tag{9.14}$$

It can be shown that the envelope of the compressed pulse output of the matched filter for this waveform has a shape represented by $(\sin x)/x$ around its main lobe and that the width of the compressed pulse at the 4 dB point is $(1/N) \times \tau$. As the duration of the transmitted pulse is N_τ, the effective pulse compression ratio $= N^2$. The waveform of the envelope of the matched filter output for $N = 50$ (50 frequencies) is shown in Figure 9.7.

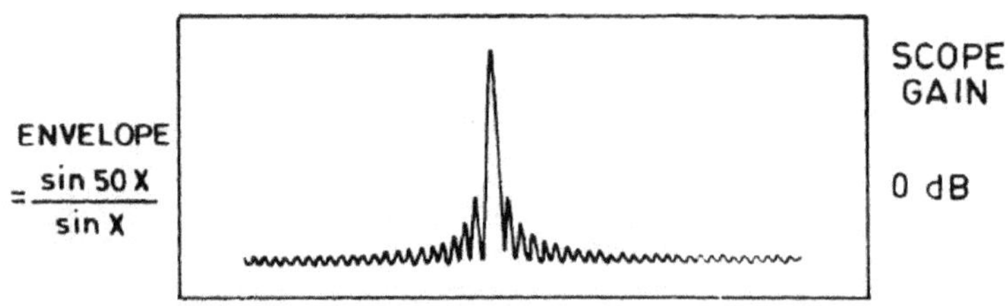

FIGURE 9.7 Waveform of the envelope of a matched filter output.

9.10 SIDE LOBE REDUCTION BY WEIGHTED AMPLITUDE OF THE FREQUENCY-CODED WAVEFORM

The side lobes in the envelope of the compressed pulse shown in Figure 9.7 can be reduced to any desired level by weighing appropriately the amplitude of the frequency-coded waveforms shown in Figure 9.6. The first side lobe level for a uniform amplitude is about 13.5 dB below the main lobe peak, the remaining side lobes decaying monotonically on neither side of the main lobe. The amplitude weighing, however, tends to widen the main lobe and reduce the peak amplitude of the main lobe, which in turn degrades the signal-to-noise ratio. A widely employed amplitude weighing is made by Gaussian amplitude modulation of a linear FM pulsed signal.

9.11 MATCHED FILTER REALIZATION FOR PULSE COMPRESSION

The output of a matched filter can be shown to be equal to the autocorrelation of its input. For a linear-stepped, frequency-modulated waveform, the matched filter can be realized by a tapped delay line acting as the autocorrelator, as shown in Figure 9.8.

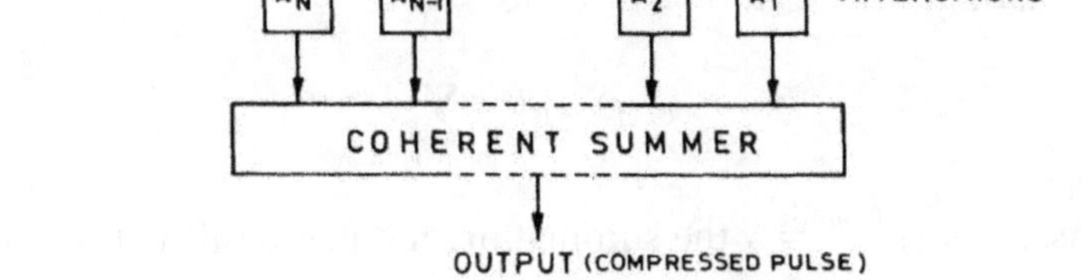

FIGURE 9.8 A tapped delay line autocorrelator.

9.12 WAVEFORM ANALYSIS OF A LINEAR-STEPPED FREQUENCY PULSE

The transmitted waveform for a linearly stepped frequency waveform can be represented by

$$s(t) = \sum_{n=0}^{N-1} [u(t - n\tau) - u(n+1)\tau] \cos(\omega_0 + n\,\Delta\,\omega)t \tag{9.15}$$

where $\omega_0 = 2\pi f_0$ = the lowest frequency in the transmission frequency steps, $\Delta\omega$ = frequency spacing = $2\pi\,\Delta f$

$$u(t) = \text{unit step function} = \begin{cases} 1, t \geq 0 \\ 0, t < 0 \end{cases}$$

N = number of frequencies in transmission.

The phase and time origin is reckoned from the leading edge of the first subpulse. The matched filter impulse response is given by

$$= \sum_{n=0}^{N-1} [u(-t - n\tau) - u(-t - \overline{n+1}\tau)] \cos(\omega_0 + n\,\Delta\,\omega)\,t \tag{9.16}$$

If the matched filter delays and filters are connected in the inverse order of that for transmission, the matched filter output can be written as

$$e_0(t) = \sum_{n=0}^{N-1} A_n^2\, e^{j(\omega_0 + n\,\Delta\omega)\,t} \tag{9.17}$$

neglecting delay to the target. A_n is the weighing function of the nth channel. Equation (9.17) is valid in the region $(N - 1)\tau < t < N\tau$. This is, in fact, the only region of interest with the assumption $\Delta f = 1/\tau$. Assuming $A_n = 1$, equation (9.17) can be written as

$$e_0(t) = e^{j\,\omega_0 t} \sum_{n=0}^{N-1} e^{jn\,\Delta\omega\,t} \tag{9.18}$$

Assuming $e^{j\,\Delta\omega\,t} = X$ the summation of equation (9.18) is given by

$$e_0 = \sum_{n=0}^{N-1} e^{jn\,\Delta\omega t} = \frac{1 - X^N}{1 - X} = \frac{1 - e^{jn\,\Delta\omega t}}{1 - e^{j\,\Delta\omega t}} \tag{9.19}$$

Combining equations (9.18) and (9.19), we have

$$e_0(t) = e^{j\left[\omega_0 + \frac{(N-1)\,\Delta\omega}{2}\right]t \left[\frac{\sin N\,(\Delta\omega/2)\,t}{\sin\,(\Delta\omega/2)\,t}\right]} \tag{9.20}$$

To take into account the autocorrelation of the rectangular segment, envelope equation (9.20) should be multiplied by the triangular function $\tau[1 - (|t|/\tau)]$ for $|t| \leq \tau$ and zero outside this region.

The frequency $\omega_0 + [(N - 1)\ \Delta\omega/2]$, included in the first term of equation (9.20), is the mean carrier frequency of the stepped frequency or comb signal. The second term determines the compressed pulse envelope. The nulls in the envelope occur at $\sin [N\ (\Delta\omega/2)t] = 0$, that is, when $N(\Delta\omega/2)t = \pm m\pi$, m being any integer.

From equation (9.20), the first side lobe peak is 13.46 dB below the main lobe for $N = 50$ and is 13.06 dB down for $N = 8$. The envelope has the $(\sin x)/x$ shape. Also, at one half the distance between the nulls, the envelope response is down by 4 dB, and the compressed pulse width is given by

$$\tau_{4dB} = \frac{1}{N\ \Delta F} = \frac{\text{Transmit envelope width}}{N^2} \qquad (9.21)$$

For the 3 dB pulse width, the pulse compression ratio would, therefore, be slightly greater than N^2. All these results are similar to that for a continuous linear FM pulse waveform for a large compression ratio.

9.13 APPLICATIONS OF FREQUENCY-CODED RADARS

Frequency-coded radars have wide applications in long-range radar systems, which are limited by the transmitter peak power. By emitting a long frequency-coded pulse, the signal energy is effectively compressed into a short duration to obtain a N-fold increase in the envelope amplitude of the compressed echo pulse, which in turn improves the signal-to-noise ratio and hence the maximum range. A recent application of frequency coded or chirp radar is in the area of detection of long-range targets like ships in the ocean by HF backscatter chirp radars, which are also called the over-the-horizon (OTH) radars. The frequency-coded HF signal propagates by reflection in the ionospheric layers to the target at ranges as great as 5000 km in deep seas. An OTH radar on 19 GHz is in operation at the University of Birmingham, United Kingdom for studies in target detection and sea waves in oceans using ionospheric propagation of the radar signals.

9.14 INTRODUCTION TO PHASE-CODED RADARS

Phase coding, like frequency coding, can also be employed for obtaining a pulse compression by a matched filter. A long pulse of duration T is divided into N subpulses, each of width τ. The phase of each subpulse may be either 0 or π radians, chosen either at random or following a certain code sequence. The output of the filter will be a spike of width τ with an amplitude N times greater than of the long pulse, with the compression ratio being $N = T/\tau = B.T$, where $B = 1/\tau$ = bandwidth of a subpulse. A phase-coded pulse compression technique is employed in a class of radars called phase-coded radar. A pseudorandom noise code (PN) is widely employed for the phase coding, apparently for the ease with which such codes may be generated and processed at the receiver by using a shift register with a feedback loop through a mod-2 adder.

9.15 PHASE CODING AND DECODING

As the output of a matched signal processor is the autocorrelation of the input signal, the criterion for selection of a good phase waveform is that its autocorrelation function should have equal side lobes. Such a code with equal side lobes is called a Barker code. An example of a Barker coded waveform and its autocorrelation are shown in Figures 9.9 (a) and (b). There are six equal time-side lobes on either side of the peak, each at level –22.3 dB below the peak.

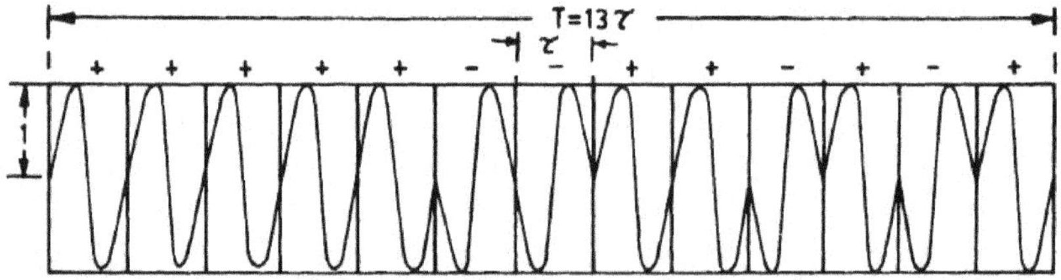

FIGURE 9.9 (a) A Barker coded waveform.

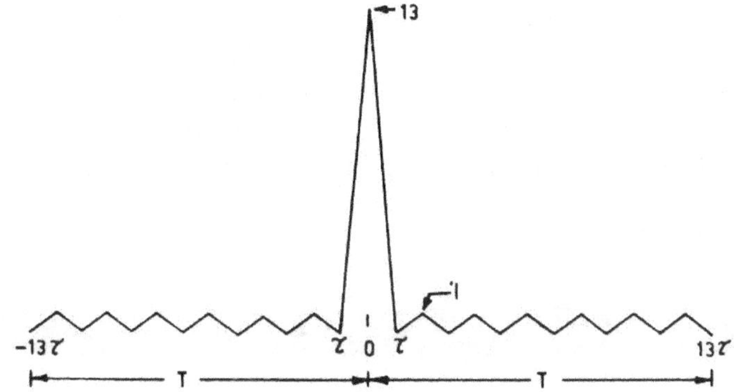

FIGURE 9.9 (*b*) Autocorrelation function.

The Barker code can be generated by a tapped delay line, as shown in Figure 9.10. The output of this coding network is fed to a biphase modulator, which is an analogue product device like a balanced modulator, to generate the desired phase-coded waveform. At the receiving end a similar product device may be used to get back the original Barker code. Also the same network can be used for both coding and decoding of the Barker codes by switching the inputs to terminals 1 and 2 respectively, the output being taken in each case from the terminal 3. Barker codes are known for lengths varying between 2 to 13. Longer lengths of code are, however, required in practice to achieve a reasonably high pulse compression ratio, and shift registers with feedback through a mod-2 adder is a very popular technique to generate the long code sequences, called pseudorandom noise codes.

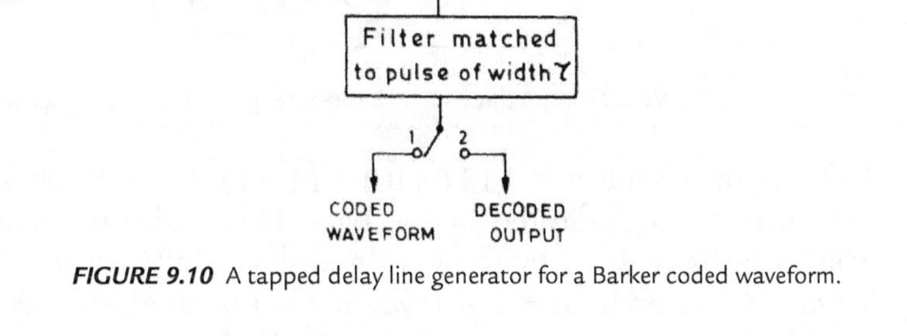

FIGURE 9.10 A tapped delay line generator for a Barker coded waveform.

A 7-bit shift register for generating a pseudorandom noise (PN) sequence of length 127 is shown in Figure 9.11 (*a*). The generation of a PN sequence may be illustrated with a truth table with reference to a shift register generator of 3 stages as shown in Figure 9.11 (*b*). If the initial state of each stage of the shift registers is 1, the truth table will be as shown in Table 9.1.

TABLE 9.1
Truth Table

Clock pulse	1	2	3
0	1	1	1
1	0	1	1
2	1	0	1
3	0	1	0
4	0	0	1
5	1	0	0
6	1	1	0
7	1	1	1

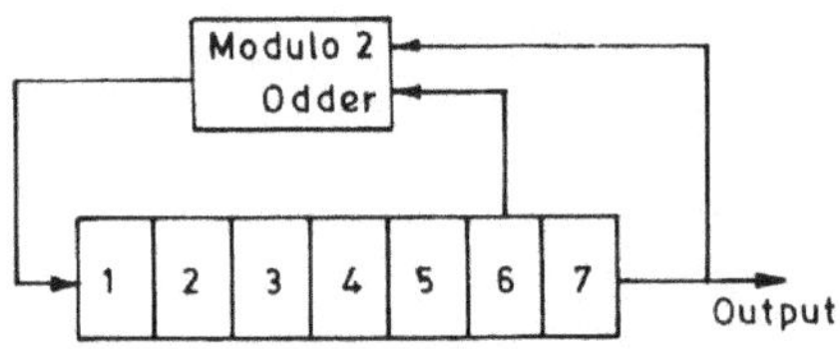

FIGURE 9.11 (*a*) A 7-bit shift register generator for a PN sequence.

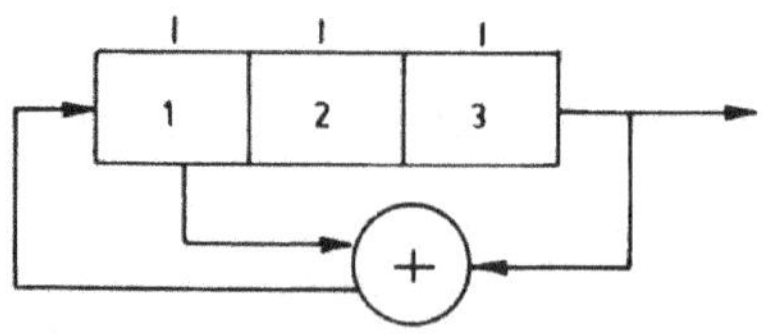

FIGURE 9.11 (*b*) A three-stage shift register generator for a PN sequence.

The output sequence is $\overline{1110100}\ \overline{1110100}$, in which length of each sequence is 7 as indicated by the bars. This is also the maximum length which can be realized by using a three-stage shift register. The maximum length, M is given by $M = 2^n - 1$, where n is the number of stages. For $n = 3$, $N = 2^3 - 1 = 8 - 1 = 7$; likewise, for $n = 7$, $N = 2^7 - 1 = 128 - 1 = 127$. Not all the sequences generated by such a shift register code generator will necessarily be maximum length. The length is, in fact, dependent on the feedback connections employed. The examples of the feedback connections required to generate maximum length sequences from an n-stage shift register PN sequence generator are shown in Table 9.2.

TABLE 9.2
Maximum Length Sequences and Examples of Feedback Connection for an *n* Stage Shift Register Generator

Number of stages, n	Length of maximal sequence $2^n - 1$	Number of maximal sequences	Examples of feedback stage connections
3	7	2	3, 2
4	15	2	4, 3
5	31	6	5, 3
6	63	6	6, 5
7	127	18	7, 6
8	255	16	8, 6, 5, 4
9	511	48	9, 5
10	1023	60	10, 7
11	2047	176	11, 9

For a large length, N, of the sequence, the peak side lobe level after matched filter processing (an autocorrelator) is approximately $1/N$ of the peak response, measured in power. For $N = 127$, the peak (side lobe level) is between −18 dB and −19.8 dB, while for $N = 255$ it varies between −21.3 dB and −22.6 dB.

9.16 BLOCK DIAGRAM OF PHASE-CODED CW RADAR

A block diagram of a phase-coded CW radar is shown in Figure 9.12. A CW signal from an RF generator is reversed in phase according to the polarity of the shift-register encoder generating a PN sequence. This phase-coded CW signal is then amplified and transmitted. In the receive channel the target echoes are mixed to get a convenient IF and demodulated. Range and Doppler prediction obtained from respective tracking circuits (not shown) allows a CW signal to appear at the output of an appropriate narrow band filter.

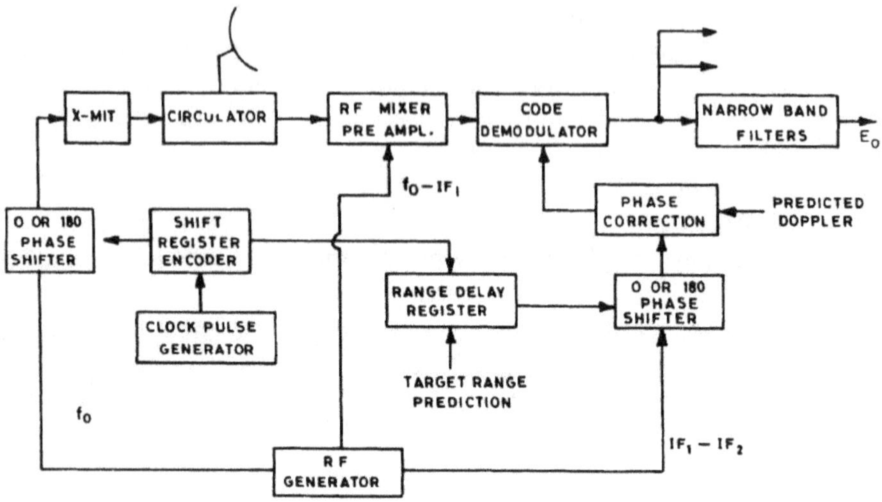

FIGURE 9.12 A block diagram of phase-coded CW radar.

9.17 DECODERS

A decoder or code demodulator for a phase-coded radar can be based on (*a*) analogue, (*b*) digital, or (*c*) cross-correlator circuits, each type having a distinctive advantage over the others. The analogue and digital types are suitable for short and long coded words respectively, but none of these types provide any Doppler coverage. The correlator type is suitable for range tracking with a limited Doppler coverage.

a. *Analogue All Range Decoders*. A block diagram of an analogue all range decoder suitable for a 7-bit Barker code is shown in Figure 9.13. For search radar in which the number of segments in the code is small, the bandwidth of the *N*-cascaded delay lines will pass a rectangular subpulse with little distortion. For this it is necessary that the overall bandwidth of *N*-cascaded stages $\geq 1/\tau$, which means that each stage must have a bandwidth $>> 1/\tau$. With a quartz or glass ultrasonic delay line, realization of such a wide bandwidth is not a problem. The progressive attenuation in the delay lines can be compensated for by varying the gain of the inverters prior to the summing amplifier. The advantage of this type of matched filter decoder over the cross-correlator shown in Figure 9.13, is that a prior knowledge of the target is not needed.

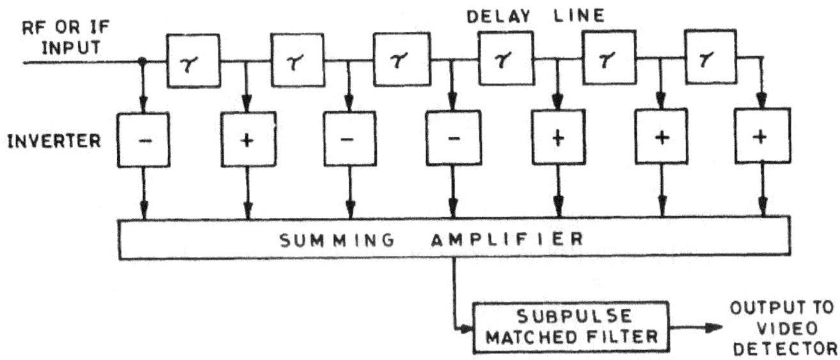

FIGURE 9.13 A cross-correlated decoder.

Another type of all range analogue decoder is based on a homodyne processor as shown in Figure 9.14. The LO for both the mixers are at the intermediate frequency to achieve homodyne processing. Their phases are, however, kept in quadrature to obtain in phase, *I* and the quadrature, *Q* components of echoes containing no carrier frequency component. The quadrature oscillator inputs are in practice derived from a single oscillator with a 90° phase shifting network at the LO input for one of the mixers. Minor Doppler shifts can be corrected by adjustment of the LO frequency to avoid degradation of the matched-filter output. The total signal may undergo any amount of phase shift due to the round trip travel between transmission and reception and as such may appear in either or both the *I* and *Q* channel outputs.

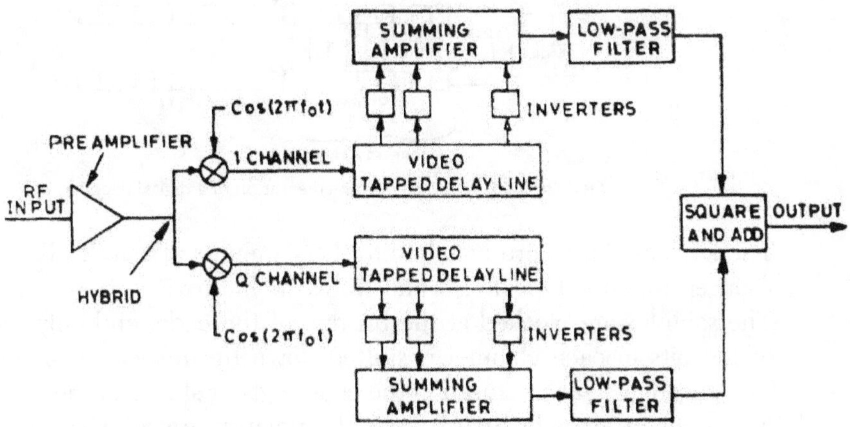

FIGURE 9.14 An all range decoder based on a homodyne processor.

The two outputs are processed by separate tapped delay line integrators and then added quadratically to obtain an output proportional to the resultant amplitude $e_0(t) = (I^2 + Q^2)^{1/2}$. This addition from both channels is necessary to avoid a 3 dB degradation in signal-to-noise ratio which would otherwise occur.

b. *Digital Decoders*. A digital version of the tapped delay line pulse compressor for all ranges can be realized for binary or quadrature (four phase) coders in a phase-coded radar. Figure 9.15 shows an all range implementation of such a digital decoder for a maximum length binary code of 255 segments. The eight-stage shift register generates the biphase code of 255 segments as shown in the upper left corner of Figure 9.15, while the *I*-*Q* mixer is at the left of the diagram.

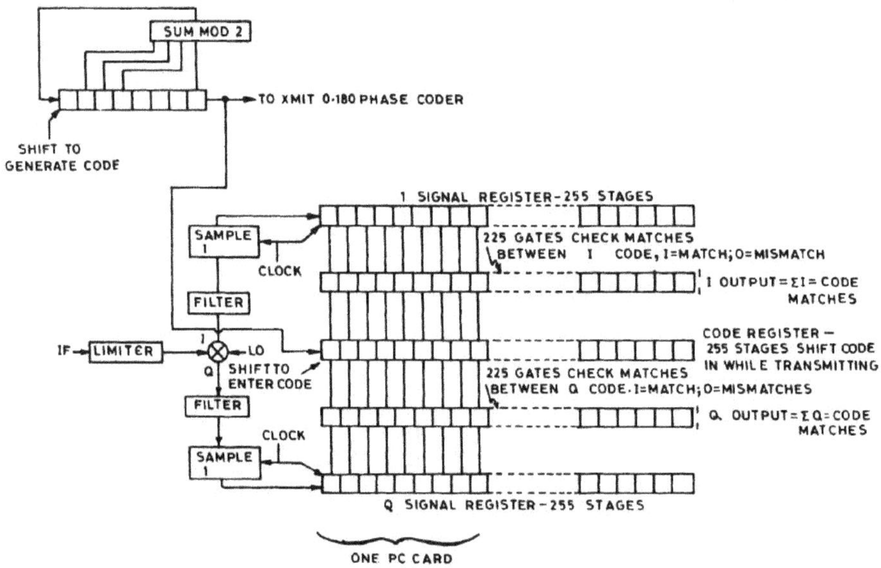

FIGURE 9.15 An all range implementation of a digital decoder.

The low pass filters are matched to the segments of a code. The IF limiter reduces impulse type noise and helps in improving the dynamic range. The samples are clocked at the bit rate of the code, and only the polarity of the bits in each channel is shifted down the respective signal register. The encoded signal is stored in the code register shown at the middle of the five series of gates in Figure 9.15, the polarity in each stage representing the polarity of the corresponding segment of the code. Any type of binary code of length 255 can, in fact, be stored in the code register prior to reception of the received echoes. This allows the code to be changed for each transmitted pulse at command, to prevent an easy detection by an enemy of the codes transmitted. Also, any shorter length code can be stored, leaving the remaining gates unused if a lower range resolution is permitted. The number of matches in the 255 gates between the signal and code register is continuously summed. For a perfect alignment of a target position and the stored word, 255 matches would appear in the I or Q summer or both. The summation of the agreements is then combined to yield the larger of $|I|$ or $|Q|$. This summation is fed to a threshold circuit, where the threshold level is predetermined by the desired false alarm rate. Target range is determined by stopping a counter when the threshold is exceeded. The broadband limiting prior to the decoder makes the false alarm rate constant (Constant False Alarm Rate, CFAR) for noise, jamming, and other uncorrelated echoes.

9.18 CROSS-CORRELATORS AND TRACKERS

Cross-correlators can be used to acquire and track the target echoes following the target motion if the range is approximately known. A block diagram of a multiple correlator, useful for negligible Doppler shifts is, shown in Figure 9.16 (*a*). Referring to Figure 9.16 (*a*), the I and Q components of the signal appear at a succession of correlator integrators, each of which corresponds to a fixed range delay. Assuming that the code generator is started when the bipolar received signal appears at the first correlator integrator, the first stage of the register will be either a plus or minus state corresponding to the first bit of the code. It will energize either A or B. The situation will be the same for the round trip delay corresponding to a multiple of code line to the correlator integrator C/i, the function of which can be explained by referring to Figure 9.16 (*b*). The block C/i will pass the bipolar video signal either directly or after inversion. When the second bit appears at the first stage of the shift register and at the first C/i, the shift register output will energize the direct or the inverter gate, depending on whether the polarity of the second bit is plus or minus. The signal applied to the integrator would thus always have the same polarity as the first bit. If the code is entirely aligned, the received voltage for all the N bits will be integrated. The I channel will also perform a similar integration over N bits. At the other integrators, cross-correlations of the codes will appear, while at the termination of the N-segment correlation code, the levels in the integrators will correspond to the side peaks of the autocorrelator waveforms.

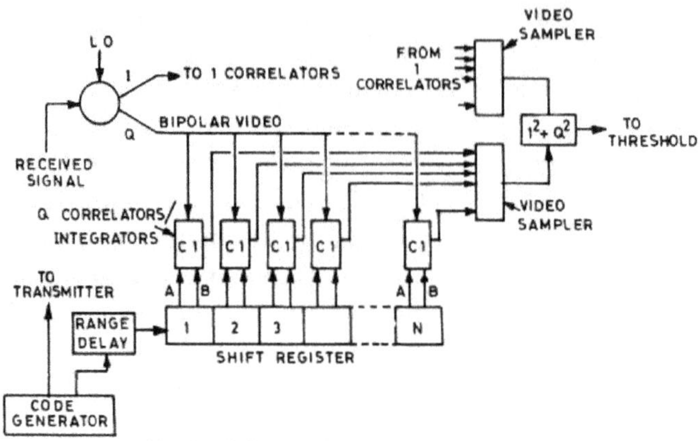

(*a*) A block diagram of a multiple correlator.

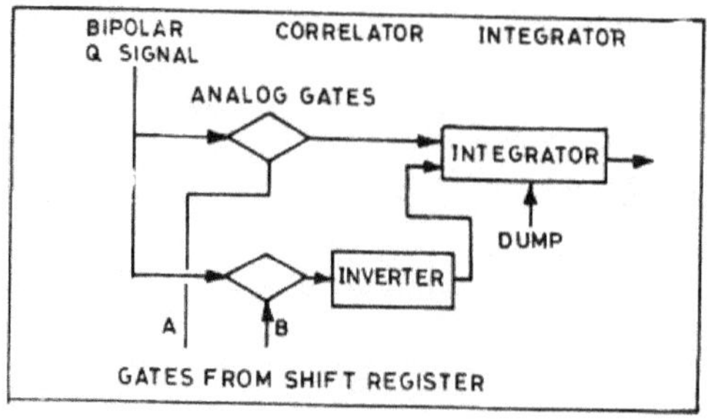

(*b*) Details of the multiple correlator.

FIGURE 9.16

The range coverage in this type of cross-correlator decoder can be increased by running the code continuously into the shift register but dumping the integrators at the end of each sequence for each stage.

9.19 RANGE TRACKERS

If the range is known to within ± one code segment, a fine range tracker can be implemented with two sets of correlators. The range delay circuit is adjusted so that first code falls between the first two *C/i* circuits. If the prediction of range is good, the first code falls exactly midway, and approximately one half of the voltage will be integrated on each circuit. For any small errors in estimation, the voltage at two sets of integrators will become different in proportion to the error, and the range-error signal can be derived from the ratio of the integrated voltages. Such a system is called an "early-late gate" or "spit-gate" range tracker.

An alternate form of range tracker is based on the range-error signal derived from the difference of the direct and inverted outputs of the two sets of correlator integrators, each of which gives the $I^2 + Q^2$ output. For the decoders in the two sets or channels separated by two code segments, this difference in the errors varies with the error in the range estimation R expressed in units of a code segment as shown in Figure 9.17 curve (*a*), while curve (*b*) is for a delay between the channels $< \tau$. The error voltage can be used to control the delay of the range-adjusting circuit by a voltage-controlled analog delay network or by its digital version using a shift register delay to form a closed negative feedback loop, called a delay lock tracking discriminator. Once the lock is achieved by a manual control of the range delay circuit it will be maintained automatically, disregarding target motion. The residual error indicated in the delay lock tracking discriminator gives an estimate of range continuously during the target motion. A simpler version of the range tracker for phase-coded radars is shown in Figure 9.18. The delayed code generated is used to switch the phase of the LO signal to decode the received signal. The received signal after amplification is split into the early and late channels, mixed to DC, and integrated. A time delay, τ/K of the order of a segment length is inserted in the early channel. The choice of the delay time τ/K is, in fact, a compromise between (*a*) the error sensitivity of the range circuit, (*b*) the pull-in or lock-in range, and (3) the efficiency or threshold level required for small signal-to-noise ratios. If the delay is made less than τ (for $K > 1$), the discriminator characteristics will have a steeper slope as in curve *b* in Figure 9.17, but will then have a small pull-in range.

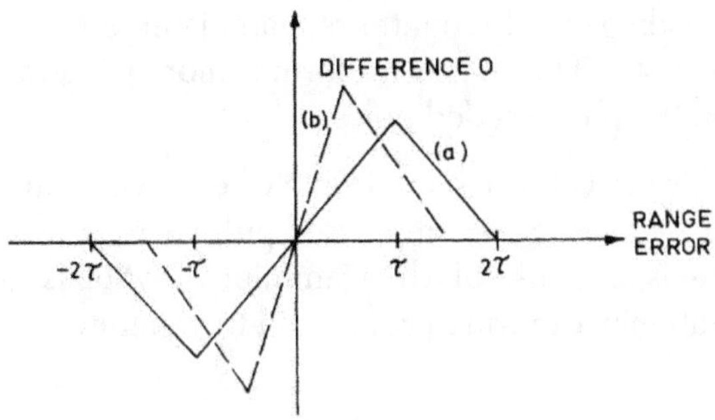

FIGURE 9.17 Error voltage variation with range for a dual channel decoder (*a*) delay between channels (*b*) delay between channels $< \tau$.

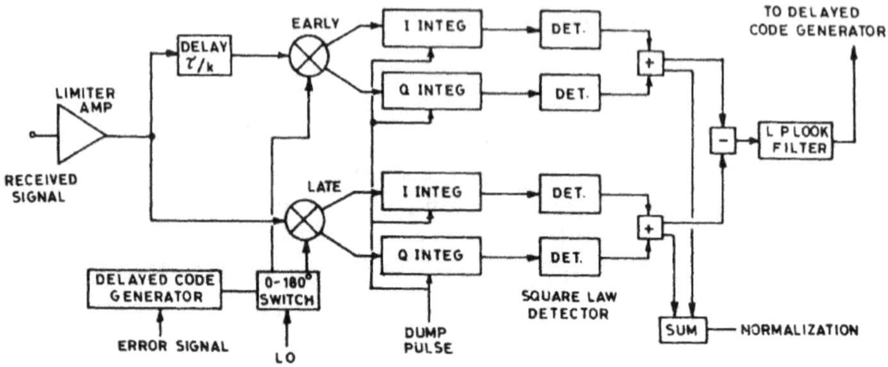

FIGURE 9.18

9.20 COMPARISON OF PHASE-CODE AND LINEAR FM PULSE COMPRESSIONS

1. The peak side lobe of the phase-coded waveform is generally lower than that in a chirp (linear FM) waveform.
2. The chirp (linear FM) is Doppler-tolerant and a single pulse-compression filter can be used. It cannot, however, provide an independent range and Doppler measurement.
3. A phase-coded radar might require a bank of continuous matched filters covering the expected range of Doppler frequencies.
4. A different phase-coded sequence can be assigned to each radar so that a number of radars can share the same spectrum.
5. In military applications, a phase-coded radar can be arranged to change the coding to help counter repeater jammers that attempt to simulate the waveform. The chirp waveform is more vulnerable to repeater jamming than the phase-coded pulse.
6. A chirp waveform is applicable when a wide bandwidth may be available and a very narrow compressed pulse is required, while the phase-coded pulse is more useful when jamming or EMC is a problem or when a long duration waveform may be used to generate a very narrow, compressed pulse.
7. The implementation of the chirp pulse compression is generally less complex than that of the phase-coded pulse.

9.21 INTRODUCTION TO MILLIMETER WAVE RADARS

The present decade is characterized by efforts to extend radar and navigational aids to the millimeter wavelength ranges. This portion of the electromagnetic spectrum has, in fact, many advantages over microwaves: (*a*) broad bandwidths, (*b*) higher spatial resolution, (*c*) low probability of interception and interference, (*d*) small size and weight of antenna and equipment, (*e*) ruggedness and reliability of systems, and (*f*) low voltage supplies. The excellent performance, precision, and maneuverability of millimeter wave systems are very attractive features for defense application in surveillance and weapon systems based on millimeter wave radars.

9.22 PROPAGATION OF MILLIMETER WAVE RADARS

A crucial initial design parameter for millimeter wave radar is the absorption coefficient for atmospheric propagation. In fact, attempts to build the first millimeter wave radar during World War II were a disappointment apparently due to the choice of the exact frequency near the water vapor molecular absorption line, at which the range was reduced seriously. This apparently inhibited the development of radar technology at millimeter waves for a long period. Besides these, J. H. Van Cleck's theoretical prediction about a strong atmospheric absorption line at 60 GHz was also classified during World War II for its military significance in developing secure communication. After World War II, the millimeter wave work continued in the United States at the National Research Laboratory (NRL) and the Bell Telephone Laboratory (BTL).

The atmospheric absorption of millimeter waves for horizontal propagation at sea level and at an altitude of 13,000 ft is shown in Figure 9.19.

This shows that the atmospheric attenuation rate is a maximum at 22, 183, and 325 GHz due to water vapor absorption lines at these frequencies and also at 60 and 19 GHz due to absorption in oxygen molecules. In between these absorption lines due to water vapor and oxygen molecules, a broad minima of attenuation rate occurs near the atmospheric window frequencies at 35, 94, 140, and 220 GHz, each at which low loss transmission through the atmosphere would be possible. The attenuation in the atmospheric gases decreases with increasing elevation angles due to a small depth of the denser lower atmosphere through which the wave will then have to pass. The total attenuation for vertical propagation is, in fact, only about a few decibels at the atmospheric window frequencies mentioned above.

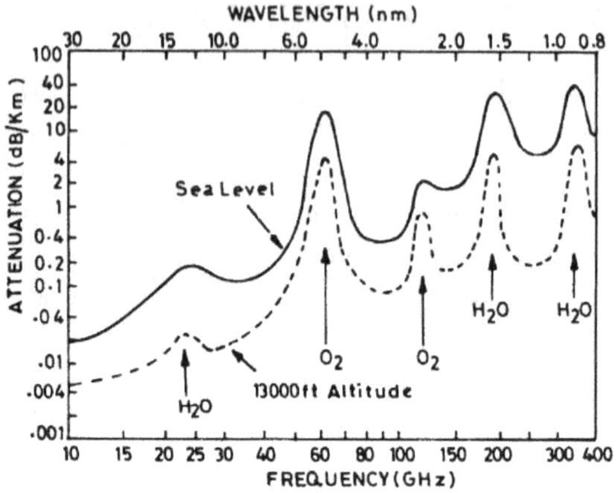

FIGURE 9.19 Atmospheric absorption of millimeter waves for a horizontal propagation path.

The millimeter wave propagation is less affected by fog, smoke, and dust compared to that of the infrared and optical wavelengths.

9.23 MILITARY RADARS

Battlefield surveillance radars at 70 GHz have been designed to cover a range up to 50 km with a beamwidth of the order of 0.2°, scan sector 30°, and scan rate of about 40 scans/sec. For a shorter range the beamwidth, scan sector, and scan rate are all made higher. The parameters of two 70 GHz radars, one for a long range and the other for a short range, are shown in Table 9.3.

TABLE 9.3
Parameters of 70 GHz Battlefield Surveillance Radars for Long and Short Ranges

Range	Long range (50 km)	Short range
Frequency	70 GHz	70 GHz
A_z beamwidth	0.2°	0.55°
Elevation beamwidth	0.3 (shaped)	3.5°
Scan rate	up to 40 scans/sec	up to 70 scans/sec
Peak power	15 kW	500 W
Pulse width	50 n sec	20—45 n sec
PRF	10 kHz	5—25 kHz

For a higher angular resolution and scan rate, higher millimeter wave window frequencies at 94 and 140 GHz would be suitable. At the higher frequencies the beamwidth will be sharper for a given antenna size. A surveillance and target acquisition radar at 94 GHz for enemy tank location and engagement was in fact developed in the United States in 1979 as shown in Figure 9.20. The parameters of this radar are shown in Table 9.4.

FIGURE 9.20 A surveillance and target acquisition radar at 94 GHz.

TABLE 9.4
Parameters of 94 GHz Surveillance and Target Acquisition Radar

Frequency	94 GHz
Beamwidth	11 m rad
Average power	0.1—0.5 W
Antenna aperture	14 in
Field of view	15° × 7.5° in wide mode and 5° × 2.5° in narrow tracking mode
Frame time	2 seconds
Target detection range	3 km with 100 m visibility
Target tracking	0.5 m rad accuracy at 2 km with 100 m visibility

9.24 ANTI-AIRCRAFT WEAPONS SYSTEMS

A short range anti-aircraft weapons system using a 94 GHz millimeter wave-directed rapid fire cannon mounted on a tank is illustrated in Figure 9.21. For a higher angular precision, a frequency of 140 GHz would be more suitable.

A comparison of the parameters of 94 and 140 GHz radars for anti-aircraft weapons systems is shown in Table 9.5.

The range performance of these 94 and 140 GHz radars can be assessed from the variation of the signal-to-clutter ratio and the signal-to-noise ratio with the range under conditions of clear atmosphere and rain as shown in Figure 9.22. It appears from the figure that ground clutter limits the range to about 4 km at 94 GHz and 2 km at 140 GHz under clear weather conditions. During rain, attenuation and scatter in raindrops further lower the range to 2 km and 1.5 km at 94 and 140 GHz bands respectively.

FIGURE 9.21 A Short range anti-aircraft weapons system.

TABLE 9.5
Parameters of 94 and 140 GHz Radars

	Frequency (GHz)	
	94	140
Attenuation rate		
In clear air	0.4 dB/km	1.5 dB/km
In rain	3.0 dB/km	3.2 dB/km
Receiver noise figure	8 dB	10 dB
System losses	4 dB	6 dB
Peak power	10 W	2.5 W
Antenna aperture diameter	10 cm	10 cm
Beamwidth	2.3°	1.6°
Antenna gain	36.9 dB	40.1 dB
Target cross-section for tank size	30 m^2	30 m^2
Pulsewidth	50 n sec	50 n sec
Matched filter bandwidth	20 MHz	20 MHz

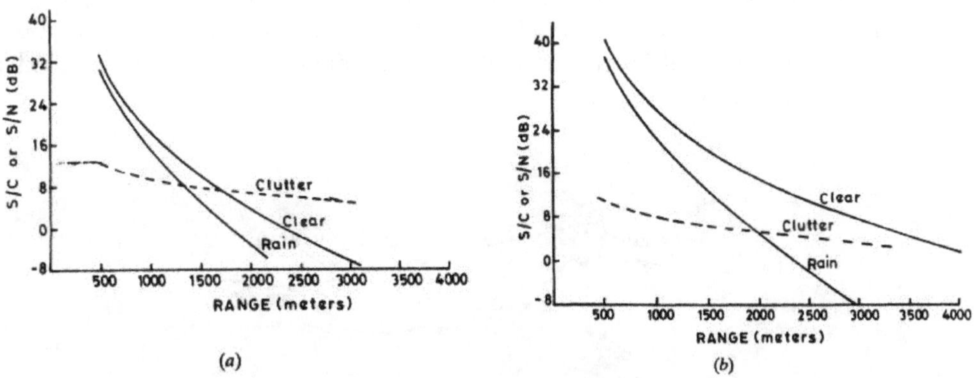

FIGURE 9.22 Range performance of (*a*) 94 and (*b*) 140 GHz radars.

9.25 MISSILE GUIDANCE AND SEEKER SYSTEMS

Millimeter wave radars are now widely employed for precision guidance of missiles. In the final phase of the attack, the missile-borne millimeter wave radar seeks for the target by scanning. Once the seeker radar gets the target within its field of view, the radar antenna beam locks on to it, and the missile subsequently tracks down to the target to hit it precisely without failure. Often during the final phase of the attack, the missile radar is switched off and replaced by a passive sensor called a radiometer. This avoids the effects of "glint" which occurs with radars at close ranges leading to erroneous tracking of the target. The passive radiometric sensor is also useful in avoiding its detection by the enemy. It can precisely track on to the target by a conical scanning of the thermal emission noise pattern of the target. The various types of missile guidance seeker and tracker systems are highlighted in the following sections.

9.26 BEAM RIDER

A millimeter wave beam rider missile guidance system is illustrated in Figure 9.23. The missile after its launch proceeds toward the antenna beam (1) and on reaching it, picks up the millimeter wave transmission along the beam. The missile electronics then guide the missile along the maximum of the beam. As the missile proceeds toward the target, the antenna beam is progressively made sharper in stages to produce beams (2) and (3) successively. This ensures precision of guidance and attack. A broad first beam facilitates acquisition of the missile into the beam near the launch site, while the sharper antenna beams during the later phases of the guidance not only ensure precision of attack but also ensure adequate strength of the signal at the missile receiver right up to the final phase for the beam rider electronics to be actuated.

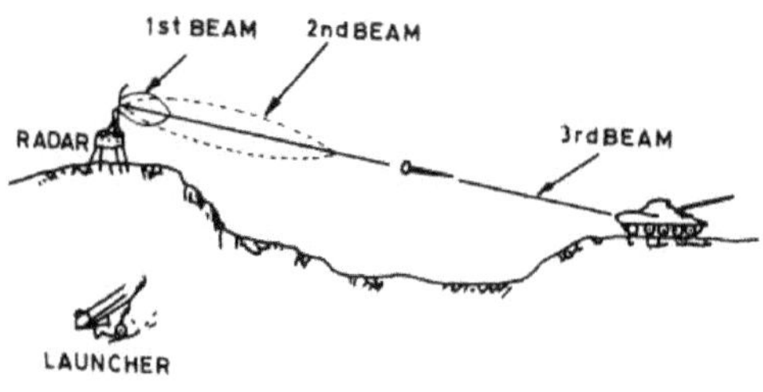

FIGURE 9.23 A mm-wave beam rider missile guidance system.

9.27 MISSILE SEEKER

A missile seeker is equipped with an onboard radar, which may be a pulse radar sensor or an frequency modulated continuous-wave (FMCW) radar sensor. The radar search beam of a seeker and its trajectory for terminal guidance are shown in Figure 9.24, while the details of a seeker search mode operation are shown in Figure 9.25. During its descent when the target approaches within the range capability of its radar sensor, the seeker antenna beam starts a scanning sequence, within the search cross range covering the target area, in a zigzag or raster scan, as shown in Figure 9.24 (*a*). At the same time the beam also executes a conical scan (Figure 9.24), until it passes through the target and locks on to it by the seeker radar electronics. The missile then proceeds or homes toward the target, tracking the target all the time (Figure 9.24 [*b*]). The zigzag search is made around the *Y*-axis, while the roll axis of the missile is vertically stabilized, as shown in Figure 9.24 (*b*). The angular error between the target line of sight (LOS) and the antenna axis of the seeker radar is used to orient the antenna by a rate stabilized LOS tracker servo system.

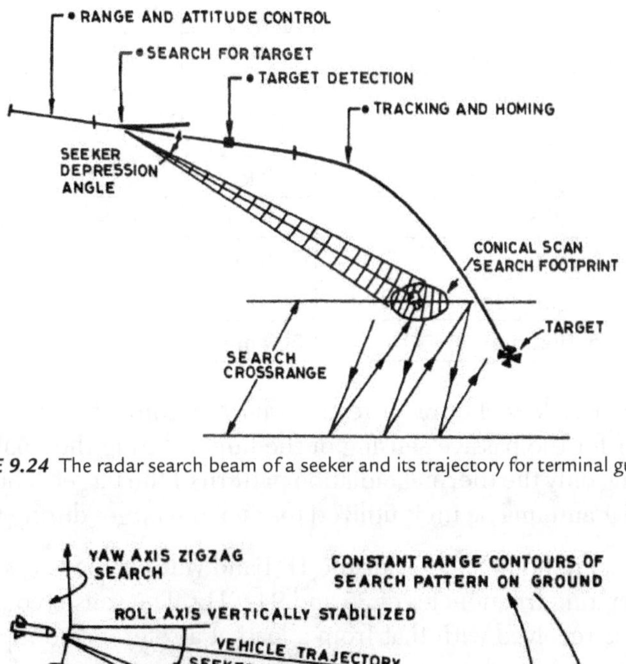

FIGURE 9.24 The radar search beam of a seeker and its trajectory for terminal guidance.

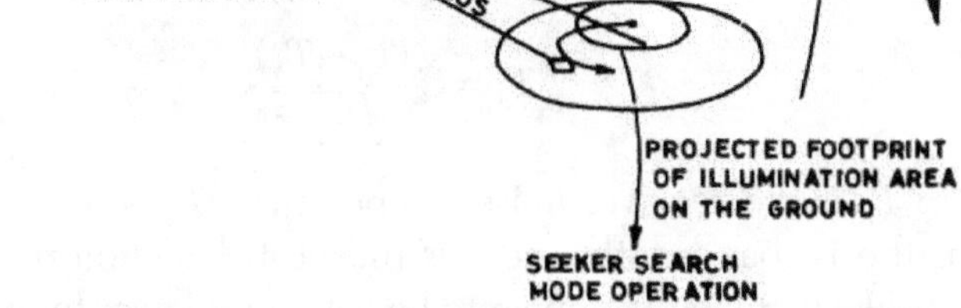

FIGURE 9.25 Details of a seeker search mode operation.

The parameters of missile seeker radars at 35 and 94 GHz bands based on a pulse radar sensor and an FMCW radar sensor are shown in Table 9.6.

TABLE 9.6

Parameters	Pulse sensor		FMCW sensor	
Frequency	35 GHz	94 GHz	35 GHz	94 GHz
Average transmitted power	10 mW	5 mW	100 mW	50 mW
Peak transmitted power	2 W	1 W	—	—
Transmitted bandwidth	300 MHz	300 MHz	500 MHz	500 MHz
Receiver IF bandwidth	300 MHz	300 MHz	200 MHz	200 MHz
Antenna diameter	16.25 cm	16.25 cm	16.25 cm	16.25 cm
Antenna efficiency	0.6	0.6	0.6	0.6
Antenna beamwidth	3.6°	1.3°	3.6°	1.3°
Transmitted pulsewidth	0.1 sec	0.1 sec	—	—
Pulse repetition frequency	50 kHz	50 kHz	—	—
Target RCS	50 m^2	50 m^2	50 m^2	50 m^2
Scanning loss	0.10	0.10	0.10	0.10
Noise figure of receiver (SSB)	9 dB	12 dB	9 dB	12 dB
Integration time	500 m sec	70 m sec	500 m sec	70 m sec

The missile seeker radar receiver may be converted to a total power radiometer for the passive sensing of the target during the final phase of attack by using only the thermal radiation pattern of the target. The conical scan of the radar antenna beam is utilized to track the target during the passive sensing.

In the radiometric sensor the IF bandwidth is made 500 MHz for both the operating frequencies of 35 and 94 GHz. The sensor compares the emission noise received with that from a load at a reference temperature. The sensitivity of the radiometer in detecting the thermal map of the target increases with the IF bandwidth and post-IF detector time constant. The minimum detectable temperature differential can be shown to be given by

$$\Delta T_A = C\frac{(T_R + T_A)}{\sqrt{B_{IF}\,\tau}}$$

where T_R is the receiver noise temperature, T_A is the antenna temperature, B_{IF} is the IF bandwidth, and τ is the post-detection time constant. Military targets like metal tanks appear to be colder (at lower temperatures) compared to the background, as it reflects the sky noise emissions at lower temperatures.

The thermal map of a typical tank for the battlefield measured by a 35 GHz radiometric sensor is shown in Figure 9.26. A radiometric sensor can also be operated in a active mode by illuminating the target with a wideband noise transmission in the millimeter waveband chosen. This increases the sensitivity of the radiometric sensing considerably. With active radiometric sensing the metal objects, like tanks, will reflect more of the incident noise power due to higher reflectivity, and would, therefore, appear to be hotter (at a higher temperature) at millimeter waves.

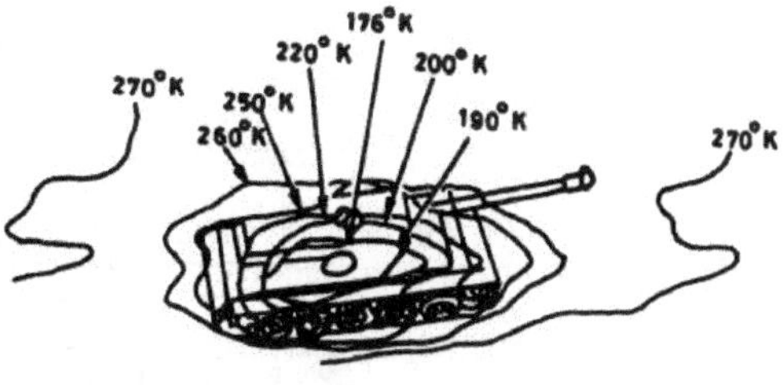

FIGURE 9.26 Thermal map of a battlefield tank measured at 35 GHz by a radiometric sensor.

9.28 CONFIGURATIONS OF MISSILE SEEKER SENSORS

Block diagrams of a pulse radar sensor, an FMCW sensor, and a radiometric sensor for a missile seeker are shown in Figures 9.27, 9.28, and 9.29 respectively. In the pulse sensor a ferrite circulator serves as the duplexer. The transmitter consists of a pulse modulated IMPATT oscillator. The active/passive selection as well as the search/track selection are made with SPDT IF switches.

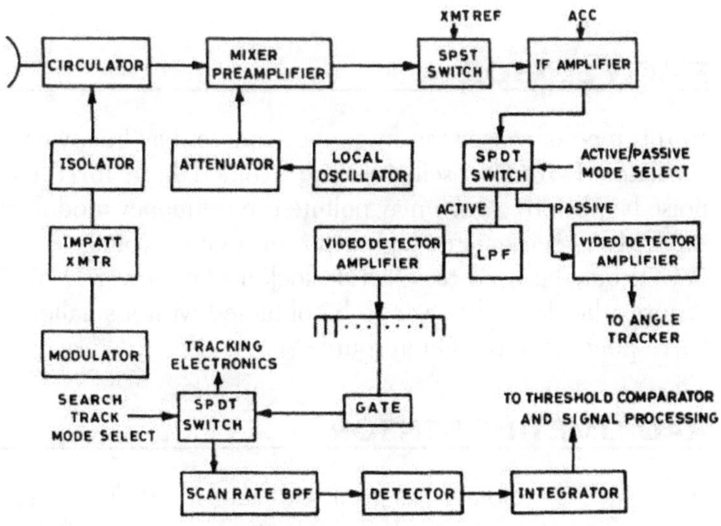

FIGURE 9.27 A pulse radar sensor.

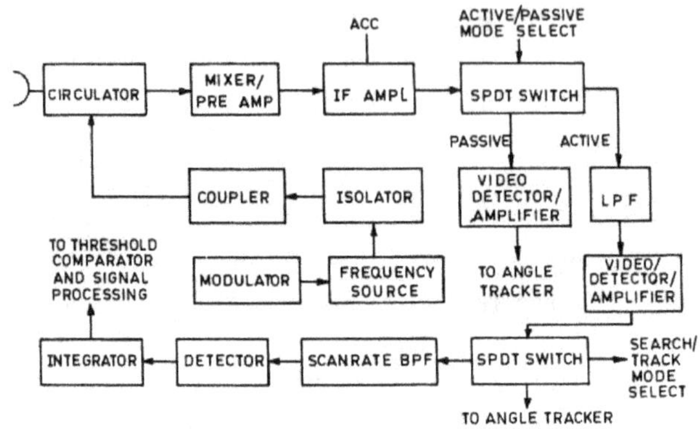

FIGURE 9.28 A pulse radar sensor.

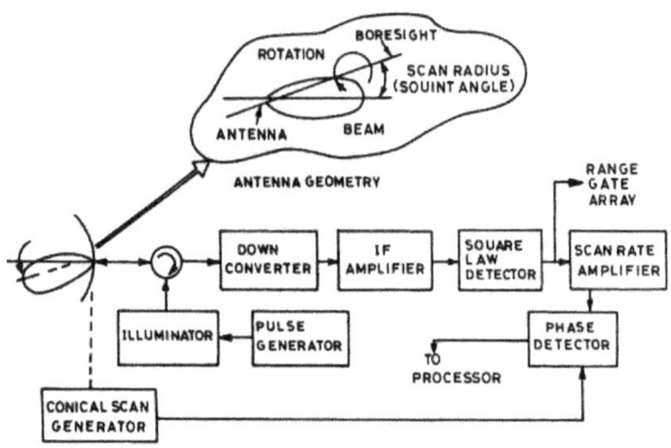

FIGURE 9.29 A radiometric sensor.

9.29 FMCW SENSOR

In this type of sensor the frequency source may be a varactor-tuned Gunn oscillator. IMPATT oscillators are avoided here for their wider spectral noise bandwidth, which may pollute the frequency modulation sidebands in the FMCW sensor. For higher power a Gunn voltage controlled oscillator (VCO) may be used to injection lock a high-power IMPATT source. This ensures the desired power to be obtained with a smaller noise bandwidth corresponding to the Gunn source.

9.30 RADIOMETRIC SENSOR

It is an active radiometric sensor with a noise transmitter radiating from the same radiometer antenna. If the noise transmitter is switched off, it becomes a passive radiometric sensor.

9.31 POWER SOURCES FOR MILLIMETER WAVE RADARS

For radar applications, the power output capabilities of millimeter wave solid-state sources like IMPATT devices decrease rapidly with increasing frequencies, as shown in Figure 9.30. The highest power devices for millimeter wave radars are the gyrotrons. The variations in the output power of gyrotrons, with frequency, are shown in Figure 9.31. More popular high-power millimeter wave sources include the magnetrons and extended interaction oscillators (EIO). It may be mentioned here that power levels even higher than that of gyrotrons have been generated by cyclotron masers driven by intense relativistic electron beams. The characteristics of various solid state and vacuum electron devices are summarized in Tables 9.7 through 9.10.

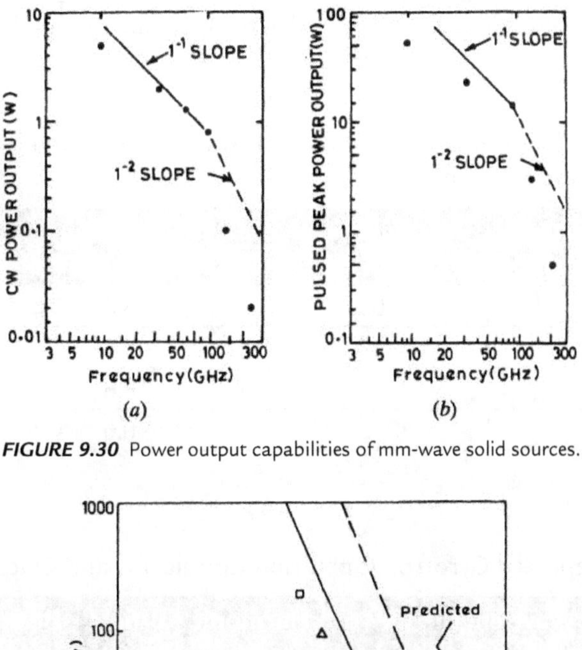

FIGURE 9.30 Power output capabilities of mm-wave solid sources.

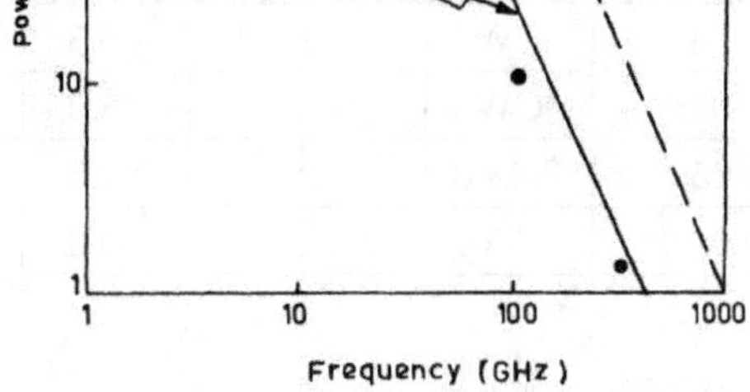

FIGURE 9.31 Variation of power output of gyrotrons with frequency.

TABLE 9.7
U.S.A. Millimeter Wave Radar Typical Transmitter Output Power (Peak)

Power source	Frequency (GHz)			
	35	95	140	220
Magnetron	50–100 kW 2–3 kW	1–6 kW	—	—
Extended interaction oscillator	200 kW	1–2 kW	200 W	60 W
Gyrotron*		20 kW	9 W	1 kW
Solid state	18 W			
(Gunn, IMPATT)		10 W	2.5	0.4 W

*Higher power levels have been reported in the literature for this device.

TABLE 9.8
U.S.S.R. Peak Power Levels from Cyclotron Masers Driven by Intense Relativistic Electron Beams

Wavelength (cm)	Peak microwave power (MW)	Accelerating voltage (MV)	Diode current (kA)
4	900	3.3	80
2	350	2.6	40
0.8	8	0.6	15
0.4	2	0.6	15

TABLE 9.9
U.S.S.R. Reported Gyrotron Operating Conditions and Output Parameters

Mode of oscillation	Wavelength (mm)	CW or pulsed	Harmonic number	B-field (kG)	Beam volts (kV)	Beam amps	Output power (kW)
TE_{021}	2.78	CW	1	40.5	27	1.4	12
TE_{031}	1.91	CW	2	28.9	18	1.4	2.4
TE_{231}	1.95	Pulsed	2	28.5	26	1.8	7
TE_{231}	0.92	CW	2	60.6	27	0.9	1.5

TABLE 9.10
U.S.A. Characteristics of Some Pulsed Millimeter EIOs

Tube type	Mech. tuning range (GHz)	Power output (peak, W)	Beam voltage WRT cathode (kV)	Anode voltage WRT cathode (kV)	Electronic tuning range (MHz)
VKV 2443	92.7–96.0	950–1770	21	12	300–360
VKT 2419	139.7–140.3	270	20	7.6	370
VKY 2429	225.5	70	21.3	8.2	400

9.32 JAMMING AND ANTI-JAMMING TECHNIQUES

Electronic warfare (EW), such as that experienced recently during the Gulf War in the Middle East, is based on a wide variety of electronic counter measures (ECM) like jamming of radars by high power radiation, electronic counter-counter measures (ECCM) like that adopted in anti-jamming or jamming-proof radars, and electronic support measures (ESM) like the wide bandwidth reconnaissance surveillance-warning receivers, often called Early Warning receivers, and direction finders. These are summarized in the following sections.

9.33 ELECTRONIC COUNTER MEASURES

The techniques for rendering the enemy radars ineffective in detecting targets are called electronic counter measures (ECM). The techniques such as jamming the enemy radars by powerful radio beams directed toward their radars are called active ECM. Passive ECM includes chaffs or needles of metal thrown out to scatter the enemy radar beams away from the real target. Other camouflage screens, false targets, and decoys (target like) are used to confuse the radar detection by the enemy.

Active ECM reduces the detection range of a radar drastically. The maximum range of a pulse radar in the absence of jamming was given by equation (3.10) as

$$r_{max} = \left[\frac{P_t G^2 \lambda^2 \sigma}{(4\pi)^3 S_{min}}\right]^{1/4} \qquad (i)$$

Assuming the receiver noise power $N_r = FkTB = NB$

where $N = FkT$ is the noise power per hertz of bandwidth, we combine it with equation (9.22) to obtain

$$\frac{S_{min}}{N_r} = \frac{P_t G^2 \lambda^2 \sigma}{(4\pi)^3 r^4 BN} \qquad (9.22)$$

In the presence of a jammer using the done way range equation, the signal from the jammer, J, received by the enemy radar to the noise ratio will be given by

$$\frac{J}{N_r} = \frac{P_j G_j G \lambda^2}{(4\pi) R_j^2 B_j N} \qquad (9.23)$$

where the subscripts for P, G, R, and B refer to parameters of the jammer while G' is the radar antenna gain in the direction of the jammer.

Taking the ratio of equations (9.22) to (9.23) we have,

$$\frac{S_{min}}{J} = \frac{P_t G^2 \lambda^2 \sigma}{(4\pi)^3 r^4 B N} \Big/ \frac{P_j G_j G^2 \lambda^2}{(4\pi)^2 R_j^2 B_j N}$$

$$= \frac{P_t G^2 \sigma R_j^2 B_j}{4\pi r^4 B P_j G_j G'}$$

$$\therefore \qquad r^4 = \frac{R_j^2 \sigma P_t G^2 B_j}{4\pi P_j G_j G' B S_{min}} \qquad (9.24)$$

The range r given by equation (9.24) in the presence of a jammer is called the burn-through range or minimum effective range of the jammer, so that for ranges greater than R, the enemy radar cannot detect targets of cross section σ or less. For a self screening jammer, for which the jammer and target for the enemy radar are collocated, we have $R_j = r$ and $G = G$, when equation (9.24) reduces to

$$r^2 = \frac{\sigma P_t G B_j^J}{4\pi P_j G_j BS_{min}} \qquad (9.25)$$

It may be noted here that in deriving equations (9.24) and (9.25) we have not assumed receiver signal processing factors and other factors which may reduce the effective jamming. Missile seekers which guide the missiles to the target use low cost airborne radar sensors with wide spectrum sources as the transmitter to reduce the target glint that would occur if the sources were of the continuous wave type. The wide bandwidth of the onboard missile seeker receiver makes it more valuable to jamming.

The effectiveness of an ECM system can be increased significantly by using high-tech intercept receivers to identify characteristics of the threat signal from the enemy radar. Identification of the threat signal from the jammer site requires wideband, high-sensitivity receivers, which determine the enemy radar frequency, location, waveforms, and so forth. Such receivers meant for early warning (EW) applications are being developed. These receivers serve as electronic support measures (ESM).

For high-power jamming, which may produce overloading and saturation of the enemy radar receivers, rather than producing simple interference, high-power sources like gyrotrons and cyclotron masers have great potential. Typical power levels of such sources are included in Tables 9.8 and 9.9, although for usual jamming interference levels, Magnetrons or Extended range Interaction Oscillators (EIO), included in Tables 9.7 and 9.10, are adequate.

Jamming by radio noise is often more effective and more difficult to overcome by the electronic counter-counter measure (ECCM) than the jamming by CW radiation. For CW jamming it may be possible to attenuate or reject the jamming signal by suitable filter at the RF or IF stages of the radar receiver, which is not possible with wideband noise jamming.

9.34 ELECTRONIC COUNTER COUNTER MEASURE

The most effective techniques of overcoming ECM produced by jamming is to use frequency agility, which changes the radar operating frequency in an unpredictable manner. The jamming producing the energy may, in fact, be overcome by frequency agility, which would force the enemy to spread the bandwidth of the noise jammer over a wide range, tending to reduce the noise power per unit bandwidth.

Another powerful technique of ECCM would be to reduce the beamwidth of the radar antenna, which is possible to achieve at millimeter waves even with an antenna diameter of moderate size. The beamwidth θ is given by $\theta = 1.22\ \lambda/D$ where D is the diameter of the antenna and λ is the wavelength. The omnidirectional antenna will have a gain of unity, and as such will receive the observed echo signal in negligible proportion.

Other ECCM techniques include: (*a*) Frequency diversity using two or more radars at different radar bands for the jamming to spread power over the entire band available to a radar, (*b*) multiple spaced radars operating at the same radar band, and (*c*) use of millimeter wavelengths.

9.35 REPEATER JAMMING AND ECCM

A radar repeater located at the target may be used to generate false echoes by delaying the received radar signal and retransmitting at a slightly later time. The delayed echo from the repeater will appear at a range different from real target, which is thus protected from attack by the enemy.

A repeater jammer can also be used to "break the lock" of a tracking radar on the target. For locking, the echo in a tracking radar is maintained between two range gates, which follow the motion of the targets after the initial lock is achieved. If the jamming signal is stronger than the echo signal, the radar tracking circuits will follow the false signal from the jammer. This is achieved by initially transmitting a single pulse and then slowly shifting the timing of its own pulse transmissions to cause apparent change in the target range.

CHAPTER 10

NAVIGATIONAL AND REMOTE SENSING RADARS

10.1 INTRODUCTION

Although radar was originally developed for detecting enemy aircraft during World War II, it was later employed extensively for navigation of aircraft and ships. Radar called airborne radar may be installed in an aircraft for this purpose. Radar called air surveillance radar (ASR) is installed at airports to obtain a panoramic view of the aircraft flying within the maximum range of the radar and also to track aircraft for air traffic surveillance and control. For guiding an aircraft in landing by supplying the pilot with positional information about the aircraft as viewed by the radar, a type of radar called precision approach radar (PAR) is also installed at the airport. Besides these, with a view to warning an aircraft with a forecast for an impending weather disturbance like a thunderstorm or a cyclonic storm, a special radar called meteorological radar (MET radar) is also installed at the airport. Ground-based radars are also employed at seaports and on ships for aiding marine navigation. The various types of radars employed in civil aviation and marine navigation are discussed in this chapter. Besides the navigational radars, a recent technique of applying radars in the remote sensing of land and sea from satellite and from the ground will also be covered in this chapter.

10.2 AIRPORT RADARS

Aircraft and ground vehicular traffic at large airports are monitored by high-resolution radars called Air Surveillance Radar (ASR) for safety controlling of air traffic in and around the airport. Guidance of aircraft to a safe landing in bad wheather is often made by a Precision Approach Radar (PAR) provided with ground control approach (GCA) systems. In addition, radar-beacons are also installed in airports for interrogation by the airbone Distance Measuring Equipment (DME), which is also a part of the Microwave Landing System (MLS), and is employed during bad weathers.

10.2.1 Air Surveillance Radar

Two different types of air surveillance radars are in use. One is the *L*-band, 200 nmi Air-Route Surveillance Radar (ARSR), while the other is the *S*-band 60 nmi Airport-Surveillance Radar (ASR) for obtaining information in the vicinity of airports to be sent to the aircraft. ASR is also called Surveillance Radar Element (SRE). The characteristics of these two types of Air Traffic Control (ATC) radars, each provided with MTI facility, are shown in Table 10.1.

TABLE 10.1
Typical Characteristics of ATC Radars

ATC Radar parameters	ARSR	ASR
Frequency	***L* band**	***S* band**
	(1250–1350 MHz)	(2700–2900 MHz)
Range	200 nmi (370 km)	60 nmi (111 km)
Peak power	5 MW	1.4 MW
Pulsewidth	2	1.6
Pulse repetition frequency	310–365 Hz	700–1200 Hz, 1040 (average)
Average power	3.6 kW	0.875 kW
Noise figure of receiver	4 dB	4 dB
Antenna size	12.8 m by 6.9 m	4.9 m by 2.7 m
Horizontal beamwidth	1.25°	1.35°
Vertical coverage	40°	30°
Antenna gain	34 dB	30 dB
Polarization	horizontal, vertical or circular	vertical or circular
Antenna rotation rate	5 RPM	12.8 RPM
MTI improvement parts	39 dB	34 dB
Blind speed	1200 knots	800 knots

The inverse fourth power variation of echo signal power in a radar with range makes small nearby targets such as birds and insects produce strong echoes appearing as clutter. This can be reduced drastically by shaping the vertical (elevation) pattern of the antenna as well as by a time dependent control of sensitivity called STC (Sensitivity Time Control).

Beam Shaping. Shaping of the elevation pattern from the idealized rectangular form as shown in Figure 10.1 (*a*) to the cosecant squared antenna pattern shown in Figure 10.1 (*b*) will effectively reduce the antenna gain at a smaller range for a constant height of flight of an aircraft, in such a way as to produce a received echo power independent of range. In this way nearby clutter from targets at the height of flight of an aircraft. However, surface clutter received at low elevation angles may still obscure the echo from a desired remote target aircraft which never flies at such low elevation angles. To reduce the surface clutter, the antenna beam is further modified by reducing the antenna gain at low elevation angles as shown in Figure 10.2 (*a*). It may be noted that the gain for an elevation greater than 30°–40° is also made negligible, for the aircraft will be very near the airport for such high elevation angles when another type of radar, the precision approach radar (PAR), takes control.

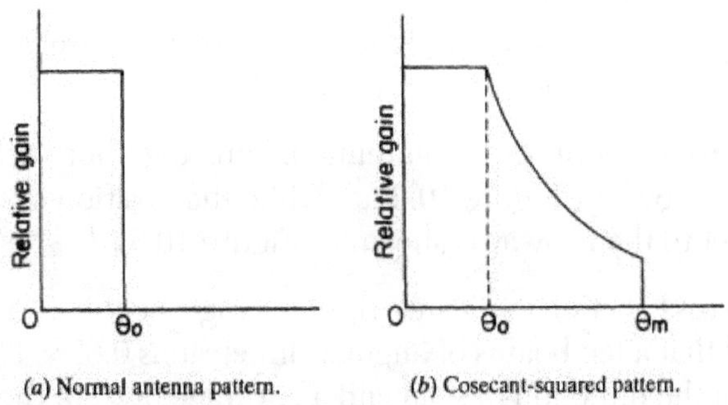

FIGURE 10.1 Idealized radar antenna patterns.

Sensitivity Time Control. In a radar set each transmitted pulse is returned back from the target and is received as an echo before the next pulse is transmitted with a delay proportional to the range of the target. Therefore, if the sensitivity or the gain of the receiver is increased progressively after the end of a transmitted pulse when the echo is expected, weaker echoes from remote objects will be amplified more, tending to make an echo signal independent of the range. The effective antenna pattern in presence of such sensitivity time control (STC) is shown in Figure 10.2 (*b*).

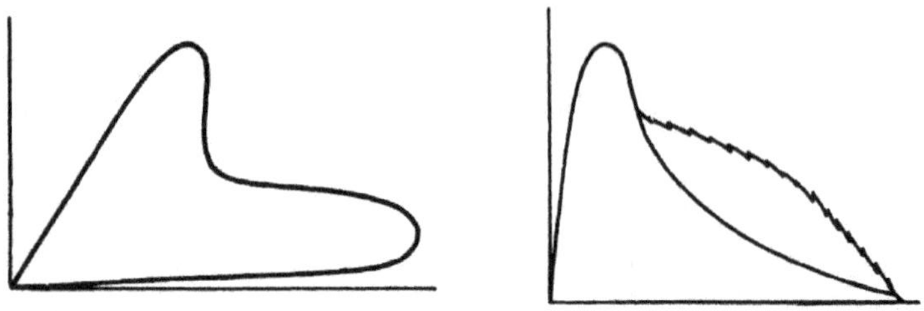

(*a*) Free space coverage diagram. (*b*) Comparison with consecant squared pattern.

FIGURE 10.2 Antenna elevation patterns for a long-range air-search radar employing STC.

10.2.2 Precision Approach Radar

A precision approach radar employs a sharp antenna beam capable of scanning both in azimuth and elevation for guiding an aircraft during its final landing phase to a point very near the runway. A typical scan width is 20° in azimuth and 7° in elevation. Two separate antennas are used for azimuth and elevation scanning. Practical considerations limit the scan width, as shown in Figures 10.3 (*a*) and (*b*). The information obtained by the PAR is displayed on two long-persistence cathode ray tubes, one displaying the range and elevation angle as in Figure 10.3 (*a*) and the other displaying the range and azimuth angle shown in Figure 10.3 (*b*). The lines of constant range are arranged to appear as straight lines instead of circles to facilitate measurement of angular position to the target.

The cross section of the antenna beams together with that of the scanned area is shown in Figure 10.3 (*c*), while the position of the PAR antenna with respect to the runway is shown in Figure 10.3 (*d*).

The cross section indicates the coverage of the PAR antenna. It may be noted that a fan beams of angular dimensions 0.6° × 4° ensures resolution of 0.6° in the direction of scan and a coverage of 4° in the perpendicular direction. Such fan beams are employed for both the azimuth and elevation scans. The steering of the beams is achieved by changing the relative phase of 209 dipoles spaced 1.92 cm apart, each one being fed from a waveguide whose dimension can be mechanically altered to produce the desired phase shift. Beam steering in phased array radars is often made by a similar technique. The advantage of the system is that the huge antenna array covering about 13 × 1.625 ft need not be steered as a whole but can be fixed permanently, while the squeezable waveguides effectively steer the beams. The mechanical problem involved in this is much simpler than the rotation of the arrays as a whole. The characteristics of a typical PAR are shown in Table 10.2.

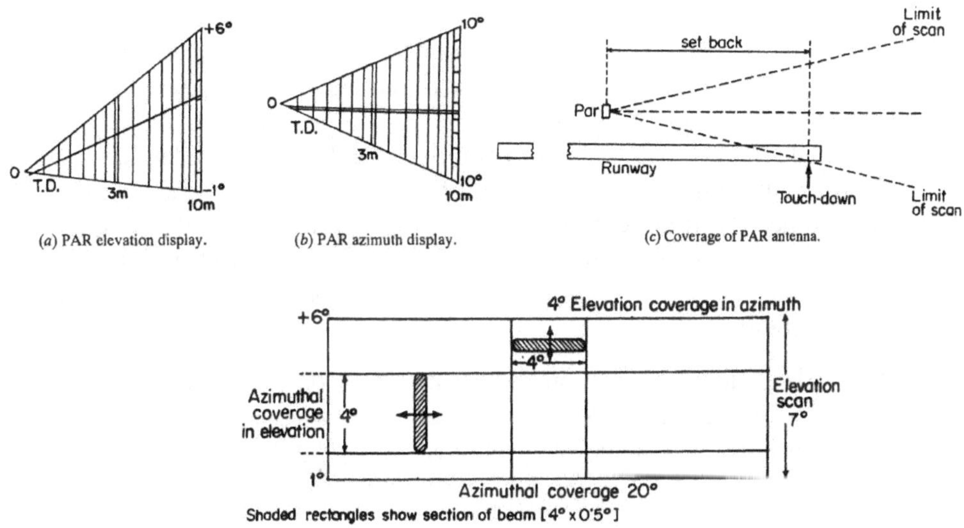

(*a*) PAR elevation display. (*b*) PAR azimuth display. (c) Coverage of PAR antenna.

(*d*) Position of PAR with respect to the runway.

FIGURE 10.3

TABLE 10.2
Typical Characteristics of PAR

Transmitted power (peak)	50 kW
Frequency	9080 MHz
Pulsewidth	0.18 sec
Pulse repetition frequency	3.825 kHz
Range discrimination	200 ft
Azimuth discrimination	0.6°
Elevation discrimination	0.6°
Range	10 nmi (18.5 km)

10.3 METEOROLOGICAL RADAR (MET RADAR)

For providing the pilot with up-to-date information about the weather around the airport, powerful meteorological radars are installed at the airport. Such radars can detect weather effects such as fog, rain, or snow, and can display advancing storm cloud formations. For this purpose, the frequencies chosen are in the *X*-band around 9 GHz, at which clutter due to raindrops, and the detectivity is about 34 dB higher than that in the *L* band, which is around 1.3 GHz. A MET radar can clearly display in PPI an advancing nor'easter or a cyclonic storm center within a range of about 300 nautical miles and can also provide the pattern of rainfall distribution over a vast area around the airport. Fog and hail are also detectable by the MET radars. Normally the scanning is made azimuthal with the horizontal antenna beam rotating all around at a certain rate to facilitate the PPI display in a long persistence CR tube. Often the range height display is needed to know the height of the clouds when the azimuthal scanning is halted, and the beam is instead steered in elevation by controlling the position of the feed horn of the parabolic dish.

Recently, a dual polarization radar capable of differentiating ice and rain using the differential reflectivity between the two has been developed and shows promise as a meteorological radar.

10.4 AIRBORNE RADARS

Airborne radars installed in aircraft are widely employed for Doppler navigation, in which the Doppler spectra of the land or sea returning from a number of directions are processed at the aircraft to obtain the velocity components of the aircraft. Other types of airborne radars include:

1. Distance Measuring Equipment (DME), which interrogates a radar beam at the airport serving as a secondary radar and measures the distance by receiving the reply pulse from the beacon,
2. FM-CW radar serving as a radio altimeter, and
3. A weather avoidance MET radar, which is a lightweight version of the airport MET radar.

10.5 DOPPLER NAVIGATION

The Doppler navigation system employs the Doppler effect to determine the velocity of the craft in a frame of coordinates fixed with respect to the aircraft. By combining this information with that obtained from a gyrocompass, the velocity with respect to the conventional earth coordinates is obtained.

Doppler Radar. For Doppler navigation a Doppler radar is carried in the aircraft. The radar directs several beams of electromagnetic waves toward the earth. The energy back scattered by the earth toward the aircraft is received and compared with the transmitted signal to ascertain the Doppler shift of frequency, which is proportional to the velocity component of the aircraft in the direction of the beam.

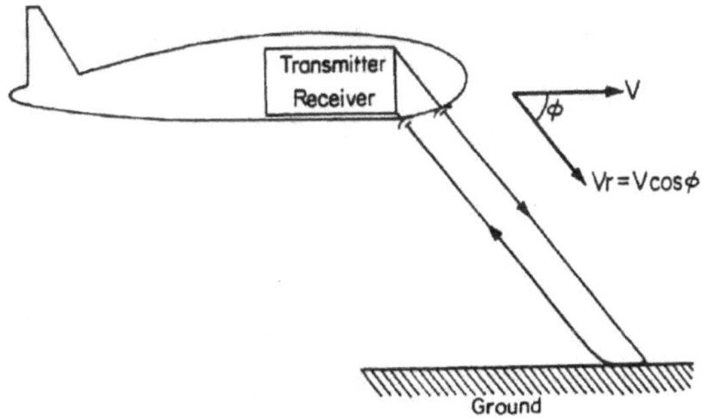

FIGURE 10.4 Doppler radar in an aircraft in level flight.

The principle of the Doppler radar is illustrated in Figure 10.4. If the velocity component of the aircraft in the direction of the beam is v_r, then for a transmitted frequency f_t, the frequency of the wave as measured at the ground, which if stationary will be $f_t\ (1 + v_r/c)$, where c is the velocity of electromagnetic waves. When this wave, reflected from the ground, is received at the aircraft, the frequency is increased further in the ratio $1 + v_r/c$. The received frequency f_r is, therefore, given by

$$\begin{aligned} f_r &= f_t\left(1+\frac{v_r}{c}\right)\cdot\left(1+\frac{v_r}{c}\right) = f_t\left(1+\frac{v_r}{c}\right)^2 \\ &= f_t\left(1+\frac{2v_r}{c}+\frac{v_r^{\,2}}{c^2}\right) \end{aligned} \tag{10.1}$$

In practice, $v_r << c$ so that $v_r^{\,2}/c^2$ may be neglected and the Doppler shift f_D is given by

$$f_D = f_r - f_t \cong f_t\frac{2v_r}{c} = \frac{2v_r}{\lambda} \tag{10.2}$$

where λ is the wavelength of the transmitted signal. The received signal at f_r is mixed with a part of the transmitted signal at f_t to produce a beat frequency signal, $f_D = f_r - f_t$. If the beam is directed backward, then the aircraft with its radar will be moving away from a fixed point on the ground, and v_r will be negative when the Doppler shift is also negative.

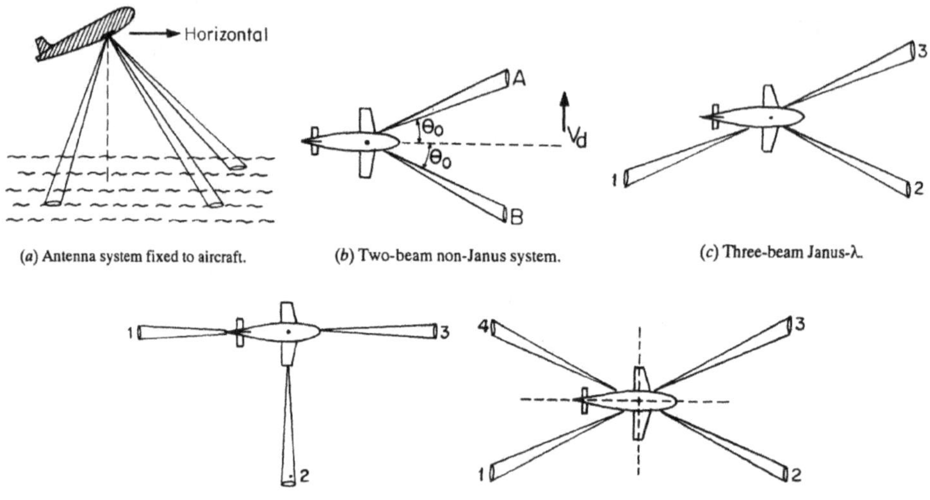

(*a*) Antenna system fixed to aircraft. (*b*) Two-beam non-Janus system. (*c*) Three-beam Janus-λ. (*d*) Three-beam Janus-T. (*e*) Four beam Janus (Janus-X).

FIGURE 10.5 Doppler radar beam Configurations.

Multibeam Configurations. A number of radar antenna beams directed simultaneously in different directions toward the ground, as shown in Figure 10.5 (*a*), allows the determination of the velocity components of the aircraft from the Doppler shifts of frequency measured at the aircraft. The plan views of the radar beams for the various multibeam configurations are shown in Figures 10.5 (*b*), (*c*), (*d*), and (*e*).

Errors in Systems. The two-beam system shown in Figure 10.5 (*b*) is subjected to errors in the presence of pitching, which causes changes in the altitude or orientation of a beam with the horizontal plane. Also a lateral drift velocity of the aircraft, as shown in the figure, causes the Doppler shift of *A* to increase while that of *B* decreases, by equal amounts, in proportion to the drift velocity. From the average of the Doppler shifts, the heading velocity of the aircraft may be obtained, while from the difference of Doppler shifts the lateral drift velocity can be derived. It may be mentioned here that a vertical velocity component changes both the Doppler frequencies, which may wrongly appear to be as due to a forward or backward velocity depending on the sign of the Doppler shift, unless the information of vertical velocity is also obtained by other navigational aids like the radio altimeter.

The Janus system, named after a Greek god who could look both ways, minimizes the errors due to change in altitude of the aircraft arising from pitch and roll. In the system, the sign of the Doppler shift of the rear beam is, in fact, opposite to that of the front beam. In level flight the two shifts are equal and their magnitudes are added during the computation of the velocity components. In the presence of a pitch angle making the front of the aircraft lower than its rear, the front beam depression angle increases, while the back beam depression decreases. This tends to decrease the magnitude of the Doppler shift of the front beam and increase that of the rear beam, tending to keep the sum of the magnitudes unaffected by the pitch angle. With a stabilized antenna, the Doppler shift due to vertical velocity cancels out in the Janus system. The vertical velocity components of the shift in the forward and backward beams are equal and cancel out if the difference is computed.

Equations for Doppler Shift. The Doppler shift with any beam may be obtained by summing the components due to each of three orthogonal velocity components of the aircraft, resolved along the direction of the beam. Assuming the antenna system to be attitude stabilized, the three components of aircraft velocity are the vertical component V_v and two horizontal components V_H and V_D, V_H being along the longitudinal axis of the aircraft. Figure 10.6 (*a*) shows the velocity components for a single beam, while Figure 10.6 (*b*) shows that for the four beams of a Janus-*X* system. From Figure 10.6 (*b*), we have the velocity components along beams 1, 2, 3, and 4 given by

$$v_1 = -V_H \cos \gamma_H + V_D \cos \gamma_D \cos \gamma_V \tag{10.3}$$

$$v_2 = V_H \cos \gamma_H + V_D \cos \gamma_D + V_V \cos \gamma_V \tag{10.4}$$

$$v_3 = V_H \cos \gamma_H - V_D \cos \gamma_D + V_V \cos \gamma_V \tag{10.5}$$

$$v_4 = -V_H \cos \gamma_H - V_D \cos \gamma_D + V_V \cos \gamma_V \tag{10.6}$$

where γ_H, γ_D and γ_V are angles of the velocity components V_H, V_D and V_V respectively with the beams. The direction cosines of the beams, for a perfect symmetry of the four beams, are given by

$$\cos \gamma_H = \cos \alpha_0 \cos \theta_0 \,;\; \cos \gamma_D = \cos \alpha_0 \sin \theta_0 \,;\; \cos \gamma_V = \sin \alpha_0$$

where α_0 is the depression angle of the beam.

From the measured Doppler shifts f_{D1}, f_{D2}, f_{D3} and f_{D4} the respective velocity components v_1, v_2, v_3 and v_4 are obtained from equation 10.2, by multiplying each by the factor $\lambda/2$. The velocity components V_H, V_D and V_V are then determined by solution of the simultaneous equations (10.3) to (10.6). The solutions are:

$$V_H = \frac{(v_2 + v_3) - (v_1 + v_4)}{4\cos\gamma_H} = \frac{\lambda}{8}\frac{(f_{D_2} + f_{D_3}) - (f_{D_1} + f_{D_4})}{\cos\alpha_0 \sin\theta_0} \tag{10.7}$$

$$V_D = \frac{(v_1 + v_2) - (v_3 + v_4)}{4\cos\gamma_H} = \frac{\lambda}{8}\frac{(f_{D_1} + f_{D_2}) - (f_{D_3} + f_{D_4})}{\cos\alpha_0 \sin\theta_0} \tag{10.8}$$

$$V_V = \frac{(v_1 + v_2 + v_3 + v_4)}{4\cos\gamma_V} = \frac{\lambda}{8}\frac{(f_{D_1} + f_{D_2} + f_{D_3} + f_{D_4})}{\sin\alpha_0} \tag{10.9}$$

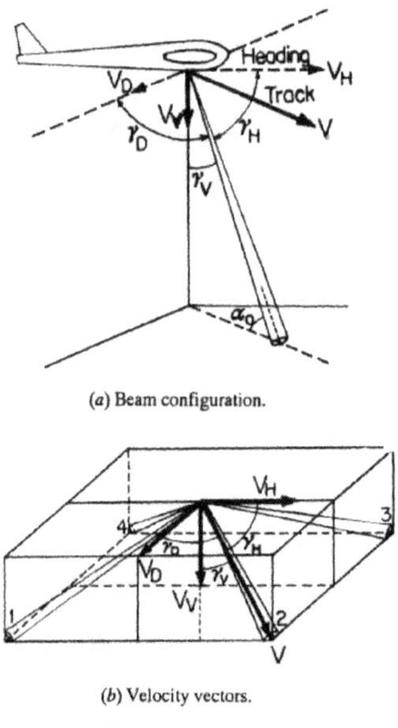

FIGURE 10.6 Geometry of one beam of a Janus X Doppler radar.

It may be mentioned here that any three of equations (10.3) to (10.6) are, in principle, sufficient to solve for V_H, V_D and V_V. However, when all four equations are considered, the solution is simple, as indicated in equations (10.7) to (10.9). Similar solutions may be obtained for the three-beam Janus-λ antenna system, using only the first three equations from equations (10.3) to (10.5).

Effect of Beam Angle on Sensitivity. The Doppler shift for a given velocity will be larger for a smaller value of γ_H, the angle between the longitudinal axis of the aircraft and the beam axis or beam centroid. However, a smaller value of γ_H makes the back scatter signal returned from the water surface smaller. A compromise value of the beam angle α_H is between 65° and 85°. The value of the angle between the longitudinal axis and the vertical plane through the beam centroid affects the sensitivity to lateral drift velocity, with larger values of θ_0 giving greater sensitivity to drift.

Heading and Track Stabilizations. An antenna system with its principal axis in the same vertical plane as the longitudinal axis or heading of the aircraft is called *heading-stabilized*, as in the previous cases considered. The horizontal components V_H and V_D are separately computed, from which the actual direction of motion or track of the aircraft is determined. The angle between the heading and the track is called the drift angle (δ) as given by

$$\tan\delta = \frac{V_D}{V_H}$$

or

$$\delta = \arc\tan\frac{V_D}{V_H} \tag{10.10}$$

If the antenna system is rotated about a vertical axis by a servo mechanism driven by the drift component V_D, the rotation will continue until the drift component becomes nil. The axis of the antenna is then oriented along the direction of the track, and the antenna is called *track-stabilized*. The track speed in this case is given by V_H only, V_D being zero.

Effect of Beamwidth on Doppler Shift. The antenna systems in practice have a finite beam width, typically 4°. Different parts of the beam within this 4° width correspond to different depression angles, causing the Doppler shifts also to be different. The various scattering elements in the area illuminated by the beam produce Doppler-shifted returns in random phase relationship to each other. The spectrum of the resulting return signal, therefore, appears like that due to a narrow band noise superimposed on the Gaussian noise background due to thermal emission from the ground and the atmosphere. To this is also added the internal receiver noise of the receiver at the aircraft, also Gaussian in nature, during the reception of the return. The spectrum of the Doppler-shifted return is typically as shown in Figure 10.7. The shape of the main peak of the spectrum is nearly Gaussian in

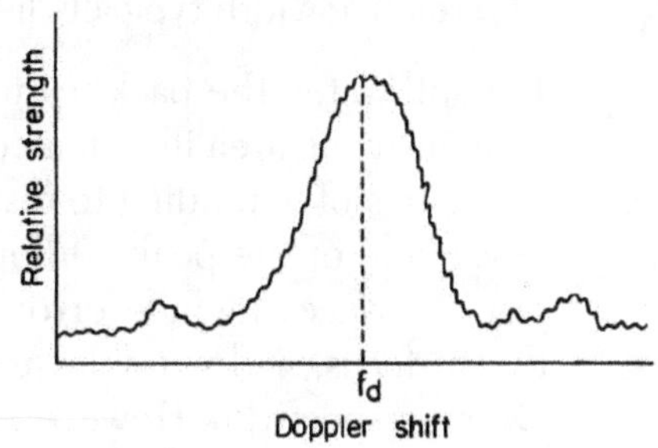

FIGURE 10.7 The Doppler spectrum of returns in a airborne Doppler radar.

nature. The amplitude of the peak is dependent on the radar parameter and the back scatter cross section of the terrain. The mean frequency of the spectral distribution is f_d, and the angle corresponding to the Doppler shift f_d defines the beam centroid.

Spectral Width. If we assume the scattering property of the terrain back along the beam to be uniform over the area illuminated by beam, and that the returns from the nearest and farthest illuminated points are nearly the same, the half power of the Doppler spectrum is given by

$$f_d = \frac{2V\cos\gamma}{\lambda} \tag{10.11}$$

If the half power beam width is $\Delta\gamma$, difference f_D between the upper and lower half-power points of the Doppler spectrum, f_{D1} and f_{D2} respectively, is given by

$$\begin{aligned}\Delta f_D = f_{D1} - f_{D2} &= \frac{2V}{\lambda}\left[\cos\left(\gamma \frac{\Delta\gamma}{2}\right) - \cos\left(\gamma + \frac{\Delta\gamma}{2}\right)\right] \\ &\cong \frac{2V}{\lambda}\sin\Delta\gamma\cdot\Delta\gamma\end{aligned}$$

The relative width of the Doppler spectrum is given by

$$\begin{aligned}\Delta f_{D_r} = \frac{\Delta f_D}{f_D} &= \frac{2V}{\lambda}\sin\gamma\cdot\Delta\gamma \Big/ \frac{2V\cos\gamma}{\lambda} \\ &= \frac{\sin\gamma\cdot\Delta\gamma}{\cos\gamma} = \tan\gamma\ .\ \Delta\gamma\end{aligned} \tag{10.12}$$

The relative width typically lies between 15% and 25%.

For still water, the back scatter cross section cannot be assumed to be uniform over the area illuminated by the beam, but it will increase for increasing values of γ, tending to distort the peak of the Doppler spectrum of the lower side of the peak. Other sources of error in Doppler navigation over water surfaces include errors due to water currents, winds blowing over the surfaces, and windblown particles, all these leading to changes in the Doppler spectrum. However, when knowing the surface wind velocity, automatic correction of the error involved may be incorporated.

10.6 DOPPLER NAVIGATION EQUIPMENT

Doppler navigation equipment is shown schematically in Figure 10.8 in which (*a*) and (*b*) illustrate the incoherent and coherent pulsed Doppler systems respectively. The second detector of the receiver is an envelope detector for the incoherent system while it is a coherent detector in the coherent system. The frequency of the local oscillator of the incoherent system is controlled by a part of the magnetron output through an automatic frequency control (AFC) circuit, in order that the drifts of the magnetron output may cancel out at the IF output of the balanced mixer. In the coherent system, on the other hand, the transmission and local oscillator frequencies are derived

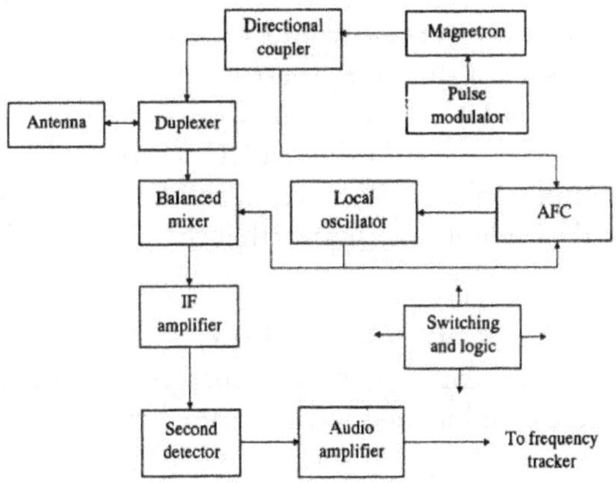

(*a*) Incoherent system.

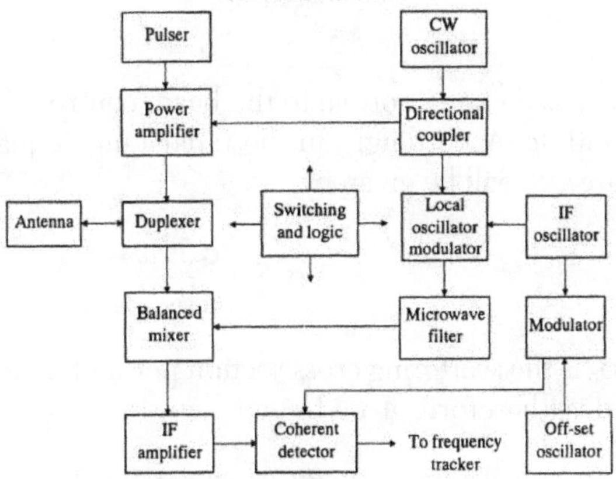

(*b*) Coherent system.

FIGURE 10.8 Pulsed Doppler systems.

from a single crystal-controlled CW oscillator, which is made to beat with a stable oscillator at intermediate frequency to synthesize the local oscillator frequency signal. In order that the coherent detector output may show up both the positive and negative Doppler shifts, the IF oscillator output is mixed with an offset oscillator to produce an output close in frequency to that of the IF, to serve as a local oscillator for the coherent detector. It may be mentioned here that the incoherent Doppler system cannot show up positive and negative Doppler shifts unambiguously.

At certain altitudes, depending on the pulse repetition frequency (PRF), the ground return of a pulse Doppler system for one pulse may arrive at or near the time of transmission of the next pulse or one of the succeeding pulses, causing gating out of the return at such times. The receiver is gated out because it is being blanked by the duplexer during the transmission of a pulse. In such a situation, the return is eliminated and the phenomenon is called an "altitude hole," which may produce serious errors at low altitudes, particularly for the high PRFs employed in the pulse Doppler systems. Altitude holes can be minimized by wobbling the PRF or by making the PRF dependent on range so that the returns do not arrive at the time of transmission of a pulse.

The Doppler Radar Range Equation. The range of a Doppler radar depends on the scattering area of the ground illuminated by a radar beam. As the beams are inclined to the vertical by an angle γ, the illuminated area is given by

$$S = \frac{r^2 w}{\cos\gamma}$$

where r^2w is the area normal to the beam centroid for a range r and angular beamwidth w. Accordingly, in the radar range equation, the effective scattering area, S_e will be given by

$$S_e = \sigma_0 S = \frac{\sigma_0 r^2 w}{\cos\gamma} \tag{10.13}$$

where σ_0 is the scattering cross section per unit area illuminated. The power received is, therefore, given by

$$P_r = \frac{P_T G_r}{(4\pi\, r^2)^2} \frac{\sigma_0 w r^2}{\cos\gamma} \cdot A_e \tag{10.14}$$

The antenna gain is related to the effective area A_e of the antenna by the relation

$$G = \frac{4\pi\, A_e}{\lambda^2},$$

while the antenna gain is also defined by

$$G = \frac{4\pi}{w}$$

combining these two relations for gain, we get

$$\frac{4\pi A_e}{\lambda^2} = \frac{4\pi}{w} \text{ or } A_e = \frac{\lambda^2}{w} \tag{10.15}$$

Substituting for A_e from equation (10.15) in equation (10.14), we get

$$P_r = \frac{P_T G_T \lambda^2 \sigma_0}{16\pi^2 r^2 \cos\gamma} \tag{10.16}$$

Dividing P_r or the signal power S by the inherent available noise power $N = kT_0BF$, we get

$$\frac{S}{N} = \frac{P_T G_T \lambda^2 \sigma_0}{16\pi^2 r^2 k T_0 BF \cos\gamma}$$

Here T_0 is the standard room temperature 290°

k is the Boltzmann constant

and B is the effective RF bandwidth as determined by the IF bandwidth.

It is interesting to note that the range equation (10.17) indicates an inverse square law dependence of (S/N) on range instead of the inverse fourth power law expected for a pulse radar detecting a target much smaller than the beamwidth. In the Doppler radar, because the target area is an extended one bigger than the beam, an increase in range also causes an increase of the illuminated area. This partly compensates for the loss of signal strength of the return from the ground for greater ranges.

The parameters of typical Doppler navigation equipment are shown in Table 10.3.

TABLE 10.3
Parameters of a Typical Doppler Navigation System

Transmitted power (peak)	1100 W
Average power	50 W
Frequency	8.7–8.9 GHz
Pulse repetition frequency (PRF)	50 kHz
Pulsewidth	0.9 μ sec
Limited speed for unambiguous range	1300 km (700 knots)
Altitude range	20,000 m
Accuracy of position	better than 1% of the distance traveled

10.7 DISTANCE MEASURING EQUIPMENT

Distance Measuring Equipment (DME) was developed in 1955 for civil aviation, along with the Tactical Air Navigation equipment TACAN. The DME is basically an airborne radar installed in the aircraft for interrogation of a secondary radar beacon of the airport, to measure the distance of the airport. In order that a number of aircraft may interrogate simultaneously a single radar beacon, the PRFs of the interrogators are made different. Each aircraft can isolate the responses of its own interrogations by using a time base triggered by its own transmitted pulse modulation. The responses produced by other aircraft will then appear as rapidly moving pulses along the time base axis, and can be easily recognized by a human observer as well as by automatic tracking systems, thus enabling isolation of the stationary pulse pattern due to the response of its own interrogations from the moving patterns due to other aircrafts.

Noise Pulse Modulation for Radar Beacons. As the transponder on the ground operates a single channel at a time, the responses due to interrogations by all the aircraft will be received by each aircraft. This causes a variation in the duty cycle due to a varying number of interrogating aircraft. The beacon transmitter is, therefore, subjected to varying loads. Also, since the receiver gain of the interrogator equipment is controlled by the average of the response pulses, the gain will be reduced drastically during busy hours when the number of interrogating aircraft is large. This may cause weak reply pulses to distant aircraft to be undetectable. To mitigate these difficulties the beacon transmitter is arranged to operate at nearly a constant duty cycle by trigger with locally generated noise pulses. However, to avoid stray responses by noise pulses, both the interrogator and the transponder operate with pulse pairs composed of two pulses spaced 12 μ sec apart. The pulse rate in the transponder is kept constant at about 3000 pulse pairs per second, and this rate is maintained even in the absence of interrogation pulses by triggering with the noise pulse pairs. This is illustrated in Figure 10.9.

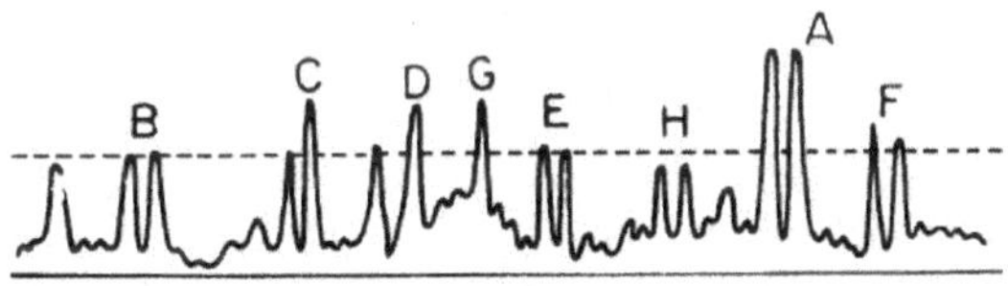

(*a*) Receiver output.

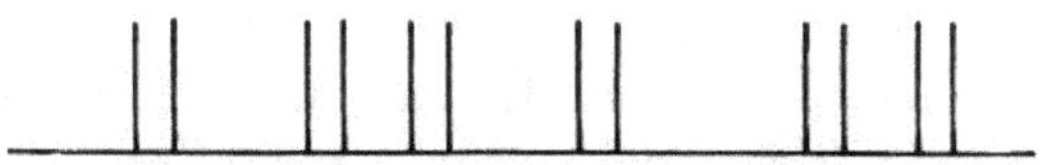

(*b*) Pulse pairs generated by the single pulses.

FIGURE 10.9 Generation of response and filter pulses in the transponder receiver.

The random noise pulses in the receiver output, in the absence of interrogation, will occur randomly. However, some of these pulses may have a spacing of 12 μ sec as in the pairs *A* and *F*, which exceed the triggering level to be recognized as a pulse pair spaced 12 μ sec apart. A single pulse is generated from each of these noise pulse pairs. Each single pulse is then converted to a pulse pair of constant amplitude, spaced 12 μ sec apart (Figure 10.9). These would falsely appear as response from the beacon. However, because of the random repetition rates of these pulse pairs, no stationary pulse pattern will be produced at the aircraft CRO having any regular time base of constant period. In the presence of interrogating pulses, the additional pulse pairs will not increase appreciably the duty cycle of the beacon in the presence of a large number of noise pulse pairs. However, those additional interrogating pulse pairs are regular in repetition rate and hence produce stationary patterns at the respective aircraft in which the time bases are triggered by their own modulation pulses for transmission.

Principle of Operation of DME. Distance measuring equipment (DME) essentially consists of two parts, an *airborne interrogator* in the aircraft and a ground-based *transponder beacon* at the airport, as shown in Figure 10.10.

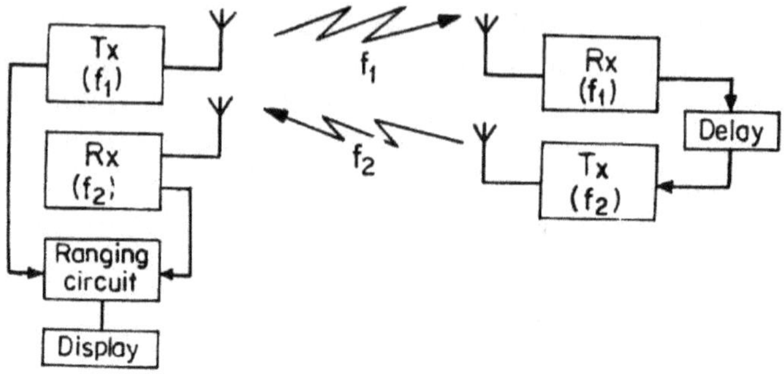

FIGURE 10.10 Components of a secondary radar.

The two parts taken together are called a secondary radar, to distinguish it from primary radars in which there is no transponder beacon, but replies or responses in the form of echoes only from passive targets are received to measure the distance of the target. In the DME, the interrogator transmits RF interrogation pulses at frequency f_1, to be received by the beacon receivers tuned to the same frequency. The receiver output, consisting of interrogation pulses as delayed by a small fixed amount, is used to modulate the beacon transmitter operating at a different frequency f_2, to avoid local feedback into the beacon receiver itself. The interrogator, on receiving the beacon transmission, measures the range by a measurement of the free space propagation delay of the received pulses from the transmitted ones.

Airborne Interrogator Equipment. The airborne DME interrogator is shown schematically in Figure 10.11. The transmitter is driven from a crystal-controlled synthesizer and frequency multiplier chain. The synthesizer can generate 126 spot frequencies around 45 MHz for transmission, after

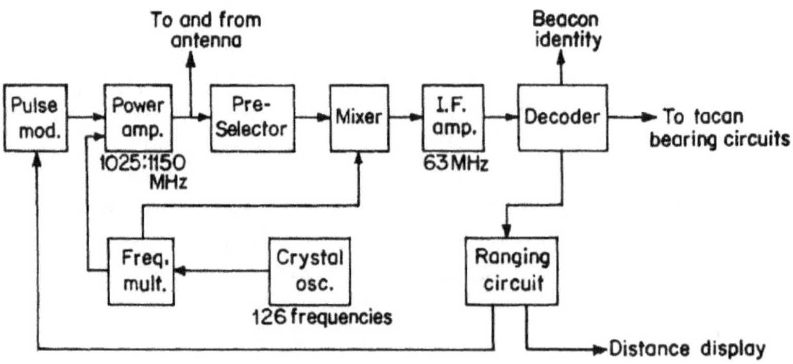

FIGURE 10.11 Airborne DME interrogator.

frequency multiplication, in the band 1025–1150 MHz. At the same time it also provides the oscillator signal to provide the 63 MHz IF signal in the receiver. The output power of the transmitter from the final power amplifier is between 50 W and 1 kW. The power amplifier is pulse modulated, the pulses being derived from a master pulse unit located in the ranging circuit unit. The IF amplifier output is demodulated and decoded in the decoder unit, the output of which is processed in the ranging circuit to obtain the delay displayed in terms of the range or distance of the beacon at the airport.

Search and Track Modes. The receiver of the interrogator can operate either in the search mode or in the track mode. In the search mode the demodulated pulse pairs due to the beacon reply signal are passed through a 20 μ sec gate, delayed after a transmitter pulse by a variable amount, increasing from 0 to 2400 μ sec. The waveforms of the search and track circuits are shown in Figure 10.12. The transmitter pulse waveform shown in (*a*) triggers the variable delay waveform shown in (*b*), which in turn, at the trailing edge, triggers the 20 μ sec gating pulse (*c*). The variable delay causes the gating pulse to move on the time axis at a rate controlled by the rate of increase of the duration of the variable delay waveform. If this rate is synchronized with the rate of occurrence of the received pulse pairs (*d*) converted to single pulses shown in (*e*), all the pulses will pass through the gate. On the other hand, if the rates differ, the number drops down drastically. Once the synchronization is obtained as indicated by a sudden increase in the number of pulses in gate output measured by a counter circuit, the system is switched to operate in track mode. The pulse operation rate is then dropped to 25 pulses per sec. If the reply falls on the early part of the gate, it advances. If it falls in the later part, the gate is delayed. In this way automatic tracking between the rates of the received and the gating pulses are achieved. A low interrogation rate is possible in the track mode since the possible change in the position of the aircraft is quite small in one interpulse period. The position of the aircraft as indicated by the delay of the gating pulse is displayed with an accuracy of 150 m.

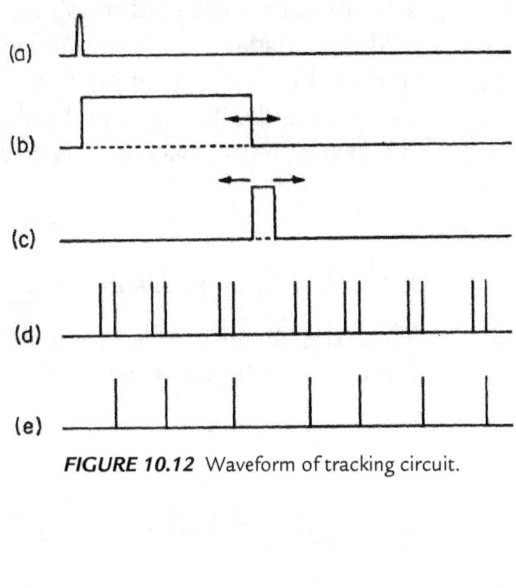

FIGURE 10.12 Waveform of tracking circuit.

Transponder Beacon. The transponder beacon equipment at the airport is similar to that of the interrogator except that no ranging circuit is required and it operates at only one pair of frequencies in place of the 126 pairs provided for transmission and reception at the interrogator. One frequency of the pair is employed for reception, while the other is for transmission from the transponder beacon. Different such pairs of frequencies are allocated for different airport radar beacons known to the pilot. He can, therefore, select his interrogator frequency pairs in conformity with that of the airport in which he desires to land. In addition, because the transponder beacon is ground based, more powerful transmitters, more sensitive receivers, and bigger antennas are used. The beacon transmits identification signals for a period of 3 sec once every 37 seconds, when the normal chain of random noise pulses are suppressed.

10.8 NAVY RADAR

A Navy Radar or Marine Radar is designed to detect targets over the sea, such as aircraft, ships buoys, icebergs, and land-sea boundaries. The ability of the radar to detect the targets on or above the sea is, however, limited by clutter interference from sea waves. Application of MTI (Moving Target Indication) will enable aircraft to be easily detected in sea clutter. For detection of ships, which usually have a large radar cross section, no MTI facility is essential and a moderate angular resolution will be adequate, while for the detection of small objects like buoys a high resolution radar free from saturation by large clutter will be needed. Besides these, the sea clutter in a Marine Radar may be minimized by a judicious choice of operating frequency. For instance, at grazing angles the sea clutter is lower at lower frequencies and with horizontal polarization. However, the elevation angle of the lowest interference lobe is given by

$$\alpha = \lambda/4h_e$$

where λ = wavelength and h_e = antenna height.

Thus, the longer wavelength involved at lower frequencies prevents targets low on the water to be detected, while a higher frequency may be preferred even though it increases sea clutter. The higher frequency allows the angular resolution to be higher, as the beamwidth is given by

$$\theta = 1.22\frac{\lambda}{D}$$

Where D is the diameter of aperture of the antenna and, therefore, shorter wavelengths at higher frequencies make the beamwidth smaller, making the beam sharper. Nevertheless, marine radars are designed both at lower and higher frequencies in the S band (10 cm) and X-bands (3 cm) respectively for the distinctive advantages in the two bands.

Scan-to-Scan Integration. The target to sea clutter ratio in marine radars can also be increased by scan-to-scan integration. At low grazing angles, the time required for the sea clutter pattern in X-band radar to become uncorrelated is about 10 ms. For periods less than this the sea can be considered "frozen" with a still surface wave pattern, and no improvement in target to clutter ratio is expected from integration. If the antenna rotation rate is as low as 20 RPM, then for a 2° beamwidth and a pulse repetition rate frequency of 3600 Hz, there will be 60 pulses returned from a target on each scan of the antenna, all occurring within a period of 1/60 sec (16 m sec) comparable to the decorrelation time of 10 ms. No improvement is, therefore, expected in the signal-to-clutter ratio from pulse-to-pulse integration over the 60 pulses. On the other hand, if the antenna rotation is speeded up to 600 RPM there will be only 2 pulses per scan, while the returns during successive scans over the target will occur at intervals of 1/10 sec (100 ms). This interval being greater than the decorrelation time, scan-to-scan integration of the successive pulse pairs will improve the target-to-clutter ratio.

Detection of Icebergs. The back scatter cross section of smooth flat ice is very small at low grazing angles on a PPI display; therefore, echoes from the ice will appear as dark areas in a much brighter background due to sea clutter. Back scattered echoes from rough ice may also be distinguished from the sea clutter by noticing that the returns from the rough ice remain stationary in a background of the changing patterns due to the sea clutter. Marine radars can also detect large icebergs having faces that are nearly perpendicular to the radar beam.

The parameters of a typical Marine Navigation Radar is shown in Table 10.4.

TABLE 10.4
Parameters of a Typical Marine Radar

Frequency	9400–9500 MHz (X-band)
Radiated power (peak)	22 kW
Pulse repetition frequency	510–160 Hz
Pulse width	0.2–0.6 ms
Antenna scan rate	20–600 RPM
Beamwidth	2°
Range	27 nmi (50 km)

10.9 REMOTE SENSING RADARS

For a remote target the angle subtended at the radar antenna may be much smaller than the angular width of the radar antenna beam. Such remote targets may not, therefore, be resolved by the radar and may even be undectable, due to the smaller ratio of echo to the clutter fields, this being proportional to the ratio of the angular extent of the target to that of the radar beam filled with the clutter. For the detection of remote targets, a very sharp radar beam free from side lobes is needed. This calls for huge antenna arrays with controlled amplitude and phases of the fields generated by each element of the array. By proper amplitude control a side lobe-free sharp beam can be generated. The technique is called pattern synthesis. For steering of the beam, a phase control of the feeds to the elements of the array is made. Such an antenna is called a *phased array antenna. Synthetic aperture radar* usually arranges the beam steering by the phase control and, therefore, may also be called a *phased array radar.* The radars for remote targets are basically remote sensing radars. When the target is very far off, the detectivity must be improved by special radar signal processing. For the very remote targets like the moon and planets, the angular motion of the target is insignificant and beam steering is no problem, although the angular resolution and detectability are outstanding problems. These radars for very remote targets are called *long-range radars.* Besides these, there is another class of radars in which the targets are neither very remote nor do they have a smaller extent. However, the radar cross section of the target may be so small that the detectability is the main problem. Several *atmospheric* radars designed for detecting and studying the *earth's lower and upper atmospheres* are examples of this type.

10.10 PATTERN SYNTHESIS

A desired radiation pattern free of side lobes may be synthesized by finding a proper distribution of current in the continuous array elements across a finite width aperture. The process is called *array synthesis,* which effectively leads to the desired radiation pattern synthesis. Pattern synthesis may in fact be obtained both for a continuous aperture antenna or an aperture filled with discrete array elements.

The electric field strength produced by a collinear array of elemental currents is given by Figure 10.13.

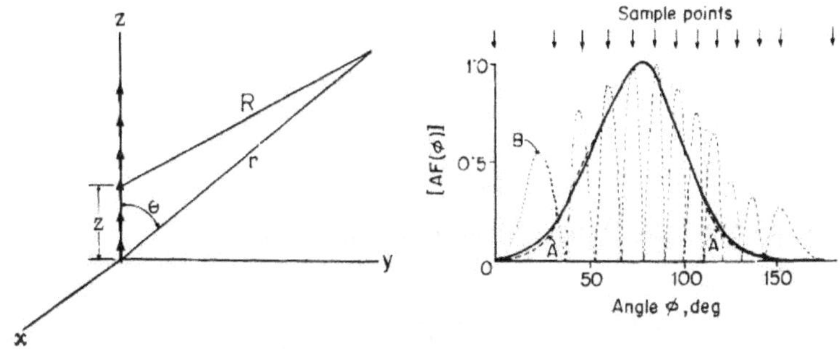

FIGURE 10.13 Geometry of a collinear array. Desired (solid curve) and synthesized (dashed) patterns are shown. The synthesized pattern is obtained (right-hand figure) by alternating signs of the Fourier coefficients.

From the left-hand figure,

$$E_\theta = jw\mu \frac{e^{-j\beta r}}{4\pi r} \sin\theta \sum_{n=0}^{N-1} I_n e^{j\beta nd\cos\theta}$$

$$= Ee \cdot Ed \tag{10.16a}$$

where $$E_e = jw\mu \frac{e^{-j\beta r}}{4\pi r} \sin\theta \tag{10.17}$$

represents the radiation pattern of an element and

$$E_d = \sum_{n=0}^{N-1} I_n e^{j\beta nd\cos\theta} = \sum_{n=0}^{N-1} e^{jn\Psi} \tag{10.18}$$

represents the array pattern due to the arraying of the elements, where $\psi = \beta d \cos = \theta$; E_d is called the array factor. Similar expressions can be deduced for broadside arrays when the factor E_e due to element pattern will only be different from that in equation (10.17) and given by

$$E_e = \frac{e^{-\beta r}}{4\pi r} \sum_{n=0}^{N-1} I_n e^{j\beta nd \cos\Omega}$$

In a *uniform array* the currents in the elements are all of equal amplitude, when the array factor is identified as a geometric series with

common ratio $e^{j\beta d \cos\theta}$. The sum of N terms of such a series is given by

$$E_d = \sum_{n=0}^{N-1} e^{j\beta nd \cos\theta}$$

$$= I \sum_{n=0}^{N-1} e^{nj\Psi} = I \frac{e^{jN\Psi} - 1}{e^{j\Psi} - 1} \tag{10.19}$$

where $I_1 = I_2 = \ldots\ldots = I_n = I$.

Equation (10.19) may be simplified to

$$E_d = I \frac{e^{jN\Psi/2}\left(e^{jN\Psi/2} - e^{-jN\Psi/2}\right)}{e^{j\Psi/2}\left(e^{j\Psi/2} - e^{-j\Psi/2}\right)}$$

$$= I\, e^{j(N-1)\Psi/2} \frac{\sin N\Psi/2}{\sin \Psi/2} \tag{10.20}$$

If the spacing d between the consecutive elements is so small they form an essentially continuous array, the summation for the array factor, given by equation (10.18), degenerates into an integral given by

$$E_d = \int I(x) e^{j\beta x \cos\eta}\, dx \tag{10.21}$$

where $\cos\eta = \sin\theta \cos\Phi$, the remote point being $P(\gamma, \theta, \Phi)$. Equation (10.21) represents the field pattern due to the continuous array. If we assume a linear progressive phase shift, the current is given by $I(x) = I_o(x)e^{jkx}$ and then

$$E_d = \int I_0(x) e^{j(\beta \cos\eta + k)x}\, dx \tag{10.22}$$

For deriving equation (10.21) we are to assume that d decreases indefinitely and N increases indefinitely in such a way that the product $d\,(N-1)$ remains a constant, L, the length of the array. The quantity nd then becomes the continuous variable, the quantity n becomes the phase shift per unit length, and I_n becomes $I(nd) = I(x)$. If the array is very large, extending on both the positive and negative directions about the origin, the limits will be $+\infty$ and $-\infty$ when we have

$$E_d = E_d(\beta\cos\,\theta) = \int_{-\infty}^{+\infty} I(x)e^{j(\beta x\cos\eta\,+\,k)\,x}\,dx$$

putting $\beta\cos\eta + k = \xi$, we get

$$E_d(\xi) = \int_{-\infty}^{+\infty} I(x)e^{-j\xi x}\,dx \tag{10.23}$$

This shows that the array factor $E_d\,(\xi)$ is the inverse Fourier transform of the current distribution $I(x)$. Conversely, if the array factor is known, the current distribution can be determined by the direct Fourier transform of the array factor $E_d\,(\xi)$. Accordingly,

$$I(x) = \frac{1}{2\pi}\int_{-\infty}^{+\infty} E_d(\xi)e^{-j\xi x}\,d\xi \tag{10.24}$$

The Fourier transform pairs in equations (10.23) and (10.24) suggest that if the array factor represents a very sharp beam, the radiation being zero outside the finite beamwidth, then the required current distribution will be infinite in extent and impracticable. Collin has shown that, if the array factor is expressible in the form

$$E_d(\xi) = \frac{\sin\dfrac{\pi\xi L}{2}}{\dfrac{\pi\xi L}{2}}\,\frac{P\left(\xi\dfrac{L}{2}\right)}{\prod\limits_{n=1}^{N}\left[\dfrac{1}{2}\xi L^2 - n^2\right]} \tag{10.25}$$

where $p(\xi L/2)$ is an arbitrary polynomial of degree $2N$, then the required $I(x)$ will be zero outside the range 0 to L. This means that the array will be finite in dimension.

Array synthesis may also be made in another way by applying mathematics similar to that used in communication theory to derive the sampling theorem. The desired array factor is sampled at an appropriate point and a corresponding space harmonic is introduced into the current distribution in such a way as to produce the correct relative field strength at the sample point. An array factor producing a single lobe is given by

$$E_d\left(\frac{\xi L}{2}\right) = e^{-c(\xi/\beta)^2}$$

The corresponding pattern is given by

$$E_d(\eta) = e^{-c\cos^2\eta}$$

The synthesized pattern is shown in Figure 10.13 (dashed curve A). The desired pattern is represented by the solid curve A. The dotted curve B shows a synthesized pattern obtained by alternating signs (phase) of the Fourier coefficients I_n.

10.10.1 Discrete Array Synthesis

The current distribution in a linear array of discrete elements required to produce a desired array factor is simpler, and equation (10.18) is appropriate for the purpose. In this case I_n is a Fourier coefficient of order n.

Chebyshev Array. In an array synthesis it is often desirable to obtain both a narrow beamwidth and a low side lobe level. However, these two are in practice conflicting requirements, and the improvement of one is associated with the deterioration of the other. C. L. Dolph showed that in such a situation, array factors derived from Chebyshev polynomials optimize the relationship between beamwidth and side lobe level. In fact, an array factor may be identified with a Chebyshev polynomial to arrange that (1) side lobe levels are all equal, (2) the main beam is a specified number of decibels above the side lobe level, and (3) the main beam has a specified angular width. Such a Chebyshev array is optimum in the sense that it produces the narrowest main beamwidth for a given side lobe level and is called a Dolph-Chebyshev array.

Unequal Element-to-Element Spacing. Studies with unequal element-to-element spacing show that (1) unequally spaced arrays having larger element-to-element spacing need not necessarily exhibit grating lobes, as do arrays having equal spacing; (2) the selection of appropriate unequal element spacing can reduce side lobe levels; and (3) preliminary designs of large arrays may be made by a statistical process, in which the element positions are described by a probability density function.

10.11 PHASED ARRAY

A directive antenna array, in which the relative phases and amplitudes of the currents at the individual elements are controlled to obtain the desired shape and direction of the radiation pattern, is called a phased array. The control of the relative phases can be used to steer a beam, while the control of relative amplitude may be used to synthesize a desired shape of the pattern with a minimum of side lobes.

Fixed (non-steered) phased array antennas were known during World War I. During World War II, the United States, Great Britain, and Germany used phased array radar in which the beam was steered by mechanically activated phase shifters. In the early 1950s the mechanically activated phase shifters were gradually replaced by electronic phase shifters, marking the beginning of a new era of electronically steered phased array radar.

10.11.1 Directional Pattern

For an array with uniform spacing and with elements fed with equal currents, the radiation pattern is given by equation (10.20). Also, if the same antenna is employed as a receiving array, with each element being connected to the receiver by the same length of cable, the resultant field will be given by a similar equation, except that the current I is replaced by the voltage induced in each isotropic antenna element. In real form, the expression may be given by

$$E_a = E\sin\left[\omega t + (N-1)\frac{\Psi}{2}\right]\frac{\sin(N\Psi/2)}{\sin(\Psi/2)} \tag{10.26}$$

where E is the amplitude of the voltage induced in each element, $\Psi = (2\pi/\lambda)$ $d\sin\theta$ and θ is the angle between the ray and the normal to the line of array. Note that in equation (10.20), we assumed θ to be the angle between the ray and the line of the array. The field intensity pattern is the magnitude of equation (10.26) given by

$$E_a(\theta) = \frac{\sin[N\pi(d/\lambda)\sin\theta]}{[\sin\pi(d/\lambda)\sin\theta]} \tag{10.27}$$

The equation shows that the pattern $E_a(\theta)$ will be zero when the numerator is zero, and occurs when

$$N\pi\left(\frac{d}{\lambda}\right)\sin\theta = \pm\, n\pi$$

where n is an integer. To find the maxima of the pattern, we note that the zero in the denominator occurs when the numerator also becomes zero and equation (10.27) assumes a 0/0 form. Differentiating both the numerator and denominator and equating each to zero, we get all the maxima of equal amplitude N, occurring when

$$\pi\left(\frac{d}{\lambda}\right)\sin\theta = \pm\, n\pi$$

that is, when

$$\sin\theta = \pm\,\frac{n\lambda}{d} \tag{10.28}$$

The maximum at sin $\theta = 0$ defines the main beam or main lobe. The other maxima are called the grating lobe or side lobe. The antenna array is in fact behaving like a diffraction grating in optics, in which the diffraction pattern results from the pattern of individual slits multiplied by a grating factor similar to the array factor for the antenna considered previously.

Equation (10.27) also shows that $E_a(\theta) = E_a(\pi - \theta)$. Therefore, an array of isotropic elements has a pattern in the rear similar to that in the front. The backward lobes are not useful in receiving the signal but are harmful for the pickup of noise and interference through the back lobes. The backward signal is usually eliminated by backing the array with some reflector. The reflector, in addition to reducing back lobe pickup, increases the signal pickup. The signal on being reflected from the reflector is received through the back lobe and may reinforce the direct pickup of the array by its main forward lobe, if the reflector is spaced from the array properly. The radiation power pattern of the array without the reflector is

$$G_a(\theta) = \frac{|E_a|^2}{N^2} = \frac{\sin^2[N\pi(d/\lambda)\sin\theta]}{N^2\sin^2[\pi(d/\lambda)\sin\theta]} \tag{10.29}$$

where N^2 in the denominator normalizes the pattern. For a spacing of $\lambda/2$ between the array elements and $\pi(d/\lambda)$ sin $\theta << 1$, equation (10.29) shows that the half-power beamwidth of the main lobe is given by

$$\theta_B \cong \frac{102}{N} \tag{10.30}$$

The first side lobe, for large N values, is 13.2 dB below the main lobe and the radiation pattern is similar to that produced by an uniformly illuminated aperture as shown in Figure 10.14. If the array elements are directive, with each having a radiation pattern $G_e(\theta)$, the resultant pattern of the array will be given by

$$G(\theta) = G_e(\theta)\cdot G_a(\theta)$$

where $G_e(\theta)$ is called the *element factor* and $G_a(\theta)$ is called the *array factor* or the *space factor.* The side lobes may be eliminated by choosing directive antenna elements of the array such that $G_e(\theta)$ drops to zero at and beyond the first side lobe. Thus for an array with element spacing $d = 2\lambda$, the side lobes occurring at ± 30° and ±90° can be suppressed if the individual elements have a beamwidth less than 60°(± 30°). It may be mentioned here that the array factor will be modified if the spacing is small enough, when the mutual coupling between the antenna elements cannot be neglected.

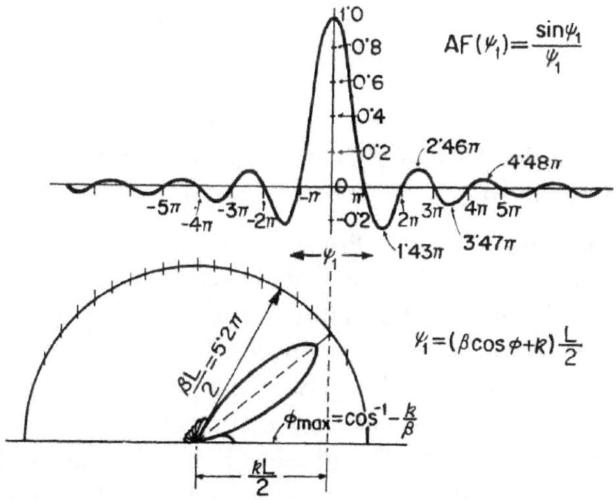

FIGURE 10.14 The radiation pattern produced by a uniformly illuminated aperture.

Two-Dimensional Array. The radiation pattern of a two-dimensional array is the product of the patterns in the two orthogonal principal planes. For a rectangular array in the ray plane, the array factor is given by

$$G = G_x(\theta_x)\, G_y(\theta_y)$$

where the subscripts x, y refer to the arrays along the X and Y axes respectively. The normalized radiation pattern or the array factor of an uniform rectangular array is given by

$$G = \frac{\sin^2[N\pi\,(d/\lambda)\,\sin\theta_x]}{N^2\sin^2[\pi\,(d/\lambda)\,\sin\theta_x]}\cdot\frac{\sin^2[M\pi\,(d/\lambda)\,\sin\theta_y]}{M^2\sin^2[\pi\,(d/\lambda)\,\sin\theta_y]} \qquad (10.31)$$

Beam Steering by Phasing. If the antenna elements of an array are all driven in phase, the main beam will occur at $\theta = 0$ corresponding to $n = 0$ as indicated by equation (10.28). The main beam will be shifted to an angle θ_0 if the phase difference between two adjacent elements is changed to $\Psi = 2\pi(d/\lambda)\sin\theta_0$. The phase at each element will in general be given by

$$\Psi_e = \Psi_c + m\Psi \tag{10.32}$$

where $m = 0,1,2, \ldots N-1$ and Ψ_c is any constant phase which may be applied to all elements. The normalized radiation pattern (array factor), when the phase difference between adjacents is Ψ, given by

$$G(\theta) = \frac{\sin^2[N\pi\,(d/\gamma)\,(\sin\theta - \sin\theta_0)]}{N^2\sin^2[\pi\,(d/\lambda)\,(\sin\theta - \sin\theta_0)]} \tag{10.33}$$

Thus the direction of the main lobe may be steered from $\theta = 0$ to $\theta = \theta_0$, for which $\sin\theta = \sin\theta_0$ and $G(\theta)$ as given by equation (10.32) attains a maximum, by applying a progressive phase shift to the elements.

Implementation of a Phased Array. A controllable progressive phase shift in the feeds of the array elements can be implemented either by a series or a parallel feed as shown in Figures 10.15 and 10.16 (*a*). In the series-fed array, the power may be fed from one end of array (Figure 10.15 [*a*]) and reaches the other end for undergoing progressive phase shift between elements, or the power is fed at the central element from which it is distributed toward the two ends. The phase shifts in this case will be positive for elements to the right and negative for the elements on the left.

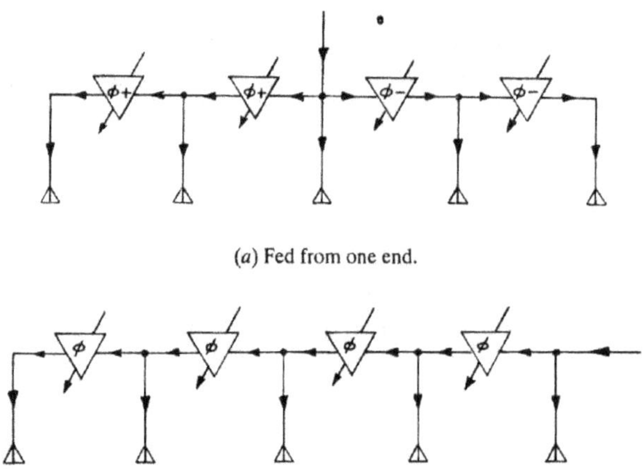

(*a*) Fed from one end.

(*b*) Center fed.

FIGURE 10.15 Series arrangements for introducing phase shifts.

Controllable Phase Shifts. The phase control for the beam steering in a phased array may be implemented in variety of ways. The simplest one is perhaps the digitally switched parallel line phase shifter as shown in Figure 10.16 (*a*). The lengths of the successive lines starting from the lowest one increases in steps of λ/16. Each of the switches may be a semiconductor diode operated with forward bias in the "on" position to present a low impedance. Pairs of diodes at the ends on a particular line are switched on one at a time, the remaining diodes being reverse biased to present a high impedance.

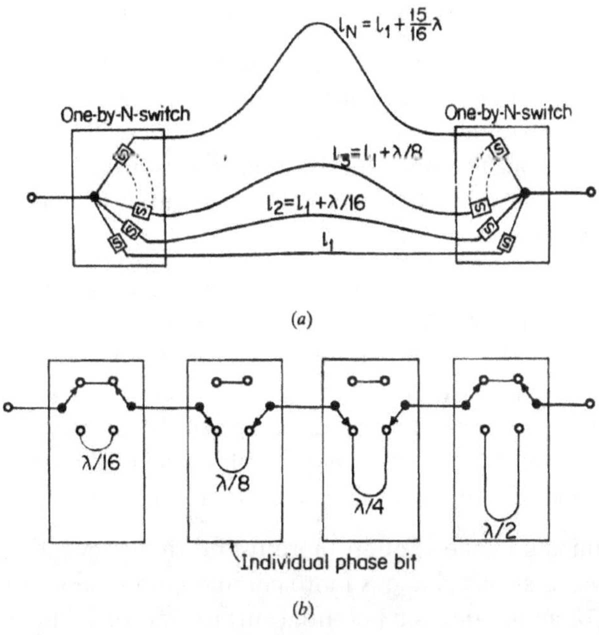

FIGURE 10.16 Digitally switched parallel line phase shifter with *N* switchable lines.

The number of switches, lines, and line lengths may be minimized by a *cascaded digitally switched phase shifter* as shown in Figure 10.16 (*b*). Each of the switching boxes can be switched between line lengths of 0 and λ/*n* to effectively introduce line lengths in steps of λ/*n* as in parallel switching.

10.11.2 Conformal and Non-Uniform Arrays

In aircraft and missiles it is often required to design directive antenna arrays spread over their surfaces so as to avoid antenna elements projecting outward from the body of vehicle significantly, thereby reducing the additional windage due to the antenna and the length of the leads from the antenna elements to the receiver. Such an array which conforms to the geometry of a nonplanar surface is called a conformal array. A nonplanar surface will introduce additional phase shifts between the array elements, and these can be corrected to obtain the desired sharp beam of the antenna with low side lobe levels. Even the spacing between the elements of an array may be forced to be non-uniform because of certain intervening structures of the body of the vehicle or for any other reason, to obtain what is called an non-uniform conformal array. With suitable phase adjustments of the array elements, it may still be possible to achieve the desired antenna pattern.

An unequal spacing can also be invoked, even in a planar array, to obtain a given beamwidth with considerably fewer elements than an equally spaced array, or to obtain a desired pattern without calling for an amplitude tapered illumination. The need for fewer elements in a non-uniform array to achieve a desired beamwidth also justifies the name "thinned array," which is often assigned to a non-uniform array.

Conformal arrays over cylindrical surfaces have been widely studied. With array elements spread uniformly over the surface, only a small aperture is switched in by a commutator, which produces a scanning beam with circular symmetry maintaining the same beam shape for all scanning angles. With the axis of the cylinder vertical, an excellent azimuthal scan may be provided in this way. At the same time the elevation of the beam can also be steered by proper phasing of the array elements along an axial plane.

A truncated cone is often more useful than a cylinder to design conformal arrays, as such a shape is more common in missiles. The nose cone-like surface of an aircraft is a potential surface for installing conformal arrays, as it would allow wide coverage, provide good aerodynamic shape, and result in larger antenna apertures than the conventional nose-cone antenna configuration. For hemispherical coverage a hemispherical conformal array may be useful, as the pattern would then be independent of beam direction.

10.11.3 Adaptive Arrays

An adaptive array adjusts the phase and amplitude of the aperture illumination by using the received echo, so as to maximize the signal-to-noise ratio. An automatic monopulse tracking radar does the job, by sensing the echo by a monopulse feed to reposition the antenna in a direction receiving the maximum signal in the sum channel.

Adaptive arrays may also be designed to minimize the clutter echoes or interference from other electromagnetic interference or noise jamming. An adaptive antenna reduces automatically the side lobes in the direction of the unwanted signal, and requires some prior knowledge of the statistical properties, direction of arrival, and waveform of the desired signal.

For an airborne MTI radar, the Doppler shift of the echo signal due to a moving target allows it to be separated from the clutter due to stationary targets exhibiting a different Doppler shift as discussed in Chapter 8. The clutter entering by side lobes can have the same Doppler shift as that due to the target of interest.

An adaptive array antenna can be arranged to place nulls in the direction of the side lobe clutter by using the side lobe clutter itself as the signal to be canceled. For this purpose, the side lobe clutter is separated from the main beam clutter by Doppler filtering, and the resulting output adaptively adjusts the aperture illumination pattern to minimize the side lobes in the direction of reception of clutter. The nulls can be adaptively placed in the direction of external noise in the coherent side lobe canceller.

10.11.4 Computer Controlled Phased Array Radar

In satellite surveillance, air defense systems, ballistic missile defense, and multifunction airborne radar, computer control of the phased array is essential. The computer control allows the amplitudes and phases of signals in the array elements to be adjusted to obtain a desired beamwidth, beamshape, and side lobe level. At the same time the scanning and tracking speeds can be optimized to suit a specific target. The computer also allows signal management by determining (*a*) the type of waveform, (*b*) the number of observations, (*c*) data rate, (*d*) power, (*e*) frequency, and (*f*) signal processing and data processing depending on the mode of operation. Besides these, the computer also provides the processed data to users, including the generation of displays, housekeeping functions of performance monitoring, fault location, data recording and simulation and, above all, the executive management of the radar. A general-purpose computer is good enough to perform all the computations and control of a phased array radar. However, a special-purpose dedicated computer may be more useful for beam steering as an integral part of the hardware, to minimize the problem of communicating a large number of phase shifter orders by the general purpose computer, which would then provide the beam steering computer with the desired elevation and azimuth angles.

10.12 REMOTE SENSING OF THE EARTH AND ITS ATMOSPHERE AT MICROWAVES

By remote sensing we mean observing from a distance or a remote location. The advent of artificial satellites has apparently opened up the possibility of observing the earth and its atmosphere from a satellite, and we call it *remote sensing*. The main advantage of remote sensing is that the observations over a wide global area can be made within a much shorter time and expenditure than that required in ground based or in situ measurement systems.

10.12.1 Atmospheric Windows for Remote Sensing

For the observations of the earth from a satellite, the intervening atmosphere must be penetrable at the observing wavelength. The "windows" of the atmosphere in the visible, infrared, and radio wavelengths are shown in Figure 10.17 (*a*). Remote sensing has been successfully made in each of the windows. However, it has been found that the radio window, particularly that covering the microwaves, is more useful, as the measurements can be made insensitive to the presence or absence of sunlight, clouds, and all but the most intense rainstorms. By "microwaves" we broadly mean here electromagnetic waves having wavelengths that range from 1 millimeter (300 GHz) to about 1 meter (300 MHz). Molecules of some constituents of the atmosphere resonate at frequencies in this region. The resonance produces narrow absorption lines for parts of the atmosphere at higher altitudes where the pressure is low. When the entire vertical thickness of the atmosphere is involved as in remote sensing, due to collisional broadening

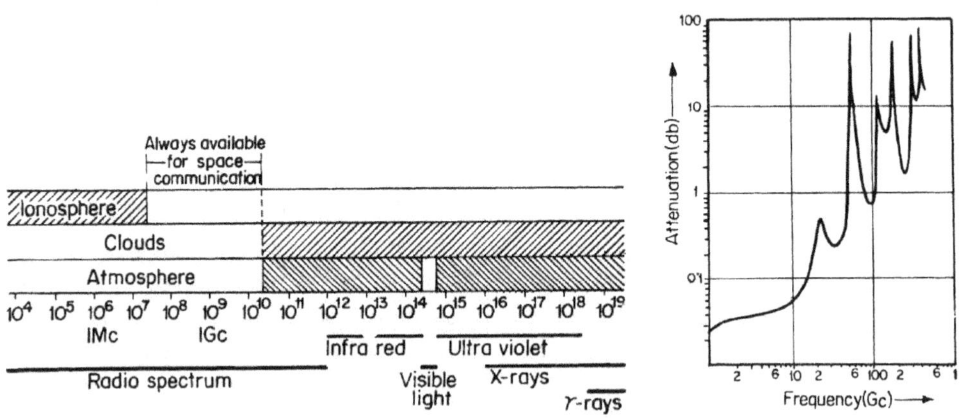

(*a*) Radio and optical windows of the atmosphere.

(*b*) Attenuation in atmospheric gases.

FIGURE 10.17

the absorption lines tend to merge, giving the attenuation spectrum in the microwaves as shown in Figure 10.17 (*b*). As good absorbers are also good emitters, the atmosphere will also have a similar emission or radiation spectrum. The remote sensing technique can be used to obtain *information of the atmosphere* as well, by studying these absorption and emission lines. The *earth's surface studies* by remote sensing depend on the measurement of the emission or reflective properties of the earth in the region of the atmospheric "windows."

10.12.2 Satellite Instruments for Microwave Remote Sensing

During the last decade microwave remote sensing instruments have been flown on the satellite series, Skylab, the Nimbus, and Geos, and later on, Seasat-A. In India, the Bhaskara and INSAT series of satellites also carried some microwave remote sensing instruments. The space verified instruments include Radar Altimeters, Scattermeters, Synthetic Aperture Radar, and Radiometers.

Radar Altimeter. A microwave pulse radar operating at 13.5 GHz and having a very short pulse length of 3 ns (10 cm range resolution) is employed as a radar altimeter mounted in the satellites. The time delay for a transmitted pulse to reach the sea surface and arrival of its reflection back at the satellite is measured at the satellite and the result telemetered to the ground by a radio telemetery link. Laser ranging with a LIDAR is used to determine the satellite orbit with a great precision. Variation of local mean sea-surface level with respect to the satellite orbit can be measured with the radar altimeter to an accuracy of 10 cm with a horizontal resolution of about 2 km.

Radar altimeter measurement may be used to study the following: (*a*) the marine geoid, (*b*) ocean tides, (*c*) currents, (*d*) eddies and circulation patterns, (*e*) sub-oceanic formations, (*f*) wave height, (*g*) raincell distribution and intensity, (*h*) sea surface wind speeds, and (*i*) the extent, age, and thickness of ice.

Scatterometer. Wind speeds over the sea surface can be measured with great precision by measuring the Doppler spectrum of a radar return from the sea surface by a radar called a wind *scatterometer.* Long pulses of length 5 ms are transmitted from the satellite at a frequency of 14 GHz. The reflected radiation is Doppler shifted due to the motion of the satellite. The degree of Doppler shift can be used to segregate returns according to their angle of incidence at the sea surface as in an airborne Doppler radar. In this way sea surface reflectivity can be measured as a function of angle in the forward and backward directions and in orthogonal polarizations. This data can be used to deduce wind speeds and directions to an accuracy of ± 10% with a spatial resolution of 50 km.

Synthetic Aperture Radar. A high resolution radar image may be obtained by a remote sensing *Synthetic Aperture Radar* (SAR), operating at 1.3 GHz. A short pulse of length 50 ns is transmitted from a large (11 m × 2.3 m) phased array antenna of the SAR producing antenna footprints, 100 km × 14 km. The reflected signal is range gated to achieve about 25 m resolution in the 100 km direction and frequency gated to achieve about 25 m resolution in the direction of satellite motion. Signals returned from a particular element of sea surface may be combined while that element is within the beam pattern, that is, for the time taken by the satellite to travel about 14 km. In this way, the effect of an antenna 14 km long can be achieved. The combined signals can be used to construct a map of surface reflectivity having a resolution of 25 m.

Synthetic aperture radar images contain a wealth of information. These images have been used to derive information about the following: (*a*) ocean pollutant distributions, (*b*) ice, (*c*) snowfield and flood water distributions, (*d*) geological surveying and resource prospecting, (*e*) land use and crop quality, and (*f*) inventory.

Radiometer. A radiometer is an instrument for passive remote sensing. It is basically a highly sensitive microwave receiver capable of measuring the thermal emissions from the earth and its atmosphere. Radiometers, so far space verified, have sensitivities of the order of 1°K. This means that the changes of thermal emission in the microwave band of reception due to change of only 1°K in the emitter is detectable by the radiometer. Dual polarizations, frequencies from 6 to 37 GHz, and spatial resolution varying from 100 to about 18 km have been employed.

Microwave radio meter measurements from satellites have been used to study the following: (*a*) vertical atmospheric water vapor and temperature distributions, (*b*) cloud top temperatures, (*c*) horizontal motion of clouds, (*d*) surface winds over the oceans, (*e*) sea surface temperature, (*f*) oil slick size and thickness, (*g*) pollution plumes, and (*h*) the extent and age of ice.

Pressure sounder. The concept of measuring atmospheric surface pressure on a global basis using an active millimeter wave sounder originated at Heriot Watt University in the early 1970s. The surface pressure charts are, perhaps, the most important database used by meteorologists for forecasting the weather. However, the necessary data is frequently inadequate or unavailable over large areas of the oceans. A review of the potential methods for remotely measuring surface pressure indicated that a satellite-borne millimeter wave radar is most suitable for the purpose.

The concept of remote pressure measurement by a *Millimeter Wave Pressure Sounder* (MPS) is based on the fact that molecular oxygen exhibits a strong absorption band spanning the 50 to 70 GHz range of frequencies. The absorption of a signal in this band by a vertical column through the atmosphere depends on the total amount of oxygen in the column. Since oxygen is a uniformly mixed constituent of dry air, a measure of the absorption by oxygen is directly related to the total weight of the column and hence the surface pressure. A signal of known strength at a frequency in the band is transmitted from the satellite-borne instrument down to the ocean surface, and the magnitude of the reflected echo is subsequently measured. A second measurement is made at a frequency outside the band. Since this second measurement essentially defines the surface reflectivity, the ratio of the received power, when normalized by the transmitter powers in the oxygen absorption band, yields surface pressure.

This simple, two-frequency measurement will, in practice, be inadequate. For the other atmospheric constituents like water vapor and other atmospheric parameters like the temperature also influence the measurement of oxygen absorption. Likewise, aerosols reduce the received signal by scattering, while the sea surface reflectivity is a function of frequency as well as the sea state. All these factors can be eliminated, however, if we make measurements at several frequencies, v_i simultaneously. A pressure index, S, can then be formed from the measured normalized received power $P(v_i)$ as given by

$$S = \pi_i P(v_i)^{\tau_i}$$

Careful selection of the operating frequencies and the indices τ_i, can make S sensitive only to pressure but very nearly independent of the other meteorological variables.

The MPS is under construction at the Jet Propulsion Laboratory. It is proposed that measurement should be made with the MPS not only at nadir but also at 14° away from nadir on either side of the sub-satellite track to provide an effective swath width of 600 km. For one measurement every 12 seconds, pressure is obtained on a grid of points separated by about 200 km within the swath. A change in surface pressure of 1 mb changes the pressure index S by 0.75%, which implies a measurement accuracy of 0.2% in each millimeter wave channel.

Microwave Limb Sounder. The Microwave Limb Sounder (MLS) is basically a passive remote sensing instrument to measure the water vapor in the upper stratosphere and mesosphere from observations of thermal emission by its rotational transition at 183 GHz (1.64 mm wavelength). A millimeter wave radiometer on an aircraft was initially employed for the purpose and was operated continuously on all the fifteen flights of the 1976 NASA CV-990 Latitude Survey Expedition. The MLS is being developed to remotely sense the composition and temperature of the Earth's stratosphere, mesosphere, and lower atmosphere. The ultimate application of the experiment is for search and monitoring of these regions of the atmosphere on a global scale from a satellite or space shuttle platform. The observed thermal emission noise depends both upon the abundance of the molecule and its temperature. By choosing the observing direction and spectral band such that the atmospheric path being observed is optically thin, the observed signal depends mainly upon the abundance of the molecule.

For limb measurements made from platforms above the atmospheric region being sensed, altitude resolution can be obtained by having an instrument with a highly directional field of view. For measurements made from platforms below the atmospheric region being sensed, as is the case for the aircraft MLS measurements, altitude resolution can be obtained by observing the shape of atmospheric emission or absorption lines.

It may be mentioned here that the term MLS signifying a Microwave Limb Sounder should not be confused with the identical term MLS to mean a Microwave Landing System, which is a modern development of an Instrument Landing System for an aircraft.

10.12.3 Remote Sensing Radars

Airborne radars meant for remote sensing of the earth and its environment from a satellite platform are called Remote Sensing Radars. Surveys of surface features of the earth over land and sea by imaging scatterometers for sensing the wind over seas and radar altimeters for the measurement of altitude of sea surface with great precision are examples of Remote Sensing Radars. Scatterometers and Altimeters have already been discussed under Section 10.12.2. Also, a type of Remote Sensing imaging radar was included there, under Synthetic Aperture Radar, which is powerful imaging radar. Further information about these radars are discussed as follows:

Side Looking Airborne Radar. Side Looking Airborne Radar (SLAR) is an imaging radar commonly used in scatterometry and altimetry. The imaging geometry of a SLAR is shown in Figure 10.18. The footprint of the radar antenna beam on the ground is elliptical in shape as shown in Figure 10.18 (*a*), since the antenna beam is looking to one side, the right side in this case, at a certain angle θ to the vertical. The along the track resolution corresponds to the width of the minor axis of the ellipse. The resolution on the surface in the orthogonal direction may appear to correspond to the switch width *S*, which is equal to the major axis of the ellipse. However, different points on the major axis are at different ranges due to the slanting look angle, θ, which is not zero. The different points along the major axis are, therefore, resolved by the range resolution capability of the radar. This is illustrated in Figure 10.18 (*b*), which is a view of the generating looking along the track. The finite pulse width τ of the radar pulse illuminates circular concentric strips in space as shown in the figure. Two consecutive strips will be spaced by *cT* where *T* is the interpulse period. The intercept of such strips with the ground is X_r which is a resolution element as shown in Figure 10.18 (*b*). The position of X_r moves toward the right due to propagation of the radar wave. Looking from an angle perpendicular to the track, the geometry appears to be as shown in Figure 10.18 (*c*) where *X* indicates the resolution element. The beamwidth θ_a is determined by the length *L* of the antenna array along the track.

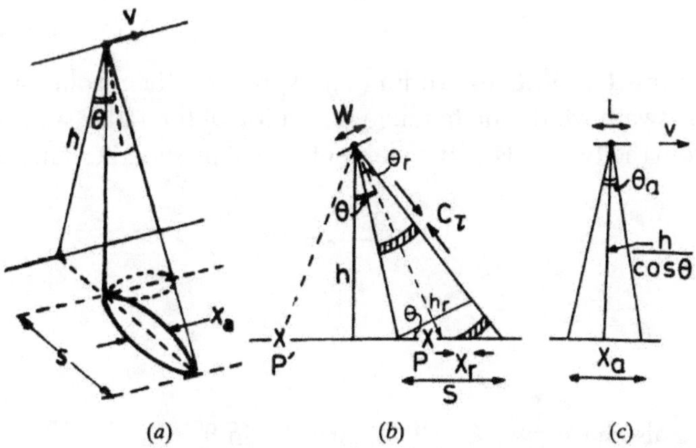

FIGURE 10.18 (*a*) Geometry corresponding to a side looking imaging radar. (*b*) Geometry in the range plane (perpendicular to the satellite track). (*c*) Geometry in the azimuth plane.

Swath Width. The width of the antenna perpendicular to the track is W as shown in Figure 10.18 (b), from which we can derive an expression for swath width S as follows:

$$\frac{S_r}{S} = \cos\theta$$

or
$$S = S_r / \cos\theta \tag{10.34}$$

For the angle between S and S_r is θ from the geometry assuming θ_r, the width of the cone in Figure 10.18 (a) and neglecting earth's curvature

Now,
$$S_r = h_r\theta_r \tag{10.35}$$

$$h_r = \frac{h}{\cos\theta} \text{ and } \theta_r = 1.22\frac{\lambda}{W} \cong \frac{\lambda}{W}$$

$$S = \frac{h_r\theta_r}{\cos\theta} = \frac{h}{\cos\theta}\cdot\frac{\lambda}{W}\cdot\frac{1}{\cos\theta}$$

$$= \frac{h\lambda}{W\cos^2\theta} \tag{10.36}$$

where θ_r, is the antenna beam width in elevation, W is the width of the antenna array, and θ is the look angle. For $\lambda = 25$ cm (1200 MHz), $h = 800$ km, $= 20°$ and $W = 2.1$ m the swath width is 92.6 km from equation (10.36).

Range Resolution. To find an expression for resolution element X_r along the swath width due to range resolution of the radar, we note that the echoes from the two ends of the element X_r will be separated in time by Δt such that

$$\frac{c\Delta t}{2} = \Delta r$$

or
$$\Delta t = \frac{2\Delta r}{c} \tag{10.37}$$

We also have
$$\Delta r = X_r \sin\theta \tag{10.38}$$

Combining equations (10.37) and (10.38)

$$\Delta t = \frac{2X_r \sin\theta}{c}$$

However, the minimum discernible time difference in a pulse radar is equal to the pulse duration τ, hence

$$\Delta t = \tau = \frac{2X_r \sin\theta}{c}$$

from which, we get

$$X_r = \frac{c\tau}{2\sin\theta} \tag{10.39}$$

The pulse duration τ is related to the B by the relation $\tau = 1/B$. Using this, equation (10.40) becomes

$$X_r = \frac{c}{2B\sin\theta} \tag{10.40}$$

Thus a pulsed signal of bandwidth B = 20 MHz will provide a range resolution element on the ground equal to 22 m for $\theta = 20°$ along the swath width.

Azimuth Resolution. The distance of two nearest separable points along the azimuth corresponds to the azimuth resolution. From Figure 10.18 (*c*) we have

$$X_a = h_r\theta_a = \frac{h}{\cos\theta}, \theta_a = \frac{h}{\cos\theta}, \frac{\lambda}{L} = \frac{h\lambda}{L\cos\theta} \tag{10.41}$$

where L is the length of the array along the track.

If h = 800 km as before, λ, = 25 cm (1.2 GHz), L = 12 m, and $\theta = 20°$, then X_a = 17.7 km. This is about two orders of magnitude larger than X_r which is only 22 m. This limitation of real aperture radars like SLAR does not allow their use in high resolution imaging for natural resources surveys. Much larger antenna dimensions have to be invoked to obtain adequate resolution both along swath width and along the track. The Synthetic Aperture Radar (SAR) discussed previously in section 10.12.2 solves the problem by artificially increasing the antenna length without the need for physical increase of the array length. More details of SAR are furnished as follows.

Synthetic Aperture Radar Geometry. The radar mounted on the satellite moves from left to right with a velocity v at a height h from the ground as shown in Figure 10.19. It was noticed earlier that in a SLAR the along-track resolution is poor, while the resolution along the swath width is adequate

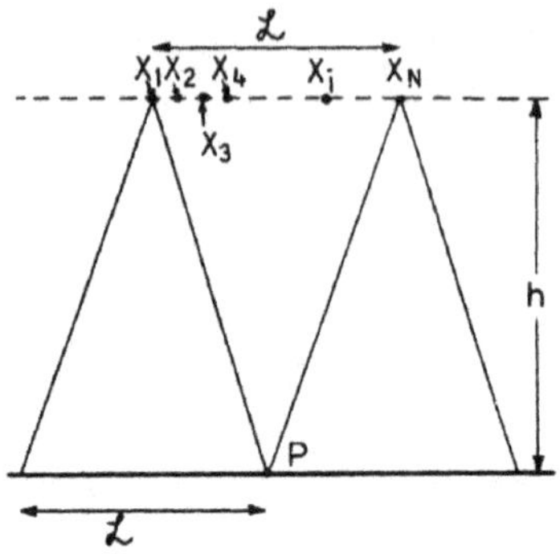

FIGURE 10.19 Geometry of a synthetic aperture array. Point *P* is visible from locations X_1, X_2 through X_N. The length of the synthetic aperture is equal to the real antenna footprint *L*.

due to range resolution by radar. The SAR improves along the track resolution as well by using successive positions of the satellite along the track marked by crosses (X) to form an array of effective length *L* instead of the physical length of the array equal to *L* only. From Figure 10.19, the footprint of the antenna over the ground, along the track, is of length *L* for a beamwidth θ_a of the antenna array of physical length *L* along the track. Therefore, we have for a SAR,

$$\mathcal{L} = h\theta_a = h\frac{\lambda}{L} \text{ as in the SLAR.} \tag{10.42}$$

However, the successive positions of the antenna between 1 and 2 is also equal to $\mathcal{L}$ where 1 and 2 corresponds to the point *P* just after coming within the beam and just before moving out of the beam respectively. If the echo signals received at the successive points are added to the phase at the point 2 after all the strong earlier echoes between points 1 and 2, the resulting antenna will effectively be an array of length L. Accordingly, the beamwidth θ of the SAR will be given by

$$\theta_s = \frac{\lambda}{2\mathcal{L}} \tag{10.43}$$

Factor 2 appears in the denominator of the two-way propagation from the path from the antenna element to the target and back for the SAR. Also, the element is moving compared to that for a stationary antenna element of a conventional array. In Figure 10.19, X_1, X_2, X_3, ..., X_n represent the positions of the radar antenna corresponding to the transmission of successive pulses. The echo received at each position is stored, and all the n pulses are then added coherently to obtain an effect similar to a linear array antenna of length X_1X_n. The distance traveled by the antenna, due to the satellite motion along the track with a velocity v, is given by $\alpha = vT = v/f_p$, where T is the interpulse period and f_p is the pulse repetition frequency. The phase difference between echoes received from two consecutive antenna positions, X_m and X_{m+1}, will be double for the two-way propagation in SAR, compared to that for one-way propagation in a usual array antenna. This is equivalent to doubling the element spacing and hence the array length by a factor of 2, which becomes $2\mathcal{L}$ instead of $\mathcal{L}$, and this was employed in equation (10.43).

Combining equations (10.42) and (10.43) we get,

$$\theta_s = \frac{\lambda}{2\mathcal{L}} = \frac{\lambda}{2h}.\frac{L}{\lambda} = \frac{L}{2h} \tag{10.44}$$

The resulting SAR footprint on the ground would be

$$X_s = h\theta_s = h\frac{L}{2h} = \frac{L}{2} \tag{10.45}$$

In contrast to this the footprint on the ground for a SLAR was shown to be

$$X_a = \frac{h\lambda}{L\cos\theta} \tag{10.45a}$$

Improvement in resolution of SAR over SLAR is therefore by a factor of M given by

$$M = \frac{X_a}{X_s} = \frac{h\lambda}{L\cos\theta}\Big/\frac{L}{2}$$

$$= \frac{h\lambda}{L\cos\theta}.\frac{2}{L}$$

$$= \frac{2h\lambda}{L_2\cos\theta} \tag{10.46}$$

with, $h = 800$ km, θ 20°, $\lambda = 25$ cm (1.2 GHz), $L = 12$ m,

$$M = 1475$$

The along track resolution for the SAR is

$$X_s = \frac{L}{2} = \frac{12}{2} = 6\,\text{m}$$

Remembering that the range resolution was 25 m, we obtain a resolution element 6 m × 25 m on the ground with the Synthetic Aperture Radar, mentioned previously.

The typical characteristics of SLAR and SAR for a 35 GHz band are shown in Tables 10.5 and 10.6 respectively.

TABLE 10.5
Characteristics of a Millimeter Wave SLAR

Parameter	Value
Peak Power output	100 kW
Transmitted frequency	34.85 GHz
Transmitted PRF	2088 pps
Pulse width	0.07 sec
Effective antenna resolution	0.1°
Receiver IF	60 MHz
Receiver bandwidth	30 MHz
Antenna polarization	Dual
Optimized Radar slant range	16 km
Optimized Radar altitude	20,000 ft
Ground speed range	130 to 300 knots
Slant range accuracy, 2	30 m
Azimuth accuracy, 2	40 m
Range resolution at 16 km start range	12 m
Azimuth resolution at 16 km slant range	21 m
Input signal dynamic range	50 dB
Power consumption, 3 phase, 400 Hz	3000 VA
Power consumption, 28 V dc	700 W

TABLE 10.6
Characteristics of the 35 GHz SAR

Parameter	Value
Frequency	33.5 GHz
Modes	Side looking SAR Spot light SAR Forward real aperture
Polarization	Horizontal (*H*) or vertical (*V*) right hand circular (RHC) or left hand circular(LHC)
Polarization, receiving	simultaneous C_0 and cross polarized
Amplitude balance	< 0.5 dB
Phase balance	< 3.0°
Absolute calibration	< 2.0 dB
Swath width	150 m
Antenna depression angle	10° to 50°
Aircraft velocity	50 to 110 m/s

10.13 CW RADAR

The phase of a continuous wave (CW) radio signal propagating through the atmosphere increases with the length of the path (phase path) traversed by the wave. When the radio signal is returned back from a target, the phase change due to the round trip path will indicate the distance of the target. The only problem in such a method of ranging of a target is that the range becomes ambiguous when the total phase shift over the round trip path exceeds 360°. Referring to Figure 10.20,

FIGURE 10.20

the round trip phase path is given by

$$\Delta\phi = \frac{2\pi}{\lambda}\cdot 2r = \frac{4\pi f_o r}{c} \tag{10.47}$$

$$r = \frac{\lambda}{4\pi}\Delta\phi \tag{10.48}$$

$$= \frac{c}{4\pi f_o}\Delta\phi \tag{10.49}$$

Maximum unambiguous range occurs at $\Delta\Phi = 2\pi$, for which $r = \frac{\lambda}{4\pi} . 2\pi = \frac{\lambda}{2}$.

For a range, λ, $\Delta\Phi$ should be 4π. However, there is no way of distinguishing between a phase difference of 2π and 4π or $2m\pi$, in usual phase measuring systems.

The unambiguous range can be extended enormously by using two CW transmissions at closely spaced frequencies, in which case the maximum unambiguous range, $r_{u\max}$, equals a half wavelength corresponding to the difference frequency. For frequencies f_1 and f_2 we have

$$r_{u\max} = \frac{(\lambda)_{f_1 - f_2}}{2} = \frac{c}{2(f_1 - f_2)} = \frac{c}{2\Delta f} \tag{10.50}$$

As f_1 is close to f_2, $(f_1 - f_2)$ is very small and, therefore, $(\lambda_{f_1 - f_2})$ must be very large, which indicates that $r_{u\max}$ must also be large.

For a moving target, the Doppler shift at each of the two frequencies may be utilized to derive a simple expression for the maximum unambiguous range as indicated in the following:

Let the two frequency transmitted signals be represented by

$$e_1 = \sin(2\pi f_1 t + \pi_1) \tag{10.51}$$

and

$$e_2 = \sin(2\pi f_2 t + \pi_2) \tag{10.52}$$

Where π_1 and π_2 are arbitrary but constant phase angles, the amplitudes of e_1 and e_2 being assumed to be unity. The echo signals are shifted by f_{d_1} and f_{d_2} at f_1 and f_2 respectively. The waveform of the Doppler shifted return signal at the two frequencies will be represented by

$$e_{R_1} = R\sin\left[2\pi\left(f_1 \pm f_{d_1}\right)t - \frac{4\pi f_1 r}{c} + \phi_1\right] \tag{10.53}$$

$$e = k\sin\left[2\pi\left(f_2 \pm f_{d_2}\right)t - \frac{4\pi f_2 r}{c} + \phi_2\right] \tag{10.54}$$

The second term within the third bracket is due the round trip phase path represented by equation (1) derived previously.

k takes account of the fall in amplitude of the return with distance of the target. The assumption that f_1 is close to f_2 suggests that the Doppler shifts f_{d_1} and f_{d_2} are close to each other, and we may write $f_{d_1} \simeq f_{d_2}$ and represent it by f_d.

At the receiver the return signals e_{R_1} and e_{R_2} are separated by two band pass filters tuned to f_1 and f_2 respectively, amplified to a level to make $k = 1$ and then arranged to beat with the respective transmitted signals at f_1 and f_2 given by equations (5) and (6) to obtain the Doppler frequency components given by

$$e_{D_1} = \sin\left(\pm 2\pi f_d t - \frac{4\pi f_1 r}{c} \right) \tag{10.55}$$

and

$$e_{D_2} = \sin\left(\pm 2\pi f_d t - \frac{4\pi f_2 r}{c} \right) \tag{10.56}$$

The + sign within the bracket indicates the Doppler shift due to an approaching target, while the -- sign indicates a shift due to a receding target.

From equations (9) and (10), the phase difference between the two Doppler frequency components is given by

$$\Delta\phi = \frac{4\pi(f_1 - f_2)r}{c} = \frac{4\pi\,\Delta fr}{c} \text{ from which}$$

$$r = \frac{c\Delta\phi}{4\pi\Delta f} \tag{10.57}$$

The maximum unambiguous range occurs when $\Delta\pi = 2\pi$, when we have

$$r_{u\max} = \frac{c.2\pi}{4\pi.\Delta f} = \frac{c}{2\Delta f} \text{ which is the same equation (10.50).}$$

From equation (10.50) we note that the closer the frequencies f_1 and f_2, the smaller Δf will be and the larger $r_{u\max}$ will be. However, the accuracy of range measurement by the two-frequency Doppler technique decreases with decreasing Δf. The rms range error r can be shown to be related to Δf through the expression

$$\Delta r = \frac{c}{4\pi\Delta f(2E/N_0)^{1/2}} \tag{10.58}$$

where E = energy of the received signal and N_0 = Noise power per unit (Hz) bandwidth. For attaining a large unambiguous range combined with a higher accuracy of range measurement (with small Δr), transmission at more than two frequencies may be used. A multiple-frequency CW radar called a tellurometer has in fact been used for accurate and unambiguous measurement of distances in surveying. A tellurometer consists of a master unit transmitting a carrier frequency of 3000 MHz (3 GHz), with four SSB (single sideband) modulated frequencies spaced from the carrier by 10,9.99,9.90, and 9.00 MHz. A remote unit at the far end of distance measurement receives the signals transmitted and retransmits after suitable amplification. Since the master and remote units are stationary, Doppler shift of the return signal will be absent and, therefore, an artificial Doppler frequency is provided by modulating the retransmitted signal at the remote unit so as to obtain a 1 kHz beat frequency by the heteordying technique at the receiver of the master unit designed for receiving the return signal. The 10 MHz difference frequency between the carrier and the SSB modulated sideband spaced 10 MHz apart determines the accuracy of measurement. The difference frequencies 1 MHz [(3000 + 10) – (3000 + 9)], 100 kHz [(3000 + 90) – (3000 + 9.9)], and 10 kHz [(3000 + 10) – (3000 – 9.99)] allow the resolution of ambiguities.

A practical tellurometer, MRA5 for instance, operating in the range 10–10.5 GHz, is capable of measuring distances between 100 m to 50 km unambiguously with an accuracy ± 0.05 m ± 10^{-5} r where r is the distance being measured, assuming the atmospheric corrections depending on meteorological conditions to be made. In case the target is stationary with no remote electronic unit, an artificial Doppler shift can be introduced at the transmitting end either by a mechanical rotation or vibration of the transmitting unit, or by electronic means.

10.14 IMAGING RADAR

The primary function of a radar is to locate a target in air or over the sea, and the maximum range over which a targent can be detected is considered to be the most important quality of the radar. However, for identification of the target the performance of the radar should be improved greatly to have a much higher angular and range resolution so that a 3D radar image can be formed and displayed on the radar screen. Such radars which can image a target are called imaging radar. The remote sensing radars like Side Looking Airborne Radar (SLAR) and Synthetic Aperture Radar (SAR), included in section 10.12.3, are examples of High Resolution Radars (HRR). However, a wide variety of special imaging radars have been developed in which the resolution of the 3D image is very high. Such a radar is called a high resolution imaging radar, or simply high resolution radar. An imaging radar may also be designed to have a 2D image, in which angular resolution in both the orthogonal planes along the central beam of the radar antenna pattern is made very high with the range resolution being kept normal to what is expected in a conventional radar. Such 2D imaging radars may be less expensive compared 3D imaging radars, but they may still allow the identification of targets from the 2D patterns. There are also applications of a high-resolution 1D radar in which only the range resolution, azimuth resolution, or elevation resolution is made very high. For instance, the precise measurement of range of a flat target with its plane perpendicular to the radar beam may be made with a 1D radar with a high-range resolution. For air traffic control where the azimuth resolution is the most important quality of the radar to avoid sideways collision between aircrafts trying to land on airports, a 1D radar with high azimuth resolution would be very useful. Similar will be the situation in radars meant for Vehicle Traffic Management System (VTMS) used to control the ship movements near ports to avoid collision. However, the 1D radars cannot identify the target like 3D or 2D imaging radars, except that the large RCS value of aircraft and ships as well as motion of the target on the display screen give an idea of the target indirectly.

10.14.1 Sources of Backscatter in Imaging Radar

The concept of RCS changes significantly for a high-resolution radar for which the radar targets and clutter observed at a given aspect to the radar are made up of individual reflection points, which may be called backscatter sources, scatters, "flare" spots, or "hot" spots of reflection on the target or clutter. Such backscatter sources for an aircraft sensed by a high-resolution radar are shown in Figure 10.21.

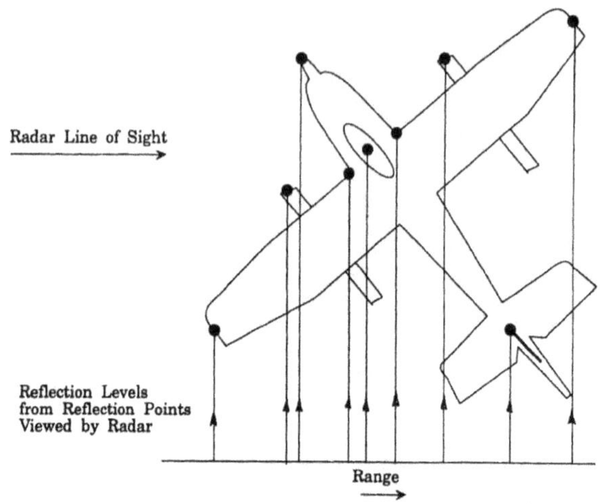

FIGURE 10.21 Radar backscatter sources.

The major backscatter sources, which are like point sources at any given target aspect, are produced by specular reflection from relatively flat portions of the surface of the target, oriented normal to the radar beam and also from corners. Besides these, "creeping-wave" reflections may also occur where the energy propagates along the surface of some part of the target to reappear after being directed toward the radar, or various resonance effects may also produce reflection to appear as pointed backscatter sources. Imaging of the target may be made using all these pointed sources of scatters on the target, each of which may have a different range and cross range, as shown in Figure 10.21.

10.14.2 Inverse Synthetic Aperture Radar (ISAR)

For imaging targets like ships, aircraft, and space objects, the technique of Inverse Synthetic Aperture Radar (ISAR) can be used. The ISAR concept can be looked upon as an improved form of the popular Synthetic Aperture Radar (SAR) described earlier in section 10.12.3. The improved form produces a higher angular resolution, making a sharper image of the target, as can be explained by comparing the geometries of SAR and spotlight SAR, as shown in Figure 10.22 (*a*) and (*b*), while Figure 10.22 (*c*) shows a variation of ISAR geometry.

Referring to Figure 10.22 (*a*) for SAR, the maximum length of the synthetic aperture is L_1 for range r_1 and L_2 for range r_2. That the synthetic aperture length is equal to the length of the footprint on the ground at range r_2

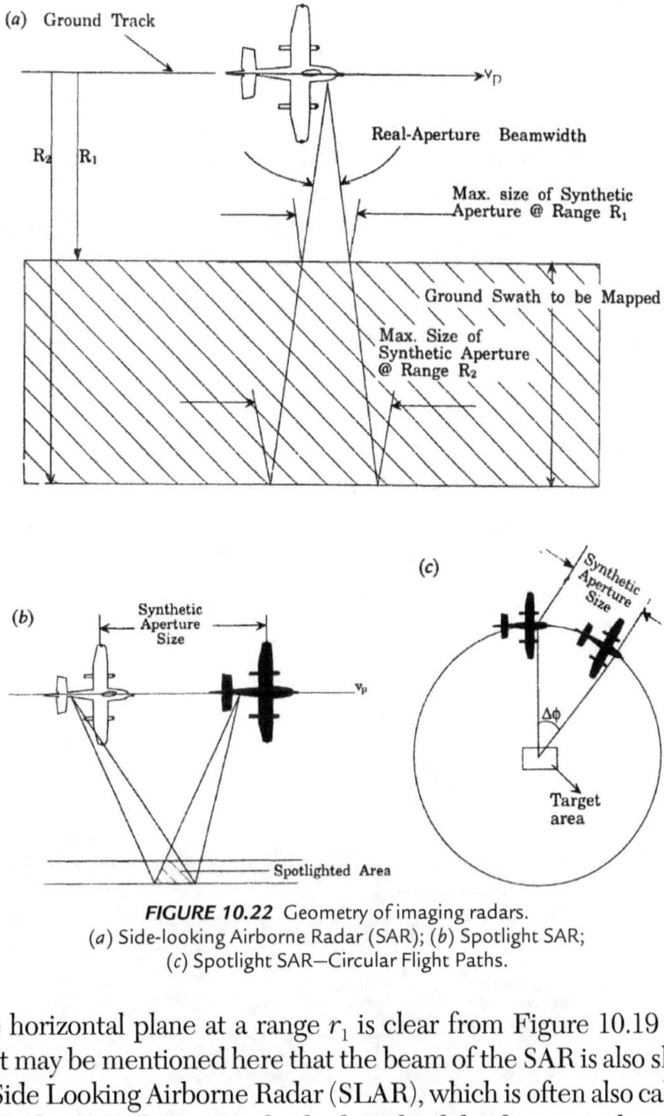

FIGURE 10.22 Geometry of imaging radars.
(*a*) Side-looking Airborne Radar (SAR); (*b*) Spotlight SAR;
(*c*) Spotlight SAR—Circular Flight Paths.

or on the horizontal plane at a range r_1 is clear from Figure 10.19 in section 10.13.2. It may be mentioned here that the beam of the SAR is also slanted like that of a Side Looking Airborne Radar (SLAR), which is often also called a Side Looking Radar (SLR). As a result, the length of the footprint along track will depend on the slant angle, which varies across the beam. The reflectivity data collected for each pulse motion of the aircraft, equal to the aperture length, is processed coherently to get a high angular resolution of a localized, target.

Referring to Figure 10.22 (*b*) for a spotlight SAR, the target is viewed by the radar beam at different orientations, unlike that in SAR for the same length of aperture motion of the aircraft. This angular information allows a higher imaging resolution. The cross-range resolution along the track is in fact determined by target dwell time rather than by the synthetic aperture length. Such a spotlight SAR may be considered to be equivalent to a radar flying in a circular path as shown in Figure 10.22 (c) over an angle $\Delta\pi$. Instead of the flying aircraft, if we could conceive of the aircraft radar to be fixed in position with the target rotating over the same angle $\Delta\pi$, we would have obtained the same type of data, which would be processed to obtain a high-resolution target image. This is called ISAR imaging, and it is widely employed to image scaled-down aircraft target models or that of ships in compact range RCS measurement systems, as shown in Figure 4.10 in Section 4.12.

It may be mentioned here that for ISAR imaging, the radar beam need not be steered to track a target during the motion of the aircraft. Instead, the beam may be kept unsteered as in SAR, and the collected data may then be processed in such a way as to obtain a result expected from ISAR imaging, with another type of processing, giving the spotlight image expected from ISAR. This is illustrated in Figure 10.23, showing the ISAR data derived

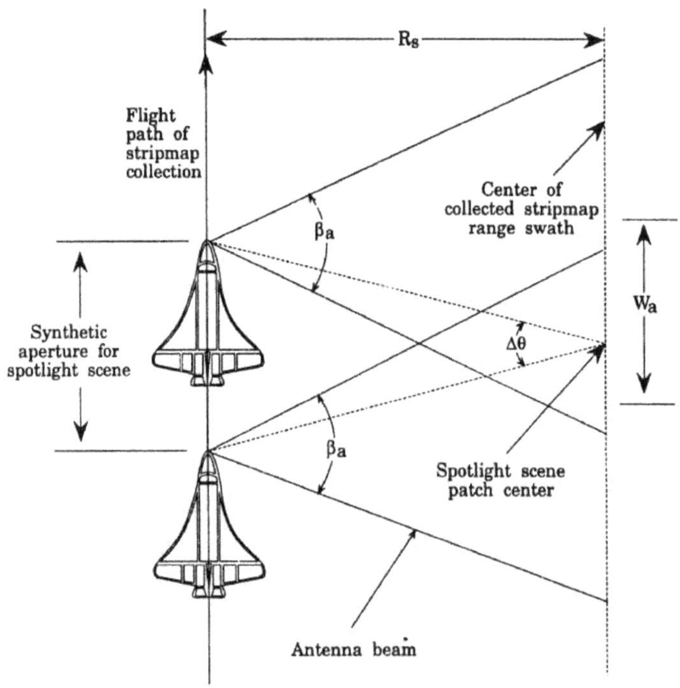

FIGURE 10.23 Illustration of allowable azimuth scene size in stripmap-to-spotlight data conversion.

from SAR data. In the figure the spotlight SAR patch center receives the radar beam along the lines joining the center to the various points between the two aircraft positions, the dotted lines indicating the boundaries of the angle $\Delta\theta$. As all these lines have different lengths, the corresponding range gate sweeping allows data for a particular line to be picked up for processing. Thus SAR data may be used and processed to obtain spotlight processing equivalent to obtain an ISAR image. It may be mentioned here that the map produced by SAR data is often referred to as a "stripmap" while the ISAR mapping is really done by spotlight data collected from SAR data, as shown in Figure 10.23. Examples of spotlight or ISAR images using stripmap data for SAR are shown in Figure 10.24 (*a*) and 10.24 (*b*). Figure 10.24 (*a*) shows the high-resolution ISAR map of the ground routes while Figure 10.24 (*b*) shows that of an aircraft whose photograph and plan view are shown in Figure 10.25.

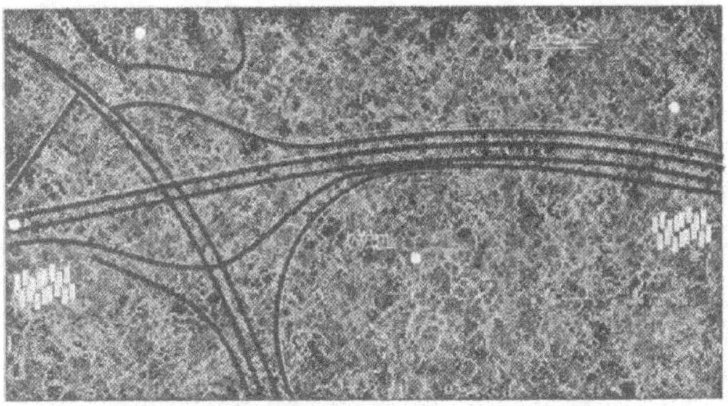

FIGURE 10.24 (*a*) Spotlight image using stripmap data and the PFA.

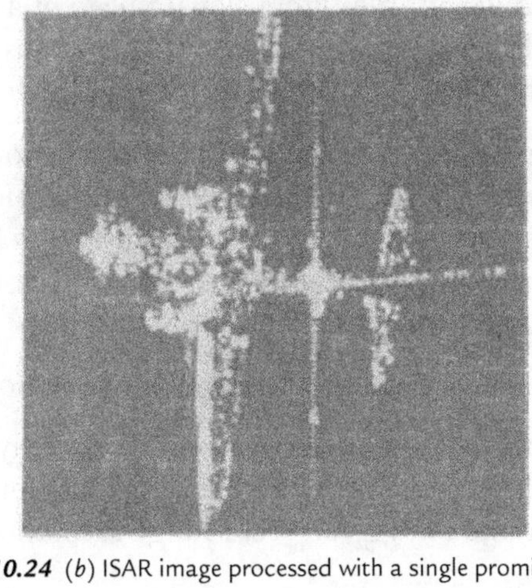

FIGURE 10.24 (*b*) ISAR image processed with a single prominent point.

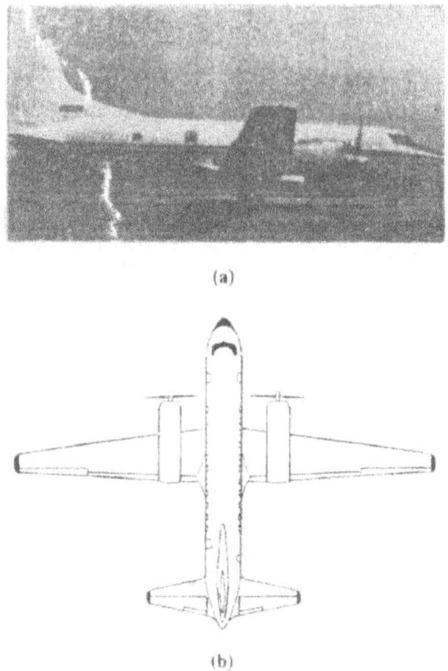

FIGURE 10.25 Convair CV-580 aircraft: (*a*) Photograph; (*b*) Plan view.

10.15 MONOPULSE RADAR IMAGING

The sum and difference channel signals of a monopulse radar may be utilized to produce a three-dimensional image of a target, free from the problems of an ISAR imaging system. The problems of ISAR imaging from the standpoint of image interpretation include the following shortcomings:

1. Distorted images result, unless the rotation rate can be determined from auxiliary data, as the cross-range resolution depends on the targets angularrotation rate. Also, during the image frame rate, defocusing may occur due to a change in the aspect rotation rate.

2. The ISAR image plane does not correspond to the true aspect of the target as the radar cannot determine the direction of target's rotation vector.

3. A long-range nonmaneuvering air target may require tens of seconds to image the target, as the target dwell time depends on the target's aspect rotation rate for a given cross-range resolution.

All these problems of ISAR imaging result from its dependence on the target's changing aspect to the radar. In spite of these problems, the ISAR technique is useful for surveillance problems requiring target identification, and methods have been found to partly resolve the problems.

The three-dimensional imaging with monopulse radar is based on the extraction of a crossed-range scatterer position from normalized monopulse error signals along a High Resolution Radar (HRR) profile of the target as shown in Figure 10.26, which indicates the generation of cross-range errors for azimuth with HRR Monopulse radar. In the figure the "hot spots" of the aircraft are used for the imaging with monopulse radar, by which the sum (Σ) channel signal is used to normalize the difference (Δ) channel signal, in which the peaks of the reflected signals from hot spots in normalized form for different ranges and bearings may be used to form an image of the hot spots, which would look like the aircraft.

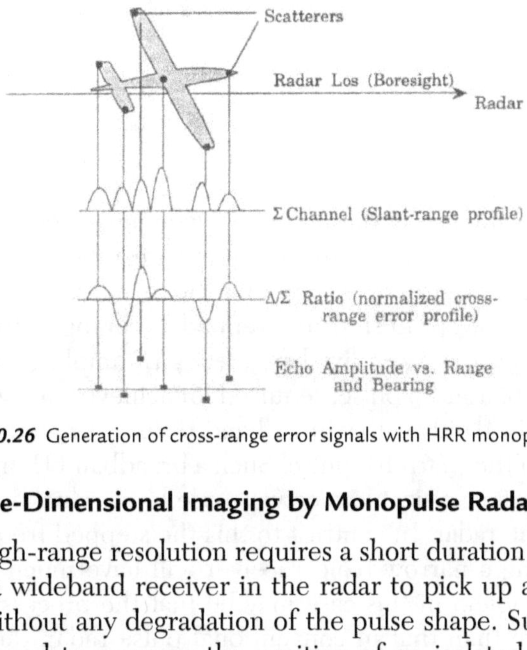

FIGURE 10.26 Generation of cross-range error signals with HRR monopulse radar.

10.15.1 Three-Dimensional Imaging by Monopulse Radar

A radar with high-range resolution requires a short duration pulse, which in turn requires a wideband receiver in the radar to pick up and process the sharp pulses without any degradation of the pulse shape. Such a wideband radar may be used to measure the position of an isolated point target in two orthogonal directions of the cross range to produce a three-dimensional (3D) image of the target. Error signals produced in the azimuth and elevation channels of the monopulse radar provide the orthogonal cross-range dimensions of the target.

10.15.2 Stepped-Frequency 3D Monopulse Radar

Instead of transmitting pulsed microwaves, it is possible to send the spectral components of the microwave pulse in steps, synthesize the pulsed return by storing the return signals in amplitude and phase for each stepped spectral component of the pulse, and then combine them at the same time slot. The spectral components of an RF pulse are shown in Figure 10.27, with the waveform of the pulse also shown at the top of the figure.

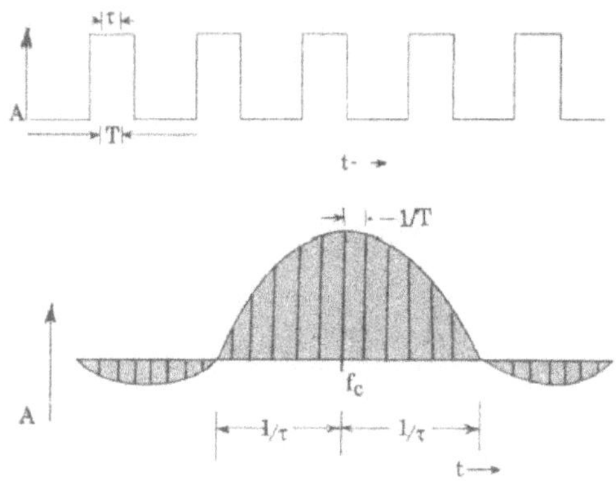

FIGURE 10.27 Repetitive pulse waveform and spectrum.

The advantage of the stepped frequency technique is that very sharp pulsed radar response from the target can be synthesized during processing of the return signal by a narrow band receiver with its frequency of reception swept in steps by stepped sweeping of the local oscillator signal using a narrow band fixed frequency IF amplifier. For conventional radar the use of a sharp pulse, required for achieving a high range resolution, can be received only with a broadband IF to cover all the wide spectral components of the sharp RF pulse. Such a broadband IF in the rader receiver will also produce a sizable receiver noise level, which may reduce the sensitivity of the radar. In contrast to this the stepped frequency radar technique requiring a narrow band receiver will have much higher sensitivity. The only problem in this case may be that the processing time required may be more than that in conventional pulse radar, due to the time taken in the synthesis of the return pulse. Also, the IF bandwidth cannot be made very narrow, as the IF amplifier has got to respond to stepped change in the frequency of reception promptly without any delay due to ringing of the narrow band circuit around each stepping of frequency. However, the overall receiver bandwidth determined by the bandwidth of the RF amplifier should be wide enough to receive all the spectral components transmitted by the radar. During synthesis of the return signal, the amplitudes of the different spectral components are modified to match the spectrum of the equivalent pulse.

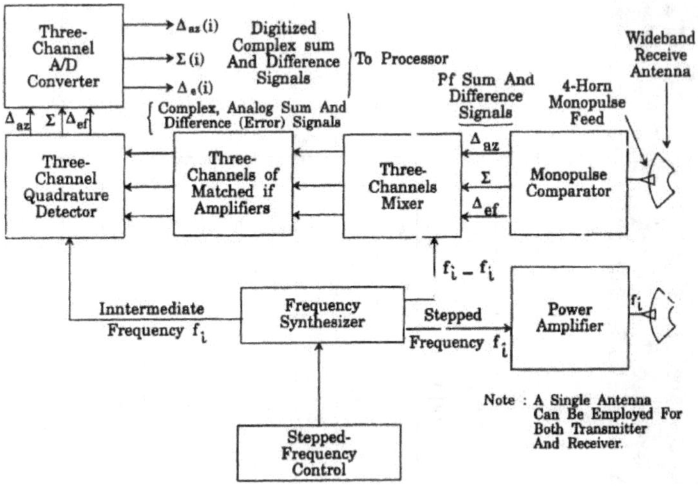

FIGURE 10.28 Generic 3D monopulse imaging radar using stepped-frequency waveforms.

A generic block diagram of a 3D monopulse radar using a stepped frequency technique is shown in Figure 10.28, in which two separate antennas are used for transmission and reception; at higher microwaves and millimetre-waves, the use of such separate dishes is not difficult as the size of the dishes will be relatively smaller at higher frequencies to achieve a reasonable sharp beamwidth. The use of separate dishes also has the advantages that good isolation between transmitting and receiving channels is possible without any circulator, and the transmission need not be stopped or blanked during reception of the return signal from the target. The continuous transmission is, in fact, required in steps at different frequencies of the stepped frequency transmission. The procuring of the return signal in a stepped frequency monopulse radar may be made by a network analyzer.

Parameters of hypothetical monopulse imaging radars at microwaves are shown in Table 10.7.

TABLE 10.7
Monopulse Imaging Radar Parameters

Parameter	Symbol	Units	Long-Range Tracking Radar	Airborne Intercept	Spaceborn Radar	Missile seeker
Antenna diameter	Da	ft	8.0 (2.4 m)	3.0 (0.91 m)	50 (15.2 m)	0.9 (0.27 m)
Wavelength	λ	cm	5.31	3.0	0.857	2
Frequency	f	GHz	5.65	10	34.29	15
Peak power	P_1	KW	2000	50	10	4
Total system	L	dB	5	13	5	10
Pulses perburst	n	—	256	128	128	128
Stepped frequency bandwidth	$n\,\Delta f$	MHz	250	100	250	250
Bursts integrated	N	—	1	200	128	20
System noise factor	F	dB	3.5	3.2	4.0	6.0
Output signal to (S/N) noise ratio (error signal)	s	dB	10	10	10	10
Pulse width	T_1	µ s	6	4	1	2
PRF	$1/T_2$	pbs	400	25.000	10.000	2500
Average power	(P)	W	4800	5000	100	20
Image frame time Range	T	sec	0.64	1.024	1.64	1.024
Resolution Δr	$s = \frac{C}{2n\Delta f}$	ft	2.0 (0.6 m)	2.0 (0.6 m)	4.9 (1.5 m)	2.0 (0.6 m)

The cross-range resolutions of the four monopulse radars listed in Table 10.7 decrease with range as shown in Figure 10.29, which shows that a cross-range resolution better than 100 ft will be realized with monopulse imaging radar for a range of 10 km.

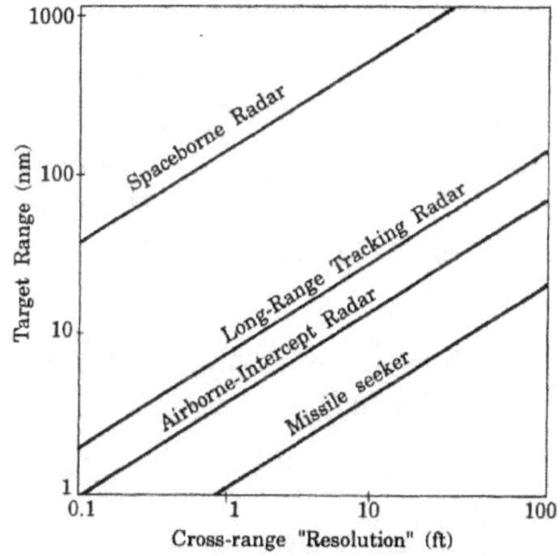

FIGURE 10.29 Monopulse radar target range versus cross-range "Resolution" (assuming spherical scatterers).

10.16 MULTIFUNCTION ARRAY RADAR (MFAR)

Multiple uses of radar for civil and military applications led to the development of the Multifunction Array Radar (MFAR) in recent years. The wide utilization of aircraft for transporting goods and passengers has, in fact, triggered the development of a sophisticated ground network of radar for air traffic control and to ensure safety in flight.

Operational concepts of radars are now focused on (*a*) accuracy of control of traffic flow in the airport terminal area, (*b*) reduced separation of the traffic routes, and (*c*) survey and transmission of important information related to flight, including risks of landing due to adverse weather conditions near the airport. These operational features need the development of high-performance radar having adequate safety, capacity, and reliability.

The safety requirements include the following:

1. Prediction and detection of dangerous wind shears, detection of turbulence, recognition of ice conditions, and detection of forms of precipitation.
2. Detection of aircraft intruding into controlled airspace, arising from traffic congestion in terminal areas.
3. Control of a number of satellite airports, outside the terminal control area, by traffic advisories. Of these items, number 1 requires multiple polarization, high update rate, high-resolution pencil beams, and clutter reduction, while items 2 and 3 require good attitude measurement accuracy, rapid track formation, a low false target rate, and accurate estimates of aircraft speed.

Regarding the requirement of high-performance capacity, it needs to be expanded due to following reasons:

1. Optimization of aircarft flow to approach more closely the theoretical throughput limit.
2. Reduction of the permitted longitudinal and lateral separation of aircraft, which determines the radar requirements for track accuracy with high update rates.

Finally, the reliability requirement is essential for operational use of radar in air traffic control systems, for which the following primary features should be provided:

1. 3D track
2. Higher angular and range measurement accuracy
3. Higher track update rate
4. Working in heavy clutter conditions
5. Rapid track formation
6. Estimation of trarget speed
7. Track performace in dense aircraft environment

The requirements of a radar regarding safety capacity and reliability are related to search and track in high-target density due to traffic congestion. In addition, the radar must be capable of operating in the environment of Electro Magnetic Interference (EMI) and should allow system integration with all auxiliary functions and data needed for integration into the radar traffic control system.

10.16.1 Multifunction Capability of Radar

The multifunction capability of radar may be defined considering search and track as the two main required functions. The search and track functions in rapid sequence independently define the multifunction capability to obtain the following specific performances:

1. Independent assignment of search and track renewal times.
2. Facility of varying the time-on-target by the radar beam, adaptively, in a wide range.
3. Coping with many geometrical target locations and distributions, keeping the search renewal time over the entire coverage solid angle.
4. Pointing of the beam for tracking independently of beam pointing required for search.
5. The search and track beam pointing with maximum flexibility, introducing significant dead time during beam switching.

The previously mentioned performance characteristics indicated in items 1–5 determine the design of the radar, particularly that of the radar computer including its software and also that of the radar antenna.

Item 1, 2, and 3 determine the array scan angle capabilities for the design of an antenna array.

Item 4 indicates that the "quantum" between two beam-pointing directions is much smaller than the beamwidth.

Item 5 indicates that the switching times of devices for beam switching are small fractions of the average time on target in each direction.

10.16.2 Design of Fixed Antenna Arrays for Search and Track

For a fixed antenna array design for search and scanning of a radar by phase shifting, the solid angle scanned Ω steer is greater than or equal to the required coverage solid angle Ω, as shown in Figures 10.30 (*a*) and (*b*). For each angular direction of the array beams of number $I = 1, 2, \ldots. N_{AR}$ the relation between Ω_{steer} and Ω is given by $\underset{1}{\Sigma} = 1\ \Omega_{steer}(i) \geq \Omega$.

The renewal time for search (t_{so}) relative to average time-on-target for each search beam pointing (t_{os}) is given by

$$\frac{t_{so}}{t_{os}} + \frac{N_s}{U_s} \qquad \text{or} \qquad t_{so} = \frac{N_s t_{os}}{U_s}$$

where N_s = number of search directions and

U_s = fraction of the radar time dedicated to search.

Likewise the renewal time for track (t_{To}) relative to average time-on-target for each track beam pointing (t_{ot}) is given by

$$\frac{t_{To}}{t_{ot}} + \frac{N_T}{U_T}$$

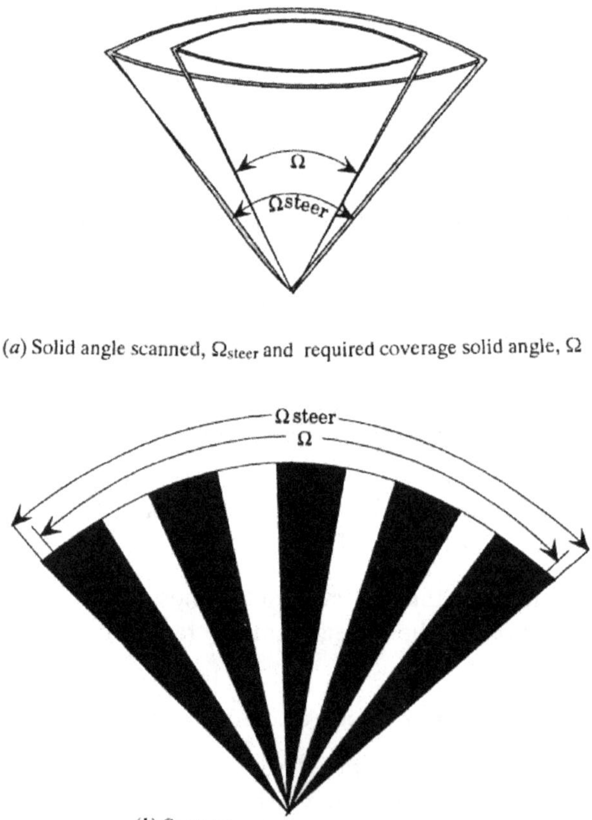

(*a*) Solid angle scanned, Ω_{steer} and required coverage solid angle, Ω

(*b*) Stepped scanning geometry

FIGURE 10.30 Scanning geometry.

where N_T is the maximum number of tracked targets in solid angle

$\Omega_T \leq \Omega$ and t_{ot} = average time-on-target for each track beam pointing.

As the radar is employed for search and track functions only, we have $U_S + U_T = 1$, indicating the limits of flexibility of MFAR.

In a radar with only search activities, $U_S = 1$ and $U_T = 0$.

We then have

$$t_{so} = N_s \,.\, t_{os} = t_{so\ min}$$

and t_{os} can be prolonged to $t_{os\ max} = \alpha$.

In a radar with only track activity, $U_T = 1$ and $U_S = 0$ and we have

$$t_{To} = N_T\, t_{ot} = t_{ro\ min}$$

and t_{ot} can be prolonged to $t_{ot\ max} = \alpha$.

The limits of t_{so} and t_{ro} for search and track respectively depend also on the radar power, and the minimum values of renewal times $t_{so\ min}$ and $t_{ro\ min}$ can be reduced, increasing the energy of the radar pulse that is equivalent to a decrease in t_{os} and t_{ot} respectively. The minimum values, however, increase when both search and track functions are required.

Design Problem. In a search and track radar find the limits for t_{so} and t_{ro}, assuming $t_{ot} = 2$ ms, $t_{os} = 3$ ms, $N_s = 1000$ and $N_T = 100$.

The limits are

$$t_{so\ min} = N_s \times t_{os} = 1000 \times 3 \text{ ms} = 3 \text{ sec}$$

$$\gamma_{to\ min} = N_T \times t_{ot} = 100 \times 2 \text{ ms} = 0.2 \text{ sec}$$

CHAPTER 10A

ATMOSPHERIC RADARS

10A.1 INTRODUCTION

RADAR, or Radio Detection And Ranging, was widely used during the war, but servicemen operating the radar would sometimes be confounded by blobs of precipitation that would block their view of the aircraft. It was only in February 1941 that radar was first used for intentionally looking at precipitation. Since then, huge advancements have been made in radar technology and its use in meteorology is widespread. At present over 150 public-access radars are in use by the National Weather Service, plus numerous radars owned and operated by private companies such as television stations. Radars are also being used on aircraft and even satellites to look down at clouds and precipitation. A radar transmits a focused high-power beam of radiation and receives a meager amount of it back from whatever the beam encounters. The antenna transmits a narrow beam from the feedhorn, which is suspended in front of the parabolic dish by struts. There are three fundamental properties of the emitted beam. These are pulse repetition frequency (PRF), transmission time, and beamwidth. The PRF is how many pulses of radiation are transmitted per second; for typical weather radars, this is around 325. The transmission time is the duration of each pulse. As the beam travels at the speed of light, the pulse length can be calculated from the transmission time. This length is also known as the spacing between "range gates" and is 1 km on average. The beamwidth describes the angular width of the emitted beam, and is typically about 1°. Combined with the pulse length, the beamwidth permits one to calculate the pulse volume. The pulse volume can be huge at long ranges, meaning that consecutive pulses will receive back scattered radiation from a large number of targets. The pulse length defines the radial resolution, while the beamwidth defines the angular resolution. The radar does more listening than talking, about 1000 times more.

More specifically, it might emit a pulse for 0.000003 seconds then listen for 0.003 seconds. So, 99.9% of the time, the radar is receiving, and 0.1 % of the time, it's transmitting.

10A.2 TERMINOLOGY

1. **Azimuth angle.** It is defined as the angle of beam with respect to north. This is illustrated in Figure 10A.1.
2. **Elevation angle.** The angle of beam with respect to the ground is called the elevation angle. It is illustrated in Figure 10A.2.
3. **Wind barbs.** It determines both wind direction and speed. In Figure 10A.3, Wind barbs point in the direction where the wind is blowing. In the figure the orientation of the wind barb indicates wind from the northeast. The term easterly means that the winds are from the east. In the example the winds are out of the northeast, or northeasterly. On the other hand, the term "eastward" means that the winds are blowing toward the east.

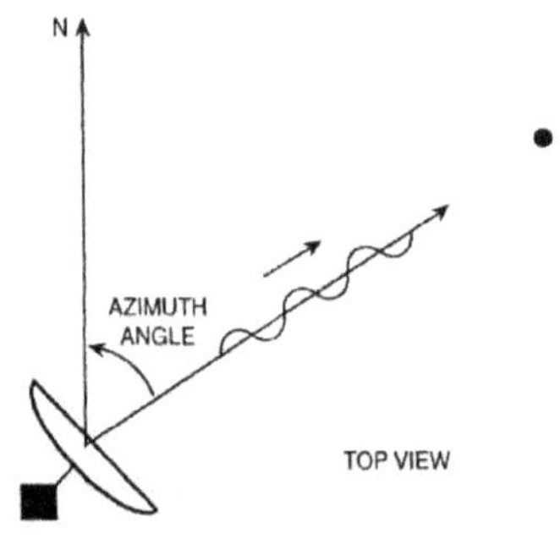

FIGURE 10A.1 Illustrating azimuth angle.

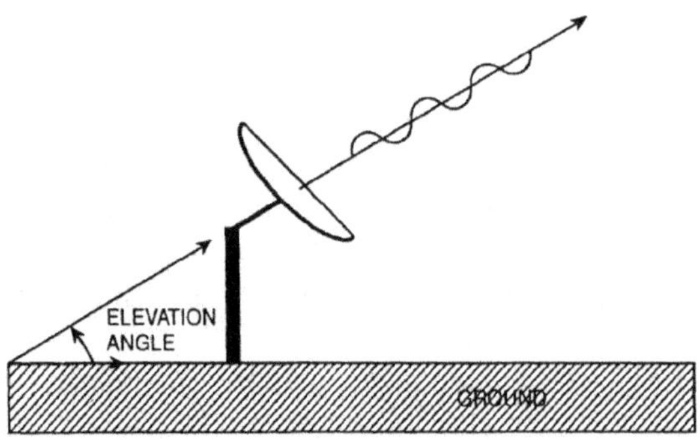

FIGURE 10A.2 Illustrating elevation angle.

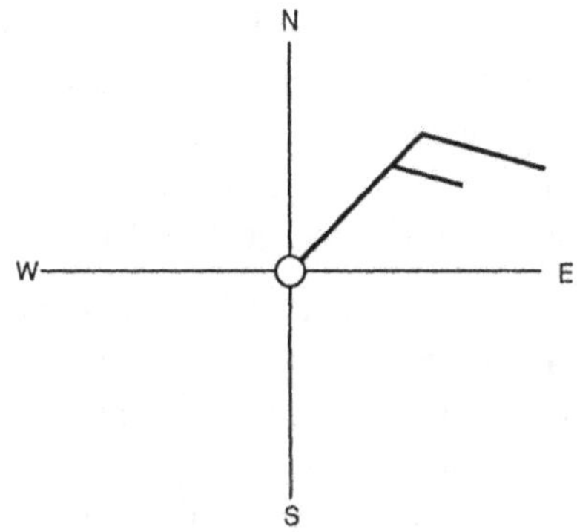

FIGURE 10A.3 Illustrating wind direction.

In Figure 10A.4 the wind speed is given in the units of "knots" (Knt). A "knot" is a nautical mile per hour.

$$1\ Knot = 1.15\ Miles\ Per\ Hour\ (MPH)$$

$$1\ Knot = 1.9\ Kilometers\ Per\ Hour\ (KM/HR)$$

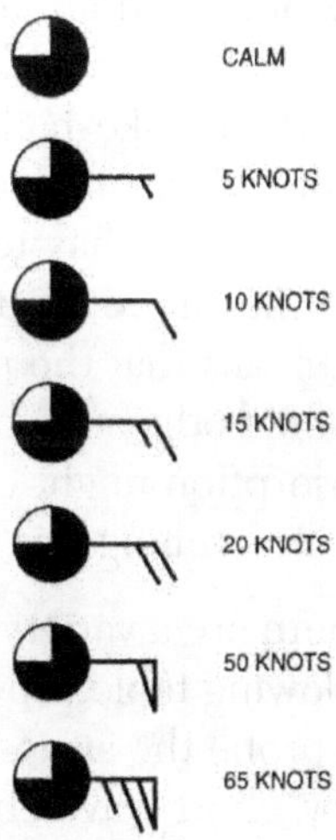

FIGURE 10A.4 Illustrating wind speed.

Each short barb represents 5 knots while each long barb is 10 knots. A long and short barb is 15 knots, simply by adding the value of each barb together (10 knots + 5 knots). If only a station circle is plotted, the winds are calm. Pennants are 50 knots. Thus the last wind example in the chart has a wind speed of 65 *knots*. (*50 knots* + *10 knots* + *5 knots*).

4. **Scattering of a radar pulse.** It means signaling back to the receiver by a target. When a pulse encounters a target, it is scattered in all directions. The signal component received back at the radar is typically much weaker than the original set from the transmitter and is called the "return signal." The larger the target, the stronger the scattered signal.
5. **Plan position indicator (PPI).** It is the scanning mode of the radar. When scanning in PPI mode, the radar holds its elevation angle constant but varies its azimuth angle. The returns can then be mapped on a horizontal plane. If the radar rotates through 360 degrees, the scan is called a "surveillance scan" while if the radar rotates through less than 360 degrees, the scan is called a "sector scan."
6. **Attenuation.** When the beam encounters a target in the atmosphere, many interesting processes take place. The raindrop will attenuate the energy by two means, namely, scattering and absorption. Upon contact with the raindrop, the energy is scattered in all directions, and hence only a very small fraction of the incident radiation is sent back in the direction of the radar. Second, the raindrop will actually absorb some of the energy. These two processes, scattering and absorption, are together known as attenuation. In addition to raindrops, atmospheric gases and small aerosols are also responsible for some absorption (too small for scattering), but particles like hail, snow, graupel, and insects are also responsible for both absorption and scattering. Attenuation becomes significant when the beam has to travel through a lot of targets and through a lot of atmosphere, that is, through heavy precipitation and long distances. It means that the power received back at the antenna is even less than it would otherwise be. Scanning through a heavy thunderstorm, gaseous absorption might cause a 5% decrease in back scattered power, while with intervening rain the hail could cause an 80% decrease.
7. **Frequencies.** There are a variety of radar frequencies used for various purposes. The following table provides the six common "bands," or frequencies, used to probe the atmosphere with radio waves. The first two are high-frequency, short-wavelength bands, which are readily attenuated by even light precipitation, making them most useful for detecting clouds and aerosols. The remaining four bands are all useful for detecting precipitation. The longer the wavelength, the less attenuation will occur. But it also cannot even "see" the smaller targets. Thus, an L-band radar will see heavy rain and hail and other things like birds and aircraft, but will not be able to see clouds, snow, or light rain. S-bands radars are widely used by the National Weather Service, as they are a fair compromise between high sensitivity and minimal attenuation.

TABLE 10A.1
Weather Radar Bands

Frequency (GHz)	Wavelength (cm)	Band
90	0.1	W (cloud)
30	1.0	K (cloud)
10	3.0	X (precip)
5	6.0	C (precip)
3	10.0	S (precip)
1.5	20.0	L (precip)

8. **Scan angle.** To keep the beam from hitting objects on the ground such as hills, buildings, and trees, the lowest scan angle used is 0.5° above horizontal. If the radar is on a flat surface and a person is 100 miles away from the radar, the beam would only be 4600' (1.4 km) over the person's head. However, as the Earth is curved, the beam deviates from the surface even more rapidly. So, at 100 miles (161 km) away, the beam is actually 9800' (3 km) over our head. In addition to the curvature of the Earth, the atmospheric lapse rate and moisture content play roles in controlling the exact path of the beam, but they are small compared to the curvature effect.
9. **Moments.** An important aspect of understanding radar image is exactly which product one is looking at. A Doppler radar can use consecutive pulses to determine motion of the targets, not just how much power they back scatter. There are three "moments of power distribution" that arise from the beam encountering targets. The 0th moment is called reflectivity. This is a very small fraction of the power that was originally transmitted into that volume. This is the product we see most often, on television or on websites, and so forth. Higher values indicate heavier precipitation. Extremely high values indicate hail, as hail is very effective at reflecting power back. The 1st moment is radial velocity and is observed by measuring the time variation of reflectivity. This product is useful in determining low-level wind shear and rotation within a thunderstorm. Finally, the 2nd moment of power distribution is known as spectrum width. This is found by measuring the time variation of radial velocity. This might be described as turbulence.

10A.3 INTERPRETATION OF THE PRODUCT

1. **Reflectivity.** The most common radar product is the reflectivity. It is a measure of how much power was scattered back to the radar from any targets. If there are no targets, no power is reflected back and there are no echoes. The term "echo" means a returned signal, whether it be from airplanes, birds, insects, snow, hail, or rain. The units of reflectivity are a convoluted form of decibels: dBZ. The higher the value of dBZ, the more power was reflected and received by the radar. Light snow is inefficient at reflecting radiation, as it might be 5-20 dBZ, while moderate rain might be 30-45 dBZ, and large hail might be around 60-75 dBZ. Radar software uses different equations to calculate dBZ from received power, depending on if it is in the deep tropics or in the mid-latitudes. This is, in fact, due to differences in moist thermodynamics, raindrop size distribution, and so on. Other features obtained by using reflectivity include smoke plumes, outflow boundaries (e.g. from thunderstorms or rain complexes), drylines (if conditions are favorable), and insects. One can easily spot mountains or hills if the radar site is near such features.
2. **Radial velocity.** There are two primary types of information one can gather from a radial velocity image. Of foremost importance is the rotation. A single radar cannot see an entire vortex of a mesocyclone, tornado, hurricane, and so forth, but rather just the components of the wind that are moving toward or away from the radar. The component of the wind moving tangentially to it will appear as very low radial velocities. A second use of radial velocity stems from the scan angle. Even at the lowest scan angle of 0.5°, the beam "sees" 8 km up in the atmosphere by the time it is 300 km away from the radar. This scan geometry provides an excellent way of getting a vertical profile of precipitation and wind. The further a storm is from the radar, the higher the level from which power is received. Also, on radial velocity images one can deduce the lower-tropospheric vertical and profile. A radial beam of power transmitted and received at an angle above horizontal acts as a proxy for a vertical wind profile.
3. **Spectrum width.** This product is available from Doppler radars, but is rarely used or shown except for specialized applications.
4. **Interpreting Doppler radar velocities.** To explain Doppler radial velocity patterns, one has to consider the geometry of a radar scan. Normally the radar beam is pointed at an elevation angle greater than zero so that the beam moves higher and higher above the surface of the earth. Due to this geometry, radar returns originating from targets near the

radar represent the low-level wind field, while returns from distant targets represent the wind field at higher levels.

On a radar PPI display, the distance away from the radar at the center of the display represents both a change in horizontal distance and a change in vertical distance. To determine the wind field at a particular elevation above the radar, one has to examine the radial velocities on a ring at a fixed distance from the radar. The exact elevation represented by a particular ring depends on the elevation angle of the radar beam.

10A.4 WEATHER SURVEILLANCE RADAR

Recent advances in technology have resulted in a new system of Doppler radars co-located with modernized weather service offices to help with unprecedented radar coverage. The new radar is the WSR-88D—the D standing for Doppler. This Doppler radar replaced the conventional non-coherent radars from 1957. The configuration of the radar control and signal processing circuits of the WSR-88D is shown in Figure 10A.5.

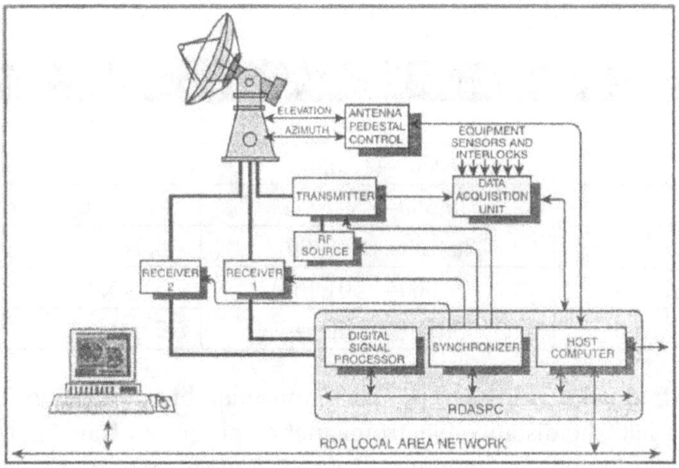

FIGURE 10A.5 Configuration of radar control and signal processing circuits of original WSR-88D.

The benefits of the WSR-88D over the WSR-57 are pointed out as follows:

1. *Improved sensitivity.* This is due to a greater amount of power transmitted and a greater ability to distinguish smaller returns. The WSR-88D's ability to detect lighter amounts of precipitation has allowed for the detection of very light precipitation and even subtle clear air boundaries.

2. *Improved resolution.* This is a function of angular beam width. The narrower the beam, the smaller the width at a given distance. This allows the WSR-88D to differentiate between objects, thereby increasing the resolution.

3. *Volume scanning.* Rather than scanning along varying azimuth angles (PPI) then stopping to scan vertically (RHI), the radar automatically scans various elevation angles while spinning around 360° of azimuth. Computers generate products based on this volume scan.

4. *Enhanced capabilities and algorithms.* Sophisticated computer programs assist the radar operator to detect various phenomena like mesocyclones and tornadoes. The radar also has a greater range of reflectivities operating in severe and non-precipitation modes.

The following is a breakdown of reflectivity (in dBZ's) corresponding to various Video Integrator and Processor (VIP) levels of the WSR-57. The WSR-88SD has fifteen different levels. The minimum detectable reflectivity for the WSR-57 was 18 dBZ whereas the minimum detectable return for the WSR-88D is –28 dBZ.

TABLE 10A.2
VIP levels vs. Reflectivity

VIP Level	Equivalent Reflectivity (Ze)	log Ze	dBZ (10 × log Ze)
1	0—(1000)	0—(3)	0—(30)
2	1000—(12,589)	3—(4.1)	30—(41)
3	12,589—(39,811)	4.1—(4.6)	41—(46)
4	39,811—(100,000)	4.6—(5)	46—(50)
5	100,000—(501,187)	5—(5.7)	50—(57)
6	501,187 or more	5.7 or more	57 or more

Maximum Range. The maximum unambiguous range, R_{max}, corresponds to half the distance electromagnetic energy can travel between pulses. The pulse repetition frequency (PRF) is a measure of how frequently the pulses are transmitted. If C is the speed of light, then

$$R_{max} = \frac{C}{2} \times \text{PRF}$$

The maximum unambiguous range is therefore inversely proportional to the PRF.

TABLE 10A.3
PRF and R_{max} for Different Radars

Radar	PRF (pulses per sec)	R_{max} (km)	R_{max} (nmi)
WSR-57	164 or 545	915 or 275	494 or 149
WSR-74 C	259	579	313
WSR-88 D	318–1304	471–115	254–62

When a second pulse is transmitted before energy from the first pulse has been received, the echo from a target beyond the maximum unambiguous range is displayed at the proper azimuth but at a distance one R_{max} closer to the radar. This "second-trip echo" exemplifies *range folding*. Echoes may be folded from third and fourth trips if the R_{max} is small enough to allow for targets to be observed at multiple intervals of R_{max} (Figure 10A.6).

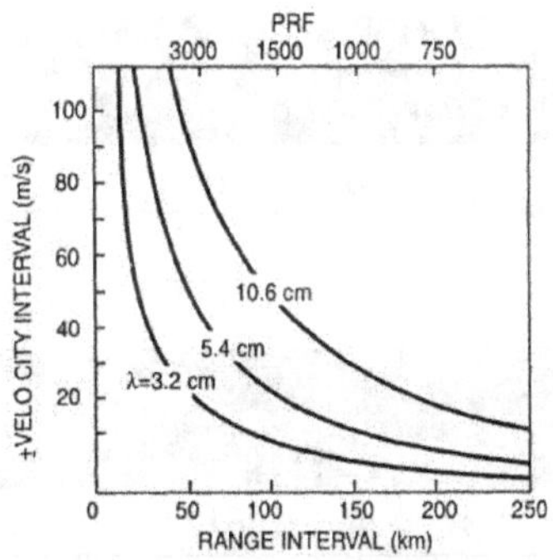

FIGURE 10A.6 Velocity interval vs. range interval or PRF.

Average Power Return. A radar can detect a small amount of power returned from a target. The more sensitive the receiver, the more a radar is able to detect smaller or less efficient scatterers. The radar needs to average returned power from successive pulses, as the power returned from pulse to pulse is highly variable. The WSR-88D typically averages power from 25 pulses to determine a representative value. A radar is able to increase its ability for detecting a target by not only changing its listening ability, but by maximizing the power it transmits. The greater the output, the greater the return.

Maximum Transmitted Power. The power output of the WSR-57 is *410,000* watts while the output of the WSR-88D is *750,000* watts. The return power is directly proportional to the transmitted power.

Antenna Gain. Gain is a measure of the antenna's ability for focusing the radiated energy. The antenna gain indicates the relative amount that the energy is focused compared to what it would be if it were an isotropic radiator. The gain of the WSR-88D is *35,481* while that for WSR-57 is *6,460*. It means that a target will be struck with over 35,000 times more energy than it would without a dish. Power received from a given target is directly related to the square of antenna gain.

Angular Beam Width. Power returned to a radar is directly related to the square of the angular beamwidth, assuming a beam narrower than 2° and at a range less than 125 *nmi*. A narrower beam improves sensitivity by focusing the outgoing power and increases the resolution by decreasing the size of the beam's cross sectional area.

TABLE 10A.4
Wavelength and Beam Width of Different Radars

Radar	Wavelength (cm)	Dish diameter (feet)	Angular beam width (degrees)
WSR-57	10.3	12	2.0
WSR-74C	5.4	8	1.6
WSR-88D	11.1	28	0.95

Range		2.0 degree beam		0.95 degree beam	
(nmi)	(km)	(feet)	(m)	(feet)	(m)
50	93	10613	3235	5041	1536
100	185	21227	6470	10082	3073
150	278	31840	9705	15124	4609
200	371	42454	12980	20165	6165
250	463	53067	16175	25206	7683

Pulse Length. The radar transmits energy for a given period of time known as the transmission time. If that is multiplied by the speed of light, a train of energy or pulse extends outward from the antenna with a pulse length, *H*. Long pulses are used for determining target locations, while short pulse lengths determine target intensity and motion characteristics. The power received from a target is directly related to pulse length. The longer the pulse length, the more energy is being transmitted. This improves the sensitivity of the radar.

TABLE 10A.5
Pulse Length of Different Radars

Radar	Transmission Time (microseconds)	Pulse Length (km)
WSR-57	0.5 or 4	0.15 or 1.2
WSR-74C	3	0.9
WSR-88D	1.57 or 4.5	0.47 or 1.35

NEXRAD WSR-88D

NEXRAD is NEXt-generation RADar, WSR-88D stands for Weather Surveillance Radar, the 88 is for 1998 (the year this technology was commissioned and implemented), and *D* is for Doppler. This means that it is capable of retrieving not only reflectivity, but also radial velocity and spectrum width. There are 158 of these radars scattered over the United States, leaving very few gaps. More populated areas are covered better, while less populated areas are likely to be covered by only one radar or may be not covered well at all. Each of them is a giant antenna enclosed in a spherical protective dome and placed at a tall tower of about *50-100'* tall to elevate it off the ground to minimize ground clutter. The fiberglass "radome" (the white sphere that encloses the antenna) has a *39 in* diameter, and the parabolic antenna inside the radome is *28'* in diameter. This *28'* antenna whirls around, gears interlocking like clockwork, belts and motors driving the whole show, and completes a full 360° scan in just 10 seconds, adjusts itself to the next higher elevation scan, and then does another 360° scan and repeats. This feat by itself would be impressive, but then one has to realize the immense of amount of genius and computing power involved with transmitting the radiation, receiving it, making sense of it, and storing it rapidly at the time of rotating.

These are S-band radars that emit a *0.75* megawatt beam and are configured to have a 460 kilometer maximum range. This range only applies to the reflectivity product; the radial velocity is much trickier to "unfold" or arrive at an unambiguous solution. Its range is half that of the reflectivity product, or 230 km.

There are two basic modes in which the NEXRADs can operate: Clear Air and Precipitation. In Clear Air mode, the radar rotates slower and performs fewer scan angles. This permits for higher resolution of fine targets like aerosol particles, insects, and snow. In precipitation mode, the radar rotates faster and performs more scan angles, sacrificing resolution for more rapid updates.

CHAPTER 11

RADARS FOR ATMOSPHERIC PROBING

11.1 INTRODUCTION TO MST RADARS

The remote sensing radars for atmospheric probing discussed in this text are confined to applications where the radar is installed in a satellite looking toward the earth and its atmosphere (Chapter 10). The space-borne design constraints of such radars do not allow high antenna apertures, transmitter powers, and high receiver sensitivity as can be easily realized by a ground based radar. Powerful special types of radars like hydrometers have in fact been built in recent years for ground-based probing of the atmosphere, like rain, hail, sleet, and snow occurring in the lower atmosphere. Most of these radars are rather expensive and have been constructed under national projects in radar meteorology in various countries. The great application potential of these studies in the field of forecasting for avionics and prediction of tropospheric propagation conditions for the VHF, UHF, microwave, and millimeter wave bands has been well recognized. Some of the important developments of such specialized radars for atmospheric probing are highlighted in the following sections.

11.2 EARLY DEVELOPMENTS

After World War II, the use of radar for research in atmospheric sciences was initiated by a number of scientists. The pioneering works of David Atlas, Louis Battan, and Roger Lithersmitte make it possible to use radar in obtaining quantitative information about the atmospheric meteorological parameters with a precision superior to that obtainable by other instruments. One can hardly conceive of an instrument which can compete with radar in furthering the advancement of knowledge in microphysics, kinematics, and the understanding of thermodynamic structures on scales ranging from a few tens of meters to several thousand kilometers. Studies on precipitation physics, mesoscale dynamics, the plantery boundary layer, and kinematic structures of the stratosphere and mesosphere have been made very successfully in recent years with these radars.

11.3 DOPPLER RADAR PROBING WITH MST RADAR

A new generation of sensitive Doppler radars has been developed recently that can be used to observe the radial compound of the wind, turbulence, and stable atmospheric layer in the troposphere (10 km), stratosphere (10-50 km), and mesosphere (50-100 km). These radars are called MST (Mesosphere, Stratosphere and Troposphere) radars. The MST radar is capable of probing the clear atmosphere over the height range of 1-100 km. These were reviewed by K. S. Gage and B. B. Balsley, NOAA in Atmospheric Technology, No.13, 1981, outlined as follows.

11.4 EVOLUTION OF MST RADAR

MST radar development is the result of decades of radio propagation experiments, radar probing of the optically clear neutral atmosphere, and radar studies of the ionosphere, by reflection as well as scattering from ionized layers of the upper atmosphere. The first upper atmospheric study based on a radar type of experiment was, perhaps, directed toward the detection of the ionized layers of the upper atmosphere using the FMCW technique followed up by a pulse technique leading to the development of a pulsed ionosonde covering the HF band 3-30 MHz. The first HF echoes from the lower non-ionized neutral atmosphere were obtained from scattering layers occurring at heights ranging from 2 to 60 km by Colwall and Friend in 1936. At that time it was not known that these scattered signals were not from any ionized layers. We now know that free charge plays no role in reflection below 40 km.

In early years, radar meteorologists sought to explain clear air echoes referred to as "angles" in terms of reflection mechanisms, as was done by Friend in 1949. Increasing sophistication of radar analysis techniques led to more detailed studies of the atmospheric structure. In 1950 Booker and

Gordon as well as Megaw developed the theoretical foundation for turbulent scattering of radio waves. Application of this theory to tropospheric radio propagation problems were studied experimentally by Chisholm and his group in 1955, Gjessing in 1964, and Lane and Solium in 1965. Powerful radars were developed during the later half of the 1960s to study the layered structure in the clear atmosphere up to the lower stratosphere using back scatter from the strong inhomogeneities of the radio refracture index. These were reviewed by K. K. Hardy in the book *Remote Sensing of the Troposphere*, edited by V. E. Derr and published by the National Oceanic and Atmospheric Administration (NOAA) in 1972. Besides these, vertically pointing FM-CW radars were used to study the fine structure of clear air reflection within the boundary layer by Gossard, Richter, and Atlas in 1970. In 1967, C. G. Little suggested that the powerful 50 MHz radar at Jicamarca, Peru, designed primarily for incoherent or thermal scatter studies of the ionosphere, might also reveal clear air structures, and the first results of this type were reported by Woodman and Guilten in 1974. They interpreted these results in terms of echoes from turbulent layers acting on temperature gradients (10–35 km) or electron density gradients (55–85 km). These successes initiated the development of several HF radar systems for neutral atmospheric research. These include the sunset radar in Colorado, Poker Flat, Alaska, SOUSY radar in Germany, and another in Platteville, Colorado.

All these VHF radars are basically coherent radars capable of measuring the Doppler spectrum of the echo from which the radial component of the wind can be determined, while the magnitude of the horizontal wind can also be deduced with certain assumptions. Such Doppler radars have been used to observe winds at all heights from the surface through the lower stratosphere. In addition, these Doppler FM-CW radars were used by R. B. Chadwick and his group in 1976 to obtain high resolution wind profiles of the boundary layer. Also, the wind profiles above the boundary layer throughout the troposphere and lower troposphere have been obtained using a variety of VHF and UHF atmospheric radar systems.

11.5 MST RADAR SYSTEMS

Optimally designed VHF systems have also been developed to extend the height range of observation through the mesosphere, thus covering the entire middle atmosphere comprised of the (*a*) Mesosphere, (*b*) Stratosphere, and (*c*) Troposphere, collectively referred to as the MST region. A list of existing and proposed MST radar facilities is shown in Table 11.1.

TABLE 11.1
MST Radar Systems for Clear Air Studies

Facility	Location	Frequency (MHz)	Average power aperture product W m^2	Beamwidth (°C)	Antenna Configuration
Jicamarca	Peru	49.9	2.0×10^{10}	1.0	Phase Doppler Array
Arecibo (S-band)	Puerto Rico	430	8.7×10^{9}	0.17	Fixed dish/Steerable feed
Poker Flat	Alaska	49.9	5.1×10^{9}	1.5	Phased dipole array
EISCAT (VHF)	N. Scandinavia	224	3.3×10^{9}	1.9×0.58	Steerable parabolic cylinder
Arecibo S-band	Puerto Rico	2380	2.0×10^{9}	0.25	Steerable dish/Steerable feed
MU^2	Japan	48	5.0×10^{8}	3	Phased dipole array
Urbana	Illinois	40.9	4.4×10^{8}	3.6×4.8	Phased dipole array
EISCAT (VHF)	N. Scandinavia	933.5	2.4×10^{8}	3.6×4.8	Phased dipole array
ALTAIR (VHF)	Kwajalein	155	1.8×10^{8}	2.8	Steerable dish
ALTAIR (UHF)	Kwajalein	415	1.6×10^{8}	1.09	Steerable dish
SOUSY	Germany	53.5	7.6×10^{7}	5	Phased Yagiuda array
Chatanika	Alaska	1290	5.8×10^{7}	0.6	Steerable dish
Millstone	Massachusetts	1290	5.8×10^{7}	0.6	Steerable dish
Swset	Colorado	40.5	9.4×10^{6}	5×9	Phased dipole array
Platteville	Colorado	49.9	4.5×10^{6}	3×3	Phased dipole array
Wallops	Virginia	430	3.4×10^{6}	2.9	Steerable dish
Defford	England	2815	3.4×10^{5}	0.33	Steerable dish
National Severe					
Storms Laboratories	Oklahoma	2851	6.5×10^{4}	0.8	Steerable dish
Tirupati	India	53	4.2×10^{10}	3	Phased Yagi array with 1024 Crossed Yagis

Of these the nine sensitive systems such as those of Jicamarca, Arecibo, Chatanika, Millstone, and EISCAT have been designed for ionospheric investigations by incoherent (thermal) scatter from electrons, while the others like the SOUSY, Swset, Platevilla, Poker Flat, Urbana, Deffeord, MU, and Tirupati radars are designed for neutral atmospheric studies in the MST or ST (Stratospheric/tropospheric) regions.

11.6 ANTENNA CONFIGURATION

At a lower VHF, phased dipole arrays having moderate steerability would be cost effective. Such an array used in the Poker Flat MST radar is shown schematically in Figure 11.1. At higher frequencies the use of steerable dishes would be more economical, as has been employed in the Chatanika incoherent radar. For larger dishes like that in the Areibo incoherent radar, a fixed dish with a steerable feed would be more economical. In general, the dish configurations are more expensive than the dipole array, but they have greater steerability and bandwidth capability.

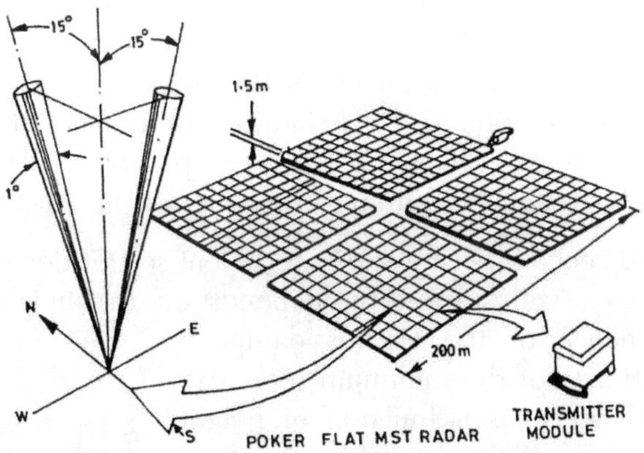

FIGURE 11.1 Antenna array in Poker flat MST radar.

11.6.1 Echoing Processes

The echoes observed on a radar may be produced by a variety of processes:

1. turbulent scatter,
2. thermal or incoherent (Thomson) scatter, and
3. Fresnel (partial) reflection or scatter.

In fact, radar echoes arise, basically by a scattering or by a reflection process from inhomogeneities in the radio refractive index, which is given by

$$n-1=\frac{3.73\times10^{-1}e}{T^2}+\frac{77.6\times10^{-6}}{T}-\frac{N_e}{2N_c}$$

where P is the atmospheric pressure in millibars, e is the partial pressure of water vapor in millibars, T is the absolute temperature in Kelvins, N_e is the number density of electrons per cubic meters, and $N_c = 1.24 \times 10^{-2} f^2$ per cubic meter is the critical plasma density in MKS units for a radio frequency f in megahertz. The first two terms express the contributions from bound electrons inherent in density fluctuations of water vapor and dry air, respectively, while the third term indicates the contributions from the electrons. For middle atmospheric heights the first two terms dominate while the third term plays a major role for ionospheric heights.

11.6.2 Turbulence and Layer Structure Measurements

At VHF, the echo intensity in MST radar roughly decreases exponentially with a height up to about 50 km. At about 50 km, the minimum echo level is determined by the thermal scatter from ambient electrons. Occasionally, however, large enhancements of the echoes may occur due to meter-scale turbulent fluctuations, which induce electron density fluctuations. Besides these, additional echo enhancements are possible at all the height ranges due to Fresnel reflection or scattering processes from horizontally stratified structures.

At UHF, echo intensity due to turbulent scatter depends on the turbulence scale size, which, in turn, depends on the height due to the height dependence of the viscous damping. An estimate of the minimum scale size of turbulence as a function of height made from MST radar data is shown in Figure 11.2. Height distribution of the minimum scale of turbulence is based on MST radar data by Rastogi and Bowhill (1976), Cunnold (1975), Roper (1977), and Gage, Green, and Vanzandt (1980). The open squares denote the maximum heights of observed atmospheric echoes for sensitive radars, obtained from Gage and Balsley, 1980.

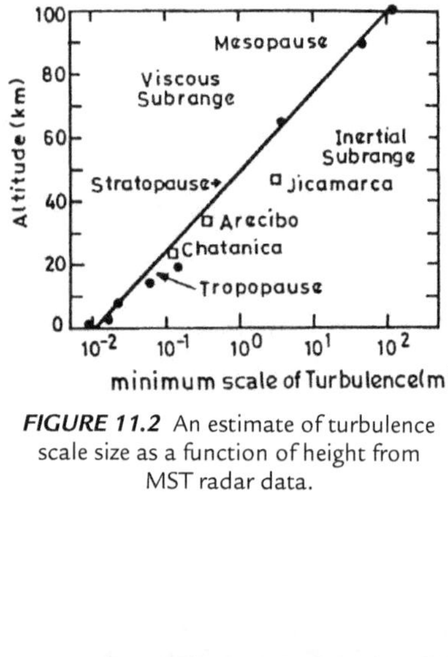

FIGURE 11.2 An estimate of turbulence scale size as a function of height from MST radar data.

11.6.3 Wind Measurement

The doppler shift in the carrier frequency of the received echoes on MST radar can be used to measure the winds in the atmosphere. Two basic techniques are available for the purpose:

1. Velocity-azimuth display (VAD) technique, which gives the mean radial velocity as a function of the antenna azimuth over a 360° rotation.
2. Multiple beam array (MAB) technique in which two or three fixed antenna beams or two or three position-switched beams are employed to determine the horizontal wind from radil wind measurements, assuming negligible vertical velocity. One of the three beams may, in fact, be directed vertically to observe any vertical component of wind. The two techniques of wind measurement are illustrated schematically in Figure 11.3.

11.6.4 Forecasting Applications

Besides the MST radars, which are basically Doppler radars in the VHF and UHF bands designed for clear air studies, Doppler radars in the C-band (4–6 GHz) have been used to obtain three-dimensional wind fields in weather forecasting applications. The National Center for Atmospheric Research (NCAR) in the United States has operated two portable C-band Doppler radars for such studies, and three Doppler radars with one vertical beam are required to determine the wind vector. However, a national network of triple Doppler radars would be rather expensive; therefore, field research programs are going on to explore the capability of a single Doppler radar network for now casting and forecasting applications.

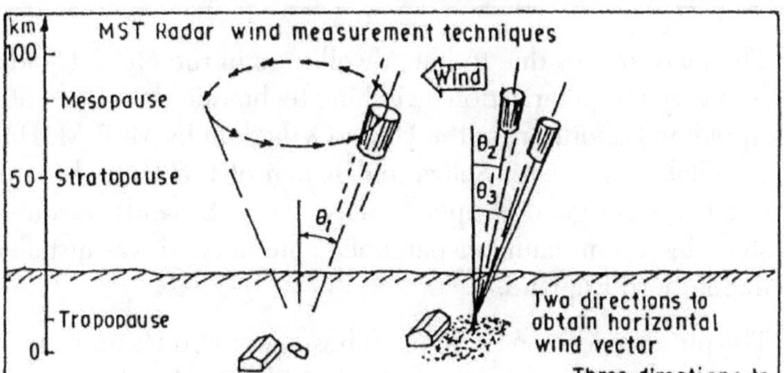

FIGURE 11.3 Two techniques of wind measurement with MST radar.

11.7 INTRODUCTION TO DUAL-POLARIZATION RADARS

A pulsed radar in which the measurements can be made in two orthogonal polarizations is called a dual-polarization radar. The reflectivity of radar targets like raindrops is often markedly dependent on the plane of polarization. A dual-polarization radar looking horizontally is capable of measuring the radar reflectivity of targets for horizontal and vertical polarization sequentially. The ratio of the two reflectivities, Z_{DR}(dB), has been used to distinguish the rain and ice clouds in a thunderstorm cell. T. A. Seliga and V. N. Bringi first introduced this concept of differential radar reflectivity, Z_{DR}(dB) in 1976 to characterize raindrop size distribution and the separations of regions of water and ice-phase hydrometers in storms.

11.8 CHILL RADAR

Seliga, Bringi, and Atkhatib made the first planned measurements of differential radar reflectivity Z_{DR}(dB) in 1977 with a slow polarization switching CHILL radar located in Oklahama, and they published the results in IEEE Transactions on Geoscience Electronics, GE-17, in 1979. An electrically actuated mechanical waveguide switch was used to alternate polarization with a long dwell time of 1.5 seconds on each polarization. The long dwell time yielded uncorrelated polarization reflectivity samples, and considerable time and spatial averaging were required to obtain accurate Z_{DR} estimates.

11.9 CHILBOLTON RADAR

The inaccuracies due to long dwell time in the CHILL radar was removed by a very fast polarization-switching technique developed at the Rutherford Appleton Laboratory in the United Kingdom by M. P. M. Hall and his group in collaboration with Seliga and Bringi of the United States in 1980. The radar was designed to operate at a 10 cm (S-band) wavelength with a fully steerable 25 m diameter parabolic antenna and was installed at Chilbolton in Southern England.

The pulse-repetition frequency chosen was 610 Hz and alternate pulse, each of 0.5 sec, which were transmitted with vertical and horizontal polarizations. A rotary vane waveguide switch coupled to a scalar feed was developed to handle the high peak power of 500 kW employed.

11.10 RADAR CEILOMETERS

Radar developed for the detection of the cloud base at airports are called a ceilometer. A radar ceilometer operating at the millimeter wave window on 35 GHz, which is capable of indicating the cloud base with a great precision, was developed at NCAR in the United States. From these indications the pilot can maneuver his aircraft to avoid the clouds.

11.11 DUAL-POLARIZATION DOPPLER RADARS

The South African Council of Scientific Research developed a dual-polarization Doppler radar at 35 GHz, developed specifically for meteorological research. The radar uses a digital pseudonoise code to modulate the transmitted wave, and two separate 1.8 m diameter antennas are used for the transmitter and receiver, with their beams crossing at the target, occurring overhead. The radar was used extensively for studies of drop size distribution in rainfall, as well for measuring the height profile of vertical air velocity in storms and developing cumulus clouds. The results obtained by F. Pasqualucei were published in the literature during the 1970s. Some of the typical results are shown in Figure 11.4. The promise of dual-polarization Doppler radars has also prompted the study of cloud microphysics, kinematics, and dynamics. The National Oceanic and Atmospheric Administration (NOAA) developed a dual-polarization Doppler radar at 35 GHz, the parameters of which are shown in Table 11.2.

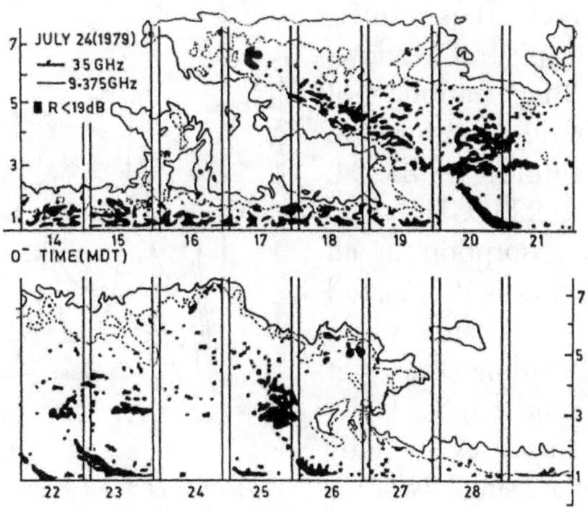

FIGURE 11.4 Some typical results of atmospheric wind parameters obtained from dual-polarization radar at 35 GHz.

TABLE 11.2
Characteristics of NOAA's Dual-Polarized 35 GHz Doppler Radar

Pulse peak power	1.5×10^5 W
Minimum detected power	–105 dB m
Antenna diameter	1.2 m
Antenna beamwidth (3 dB)	0.5°
Polarization of transmitter	Linear or circular
Polarization of receiver	Dual circular or dual linear
Maximum range	10 km
Pulse duration	0.3 sec.
Interpulse period	600 μ sec.

Dual-polarization doppler radars have also been proposed at frequencies higher than 35 GHz, with the expectation that a shorter wavelength would increase the radar reflectivity. A 94 GHz dual polarization Doppler radar has in fact, been developed recently using technology similar to that for the 35 GHz system. The wavelength selection for a radar operates in the earth's atmosphere; however, it is restricted to spectral regions where absorption by atmospheric gas is low. The millimeter wave absorbtion spectrum is shown in Figure 11.5. At frequencies beyond the water vapor line at 22.235 GHz, the spectrum is dominated by three strong absorption regions: (*a*) a 60 GHz rotational O_2 line, (*b*) another O_2 line at 118 GHz, and (*c*) a strong water vapor line at 183 GHz. In between these lines, there are spectral windows where the minima of atmospheric absorption occur, centering on 35, 94, 140, and 220 GHz. The residual absorption in all the windows is dominated by skirts of water vapor lines, including those at 22.234 and 183 GHz, and it increases steadily with increasing window frequency. This is clearly noticeable in Figure 11.5.

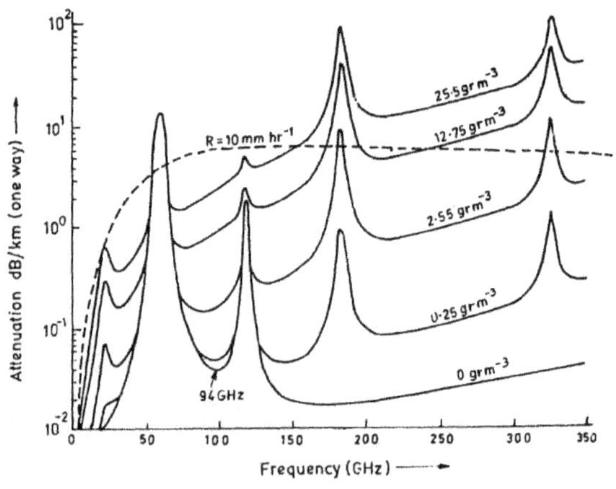

FIGURE 11.5 Atmospheric gas absorption spectrum at the ground in various humidity conditions indicated by the specific humidity values. Absorption for a 10 mm h^{-1} rain is also indicated.

11.11.1 Cloud and Precipitation Radar Reflectivity at Millimeter Waves

At millimeter waves, the drop size of cloud and precipitation particles may be comparable to wavelength when the Mie scattering from the drops have to be considered, particularly at higher millimeter waves.

In the Rayleigh region, the back of dia scattering cross section of a spherical drop dielectric constant K is $\sigma = \pi^5 D^6 \mid K \mid /\lambda^4$ so that cloud radar reflectivity η is given by,

$$\eta = (\pi^5 \mid k \mid^2 /\lambda^4) \int v(D)\, dD \tag{11.1}$$

where η is in cm^{-1} and D and λ are in cm. $\mid K \mid^2$ is shown in Table 11.3 for 94 GHz.

TABLE 11.3
Index of Refraction of Liquid Water Absorption and Scattering Terms at 94 GHz

Temperature	n'	n''	$I_m(-K)$	α_a dB/km/(gm/m^3)	$\|k\|^2$
0°C	3.84	1.48	0.1938	4.80	0.711
15°C	3.21	1.79	0.1564	4.0	0.787
20°C	3.41	2.02	0.1470	3.64	0.828
23°C	3.05	2.05	0.1402	3.47	0.832

For non-Rayleigh scatters, like drops having a diameter longer than 0.5 mm at 94 GHz, the Mie scattering cross section must be computed. The radar

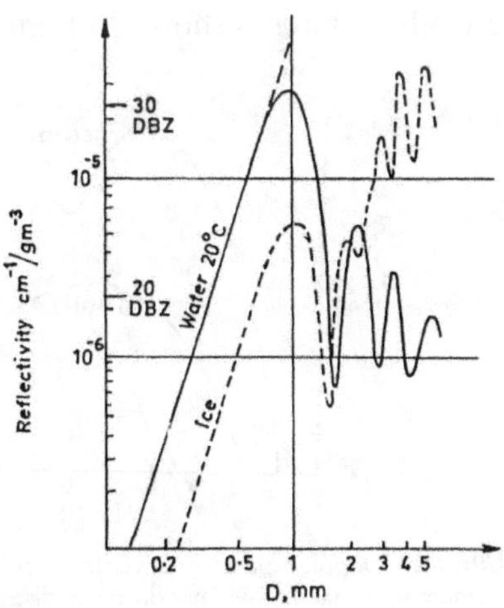

FIGURE 11.6 Radar reflectivity contribution based on Mie scattering functions by particles of diameter D, per grams times cubic meters of water content, as a function of D. The same calculations for ice spheres are also indicated.

reflectivity in practice may be considered as due to an integral one all particle sizes, encompassing both Rayleigh and Mie regions, thereby involving the drop size distribution. The estimated back scattering cross section of water and ice spheres at 94 GHz for various diameters of the spheres is shown in Figure 11.6 The presence of deep Mie scattering oscillation of the radar reflectivity within the rain drop size range may provide an opportunity to identify raindrop size by remote sensing of the drops with a Doppler radar looking vertically upward against the falling raindrops. The velocity of falls *V* of a raindrop increases with drop diameter, and hence the Doppler shift of frequency will be longer for bigger drops, and the Doppler spectrum or density distribution of radar reflectivity will resemble the drop size distribution and as such can be utilized to estimate the drop size distribution of rain and ice. A theoretical evaluation of the Doppler spectrum observable at vertical incidence in a mm/h rainfall assuming a Marshall Polmer drop size distribution is shown in Figure 11.7 for frequencies 9.3 and 94 GHz. The ratio *r* between the Doppler spectra at 94 and 9 GHz is a function of drop size only, but is independent of drop size distribution. This ratio *r* is plotted against the drop size as shown in Figure 11.8, both for raindrops and ice of hail particle. The figure exhibits that for particle diameters exceeding 1 mm, the ratio is considerably higher for ice particles than that for water drops, and this may be utilized to differentiate between raindrops and hailstorms. The vertical profile, that is, the highest distribution of the radar reflectivity and that of the fall velocity of the drop or particle derived from the mean Doppler shift observed with a 94 GHz Doppler radar at different times and location in the United States is shown in Figure 11.9.

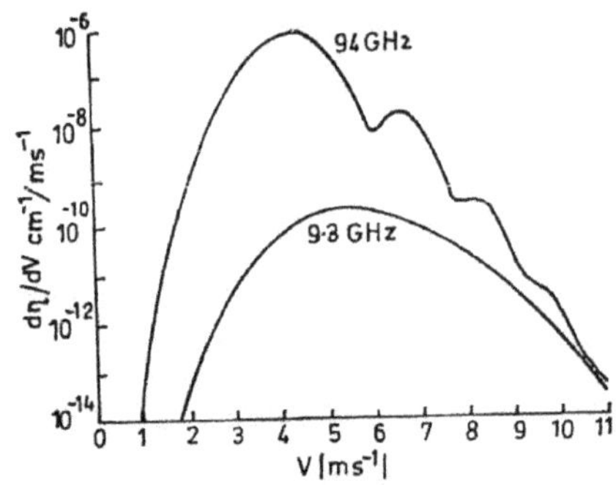

FIGURE 11.7 Doppler spectrum at vertical incidence for a 1-mm/h rainfall assuming a Marshall-Palmer drop size distribution, calculated at 94- and 9.3-GHz radar frequency.

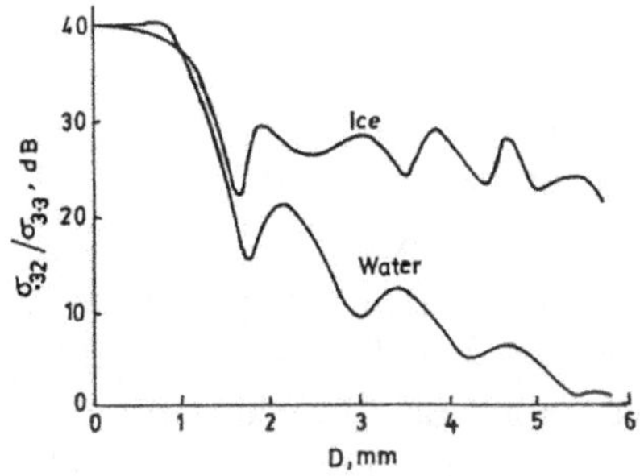

FIGURE 11.8 Ratio between Doppler spectra at 94- and 9.3-GHz wavelength as a function of particle diameter. This ratio is effectively represented by the ratio between radar cross sections of the two wavelengths. The results are also shown for ice spheres.

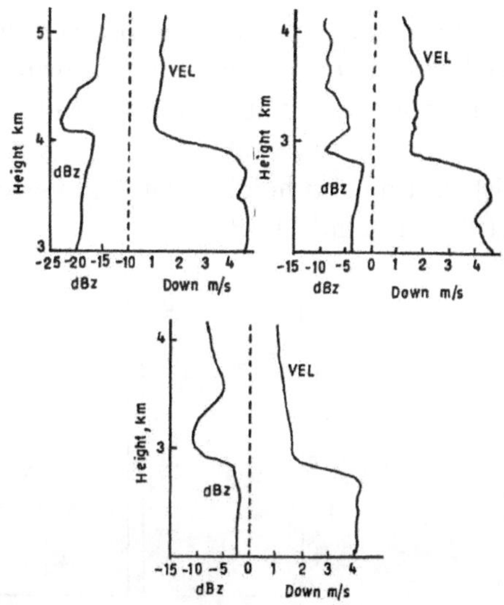

FIGURE 11.9 Vertical profiles of radar reflectivity and mean Doppler observed at 94 GHz through the melting band in stratiform rain at different locations and times (*a*) July, 2,1985, Miami, FL; (*b*) February 6,1987, Miami, FL; (*c*) May 15,1987, Boston, MA.

It may be mentioned here that both the 35 and 94 GHz doppler radars are of the dual-polarization type and as such are capable of distinguishing the rain and ice clouds by using the differential radar reflectivity for the orthogonal polarizations arising from the non-spherical shape of the raindrops in just the same way as the dual-polarization S-band Radar indicated in Art. 11.6. However, the range capability of the millimeter wave

Doppler radar will be to that based at the S-band, due to the sizable rain and atmospheric attenuation coupled with limited transmitter powers available at millimeter waves. The range resolution as well as the Doppler shift are, however, much higher at millimeter waves. With the advent of long-life transmitters, types like extended interaction oscillators, EIO at millimeter waves, and dual-polarization Doppler radars have been developed at 35 and 94 GHz that are capable of increasing up to a height of 10 km for (a) use as a ceilometer at an airport in making the cloud pattern, (b) to increase the drop size distribution of raindrops and hailstorms as a function of height, and (c) to distinguish between rain and ice particles. Higher transmitter power at millimeter waves is now becoming increasingly available, such as with the Gyrotron and other devices as indicated in Tables 9.7 to 9.10 in Chapter 9. Vertical profiles of radar reflectivity and fall velocity of rain and ice particles during straight rain observed with 94 GHz Doppler radar. Observation during a developing thunderstorm in South Africa with a dual-polarization Doppler radar at 35 GHz developed at the National Center of Atmospheric Research (NCAR), U.S., provided interesting results about drop size distribution and height profile of vertical air velocity as shown in Figures 11.10, 11.11, and 11.12.

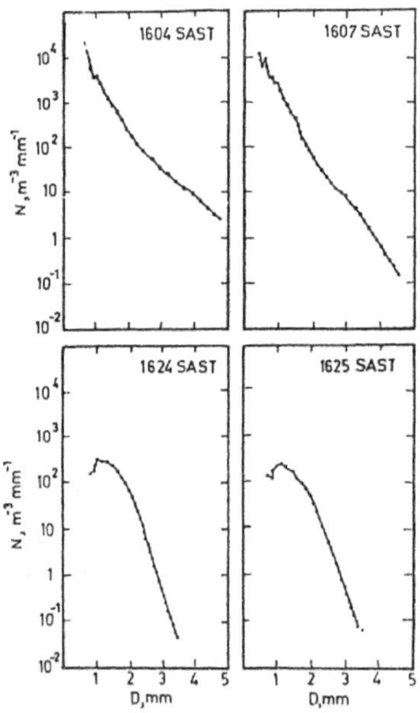

FIGURE 11.10 Drop-size distributions measured in a squall line with a vertically pointing 35 GHz radar. The height of the measurement is 19 *m* AGL.

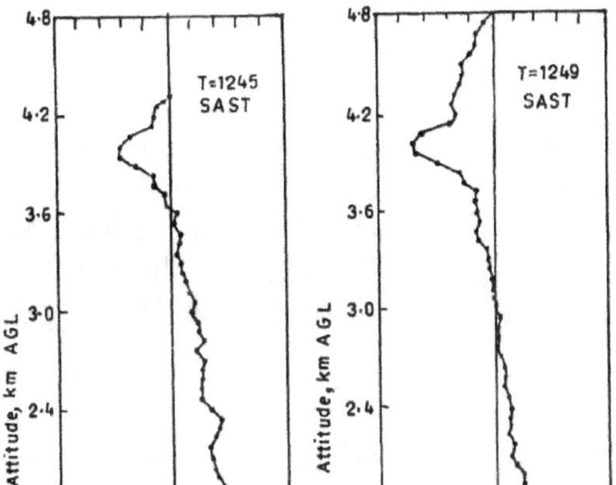

FIGURE 11.11 Height profiles of vertical air velocity measured with the vertically pointing 35 GHz Doppler radar. Note the presence of penerative downdraft between 3.6 km and the cloud top at 12:45 SAST and between 3 and 4.8 km at 12:49 SAST.

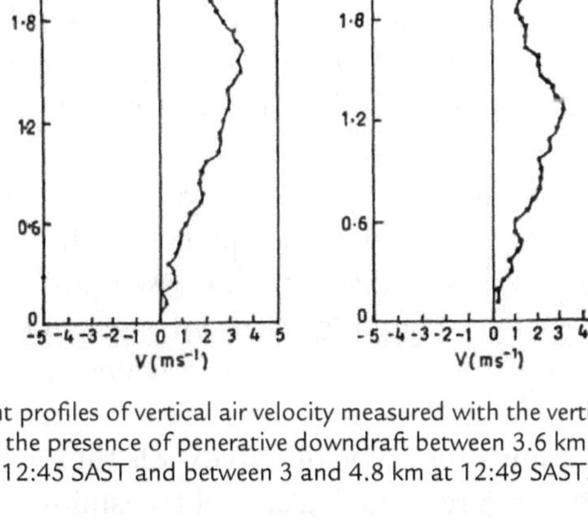

FIGURE 11.12 Height profiles of vertical air velocity in a storm. (SAST is South African Standard Time).

11.11.2 Configuration of Dual-Polarization Doppler Radars for Millimeter Waves

A Doppler radar requires a coherent or phase stable oscillator based on a phase locked source both for the transmitter and receiver local oscillator in order that the measured Doppler shift is not contaminated by frequency or phase drifts in the radar system. A coherent instrumentation pulse radar similar to that used for RCS measurement (included in Section 4.13.4 in Chapter 4) is employed for this purpose. To incorporate the dual-polarization facility in the system, the antenna feed is made of a circular over moded waveguide or a circular horn with the output fed to an orthomode transducer to divide the output into two orthogonal polarizations, horizontal/vertical (dual linear), or left circular/right circular (dual circular). In the latter case a linear to circular polarizer will have to be added.

A block diagram of a 35 GHz dual-polarization Doppler radar is shown in Figure 11.13. In this the transmitter is a 35 GHz magnetron. The frequency chirps of the magnetron will be without any affect, as it is used to derive the local oscillator frequency by mixing it with a stable local oscillator (STALO), in the same way as that in the MTI configuration shown earlier in Chapter 8, Figure 8.3 (*b*). The IF_1 and IF_2 carry information about the received signals for the two orthogonal polarizations, and these are processed further by coherent detectors using a 30 MHz reference signal also derived by mixing a sample of the magnetron frequency with the STALO frequency. The antennas are parabolic dishes with Cassegrain feed for both the transmitter and receiver. The coupling between such antennas placed side by side would be very small for such dishes which produce very narrow antenna beam widths at millimeter waves with minimal side lobes. The use of a circulator or any other type of duplexer is thus avoided. However, to protect the receiver from residual leakage of transmitter RF power to the low noise mixer amplifier, some form of fast limiter like a PIN diode limiter may have to be used at the receiver input, particularly if the transmitter peak power is very high. At the same time, the IF amplifier has to be blanked by a gating pulse during the transmission period to prevent saturation and long recovery time. All these features are, however, typical of an instrumentation radar as described in Chapter 4, in connection with the RCS measurement system.

A block diagram of a 94 GHz dual-polarization Doppler radar is shown in Figure 11.14. In this an extended range interaction oscillator, or EIO, is used as a transmitter. Refering to the figure, it is clear that any drift of frequency or phase of the EIO is sampled by the mixer $M_1$and produces a similar drift in COHO, so that this drift and the drift in the received signal may be the same. This will maintain the phase relationship of the two inputs of each of the coherent detectors D_1, D_2, D_3, and D_4. The four outputs I_V, O_V, I_H, and O_H can be used to obtain the result and echo signal, and its Doppler shift for both horizontal and vertical polarizations. The echo signal is indicative of the

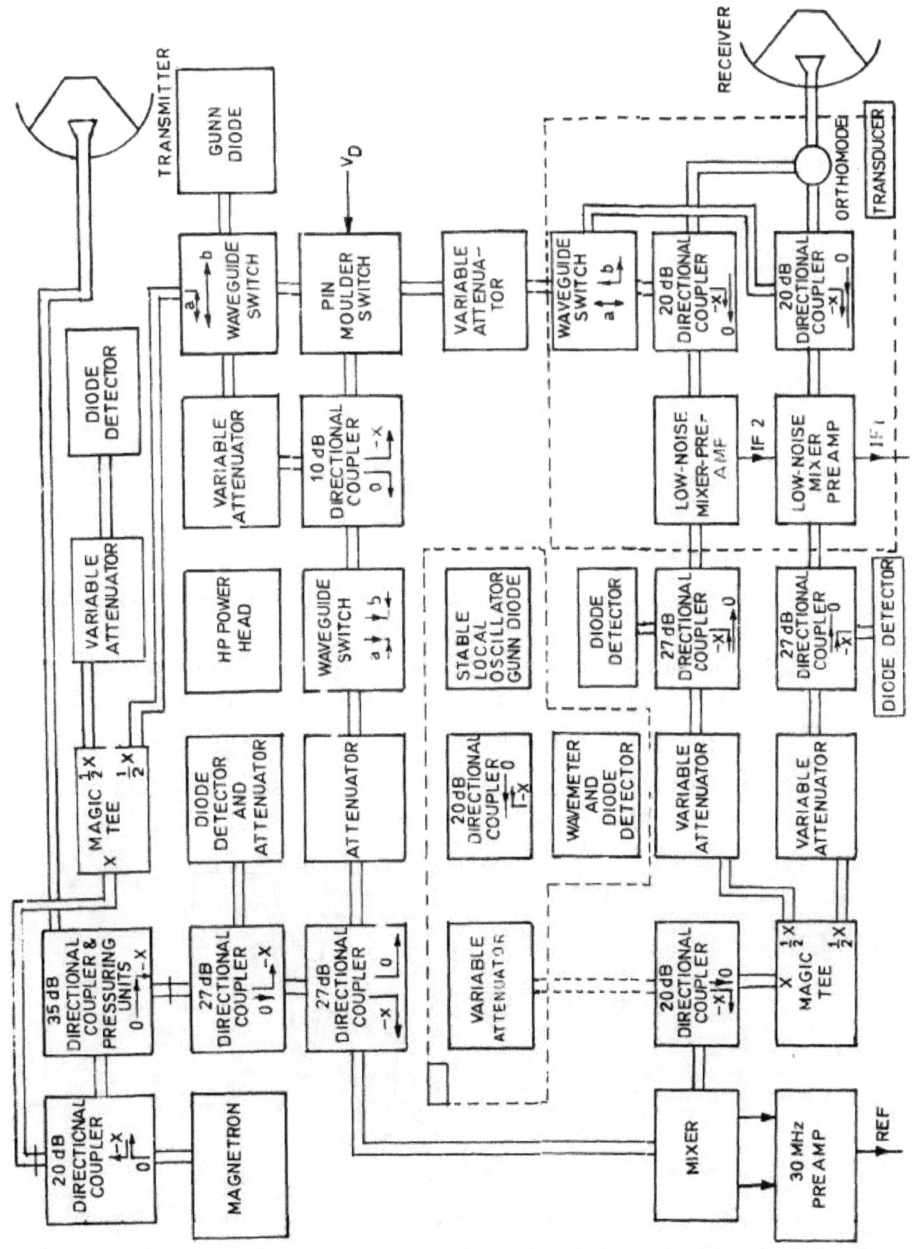

FIGURE 11.13 Block diagram of 35 GHz dual-polarization Doppler radar.

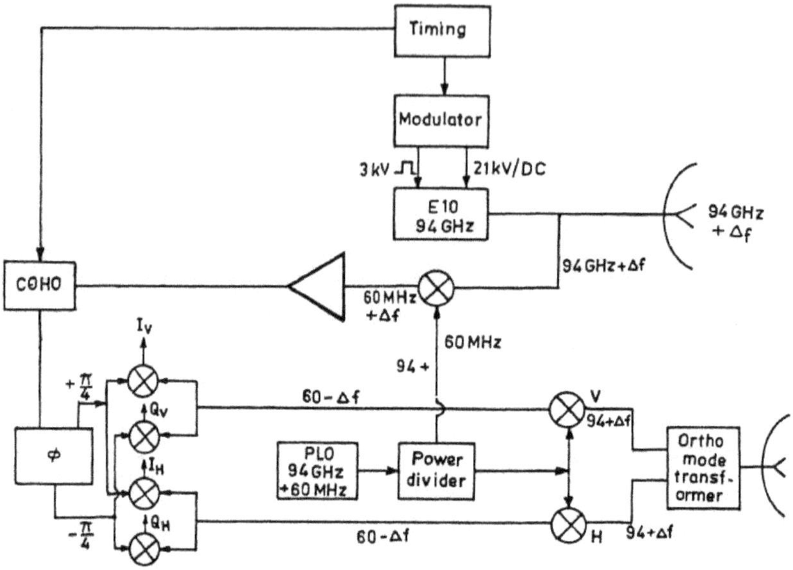

FIGURE 11.14 Block diagram of dual-polarization Doppler radar at 94 GHz.

back scattering cross sections, while the Doppler shift is proportional to the radial velocity, which is nearly proportional to the drop diameter, *D*. Thus, the Doppler spectrum is indicative of the drop size distribution.

A theoretical estimate of the back scattering cross section of spherical liquid water drops in the Mie scattering region at 94 GHz is shown in Figure 11.15.

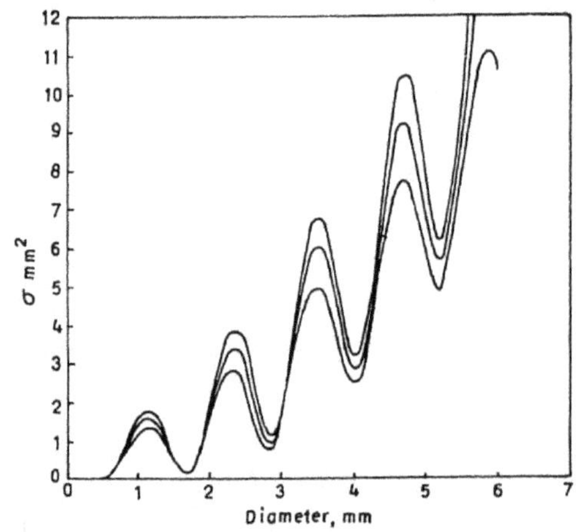

FIGURE 11.15 Back scattering cross section of spherical liquid water drops as a function of their diameter at 3.2 mm wavelengths and for 0, 10, and 20° temperatures.

The value of σ exhibits successive maxima and minima with increasing drop diameters. The rate of change of radar reflectivity radial velocity, *dx/dy*, measured with a 94 GHz Doppler radar during 15 millimeter/hr rain, plotted as a function of vertical velocity, is shown in Figure 11.16, the dashed line indicating the predicated spectrum based on a Mie back scattering cross section and an exponential drop size distribution. The most noticeable feature is the spectral dip at 6.2 m/s. The function at the top of the figure is the cross correlation between the two spectra plotted as a function of velocity lag. Figure 11.17 shows the rate of change of radar reflectivity with drop diameter observed with the 94 GHz Doppler radar plotted as a function of drop diameter, the dashed line indicating the predicated spectrum based on the Mie function and an exponential distribution. The agreement is remarkable between $D = 0.5$ mm and $D = 3$ mm.

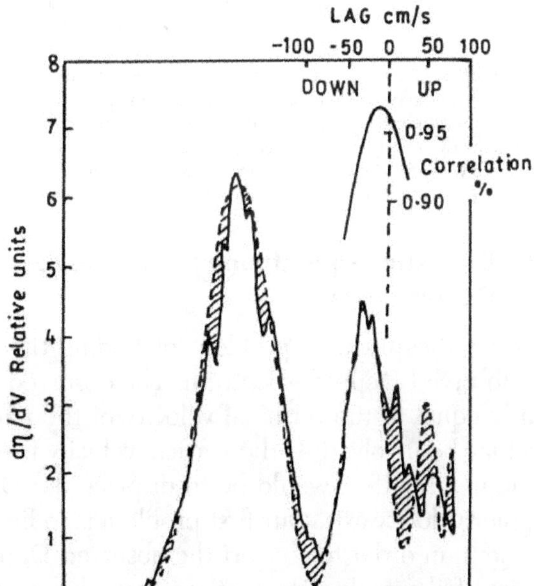

FIGURE 11.16 Radial velocity (Doppler) spectrum observed in a 15 mm/hr rain at vertical incidence together with a predicted spectrum (dashed line) based on Mie back scattering cross section and an exponential dropsize distribution with A adjusted for best fit. The shaded area shows the difference between the observed and predicted spectrum. The most noticeable feature is the spectral dip at 6.2 m/s. The function at the top of the figure is the cross correlation between the two spectra as a function of a velocity lag. The correlation peak (0.965) is obtained for a very small velocity lag (less than 5 cm/s).

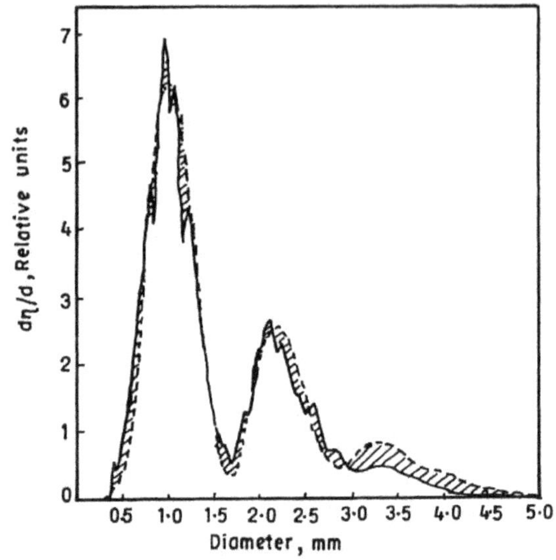

FIGURE 11.17 Spectrum as a function of equal raindrop diameter *D* intervals. Both the observed spectrum and a predicted spectrum (dashed line) based on the Mie function and an exponential dropsize distribution are shown. The shaded area highlights the small difference between the two functions. The agreement is remarkable between $D = 0.5$ to $D = 3$ mm, but for diameters larger than approximately 3 mm the observed spectrum falls off more rapidly.

11.11.3 Estimating Updraft and Particle Size Spectra from Doppler Data

We consider the practical problem of finding the particle size spectrum from an observed Doppler spectrum. The observed Doppler velocity is not necessarily equal to the terminal velocity of the particle, since vertical air motions may be involved. If the vertical velocity is uniform throughout the pulse volume, its effect would be to displace the Doppler spectrum along the Doppler velocity axis. Our first problem is to find the magnitude of this displacement, in order to convert the observed Doppler velocities to particle terminal fall speeds. Two methods have been used to accomplish this: (*a*) lower bound method, and (*b*) Roger's method.

In the first method it is assumed that particles of a certain minimum size and, therefore, minimum terminal velocity, are present in large enough numbers to be just detectable. The minimum fall speed is usually taken as 1 m sec^{-1}. This means that the (algebraically) largest Doppler velocity (taken positive away from the radar) is due to particles of terminal fall speed 1 m sec^{-1}. Therefore, the largest Doppler velocity in the observed spectrum is taken to correspond to particles of fall speed 1 m sec^{-1}. This information immediately gives the magnitude and sign of the vertical air velocity, and enables conversion of the Doppler to terminal velocities. Difficulties and inaccuracies arise in the application of this method due to noise in the data, non-uniform air motion, and turbulence. Because of these factors, the Doppler spectrum does not show a sharp cutoff. In a variant of this method (Battan and Theiss), the point at which the Doppler power falls to a certain level below its peak value (usually 10 dB) is taken as corresponding to the particle of minimum detectable size. Referring to Figure 2 (*a*), we see that for an exponential particle size distribution, the Doppler power falls to 10 dB below its peak value at terminal velocities of 2.5, 3.0, 4.4, and 5.6 m/sec for rainfall rates of 1, 2, 10, and 30 mm/hr respectively. This example brings out the drawbacks of this method. An inherent drawback of the previous method is that the assumption that particles of a certain minimum size are present in the pulse volume is not always valid, especially in situations where particle sorting takes place as a result of wind shear.

The second method depends upon assuming a relationship between the radar reflectivity and the mean Doppler velocity under no vertical air motion. Rogers (1964) found this relation by assuming that the particle size distribution is that given by Marshall and Palmer. For this distribution, we have

$$\overline{v} = 3.8\,Z^{1/14}$$

where $\overline{v}$ is in m sec^{-1} and Z is in mm^6/m^3. The observed mean Doppler velocity will in general be different from the $\overline{v}$ computed from the measured Z using the previous equation, the difference being attributable to the vertical air motion. Thus the vertical air velocity can be deduced, and the Doppler velocities can be converted to particle fall speeds. This method suffers from the drawback that it is necessary to make an assumption regarding the particle size distribution.

Once the Doppler velocities have been converted to particle fall speeds using either of the previous methods, the particle concentration can be computed from

$$Z'(v) = D^6 N(v)$$

where $Z'(v)\,dv$ and $N(v)\,dv$ are the contribution of particles of fall speed v to $v + dv$ to the radar reflectivity factor and the particle concentration, and D is the diameter of the particle of fall speed v (Rayleigh approximation has been assumed). In terms of $N(D)$, the particle concentration per unit diameter interval, we have

$$N(D) = \frac{Z'(v)}{D^6}\frac{dv}{dD} \tag{11.2}$$

This shows how the number concentration density can be derived from the $Z'(v)$ spectrum. As we know, however, there will be a minimum detectable power level for any radar P_{min}, and a corresponding minimum detectable power density $S(f)$ such that

$$P_{min} = \int S(f)\,df = CZ_{min} / r^2$$

where the last form on the right is the radar range equation and C includes all the radar parameters. Accordingly there will be minimum detectable $S(f)$ over the entire acceptable frequency band of the radar such that

$$P_{min} = S_{min}(f)[f_{max} - 0]$$

where $f_{max} = PRF/2$ or PRF depending upon the system. Since $S(f)df = Z'(v)$ dv, there will also be a corresponding minimum $Z'(v)$ which is independent of v. Calling this minimum detectable value Z'_{min} we then have from equation 11.2 the minimum detectable number concentration density

$$N_{min}(D) = \frac{Z'_{min}}{D^6}\frac{dv}{dD}$$

Values of $N_{min}(D)$ depend upon radar sensitivity and range.

There will usually be some error in estimating the updraft. It is, therefore, of some value to estimate the error in $N(v)$ due to a given error in the vertical air velocity. Assuming that there is no error in $Z'(v)$ we have

$$\frac{\Delta N_v}{N_v} = -\frac{6\Delta D}{D}$$

If there is an error ΔU in estimating the updraft, the error in the particle terminal velocity will also be ΔU. The resulting error in the diameter is

$$\Delta D = (dD / dv)\Delta U$$

Assuming $$v = aD^b$$

We get, $$\frac{\Delta N_v}{N_v} = \frac{6\Delta U}{bV}$$

Which may also be written as

$$\Delta \log_{10} N_v = 2.606\frac{\Delta U}{bV}.$$

11.12 RAINFALL MEASUREMENT WITH METEOROLOGICAL RADAR

The difficulties of rainfall measurement over a wide area are well recognized. Due to the small spatial variability of rain, a network of gauges has severe limitations in sampling representatively. On the other hand, radar suffers from some basic uncertainties, primarily due to the fact that the measured quantity, namely, the equivalent radar reflectivity factor, is not uniquely related to the rainfall rate. Although radar has been used experimentally for over thirty years for measuring precipitation, operational implementation has been slow. Now we find that the data available are under utilized, and both misunderstanding and confusion exist about the accuracy of the results. However, implementation of digital recording and processing of weather radar data in recent years has largely removed this obstacle. For operational forecasting of river flow and flash floods, dense rain gauge observations are desirable, no doubt, but their installation has not been practical. Thus there has been a considerable interest to use weather radar, as it provides spatially and temporally continuous measurements available immediately in one location. Both scatter and attenuation of microwaves are considered in the precipitation estimated. Presently reflectivity data are considered practical for operational measurement of rainfall over large areas.

11.12.1 Theoretical Consideration

The back scattered radar power due to precipitation particles is directly proportional to the sixth power of particle diameter (D^6) per unit volume illuminated by radar beam. If N_i denotes the number of drops in a unit volume of air with diameter D_i then the radar reflectivity factor Z can be mathematically expressed as

$$Z = \sum_{i=1}^{\infty} N_i D_i^6 = \int_0^{\infty} N(D) D^6 dD \tag{11.2a}$$

where $N(D)$ is the number of drops with diameters between D and $(D + dD)$ in a unit volume of air. If the vertical air motions are absent, then the rainfall rate R is related to D by an equation

$$R = \frac{\pi}{6} \int_0^{\infty} N(D) D^3 V_t(D) dD \tag{11.2b}$$

where $V_t(D)$ represents the terminal velocity of a drop of diameter D that is approximated by

$$V = 1400\, D^{1/2} \tag{11.3}$$

Substituting the Marshall-Palmer (1948) exponential drop size distribution into equations (11.2a) and (11.2b) and using the empirical relation between V_t and D_3, one may get a relation between Z and R which is of the form

$$Z = AR^b \tag{11.4}$$

Hence, if the drop size distributions are exponential and also the vertical air motions are small relative to the drop terminal velocities, then no fundamental limitation arises for the accurate estimates of rainfall using radar. But in practice, the drop size distribution is rarely known, and it varies in time and space. Furthermore, the vertical air motions are frequently of the same magnitude as the terminal velocities, particularly in cases of thunderstorms. Thus the Z-R relation is not a unique one, and we are compelled to rely on average empirical relations. Battan (1973) has presented a comprehensive list of the Z-R relationship as established by different investigators. A frequently used relation is of the form

$$Z = 200\, R^{1.6} \tag{11.5}$$

11.12.2 Determination of Radar Parameters

The average power received P_r at range r can be expressed as

$$P_r = \frac{P_t G^2 \lambda^2 h \theta^2}{512\pi^2 r^2} \sum_{vol} \sigma_i \tag{11.6}$$

where P_i = peak transmitted power in the pulses

G = antenna gain

h = pulse length in space

λ = wavelength

θ = beam width between half power points

and r = range

Again $$\sigma_i = \frac{\pi^5}{\lambda^4} |k|^2 D_1^{\,6} \tag{11.7}$$

where $|k|^2$ is a function of the dielectric constant of the targets and is approximately 0.93 for water and 0.179 for ice in the microwave band. By equations (11.6) and (11.7)

$$P_r = \frac{P_t G^2 h \theta^2}{512 r^2} \frac{\pi^3}{\lambda^2} |k|^2 \sum_{vol} D_t^6$$

$$= \frac{P_t G^2 h \theta^2 \pi^3 |k|^2}{512 r^2 \lambda^2} Z \tag{11.8}$$

The quantity $\sum_{vol} D_i^6$ commonly designated by Z is known as the reflectivity factor.

In practice the gain G is realized only at the center of the beam, and the intensity tapers off with increasing axial angle. If the antenna is a circular paraboloid, the distribution of intensity in the main beam is closely approximated by the Gaussian function of the axial angle α and is expressed as

$$G(\alpha) = G \exp = [4(\text{In } 2)\,\alpha^2 / \theta^2] \tag{11.9}$$

An effective beam width θ_e can be found by integrating the actual beam intensity pattern for two-way transmission out to some axial angle α_1 which is considered large enough to include all the scatterers contributing to the signal. Then,

$$G^2 \left[\frac{\theta_e}{2}\right]^2 = \int_0^{\alpha} G^2(\alpha) 2\alpha \, d\alpha \tag{11.10}$$

Integration of equation (11.10) yields

$$\theta_e^{\,2} = 0.69\,\theta^2 \text{ for } \alpha_1 = \frac{3\theta}{4}$$

and

$$\theta_e^{\,2} = 0.72\,\theta^2 \text{ for } \alpha_1 \geq \theta$$

So the choice of α_1 is not critical, and θ_e^2 can be set equal to 0.7 θ^2 with very little error. For a conical beam,

$$P_r = \frac{P_t G^2 h \theta^2 \pi^3 |k|^2}{730\, r^2 \lambda^2} Z \tag{11.11}$$

In log form the previous equation reduces to

$$Z_c\,(\text{dBz}) = P_r\,(\text{dBm}) - 10 \log P_t + 20 \log r + C \tag{11.12}$$

where $$Z_e\,(\text{dBz}) = 10\log Z_e(\text{mm}^6\text{m}^{-3}) \tag{11.13}$$

and $$C = \text{radar constant}$$

The power transmitted P_t is not included in the constant, as it varies somewhat with transmitter age, and so on, and should be monitored regularly. In equation (11.11) h and λ can be accurately determined with standard test instruments. Beam patterns at the desired frequencies are usually measured at the factory, and the beam width can be accurately obtained directly from them. By comparing with a standard horn, the gain is also measured in the factory, but the value is not adequate for the field. Two-way losses like waveguides, rotary joints, and random losses must also be taken into consideration.

11.12.3 Reflectivity Factor

The most important parameter that is measured by using meteorological radar is the reflectivity of the scattering volume. From a knowledge of reflectivity, by using suitable empirical relations, one may deduce useful meteorological quantities like rainfall rate and liquid water content. Moreover, severe storms, hail storms in particular, can often be identified by their high reflectiveness.

In order to determine reflectivity, the quantity which is to be measured is the power received. From the average power received P_r and the radar equation, the volume reflectivity η can be calculated. If it is then normalized for wavelength, we can get reflectivity factor Z_e. By definition Z is equal to the summation over the sixth power of the drop diameters, but when it is obtained from radar measurements with the radar equation, it is called equivalent Z and is denoted as Z_e. The accuracy of Z_e depends on how accurately one can measure P_t and also how well known the parameters are in the radar equation.

11.12.4 Signal Averaging

The power received from a radar volume at any instant (P_r) depends mainly on the configuration of the scatterers at that moment. By averaging P_r either in time or in range or both, the desired quantity P_r can be achieved. In general, the time integration is obtained by digital sweep integrators. The signal from the receiver (a log receiver usually, for accommodating the large dynamic range involved) is digitized in a number of range bins, a few hundred generally, and finally summed over a number of transmitted pulses of the order of 16 or 32. Range averaging is obtained either digitally by combining range bins or analoguewise by some sort of smearing (e.g., RC or tapped delay line).

Each integration cycle should be completed in about a beamwidth, but the antenna should not be turned too slowly. In practice thirty or forty samples are averaged, which results in a standard deviation of less than a dB in the fluctuations of the signal. But for an unaveraged signal, the standard deviation is about ± 6 dB. The performance of the receiver also affects the measurement of P_r. One source of variation is drift in the DC level. It is important to note that this should either be held to a few millivolts or to be compensated by some means.

11.12.5 Antenna Gain

There are two ways of measuring antenna gain in the field:

1. to measure the back scatter from a standard target, and
2. to measure the power received by a standard gain horn in the far field. The 'horn' method is superior to the standard target method due to its steadiness and reliability of measurement and also its error-canceling characteristics. If P_{ff} is the power received by an antenna of gain G_r then in the far field of an antenna of gain G_t and transmitting power P_t, one can write

$$P_{ff} = \frac{P_t G_t G_r \lambda^2}{(4\pi r)^2} \tag{11.14}$$

P_t is measured through a directional coupler at the same place where the receiver is calibrated by the signal generator, G_r is precisely known as the receiving antenna in a calibrated standard gain horn, and P_{ff} is measured with the same power meter used to measure P_t. From an accurate knowledge of wavelength and range, the effective gain G_t can be computed.

11.12.6 Radar Rainfall Adjustment

Adjustments of the radar estimates can be made either by:

1. changing the Z-R relationship or
2. keeping the Z-R relationship fixed and using rain gauge observations. In the simplest calibration technique, a number of gauges (N) are utilized

and a multiplicative adjustment factor (F), the ratio of gauge derived (G), and radar indicated rainfall (R) are computed by

$$F = \frac{\sum_{i=1}^{N} G_t}{\sum_{i=1}^{N} R_i} \tag{11.15}$$

or by using,

$$F = \frac{1}{N} \sum_{i=1}^{N} \frac{G_i}{R_i} \tag{11.16}$$

In equation (11.15) observations received a weight proportionate to depth while in equation (11.16) all gauge-radar comparisons have equal weight. Indeed, the relationship selected for converting reflectivity to rainfall is of little importance and has negligible impact on the corrected radar depth estimate.

11.12.7 Factors Affecting Rainfall Measurement

Numerous raindrop-size measurements have shown that drop sizes vary significantly from storm to storm and also within storms. This variability introduces uncertainty into the Z-R relation, and some meteorologists assume it to be the major cause of uncertainty in radar rainfall measurements. In addition, the following factors have significant effects, at least as much as those arising from variations in drop sizes.

1. *Vertical Variations in Reflectivity.* The spatial averages at several stages in the recording and processing of radar data tends to depress the peak reflectivity in showers of small dimension. There is evidence also that Z values and the corresponding values of Z_e are sometimes reduced by downdrafts associated with rainfall shafts.
2. *Horizontal Variations in Rainfall.* Small-scale variability of precipitation in the horizontal plane creates some complications. Reflectivity gradients as great as 5 dB km^{-1} are not particularly unusual, and they often occur in the vicinity of heavy rain.
3. *Time interval between Radar Maps.* When the time interval between radar maps is larger than the time necessary for small intense echoes to move from one grid point to the next, then the time integrations of R will be in error unless the radar patterns are advected along the motion vectors.

4. *Sensitivity of the Radar.* A radar may fail to detect light rain at large ranges. For example, it might indicate a 2 hr accumulation of 5 mm due to a short intense shower but fail to detect similar accumulation which occurred as steady light rain. Clearly an adjustment based on either of the previous situations would be inappropriate for the other.

11.12.8 Conclusions

Measurement of reflectivity by the attenuation of the signal due to precipitation. Attenuation mostly affects the shorter wavelengths, so a wavelength which is not significantly attenuated by the precipitation to be measured should be chosen. If light rain is involved, wavelengths as short as 3 cm may be satisfactory, but for the measurement of heavy rain and extended areas, a wavelength of 10 cm is recommended.

Various factors as discussed here contribute to some discrepancies in radar measurements. A more comprehensive analysis is desirable to achieve a better physical understanding and a quantitative assessment of the factors.

Since the physical phenomena involved in precipitation are complex and variable, they should be analyzed in depth so that they can properly be taken into account in attempts to develop objective monitoring techniques and useful operational procedures.

In order to adjust the radar, it is desirable that frequent comparisons be made between rainfall rates deduced from radar measurements and those measured with rain gauges.

Radar meteorology, being largely an observational science, is sensitive to the quantity and cost of the means for making measurements. Recent rapid advances in digital electronics should, therefore, have a significant impact on the field. As part of the weather radar systems, many installations now use digital integrators and some incorporate mini-computers. As a result, accuracy of measurement has been greatly increased. A computer can also eliminate the necessity of some standard radar components like the specialized displays (e.g., A-scope, PPI), timing circuitry, and so forth. Its ability to store and integrate the signal over times and areas has obvious hydrological applications. Radar observations from satellite platforms are now considered as further possibilities to establish detailed rainfall patterns in the near future, particularly over the tropical oceanic regions and in the maritime continental area.

11.13 WIND MEASUREMENT BY DUAL BEAM INCOHERENT RADAR

The conversion from relative velocity u to fluctuation frequency F is through the Doppler relation $F = 2u/\lambda$. These fluctuation spectra show a secondary peak at a beat frequency $\hat{F} = \left|\bar{f}_1 - \bar{f}_2\right| = 2\left|\bar{v}_1 - \bar{v}_2\right| / \lambda$, where $\bar{f}_1$ (or $\bar{v}_1$) and $\bar{f}_2$ (or $\bar{v}_2$) are the mean Doppler frequencies (or velocities) on the axes of the respective beams, provided that the beam spacing is sufficiently large. In the situation in which the beams are tilted by angles $\pm\,\delta$ from the normal to the wind vector, $\bar{v}_1 = v_0 = -\bar{v}_2$ and so

$$\hat{F} = 4v_0 / \lambda \cong 4\,W \sin\delta = 4W\delta$$

where W is the speed of the crosswind, and the last identity on the right is applicable to small angles δ. This secondary peak occurs in the fluctuation spectrum under the condition that

$$k = \left|\bar{v}_1 - \bar{v}_2\right| / \sigma_v \geq 4$$

or for the symmetrical case

$$k = 2\,v_0 / \sigma_v \geq 4$$

where σ_v is the standard deviation of the Doppler spectrum in either beam (the σ_v' s in both beams were assumed equal). For $k = 4, \hat{F} / \Sigma = \hat{u} / \sigma_v = 3.9$, where $\Sigma = 2^{1/2}(2\sigma_v / \lambda)$ is the standard deviation of the single beam fluctuation spectrum, and $\hat{u} = \lambda\hat{F} / 2$ (that is, u is the relative velocity corresponding to any beat frequency F). Thus when

$$k = \frac{2v_0}{\sigma_v} = 4, \frac{\hat{u}}{\sigma_v} = 3.9, \text{ or } \frac{\hat{u}}{2v_0} = \frac{3.9}{4.0} = 0.975$$

In other words $\bar{u}(\text{or } \hat{F})$ is a reasonably accurate measure of the relative Doppler velocities $(2v_0)$ or $\left|\bar{v}_1 - \bar{v}_2\right|$ between the two beam axes as long as the beams are sufficiently wide apart. In the symmetrical beam case where $v_0 = W \sin\delta$, the requirement that $k \geq 4$ implies that $2\,W \sin\delta/\sigma_v \geq 4$ or $\sin\delta \geq 2\sigma_v/W$.

The previous discussion applies to the identification of the crosswind W from the secondary peak in the dual beam fluctuation spectrum. However, the variance of the relative velocity spectrum s^2 (or of the corresponding fluctuation spectrum $\Sigma^2 = (4/\lambda^2)s^2$) in the symmetrical beam case is

$$s^2 = 2(\sigma_v^{\ 2} + \delta^2 W^2)$$

In terms of the fluctuation spectrum, then

$$\Sigma_{1,2}^{\ \ 2} = (8/\lambda^2)(\sigma_v^{\ 2} + \delta^2 W^2)$$

where the subscript 1, 2 has been added to signify the variance of the dual beam case. Of course, for either beam individually $s^2 = 2\sigma_v^{\ 2}$ and

$$\Sigma_1^{\ 2} = \Sigma_2^{\ 2} = (8/\lambda^2)\,\sigma_v^{\ 2}$$

A subtraction gives,

$$(\Sigma_{1,2}^{\ \ 2} - \Sigma_1^{\ 2}) = (8/\lambda^2)\,\delta^2 W^2$$

which allows the determination of the crosswind W from measurement of the dual beam variance $\Sigma_{1,2}^{\ \ 2}$ the single beam variance $\Sigma_1^{\ 2}$ and the known beam spacing. More generally, of course $\delta W = |\overline{v}_1 - \overline{v}_2|/2$ so that we can write,

$$\Sigma_{1,2}^{\ \ 2} - \Sigma_1^{\ 2} = (2/\lambda^2)[\overline{v}_1 - \overline{v}_2]^2$$

In this form the differences in the dual and single beam variances allow the measurement of the difference in mean Doppler velocities between the two beams regardless of their orientation relative to the winds.

The basic assumptions made in the previous derivations were that: (*a*) the two beams are identical in shape and gain, (*b*) the individual beam patterns are Gaussian, (*c*) the volumes illuminated had equal reflectivity, and (*d*) the spectrum variances in the individual beams were equal.

11.13.1 The Generalized Dual Beam Fluctuation Spectrum

Let the Doppler spectrum returned on beam 1 be represented by $\psi s_1(f)$ and that on beam 2 by $(1 - \psi)\ S_2(f)$, where ψ is the fraction of the total power returned on beam 1 and the remainder $(1 - \psi)$ is that received through beam 2. When the corresponding signals returned simultaneously are added in the RF for IF stages of the radar, the resulting normalized Doppler spectrum is

$$S_{1,2}(f) = \psi_1 S(f) + (1-\psi)S_2(f) \qquad (11.17)$$

The corresponding fluctuation spectrum is then the auto-correlation of $S_{1,2}(f)$ with itself or

$$S_p(f) = \int_{-\infty}^{\infty} S(f)\, S\,(f+F)\, df \tag{11.18}$$

Substituting (*A*) into (*B*), we get

$$\begin{aligned} S_p(f) = \int_{-\infty}^{\infty} [\psi^2 S_1(f)\, S_1(f+F) + (1-\psi^2) S_2(f)\, S_2(f+F) \\ +\psi(1-\psi)\, S_1(f) S_2(f+F) + \psi(1-\psi) S_2(f)\, S_1(f+F)]\; df \end{aligned} \tag{11.19}$$

It is then necessary to select the analytical forms of $S_1(f)$ and $S_2(f)$ and carry out this rather cumbersome integration. This is essentially with the simplifications that $\psi = 1/2$ (equal power in both beams) and $S_1(f)$, $S_2(f)$ are both Gaussian with equal variance. However, the problem is greatly facilitated by first finding the auto-correlation function $G(\tau)$ of the dual beam spectrum.

Now, we have

$$\rho\,(\tau) = G(\tau)\, G^*(\tau)$$

and finally taking the inverse Fourier transform of $\rho(\tau)$ to get

$$S_p(F) = \int_{-\infty}^{\infty} \rho\,(\tau) \exp(-j\, 2\pi F \tau)\, d\tau$$

We now proceed by first assuming both $S_1(f)$ and $S_2(f)$ to have Gaussian forms with means $\bar{f}_1$, $\bar{f}_2$ and variances $\sigma_1^{\;2}, \sigma_2^{\;2}$ respectively. We know that

$$G(\tau) S(f) \exp(j\, 2\pi\, f \tau)\, df$$

We have the Fourier transform pair

$$e^{-at^2} \leftrightarrow \sqrt{\frac{\pi}{a}} \exp\,[-\omega^2 / 4a]$$

Or writing $\omega = 2\pi\, f$ and $a = 2\pi^2\sigma^2$

$$e^{-at^2} \begin{cases} \leftrightarrow \sqrt{\dfrac{\pi}{a}} \exp\,[-\pi^2 f^2 / a] \\ \leftrightarrow \dfrac{1}{\sqrt{2\pi}\,\sigma \exp\,[-f^2 / 2\sigma^2]} \end{cases}$$

where $a = 2\pi^2\sigma^2$

Thus the auto-correlation function corresponding to a normalized Gaussian Doppler spectrum centered at $f = 0$ is $G(\tau) = \exp[-2\pi^2\sigma^2\tau^2]$

By the frequency shift theorem, a shift in the Doppler spectrum to a center at frequency f_1 simply introduces a phase shift in $G(\tau)$, so if

$$\frac{S(f)}{\frac{1}{\sqrt{2\pi}\sigma}\exp\left[-\frac{(f-f_1)^2}{2\sigma^2}\right]} \leftrightarrow \frac{G(\tau)}{\exp[j\,2\pi\, f_1\tau]\exp[-2\pi^2\sigma^2\tau^2]}$$

Alternatively, the shifted Gaussian on the left hand side may be regarded as the convolution of the unshifted Gaussian, and a Dirac delta function at $f=f_1$. The right hand side of the equation is then the product of the inverse Fourier transforms of the Dirac delta function $f=f_1$ and the transform of the unshifted Gaussian.

Taking the FT of the dual lobe spectrum, we find

$$G(\tau) = \int_{-\infty}^{\infty}[\psi S_1(f) + (1-\psi)S_2(f)]\exp[+j2\pi f\tau]\,df$$
$$G(\tau) = G_1(\tau) + G_2(\tau)$$

where $G_1(\tau)$ and $G_2(\tau)$ are the auto-correlation functions of the individual beam spectra respectively. Now

$$\begin{aligned}\rho(\tau) &= [G_1(\tau)+G_2(\tau)][G_1*(\tau)+G_2*(\tau)]\\ &= G_1(\tau)G_1*(\tau)+G_2(\tau)G_2*(\tau)+G_1(\tau)G_2*(\tau)+G_2(\tau)G_1*(\tau)\end{aligned}$$

Let us evaluate the four terms in the previous equation, making use of the transform pair

Term 1 [$\rho_1(\tau)$ *for beam 1*]

$$G_1(\tau)\,G_1*(\tau) = \psi^2\exp[-4\pi^2\sigma_1{}^2\tau^2]$$

The inverse FT gives

$$\psi^2\int S_1(f)S_1(f+F)\,df = \frac{\psi^2}{2\sqrt{\pi}\sigma_1}\exp[-F^2/4\sigma_1{}^2]$$

Term 2 [$\rho_2(\tau)$ *for beam 2*]

$$G_2(\tau)G_2*(\tau) + (1-\psi)^2\exp[-4\pi^2\sigma_2{}^2\tau^2]$$

The inverse of FT gives

$$(1-\psi)^2 \int S_2(f)S_2(f+F)\,df = \frac{(1-\psi)^2}{2\sqrt{\pi}\sigma_2}\exp[-F^2/4\sigma_2{}^2]$$

Term 3

$$\begin{aligned} G_1(\tau)G_2*(\tau) &= \{\psi(1-\psi)\exp[j2\pi f_1\tau]\exp[-2\pi^2\sigma_1{}^2\tau^2]\}X \\ &\quad \{\exp[-j2\pi f_2\tau]\exp[-2\pi^2\sigma_2{}^2\tau]\} \\ &= \psi(1-\psi)\exp[j2\pi(f_1-f_2)\tau]\exp[-2\pi^2(\sigma_1{}^2+\sigma_2{}^2)\tau^2] \end{aligned}$$

The inverse FT gives

$$\begin{aligned} &\psi(1-\psi)\int S_2(f)S_1(f+F)\,df \\ &= \frac{\psi(1-\psi)}{\sqrt{2\pi(\sigma_1{}^2+\sigma_2{}^2)}}\exp[-(F-\Delta)^2/2(\sigma_1{}^2+\sigma_2{}^2)] \end{aligned}$$

where $\Delta = -(f_1 - f_2)$. Physically this corresponds to the fluctuation spectrum resulting from the beats between the Doppler spectrum of lobe 2 with those from lobe 1. Clearly this beat frequency spectrum is maximized when lobe 1 is shifted by an amount $F = \Delta = -(f_1 - f_2)$ so that the maximum in the fluctuation spectrum is centered at $F = \Delta$.

Term 4

$$G_1*(\tau)\,G_2(\tau) = \psi(1-\psi)\exp[j2\pi(f_2-f_1)]\exp[-2\pi^2(\sigma_1{}^2+\sigma_2{}^2)\tau^2]$$

with Fourier transform

$$\begin{aligned} &\psi(1-\psi)\int S_1(f)S_2(f+F)\,df \\ &= \frac{\psi(1-\psi)}{\sqrt{2\pi(\sigma_1{}^2+\sigma_2{}^2)}}\exp\left[-\frac{(F+\Delta)^2}{2(\sigma_1{}^2+\sigma_2{}^2)}\right] \end{aligned}$$

We have for the auto-correlation function $\rho(\tau)$ and its Fourier transform, the fluctuation spectrum, $S_p(F)$:

$$\begin{aligned} \rho(\tau) &= \psi^2\exp[-4\pi^2\sigma_1{}^2\tau^2] + (1-\psi)^2\exp[-4\pi^2\sigma_2{}^2\tau^2] \\ &\quad + \psi(1-\psi)\exp[j2\pi(f_1-f_2)\tau]\exp[-2\pi^2(\sigma_1{}^2+\sigma_2{}^2)\tau^2] \\ &\quad + \psi(1-\psi)\exp[-j2\pi(f_1-f_2)\tau]\exp[-2\pi^2(\sigma_1{}^2+\sigma_2{}^2)\tau^2] \quad [11.20] \end{aligned}$$

$$S_p(F) = \frac{\psi^2}{2\sqrt{\pi}\sigma_1} \exp[-F^2/4\sigma_1^{\,2}] + \frac{(1-\psi)^2}{2\sqrt{\pi}\,\sigma_2} \exp[-F^2/4\sigma_2^{\,2}]$$
$$+ \frac{\psi(1-\psi)}{\sqrt{2\pi(\sigma_1^{\,2}+\sigma_2^{\,2})}} \exp[-(F-\Delta)^2/\{2(\sigma_1^{\,2}+\sigma_2^{\,2})\}]$$
$$+ \frac{\psi(1-\psi)}{\sqrt{2\pi(\sigma_1^{\,2}+\sigma_2^{\,2})}} \exp[-(F+\Delta)^2/\{2(\sigma_1^{\,2}+\sigma_2^{\,2})\}] \qquad [11.21]$$

It will be noted that the last two terms of the equations represent Gaussian spectra centered at $F = +\Delta$ and $F = -\Delta$ respectively. Since in the case of the measurable fluctuation spectrum, negative F's are indistinguishable from positive ones, the "power" appearing at $-F$ will be measured at $+F$, so the measured spectrum is defined for positive F only, and may be expressed as

$$S_p(F) = \frac{\psi^2}{\sqrt{\pi}\sigma_1} \exp[-F^2/4\sigma_1^{\,2}] + \frac{(1-\psi)^2}{\sqrt{\pi}\,\sigma_2} \exp[-F^2/4\sigma_2^{\,2}]$$
$$+ \frac{2\psi(1-\psi)}{\sqrt{2\pi(\sigma_1^{\,2}+\sigma_2^{\,2})}} \exp\left[-\frac{(F-\Delta)^2}{2(\sigma_1^{\,2}+\sigma_2^{\,2})}\right]$$
$$+ \frac{2\psi(1-\psi)}{\sqrt{2\pi(\sigma_1^{\,2}+\sigma_2^{\,2})}} \exp\left[-\frac{(F+\Delta)^2}{2(\sigma_1^{\,2}+\sigma_2^{\,2})}\right], F > 0$$

where $\Delta = |f_1 - f_2|$

It is interesting to interpret the various terms physically. This is done in Figure 11.18.

Term I represents the auto-correlation integral of spectrum 1 with itself and so term I represents the fluctuation spectrum of the signals received via beam 1 alone, with variance $\Sigma_1^{\,2} = 2\,\sigma_1^{\,2}$. Similarly, Term II of (C) represents the auto-correlation integral of spectrum 2 with itself, so term II of (D) corresponds to the

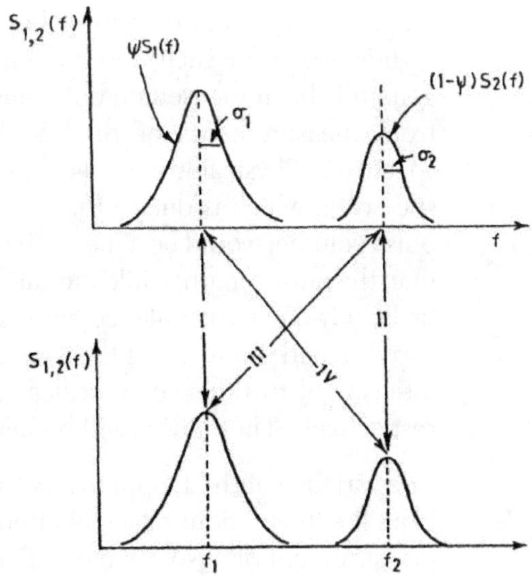

FIGURE 11.18 Physical interpretation of a dual volume fluctuation spectrum.

fluctuation spectrum received via beam 2 alone with variance $\Sigma_2^{\,2} = 2\,\sigma_2^{\,2}$. Now term III of ($C$) represents the cross-correlation integral of spectrum 1 with the displaced version of spectrum 2. Since $S_1(f)$ has a maximum value at f_1 and $S_2(f)$ has a maximum at f_2, the integral of the product $S_1(f)S_2(f+F)$ will be a maximum when $(f_2 + F) = (f_1)$ or when $F = (f_1 - f_2) = \Delta$. Since both spectra are Gaussian, their variances add and the corresponding fluctuation spectrum has variance $= 2(\sigma_1^{\,2} + \sigma_2^{\,2})$. Thus, this component of the fluctuation spectrum may also be written down virtually by inspection, noting that it is also Gaussian and centered at $F = \Delta$

$$S_p(F)\,(\text{Term III}) = \frac{\psi(1-\psi)}{\sqrt{2\pi(\sigma_1^{\,2}+\sigma_2^{\,2})}}\exp\left[-\frac{(F-\Delta)^2}{2(\sigma_1^{\,2}+\sigma_2^{\,2})}\right]$$

Similarly, Term IV of (C) is the auto-correlation integral of $S_2(f)$ with the displaced version of $S_1(f)$. Since spectrum 2 is maximized at f_2, and spectrum 1 is maximized at f_1, the integral of the product is a maximum where $f_1 + F = f_2$, or $F = (f_2 - f_1) = -\,\Delta$. Thus, Term IV of the fluctuation spectrum looks just like Term III with the exception that it is symmetrical with respect to $(F + \Delta)$.

11.14 TURBULENCE MEASUREMENT

It may be possible to deduce the spectrum of air velocities from radar observations of chaff or snowflakes at horizontal incidence. Indeed, both Doppler and incoherent radar have been used to estimate the spectrum of turbulence from such observations. In such measurements, attention has generally been focused on obtaining estimates of the intensity of turbulence from measurements of the mean and variance of the Doppler power spectrum. Physically, one may anticipate that the variance of the Doppler spectrum, which is due to the spread in velocities of the scatterers within a pulse volume, would be a measure of turbulent air velocities on scales smaller than the pulse length, while the variations of the mean Doppler velocity would be largely due to turbulence on scales larger than the pulse length. Thus, the variance and the mean of the Doppler spectrum would give estimates of the intensity of turbulence on scales smaller and larger than the pulse volume respectively. These ideas will be made precise in the next section.

The variance of the Doppler spectrum can be measured by incoherent radar from the fluctuations of signal intensity. A Doppler radar is necessary for the measurement of absolute mean Doppler velocity.

11.14.1 Mean and Variance of the Doppler Velocity, and the Spectrum of Turbulent Velocity

The air velocity will be denoted by V, and ensemble averages and averages within the pulse volume will be denoted by <> and – respectively. With this notation the variance σ^2 of the turbulent velocity may be written:

$$\sigma^2 = <V^2> - <V>^2 \tag{11.22}$$

The variance σ_d^2 of the Doppler spectrum may be written:

$$\sigma_d^2 = \overline{V}^2 - (\overline{V})^2 \tag{11.23}$$

and the variance, σ_v^2, of the mean Doppler velocity $(\overline{V})$ may be written

$$\sigma_v^2 = <(\overline{V})^2> - <\overline{V}>^2 \tag{11.24}$$

Now the operations < > and – commute, therefore

$$<\overline{V}> = <\overline{V}> = <V> \tag{11.25}$$

Using (11.25), (11.24) may be written

$$\sigma_v^2 = <\overline{V}^2> - <V^2> \tag{11.26}$$

Taking the ensemble average of (11.23), we have

$$<\sigma_d^2> = <\overline{V}^2> - <\overline{V}^2> \tag{11.27}$$

Once again using the commutation property, we have

$$<\sigma_d^2> = <V^2> - <\overline{V}^2> \tag{11.28}$$

Adding (11.26) and (11.28), and comparing with (11.22), we have

$$\sigma^2 = \sigma_v^2 + <\sigma_d^2> \tag{11.29}$$

which says that the variance of the turbulent velocity is equal to the sum of the variances of the mean velocity over the pulse volume, and the variance of the velocity within the pulse volume. This is a general result independent of the shape or size of the pulse volume.

We now consider the relationship of the spectral density fraction of $\overline{V}$ to that of V. To simplify the treatment, consider a one-dimensional case first.

Let the coordinate distance along this dimension be measured by x, and let the pulse of length a be centered at x, then

$$\bar{V}(x) = \frac{1}{2}\int_{x-a/2}^{x+a/2} V(x)\,dx \tag{11.30}$$

where $V(x)$ is the radial velocity. This equation may be written as

$$\bar{V}(x) = \int_{-\infty}^{+\infty} G_a(x-X)\,V(X)\,dX \tag{11.31}$$

where

$$G_a(x-X) = \begin{matrix} 1/a, & |x-X| < a/2 \\ 0, & |x-X| > a/2 \end{matrix} \tag{11.32}$$

Now, according to (11.31), $\bar{V}(x)$ is simply the convolution of the gate function $G_a(X)$ and the radial velocity $\bar{V}(X)$; therefore, the Fourier transform of $\bar{V}(x)$ is simply the product of the Fourier transforms of $G_a(X)$ and $V(X)$ The Fourier transform of the gate function $G_a(X)$ is known to be $\sin(k\,\pi\,a)/(k\,\pi\,a)$, k being the wave number. Let the Fourier transform of $V(X)$ be denoted $F_v(k)$, and that of $\bar{V}(X)$ be denoted $F_{\bar{v}}(k)$, then

$$F_{\bar{v}}(k) = F_v(k).\frac{\sin(k\pi a)}{(k\pi a)} \tag{11.33}$$

In terms of the power spectral density S, we have

$$S_{\bar{v}}(k) = S_v(k).\left(\frac{\sin k\pi a}{k\pi a}\right)^2 \tag{11.34}$$

Equation (11.34) shows that it is possible to infer the power spectrum of V from the measurable power spectrum of $\bar{V}$. However, note that at large k (small length scale), the factor $\sin(k\pi a)/(k\pi a)$ is small, and a small error in $S_{\bar{v}}(k)$ is amplified. Figure 11.19 shows a graph of the function $[\sin(k\pi a)/(k\pi a)]^2$. Note that the factor $[\sin(k\pi a)/(k\pi a)]^2$ has fallen by 10 dB at $k\pi a \approx 2$, or $L = 1/k = \frac{\pi a}{2} \approx 1.5a.$ Thus, at a scale of about one and a half times the pulse length, any error in estimating $S_{\bar{v}}$ will mean an error in S_v ten times as large.

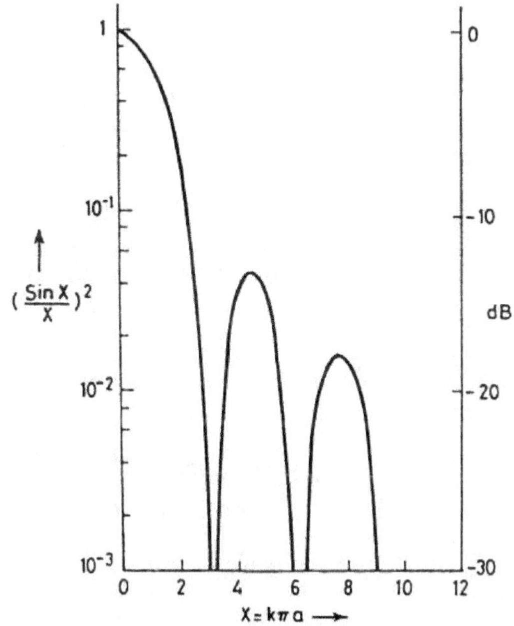

FIGURE 11.19

From Parseval's theorem, we also have

$$\sigma_{\bar{v}}^{\ 2} = \int_{-\infty}^{+\infty} S_v(k)\left(\frac{\sin k\pi a}{k\pi a}\right)^2 dk = 2\int_0^{\infty} S_v(k)\left(\frac{\sin k\pi a}{k\pi a}\right)^2 dk \qquad (11.35)$$

Also

$$\sigma^2 = 2\int_0^{\infty} S_v(k)\,dk \qquad (11.36)$$

Thus, the variance of the mean velocity $\bar{V}$ gives the variance of the velocity V itself but for the weighting factor $(\sin k\pi a/k\pi a)^2$. Now this factor is nearly unity at small k, or scales large compared to the pulse length a, and falls to small values of large K, or scales comparable to and smaller than the pulse length. Therefore, the variance of the mean velocity is essentially a measure of the kinetic energy in scales of motion larger than the pulse length. A reference to equation (11.29) now shows immediately that the variance of the Doppler spectrum is essentially a measure of the kinetic energy of turbulence in scales comparable to and smaller than the pulse length. From equations (11.29) and (11.35), we have the useful result:

$$< \sigma_d^2 > = 2\int_0^\infty S_v(k)\left[1-\left(\frac{\sin k\pi a}{k\pi a}\right)^2\right]dk \tag{11.37}$$

The extension of the previous results to consider sampling in three dimensions is straightforward. Instead of the one-dimensional gate function, we use a gate function which is constant within the pulse volume and zero outside it. In equation (11.33), we shall now get the Fourier transform of the three-dimensional gate function instead of that of the one-dimensional gate function. If the sampled region is a rectangular parallelepiped of sides, a, b, and c, then instead of (11.33), we shall have

$$F_{\bar{v}}(k) = F_v(k)\left(\frac{\sin k_a\pi a}{k_a\pi a}\right)\left(\frac{\sin k_b\pi b}{k_b\pi b}\right)\left(\frac{\sin k_c\pi a}{k_c\pi a}\right) \tag{11.38}$$

where k_a, k_b, and k_c are the wave numbers along the dimensions a, b, and c respectively.

11.15 DOPPLER RADAR AND ITS APPLICATIONS TO METEOROLOGY

11.15.1 Doppler Radar Techniques

Let us consider a radar viewing a target at range r. The electric field intensity scattered back to the radar will have the form,

$$A_s = A_0 e^{-i(\omega_0 t - 4\pi r/\lambda)} = A_0 e^{-i\psi(t)} \tag{11.39}$$

Equation (11.39) represents wave propagating toward increasing r of the signal. Clearly $\psi(t)$ represents the phase of the received signal as a function of time. The corresponding angular frequency of this signal is

$$\omega = \frac{d\psi}{dt} = \omega_0 - \frac{4\pi}{\lambda}\frac{dr}{dt} = \omega_0 - \frac{4\pi}{\lambda}v_r \tag{11.40}$$

where dr/dt is the radial velocity v_r. The returned signal returns at the carrier frequency ω_0 only when the target is fixed. Otherwise the echo angular frequency is Doppler shifted by an amount

$$\omega_d = 2\pi f_d = -4\pi v_r/\lambda \tag{11.41}$$

or

$$f_d = -2v/\lambda \tag{11.42}$$

The Doppler shift is negative when v_r is positive and the target is moving away from the radar.

In the case of continuous wave (CW) radar, the transmitted signal corresponds to a single spectral line at frequency f_0, while the returned signal corresponds to a spectral line at $(f_0 + f_d)$.

In general the signal of equation (11.39) may be represented by

$$A_s = a(t)e^{-i\omega_0 t} \tag{11.43}$$

However, what we measure is really a Cartesian component of the signal A_S or

$$A(t) = R(A_s) = R\{a(t)e^{-i\omega_0 t}\} \tag{11.44}$$

where "R" denotes the real part, and

$A(t) \equiv$ the real signal

$a(t) \equiv$ complex amplitude associated with the real signal.

In the case of a single scatterer moving at velocity, v_i and producing Doppler angular shift ω_i

$$a(t) = a_i e^{-i(\omega_i t + \phi_i)} \tag{11.45}$$

where ϕ_i is the initial phase position of the scatterer. With many scatterers and associated Doppler shifts,

$$a(t) = \sum_i a_i e^{-i(\omega_i t + \varphi_i)} \tag{11.46}$$

In the case of a single scatterer, then,

$$A(t) = R\{a_i e^{-i\omega_i t} e^{-t\omega_0 t}\} = a_i \cos(\omega_0 + \omega_i)t \tag{11.47}$$

ignoring the initial constant phase ϕ_i.

The Heterodyning Process and Simple CW Doppler Radars

The heterodyning or mixing process corresponds to the multiplication of the two signals. In the simplest CW Doppler radar such as that shown in Figure 11.20, the echo signal represented by Equation 11.47 is heterodyned against a reference signal $k \cos \omega_0 t$ having the same frequency as the carrier, so that the output of the detector (mixer) produces a signal

$$\begin{aligned} A_m(t) &= a_i \cos(\omega_0 + \omega_i)t \; k \cos(\omega_0 t) \\ &= \frac{a_i k}{2}[\cos \omega_i t + \cos(2\omega_0 + \omega_i)t] \end{aligned} \tag{11.48}$$

with frequencies corresponding to the sum and difference of the incoming and carrier signals. The heat frequency amplifier in Figure 11.20 (*a*) is designed to respond only in the audio range (or range covering all the desired Doppler frequencies) with the response curve of Figure 11.20 (*b*). Accordingly, the high frequency component of Equation 11.48 with frequency $(2\omega_0 + \omega_i)$ is blocked by this amplifier, and the final output to the indicator is a signal with amplitude directly proportional to a_i (scaled by $k/2$) and frequency identical to the Doppler angular frequency ω_i.

It should be noted that the beat frequency amplifier response curve cuts off at frequencies close to zero in order to reject signals from fixed targets. Figure 11.20 is in essence the design of a police Doppler speedometer. The receiver is called a "homodyne" receiver or superheterodyne with zero IF.

The crucial point is that the detection of the Doppler frequency shift requires a heterodyning or beating of the echo signal with a reference signal having the frequency of the carrier so that the *difference* frequency is identically the Doppler shift. In the case illustrated previously, the amplitude spectrum of $f(t)$ is a line at $(\omega_0+\omega_i)$, which, when heterodyned with ω_0, produces a spectrum having two lines at ω_i and $\omega_0+\omega_i$, each with half the amplitude.

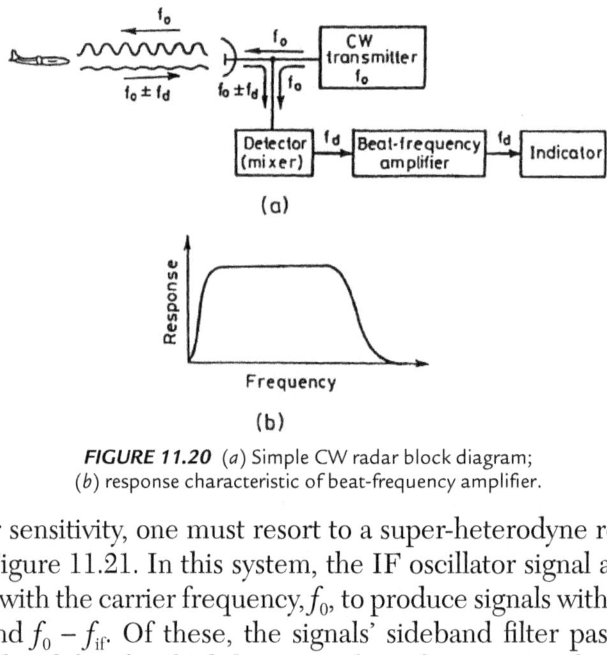

FIGURE 11.20 (*a*) Simple CW radar block diagram; (*b*) response characteristic of beat-frequency amplifier.

For greater sensitivity, one must resort to a super-heterodyne receiver such as that in Figure 11.21. In this system, the IF oscillator signal at frequency, f_{if}, is mixed with the carrier frequency, f_0, to produce signals with frequencies f_0, $f_0 + f_{\text{if}}$ and $f_0 - f_{\text{if}}$. Of these, the signals' sideband filter passes only the upper side band $f_0 - f_{\text{if}}$, which becomes the reference signal against which the incoming signal at $f_0 \pm f_d$ is beat in the receiver mixer.

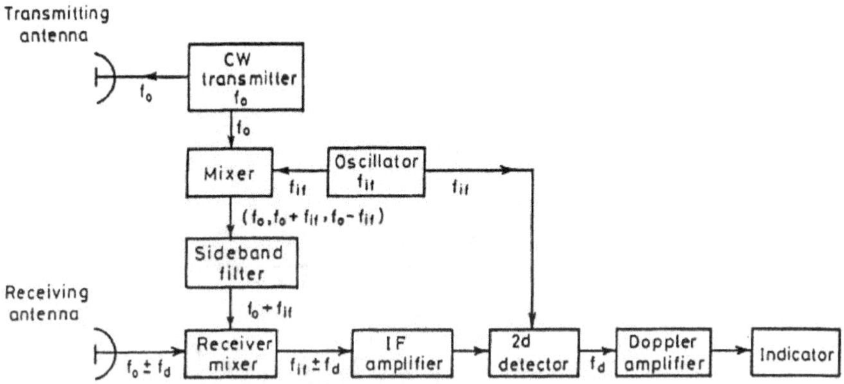

FIGURE 11.21 Block diagram of CW Doppler radar with non-zero IF receiver, sometimes called a sideband superheterodyne.

Detection of Sign of Doppler Shift

The low frequency "line" of the spectrum at the output of the heterodyne mixer as represented by Equation 11.48 is a signal

$$A_{ml} = \frac{a_i k}{2} \cos(\omega_i t) \tag{11.49}$$

which is the cosine of a phasor rotating at angular Doppler frequency θ. Clearly since $\cos(\omega_i t) = \cos(-\omega_i t)$, the output of the mixer provides no information as to the sign of the Doppler shift. To obtain the sign one must resort to a technique such as that illustrated in Figure 11.22. There the reference signal from the CW transmitter goes to two mixers, one directly and the other through a 90 degree phase shifter. By analogy to Equation 11.48, the low frequency output of channel A is the same as the first term in Equation 11.48.

$$E_A = E_o \cos(\pm\, \omega_i t) \qquad [11.50]$$

while the output of channel *B* is

$$E_B = E_o \cos(\pm\, \omega_i t + \pi/2) \qquad [11.51]$$

where we have allowed the Doppler shift to take either sign. For positive Doppler shifts, then

$$E_A(+) = E_o \cos(\omega_i t) \qquad [11.52]$$

$$E_B(+) = E_o \cos(\omega_i t + \pi/2) \qquad [11.53]$$

that is, channel B leads channel A by $\pi/2$.

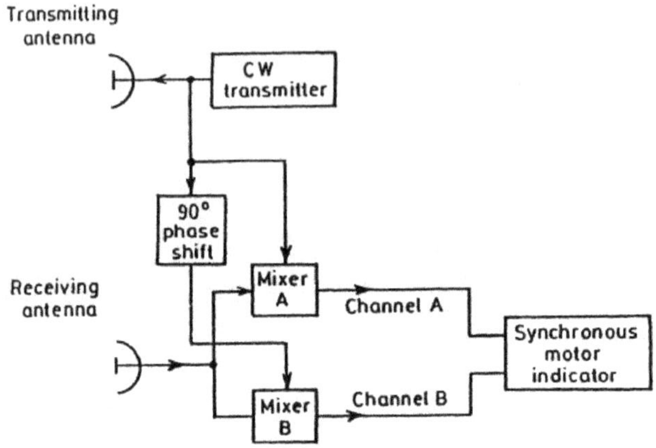

FIGURE 11.22 Measurement of Doppler direction using a synchronous, two-phase motor.

In contrast, for negative Dopplers,

$$E_A(-) = E_o \cos(-\omega_i t) = E_o \cos \omega_i t$$

$$E_B(-) = E_o \cos\left[-\omega_i t + \frac{\pi}{2}\right] = E_o \cos\left[\omega_i t - \frac{\pi}{2}\right] \quad [11.54]$$

that is, channel B lags channel A by $\pi/2$.

Thus, the sign of the Doppler shift is determined by whether the output of channel *B* leads (+ Doppler) or lags (– Doppler) that of channel A. In Figure 11.22, the outputs are applied to a two-phase synchronous motor whose direction is the direction of target motion and with rotation rate equal to the speed.

Pulse-Doppler Radar

While there are a variety of reasons for the use of pulse-Doppler radars instead of CW radars, by far the most important one in the case of weather is the need to obtain a measurement of the target's range. CW radars may also be used for combined ranging of the velocity determination by means of frequency modulation (FM) techniques.

The basic elements of pulse-Doppler radar are illustrated in Figure 11.23, which also shows a simple CW radar by way of direct comparison. There a CW oscillator operating at the transmitter frequency f_t, feeds a power amplifier (PA) to raise the transmitted power levels to that required for sufficient sensitivity, and a pulse modulator turns the PA on for the pulse duration, τ, at

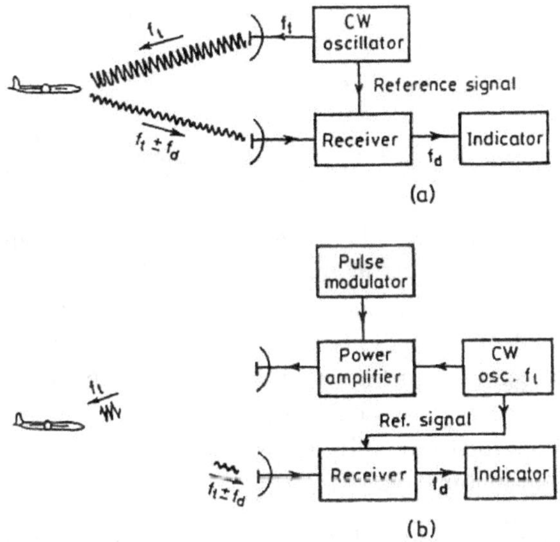

FIGURE 11.23 (*a*) Simple CW radar; (*b*) pulse radar using Doppler information.

the desired PRF. The CW oscillator also provides a *coherent* reference signal to the receiver. By *coherent*, we mean that the phase of the reference signal for heterodyning in the receiver is identical to that of the transmitted signal and is preserved at least until all the echoes are returned. This kind of coherent reference signal is the distinguishing feature of coherent pulse-Doppler or MTI (moving target indication) radars.

If the CW oscillator voltage is represented as $A_1 \sin 2\pi f_t t$, where A_1 is the amplitude and f_t the carrier frequency, the reference signal is

$$V_{ref} = A_2 \sin 2\pi f_t t \tag{11.55}$$

and the Doppler-shifted echo-signal voltage is

$$V_{echo} = A_3 \sin\left[2\pi(f_t \pm f_d)t - \frac{4\pi f_t R_o}{c}\right] \tag{11.56}$$

where A_2 = amplitude of reference signal
A_3 = amplitude of signal received from a target at a range R_0
f_d = Doppler frequency shift
t = time
c = velocity of propagation

The reference signal and the target echo signal are heterodyned in the mixer stage of the receiver. Only the low-frequency (difference-frequency) component from the mixer is of interest and is a voltage given by

$$V_{diff} = A_4 \sin\left(2\pi f_d t - \frac{4\pi f_t R_o}{c}\right) \tag{11.57}$$

Note that equations (11.55) to (11.57) represent since-wave carriers upon which the pulse modulation is imposed. The difference frequency is equal to the Doppler frequency f_d. For a stationary target the Doppler frequency shift f_d will be zero; hence, V_{diff} will not vary with time and may take on any constant value from $+A_4$ to $-A_4$, including zero. However, when the target is in motion relative to the radar, f_d has a value other than zero, and the voltage corresponding to the difference frequency from the mixer (Equation 11.57) will be a function of time.

An example of the output from the mixer when the Doppler frequency f_d is large compared with the reciprocal of the pulse width is shown in Figure 11.24 (*b*). The Doppler signal may be readily discerned from the information contained in a single pulse. If, on the other hand, f_d is small compared with the reciprocal of the pulse duration, the pulses will be modulated with an amplitude given by Equation 11.57 (Figure 11.24 [*c*]) and many pulses will be needed to extract the Doppler information. The case illustrated in Figure 11.24 (*c*) is more typical of aircraft-detection radar, while the waveform of Figure 11.24 (*b*) might be more applicable to a radar whose primary function is the detection of extraterrestrial targets such as ballistic missiles or man-made satellites. Ambiguities in the measurement of Doppler frequency can occur in the case of the discontinuous measurement of Figure 11.24 (*c*), but not when the measurement is made on the basis of a single pulse. The video signals shown in Figure 11.24 are called *bipolar*, since they contain both positive and negative amplitudes.

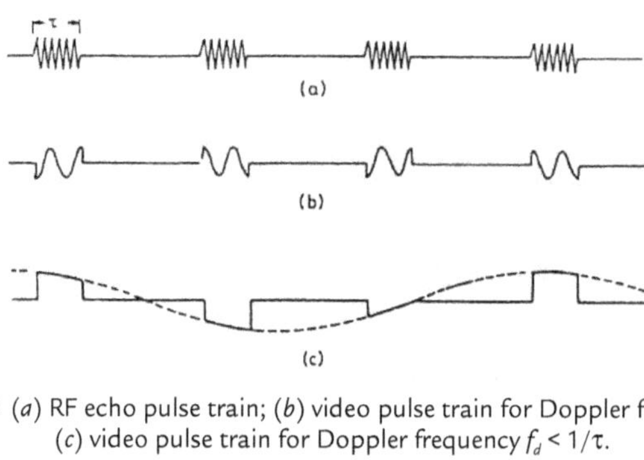

FIGURE 11.24 (*a*) RF echo pulse train; (*b*) video pulse train for Doppler frequency $f_d > 1/\tau$; (*c*) video pulse train for Doppler frequency $f_d < 1/\tau$.

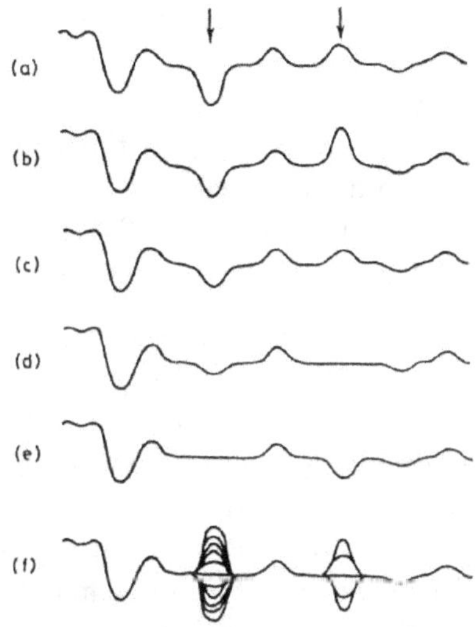

FIGURE 11.25 *(a–e)* Successive sweeps on an MTI radar A-scope display (echo amplitude as a function of time); *(f)* superposition of many sweeps; arrows indicate position of moving targets.

Moving targets may be distinguished from stationary targets by observing the video output on an A-scope (amplitude vs. range). A single sweep on a an A-scope might appear as in Figure 11.25 (*a*). This sweep shows several fixed targets and two moving targets indicated by the two arrows. On the basis of a single sweep, moving targets cannot be distinguished from fixed targets. (It may be possible to distinguish extended ground targets from point targets by the stretching on the echo pulse. However, this is not reliable means of discriminating moving from fixed targets, since some fixed targets can look like point targets, e.g., a water tower. Also, some moving targets such as aircraft flying in formation can look like extended targets.) Successive A-scope sweeps (pulse-repetition intervals) are shown in Figure 11.25 (*b* or *e*). Echoes from fixed targets remain constant throughout, but echoes from moving targets vary in amplitude from sweep to sweep at a rate corresponding to the Doppler frequency. The superposition of the successive A-scope sweep is in Figure 11.25 (*f*). The moving targets produce, with time, a "butterfly" effect on the A-scope.

We proceed now to the essential design of a modern coherent pulse Doppler radar as shown in Figure 11.26. We start with the so-called "Coho" or coherent local oscillator, which is a highly stable oscillator selected to operate at the lowest intermediate frequency (IF) between the low frequency Doppler shifts and the microwave carrier. The output frequency, f_c, of the Coho (usually 30 or 60 MHz), is mixed with that of the Stalo (stable local oscillator) operating at f_l in Mixer 1. Since we may anticipate down-conversions to lower frequencies by heterodyning in the receiver, it is clear that we wish to use the *sum* frequency $fc + f_l$ at the output of Mixer 1 for transmission. The difference frequency is rejected, and the signal at $f_l + f_c$ is amplified strongly by the Klystron Amplifier so that the transmitted frequency $f_t = f_l + f_c$. Since f_t is a microwave frequency, say 3000 MHz, it is clear that f_l is also in the microwave region but in an amount f_c smaller than f_t.

The back scattered signal returns at frequency $f_t \pm f_d = f_l + f_c \pm f_d$,where f_d is the Doppler frequency shift and is heterodyned with a small part of the Stalo output signal in Mixer 2. Now we use the difference frequency output of the mixer, since we ultimately want to recover the low frequencies. The selected output of Mixer 2 is thus at frequency $f_c \pm f_d$. Since f_c was chosen to be the lowest IF frequency, and $\pm f_d$ is small relative to f_c, the IF amplifier amplifies these signals to provide signals of unchanged frequency but of the required strength. The final down-conversion in frequency is done by heterodyning the IF amplifier output signal at $f_c \pm f_d$ with the Coho signal f_c in the Phase Detector (i.e., another mixer) to provide the difference frequency $\pm f_d$, the desired Doppler shifts.

One may also obtain the standard unipolar A-scope display of signal intensity (αA^2) versus range by square-law detection of the output of the IF amplifier in the video detector. The corresponding A-scope display is shown schematically by waveform B in Figure 11.26. The incoherent "A^2" display contains no phase information, since it has been derived without beating against the reference Coho frequency. Another type of unipolar A-scope with signal amplitude displayed linearly may be obtained by clipping the negative going portions of the bipolar video so that only the upper half of waveform A in Figure 11.26 is shown.

In moving target indicator (MTI) radars which are required only to discriminate moving targets from ground clutter, the bipolar video during one pulse repetition period is stored and subtracted from that returned during the next successive period. In this way, fixed targets producing no "butterfly" fluctuation from one pulse to the next are cancelled out, and only the fluctuating bipolar signals corresponding to moving targets are passed to the PPI scope.

The incoming signals may be traced through the receiver with the aid of Figure 11.27, where it is understood that all heterodyning operations in the receiver exploit the *difference* between the heterodyned frequencies. It will be noted that the analytical form of the signals correspond to that from a

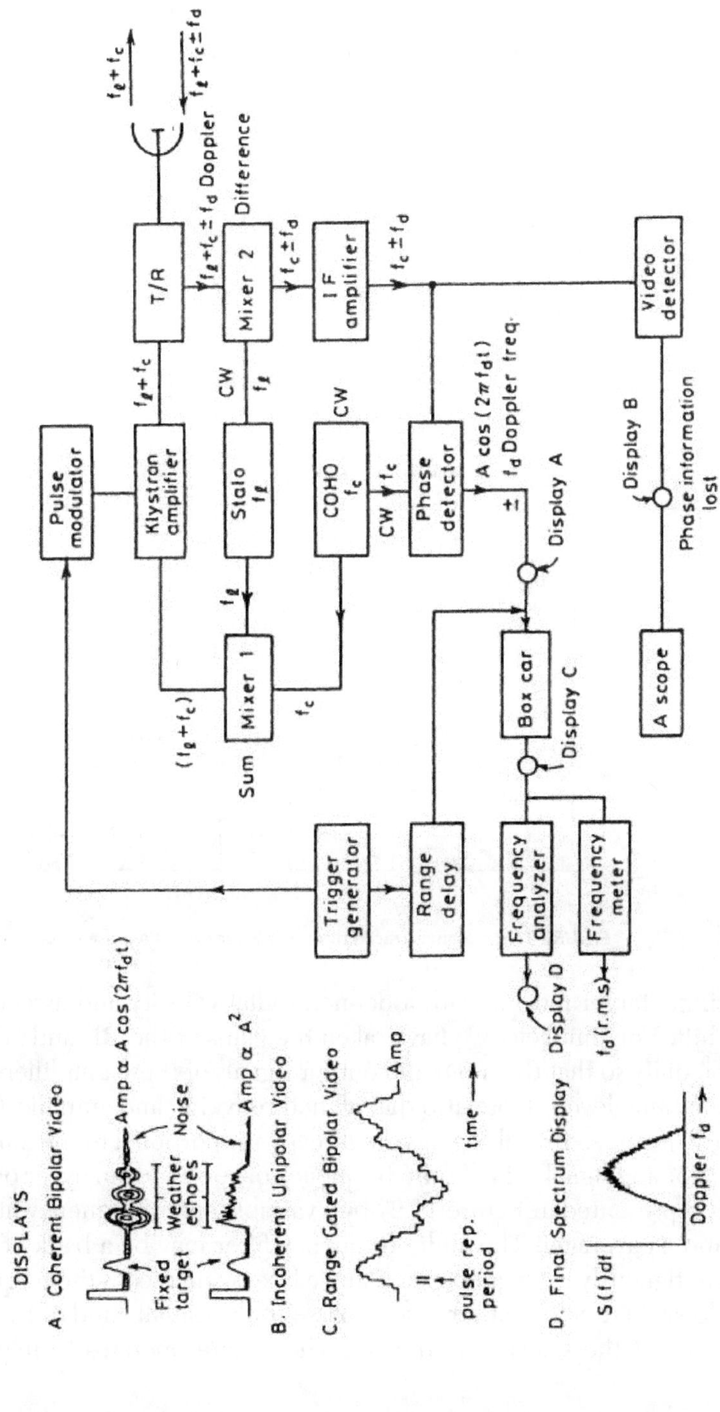

FIGURE 11.26 Completely coherent pulse Doppler radar.

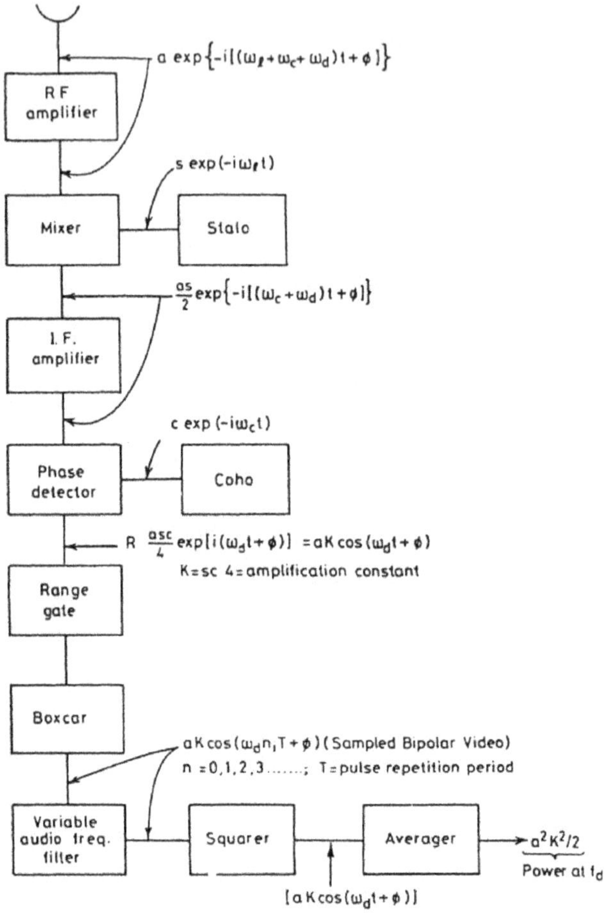

FIGURE 11.27 Signals traced through the receiver of a pulse Doppler radar.

single target moving with a defined radial velocity and associated Doppler shift. For simplicity we have taken the gains of the RF and IF amplifiers to be unity so that the input and output signals of these amplifiers are shown at the same levels. Note also that all heterodyning and amplification processes are *linear*, so that all signals are directly proportional to the amplitude of the incoming signal. Finally, the frequency or spectrum analyzer of Figure 11.26 is represented in Figure 11.27 by a variable audio frequency filter, a squarer, and an averager. The audio frequency filter may be a bank of filters spaced contiguously in frequency or a single filter which scans the frequency domain slowly in time. In either case, it passes only a small band of frequencies centered at the selected frequency. The square operates to make the signals

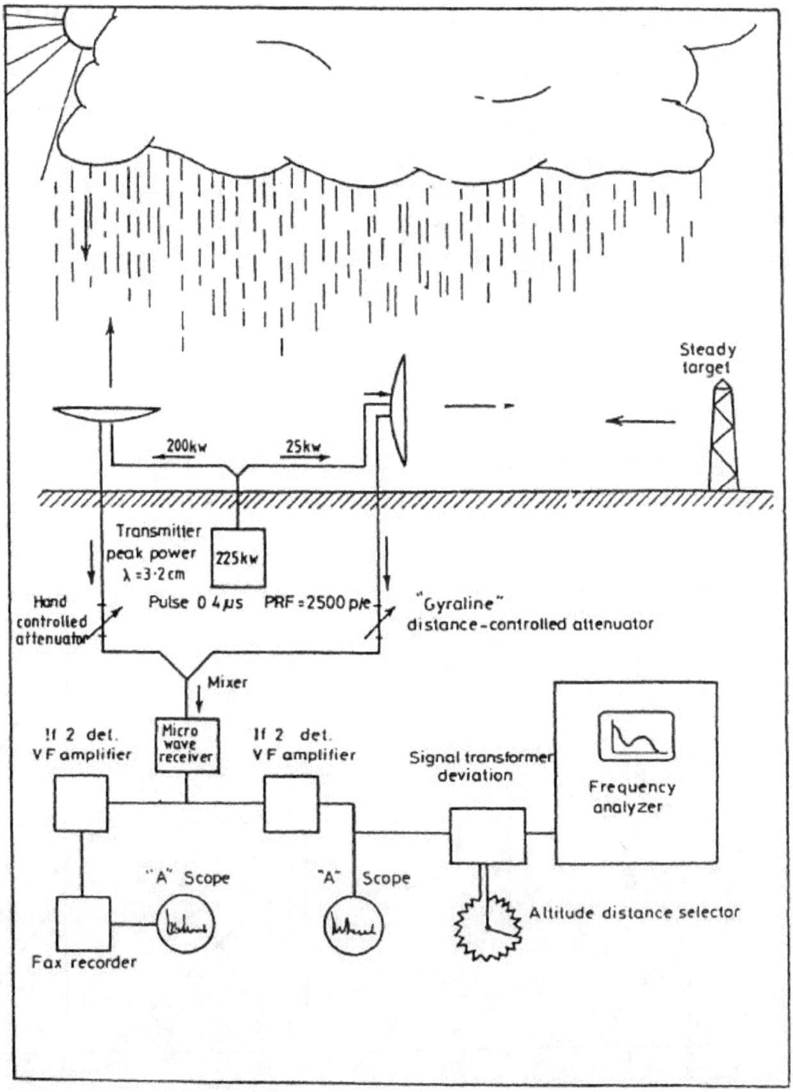

FIGURE 11.28 Pulse Doppler radar installation schematic diagram.

unipolar and the averager takes a time average of the squared signal so that its DC output is directly proportional to the power in the Doppler signal passed by the audio filter. In this way, the last three elements in the system provide values of $(a_i^2/2)$, the average power at each Doppler frequency f_i. The factor K^2 is a scaling or calibration coefficient. A schematic diagram of pulse Doppler radar installation is shown in Figure 11.28.

11.16 EXAMPLES OF DOPPLER AND FLUCTUATION SPECTRA

Figure 11.29 gives six examples of normalized Doppler power spectra, $S_1(f)$, and their corresponding fluctuation spectra $S_p(F)$. To facilitate comparison, the fluctuation spectra have been normalized and replotted in Figure 11.30. The normalization has been carried out by dividing the fluctuation frequency F by the standard σ_f, of the Doppler power spectrum and multiplying $S_p(F)$ by σ_f so as to conserve the area under the curve. This is to say that the area under the $S_p(F)$ *vs*. F curve is equal to that under the $\sigma S_p(F)$ vs. F/σ_f curve.

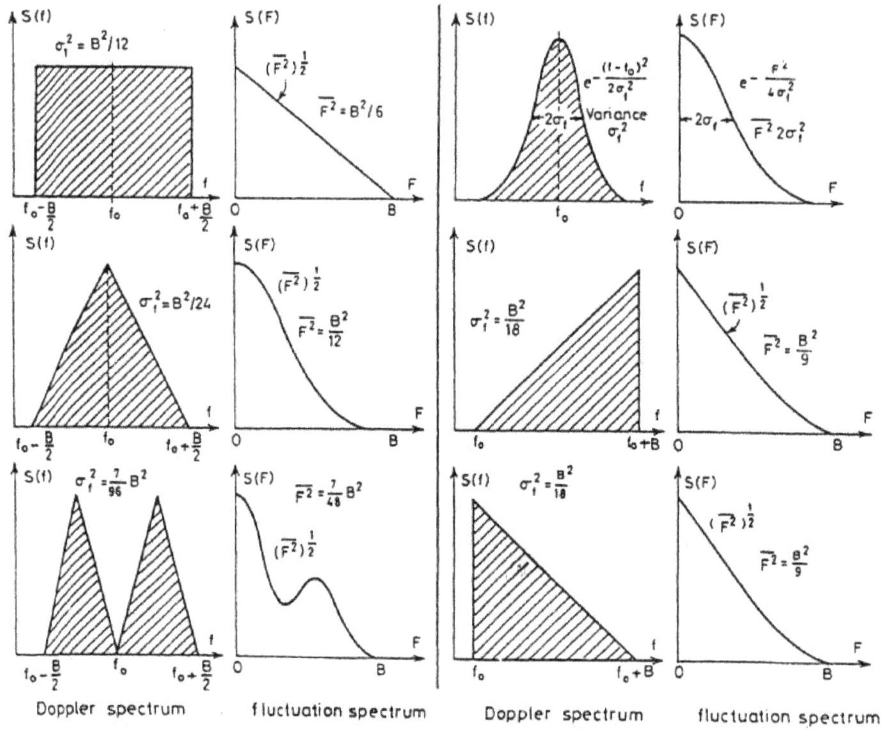

FIGURE 11.29 Doppler and fluctuation spectra pairs (after Atlas, 1964).

The auto-correlation function for the intensity, $\rho(\tau)$, is plotted in Figure 11.31. The lag τ has been normalized by multiplying by the standard deviation, σ_f of the Doppler spectrum. It is seen that for all the spectra $\rho(\tau)$ falls to about 0.1 at $\tau \cong 1(4\sigma_f)$. Moreover, except for certain details, $\rho(\tau)$ does not depend too strongly on the Doppler spectrum; consequently, any of the spectra may be used as a model. Since the Gaussian spectrum is easy to handle analytically, it will be taken a model spectrum in the discussions of the following section.

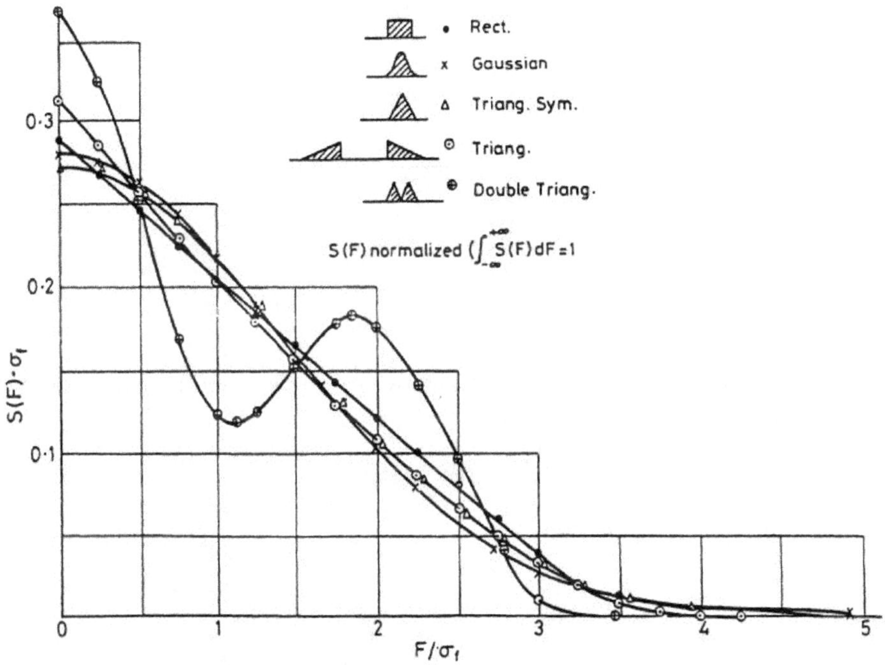

FIGURE 11.30 Normalized fluctuation spectra (after Lhermitte, 1963).

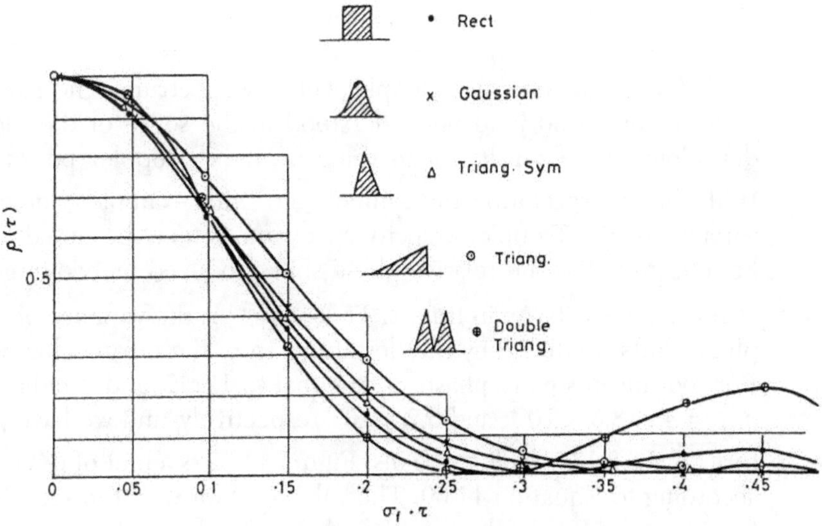

FIGURE 11.31 Auto-correlation function of signal amplitude for various Doppler spectra (after Lhermitte, 1963).

Auto-correlation function, time to independence, and independent samples

The normalized Gaussian Doppler power spectrum

$$S_1(f) = \frac{1}{\sqrt{2\pi}\sigma_f} \exp[-(f - f_o)^2 / 2\sigma_f^{\,2}] \tag{11.58}$$

has the fluctuation spectrum

$$S_p(F) = \frac{1}{\sqrt{4\pi}\sigma_f} \exp[-F^2 / 4\sigma_f^{\,2}] \tag{11.59}$$

which is also Gaussian and has twice the variance of Equation 11.58. The corresponding auto-correlation function of intensity is

$$\rho(\tau) = \exp[-(2\pi\sigma_f\tau)^2] \qquad [11.60(a)]$$

$$= \exp[-(4\pi\sigma_v\tau/\lambda)^2] \qquad [11.60(b)]$$

$$= \exp[-\overline{(\Delta\phi)^2}] \qquad [11.60(c)]$$

where $$\Delta\phi = 4\pi.\Delta v.\tau / \lambda \tag{11.61}$$

Equation 11.59 follows by taking the auto-correlation integral of 11.58. Equation 11.60 follows from the property that the auto-correlation function and the fluctuation spectrum are a Fourier transform pair. The quantity $\sigma_v^{\,2}$ is the variance of the Doppler velocity spectrum. The bar over $(\Delta\phi)^2$ in Equation 11.60 is to be understood in the sense of the mean square deviations of ϕ from its mean value over the Doppler power spectrum. With this interpretation the symbol $\overline{(\Delta v)^2}$, for example, would equal the variance of the Doppler velocity spectrum. It may be noted that $\Delta\phi$ can be interpreted as the relative phase shift produced by two targets moving at relative velocity Δv in time τ. Thus, $\overline{(\Delta\phi)^2}]$ is the variance of the relative phase shifts produced by the deviations from the mean velocity in time τ. For root mean square phase shifts equal to 1, $\pi/2$, and π radians, we have $\rho(\tau) = 1/e$, 8.5×10^{-2}, and 5.2×10^{-5} respectively, and we have $\rho(\tau) = 0.01$ for $\overline{(\Delta\phi)^2}] = 2.15 = 0.68\,\pi$ radians. Figure 11.32 is a plot of $\rho(\tau)$ *vs*. $\overline{(\Delta\phi)^2}]^{1/2}$ according to Equation 11.60. The value of τ for $\sigma_v = 1$ m sec^{-1} is indicated for selected values of the wave length λ.

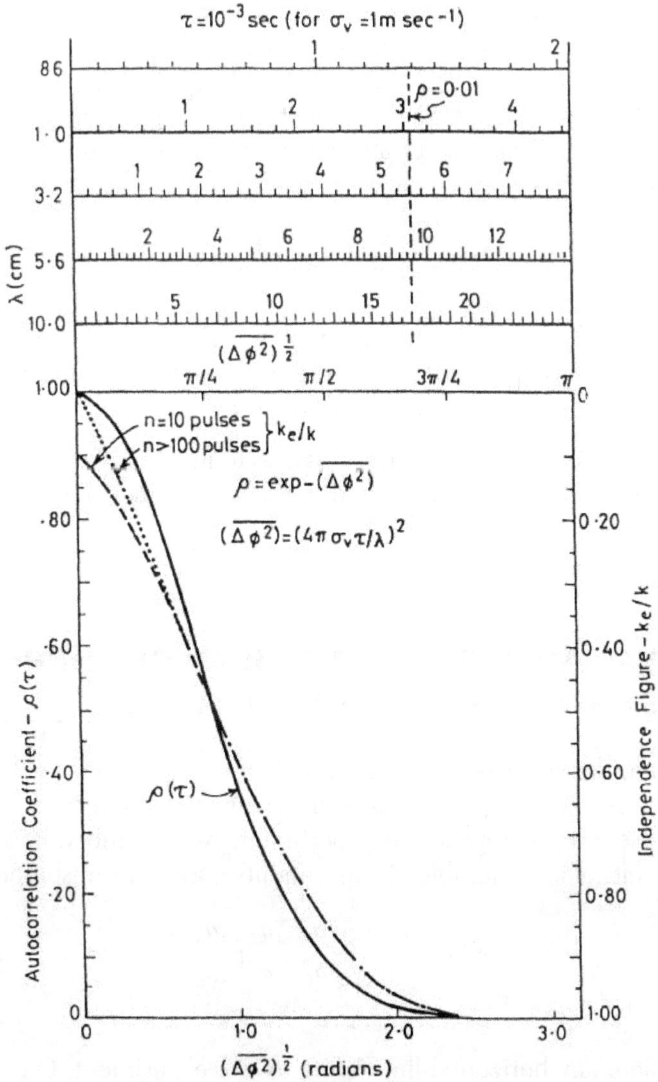

FIGURE 11.32 The auto-correlation function of signal intensity fluctuations for a Gaussian Doppler power spectrum. The independence figure, k_e/k, is also shown for 10 and 100 pulses. (After Atlas, 1964).

The signal amplitude has a Rayleigh probability density, and because of this it is necessary to take the average of a number of samples in order to estimate the values of meteorologically significant parameters with acceptable accuracy. The confidence level of the estimate was found as a function of the number of samples. In this development, it was assumed that the samples of which the averages are taken are independent. Assuming now that the signals may be regarded as effectively independent when $\rho(\tau) = 0.01$, it is seen from Equation 11.59 that the samples should be spaced at least $\tau_{0.01}$ apart, where

$$\tau_{0.01} = \frac{1.71\lambda \times 10^{-3}}{\sigma_v} \sec \qquad (11.62)$$

(σ_v and λ are in m sec^{-1} and cm respectively). The time to independence is thus directly proportional to the wavelength of the radiation and inversely proportional to the standard deviation of the velocity-spectrum.

For correlated samples, it is possible to find the effective number k_e of independent samples, which yield the same accuracy of estimate as k correlated samples.

11.17 VARIANCE OF SAMPLE MEAN OF CORRELATED SAMPLES INDEPENDENCE FIGURE

Consider an experiment in which the sequence of values $a_1^{(1)}$, $a_2^{(1)}$,... is observed (in the radar case, these may be intensity values from pulse to pulse). If we repeat the experiment, we get the values $a_1^{(2)}$, $a_2^{(2)}$,... The results of an ensemble of experiments may be represented by the array:

$$\begin{array}{l} a_1^{(1)}, a_2^{(1)}, a_3^{(1)}, \ldots \\ a_1^{(2)}, a_2^{(2)}, a_3^{(2)}, \ldots \\ a_1^{(3)}, a_2^{(3)}, a_3^{(3)}, \ldots \end{array}$$

where any horizontal line refers to one experiment. Our purpose is to find the accuracy of the estimate of the ensemble average from the average of k successive observations in one experiment. When discussing one experiment, the superscript on a will be dropped. Averages of k values in one experiment (i.e., along a row) will be called sample averages or means; averages over the ensemble (i.e., along a column) will be denoted by angular brackets. Stationarity and ergodicity will be assumed throughout.

Consider the sample mean m of k samples values a_i, $i = 1, 2, ..., k$.

$$m = (a_1 + a_2 + ... + a_k)/k \tag{11.63}$$

The (ensemble) average square value of the sample mean is

$$\begin{aligned} < m^2 > &= < (\Sigma a_i)(\Sigma a_j) > /k^2 \\ &= [< \Sigma a_i^2 > + < \sum_{i \neq j}\Sigma a_i a_j >] / k^2 \end{aligned} \tag{11.64}$$

If the samples are independent, then

$$< a_i a_j > = < a_i >< a_j > = M^2 \tag{11.65}$$

where M is the population mean. Substituting Equation 11.65 into 11.64, and remembering that there are $k(k–1)$ terms in the double summation on the right hand side of Equation 11.64 and that the operations Σ and $< >$ commute, we get

$$< m^2 > = [\Sigma < a_i^2 > + k(k-1)M^2] / k^2 \tag{11.66}$$

Also, by definition of the variance, σ_i^2, of the a_i, we have

$$<a_i^2> = \sigma_a^2 + M^2 \tag{11.67}$$

From Equations 11.66 and 11.67, we have

$$<m^2> = [\sigma_a^2/k + M^2] \tag{11.68}$$

Now the variance of the sample mean is given by

$$\sigma_m^2 = <m^2> - M^2 \tag{11.69}$$

From Equations 11.68 and 11.69, we get

$$\sigma_m^2 = \sigma_a^2/k \tag{11.70}$$

This is a well-known result which holds for uncorrelated samples.

Let us now consider the case of correlated samples. Then instead of Equation 11.65, we have

$$< a_i a_j > = R(|i - j|T) \tag{11.71}$$

where R denotes the auto-covariance function of the a_i (without the mean subtracted out), and successive samples are supposed to be spaced by T, the pulse repetition period. (As already stated, stationarity and indeed ergodicity is assumed throughout.) Now in the double summation of Equation 11.64

$$\sum_{i\neq j}\sum a_i a_j$$

there are two terms with a lag $(k-1)\,T$, 4 terms with a lag $(k-2)\,T$, 6 terms with a lag $(k-3)\,T$ ….., and $2\,(k-1)\,T$ terms with a lag $[k-(k-1)]\,T$ or T (Figure 11.33).

Hence,
$$<\sum_{i\neq j}\sum a_i a_j> = \sum_{i\neq j}\sum <a_i a_j>$$
$$= 2.1R[(k-1)T] + 2.2R[(k-2)T] + \ldots + 2(k-1)R[T]$$
$$= 2\sum_{j=1}^{k-1}(k-j)R(jT) \tag{11.72}$$

Substituting from Equations 11.67 and 11.72 into Equation 11.64 we get

$$<m^2> = \left[k(M^2+\sigma_a{}^2) + 2\sum_{j=1}^{k-1}(k-j)R(jT)\right]\Big/k^2 \tag{11.73}$$

Using Equation 11.69, we have

$$\sigma_m{}^2 = \sigma_a{}^2/k - (k-1)M^2/k + 2\sum_{j=1}^{k-1}(k-j)R(jT)/k^2 \tag{11.74}$$

Now $\sum_{j=1}^{k-1}(k-j) = (k-1)+(k-2)+\ldots+1$

$$= k(k-1)/2$$

Thus, the second term in Equation 11.74 may be written as

$$\frac{(k-1)M^2}{k} = \frac{2M^2}{k^2}\left[\frac{k(k-1)}{2}\right] = \frac{2}{k^2}M^2\sum_{j=1}^{k-1}(k-j)$$

Hence, Equation 11.74 may be written as

$$\Sigma_m{}^2 = \left[\sigma_a{}^2 + 2\sum_{j=1}^{k-1}\{(k-j)[R(jT)-M^2]\}/k\right]\Big/k \tag{11.75}$$

Let us now introduce the auto-correlation function $\rho(\tau)$:

$$\rho(\tau) = \frac{R(\tau)-M^2}{R(0)-M^2} \tag{11.76}$$

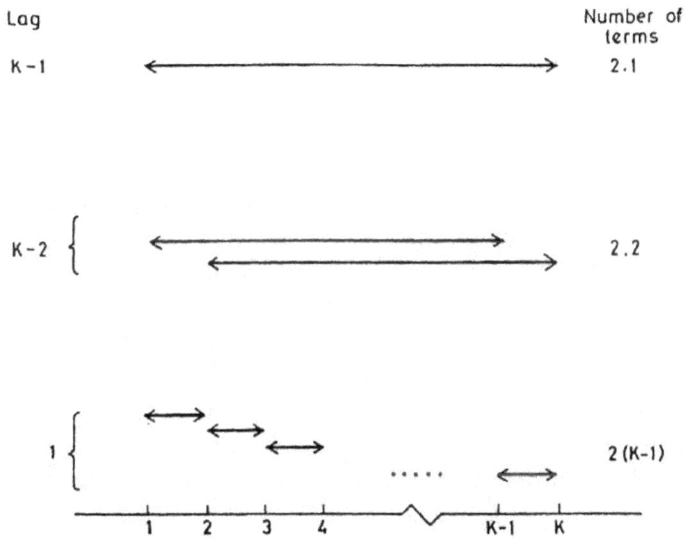

FIGURE 11.33 Number of terms having various lags.

But
$$R(0) = \sigma_a^{\,2} + M^2 \tag{11.77}$$

Hence,
$$\rho(\tau) = [R(\tau) - M^2]\sigma_a^{\,2} \tag{11.78}$$

Substituting Equation 11.78 into Equation 11.75, we have

$$\sigma_m^{\,2} = \frac{\sigma_a^{\,2}}{k}\left[1 + \frac{2}{k}\sum_{j=1}^{k-1}(k-j)\rho(jT)\right] \tag{11.79}$$

Now, let the effective number of samples, k_e, be defined as the number of independent samples which yield the same variance as the k correlated samples. By Equations 11.70 and 11.79, this requires that

$$\sigma_a^{\,2}/k_e = \left[1 + 2\sum_{j=1}^{k-1}(k-j)\rho(jT)\right]\sigma_a^{\,2}/k \tag{11.80}$$

or
$$\frac{k}{k_e} = (1 + 2/k)\sum_{j=1}^{k-1}(k-j)\rho\, jT \tag{11.81}$$

If the samples are independent $\rho(jT) = 0$, and $k = k_e$. Numerical values of the independence figure (k_e/k) are given in Figure 11.32 for a Gaussian Doppler spectrum. It will be seen that for $k > 10$ a good approximation for the independence figure is

$$k_e / k = 1 - \rho(T) \tag{11.82}$$

From Equation 11.82 it is seen that for independent samples ($\rho = 0$), $k_e = k$, and for perfectly correlated samples ($\rho = 1$), and $k_e = 0$. For intermediate values of ρ, k_e varies approximately linearly between these two extremes. It is helpful to think of the observation as being taken by a moving pulse. If the two positions of the pulse overlap completely, then the samples may be regarded as completely dependent; on the other hand if the pulses do not overlap at all, the samples may be regarded as completely independent. In between we have varying degrees of overlap, and the degree of independence may be regarded as varying linearly from 0 when the pulses are coincident to 1 and are just adjacent.

For any given sampling rate T, the independence figure approaches unity as the width of the Doppler spectrum increases. A knowledge of the width of the Doppler spectrum is thus useful in finding the level of confidence of the estimate of mean intensity found by averaging equally spaced observations.

11.18 MEASUREMENT OF VARIANCE OF DOPPLER SPECTRUM: R-METER QUALITATIVE CONSIDERATIONS

Consider a continuous time record of the intensity, I (Figure 11.34). Rate the time at which the intensity that crosses a prescribed threshold, L, may be anticipated to increase with the speed of the fluctuations of the intensity. Now the fluctuation of intensity is due to changes in the relative phase positions of the scatterers, and therefore the speed of the fluctuations is expected to increase with the variance of the Doppler spectrum. Thus, the rate at which the intensity crosses a given threshold may provide an estimate of the variance of the Doppler power spectrum. This is the basic idea behind the R-meter method of measuring the variance of the Doppler power spectrum.

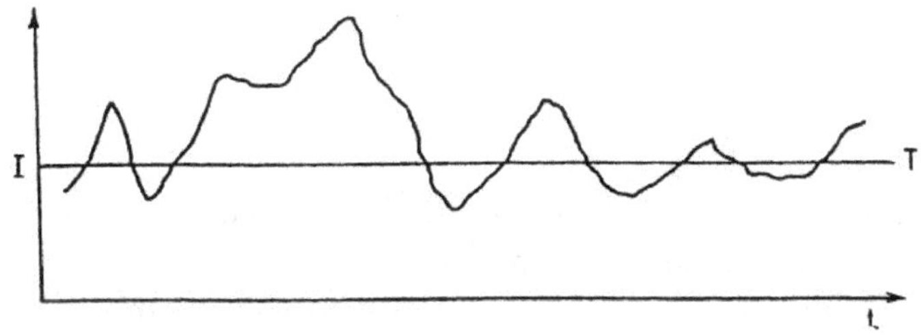

FIGURE 11.34 Threshold crossing.

Rate of Crossing.

Consider a random function ξ of time t. Let $\dot{\xi} \equiv \partial\xi / \partial t$, and let $P(\xi, \dot{\xi})$ be the joint probability density of ξ and $\dot{\xi}$, that is, $P(\xi,\dot{\xi})d\xi d\dot{\xi}$ is the probability that ξ and $\dot{\xi}$ lie in the intervals ξ to $\dot{\xi}$ + $d\xi$, ξ to $\dot{\xi}$ + $d\dot{\xi}$. It is required to find the rate at which ξ crosses any prescribed level which may be taken as zero without loss of generality. Consider the portion of one ξ – curve between t and $t + dt$ (Figure 11.35). If dt is sufficiently small, the curve may be regarded as straight with slope $\dot{\xi}$. If ξ is the cross zero in the positive direction during the period dt, we must have ξ < 0. Further, the increment of ξ during dt at the rate $\dot{\xi}$ which is equal to $\dot{\xi}$ dt must exceed ξ. Hence we must have

$$-\dot{\xi}dt < \xi < 0 \qquad (11.83)$$

in order that a zero crossing in the positive direction occur during t to $t + dt$. The probability of this happening is

$$\int_0^\infty d\dot{\xi} \int_{-\dot{\xi}dt}^0 P(\xi,\dot{\xi})\,d\xi \cong dt \int_0^\infty \dot{\xi}P(0,\ \dot{\xi})\,d\dot{\xi} \qquad (11.84)$$

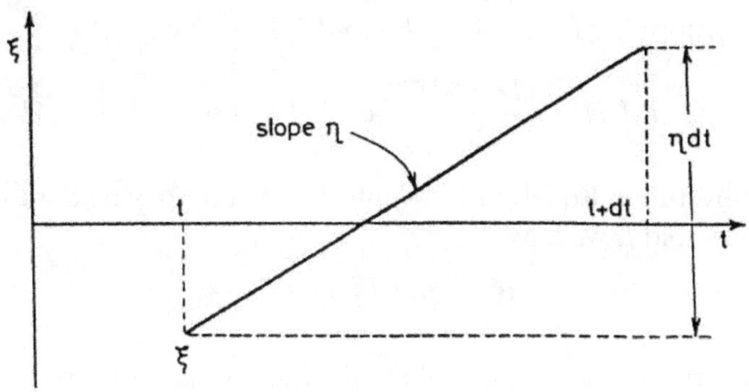

FIGURE 11.35 Zero crossing in the positive direction.

Hence, the rate of crossing zero in the positive direction is given by

$$W_+ = \int_0^{\infty} \dot{\xi} P(0,\dot{\xi})\, d\dot{\xi} \tag{11.85}$$

If the rate of crossing any threshold L in the positive direction is required, the 0 in $P(0,\dot{\xi})$ should be replaced by L. Similarly, it may be shown that the rate of crossing in the negative direction is given by

$$W_- = -\int_{-\infty}^{0} \dot{\xi} P(0,\dot{\xi})\, d\dot{\xi} \tag{11.86}$$

and the total rate of crossing irrespective of direction is given by

$$W = \int_{-\infty}^{+\infty} |\dot{\xi}|\, P(0,\dot{\xi})\, d\dot{\xi} \tag{11.87}$$

We shall now apply Equation 11.85 to our problem. For this purpose, we identify ξ with the intensity I. We now need the joint probability density of I and dI/dt. We have after suitable change of notation

$$P(I,\delta) = \frac{\sqrt{I}}{\sqrt{\pi} <I>^{3/2}\, \sigma_\omega} \exp[-I/<I>].\exp[-I\delta^2/(<I>\sigma_\omega{}^2)] \tag{11.88}$$

where

$$\delta = \frac{1}{2I}\frac{dI}{dt} \tag{11.89}$$

and σ_ω is the variance of the Doppler power spectrum in terms of ω. In terms of f

$$\sigma_\omega = 2\pi\sigma_f \tag{11.90}$$

Transforming Equation 11.88 to $P(I, I)$,we have

$$P(I,I) = \frac{[\pi I <I>^3]^{-1/2}}{2\sigma_\omega} \exp[-I/<I>].\exp[-I^2/(4I<I>\sigma_\omega{}^2)] \tag{11.91}$$

Substituting Equation 11.91 into 11.85 and carrying out the integration with threshold I, we have

$$W_+ = 2\sigma_f \sqrt{\pi}(I/<I>)^{1/2} \exp[-I/<I>] \tag{11.92}$$

In terms of the variance of the fluctuation spectrum $\sigma_F = \sqrt{2}\,\sigma_f$, we have

$$W_+ = (I/<I>)^{1/2} \sqrt{2\pi}\sigma_f \exp[-I/<I>] \tag{11.93}$$

For level crossing in both directions we have

$$W = 2W_+ = 4\sqrt{\pi}.\sigma_f[I/<I>]^{1/2}\exp[-I/<I>] \tag{11.94}$$

If the threshold is set at $I = <I>$, the rate of crossing in both directions is given by, W_o, where

$$W_o = 4\sqrt{\pi}\,\sigma_f / e \tag{11.95}$$

In terms of the standard deviation of the Doppler velocity,

$$\sigma_f = \frac{2\sigma_v}{\lambda} \tag{11.96}$$

We have

$$W_o = 8\sqrt{\pi}\,\sigma_v / (\lambda e) \tag{11.97}$$

If the threshold is set at $<I^{1/2}>$, we have

$$W_o = \frac{4\pi\sigma_v}{\lambda}\exp(-\pi/4) = \frac{5.73\sigma_v}{\lambda} \tag{11.98}$$

It is seen that the right hand sides of Equations 11.92, 11.93, and 11.94 for the rate of crossing a threshold I are essentially the same (except for a factor) as the expression for the Rayleigh probability density of the amplitude $I^{1/2}$. This is illustrated in Figure 11.36.

The previous equations confirm the qualitative considerations which indicated that the rate of level crossings should increase with the variance of the Doppler spectrum. That the rate of level crossing is directly proportional to the standard deviation σ_f may also be seen very simply as follows. Suppose that all frequencies in a given Doppler spectrum are doubled so that the form of the spectrum remains the same but σ_f is doubled. Now by the scaling property of Fourier transforms, this corresponds to compressing the time scale by a factor of two. Hence, the rate of level crossings is doubled.

In the previous derivations, a continuous sample of I was considered. In radar practice, the signal is boxcared. The presence of noise added to the signal also modifies the crossing rate.

Zero-Crossing Rate.

The rate at which the intensity crosses a prescribed level was considered in the previous section. We will now consider the rate at which one component of the signal amplitude crosses the prescribed level, and in particular the rate at which it changes sign or crosses zero, the so-called zero-crossing rate. In practice, the zero-crossing rate can be determined from the bipolar video presentation of the signal.

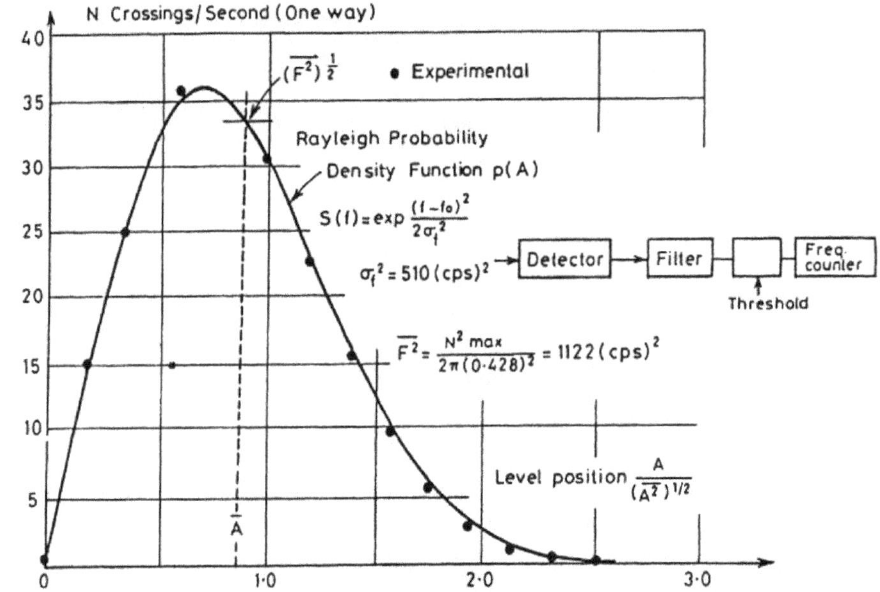

FIGURE 11.36 Rate of crossing, one way. The ordinate is the rate of crossing, the abscissa is the threshold level in units of </>$^{1/2}$.

11.19 DOPPLER SPECIAL CHARACTERISTICS OF PRECIPITATION

In this section, we shall deal with the characteristics of the Doppler spectra of precipitation viewed by a vertically pointing radar. In particular we shall be concerned with the information which can be inferred concerning the nature of the particle size distribution, the precipitation intensity, and other relevant factors. We shall also deal with a method of inferring vertical velocities from the Doppler spectra.

In previous sections we have generally dealt with the normalized (unit area) power density spectrum $S_1(f)$, where we used the subscript 1 to designate the normalization $\int S_1(f)df = 1$. We have also noted that the total power under the actual spectrum corresponds to the average echo power

$$\overline{P}_r = \overline{I(t)} = \overline{a(t)a*(t)} = \int_{-\infty}^{\infty} S(f)df \tag{11.99}$$

so that

$$\frac{d\overline{P}_r(f)}{df} = S(f) \tag{11.100}$$

This simply emphasizes the point that $S(f)$ is the power density spectrum and that a measurable amount of power occurs only in some finite frequency interval f to $f + df$. Obviously the normalized density spectrum is given by

$$S_1(f) = \frac{S(f)}{\infty S(f)df} = \frac{S(f)}{\overline{P}_r} \tag{11.101}$$

In what follows we shall generally be concerned with $S(f)$ rather than $S_1(f)$. Also, for simplicity, we shall write $\overline{P}_r = P$.

Since $f = 2v/\lambda$ where v is the radial velocity, it is also clear that there is a linear transformation such that

$$S(v)\,dv = S(f)\,df \tag{11.102}$$

so that $S(v) = S(f)(df/dv) = S(f)\ 2/\lambda$. $S(v)$ is the doppler velocity spectrum and is frequently used interchangeably with $S(f)$, the frequency spectrum. Moreover, since $\overline{P}_r$ is proportional to reflectivity, η, through the radar equation for distributed targets, it will frequently be convenient to write

$$\frac{dP}{dv}dv = S(v)\,dv = C\frac{d\eta(v)}{dv}dv = C\eta'(v)dv \tag{11.103}$$

where the constant C contains all the parameters in the radar equation including the range dependence. Here $\eta' = \frac{d\eta}{dy}$ is the spectral density of reflectivity as a function of Doppler velocity.

A precautionary note must be added since η generally refers to the cross section per unit volume and may vary through the pulse volume, while $S(v)\ dv$ represents the distribution of average echo power with v from the entire pulse volume. Thus, Equation 11.103 is rigorous only if $\eta(v)$ is uniform throughout the pulse, otherwise $\eta(v)\ dv$ must be thought of as the average reflectivity contributed by scatterers in the velocity interval v to $v + dv$. Clearly $\eta'(v)$ is the reflectivity velocity or Doppler spectrum.

Now the reflectivity is given by

$$d\eta(v) = \eta'(v)dv = N(v)\sigma(v)dv \tag{11.104}$$

where $N(v)\ dv$ is the number of particles per unit volume in velocity interval v to $v + dv$ of cross section $\sigma(v)$. Moreover, if there is a well-defined relationship between particle fall speed and diameter, as is the case with rain in the absence of turbulence, then

$$N(v)dv = N(D)dD \tag{11.105}$$

If, in addition, there is a known relationship between particle cross section σ and D, we may substitute $\sigma(D)$ for $\sigma(v)$, and using Equation 11.66 in Equation 11.106, we get

$$\eta' dv = \frac{d\eta(v)}{dv}dv = \left[N(D)\sigma(D)\frac{dD}{dv} \right] dv \tag{11.106}$$

where the quantity in brackets is the Doppler reflectivity density spectrum. In the case of Rayleigh scatterers in which $\sigma(D) = kD^6$, the quantity

$$N(D)\,D^6\ dD = \frac{dZ(D)}{dD}dD = Z'dD \tag{11.107}$$

where Z is the well-known reflectivity factor. We may therefore write

$$Z'dv = \frac{dZ(v)}{dv}dv = [N(D)\,D^6\,(dD\,/\,dv)]dv \tag{11.108}$$

Where $Z' = \dfrac{dZ}{dv}$ is the Doppler spectrum of reflectivity factor density. This is the form which we shall find most useful in dealing with the Doppler spectra of rain and snow.

General Form of the Z′(v) Spectrum for Exponential Size Distributions.

The number-size distribution of rain and snow may be expressed as,

$$N(D) = N_o \exp[-\Lambda D] = N_o \exp[-\alpha\, D\,/\,D_o] \tag{11.109}$$

where $N(D)$ is the number concentration per unit volume per unit size interval of particles of *melted* diameter D, and N_o, and Λ have the values given as follows

	$N_o(\text{cm}^{-4})$	$\Lambda = \alpha\,/\,D_o(\text{cm}^{-1})$	
Rain	8×10^{-2}	$41\,R^{-0.21}$	[11.110(*ab*)]

(after Marshall and Palmer, 1948)

$$\text{Snow } 2.5\times10^{-2}R^{-0.94}\ 22.9R^{-0.45} \qquad [11.110(ab)]$$

(after Sekhon and Srivastava,1969)

where R is in mm/hr.

The parameter $\alpha = \Lambda D_o$ where D_o is the median volume diameter has an asymptotic value of 3.67 when the maximum diameter in the size spectrum $D_m \rightarrow \infty$. In practice $\alpha = 3.67$ for $D_m \geq 2.5\ D_o$.

Although no simple form of the fall speed-diameter relationship is valid for the entire range of particle sizes, the use of the approximation

$$V_D \cong a\,D^b \qquad (11.111)$$

greatly facilitates tractable analytical solutions. The values of a and b are as follows:

	a (cm$^{1/2}$sec^{-1})	b
Rain (Spilhaus, 1948)	1420	0.5
Rain (Liu and Orville, 1968)	2115	0.8
Aggregate Snow (Langleben, 1954)	207	0.31

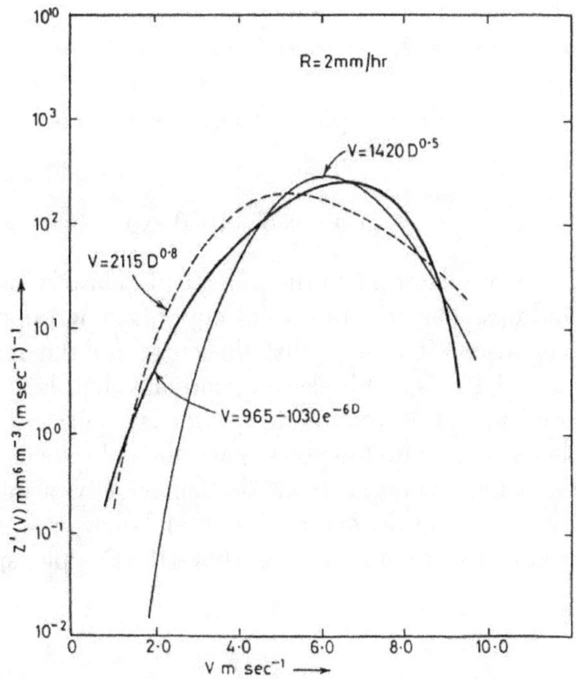

FIGURE 11.37 Radar reflectivity factor density vs. Doppler velocity for Marshall-Palmer distribution for $R = 2$ mm/hr. The fall velocity equations used are as indicated.

From Equation 11.111, we obtain

$$\frac{dD}{dv} = \frac{D^{(1-b)}}{ab} \tag{11.112}$$

Substituting Equation 11.112 into Equation 11.108 we obtain the general reflectivity Doppler spectral density

$$Z'(v) = \frac{dZ(v)}{dv} = N(D)D^{7-b}/ab \tag{11.113}$$

and using Equation 11.109 in Equation 11.113, we get

$$Z'(v) = \frac{N_o D_{(7-b)}}{ab} \exp[-\Lambda D] \tag{11.114}$$

explicitly in terms of v, we get after some manipulation

$$Z'(v) = \frac{N_o V^{(7/b-1)} \exp[-\Lambda(v/a)^{1/b}]}{a^{7/b} b} \tag{11.115}$$

Using Equation 11.110 for rain, plots of $Z'(v)$ are represented in Figure 11.37 for the cases of both $a = 1420$, $b = 0.5$ (solid curves) and $a = 2115$, $b = 0.8$ (dashed curves) for a rainfall rate of 2 mm hr^{-1}. In order to obtain more realistic plots of the Doppler reflectivity spectra for rains, we have also used the equation

$$v = 965 - 1030 \exp(-6D) \tag{11.116}$$

which is a very close fit to the fall speed data of Gunn and Kinzer (1949) for raindrops. The resulting $Z'(v)$ curve is superimposed on Figure 11.37 for comparison. It is seen that the curve for the realistic fall velocity in Equation 11.107 is highly skewed, and falls sharply at the high velocity end compared to the curves for the power law fall velocity rules. This is not unexpected since with the power laws, the fall velocity continues to increase with diameter, whereas actually the fall velocity of raindrops tends to level off with increasing diameter. The latter behavior is well approximated by Equation 11.107. Figure 11.38 (a) shows the Doppler spectra for $R = 1.0$, 2.0,

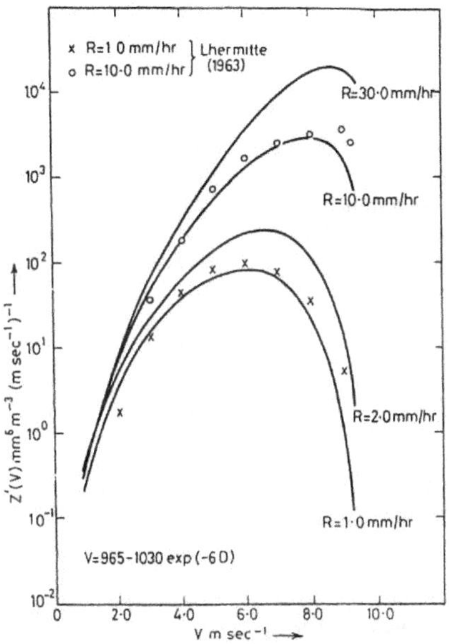

FIGURE 11.38 *(a)* Radar reflectivity factor density vs. Doppler velocity for the Marshall-Palmer distribution for $R = 1$, 2, 10, and 30 mm/hr using the fall soeed equation, $v = 965 - 1030 \exp(-6D)$. Values computed by Lhermitte (1963) using Gunn-Kinzer velocity data are also shown.

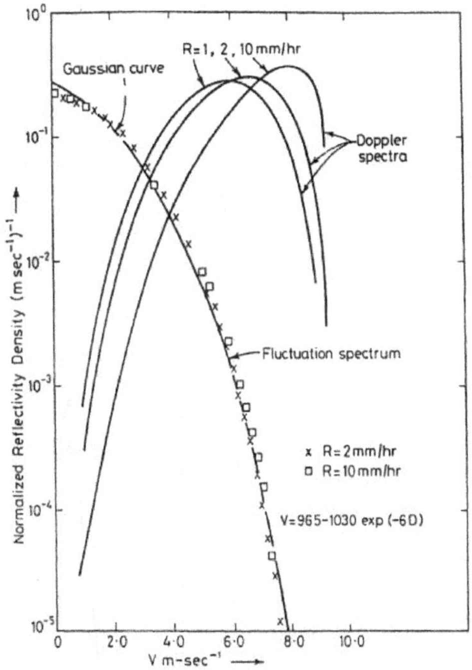

FIGURE 11.38 *(b)* Normalized Doppler and fluctuation spectra for three rainfall rates. The fluctuation spectra are very well approximated by a Gaussian curve.

10.0, and 30.0 mm/hr, computed using the fall velocity (Equation 11.107). Values computed by Lhermitte (1963) using the *actual* Gunn-Kinzer velocity data are also shown. The agreement between the two is good. In Figure 11.38 (*b*), the normalized Doppler and fluctuation spectra computed using the fall velocity (Equation 11.107) are shown. It is seen that the fluctuation spectra are practically independent of the rainfall rate, and further are very well represented by a normal curve. In Figure 11.39, the Doppler spectra for snow are shown for three snowfall rates. For this case, the fall speed law $v = 207\ D^{0.31}$ was used, which holds for aggregate snowflakes. The Doppler spectra for snow are, of course, very much narrower compared with those for rain.

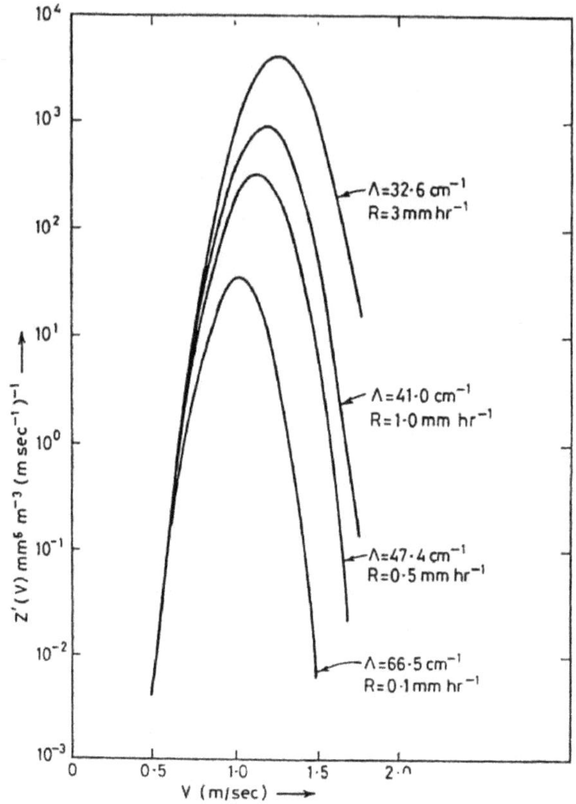

FIGURE 11.39 Doppler spectra for the Marshall-Palmer distribution using the fall velocity equation for aggregate snow.

The Mean Doppler Velocity and Variance of the Spectrum.

The nth moment of the Doppler reflectivity spectrum may be expressed

$$\overline{v^n} = \int_0^\infty v^n \frac{dZ}{dv} dv \Big/ \int_0^\infty \frac{dZ}{dv} dv \tag{11.117}$$

we have $$\frac{dZ}{dv}\frac{dv}{dD} = \frac{dZ}{dD} = N(D)D^6 \frac{dD}{dv}\frac{dv}{dD} = N(D)D^6$$

Substituting $v^n = a^n D^{bn}$ and $N(D) = N_o e^{-\Lambda D}$

$$\overline{v^n} = \frac{\int a^n D^{(6+bn)} e^{-\Lambda D} dD}{\int D^6 e^{-\Lambda D} dD} = \frac{a^n \Gamma(7+bn)\Lambda^7}{\Lambda^{7+bn}\Gamma(7)}$$

$$\overline{v^n} = \frac{a^n \Gamma(7+bn)}{\Lambda^{bn}\Gamma(7)} \tag{11.118}$$

Thus, the mean Doppler velocity (*i.e.*, *n*=1) is

$$\bar{v} = \frac{a\Gamma(7+b)}{\Lambda^b\Gamma(7)} = \frac{a}{\Lambda^b}F_1(\Gamma,b) \tag{11.119}$$

and the second moment is

$$\overline{v^2} = \frac{a^2}{\Lambda^{2b}}\frac{\Gamma(7+2b)}{\Gamma(7)} = \frac{a^2}{\Lambda^{2b}}[F_2,(\Gamma,b)]^2 \qquad [11.120(a)]$$

The rms Doppler velocity is therefore

$$\overline{(v^2)}^{1/2} = \frac{a}{\Lambda^b}\left[\frac{\Gamma(7+2b)}{\Gamma(7)}\right]^{1/2} = \frac{a}{\Lambda^b}F_2(\Gamma,b) \qquad [11.120(b)]$$

Finally, the variance of the Doppler spectrum,

$$\sigma_v^{\,2} = \overline{v^2} - (\bar{v})^2 = \frac{a}{\Lambda^{2b}}\left\{\frac{\Gamma(7)\Gamma(7+2b)-[\Gamma(7+b)]^2}{[\Gamma(7)]^2}\right\} \tag{11.121}$$

$$\sigma_v = \frac{a}{\Lambda^b}\left\{\frac{\Gamma(7)\Gamma(7+2b)-[\Gamma(7+b)]^2}{[\Gamma(7)]^2}\right\}^{1/2} = \frac{a}{\Lambda^b}F_3(\Gamma,b) \tag{11.122}$$

It will be recalled that $\Gamma(k+1) = k\Gamma(k) = k\,!$. Thus $\Gamma(7)\ 6\,! = 720$. Since $\bar{v},\overline{(v^2)}^{1/2}$, and σ_v all involve the coefficient a/Λ^b multiplied by functions of Γ and *b*, we have evaluated the required functions in Table 11.4 for the various values of *b* with which we are concerned.

TABLE 11.4
The Functions of Γ and *b* Involved in $\bar{v},(\bar{v}^2)^{1/2}$ and σ_v for Selected Values of *b*

b	$\Gamma(7+b)$	$\Gamma(7+2b)$	$F_1(\Gamma,b)$	$F_2(\Gamma,b)$	$F_3(\Gamma,b)$
0.31	1290	2360	1.79	1.81	0.260
0.5	1870	5040	2.60	2.65	0.504
0.8	3380	17290	4.69	4.90	1.406

Using the values in Table 11.4, we then get the following relations

$$\bar{v} = (1.79; 2.60; 4.69)a/\Lambda^b \tag{11.123}$$

$$\overline{(v^2)^{1/2}} = (1.81;\ 2.65;\ 4.90)a/\Lambda^b \tag{11.124}$$

$$\sigma_v = (0.26; 0.50; 1.41)a/\Lambda^b \tag{11.125}$$

for $b = 0.31$, 0.5, and 0.8.

Now we may obtain explicit expressions for these functions corresponding to the various values of a and b of interest in the case of rain and snow. The resulting equations are listed in Table 11.5.

Using the realistic fall velocity equation,

$$v = \alpha_1 - \alpha_2 \exp(-6D)$$

where $\alpha_1 = 965$, $\alpha_2 = 1030$

we have

$$\bar{v} = \int e^{-\Lambda D} D^6 [\alpha_1 - \alpha_2 e^{-6D}] dD \Big/ \int e^{-\Lambda D} D^6 dD$$

$$= \left[\alpha_1 \frac{\Gamma(7)}{\Lambda^7} - \alpha_2 \frac{\Gamma(7)}{(\Lambda+6)^7}\right] \Big/ [\Gamma(7)/\Lambda^7]$$

$$= \alpha_1 - \alpha_2 \left(\frac{\Lambda}{\Lambda+6}\right)^7 \qquad [11.126(a)]$$

or

$$\bar{v} = 965 - 1030\ [\Lambda/(\Lambda+6)]^7 \qquad [11.126(b)]$$

$$\overline{v^2} = \int e^{-\Lambda D} D^6\ [\alpha_1 - \alpha_2 e^{-6D}]^2\ dD \Big/ \int e^{-\Lambda D} D^6\ dD$$

$$= \frac{\Lambda^7}{\Gamma(7)} \int e^{-\Lambda D} D^6 [\alpha_1^2 + \alpha_2^2 e^{-12D} - 2\alpha_1\alpha_2 e^{-6D}] dD$$

$$= \alpha_1^2 + \alpha_2^2 \left(\frac{\Lambda}{\Lambda+12}\right)^7 - 2\alpha_1\alpha_2 \left(\frac{\Lambda}{\Lambda+6}\right)^7 \qquad [11.126(c)]$$

From Equations [11.126 (a) and (c)], we have

$$\sigma_v^2 = \overline{v^2} - (\overline{v^2})^2 = \alpha_2^2 \left[\left(\frac{\Lambda}{\Lambda+12}\right)^7 - \left(\frac{\Lambda}{\Lambda+6}\right)^{14}\right] \qquad [11.127(a)]$$

or

$$\sigma_v = 1030 \left[\left(\frac{\Lambda}{(\Lambda+12)}\right)^7 - \left(\frac{\Lambda}{(\Lambda+6)}\right)^{14}\right]^{1/2} \qquad [11.127(b)]$$

TABLE 11.5
Equations for $\overline{v}, (\overline{v^2})^{1/2}$, σ_v, v^*, v_m and v_0 (all in m sec^{-1}) for Rain and Snow

Parameter	Rain I	Rain II	Snow	Equation No.
$a(\mathrm{cm}^{1/2}\ \mathrm{sec}^{-1})$	1420	2115	207	
b	0.5	0.8	0.31	
$\overline{v}$	$(36.92)\ \Lambda^{-0.5}$	$(99.2)\ \Lambda^{-0.8}$	$(3.71)\ \Lambda^{-0.31}$	(11.123)
$(\overline{v^2})^{1/2}$	$(37.63)\ \Lambda^{-0.5}$	$(103.6)\ \Lambda^{-0.8}$	$(3.75)\ \Lambda^{-0.31}$	(11.124)
σ_v	$(7.16)\ \Lambda^{-0.5}$	$(29.74)\ \Lambda^{-0.8}$	$(.54)\ \Lambda^{-0.31}$	(11.124)
v^*	$(36.2)\ \Lambda^{-0.5}$	$(91.0)\ \Lambda^{-0.8}$	$(3.7)\ \Lambda^{-0.31}$	(11.131)
v_m	$(36.6)\ \Lambda^{-0.5}$	$(96.4)\ \Lambda^{-0.8}$	$(3.7)\ \Lambda^{-0.31}$	(11.133)
v_0	$(27.3)\ \Lambda^{-0.5}$	$(59.9)\ \Lambda^{-0.8}$	$(3.1)\ \Lambda^{-0.31}$	(11.136)
$v_m \sqrt{v}$	0.99	0.97	1.00	
v_m / v^*	1.01	1.06	1.00	
$v^* \sqrt{v}$	0.98	0.97	1.00	
$\overline{v} / v_o$	1.35	1.66	1.20	

The Peak of the Doppler Spectrum.

It will sometimes be simpler to observe or measure the velocity at which the Doppler spectrum is peaked. Thus, let us calculate v_* corresponding to the spectral peak. From Equation 11.114 we have for the reflectivity density Doppler spectrum

$$Z'(v) = \frac{N_o D^{7-b}}{ab} e^{-\Lambda D}$$

The peak of the spectrum thus occurs at a v_* corresponding to a diameter D_* such that

$$\frac{dZ'}{dv} = \frac{dZ'}{dD}\frac{dD}{dv} = \frac{dD}{dv}\frac{N_o}{ab}\left[(7-b)D_*^{6-b}e^{-\Lambda D*} - \Lambda D_*^{7-b}e^{-\Lambda D_*}\right] = 0 \quad (11.128)$$

or

$$(7-b)D_*^{(6-b)} = \Lambda D_*^{(7-b)}$$

or

$$D_* = (7-b)/\Lambda \quad (11.129)$$

To obtain the corresponding velocity, use Equation 11.73 with $v_{\circ} = a\, D_{\circ}^{b}$ so that

$$v_{\circ} = \frac{a}{\Lambda^{b}}(7-b)^{b} \tag{11.130}$$

Clearly, like $\bar{v}$, $(\overline{v^2})^{1/2}$, and σ_v, $v_{\circ}$ is related to a/Λ^b through another proportionality constant. The $v_{\circ} - \Lambda$ relations corresponding to the various values of a and b for rain and snow are also listed in Table 11.5 as in Equation 11.131.

The Median Velocity (and Diameter) of the Doppler Spectrum.

The median velocity, v_m, of the Doppler spectrum is defined by

$$\int_0^{v_m} Z'(v)\, dv = \int_{v_m}^{0} Z'(v)\, dv$$

or by

$$\int_0^{\infty} Z'(v)\, dv = 2\int_0^{v_m} Z'(v)\, dv$$

Since $Z'(v)\, dv = Z'(D)\, dD$, and $Z'(D) = N_o e^{-\Lambda D}\, D^6$, we may also write

$$N_o \int_0^{\infty} D^6 e^{-\Lambda D} dD = 2N_o \int_0^{D_m} D^6 e^{-\Lambda D} dD$$

where D_m is the median of the reflectivity spectrum. Making the substitution $\Lambda D = x$, $D^6 = (x/\Lambda)^6$ and dc/Λ, we then get

$$\Lambda^{-7}\int_0^{\infty} x^6 e^{-x} dx = \Gamma(7)\Lambda^{-7} = 2\Lambda^{-7}\int_0^{x_m} x^6 e^{-x} dx$$

where x_m=ΛD_m. This then reduces to

$$6!/2 = \int_0^{x_m} x^6 e^{-x} dx$$

$$= -\left\{ 6!e^{-x}\left[\frac{x^6}{6!}+\frac{x^5}{5!}+\frac{x^4}{4!}+\frac{x^3}{3!}+\frac{x^2}{2!}+x\right]\right\}_0^{x_m}$$

or

$$\frac{6!}{2} = e^{-x_m}[x_x^{\ 6} + 6x_m^{\ 5} + \ldots 720x_m + 720]$$

This has the solution

$$x_m = \Lambda D_m = 6.669 \tag{11.131}$$

Substituting $v_m = aD_m{}^b$ we also gets

$$v_m = \frac{a}{\Lambda^b}(6.669)^b \tag{11.132}$$

Once more we see that this velocity parameter of the spectrum, like all the others, is related to $a/\Lambda b$ through another proportionality constant. The relations $v_m - \Lambda$ for the various values of a and b are also shown in Table 11.5 as in Equation 11.132.

It is of interest to see how v_m is related to $\overline{v}$ and v_* of the spectrum. Therefore, the ratios $(v_m/\overline{v}), (v_m/v*)$, and $(v*/\overline{v})$ are also presented in Table 11.5. It is evident that all these ratios are very close to unity regardless of the values of a and b in the relationship $v = aD^b$ or Λ. Of course, when the mean peak and median values of a distribution are identical, the distribution is symmetrical. Although this is essentially the case for the Doppler spectra of any exponential size spectrum with $v = aD^b$, we have already seen in Figure 11.37 that the actual Doppler spectra becomes asymmetrical at the larger precipitation rates when a more realistic fall velocity law is used.

Finally, it is of interest to compare the various velocity parameters of the Doppler spectra to v_o, the fall velocity corresponding to the median volume diameter, D_o, in the size distribution of liquid water content. In general, $D_0 = \alpha/\Lambda$ so that

$$v_o = \frac{a}{\Lambda^b}\alpha^b \tag{11.133}$$

When the maximum diameter in the size spectrum $D_m > 2.5$, we have already indicated that $\alpha = 3.67$. Accordingly, we may write

$$v_o = \frac{a}{\Lambda^b}(3.67)^b \tag{11.134}$$

The corresponding v_0–Λ relations are also presented in Table 11.5 as in Equation 11.135. Table 11.5 also lists the associated ratios of $(\overline{v}/v_o)$ and shows that the mean Doppler velocity is 1.35, 1.66, or 1.20 times the fall speed of the median volume particle, depending upon whether the value of b is 0.5 or 0.8 or 0.31 respectively.

11.20 RELATIONS OF THE DOPPLER SPECTRAL VELOCITY PARAMETERS TO THE VARIOUS CHARACTERISTICS OF THE PARTICLE SIZE DISTRIBUTION RELATIONS TO D_o

Since all the velocity parameters of the Doppler spectrum have been calculated on the assumption of an effectively infinite maximum diameter in the size spectrum, we proceed accordingly. Thus, we have $\Lambda = 3.67/D_o$ and $\Lambda^{-b} = D_o^{\,b}/(3.67)^b$. The appropriate values of $(3.67)^{-b}$ are as follows:

b	0.50	0.8	0.31
$(3.67)^{-b}$	0.52	0.35	0.67

Thus, each of the velocity parameters in Table 11.5 may be expressed in terms of D_o simply by replacing Λ^{-b} by $D_o^{\,b}$ and multiplying the coefficients by the appropriate value of $(3.67)^{-b}$. Plots of all the velocity parameters in Table 11.5 versus both Λ and D_o are presented in Figure 11.40.

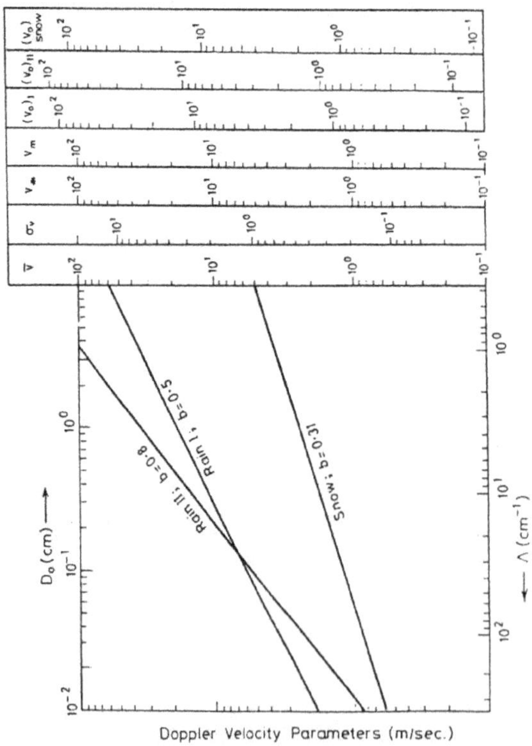

FIGURE 11.40 Doppler velocity parameters vs. Λ and D_o for Marshall-Palmer distribution using the three power law fall velocity equations.

Relations to Precipitation Rate.

To determine the corresponding relations to the precipitation rate, we simply have to substitute Equations 11.110 (a) and 11.110 (b) which are

$$\Lambda = 41\, R^{-0.21} \text{ for rain}$$

and

$$\Lambda = 22.9\, R^{-0.45} \text{ for snow}$$

in the various equations of Table 11.5. Note that all the velocity parameters are independent of N_o. Thus,

$$\Lambda^{-b} = (41)^{-b}\, R^{0.21b} \text{ (rain)} \qquad (11.134a)$$

$$\Lambda^{-b} = (22.9^{-b})\, R^{0.45b} \text{ (snow)} \qquad (11.134a)$$

The coefficients and exponents in these relations are as follows:

	b	0.5	0.8	0.31
Rain	(41^{-b})	.156	.051	–
	$0.21b$	.105	.168	–
Snow	22.9^{-b}	–	–	.379
	$0.45b$	–	–	.140

All the velocity parameters of the Doppler spectrum are plotted in Figure 11.41 as a function of the precipitation rate.

Relations to the Reflectivity Factor, Z.

Since the reflectivity factor for an exponential size distribution is given by

$$Z = \int_0^\infty N_o D^6 e^{-\Lambda D} = \frac{N_o}{\Lambda^7} \int_0^\infty x^6 e^{-x} dx$$

where $x = \Lambda D$, we find as before

$$Z = 6!\, N_o / \Lambda^7 = 720\, N_o / \Lambda^7 \qquad (11.135)$$

or

$$\Lambda^{-1} = Z^{1/7} / (N_o 720)^{1/7} \qquad (11.136)$$

Thus,

$$\Lambda^{-b} = Z^{b/7} / (N_o 720)^{b/7} \qquad (11.137)$$

For rain

$$N_o = 0.08 \text{ cm}^{-4} \qquad [11.138(a)]$$

and for snow

$$N_o = 0.025\, R^{-94} (\text{cm}^{-4}) \qquad [11.138(b)]$$

making use of Z-R relation, i.e.,

$$Z = 1780\, R^{2.21} (\text{Z in mm}^6/\text{m}^3)$$

or

$$Z \times 10^{-12} = 1780\, R^{2.21} (\text{Z in cm}^3) \qquad (11.139)$$

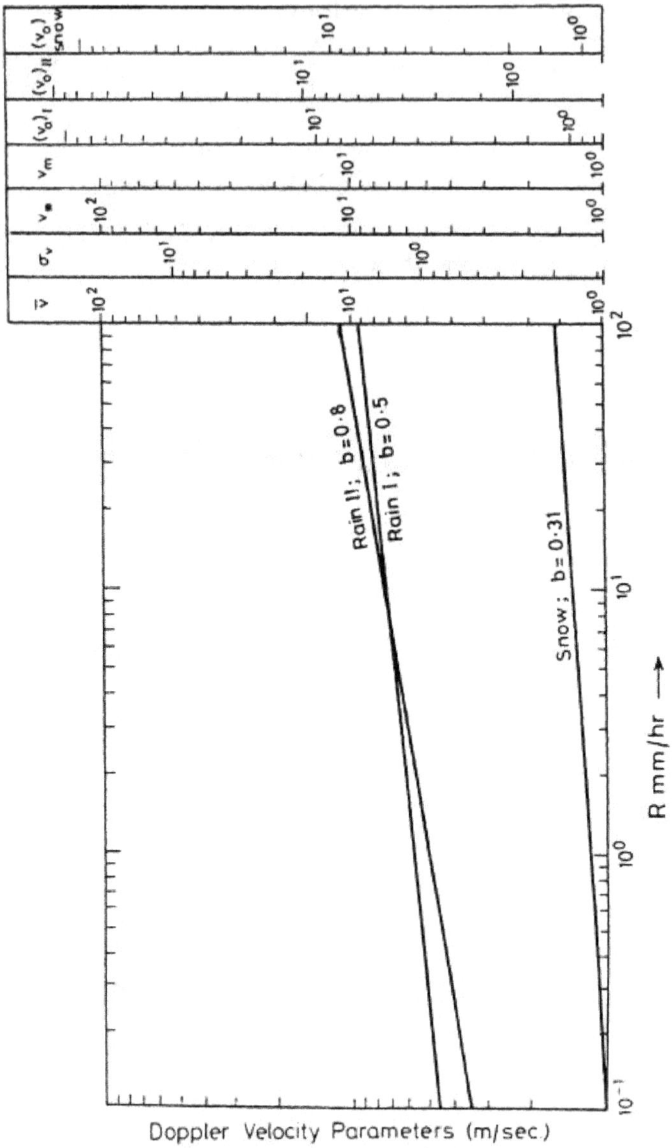

FIGURE 11.41 Similar to Figure 11.40 except that the Doppler velocity parameters are now plotted as a function of the rainfall rate, *R*.

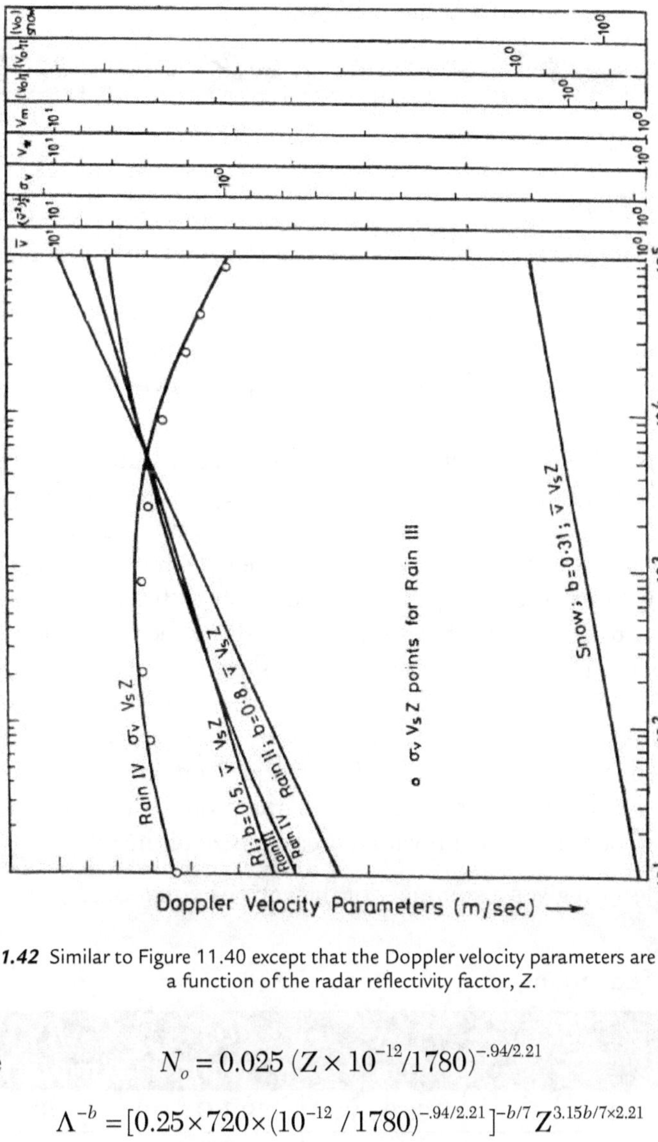

FIGURE 11.42 Similar to Figure 11.40 except that the Doppler velocity parameters are now plotted as a function of the radar reflectivity factor, *Z*.

we have

$$N_o = 0.025\,(Z \times 10^{-12}/1780)^{-.94/2.21} \tag{11.140}$$

Thus

$$\Lambda^{-b} = [0.25 \times 720 \times (10^{-12}/1780)^{-.94/2.21}]^{-b/7} Z^{3.15b/7\times 2.21}$$

$$= [2.272 \times 10^{6}]^{-b/7} Z^{3.15b/15.47} \tag{11.141}$$

so the appropriate coefficients and exponents are as follows:

b	0.5	0.8	0.31
$b/7$	.071	.114	–
$3.15b/7 \times 2.21$	–	–	.063
$(720N_o)^{-b/7}$	.75	.63	–
$[0.25 \times 720 \times (10^{-12}/1780)^{-.94/2.21}]^{-b/7}$	–	–	.523

Therefore, the relations of the various velocity parameters to Z for rain may be obtained simply by replacing Λ^{-b} in Table 11.5 by $Z^{b/7}$ and multiplying the coefficient by $(720\,N_o)^{-b/7}$. It is to be noted that the resulting relations are functions of Z in units of cm^3. We have also superimposed curves of $\bar{v}$ and σ_v computed by use of the Gunn and Kinzer fall speed law, and also the values computed according to Equations 11.126 and 11.127. Note that the standard deviation computed according to the realistic fall velocities at first increases with Z (or rainfall rate), attains a maximum, and then decreases again. This is because the fall velocity of raindrops tends to level off with increasing size, consequently as the relative proportion of the bigger drops increases with increasing rainfall rate, the Doppler spectra tend to be skewed with a large fraction of the total power at high Doppler velocities. This results in a decrease of standard deviation with increasing Z or *R*. In contrast, it will be seen that with power law fall speed equations, the standard deviation is a monotonie increasing function of Z or *R*. This is because in the case of the power law fall speed equation, the fall speed does not level off with increasing size of the drop.

Table 11.6 gives a summary of the various equations.

TABLE 11.6
Summary of Equations

		Λ (cm^{-1})	D_o (cm)	R (mm/hr)	Z (mm^6/m^3)
$\bar{v}$	R_I	$36.9./\Lambda^{.5}$	$19.3\,D_o^{.5}$	$5.8\,R^{11}$	$3.8\,Z^{.07}$
	R_{II}	$99.2/\Lambda^{.8}$	$35.1\,D_o^{.8}$	$15.5\,R^{13}$	$2.7\,Z^{11}$
	Snow	$3.71/\Lambda^{.31}$	$2.5\,D_o^{.31}$	$0.91\,R^{14}$	$0.8\,Z^{.06}$
$(\overline{v^2})^{1/2}$	R_I	$37.6/\Lambda^{.5}$	$19.6\,D_o^{.5}$	$5.9\,R^{.11}$	$3.9\,Z^{.07}$
	R_{II}	$103.6/\Lambda^{.8}$	$36.6\,D_o^{.8}$	$16.2\,R^{.13}$	$2.8\,Z^{.11}$
	Snow	$3.8/\Lambda^{.31}$	$2.5\,D_o^{.31}$	$0.9\,R^{.14}$	$0.8\,Z^{.06}$

		Λ (cm^{-1})	D_O (cm)	R (mm/hr)	Z (mm^6/m^3)
σ_v	R_I	$7.2/\Lambda^{.5}$	$3.8\,D_o^{5}$	$1.1\,R^{.11}$	$0.5\,Z^{.07}$
	R_{II}	$29.7/\Lambda^{.8}$	$10.5\,D_o^{.8}$	$4.6\,R^{.13}$	$0.7\,Z^{11}$
	Snow	$0.5/\Lambda^{.31}$	$0.4\,D_o^{.31}$	$0.1\,R^{.14}$	$0.2\,Z^{.06}$
$V_\circ$	R_I	$36.2/\Lambda^{.5}$	$18.9\,D_o^{.5}$	$5.7\,R^{.11}$	$3.8\,Z^{.07}$
	R_{II}	$91.0/\Lambda^{.8}$	$32.2\,D_o^{.8}$	$14.2\,R^{.13}$	$2.4\,Z^{.11}$
	Snow	$3.7/\Lambda^{.31}$	$2.5\,D_o^{.31}$	$0.9\,R^{.14}$	$0.8\,Z^{.06}$
v_m	R_I	$36.6/\Lambda^{.5}$	$19.1\,D_o^{.5}$	$5.7\,R^{.11}$	$3.8\,Z^{.07}$
	R_{II}	$96.4/\Lambda^{.8}$	$34.1\,D_o^{.8}$	$15.1\,R^{.13}$	$2.6\,Z^{.11}$
	Snow	$3.7/\Lambda^{.31}$	$2.5\,D_o^{.31}$	$0.9\,R^{.14}$	$0.8\,Z^{.06}$
v_o	R_I	$27.3/\Lambda^{.5}$	$14.3\,D_o^{.5}$	$4.3\,R^{.11}$	$2.8\,Z^{.07}$
	R_{II}	$59.9/\Lambda^{.8}$	$21.2\,D_o^{.8}$	$9.4\,R^{.13}$	$1.6\,Z^{.11}$
	Snow	$3.1/\Lambda^{.31}$	$2.1\,D_o^{.31}$	$0.8\,R^{.14}$	$0.7\,Z^{.06}$

11.21 REAL-TIME DOPPLER RADAR DATA PROCESSING AND DISPLAY

Doppler radars have been used in atmospheric research, where the inherent capability of these radars has provided a valuable tool in the study of storm dynamics. Further, the widths of the Doppler frequency spectra can be obtained as a measure of velocity dispersion. Such Doppler data is used effectively in real time to detect severe weather and issue advisories. Doppler radar technology has evolved to the point of transfer to the wide-scale, real time operational environment. This application implies development and implementation of complex signal processing techniques to produce easily interpretable outputs for the observer.

At the National Severe Storms Laboratory (NSSL), a powerful data processing system was developed to produce displays of Doppler radar signal parameters in real time. This system generates simultaneous displays of reflectivity, velocity, and spectrum width on raster-scan color monitors. In addition, a high resolution black and white monitor is used for various special purpose displays.

The maximum unambiguous range and velocity, called Nyquist Velocity, are related by the equation

$$R_m V_m = \frac{c\lambda}{8} \tag{11.142}$$

where c is the electromagnetic propagation constant and λ is the wavelength. For conventional centimeter-band Doppler radars, the range-velocity ambiguity product is not too large to allow observation of severe convective storms without some obscuration due to overlaid echoes.

The radar system has three hard-wired digital signal processors to provide estimates of signal intensity ($\hat{I}$), Doppler velocity ($\hat{V}$), and velocity spectrum width ($\hat{W}$) at 762 locations along the radar radial.

The intensity integrator operates on the returned signal at the output of a logarithmic receiver. This device uses an exponential time window that performs no range integration in the normal mode and a linear range averaging over 600 meters in the intensity range expanded mode.

The velocity processor provides an estimate of the mean radial velocity by the complex covariance technique operating on the inphase (I) and quadrature (Q) components of the complex video signal at the output of the phase demodulator. This device uses uniformly weighted echo samples, and it does not perform range averaging. The sample pairs are contiguous in the normal mode, but occur in groups of eight in the expanded mode. The spectrum width processor estimates the velocity spectrum width by a complex correlation scheme. This device is associated with the velocity processor, sharing the same set of operator options and the same circuitry for many operations.

The outputs from the primary processors are multiplexed into a block of data corresponding to a radial of radar data. Housekeeping information, for example, radar control parameters and selected options, are also included in every block. These records are written on magnetic tape and introduced to the computer terminals.

11.22 STATISTICAL PROPERTIES OF DUAL-POLARIZED RADAR SIGNALS

1. A pulsed, meteorological radar illuminates a radar resolution volume which depends on range, antenna beam widths, and also on the transmitted pulse width. Within this volume hydrometeors are randomly positioned and constitute a random medium from which radar

measurements can be obtained. Radar measurements of reflectivity, and differential reflectivity. Z_{DR}, involve estimation of mean back scattered power $(\overline{p}_h)$ and $(\overline{p}_V)$, where $\overline{P}_H$ and $\overline{P}_V$ are the mean back scattered powers at horizontal and vertical polarizations respectively. Proper measurement of Z_{DR} is important for rain and ice melting studies. Many radar systems employ alternate polarization sampling on a pulse-to-pulse basis to estimate Z_{DR}.

We first discuss the Z_{DR} variance for logarithmic receivers and then present an algorithm to simulate samples of Z_{DR} from dual polarized time series. These simulated signals offer great potential for studying the fluctuations of derived quantities from Z.

2. **Z_{DR} Using Logarithmic Receivers.** Many meteorological radars are equipped with logarithmic receivers, and it has been established that the variance of reflectivity estimates from log receivers are higher than those from square law receivers. Here we compute the mean and variance of Z_{DR} estimates from log receivers.

A log receiver estimate of Z_{DR} is defined as,

$$\hat{Z}_{DR}(dB) = \frac{1}{M} 10 \log \left[\frac{p_H^i}{p_V^i} \right] \tag{11.143}$$

where M is the number of sample pairs, and p_H^i and p_V^i are the ith power samples at H and V polarization respectively. Taking expectation on both sides of equation (143) we get

$$\varepsilon(\hat{Z}_{DR}) = \frac{4.343}{M} \left\{ \sum_{i=1}^{m} \varepsilon\left[\ln(p_h^i)\right] - \sum_{i=1}^{m} \varepsilon\left[\ln(p_v^i)\right] \right\}$$

where $\varepsilon(\hat{Z}_{DR})$ stands for statistical expectation. p_H^i and p_V^i are distributed exponentially, and it can be shown that

$$\varepsilon \ln\,(p_{H.V}^i) = \ln\,(\overline{p}_{H.V}) - r \tag{11.144}$$

where r is Euler's constant and p_H and p_V are the mean back scattered powers at horizontal and vertical polarizations respectively.

Substituting (144) in (143) we get

$$\varepsilon\,(\hat{Z}_{DR}) = \overline{Z}_{DR} = 10 \log \left(\frac{\overline{p}_H}{\overline{p}_V} \right) \tag{11.145}$$

This shows that the log receiver estimate of Z_{DR} is unbiased. The variance of $\hat{Z}_{DR}$ can be expressed as

$$\mathrm{var}(\hat{Z}_{DR}) = \frac{18.86}{M^2}\left\{\mathrm{var}\sum_{i=1}^{M}\ln(p_H^i) + \mathrm{var}\sum_{i=1}^{M}\ln(p_V^i)\right\}$$

$$-2\,\mathrm{cov}\left[\sum_{i=1}^{M}\ln(p_H^i), \sum_{i=1}^{M}\ln(p_V^i)\right] \quad (11.146)$$

Where $\mathrm{var}\left[\sum_{i=1}^{M}\ln p_{H,V}^i\right] = m = -\sum_{(M-1)}^{(M-1)}(M-|m|)\,\mathrm{cov}\left[\ln p_{H,}^i \ln p_H^{i+m}\right]$ (11.147a)

and $\mathrm{cov}\left[\sum_{i=1}^{M}\ln[p_H^i], \sum_{i=1}^{M}\ln[p_V^i]\right] m = -\sum_{(M-1)}^{(M-1)}(M-|m|)$

$\mathrm{cov}\left[\ln p_{H,}^i \ln p_V^{i+m}\right]$ (11.147b)

it can be shown that $\mathrm{var}\left[\ln p_{H,V}^i\right] = \frac{\pi^2}{6}$. Substituting equation (11.147) into equation (11.146) we get

$$\mathrm{var}\,\hat{Z}_{DR} =$$

$$\frac{18.86}{M^2}\cdot\frac{\pi^2}{6}\left[m = -\sum_{(M-1)}^{(M-1)}(M-|m|)\left\{\rho_{HH}^{\log}(m) + \rho_{VV}^{\log}(m) - 2\rho_{HV}^{\log}(m)\right\}\right]$$

where $\rho^{\log}$ is the correlation between log of the samples.

Assuming $\rho_{HH} = P_{VV}$, we get

$$\mathrm{var}\,\hat{Z}_{DR} =$$

$$\frac{37.72}{M^2}\cdot\frac{\pi^2}{6}\left[m = -\sum_{(M-1)}^{(M-1)}(M-|m|)\left\{\rho_{HH}^{\log}(m) + \rho_{VV}^{\log}(m) - 2\rho_{HV}^{\log}(m)\right\}\right] \quad (11.148)$$

Equation (148) gives the expression for the variance of $\hat{Z}_{DR}$ for simultaneous sampling. For alternate sampling $\rho_{HH}(m)$ becomes $\rho_{HH}(2m)$ and $\rho_{HV}(2m)$ becomes $\rho_{HV}(2^{m+1})$. Hence we obtain,

$$Var\,(\hat{Z}_{DR}) = \frac{32.72}{M^2}\cdot\frac{\pi^2}{6}\left[(m-|m|)\left\{\rho_{HH}^{\log}(2m) - \rho_{HV}^{\log}(2m+1)\right\}\right] \quad (11.149)$$

It is a well-known approximation to assume a Gaussian form for the auto-correlation function of the complex signals from which $P_{H,V}$ are derived. The relationship between the correlation of the complex signals is given by

$$\rho^{\log} = \frac{6}{\pi^2} \sum_{m=1}^{\infty} \frac{|\rho|^{2m}}{m^2} \tag{11.150}$$

Using equation (11.150) and the approximation used by Sachidananda and Zrnic (1986),

$$\rho_{HV}^{p}(m) \simeq \rho_{HH}^{p}(m) \cdot \rho_{HV}^{p}(0) \tag{11.151}$$

where the superscript "p" refers to power samples, we can compute the variance in Z_{DR} estimates for log receivers. In the presence of noise the correlation between the p_H and p_V samples is reduced, and it can be obtained as

$$\rho_{\text{noise}}^{p} = \rho_{\text{no noise}}^{p} \Big/ \left(1 + \frac{1}{SNR}\right)\left(1 + \frac{\sigma_{DR}}{SNR}\right) \tag{11.152}$$

where SNR is the signal-to-noise ratio and σ_{DR} is Z_{DR} in linear scale.

The standard deviation in $Z_{DR}(\sigma_{ZDR})$ as a function of number of sample pairs used for alternate sampling, for $\left|\rho_{HV}^{(0)}\right|^2 = 0.99$ and 0.995, respectively for Gaussian Doppler spectrum with varying spectral width, pulse repetition time, of 1 ms and wavelength = 10 cm. These curves can be compared with similar curves for a square law receiver given by Sachidananda and Zrnic (1986). σ_{ZDR} has the same features of variability with spectral width and number of sample pairs.

However, the values of σ_{ZDR} for log receivers are higher than those for square law receivers by a factor of approximately 2. This suggests that a square law receiver is preferable to a log receiver in the context of Z_{DR} estimation.

It is useful to know the covariance (and hence correlation) between Z and Z_{DR} so that it can be used in computing the variance of functions of Z and Z_{DR}. Using the same techniques outlined for computing σ_{ZDR}, we get

$$\begin{aligned} \text{cov}[Z, Z_{DR}] &= \frac{18.86}{M^2} \text{cov}\left[\sum_{i=1}^{M} \ln p_H^i, \sum_{i=1}^{M} \ln\left[\frac{p_H^i}{p_v^i}\right]\right] \\ &= \frac{18.86}{M^2} \text{cov}\left[\sum_{i=1}^{M} \ln p_H^i, \sum_{i=1}^{M} \ln p_H^i\right] \\ &\quad - \frac{18.86}{M^2} \text{cov}\left[\sum_{i=1}^{M} \ln p_H^i, \sum_{i=1}^{M} \ln p_v^i\right] \end{aligned} \tag{11.153}$$

$$= \frac{18.86}{X^2} \cdot \frac{\pi^2}{6} \left[m = -\sum_{(M-1)}^{(M-1)} (M - |m|) \left\{ \rho_{HH}^{\log}(m) - \rho_{H\upsilon}^{\log}(m) \right\} \right] \quad (11.154)$$

We can see from equation (11.154) that cov [Z, Z_{DR}] is half the variance of Z_{DR}. Hence the correlation coefficient ρ is given by,

$$\rho = \frac{1/2\sigma^2_{ZDR}}{\sigma_Z \cdot \sigma_{ZDR}} \quad (11.155)$$

where σ_z is the variance of the Z estimate.

Equation (13) simplifies to $\rho_{Z.Z_{DR}} = \frac{1}{2}\left(\frac{\sigma_{ZDR}}{\sigma_Z}\right)$

which typically takes values around 0.08. This indicates that, for all practical purposes, we can consider the Z and Z_{DR} estimates from log receivers to be uncorrelated.

3. **Simulation of Dual-Polarized Signals.** Simulation has always been a powerful method in statistics to handle analytically interactable quantities. Z_{DR} is used in many situations in non-analytical form and it would be useful to have some simultation scheme to simulate Z_{DR} samples.

Let *H*, *V* be the marginals of the bivariate complex process that describes the dual-polarized complex signals. The time correlation of these signals is described by their Doppler spectrum. The correlation between *H* and *V* signals at a given time is described by the type of scatterers, and their shape and orientation distributions. For raindrops this correlation depends on drop shape, oscillation, and canting angle distributions, and this has been shown to be very high, that is, $|\rho_{HV}|^2 > 0.99$,

where ρ_{HV}, is the correlation between *H* and *V* signals.

The *H* and *V* signals can be written as,

$$H, V = X_{h,\upsilon} + iY_{h,\upsilon} \quad (11.156)$$

(X_h, Y_n, X_υ, Y_υ) are Gaussian and have a hermitian symmetric autocovariance function. They are described by the classical 2-D random walk problem.

The *H* and *V* complex signals satisfy the correlation structure, and they form a complex bivariate Gaussian process.

For the previous signals we can define the spectral estimates as follows:

$$\frac{1}{2}n^{1/2}(a_{1k}+b_{1k})=\frac{1}{n^{1/2}}\sum_{t=1}^{n}H_t e^{i\omega_k t} \tag{11.157}$$

$$\frac{1}{2}n^{1/2}(a_{2k}+b_{2k})=\frac{1}{n^{1/2}}\sum_{t=1}^{n}V_t e^{i\omega_k t}$$

Where $\omega_k=\frac{2\pi k}{n}; k=0,\ 1,\ \ldots\ldots,\ n-1$ and $1=\sqrt{-1}$.
It can be shown for real bivariate Gaussian time series that asymptotically,

$$\vec{r}_k=2^{-1/2}n^{-1/2}(a_{1k},b_{1k},a_{2k},b_{2k})$$

is distributed as a multivariate normal random vector with zero mean and covariance matrix given by,

$$\varepsilon\{\vec{r}_k\,\vec{r}_k\}=2\pi \begin{bmatrix} f_{11}(\omega_k) & 0 & c_{12}(\omega_k) & q_{12}(\omega_k) \\ 0 & f_{11}(\omega_k) & -q_{12}(\omega_k & c_{12}(\omega_k) \\ c_{12}(\omega_k) & -q_{12}(\omega_k) & f_{22}(\omega_k) & 0 \\ q_{12}(\omega_k) & c_{12}(\omega_k) & 0 & f_{22}(\omega_k) \end{bmatrix}+0(n^{-1}) \tag{11.158}$$

where $f_{12}(\omega_k)=c_{12}(\omega_k)-iq_{12}(\omega_k)$ and $f_{11}\cdot f_{12}$ are the spectra of H, V, respectively, and f_{12} is the cross-spectrum.

Also $E\ \{r_k\ r_j^t\}=0(n^{-1})$, for $j\neq k$. The previous results indicate that the spectral estimates of H and V defined by equation (11.157) are asymptotically independent when $j\neq k$ and the correlation structure is completely known in terms of the spectra. This result can be extended for multivariate signals.

$$f_{11}(f)=\frac{S_0}{\sqrt{2x\sigma_f}}\exp\left\{\frac{(f-\bar{f})^2}{2\sigma_f^2}\right\} \tag{11.159}$$

where $f=\frac{\omega}{2\pi},\sigma_f=\frac{2\sigma_v}{\lambda}$ and S_0 is the mean signal power at horizontal polarization. The spectrum of the V signal is assumed to be the same as that of H except that S_0 is lower by the mean Z_{DR}. With no loss of generality we can assume the mean velocity to be zero if we are interested in simulating only the power signals. Regarding the cross-spectrum, it can be obtained by making the assumption $\rho_{HV}(n) = \rho_{HH}(n)\rho_{HV}(0)$ where $\rho_{HV}(0)$ is the correlation between H and V signals at the same time. This assumption implies that the Doppler spectrum is the same for each size category, and that the joint distribution can be written as a product of two marginal distributions. With this assumption we get

$$f_{12}(f) = f_{11}(f) \cdot \rho_{HV}(0).$$

Noise can easily be incorporated into the simulations by adding a constant to both the co-spectral components according to the required signal-to-notes ratio.

$\rho_{HV}(0)$ in rainfall is very high, typically greater than 0.95.

The simulation algorithm is as follows:

1. Generate X with 4 iid N(0.1) components.
2. Factor the positive definite matrix A described by equation (11.158) using the Cholesky factorization, $A = LL^T$.
3. Compute the random vector XL^T with covariance matrix A.
4. Repeat (1) through (3) for all values of ω_k: $k = 0, 1, 2, \ldots. n - 1$.
5. Form the complex vectors $(a_{1k} + ib_{1k})$ and $(a_2k + ib_{2k})$
6. Compute the time sequence by performing a discrete Fourier inverse of the spectral sequences.
7. Obtain the power sequence by squaring the magnitude of the complex time sequence.

The previous simulation procedure has been tested for distributional, auto-, and cross-covariance properties.

We can use the algorithm to simulate bivariate power samples P_H and P_V for simultaneous sampling. Alternate polarization sampling can be simulated by selecting alternate samples starting with the P_V sequences and time lagging the start of the P_H sequence by one count. From the derived time series Z_{DR} samples can then be computed. The Z_{DR} variance obtained from simulations match with those obtained analytically for both log and square law receivers.

Log receivers are used in many meteorological radar systems, and the estimates of Z from log receivers are more noisy than those from square law receivers. We have shown that estimates of Z_{DR} from log receivers also have a higher standard of deviation than those from square law receivers by approximately a factor of 2. Z_{DR} and Z estimates are nearly uncorrelated, with the correlation coefficient as low as 0.08.

Simulation of Log Receiver Method. Figure 11.43 shows the block diagram for the log-linear conversion technique. When E is given as the radar input signal of the precipitation echo, the signal amplitude at the output of the logarithmic receiver is given by

$$E_{\log} = \frac{2^N - 1}{\log 2^M} \log(E+1) \tag{11.160}$$

where N is the A-D converter's bit value, and M(bit) the dynamic range of the input signal. Now if $E_{max} = 2^M - 1$ and $E_{min} = 0$, then

$$2^M = \frac{E_{max} + 1}{E_{min} + 1} \tag{11.161}$$

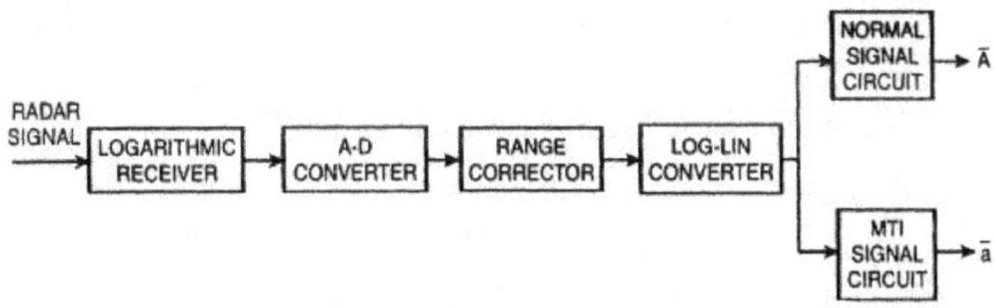

FIGURE 11.43 A block diagram showing radar signal processing using the linear characterized signal converted from the output amplitude of a logarithmic receiver.

The digital signal is converted from the log form to the linear form at the log-lin conversion circuit as follows:

$$(E_0 + 1) = 2^Q.(E+1)^{L/M} \tag{11.162}$$

where the L bit shows the dynamic range of the conversion circuit. If L is put equal to the value of M, a linear signal can be obtained.

$$(E_0 + 1) = 2^Q.(E+1) \tag{11.163}$$

If the range correction is carried out and the resultant dynamic range is $M_R(< M)$, the circuit configuration can be made as $L = M_R$, and the total bit number at the output of the log-lin conversion circuit can be decreased.

Multiple 2^Q is used to improve the reproducibility due to the inferiority of the quantization resolution in the smaller amplitude region. Therefore, the significant figure of E_o is increased by 2^Q and the accuracy of the reproducibility is also improved. Of course, the order is kept constant regardless of the value of 2^Q. Accordingly, equation (11.163) is rewritten in the form of $E_o = E$, where the figure has the quantization resolution of $1/2^Q$.

In the case of normal signal processing, though the negative error is rather small in the small amplitude region, it will increase the chance to miss the radar echo in the radar noise. Therefore, in order to improve the error to the inherent error due to the characteristics of the logarithmic receiver, it is necessary to add $Q = 5$ bit to the bit number M of the dynamic range of the input signal. However, the error can be put within –0.3 dB (0.965) by adding $Q = 2$ bit at least.

In the case of MTI signal processing, the decrease of $\overline{A}_T$ tends to increase a positive error, and since the amplitude difference of the adjacent pulses is processed, there is no significant difference in the characteristics. From equation (11.164), R_T is 0.5 mm/hr at $\overline{A}_T = 2.1$, but the measured value R is 0.7 mm/hr. Therefore, the absolute error is only 0.2 mm/hr, so it will not bring about any serious problem.

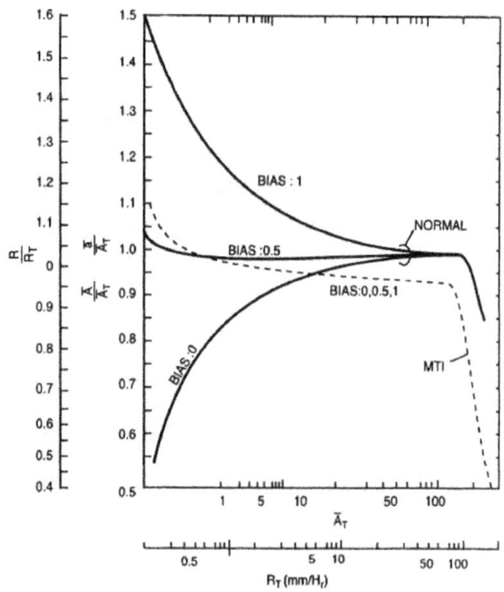

FIGURE 11.44 The input-to-output characteristics of the conventional and MTI precipitation echo measurements as the parameter of bias.

The amplitude of the output signal is affected by the DC setting level at the A-D conversion, and the magnitude is given by 10 to the $\left(\frac{\text{B.M. log 2}}{2^N - 1}\right)$ power, where B is the deviated quantity. For example, $B = +1$ corresponds to the deviation of one quantization resolution in the forward direction of the input signal's amplitude. Therefore, for $B = \pm 1$, and ± 2, the output amplitude will vary by $\pm$ 2.2 % and $\pm$ 4.3 % in the linear region respectively.

$$\left(\frac{\overline{A}}{2^N - 1}\right)^2 = \left(\frac{R}{R_{max}}\right)^\beta \tag{11.164}$$

where N is the bit number of the A-D converter, R_{max} is the maximum rainfall intensity corresponding to $(2^N - 1)$ or $\overline{A}_{max}$, and β is the exponent of A-R relationship. In this simulation, R_{max} and β are set in 200 mm/hr and 1.6 respectively.

Concerning the normal signal for bias = 0, which is a normal DC setting level, the output tends to decrease as the input decreases. For bias = + 1, a positive error tends to increase too much in the smaller input level, and when the bias value is examined in every 0.5 level, it seems that the bias + 0.5 level yields the minimum measurement error. Meanwhile, the downward curve beyond $\overline{A}_T \approx 100$ is caused by the saturated characteristics of the signal amplitude.

As for the MTI signal, the characteristics $(\overline{a} / \overline{A}_T)$ are not dependent on the value of bias; the reason is that, since MTI signal processing uses the amplitude difference of the adjacent pulses, the dependency to the value of bias can be neglected in this case.

In this computer simulation, the ADC's bit number N is set in 8. Now we consider that how to use Figure 11.44 characteristics, when the bit number changes from $N = 8$ bit. If the bit increases by ΔN, the linear characteristic region in the figure will expand by ΔN bit, and the saturated region moves toward the larger value of $\overline{A}_T$ by the same amount. Accordingly, the rainfall measurement accuracy in the small signal level is obtained at the scale of $(2^{N+\Delta N} - 1).(R/R_{max})^{\beta/2}$. For example, in the use of $N = 8$ bit, at $R_T = 1$ mm/hr, $\overline{A}_T = 3.68$ and R/R_T is given by 0.85 (bias = 0). Now if 2 bit for ΔN is added, then R/R_T is improved to 0.95 as $\overline{A}_T$ is given by 14.76. Furthermore, when Figure 11.44 is used in the saturated region, the mean value corresponding to $N + \Delta N$ bit again has to be converted to that at N bit, because R_{max} is kept in a given value in spite of the value of the ADC's bit in this simulation.

11.23 RADAR CHARACTERISTICS OF PRECIPITATION ECHO DUE TO RECEIVERS

In the quantitative rainfall measurement by radar, the logarithmic receiver is used to cover the rain echo signal's wide dynamic range. The noncoherent MTI weather radar has also been put into practice. It has become necessary to use the linear characterized signal for MTI signal processing. There are two ways to yield such a linear signal. One is to use directly the linear receiver and the other is to get that from the log-linear conversion at the output of the logarithmic receiver.

LINEAR RECEIVER METHOD.

Simulation.

Figure 11.45 shows the block diagram of signal processing by the linear receiver. To process the linear signal of the wide dynamic range as much as 50 dB, the amplitude information of the orthogonal phase detector can be used.

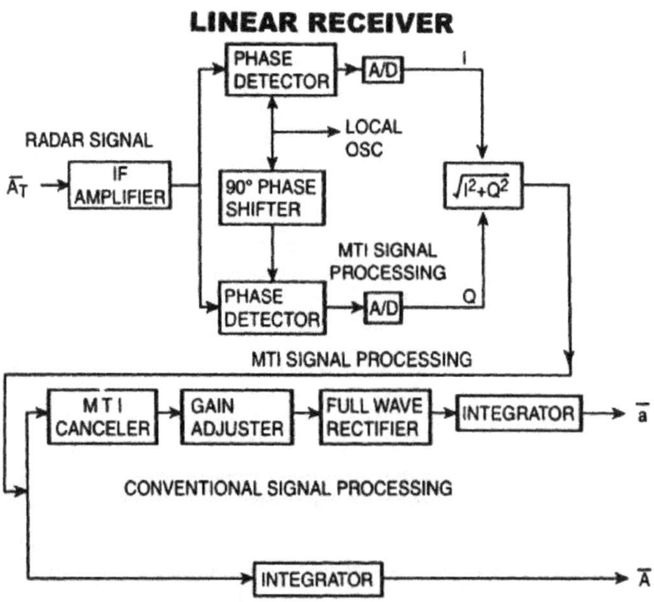

FIGURE 11.45 A block diagram showing radar signal processing using a linear receiver.

In Figure 11.45, $\overline{A}_T$ is the actual mean amplitude of precipitation echo signal in which Rayleigh distribution is valid. In conventional signal processing, the measured mean amplitude is shown by $\overline{A}$ as a normal signal. In MTI signal processing, the MTI mean amplitude $\overline{a}$ is expressed as follows:

$$\overline{a} = \sqrt{\frac{\pi}{4-\pi} k^2 A^{-2}\{1 - f(\rho)\}} \tag{11.165}$$

From this equation, it is found that there are two factors which contribute to the MTI measurement error. One is that the amplitude distribution is deviated from the Rayleigh distribution. If the Rayleigh distribution is valid, since $k^2 = (\sigma / \overline{A})^2$ is given by $(4-\pi)\pi$, where σ is the standard deviation of the amplitude distribution, equation (11.165) becomes

$$\overline{a} = \overline{A}\sqrt{1 - f(\rho)} \tag{11.166}$$

The other is the correlation coefficient ρ of the signal amplitude of $A(T)$ and $F(\rho)$ is the function of ρ decided by the canceler used.

Figure 11.44 shows the input-to-output characteristics in normal and MTI signal processing. The ADC's bit is 8 and the pulse number is 1024, which is used to get the mean value. The solid lines show the mean ratio of the measured value $\overline{A}$ to the actual value $\overline{A}_T$. The broken lines show the MTI mean value to the actual mean value. Those curves are drawn as the parameter of "Bias." The term "Bias" is the deviation of the A-D converter's DC setting level.

In the figure, R is the measured rainfall intensity and R_T is the actual one. There is the following relation between R (or R_T) and the signal amplitude $\overline{A}$ (or $\overline{A}_T$, $\overline{a}$) from the radar equation.

11.24 ADVANCED METEOR WIND OBSERVATIONS USING MST RADARS

MST radars in the VHF band have a great potential in meteor echo observations due to its high transmitting power. The meteor measurement can be conducted throughout a day and compensate the turbulent echo measurement in the mesosphere, which is limited to daylight hours only. The MU radar of Kyoto University is one of these radars and has been applied to meteor studies by utilizing its very high versatilities, and various atmospheric phenomena such as inertial gravity waves, planetary waves, and atmospheric tides have been successfully studied. The MU radar was recently renewed, and its signal processing system is upgraded from a 4 analog receiver system to a 25 digital receiver system.

Observation. In order to improve the time and spatial resolutions of MU radar meteor wind the temperature measurements and to increase the number of received meteor echoes, the following techniques are being used:

1. Online post beam steering to increase the signal-to-noise ratios (SNRs) and facilitate the detection of weaker meteor echoes. This has not been tried with the old MU radar system because of its limited CPU power. When using a 4-antenna interferometer, which has been used so far, the SNRs are expected to be improved by 4 times (6 dB).
2. Radio interferometer consisting of 25 receivers. The SNRs can be further boosted by 25/4 times (about 8 dB).

To develop the online post-beam steering algorithm, meteor echo observations are conductcd, and the observation parameters were almost identical except for the major difference that all the raw data were stored for the offline analysis and development.

Analysis and Initial Result Offline post-beam steering is conducted. Every few hundred seconds the phase differences between the 4 receiver outputs are estimated and are subtracted from the 4 outputs to synthesize the coherently integrated time series. This beam steering method does not consume much of the CPU power and is expected to be applied to new online meteor detection software easily. Data quality control criteria are also tuned; the echo quality check is rather strict and only echoes with excellent quality are used for further analysis. Although the development is still in its initial stage, the number of meteor echoes which can be used for the wind analyses will be increased from the current 10^4/day to around 4×10^4/day.

Discussions. If the observations using all twenty-five receivers are successfully conducted, the number of meteor echoes is further increased and is expected to be nearly 10^5/day, which is about 10 times as many as the current observation. The time and spatial resolutions of the observed wind and temperatures will be significantly improved.

CHAPTER 11A

RADAR MEASUREMENT OF RAIN INTENSITY, DROP SIZE, AND THE RELATED ATTENUATION OF RADIO WAVES

11A.1 INTRODUCTION

The digital revolution has been successfully exploited in a new generation of radars with full digitized outputs to facilitate the quantitative use of radar information. It is a complementary source of information of rain gauges and is useful for gathering information in conjunction with other relevant meteorological and hydrological information by means of a multi sensor approach. Besides quantitative information the radar is ideal for getting qualitative images of the atmosphere. In this chapter radar measurements of rainfall have been taken into account, emphasizing radar reflectivity from which the most probable drop size diameter and rain rate have been calculated. Taking the frequencies 11 GHz to 14 GHz separately and 0°C isotherm height as 5.0 km, the effective rain heights are evaluated. The attenuation of radio waves due to rain are then determined for different rain rates corresponding to different γ-values chosen.

The time variant fading in microwave communication has resulted in a closer look on fade dynamics and modeling, which is intimately related to the equivalent subjects in rainfall rate and rain intensity at a particular location. The increasing demand on information about duration of exceedances of rain rates and associated intensity has a very important impact on the statistics of durations and time intervals. Since each geographical area corresponds to its own distribution, they must be selected on the basis of measurements in such a way as to be as representative as possible and to be carried out at stations of a well-defined rainfall climate. The purpose of the chapter is to provide statistics of rain rate and rain intensity from the available data, particularly in the tropics. For this purpose we have elaborately considered the existing models of rain rate and rain intensity with a critical analysis of the various parameters involved.

11A.2 RADAR MEASUREMENTS OF RAINFALL

For the purpose of radar measurement of rainfall, the radar antenna is utilized for emitting a short pulse of electromagnetic radiation in a predetermined direction. A small fraction of this energy is reflected by the targets to the radar antenna. The back scattered mean power P_r received by the radar is proportional to the reflectivity factor Z, provided the scattering particles are considerably smaller than the wavelength and are spherical in shape. The received power is also proportional to the radar constant C which includes the emitted power. It is again inversely proportional to the square of the target distance r^2 and the square of the one-way atmospheric attenuaton L^2_{Atm}. In fact, the radar constant C reflects the radar properties like the emitted power, 3-dB beam shape, pulse length, antenna gain, attenuation of the radar hardware, and so on. The distortion of the beam by the radome is largely neglected in practice, and some of these values are calibrated regularly. The radar reflectivity factor Z is a purely meteorological quantity and is independent of any radar property. As the variations in radar reflectivity may span several orders of magnitude, it is preferable to use a logarithmic scale. The logarithmic radar reflectivity is expressed in units of dBZ. The radar equation becomes accurate under specific assumptions like the physical properties of the target, uniform factor Z through the sample volume, beam filling with the randomly scattered precipitation particles, and so forth, and so a quantity called the *effective radar reflectivity* Z_e is used in practice, which is defined as the summation per unit volume of the sixth power of the diameter of spherical water drops in the Rayleigh scattering region.

11A.3 THEORETICAL APPROACH ON RAIN RATE VIS-À-VIS RADAR REFLECTIVITY

The integral equation that relates radar reflectivity factor (Z) in dBz, effective radar reflectivity factor (Z_e) in $mm^6\ m^{-3}$, rain drop (D) in mm, drop size distribution $N(D)$ in $m^{-3}\ mm^{-1}$, rain rate (R) in mm hr^{-1}, back scattering cross section (σ_b) in m^2, terminal velocity of rain drop (V) in m s^{-1}, wavelength of operating radar (λ) in m, and dielectric constant of water (K_w) is expressed as follows:

$$Z = \int_0^\infty D^6 N(D)\, dD \tag{11A.1}$$

$$Z_e = \frac{10^{18}\lambda^4\ \pi^{-5}}{k_w^{\ 2}} \int_0^\infty \sigma_b(D) N(D)\, dD \tag{11A.2}$$

$$R = 6\pi \times 10^{-4} \int_0^\infty D^3 N(D) V(D)\, dD \tag{11A.3}$$

In equation (11A.1) the radar reflectivity or the radar cross section per unit volume is the scattering cross section, which is a function and dimension of the scatterer. Assuming an average rain drop size distribution attributed to Marshall-Palmer, the resultant expression of the reflectivity can be modified to get the effective reflectivity, Z_e. The formulation rain rate R is a function of rain drop diameter, number of rain drops, and terminal velocity of rain drops. The terminal velocity $V(D)$ is expressed in terms of the empirical relations for different rates, and it is taken in terms of Stoke's law. The terminal velocity function is obtained by determining the constant of the equation.

If the effects of updraft and downdraft wind, turbulence, and raindrop interaction are neglected, the stationary rain rate R in mm h^{-1} is related to the raindrop size distribution $N_v(D)$ according to the relation

$$R = 6\pi \times 10^{-4} \int_0^\infty 3D v(D) N_v(D)\, dD \tag{11A.4}$$

Marshall and Palmer [10] proposed a simplified negative exponenetial parameterization for the raindrop size distribution $N_v(D)$ as a fit to filter-measurements of raindrop size spectra if the rain rates lie between 1 and 23 mm h^{-1},

$$N_v(D) = N_0 \exp(-\Lambda D) \tag{11A.5}$$

where $N_0 = 8.0 \times 10^3$ and that is decreased with increasing rain rate R according to the power law. Through the filter paper raindrop size measurements to which the Marshall-Palmer parameterization was adjusted (Figure 11A.1), it has been noted that for a realistic representation of averaged raindrop size distribution, a higher rain rate beyond 23 mm h^{-1} is appropriate.

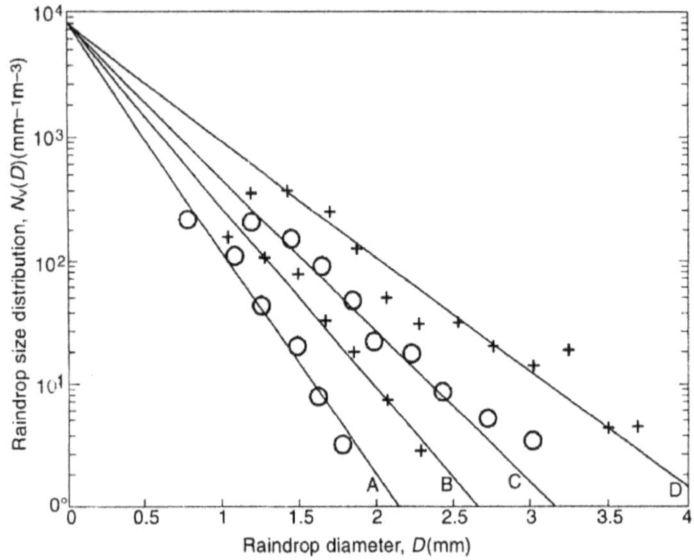

FIGURE 11A.1 Experimental raindrop size distribution (circles, crosses) present in a volume of air and fitted exponential parameterization (solid lines) for different rain rates represented as A : 1.0 mm h^{-1}; B : 2.8 mm h^{-1} ; C : 6.3 mm h^{-1} ; D : 23.0 mm h^{-1}.

11A.3.1 Empirical Rain Rate—Radar Reflectivity Relationships

It is possible to derive Z-R relationships through regression analysis. Such relationships generally follow power laws of the form $z = aR^b$, where a and b are coefficients that may vary from location to location and also from one season to another. These coefficients reflect the climatological behavior and type of rainfall of a particular location or season for which they are evaluated. Battan quoted a list of 69 such empirical power law Z-R relationships for different climatological conditions in various parts of the globe. Figure 11A.2 (a) reveals these relationships in a single plot. For equilibrium, rainfall conditions which have been observed during steady tropical rain, are also included in Figure 11A.2 (b).

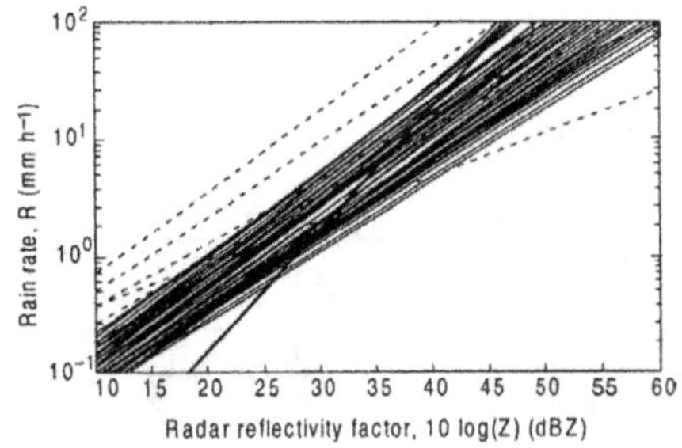

(*a*) The 69 power law *Z-R* relationships, including five deviating relationships as indicated by dashed lines, four of which have pre factors *a* significantly smaller than 100 and one of which has an exponent *b* as high as 2.87. The bold line in the figure represents the linear relationship as proposed by List.

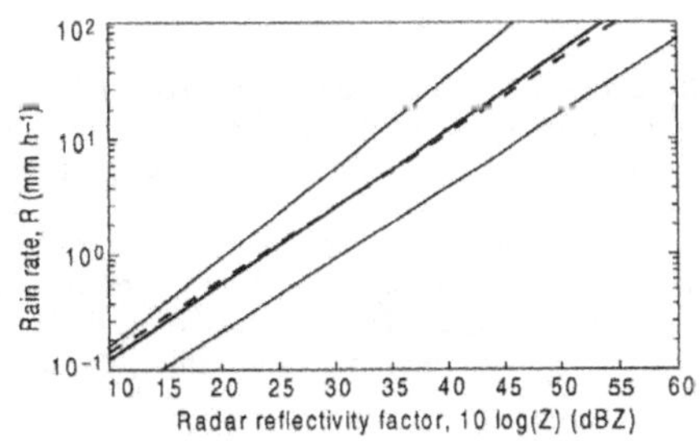

(*b*) The bold solid line represents the mean of Battan's relationships, while the bold dashed line represents the reference relationship of Marshall et al. All the individual relationships lie in between the thin solid lines as shown on either side in the form of an envelope.

FIGURE 11A.2 Radar reflectivity *versus* rain rate plot.

Figure 11A.3 shows that there is a considerable variability in the coefficients of these *Z-R* relationships owing to varying rainfall conditions; there appears to be a well-defined envelope comprising most of the relationships developed. Considering the geometric mean of the individual prefactors *a* and the arithmetic mean of the exponents *b* corresponding to averaging the linear relationships, an attempt has been made by Battan that leads to the mean power log Z-log *R* law relationship:

$$Z = 238R^{1.5} \tag{11A.6}$$

The relationship has been compared in Figure 11.2 (*b*) with the well-known Marshall-Palmer *Z-R* relationship:

$$Z = 200R^{1.6} \tag{11A.7}$$

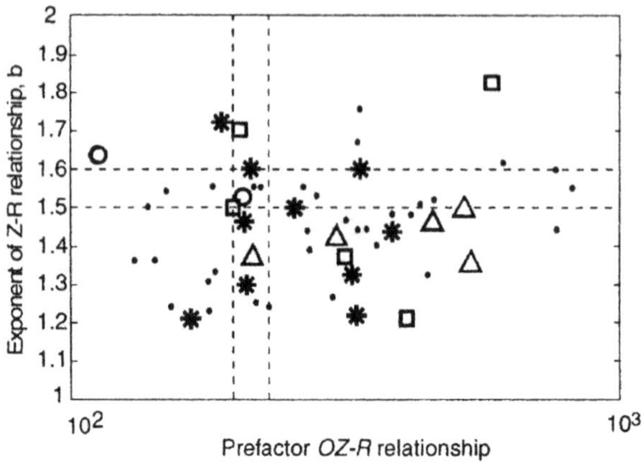

FIGURE 11A.3 The coefficients *a* and *b* of the 69 power law *Z-R* relationships $z = aR^b$ (where *Z* is expressed in $mm^6\ m^{-3}$ and *R* in $mm\ h^{-1}$) as quoted by Battan. The rainfall types considered are: thunderstorms (triangles); showers (squares); widespread/stratiform (stars); orographic (circles); and no unambiguous identification possible (dots). The vertical dashed lines represent the reference relationship $Z = 200R^{1.6}$ while the horizontal dash-dotted lines correspond to Marshall and Palmer's relationship $Z = 237R^{1.5}$, which is nearly equal to the mean of Battan's relationships, $Z = 238R^{1.5}$.

When rain data of tropical countries, like India, are taken into account, it appears that the correspondence is very close, particularly for rain rates in between 1 to 50 mm h^{-1}. In fact, from a close scrutiny we find that it is valid not only for tropical rain data but also for different types of rainfall in many parts of the globe.

In order to verify whether the coefficients *a* and *b* of the power law *Z-R* relationships systematically vary corresponding to different types of rainfall, we may consider Figure 11A.3 showing a plot of *b* versus *a* for Battan's 69 *Z-R* relationships. According to Battan it is possible to associate 25 of these *Z-R* relationships unambiguously with a particular type of rainfall. An interpretation of the coefficients *a* and *b* in terms of the parameters of the corresponding raindrop size distributions in a particular location may help to explain their variability.

11A.3.2 Drop Size Distribution from Radar Reflectivity

The rain characteristics, particularly rain rate and drop size distribution, are important for estimating attenuated radio waves due to rain. It is very difficult to get the exact drop size distribution, but the rain rate can be measured accurately by rain gauge. With the help of radar reflectivity measurements from radar data located at Kolkata (22.39°N, 88.27°E), drop size distribution was being estimated for different rain rates. A comparative study of drop size distributions obtained from reflectivity measurements and also from other techniques observed that the reflectivity measurement is consistent at high rain intensities. The most probable drop size diameter and rain rate have been calculated by

$$Z_e = \frac{48 \times 3.3978 \times N_T}{A^3} \tag{11A.8}$$

and

$$R = \pi^{\frac{3}{2}} \times 45 \frac{N_T}{4A^{5/2}} \times 10^{-3} \tag{11A.9}$$

and is illustrated in Figure 11A.4. From the figure it is clear that the most probable rain drop size varies exponentially with radar reflectivity and also with rain rate.

The spatial density distribution of scatterers at rain rates ~49 mm h^{-1} and 75 mm h^{-1} have also been estimated and are plotted in Figure 11A.5. The theoretical model derived by using the measured rain drop sizes over the tropical sites also shows the same kind of variation. In fact investigation reveals that for different rain rates, different *Z-R* relations give accurate results.

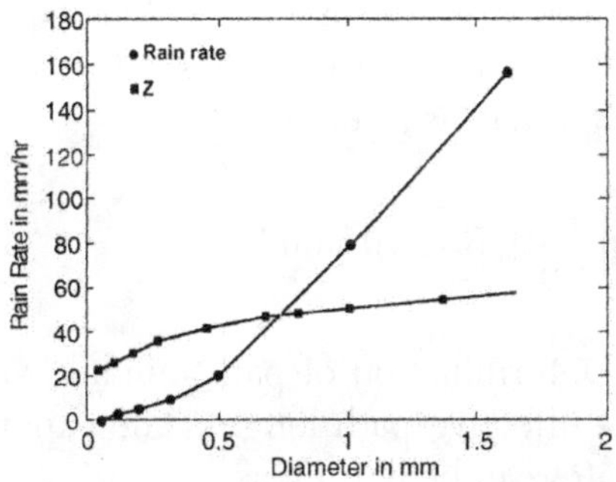

FIGURE 11A.4 Variation of most probable rain drop size with radar reflectivity and rain rate.

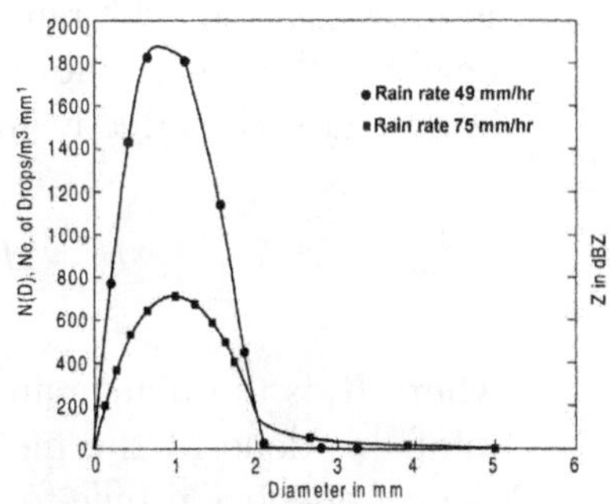

FIGURE 11A.5 Rain drop size distribution at ~49 mm h^{-1} and 75 mm h^{-1} rain rate.

11A.4 ATTENUATION OF RADIO WAVES DUE TO RAIN

Knowledge of rain attenuation is essential for design of a reliable communication link above 10 GHz. The point rainfall rates and the vertical and horizontal structures of rain are important parameters to be considered for estimating rain attenuation.

The relation between a specific attenuation and rain rate can be expressed as:

$$\alpha(R) = aR^b \tag{11A.10}$$

where a and b are two constants. These constants depend on the frequency and microstructure of the rain. The value of a and b can be estimated from

$$\begin{aligned} a(f) &= 4.21\times10^{-5}\times f^{2.42}, \text{ for } 2.9 \le f < 54\,\text{GHz} \\ &= 4.09\times10^{-2}\times f^{0.699}, \text{ for } 54 \le f \le 180\,\text{GHz} \end{aligned} \tag{11A.11}$$

and

$$\begin{aligned} b(f) &= 1.41\times f^{-0.0779}, \text{ for } 8.5 \le f \le 25\,\text{GHz} \\ &= 2.63\times f^{-0.272}, \text{ for } 25 \le f \le 164\,\text{GHz} \end{aligned} \tag{11A.12}$$

Where f represents the frequency. Knowing the rain rate profile, $R(l)$, along the propagation path, L, the total attenuation can be determined by integrating the incremental attenuation. The total attenuation can be written as

$$A(R_0) = \int_0^L \alpha[R(l)]dl \tag{11A.13}$$

where $R(l)$ represents the rain rate along the path and R_0 is the rain rate at $l = 0$, that is, the point rain rate at $l = 0$ at the end of the path.

For low rain rates the rain height is nearly constant [22] and is equal to the height of the 0°C isotherm. But the rain height increases with the increase in rain rate. The effective rain height (H_e) is given by

$$H_e = H_i + \log\left(\frac{R_2}{10}\right); R_0 \ge 10 \text{ mm h}^{-1} \tag{11A.14}$$

where R_0 is the point rain rate. Determination of path rainfall from point rainfall is done by the method of effective path length. The exponentially shaped effective profile for rain rates can be written as

$$R(Z) = R_0\, e^{-\gamma \ln\left[\frac{R_0}{10}\right] Z} \quad \textit{for } R_0 \ge 10 \text{ mm h}^{-1} \tag{11A.15}$$

where Z is the horizontal distance along the path; γ is a parameter for controlling the rate of decay of the rain profile. The path averaged rain rate for path length, D, can be calculated from:

$$R_{ave} = \frac{1}{D}\int_0^D R(Z)dz = \frac{R_0\left(1-e^{\gamma \ln\left(\frac{R_0}{10}D\right)}\right)}{\gamma \ln\left(\frac{R_2}{10}\right)D}; \text{for } R_0 \geq 10 \text{ mm h}^{-1} \quad (11A.16)$$

The total attenuation due to a point rain rate, R_0, is computed from the effective rain profile. The attenuation due to path averaged rainfall is attained from

$$A(R_0) = R_0^b \frac{1-o^{-\gamma b \ln\left[\frac{R_0}{10}\right]Lcost}}{\gamma\, b \ln\left(\frac{R_0}{10}\right)\cos\theta}; \text{for } R_0 \geq 10 \text{ mm h}^{-1} \qquad (11A.17)$$

where $L = \frac{H_e - H_0}{\sin\theta}$, and H_0 is the station height from the earth's surface.

Using equations (11A.11) and (11A.12) the values of a and b are calculated for 11 GHz, 12 GHz, 13 GHz, and 14 GHz. The effective rain heights (H_e) over Kolkata for different rain rates are next calculated using equation (11A.14), considering the 0°C isotherm heights as 5.0 km as an annual average. This value of H_e is necessary for estimating the value of L, which is needed for the calculation of attenuation, using equation (11A.17). The elevation angle we have considered for the present study is ~56° as most of the stations in India have an elevation angle of ~56° with a geostationary satellite. Taking the frequencies 11 GHz, 12 GHz, 13 GHz, and 14 GHz separately and choosing $\theta = 56°$ and 0°C isotherm height H_i = 5.0 km, the effective rain heights, H_e are determined. The attenuation values are calculated by using equation (11A.17) for different rain rates assigning different γ-values.

Attenuation due to rain obtained by using the concept of rain path profile at 11 GHz, 12 GHz, 13 GHz, and 14 GHz for the elevation angle 56° with rain height 5.0 km is presented in Figure 11A.6 (*a*), (*b*), (*c*), and (*d*). The attenuation is found to vary significantly with the γ -values. Also the attenuation curves are closer when the γ-values are of low order. This means that the attenuation increases at a lower rate as the γ-value decreases. This attenuation behavior is found to be more prominent for higher rain rates.

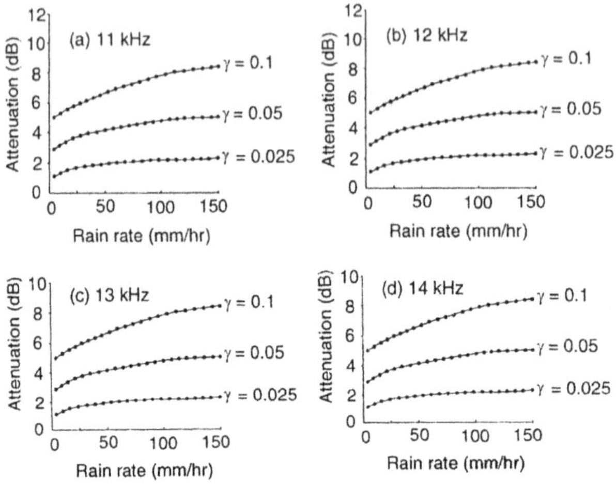

FIGURE 11A.6 Variation of attenuation with rain rates for different values of γ.

It is well-known that the radar beam is slightly curved downward in a standard atmosphere, which reduces the height of the beam above the ground; hence, one can measure the reflectivity at a lower height than in a homogeneous atmosphere. In an analysis the physical quantity to be averaged is the rain intensity, while the primarily measured quantity is the reflectivity, which is related non-linearly to the rain intensity. For this reason averaging in reflectivity introduces a bias which increases with the homogeneity of the reflectivity field and with distance from the radar. In order to allow the reflectivity of a single scatter to be calculated by using the Rayleigh approximation, the scatterer has to be small compared to the Wavelength; that is, the diameter should be 10 to 16 times smaller than the wavelength. It thus appears that the laragest particle should be smaller than 10 mm, 5 mm, and 3 mm for S-, C-, and X-band radars respectively.

In tropical climates, like India, the raindrops of 8 mm are frequently observed, while the frozen hydrometeors may be even larger than 10 mm. These large particles reflect much less than considered by Rayleigh approximation, and so they are estimated to be smaller than they are in reality. In fact, these large particles are not suitably described by the normal drop size distributions that follow the *Z*-*R* relation. Besides, other errors that may be present in the estimation of radar precipitation include reflections received from a variety of non-meteorological targets like aircraft, chaff, insects, birds, solar radiation, and so on.

Radar radiation is mainly attenuation by atmosphereic gases and hydrometeors. Attenuation is largely dependent on the wavelength. However, S-band radar (of about a 10 cm wavelength) is much less affected by attenuation then X-band radar (approx. a 3 cm wavelength). The static part of attenuation is corrected for by the radar signal processor. The processor uses a user-defined value for the attenuation coefficient to correct the data.

The more problematic part of the attenuation is of course due to the precipitation itself. It may be pointed out that the CCIR model does not hold true for lower elevation angles and also is not valid for vertical propagation. It is also noted from the profile of γ-values that higher rain rates contribute more toward attenuation.

11A.5 STATISTICAL MODEL FOR RAIN RATES

For the statistical distribution of rain rates some models have been proposed under different climatic conditions. Out of those the long-normal distribution is a good approximation for the low rain rates but overestimates largely the high rain rates corresponding to very low time percentages. The gamma distribution is better for rain rates > 15 mm h^{-1}, in temperature climates. But the data from tropical regions reveal that the gamma distribution in these zones itself overestimates the high rain rates. For the low rain rates the curvature is upward, while it is downward for the high rain rates. It would be desirable to have a statistical model that is valid in any climate. One can consider that in any climate the curve representing R versus $\log P$ has an inflexion, and the rain rate is always smaller than a maximum value R_M which is dependent on the rain climate. Accordingly, a function can be chosen conveniently to prevent the introduction of a parameter representing the percentage of time without rain. The function $e^{-1/x}$, which starts slowly from zero, is appropriate from this point of view. The model is then represented by the equation

$$P = \exp\left[-\frac{\alpha}{\left(\ln\frac{R_M}{R}\right)\beta}\right] \tag{11A.18}$$

P in equation (11A.18) being the probability that the value R be exceeded; α and β are two positive parameters.

There is no mathematical restriction on the values. The usual values of α, β, and R_M are found in the following ranges: α: 5 to 50, β : 0.5 to 1, R_M: 250 to 5000 mm h^{-1}.

Equation (11A.18) can then be written as

$$R = R_M \exp - \left[-\frac{\alpha}{\ln P} \right]^{\frac{1}{\beta}} \tag{11A.19}$$

Choosing different values of α, β, R, and R_M, we have calculated the probability P against the rain rate. The curves so obtained by using the model are plotted in Figure 11A.7. It appears from the figure that for tropical climates the value of the exponent β is appropriate to a value of 0.5 as we have used for the present computation. However, for temperate climates β should be taken as 1 as reported by others using the same model.

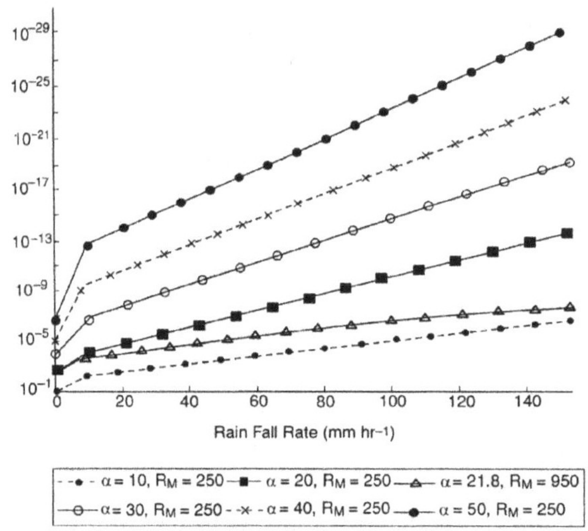

FIGURE 11A.7 Comparison between the measurement and the model.

11A.6 STATISTICAL MODEL FOR RAINFALL INTENSITY

The log-normal law can provide a correct representation of low rain intensities, but it overestimates the high intensities corresponding to low time percentages. On the other hand the gamma law is appreciable to reflect rainfall intensities above about 12 to 20 mm h^{-1} in temperature climates but is unsuitable for low rainfall intensities.

A mathematical function can be so selected that a point of inflexion is obtained in the graphical representation in the coordinates *R/log P.* This function should be probability law, and it would be useful if the function could be inverted for either R versus P or P versus R. There are models which meet either all or some of these conditions. One very appropriate model applicable to all climates is represented by the relation (Boithias Model)

$$P = \exp\left[-\frac{\alpha}{\left(\ln\frac{R_M}{R}\right)\beta}\right] \tag{11A.20}$$

Inverting relation (4.3) we get

$$R = R_M \exp - \left[-\frac{\alpha}{\ln P}\right]^{\frac{1}{\beta}} \tag{11A.21}$$

This relation is based on three parameters: R_M, α, and β, where R_M represents the maximum rainfall intensity for P tending toward zero. The coordinates of the point of inflexion can be expressed as

$$P_i = \exp\left[-\alpha\left(\frac{\beta}{\beta+1}\right)^{\frac{1}{\beta}}\right] \tag{11A.22}$$

$$R_i = R_M \exp\left[-\left(\frac{1+\beta}{\beta}\right)\right] \tag{11A.23}$$

No mathematical restriction is there on the three parameters, and P always lies between 0 and 1. In practice for a tropical country, the exponent β may be selected as equal to 1 and the relation comes to the form:

$$\ln P \ln \frac{R}{R_M} = \alpha \tag{11A.24}$$

and then the coordinates of the point of inflexion are

$$P_i = \exp\left(-\frac{\alpha}{2}\right) \tag{11A.25}$$

$$R_i = 0.135\, R_M \tag{11A.26}$$

If the maximum rain rate R_M and probability at point of inflexion P_i are adopted, we get

$$\ln P . \ln \frac{R}{R_M} = 2 \ln P_i \tag{11A.27}$$

Again if we take the parameters R_M and $P(1)$ (Probability that 1 mm h^{-1} will be exceeded), we have the expression

$$\ln P\left(1-\ln\frac{R}{R_M}\right)=2\ln P(1) \tag{11A.28}$$

11A.6.1 Determination of the Parameters R_M, α and β

To represent a statistical distribution, the determination of the parameters of a function may have different objectives:

1. to obtain these parameters for each experimental distribution to check that the representation is possible in all cases using the function, and
2. to utilize these parameters at a given point on the earth where no measurement is available.

We shall consider the first point taking the value of β as 1 for most distributions. This introduces a degree of simplification, as the function will then depend only on two parameters R_M and α.

These two parameters are determined by the measurement point R_1, P_1 and R_2, P_2 as, for example, the rainfall intensities for 0.01% and 0.1% of time. We can write

$$\ln R_M=\frac{\ln P_2\ln R_2-\ln P_1\ln R_1}{\ln\frac{P_2}{P_1}} \tag{11A.29}$$

$$\alpha=\frac{\ln P_1\ln P_2}{\ln\frac{P_2}{P_1}}\ln\frac{R_1}{R_2} \tag{11A.30}$$

By equations (11A.29) and (11A.30) we get,

$$\frac{\ln R_M}{\alpha}=\frac{\frac{\ln R_2}{\ln P_1}-\frac{\ln R_1}{\ln P_2}}{\ln\frac{R_1}{R_2}} \tag{11A.31}$$

The function itself may then be written as

$$\ln P\left[\frac{1}{\ln P_1}\ln\frac{R}{R_2}-\frac{1}{\ln P_2}\ln\frac{R}{R_1}\right]=\ln\frac{R_1}{R_2} \tag{11A.32}$$

Due to rainfall variability it is preferable to find the parameters on the largest possible number of R_i, P_i pairs. For the purpose, we can present graphically, in a simple way, by plotting *log R* versus 1/*log P*. The straight line for the adjustment of the points so found may provide both R_M (ordinate at origin) and α (slope of the straight line).

11A.7 R_M AND α AS CALCULATED FROM CCIR RAINFALL DATA

Table 11A.1 shows the values of R_M, α and *log* R_M/α calculated from available rainfall data of the CCIR data bank. The adjustment was optimized in between 1%, .001% ($10^{-5} < P < 10^{-2}$). The table shows that the values of R_M and α vary within wide limits, even in the same climatic region. On the contrary the quantity *log* R_M/α did not change much and appears to characterize the climate region. It is around 0.3 in India as evident from the table. For β = 1, the quantity *log* R_M/α is directly related to the probability that the rainfall will be above 1 mm h^{-1} in the region chosen. We have

$$\frac{\ln R_M}{\alpha} = -\frac{1}{\ln P(1)} \tag{11A.33}$$

This model of rainfall intensity is applicable to all rainfall climates throughout the world and can serve from the standpoint of both radio meteorology and applications connected with the forecasting of the various effects of rainfall.

TABLE 11A.1
Determination of the Parameters R_M and α (Value of β for Stations Number 1, 2, and 3 are Taken as 2 and for the Rest Stations as 1)

Station	Observing Station	R_M	α	$\frac{\log R_M}{\alpha}$	β
1	Pading	270	7.14	0.78	2
2	Miami	420	10.50	0.58	2
3	Brazzaville	470	11.40	0.54	2
4	Osaka	540	21.20	0.30	1
5	Darwin	570	25.70	0.30	1
6	Douala	600	14.80	0.43	1
7	Belem	640	16.40	0.39	1
8	Papua-New Guinea	730	19.80	0.33	1

Station	Observing Station	R_M	α	$\frac{\log R_M}{\alpha}$	β
9	Innisfail	745	14.80	0.45	1
10	Manaus	750	18.00	0.37	1
11	Kjeller	810	27.80	0.24	1
12	Rio de Janeiro	950	22.00	0.31	1
13	New Delhi	950	21.80	0.31	1
14	Yamaguchi	1220	24.40	0.29	1
15	Ogasawara	1350	28.10	0.26	1
16	Darmstadt	1400	38.10	0.19	1

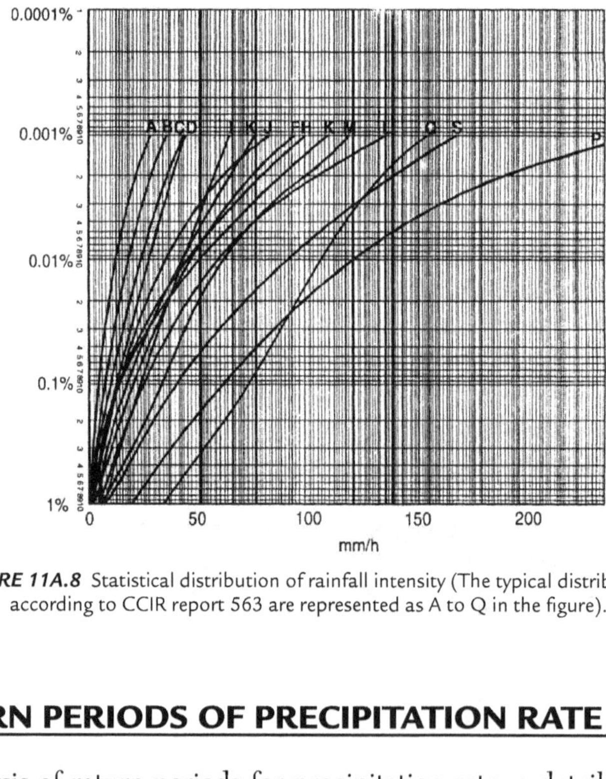

FIGURE 11A.8 Statistical distribution of rainfall intensity (The typical distributions according to CCIR report 563 are represented as A to Q in the figure).

11A.8 RETURN PERIODS OF PRECIPITATION RATE

In an analysis of return periods for precipitation rate, a detailed statistical study is made of the three types of intervals between rain rate exceedances: (*a*) return time intervals I_i to the exceedance of a given R irrespective of whether it happens within the current event or at some future one, (*b*) intra exceedance I_a or return period to the exceedance of a given R within the same rain event, and (c) inter exceedance I_e or return period to the exceedance of R at the following rain event. Figure 11A.9 illustrates durations $D(R)$ of exceedance of a threshold R. The intervals can be analyzed in terms of annual averages $<N_i>$, $<N_e>$, $<N_a>$, mean $<I_i>$, $<I_e>$, $<I_a>$, and standard deviations σ_i, σ_e, σ_a, Long-term statistics of precipitation rate return periods in the context of microwave communication was reported by Burgueno et al. Their reported annual mean intervals are illustrated in Figure 11A.10. The figure shows that both the annual average number of events and the mean can be well fitted by

number of events $$\log N = aR \mid b \quad (11A.34)$$

mean $$<1>, <D> = aR^b \quad (11A.35)$$

The coefficients a and b are summarized in Table 11A.2.

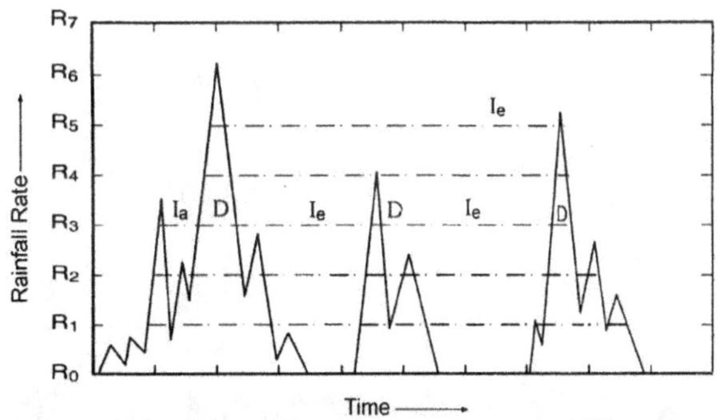

FIGURE 11A.9 An illustration of the concept of intra exceedance I_a(...), duration of an exceedance D(–), and inter exceedanace I_e(-.-.-.).

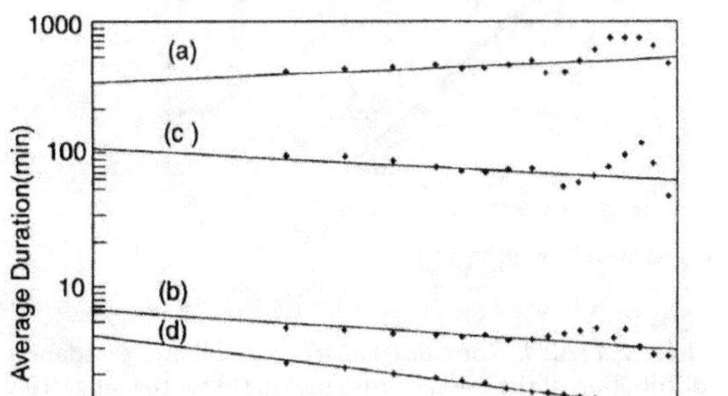

FIGURE 11A.10 Average duration of return periods (–) best fit. Inter $<I_e>$ (a), intra $<I_a>$ (b) Interval $<I_i>$ (c), and duration of exceedance $<D>$ (d).

TABLE 11A.2

Coefficients *a* and *b* for the Best Fit Modeling (Equations 11A.34, 11A. 35)

	a	*b*
$\langle N_e \rangle$, $\langle I_e \rangle$	-1.430×10^{-2}, 247.66	1.408, $+7.21 \times 10^{-2}$
$\langle N_a \rangle$, $\langle I_a \rangle$	-1.050×10^{-2}, 9.73	1.829, -0.184 ($\sim -1/5$)
$\langle N_i \rangle$, $\langle I_i \rangle$	-1.105×10^{-2}, 221.70	1.952, -0.313 ($\sim -1/3$)
$\langle N \rangle$, $\langle D \rangle$	-8.960×10^{-3}, 12.23	2.012, -0.469 ($\sim -1/2$)

It appears from Table 11A.2 that the intra exceedances I_a are longer and have greater dispersion σ_a than the duration of single exceedances D. Further, it is noted that as the rate R increases, the duration $\langle D \rangle$ decreases more rapidly than the return intervals, and the ratio $\frac{\langle I_a \rangle}{\langle D \rangle}$ ranges between 1.8 and 3 as R increases from 10 to 100 mm h^{-1}. It can be seen that

$$\frac{\langle I_a \rangle}{\langle D \rangle} = 0.795R^{0.285} \tag{11A.36}$$

$$\frac{\langle I_e \rangle}{\langle D \rangle} = 20.25R^{0.541} \tag{11A.37}$$

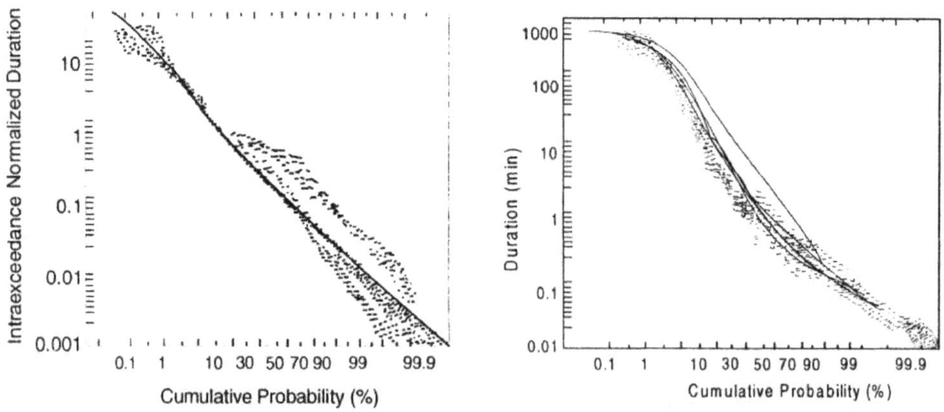

(*a*) (*b*)

FIGURE 11A.11 Normalized distribution of intra excedances (*a*) and the distribution of the overall return period (*b*) for the range 10-200 mm h^{-1}.

These results can be implemented to derive the $\langle I \rangle$s from the knowledge of $\langle D \rangle$s. The statistical analysis of both intra and inter exceedances has been seen to be log normal, though in the analysis of the overall set of return periods, the intra and inter exceedances (known as the intervals I_i) exhibit a distribution showing 2 ranges of different slopes consistent with the two types of distributions. As a result no universal distribution of return periods $P\left(\frac{I_i}{<I_i>}|R\right)$ existed. The distribution of the normalized variable $x_a(R) = \frac{I_a}{<I_a>}$ when plotted in log normal probability appears as in Figure 11A.11 (*a*), while the cumulative distribution, when all the observational data are taken into consideration, looks like Figure 11A.11(*b*).

11A.9 RAIN RATES IN THE TROPICS MEASURED BY RAIN GAUGES

The prediction of rain attenuation involving rain rate and rain drop size distribution is equally important since the size and shape of the drops may introduce scattering of electromagnetic waves. In fact, the prediction of rain attenuation and rain drop size distribution has high uncertainty owing to variation of non spherical rain drops at different locations and associated variations of the rainfall characteristics. In this section, we have critically examined the published data obtained by using three rapid response rain gauges with 10s integration time. The stations we have chosen are at three

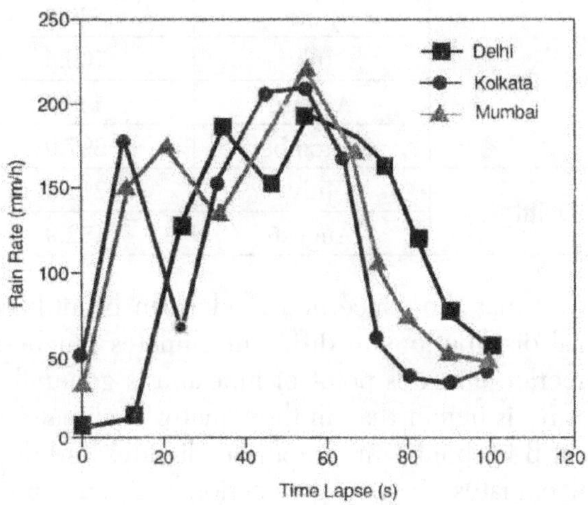

FIGURE 11A.12 Rain rates by rain gauges at three locations in the tropics.

widely different locations in India, namely, Kolkata (22.39°N, 88.27°E), Delhi (28.32°N, 77.12°E), and Mumbai (19.04°N, 72.06°E). In the observational part, the rainwater was collected in a standard funnel and was converted into equal size drops maintained by a capillary tube 3 mm in diameter. The number of drops collected every 10 seconds was counted electronically and thereafter converted into corresponding rain rates in mm h^{-1}. Figure 11A.12 reveals the characteristic variations of the rain rates as measured by rain gauges including the non-rainy months. It appears from the figure that out of the three stations, the rain rate over Delhi is lowest and highest at Mumbai. As the rainfall data reported by the India Meteorological Department (IMD) on a monthly basis includes the non rainy period also, it becomes difficult to get an instantaneous rain rate. The monthly total rainfall reported by the IMD can be converted into mm h^{-1} by dividing the total reported rainfall in mm divided by 720 (24 by 30 hours). The values so obtained are presented in Table 11A.3.

TABLE 11A.3
Reported Monthly Rainfall and the Rain Rate for Kolkata, Mumbai, and Delhi

Station	Worst months	Monthly total rainfall in mm	Rain rate in mm h^{-1}
Kolkata	June	300.6	0.418
	July	306.3	0.404
	August	289.7	0.389
Mumbai	June	520.3	0.723
	July	709.5	0.954
	August	439.3	0.590
	September	297.0	0.413
Delhi	July	211.1	0.284
	August	172.9	0.232

It appears that the concerned model can be utilized largely to represent statistical distributions in different climates but under certain conditions. In temperate climates point of inflexion is generally not reached, and the value of R_M is higher than in the equatorial climate. Again, the value of the exponent β is about 1 in temperate climates and decreases to about 0.5 in tropical climates. In this connection we can consider the Capsoni and Paraboni model from which we have,

$$P(R) = P_0 \left(\ln \frac{R_M}{R} \right)^n \quad (11A.38)$$

or inverting,

$$R = R_M \exp - \left(\frac{P}{P_0} \right)^{\frac{1}{n}} \quad (11A.39)$$

[P and P_0 may be either probabilities or time percentages]

This simple relation has certain drawbacks. It reflects that when P tends to zero R_M represents the maximum rainfall intensity, and P_0 represents the probability corresponding to point of inflexion of curve R. This value being very low in temperature climates, the main part of the curve becomes concave toward the top, while for equatorial climates the point of inflexion remains in the main part. The probability at the point of inflexion is $R_M/e = 0.368\ R_M$. This means that the function cannot be properly adapted to certain measured distribution, especially in equatorial climates. The value of R_M is extremely high in temperate regions, where the statistical distribution has a very steep slope owing to the occurrence of very violent but short-lived storms. Clearly, in this case, the useful part of the distribution remains well below the point of inflexion. With decreasing latitudes, the value of R_M tends to drop, but the useful part of the curve reaches and even passes the point of inflexion. Finally, it is found that for some stations in low latitude, a value of β must be chosen equal to 2 in order to ensure the proper representation of that part of the distribution which corresponds to high time percentages. However, the value of $\beta = 1$ is suitable for representing almost all measured distributions. It would not seem necessary to use intermediate values of β between 1 and 2 owing to the variability of rain fall distribution.

CHAPTER 12

DIRECTION FINDING

12.1 INTRODUCTION

A radio navigational aid which is based on the measurement of the direction of arrival of a radio wave is called a *direction finder.* A direction finder really determines the vertical plane of travel of the radio wave transmitted from a fixed location on the ground or from a moving vehicle in the sea or air. For a ground-based transmission the navigator in a moving vehicle directly obtains the information about the vertical plane of travel or bearing of the vehicle by direction finding, while for transmission from the vehicle, the bearing information obtained at the ground is transmitted to the navigator by radio communication. The radio direction finders were perhaps the earliest form of radio navigational aids and are still widely employed in radio navigation in a variety of forms. In the simplest form, for instance, the direction of radio transmission from a port may be determined from an approaching ship by a directional antenna for reaching the port. An aircraft approaching an airport needs greater precision in direction finding, for which a specialized antenna is employed either in the aircraft or at the airport. Those direction finders are specially called homing systems or radio ranges. In all direction finders, the determination of direction is made by utilizing the directional property of a *loop antenna* or its modified form.

12.2 LOOP ANTENNA

A rectangular *loop antenna* made of a conducting wire bent into the form of a rectangle ABCD of height and width is as shown in Figure 12.1. The

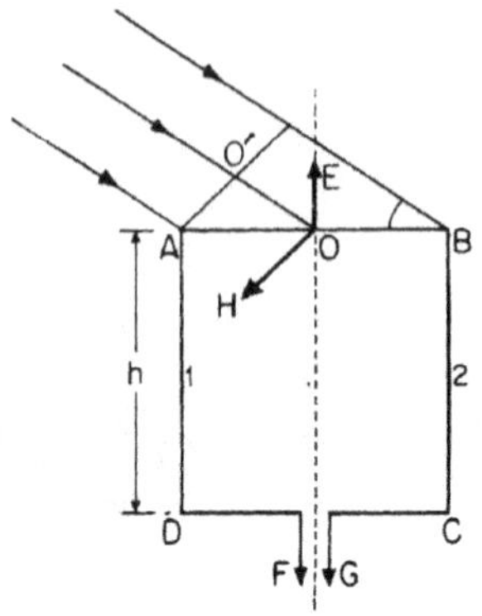

FIGURE 12.1 Loop antenna as a direction finder.

loop is placed in the electromagnetic fields of the radio wave from a distant transmitter. If the electric field is vertical, as shown in the figure, then at the center of the loop, E is given by

$$E = E_0 e^{j\omega t} \tag{12.1}$$

At A the radio wave will travel a distance shorter by OO' for which the phase of wave at A will lead by $\frac{2\pi}{\lambda}.OO' = \frac{2\pi l}{\lambda 2} \cos \Phi$, where $AB = l$ and ϕ is the angle between the direction of propagation of the wave and the plane of the loop antenna. The field at A will, therefore, be given by

$$E_A = E_0\ e^{j\left(\omega t + \frac{2\pi}{\lambda}.\frac{t}{2}\cos\phi\right)}$$

Likewise, the field of the radio wave at B, which travels a distance longer by OO' will be represented by

$$E_B = E_0\ e^{j\left(\omega t + \frac{2\pi}{\lambda}.\frac{l}{2}\cos\phi\right)}$$

If $h < < \lambda$, the voltages induced in conductors 1 and 2 due to the fields E_A and E_B respectively given by

$$E_1 = E_A\ h = E_0\ e^{j\left(\omega t + \frac{2\pi}{\lambda}.\frac{l}{2}\cos\phi\right)}$$

and $bE_2 = E_B h = E_0\ he^{j\left(\omega t + \frac{2\pi}{\lambda}.\frac{l}{2}\cos\phi\right)}$. As the electric field is vertical, no voltage is induced in the horizontal parts AB and CD of the loop antenna. The resultant output voltage of the loop between F and G is given by

$$E_f = E_1 - E_2$$

$$= E_0\, he^{j\left(\omega t + \frac{2\pi}{\lambda}\cdot\frac{l}{2}\cos\phi\right)} - E_0\, he^{j\left(\frac{2\pi}{\lambda}\cdot\frac{l}{2}\cos\phi\right)}$$

$$= E_0\, he^{j\omega t}\left[e^{j\left(\frac{2\pi}{\lambda}\cdot\frac{l}{2}\cos\phi\right)} - e^{-j\left(\frac{2\pi}{\lambda}\cdot\frac{l}{2}\cos\phi\right)}\right]$$

$$= 2j\, E_0\, h\left[\frac{e^{j\left(\frac{2\pi}{\lambda}\cdot\frac{l}{2}\cos\phi\right)} - e^{-j\left(\frac{2\pi}{\lambda}\cdot\frac{l}{2}\cos\phi\right)}}{2j}\right]$$

Since $$\frac{e^{j\theta} - e^{-j\theta}}{2j} = \text{Sin}\,\theta$$

$\therefore$ $$E_f = 2j\, E_0\, h \sin\left(\frac{2\pi}{\lambda}\cdot\frac{l}{2}\cos\phi\right)e^{j\omega t}$$

$$= 2E_0\, h \sin\left(\frac{\pi}{\lambda}l\cos\phi\right)e^{j\left(\omega t + \frac{\pi}{2}\right)} \quad (12.2)$$

If $\pi l/\lambda << 1$, since $\cos\phi$ lies between + 1 and – 1, $\pi l/\lambda \cos\phi << 1$ and $\sin(\pi/l/\lambda \cos\phi) \cong (\pi/\lambda)l\cos\phi$.

Using this approximation, the loop equation reduced to

$$E_f = 2\pi.\, E_0.\frac{hl}{\lambda}\cos\phi\, e^{j(\omega t + \pi/2)}$$

$$= \frac{2\pi E_0 A}{\lambda}\cos\phi\, e^{j(\omega t + \pi/2)} \quad (12.3)$$

If there are N turns of the loop, the induced voltage will be N fold, as the voltages induced in all the turns add up acting in series. Accordingly, we have

$$E_N = NE_f = \frac{2\pi\, E_0\, AN}{\lambda}\cos\phi\, e^{j(\omega t + \pi/2)} \quad (12.4)$$

since $hl = A$, the area of the loop antenna. Comparing equation (12.1) with equations (12.2), (12.3), or (12.4), we note that the induced voltage is leading by $\pi/2$ relative to the field at the center. This means that the instantaneous loop voltage will be at its maximum when the field is passing through zero. Equation (12.4) shows that the amplitude of the loop voltage is given by

$$|E_f| = \frac{2\pi E_0\ AN}{\lambda} \cos\phi = b\cos\phi \tag{12.4 a}$$

where $b = \dfrac{2\pi E_0\ AN}{\lambda}$.

A plot of equation (12.4 *a*) is shown in Figure 12.2, which also indicates that in the plan view of the loop, shown dotted, E_f is positive. Beyond these limits E_f becomes negative, and E_f is thus positive $-\pi/2 \le \phi \le \pi/2$ in the right half plane and negative in the left half plane.

A negative amplitude of E_f can be interpreted as a reversal of the phase of the radio frequency voltage E_f. For,

$$\begin{aligned} -E_0\ e^{j\omega} &= E_0\ (\cos\pi + j\cos\pi)\ e' = E_0\ e^{j\pi}\ e^{j\omega t} \\ &= E_0\ e^{j(\omega t + \pi)} \end{aligned}$$

A plot of the waveform of E_f as a function of ϕ is shown in Figure 12.3 when the loop is rotated at a frequency $\omega_r = \phi(i)$, $\phi(t)$ being the angular rotation in a time t. An envelope detector cannot recognize the reversal of the phase of

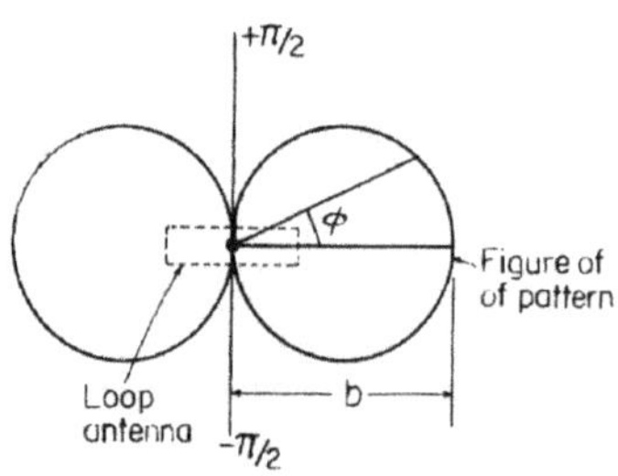

FIGURE 12.2 Directional pattern of a loop antenna.

E_f, and the detected voltage E_D will depend on ϕ as shown in Figure 12.3 (*b*). Thus, maxima of E_D occurs at 0 or π while minima or nulls is at $\pi/2$ or $3\pi/2$. Such a direction finder determines only the direction of the ratio wave but cannot recognize the sense, whether it is 0 or π, or whether it is $\pi/2$ or $3\pi/2$. It may be noted here that the rate of change of E_D with ϕ is maxima at the minimum of E_D at $\phi = (2n+1)\,\pi/2$ and therefore, the accuracy of direction finding

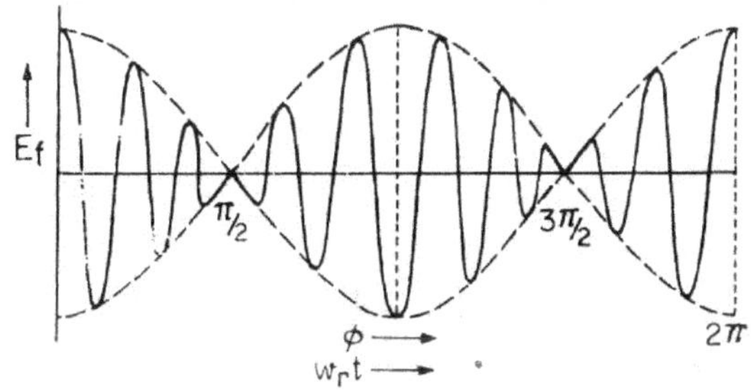

(*a*) Reversal of phase of loop voltage around the direction of null, $\phi = \pi/2, 3\pi/2, ...$

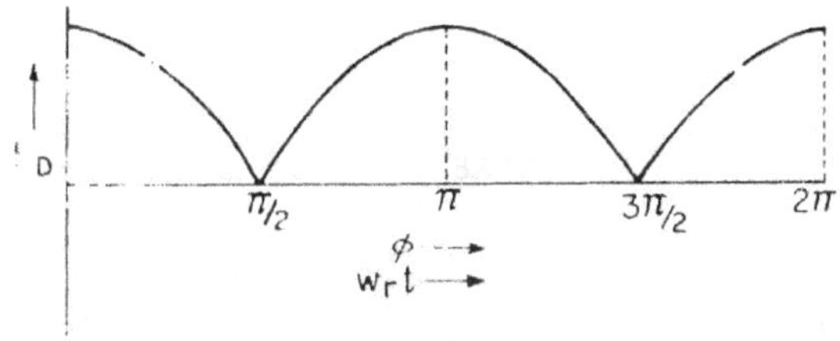

(*b*) Null of the envelope voltage at $\phi = \pi/2, 3\pi/2, ...$

FIGURE 12.3

will be highest if the nulls are used in direction finding. In practice, the loop may be steered about a vertical axis through the center of the loop in search of the null. The orientation of the plane, which is perpendicular to the plane of the loop, then gives the bearing. For determining the sense as well as the direction, we may combine the loop output with a voltage of the same frequency and with an amplitude and phase not dependent on the bearing. Such an arrangement is called a *sense finder*.

12.3 SENSE FINDER

A typical sense finder is illustrated in Figure 12.4. If h is the effective height of the vertical antenna placed at the center of the loop antenna, the voltage induced in it will be $Eh = E_0\ h^{just}$ and this is independent of the bearing ϕ since the azimuthal pattern of the vertical antenna is omnidirectional. The induced voltage drives a current $i = E_0 h / R\ e^{j\omega t}$ through the coil L_1, since L_1 and C in association with the antenna reactance is adjusted to make the net

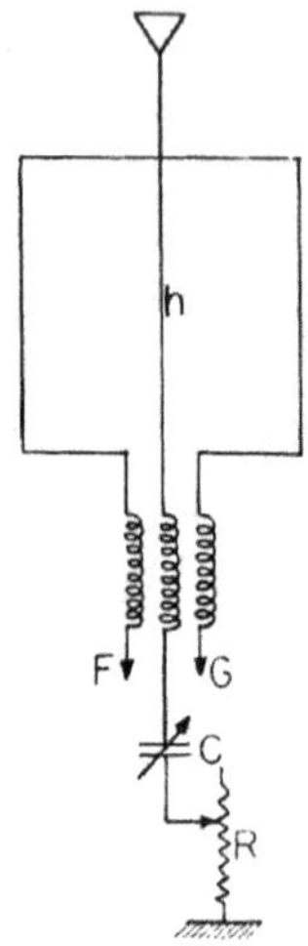

FIGURE 12.4 Sense finder antenna system.

reactance in the circuit zero. The voltage induced in the loop circuit due to this current is given by

$$E_v = -j\,\omega(-M)\,i$$
$$= j\,\omega(M)\frac{E_0\,h}{R}\,e^{j\omega t} = \frac{\omega M E_o h}{R}\,e^{j(\omega t+\pi/2)}$$

where – M is the mutual inductance between L_1 and the inductance L_2, L_3 in the loop circuit.

The net voltage across the terminals F and G of the loop of N turns is, therefore, given by

$$E_s = E_v + E_N$$
$$= \frac{\omega M E_0\,h}{R}\,e^{j(\omega t+\pi/2)} + \frac{2\pi\,E_0\,AN}{\lambda}\cos\phi\,e^{j(\omega t+\pi/2)}$$
$$= \left(\frac{\omega M E_0\,h}{R} + \frac{2\pi\,E_0\,AN}{\lambda}\cos\phi\right)e^{j(\omega t+\pi/2)} \tag{12.5}$$
$$= (a + b\cos\phi)\,e^{f(\omega t+\pi/2)} \tag{12.6}$$

where $$a = \frac{\omega M E_0\,h}{r} \text{ and } b = \frac{2\pi\,E_0\,AN}{\lambda}$$

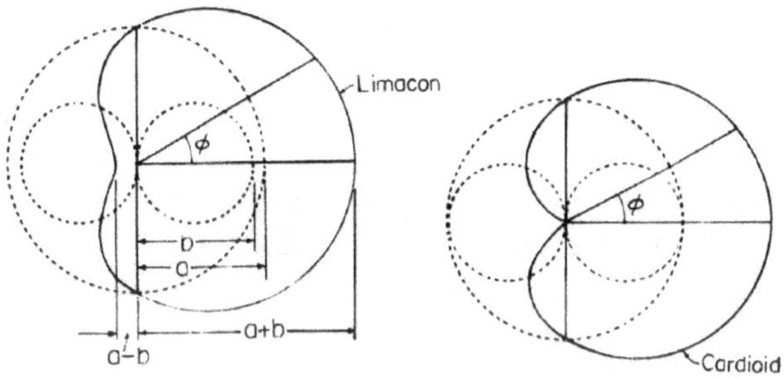

(*a*) Limacon pattern for $a > b$. (*b*) Cardioid pattern for $a = b$.

FIGURE 12.5

The amplitude of the sense finder voltage E_s is given by

$$|E_s| = a + b\cos\phi \tag{12.7}$$

Equation (12.7) represents a limacon as shown in Figure 12.5 (*a*) for $a > b$ and a cardioid for $a = b$ as shown in Figure 12.5 (*b*). In either case the pattern can distinguish between the directions with $\phi = 0$ and $\phi = \pi$. The difference in $|E_s|$ for these two directions is $2b$ in each case; the ratios of $|E_s|$ for the two directions is, however, $a + b/a - b$ and α for the limacon and cardioid respectively. In practice R may be adjusted to make $a = b$ to realize the cardioid pattern.

12.4 INCREASING THE SENSITIVITY OF A DIRECTION FINDER

The sensitivity of a direction finder depends on the amplitude of the detected loop voltage $|E_D|$ as shown in Figure 12.6. Now $|E_D| = b = 2\pi E_0 An/\lambda$ if the detector efficiency is unity. An increase of b will not only increase the amplitude of detected output voltage $|E_D|$ but will also increase the output at a given angle away from the null, proportionately, thus facilitating the detection of the null. In the presence of inherent noise of the receiving system of the direction finder, the null will be uncertain. The detected voltage cannot be recognized unless it exceeds the noise voltage of the direction finder, and the null is drowned in noise for

$$\frac{\pi}{2} - \phi_{min} < \phi < \frac{\pi}{2} + \phi_{min}$$

or

$$\frac{3\pi}{2} - \phi_{min} < \phi < \frac{3\pi}{2} + \phi_{min}$$

as shown in Figure 12.6 curve 1.

The bearing determination using the null is therefore uncertain by $\pm\ \phi_{min}$. If b is increased, we get curve 2 for $|E_D|$ and ϕ_{min} is reduced as shown in the figure. Thus, in the presence of noise the accuracy of bearing determination increases with the increase of b. Alternatively, for a specified accuracy as indicated by ϕ_{min}, the weaker radio waves may be detected by an increase in the value of b. The sensitivity of the direction finder is thus increased by an increase of b, the loop constant.

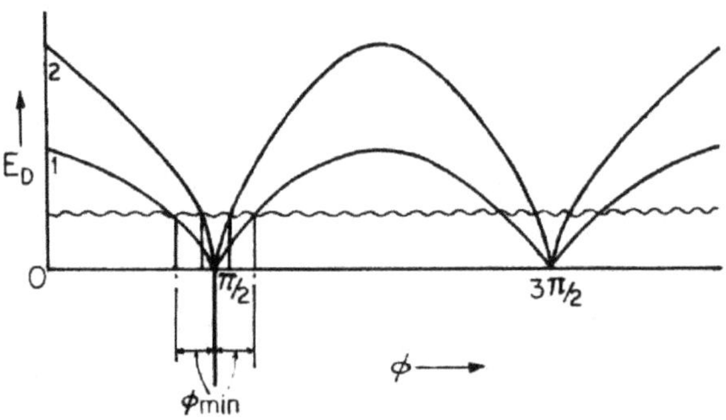

FIGURE 12.6 Receiver noise affecting precision of direction finding, ϕ_{min}.

12.5 METHODS OF INCREASING LOOP VOLTAGE

12.5.1 Increase of A

Considering the equation $b = (2\pi E_0 AN)/\lambda$, we note that b can be increased by increasing the area A of the loop. However, A cannot be increased beyond a certain limit at which the physical dimension of the antenna becomes too large to facilitate steering. The area can, in fact, be increased in a variety of ways. The height h or the width l or both h and l may be increased for an increase of A. However, for the condition $\pi d/\lambda << 1$ assumed in deducing equation (12.3) from equation (12.2), we cannot increase l beyond a limit to meet this condition as shown in loop 1 of Figure 12.7. We also cannot increase h beyond a limit as shown in loop 4, for which the induced voltage in the vertical arms will not be obtained from Eh, unless $h << \lambda$ as well. For h comparable to λ the current in the vertical arm will not be uniform, and the induced voltage will be less than Eh. For such cases, therefore, even equation (12.2) cannot be used. Within the limits of l and h as indicated previously, equation (12.3) will be valid and b will increase with A as shown by the solid line in Figure 12.8, the dotted part indicating the region of invalidity of equation (12.3).

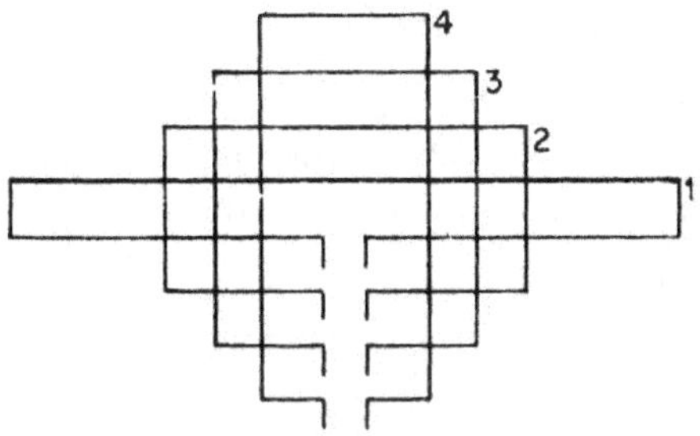

FIGURE 12.7 Rectangular loop antennas of various width to height ratios.

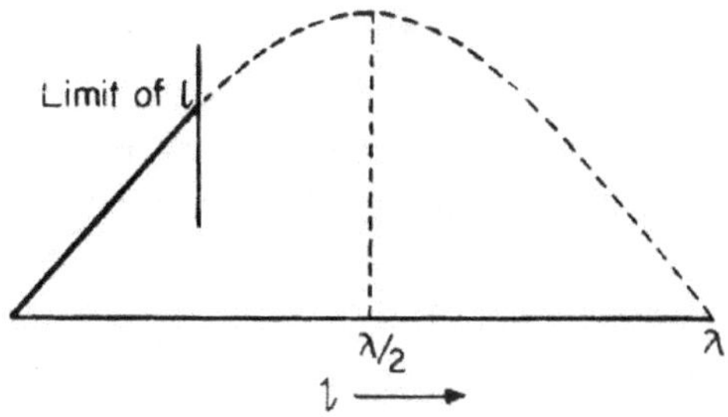

FIGURE 12.8 Variation of loop output voltage with the width of the loop.

12.5.2 Increase of N

Because b is proportional to the number of turns N, the largest value of N may be employed. Increase of N, however, leads to a width of the loop antenna. The width will lead to a loop output even in the direction of null as illustrated in Figure 12.9 (a). For an incident radio wave in the direction of the arrow along the axis of the loop antenna, the induced voltages in the loops 2, 3, and so on will be progressively delayed relative to that in loop 1. This is called the *frame effect* arising from the frame-like appearance of the loop antenna. The frame effect can be minimized by multiplying the number of turns N without adding to the width as shown in Figure 12.9 (b). This is called a pancake loop

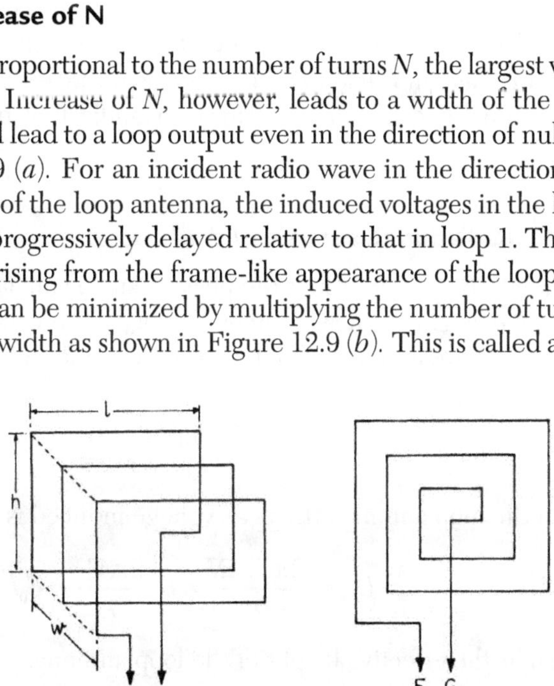

(a) Frame effect in loop antenna. (b) Pancake loop antenna.

FIGURE 12.9

antenna. In this, however, the individual turns are of different areas, and this should be taken into account in deducing the loop voltage. If the successive areas are A_1, A_2,...A_n, the effective area will be given by

$$A = \sum_{n=1}^{n} A_n.$$

12.5.3 Tuning of the Loop to Increase Loop Voltage

A remarkable increase in b can be obtained by a tuning of the loop antenna to the frequency f of the radio wave. If the inductance of the loop is L, then a capacitance C connected to its terminals F and G will resonate the loop if

$$C = \frac{1}{4\pi^2 f^2 L}$$

At resonance, the voltage across the capacitance will be Q times the induced loop voltage E_f, where $Q = \omega L/R$, R being the loss resistance of the loop. Q is typically 10 to 100. The loss resistance is given by $R_L = R_0 + R_r$, where R_0 is the ohmic resistance in the conductor and R_r is the radiation resistance. It should be noted that the tuning of the loop will not really increase the induced voltage in the loop, but it increases the available voltage across the loop terminal because of the voltage magnification by the resonant circuit formed.

12.5.4 Radiation Resistance of the Loop

To estimate R_r, consider first the peak induced voltage E_m in a straight wire of effective height $h_f \leq \lambda$ as given by $E_m = E_0 h_f$, h_f is called the effective height of the short wire antenna. The radiation resistance of the wire is given by

$$R_r = 80\pi^2 \left(\frac{h_f}{\lambda}\right)^2$$

Now in the loop antenna, the peak voltage induced is given by

$$E_{fm} = \frac{2\pi E_0 AN}{\lambda} = E_0 \frac{2\pi AN}{\lambda} = E_0 h_f$$

where h_f is the effective height of the loop antenna.

The radiation resistance of the loop antenna will then be obtained from the equation given by

$$R_r = 80\pi^2 \left(\frac{h_f}{\lambda}\right)^2$$

$$= 80\,\pi^2 \left(\frac{2\pi AN}{\lambda}/\lambda\right)^2 = 80\,\pi^2 . 4\,\pi^2 . \left(\frac{AN}{\lambda^2}\right)^2$$

$$= 320\,\pi^4 \left(\frac{AN}{\lambda^2}\right)^2 \qquad (12.7a)$$

If $A = 1$ sq. m, $N = 10$, $\lambda = 100$ m, then

$$\left(\frac{AN}{\lambda^2}\right)^2 = \left(\frac{1\times 10}{10^4}\right)^2 = 10^{-6}$$

and $\quad R \cong 320 \times 100 \times 10^{-6} = 3.2 \times 10^{-2}\ \Omega$

Thus, the radiation resistance R of the loop is negligible compared to R_0, the ohmic resistance which is typically a few ohms, and $R_L = R_o + R_r \cong R_o$.

12.5.5 Increasing the Q of the Loop

The Q of the loop may be increased by maximizing $\omega L/R_o$. For a given length of the wire the ohmic resistance R_o is fixed, and L will be maximum for a maximum area of the coil formed with the wires. For a rectangular coil of height h and breadth L, if the perimeter, given by $p = 2(l + h)$, is to remain constant, R_o will be constant and the area A will be maximum when $A = h.l = \sqrt{\frac{(h+l)^2 - (h-l)^2}{4}}$ is a maximum, that is, when $h = l$. Thus, a square loop will have the maximum area for a given length of the wire and hence will have the maximum Q. Likewise, if the loop is elliptical in shape, then for a given length of the wire, the area will be largest when the major and minor axes are of equal length, that is, when the ellipse degenerates into a circle. A circular wire, in fact, has an area greater than that of any other shape of the loop, for a given length of the wire. Such a loop is, therefore, sought for the highest Q of the loop antenna.

12.5.6 Use of Ferrites to Increase Loop Voltage

The effect of a highly permeable core such as a ferrite on the loop voltage can be investigated by referring to Figure 12.10. The magnetic field component of the incident radio wave will induce a voltage in the loop AB due to the varying magnetic flux linkage in the coil of the loop. The magnetic field component along the normal ON is $-H\cos\phi$. If A is the area of the loop and N the number of turns, then the flux linkage is given by

$$\psi = \mu AN\,(-H)\cos\phi \qquad (12.8)$$

where ϕ is the angle between the plane of the loop and the direction of the radio wave, which is equal to the angle between H and the normal to AB, NON', and μ is the permeability of the medium.

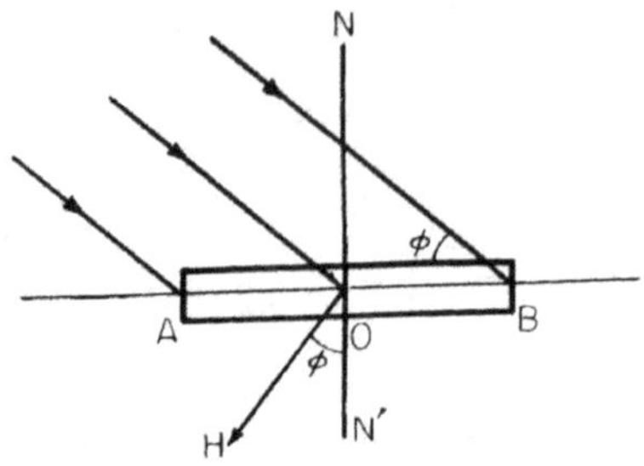

FIGURE 12.10 Induced voltage in a loop antenna due to rate of change of magnetic flux linkage.

Putting $H = H_0 e^{j\omega t}$ in equation (12.8), we get

$$\psi = -\mu ANH_0 \cos\phi\, e^{j\omega t} \tag{12.9}$$

The induced voltage is, therefore, given by

$$\begin{aligned} E_f &= -\frac{d\psi}{dt} = j\,\mu\, AN\,\omega\, H_0 \cos\phi\, e^{j\omega t} \\ &= \mu\, AN\omega H_0 \cos\phi\, e^{j(\omega t + \pi/2)} \end{aligned} \tag{12.10}$$

Now E_0 and H_0 are related through

$$\frac{E_0}{H_0} = Z_0 = \sqrt{\frac{\mu_0}{\varepsilon_0}} \tag{12.11}$$

where μ_0 and ε_0 are the permeability and permittivity of the free space respectively. Also

$$\omega = 2\pi f = \frac{2\pi c}{\lambda} = \frac{2\pi}{\lambda}\cdot\frac{1}{\sqrt{\mu_0\varepsilon_0}} \tag{12.12}$$

Since $$c = \frac{1}{\sqrt{\mu_0\varepsilon_0}}$$

Putting equations (12.11) and (12.12) in equation (12.10) and noting that $\mu = \mu_r\, \mu_0$, μ_r being the relative permeability, we get

$$\begin{aligned} E_f &= \mu_r\mu_0 \frac{2\pi}{\lambda}\frac{1}{\sqrt{\mu_0\varepsilon_0}} \times \frac{E_0}{\sqrt{\mu_0/\varepsilon_0}} A\cos\phi\, e^{j(\omega t + \pi/2)} \\ &= \mu_r \frac{2\pi E_0 AN}{\lambda} \cos\phi\, e^{j(\omega t + \pi/2)} \end{aligned} \tag{12.13}$$

For $\mu_r = 1$, valid for the medium being a vacuum, equation (12.13) becomes identical with the loop equation (12.4). The effect of a permeable medium is thus to increase E_f by μ_r. For a ferrite rod inserted inside the loop coil, μ_r is typically 100 to 1000. In this case, the presence of the ferrite concentrates the field inside the coil, thus increasing the induced voltage. The ferrite, in fact, affects the distribution of the field in space around it, causing the fields over a considerable region to bend and pass through it.

12.5.7 Increasing Loop Voltage by Large Fixed Crossed Loops: Bellini Tosi Direction Finder

The limitation arising from the difficulty of steering a large loop can be avoided by using two fixed loops with their planes at right angles as shown in Figure 12.11 (*a*). The voltages induced in these loops may be used to produce a resultant magnetic field, the direction of which depends on ϕ. A small search coil may be steered with its plane lying on the plane of the resultant magnetic field to produce a null at its terminals *FG*. In this way the fixed loops which is placed in the incident radio wave need not be steerable and, therefore, can be made large to increase the area *A* and hence the induced voltage.

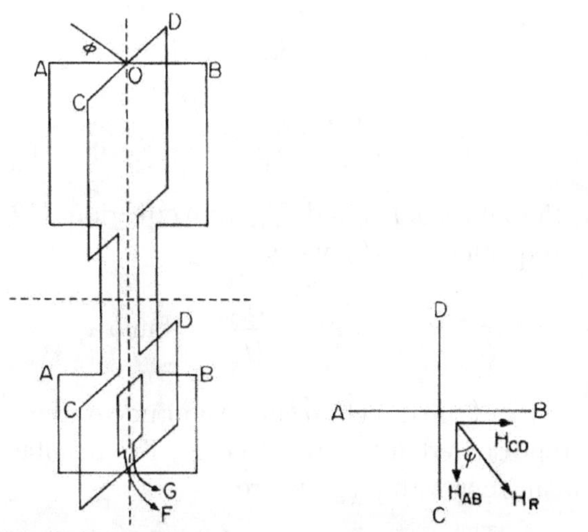

(*a*) Bellini-Tosi direction finder antenna system.

(*b*) Plane view of the Bellini-Tosi antenna showing the direction of the resultant magnetic field.

FIGURE 12.11

The voltage induced in loop *AB* is given by

$$E_{AB} = \frac{2\pi E_0 AN}{\lambda} \cos\phi \, e^{j(\omega t + \pi/2)} \tag{12.14}$$

while that in the crossed, or orthogonal, loop *CD* is given by

$$E_{CD} = \frac{2\pi E_0 AN}{\lambda} \sin\phi \, e^{j(\omega t + \pi/2)} \tag{12.15}$$

Since the angle between *CD* and the direction of the radio wave is $(\pi/2 + \phi)$. Each of these induced voltages are used to drive current through the respective loop circuits which are given by

$$i_{AB} = \frac{E_{AB}}{R}$$

and

$$i_{CD} = \frac{E_{CD}}{R}$$

assuming the individual loop circuits to be tuned to resonance at the frequency *f* of the incident radio wave by inserting suitable reactances in the circuits, *R* being the total resistance of each loop circuit. Due to the flow of the current, the magnetic fields due to *AB* and *CD* are in the ratio of the respective currents and, therefore, we have

$$\frac{H_{C'D'}}{H_{A'B'}} = \frac{i_{C'D'}}{i_{A'B'}} = \frac{E_{CD}}{R} \Big/ \frac{E_{AB}}{R} = \frac{E_{CD}}{E_{AB}} \tag{12.16}$$

Putting the values of E_{AB} and E_{CD} from equations (12.14) and (12.15) respectively, in equation (12.16), we get

$$\frac{H_{C'D'}}{H_{A'B'}} = \tan\phi \tag{12.17}$$

A plane view of the direction and magnitude of the magnetic fields $H_{C'D'}$ and $H_{A'B'}$ is represented in Figure 12.11 (*b*). The resultant magnetic field H_R will make an angle ψ with $H_{A'B'}$, where

$$\tan\psi = \frac{H_{C'D'}}{H_{A'B'}} = \tan\phi \text{ [from Equation (13.17)]}$$

$$\therefore \quad \tan\psi = \tan.\ \phi \text{ or } \psi = \phi \tag{12.18}$$

Thus, the direction of the resultant magnetic field makes the same angle with one loop $C'D'$ or CD as the incident radio wave makes with the AB. If the search coil is rotated to a position with its plane lying on that of H_R, a null will be indicated at its terminal F and G. The crossed loop $A'B'$ and $C'D'$ can be small and kept inside the deck of a ship, not exposed to the radio wave, while the large crossed loops AB and CD are kept high up above the deck exposed to the incoming radio wave. Such a direction finder is called a Bellini Tosi Direction Finder after its inventor. It may be mentioned here that H_R is given by

$$\begin{aligned} H_R &= \sqrt{H_{A'B'}^2 + H_{C'D'}^2} = k\sqrt{i_{AB}^{\ 2} + i_{CD}^{\ 2}} \\ &= k\frac{2\pi E_0 AN}{\lambda R} e^{j(\omega t + \pi/2)} \sqrt{(\cos^2\phi + \sin^2\phi)} \\ &= k\frac{2\pi E_0 AN}{\lambda R} e^{j(\omega t + \pi/2)} \\ &= H_{RO}\, e^{j(\omega t + \pi/2)} \end{aligned} \tag{12.19}$$

k being a constant of proportionality for the relation between current and magnetic field.

12.6 ERRORS IN DIRECTION FINDING

The errors in direction finding may arise either from (1) the direction finding equipment of from (2) anomalous radio propagation effects.

12.6.1 Errors Due to Direction Finding Equipment Antenna Effect

Connecting a loop antenna to an amplifier of the direction finder equipment may introduce an unbalance, producing a loop output even in the null position. In Figure 12.12 (a), a loop antenna is shown connected to a FET amplifier. Note that a large voltage is induced in each of the vertical arms of the loop acting as the antenna. The pickup in the vertical parts 1 and 2 of the loop antenna, act as vertical antennas, and the unequal impedance from 1 and 2 to ground will lead to a large voltage at G, even in the position of null. The voltage induced in 2 appears at G while that at F is earthed. The presence of induced voltage even in the null position due to the unbalance in the impedance to earth between the loop terminals is called an *antenna effect* since it arises from a vertical arm of the loop acting as an antenna.

The pickup in the vertical part is, however, omnidirectional, but is in quadratic with the loop voltage. The effect is therefore observed only as a broadening of the null. If the vertical antenna pickup is shifted in phase by 90° then the loop output will be in phase or in antiphase with it, resulting in one of the figure eight loops being bigger than the other. The net effect in this will be to produce nulls which are no longer 180° apart. The unbalance can be reduced by earthing the electrical midpoint of the loop as shown in Figure 12.12 (*b*). A residual unbalance due to the input capacitance C_i at *G* may be removed by connecting a suitable compensating capacitance at *F* as shown as dotted lines in Figure 12.12 (*b*). The antenna effect can thus be removed. However, the voltage available for amplification is between *G* and the earth, which will be half of the voltage induced in the loop. An alternative way to obtain a balance is to connect an amplifier with a balanced input as shown in Figure 12.12 (*c*).

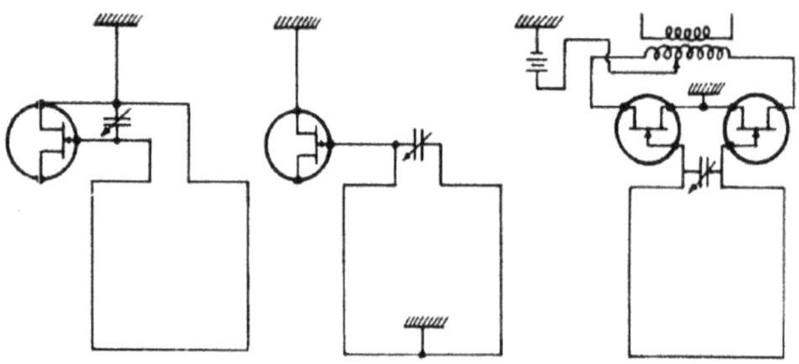

(*a*) Antenna effect in loop direction finder. (*b*) Center tapped loop to reduce antenna effect. (*c*) Balanced output of loop to eliminate antenna effect.

FIGURE 12.12

Shielded loop. Any unbalance which may crop up due to the unsymmetrical disposition of the loop with respect to the earth or due to unequal capacitance to the ground for the arms 1 and 2 due to the proximity of the operator can be removed by electrostatic shielding of the loop. However, a total shielding of the conductors of the loop by a metallic tube as shown in Figure 12.13 (*a*) will also shield the radio wave completely from the loop. For then the shield itself will act as a loop antenna, and the current driven in the shield by the induced voltage will produce a re-radiated field which will appose the primary field to which it is due following Lenz's law. An electrical break is, therefore, introduced in the shield circuit, usually at the midpoint of the loop horizontal arm, with the electrical midpoint being earthed as shown in Figure 12.13 (*b*). If the capacitance between the air gap of the shield is increased to tune the loop circuit near resonance, it can be shown that the re-radiated field may add up with the primary field tending to increase the induced voltage in the loop. The increase will, in fact, be maximum when the shield circuit is tuned to a frequency about times the frequency of the radio wave. The rather large capacitance between the air gap needed for the purpose is obtained by introducing the shield at the open ends, one into the other, as shown in Figure 12.13 (*c*), with an insulating spacer in between them.

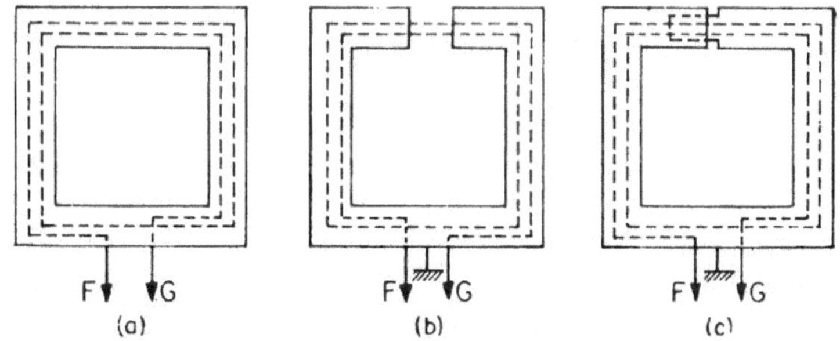

FIGURE 12.13 (*a*) Screened loop without break in screen (*b*) Screened loop with break in screen to eliminate eddy current (*c*) Screened loop with mutual foldover at the break point of the screen to increase capacitance across the break.

12.6.2 Errors Due to Anomalous Propagation Effects

Night effect. In a vertical loop antenna, no voltage is induced in the top and horizontal conductors when the loop is placed in a vertically polarized radio wave field. At night, however, the radio wave is propagated efficiently by reflection in the ionosphere. This reflected component, in general, undergoes a rotation of the plane of polarization or may become elliptically polarized, and a horizontal component of electric field appears which induces a voltage in the loop even in the position of null. This is called *night effect*, which is illustrated in Figure 12.14. For a horizontal ray no voltage is induced in *AB* or *CD* if the electric field is vertical. Even when the field is not vertically polarized, its horizontal component will arrive at *AB* and *CD* at the same instant and there will be no resultant induced voltage due the horizontal components. For a down coming ray, if the electric field is in the vertical plane of travel of the radio wave, its horizontal component will be perpendicular to *AB* or *CD* and will not induce any voltage in it. If the plane of polarization of the downcoming wave is rotated to *ER* or if it is elliptically polarized, in either case, the horizontal component becomes parallel to *AB* or *CD* and since the wave arrives at *AB* earlier to that at *CD*, a resultant voltage will be induced due to the pickup in the horizontal parts even in this position of null of the loop antenna.

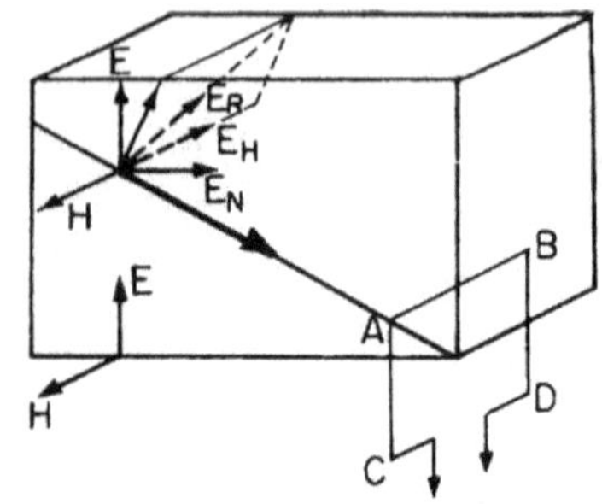

FIGURE 12.14 Night effect geometry in loop direction finding.

Bearing Error Due to Night Effects. Consider a downcoming ray incident at an angle β to the vertical on a loop oriented with its plane in the *yz* plane which is perpendicular to the vertical plane of travel *xz* as shown in Figure 12.15. If the electric field component $E = E_0 e^{j\omega t}$ is rotated by an angle α from the vertical plane of travel *XZ*, as shown on the circular plane *S* perpendicular to the ray, the magnetic field component $H = H_0 e^{j\omega t}$ will also be on *S* making angle α with the *Y*-axis.

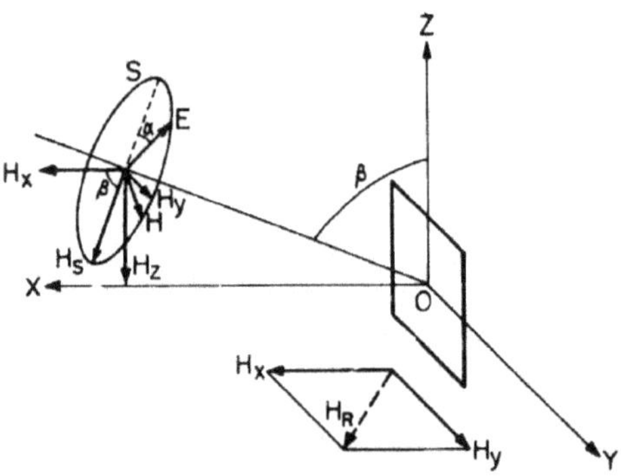

FIGURE 12.15 Night effect geometry in Cartesian coordinate.

Accordingly, we have

$$H_y = H \cos \alpha \tag{12.20}$$

and H_s, which is perpendicular to H_y but in the *S* plane is given by

$$H_s = H \sin \alpha \tag{12.21}$$

H_s can be further resolved into H_x and H_s given by

$$H_x = H_s \cos \beta \tag{12.22}$$

Since the angle between H_x and H_s is equal to the angle between their perpendiculars, that is, between the Z-axis and the ray direction. The other resolved part of H_s is given by

$$H = -H_s \sin \beta \tag{12.23}$$

Putting equation (12.20) in equations (12.21) and (12.23), we get

$$H_x = H \sin \alpha \cos \beta \tag{12.24}$$

and

$$H_z = -H \sin \alpha \sin \beta \tag{12.25}$$

The component H_Y will produce no flux linkage in the loop and, therefore, no voltage is induced. The components H_x and H_y combine to produce H_r at an angle δ to the plane of the loop xz, given by

$$\tan \delta = \frac{H_x}{H_y}$$

$$= \frac{H \sin \alpha \cos \beta}{H \cos \alpha} = \tan \alpha \tan \beta \tag{12.26}$$

from equations (12.20) and (12.24).

If the loop is rotated through δ, H_r will produce no flux linkage and hence a null will be obtained. The bearing error is thus given by equation (12.25). In practice, a direction finder may show an error greater than δ due to limitations of the instrument. To compare various instruments, therefore, we test these with a standard wave which is defined to be a wave having $\alpha = \beta = \pi/4$. From equation (12.25) the error in bearing due to a standard wave is given by $\tan \delta = \pm\, \pi/4 \cos \pi/4$ which gives $\delta = \tan^{-1} 1/\sqrt{2} \cong 35.5° = 1.1/\sqrt{2}$.

It may be mentioned here that for an elliptically polarized downcoming ray, H_x and H_y will not be in phase, and then equation (12.25) will not be valid. The effect is then observed as a poorly defined null. Since the error in bearing during a night effect arrives from abnormal polarization of the downcoming ray, it is also called a polarization error.

Adcock Antenna to Eliminate Night Effect. The pickup in the horizontal parts of a loop antenna being responsible for night effect, it can be avoided by eliminating the top and shielding the bottom horizontal portions as shown in Figure 12.16 (*a*). Such an antenna is called an Adcock antenna after its inventor. The shielding can also be achieved by burying the bottom part within the ground. Due to the *U*-shaped appearance it is called a U-type

Adcock antenna. Instead of shielding the bottom, a balanced pair of leads may be taken from the center of each of the vertical parts from an H-type Adcock antenna as shown in Figure 12.16 (b).

The open ends of the vertical parts make the voltages induced in each less effective in inducing current through the elements. The current driven by the induced voltage will be maximum at the feeder end and zero near the open end, making a distribution current along the wire as shown as dotted lines in Figures 12.16 (a) and (b). As a result, the effective height of each vertical element is about half of the actual height, leading to an induced voltage across FG half as large compared to a loop of single turn having the same height and width as the Adcock antenna. The pickup in an Adcock is therefore considerably lower than a loop antenna of similar dimension but having a number of turns in it. To increase the pickup, the lengths of the vertical parts are increased and operated in the VHF band, at which the pickup will be large even in a short length of wire. In Bellini-Tosi direction finders, the night effect can also be eliminated by employing Adcock versions of the crossed loops above the deck as shown in Figure 12.16(c).

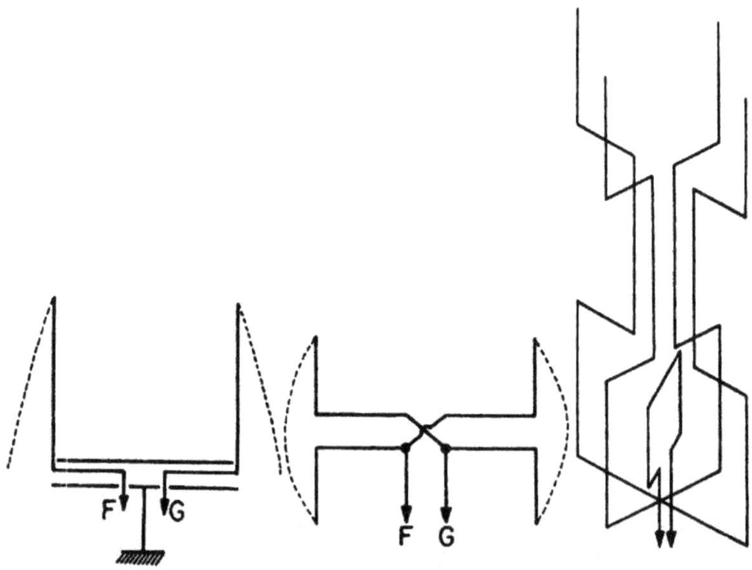

(a) Adcock antenna to remove night effect.

(b) Adcock antenna with center fed dipole elements.

(c) Adcock version of Bellini-Tosi direction finder.

FIGURE 12.16

Shore Effect. Refraction of the ground wave at the coastal boundary between land and sea may introduce errors in bearing in direction finding as shown in Figure 12.17. A ground wave traveling from the transmitter T at the port will be refracted due to different refractive indices of land and sea, reaching the direction finder in the ship. The apparent position of the transmitter will be T, leading the bearing error of δ as shown. This is called the shore effect. The refractive index of earth's surface, through which ground wave propagates, is given by

$$\varepsilon' = \varepsilon + \frac{\sigma}{j\omega}$$

where σ and ε are the conductivity and dielectric constants of the material of the earth. For land, typical values of σ and ε are 10^{-2} mho/m and 15 respectively, while those for sea are 5 mho/m and respectively. It may be noted that the bearing errors due to shore effect will be nil if the coastal boundary is semi-circular with the transmitter at the center of the circle.

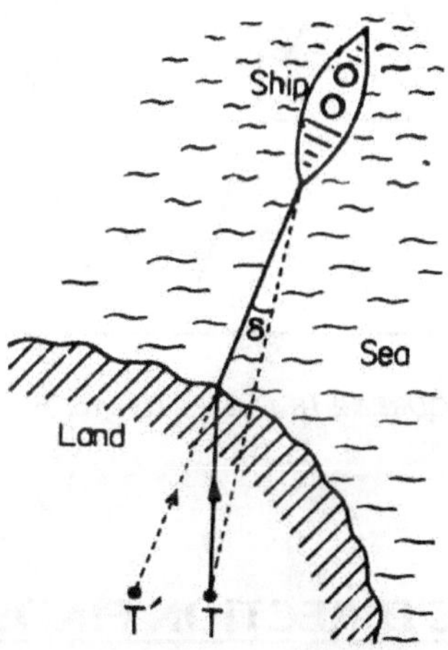

FIGURE 12.17 Refraction effect near the seashore.

Site Error. The presence of metallic conductors or objects near a direction finder may alter the directional pattern. Voltages are, in fact, induced in such neighboring objects at the site and drive current through them to produce *re-radiated fields*, which add up with the primary field to produce a new distribution of fields in the neighborhood of the loop antenna. The apparent bearing will then be different from the true one, the error in bearing being called site error. In ships, the sticks, poles, and rings act as re-radiators. Their distances being small the direction finder picks up the induction and electrostatic fields are re-radiated. If the direction finder antenna is on the center line of the ship, the error in bearing due to re-radiators becomes zero in four directions and varies with direction as shown in Figure 12.18. This is called quadrantal error.

The quadrantal error can be reduced greatly by using a shield loop, since the induction and electrostatic components of the re-radiated field which dominate can be effectively shielded from the loop inside the shield. Nevertheless, some residual quadrantal error is inevitable and needs calibration of bearing.

Quadrantal error may occur in aircraft direction finders as well. Here the wings, engines, propellers, and other parts of the aircraft act as re-radiators. Some mechanical systems are used to supply the corrections in bearing in the presence of quadrantal error.

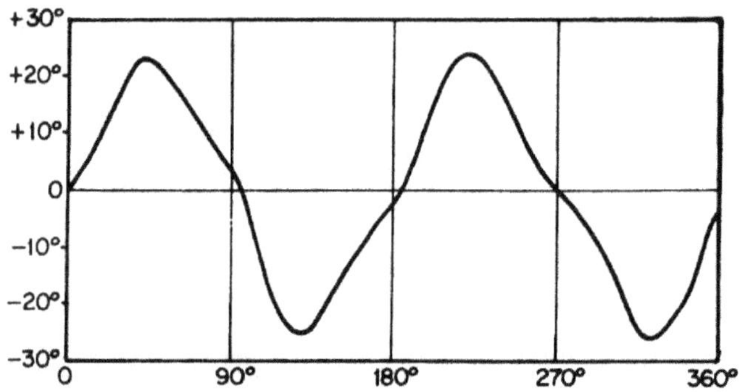

FIGURE 12.18 Quadrantal error in direction finding on ships.

12.7 AUTOMATIC DIRECTION FINDERS

Direction finding can be made fully automatic requiring no human operator to rotate the loop and find the bearing. Automatic direction finders are particularly useful for aerial navigation, where the motion of the vehicle is very fast and, therefore, the bearing must be determined in a very short time, either from the ground and sent to the vehicle by radio communication or directly from the vehicle. The ground-based automatic finders are often made very complicated to make the system in the aircraft rather simple. These topics are covered in Chapter 13 on Radio Range.

12.7.1 Airborne Automatic Direction Finder Radio Compass

In aircraft an automatic direction finder called a *radio compass* is very widely employed. A schematic diagram of the radio compass is shown in Figure 12.19. Two crossed loop antennas feed a goniometer coil which can

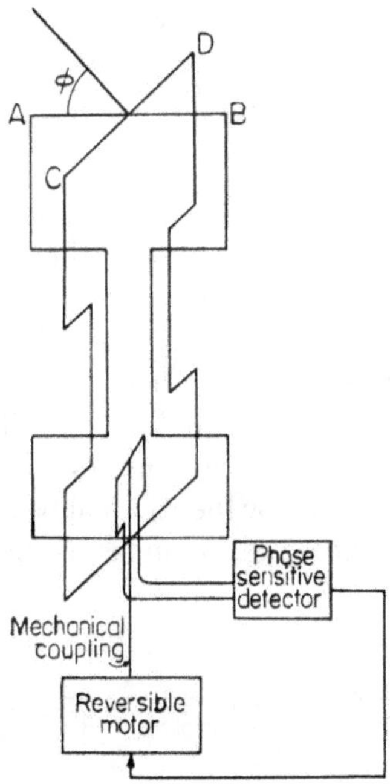

FIGURE 12.19 Automatic direction finder based on a Bellini-Tosi system for ships.

be looked upon as a search coil of a Bellini-Tosi direction finder shown earlier in Figure 12.11. Accordingly the direction of the resultant magnetic field H makes an angle ϕ with the plane of the loop CD. If the goniometer coil makes an angle $\Delta\phi$ with direction of the H_ϕ then the flux linkage B is given by

$$B = H_{0\phi} \sin \Delta\phi \, e^{j(\omega t + \pi/2)}$$

where $H_{0\phi}$ is the peak amplitude of $H\phi$. Voltage induced in the goniometer coil is therefore

$$E = -\frac{dB}{dt} = -\frac{d}{dt}[H_{0\phi} \sin \Delta\phi \, e^{j(\omega t + \pi/2)}]$$

$$= H_{0\phi} \sin \Delta\phi \, (j\omega) \, e^{j(\omega t + \pi/2)}$$

$$= -j\omega H_{0\phi} \sin \Delta\phi \, e^{j(\omega t + \pi/2)}$$

$$= \omega \sin \Delta\phi \, H_{0\phi} \, e^{j(\omega t + \pi/2)} \, e^{-j\frac{\pi}{2}}$$

$$= \omega \sin \Delta\phi \, N_{0\phi} \, e^{j\omega t}$$

$$= A_0 \sin \Delta\phi \, e^{j\omega t}$$

$$\cong A_0 \, \Delta\phi \, e^{j\omega t} \qquad (12.27)$$

if $\Delta\phi$ is small.

The voltage of the goniometer coil is detected by a phase-sensitive detector, and the output is used to rotate the goniometer in a direction tending to reduce $\Delta\phi$ to zero by a reversible motor fed with the detected voltage. Equation (12.27) shows that the phase of E_G is dependent on the sign of $\Delta\phi$ and, therefore, the sign of the detected voltage corresponds to that of $\Delta\phi$. For $\sin(-\Delta\phi) = -\sin\Delta\phi = \sin\Delta\phi \, e^{-j(\pi/2)}$. The motor will rotate the goniometer coil, tending to reduce $\Delta\phi$. The system thus becomes a closed loop servo control system actuated by the residual bearing error of the goniometer coil. A block diagram of the radio compass is shown in Figure 12.20.

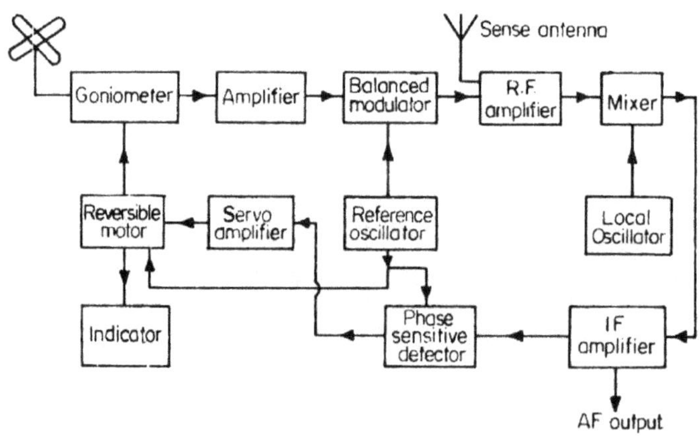

FIGURE 12.20 Automatic direction finder for aircraft.

Goniometer output E_G is amplified and fed to a balanced modulator, which produces the product of E_G with an oscillator output at frequency ω_r supplying a reference phase. Accordingly, the output of the balanced modulator is given by

$$\begin{aligned} E_B &= E_G\ E_R \\ &= A_0 \sin \Delta\phi\ e^{j\omega t} . E_r \cos \omega_r t \\ &= A_0 E_r \sin \Delta\phi \cos \omega_r\ t\ e^{j\omega t} \end{aligned} \tag{12.28}$$

To this is added the voltage induced in the vertical antenna of effective height h_f, the pickup of which is independent of the azimuthal direction of arrival of the radio wave due its omnidirectional pattern. The net voltage at the input of the *RF* amplifier is given by

$$\begin{aligned} E_I &= E_B + h_f E_0\ e^{j\omega t} \\ &= (E_r A_0 \sin \Delta\phi \cos \omega_r\ t + h_f E_0)\ e^{j\omega t} \\ &= h_f E_0 \left[1 + \frac{E_r A_0 \sin \Delta\phi}{h_f E_0} \cos \omega_r t \right] e^{j\omega t} \end{aligned} \tag{12.29}$$

This represents an amplitude modulated signal modulated at frequency of the reference oscillator and having an amplitude proportional to sin $\Delta\phi$. For $\Delta\phi$ lying between 0 and π, sin $\Delta\phi$ is positive and the demodulated signal at ω_r will be in phase within the reference signal. For $\pi < \Delta\phi < 2\pi$ is negative and the demodulated signal will have negative amplitude, which indicates that its phase is reversed. The reference signal and the demodulated signal are fed to a pair of coils at right angles with a 90° phase shift between them to produce a rotating magnetic field at ω_r for a synchronous motor, the direction of rotation being controlled by the sign of $\Delta\phi$. The motor is coupled mechanically to the goniometer rotor so as to turn it in a direction to reduce $\Delta\phi$, tending to orient the plane of the goniometer coil along the direction of the resultant magnetic field $H\phi$. The position of the goniometer rotor then indicates the bearing. To facilitate the generation of the rotating magnetic field, the amplitude of phase-sensitive detector output may be limited to a constant level, so that the two coils of the motor may be fed with gradrature voltage of equal amplitudes.

A radio compass operates in the *LF*/*MF* bands between 200–3000 kHz. It is universally employed as a reliable aid for the homing of aircrafts. It fails to indicate direction only during interference due to lightning discharges appearing as atmospherics. Also, in the presence of lateral winds causing a drift of the aircraft perpendicular to the direction of motion, the radio compass being always pointed toward the airport will guide the aircraft along a curved path to the airport. In modern versions the antennas employ ferrite loops mounted on the fuselage so that no additional air drag may be introduced.

Calculate the maximum voltage induced in the loop antenna having

Number of turns	$N = 10$
Quality factor	$Q = 100$
Frequency of operation	$= 300$ kHz
Area of the loop	$A \simeq 1\ \text{m}^2$
and field strength	$E_0 = 3.146\ \mu\text{V/m} = \pi\ \mu\text{V/m}$

$$E_V = \frac{2\,\pi\,E_0\,AN\,Q}{\lambda}$$

$$= \frac{2\,\pi\,E_0 \times 1 \times 10 \times 100}{1000}$$

$$\therefore\ \lambda = \frac{c}{f} = \frac{3 \times 10^8}{300 \times 10^3}\ \text{m} = 1000\ \text{m}.$$

$$= 2\,\pi\,E_0$$

$$= 2\,\pi \times \pi\ \mu\text{V}$$

$$\simeq 2 \times 10\ \ \mu\ \text{V, assuming } \pi^2 \simeq 10$$

$$= 20\ \mu\ \text{V}.$$

CHAPTER 13

AIRCRAFT HOMING SYSTEM AND INSTRUMENT LANDING SYSTEM

13.1 INTRODUCTION

The guidance of an aircraft toward an airport is called homing. In busy airports, where a large number of aircrafts are scheduled to arrive simultaneously, the pilot of each aircraft must known its own bearing in flight, with precision. Otherwise, such a simultaneous approach to an airport by a number of aircraft may lead to a collision between two adjacent aircraft. A "radio range" system is designed to provide precision bearing information to an aircraft with a simple antenna system on it. In the presence of lateral wind drag of the aircraft, any deviation from the correct course will be indicated, allowing the pilot to correct the course, guiding it along a straight path. A radio compass, however, always points toward the airport, irrespective of its bearing, and causes the aircraft to approach the airport in a curved path in the presence of lateral wind drag. After approaching closely to the airport, the landing of an aircraft is often aided by radio aids called instrument landing systems (ILS). This is particularly useful during poor visibility conditions and at night.

13.2 SWITCHED CARDIOID HOMING SYSTEM

The precision and range of a radio compass can be increased greatly by employing a switched cardioid antenna pattern for reception of the beacon signal at the aircraft. Figure 13.1 shows how the two switched positions 1 and 2 of the pattern enable one to receive equal amplitude signals at the two switched positions when the beacon is along the intersection of the two patterns. This defines an equisignal course, and the technique is called lobe switching. Any deviation of the aircraft from the equisignal course, + ϕ, will lead to detected output waveforms as shown in Figure 13.2 (*a*). If the polarity of the detected signal is reversed whenever the pattern is switched to pattern 2, the waveforms will have an average which is zero for correct course (III), as in Figure 13.2 (*b*) positive for course I as in Figure 13.2 (*c*), and negative for course II. Accordingly, a center zero meter indicating the average signal will directly show the course followed by the aircraft.

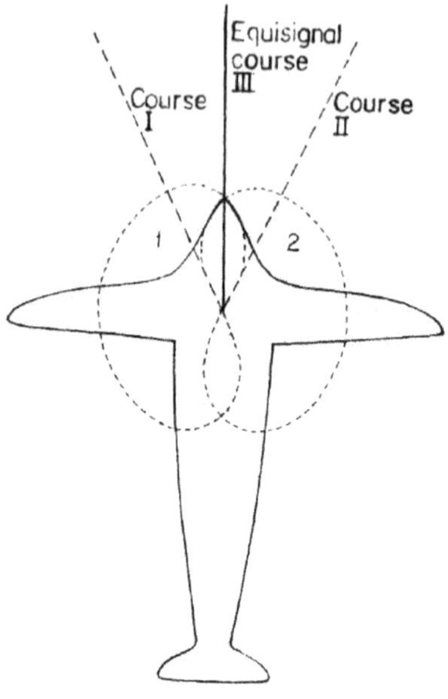

FIGURE 13.1 Switched cardioid direction finding.

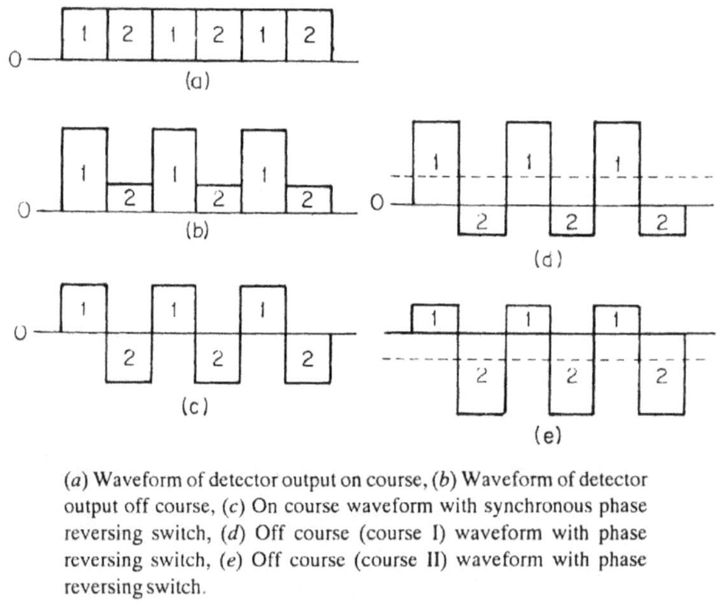

(*a*) Waveform of detector output on course, (*b*) Waveform of detector output off course, (*c*) On course waveform with synchronous phase reversing switch, (*d*) Off course (course I) waveform with phase reversing switch, (*e*) Off course (course II) waveform with phase reversing switch.

FIGURE 13.2

The antenna system for generating the switched cardioid pattern is shown in Figure 13.2 (*d*). The combination of a loop and vertical antenna generates a pattern given by equation (12.10) which is of the form $E_t = (a \pm b \cos \theta)$, the sign of *b* depending on the sign of the mutual inductance coupling between the vertical and the loop antennas. This sign can be reversed by reversing the polarity L_1 by the commutator, which can be an electronic commutator as well.

The advantage of the switched cardioid system is that two signals of rather large amplitudes are effectively compared for equality for orienting the aircraft to the correct course. The equisignal course is, therefore, much less affected by noise which, however, limits the accuracy of the null in a simple loop direction finder. Also, for a given noise level the lobe switching will provide a higher accuracy bearing. The lateral wind drag on the aircraft cannot be sensed by pilot with the switched cardioid homing system, as is also the case with a loop direction finder. For the error in bearing resulting from lateral drag, the pilot will be prompted to reorient the aircraft by looking at the radio compass pointer for regaining the equisignal course. This causes the aircraft to be guided along a curved path being always oriented toward the airport.

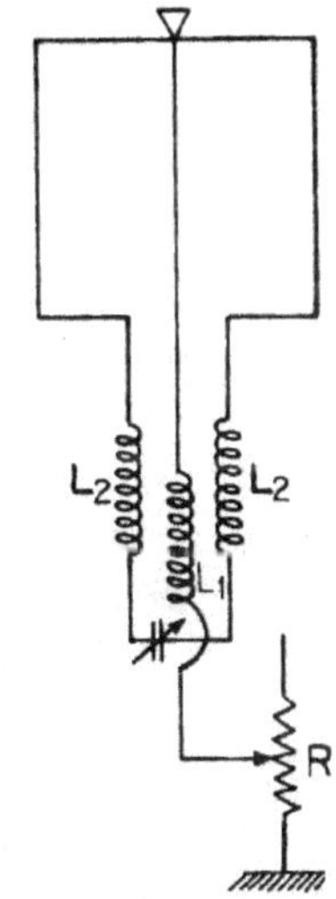

FIGURE 13.3 Sense finder antenna system.

13.3 FOUR COURSE RADIO RANGE

The higher accuracy of the lobe switching technique is utilized in a four course radio range. Switching between two loop antenna patterns at the transmitter end at airport can be employed to generate four equisignal courses. In that case is the aircraft antenna may be a simple loop or dipole. Such a system is called a four course homing system.

The four equisignal courses will be along directions determined by the patterns of the two loop antennas. Referring to Figure 13.4 (*a*), two loop antennas, L_1 and L_2, of equal area with their planes at right angles, will have patterns 1 and 2 respectively, when fed with equal powers. The lines joining the points of intersection of the two patterns define the equisignal courses, which are a long ON_1, ON_2, ON_3, and ON_4, spaced 90° apart from each other. An aircraft flying along any one of these courses will pick up signals of equal strength for the two switched positions of the antennas L_1 and L_2.

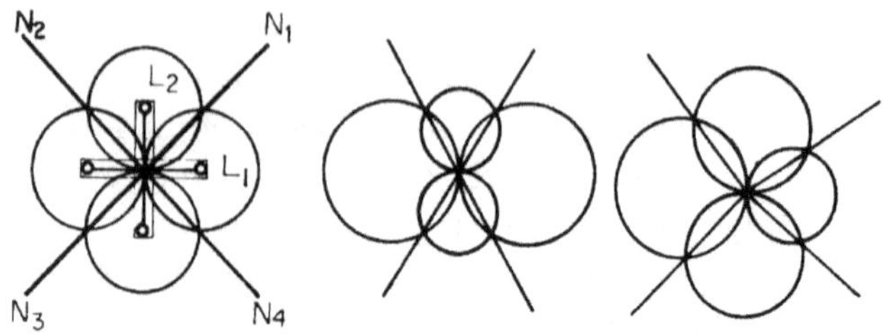

(*a*) Four equispaced courses.

(*b*) Courses with unequal angular spacing.

FIGURE 13.4

To detect the equisignal course from the aircraft, the transmitter is modulated by a 1020 Hz tone and is switched on to the transmitting loops L_1 and L_2 alternately in a particular sequence to produce the codes for the letter A and N, for transmissions from L_1 and L_2 respectively. The codes being complimentary, a steady 1020 Hz tone will be heard in the receiver of an aircraft in the equisignal course. Any deviation from the equisignal course will make the letter A or N detectable, depending on whether the aircraft is on the L_1 or L_2 side of the equisignal course respectively. Considering that the ears cannot distinguish changes of amplitude of the tone is below a certain limit, the equisignal course will not be sharp, the accuracy being limited to 3°. For adjustment of the exact direction of the course, the power fed to one of the loops may be increased to increase the size of its pattern, as shown in Figure 13.4 (*b*).

In practice, the loops can be replaced by four vertical tower antennas at the corners of a square, when each diagonal pair generates a figure of eight patterns. A fifth antenna at the center of the square transmits at a frequency differing by 1020 Hz from the transmissions of the diagonal pairs. This produces a beat note of 1020 Hz at the aircraft receiver. Thus, none of the transmissions needs the 1020 Hz modulation which was introduced in the crossed loop system. This system is called SRA (Simultaneous Range Adcock). The radiation from the central tower can be modulated to transmit weather information to the aircraft, and can further be employed for radio compass operation at the aircraft.

The four course radio ranges operate in the LF/MF range of 200–400 kHz and provides a course over a range of about 200 km. The sources defined may be followed by an aircraft unaffected by lateral wind drag. The disadvantages include (*a*) fatigue caused by listening to tones, (*b*) ambiguity of course of which the number is also limited to four, and (*c*) interference due to atmospherics. All these disadvantages were removed in a system called VHF omnirange, which effectively replaced the four course radio range.

13.4 OMNIDIRECTIONAL RANGES

An omnidirectional range is capable of providing course information to an aircraft flying at any azimuthal angle or bearing. The system is often called omnirange, although the term range may be a misnomer. In most radio ranges, the range of the aircraft is not determined at all. The most widely employed omnirange operates in the VHF band between 108–136 MHz, and it is then called a VHF omnirange or VOR.

13.4.1 VHF Omnirange (VOR)

In a VHF omnirange a VHF transmitter at the airport radiates from an antenna similar to that of a sense finder generating a limacon pattern as shown in Figure 13.5 (*a*). The pattern is rotated at 30 Hz revolutions per second either by mechanical rotation of the antenna or by electrical means. The figure shows successive orientations of the limacon pattern during one complete revolution. Due to the pattern rotation, the signal received in an aircraft flying along a particular direction will undergo a periodic variation at the frequency of the pattern rotation. The waveforms of the signal strength variation observed in four aircraft flying from the north, south, east, and west will be as shown in Figure 13.5 (*b*), from which it is evident that the phase of the waveform depends on the bearing. To facilitate measurement of the phase of the aircraft, the transmission is interrupted whenever the maximum of the pattern is due north. The dotted lines in Figure 13.5 (*b*) indicate these periodic interruptions. By noting the position of the interruption in relation to the maximum of the pattern, the phase of the signal strength variation can be determined from which the bearing information can be obtained. The interruption is thus supplying reference phase information which is independent of the direction of flight of the aircraft.

The interruption of transmission to provide a reference phase leads to the generation of sidebands covering much wider RF band widths than that due to the 30 Hz pattern rotation alone. The sidebands may interfere with other VHF equipment near the airport. In an improved system the reference phase is provided by amplitude modulation of an omnidirectional transmission by a 30 Hz reference phase signal. Also, the rotation of the limacon pattern is obtained by electrical means.

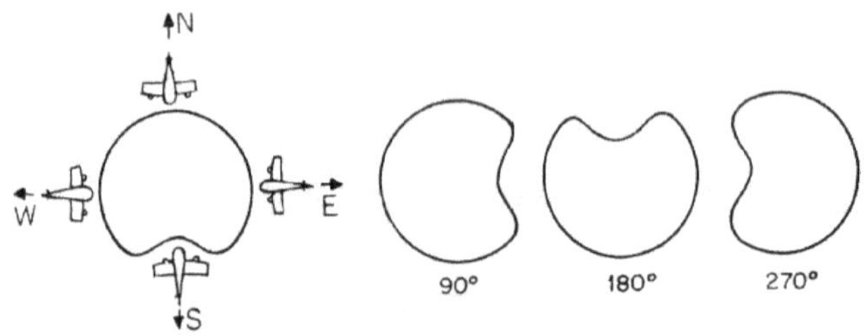

(*a*) Rotating limacon pattern of VOR.

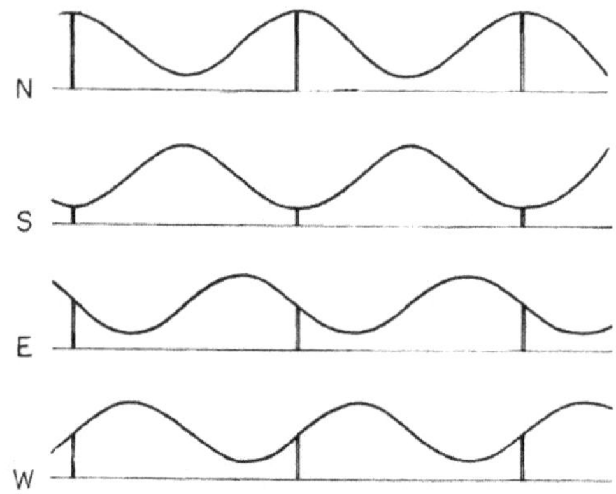

(*b*) Waveform of detector output in aircraft flying from the North, South, East, and West.

FIGURE 13.5

Electrical Pattern Rotation. The pattern rotation by a stationary antenna system may be achieved by electrical means as shown in Figure 13.6. Five vertical tower antennas generate a limacon radiation pattern as follows: the antenna pairs 1–3 and 2–4 produce two crossed figures of eight patterns, as in Adcock pairs. For equal power fed to the pairs, the radiation pattern will be represented by

$$|E_{1\text{-}3}| = E_L \cos\phi$$

$$|E_{2\text{-}4}| = E_L \sin\phi$$

where $|E_{1\text{-}3}|$ and $|E_{2\text{-}4}|$ are the amplitudes of the fields at a given distance r from the center O, due to the pairs 1–3 and 2–4 respectively, and E_L is the constant for a pair at the distance r and ϕ is the direction of an aircraft receiving the signal. Now if the power fed to each pair is amplitude-modulated at 30 Hz with modulation signals to pair 1–3, 90° out of phase, or in quadrature with that of pair 2–4, we get

$$|E_{1-3}| = E_L \cos pt \cos \phi$$

$$|E_{2-4}| = E_L \sin pt \sin \phi$$

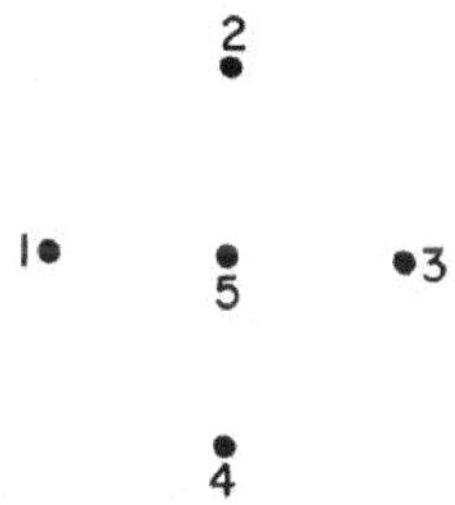

FIGURE 13.6 Antenna configurations in VOR for generation of a rotating limacon radiation pattern.

if we assume $E_L \cos pt$ and $E_L \sin pt$ to represent the modulating signals to the antenna pairs 1–3 and 2–4 respectively. The resultant field will then be

$$\begin{aligned} |E| &= |E_{1-3}| + |E_{2-4}| \\ &= E_L(\cos pt \cos\phi + \sin pt \sin\phi) \qquad (13.1) \\ &= E_L \cos(pt - \phi) \end{aligned}$$

This represents a figure of eight pattern rotating at a frequency f_p, where $p = 2\pi f_p$.

Now, if the fifth tower antenna is also energized with a constant power, the radiated resultant field due to the five antennas will be represented by

$$\begin{aligned} |E_r| &= |E_5| + |E_{1-3}| + |E_{2-4}| \\ &= k + E_L\, cos\,(pt - \phi) \qquad [\text{since } |E_5| = k, \text{ a constant}] \end{aligned}$$

This is of the form $|E_r| = a + b\, cos\,(pt - \phi)$, which represents a rotating limacon pattern as shown in Figure 13.4. If the pattern rotates through an angle in a line t, then for a rotation frequency, $f_p = p/2\pi$, $\theta = pt$.

Angular deviation of the direction of flight of an aircraft from the direction of maximum, at a time t, is $\theta - \phi = pt - \phi$. Therefore, the position of the pattern at a time t is represented by

$$|E_r| = a + b \cos(pt - \phi)$$

Introduction of a Reference Phase. The reference phase could be introduced by modulating the constant amplitude term a, which is due to the omnidirectional radiation from the central tower, with a 30 Hz reference signal using 30% modulation of the VHF carrier. However, in that case, the 30 Hz reference phase will not be distinguishable from the 30 Hz modulation due to pattern rotation. The term a is therefore first modulated by a signal of frequency f'_p = 9960 Hz, that can be easily separated from the 30 Hz signal due to pattern rotation by filtering during reception at the aircraft. The 9960 Hz signal is further frequency modulated by the 30 Hz reference signal. The resulting signal is then represented by

$$|E_r| = a\,[1 + m \cos(p't + f_a/f_p \sin pt)] + b \cos(pt - \theta)$$

where $p' = 2\pi f'_d$.

The 30 Hz reference phase is thus introduced as a frequency modulation of a subcarrier at 9960 Hz, which in turn amplitude modulates the term "a." The frequency deviation f_d is 480 Hz.

Recovery of Reference Phase and Measurement of Bearing. The structure of the receiver at the aircraft for the measurement of bearing is as shown in Figure 13.7. The incoming signal received by the antenna is fed to a superheterodyne receiver tuned by the local oscillator frequency to the frequency of the incoming VOR signal at VHF. The detected output is divided into two parts which are passed through bandpass filters tuned to f_p and f'_p, which are 30 Hz and 9960 Hz respectively. The output at f_p from the filter tuned to f_p is due to pattern rotation, while the output of f'_p from the other filter is fed to an FM demodulator to recover the reference phase signal at f'_p. The difference of phase between those two 30 Hz outputs, as $\cos(pt - \theta)$ due to pattern rotation and $\cos pt$ due to reference phase signal, is measured by a direct reading phase meter. In this part the two signals are first limited to equal amplitude square wave signals. The difference between these square waves will be zero for common areas between them, and the output from the subtractor will be proportional to the phase difference ϕ. Averaging the difference output will in fact produce a DC signal proportional to the phase difference ϕ, which is really the bearing of the aircraft.

Antenna Pattern with Horizontal Polarization. Operation of VOR in the VHF band ensures a low radio interference level. The interference level can be minimized further by employing horizontal polarization instead of vertical polarization realized by tower antennas. The polarization of most of

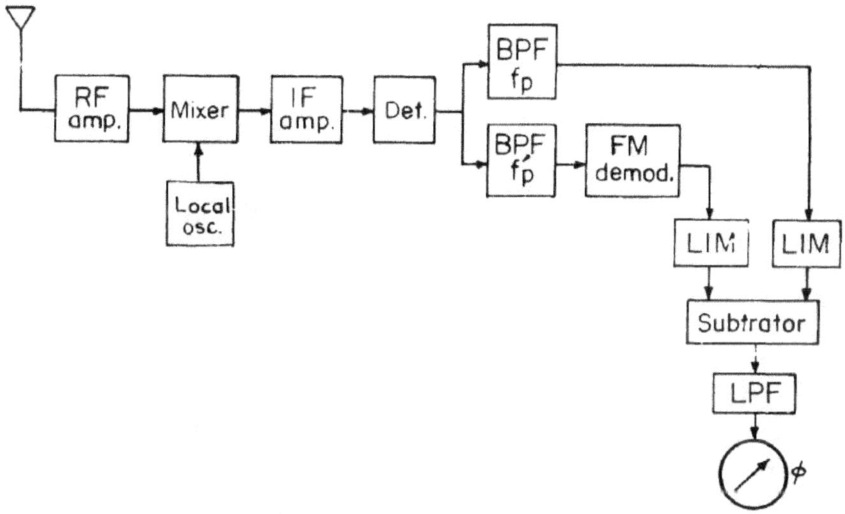

FIGURE 13.7 On board receiving equipment for VOR.

the interfering sources is predominantly vertical, and a horizontally polarized receiving antenna will pick up less interference. In fact, this prompted TV broadcasters to employ horizontal polarization. To radiate with a horizontally polarized rotating limacon pattern in a VOR, each of the five elements constituting the antenna system is replaced by an antenna with an omnidirectional pattern but radiating horizontally polarized waves. This can be achieved by using a magnetic dipole in place of the electric dipole which was realized with a tower antenna.

A single turn of wire, small in dimension compared to wavelength, λ, constitutes a magnetic dipole. When fed from an electrical source, vertical magnetic fields appear if the plane of the coil is horizontal. The instantaneous polarity of the field E_1 and the current (i) causing it are shown in Figure 13.8 (a). The polarities alternate at the frequency of the source. The electric field from such an antenna will be horizontally polarized. The connecting leads are closely spaced in the form of a transmission line so that any radiation from the two leads tend to cancel each other out and the radiation loss from the leads is negligible. A pair of magnetic dipoles producing a figure eight pattern with horizontal polarization is shown in Figure 13.8 (b). Note the opposite direction of instantaneous current flow to produce opposite magnetic polarity of the two loops required to produce zero fields at right angles to the line joining their centers.

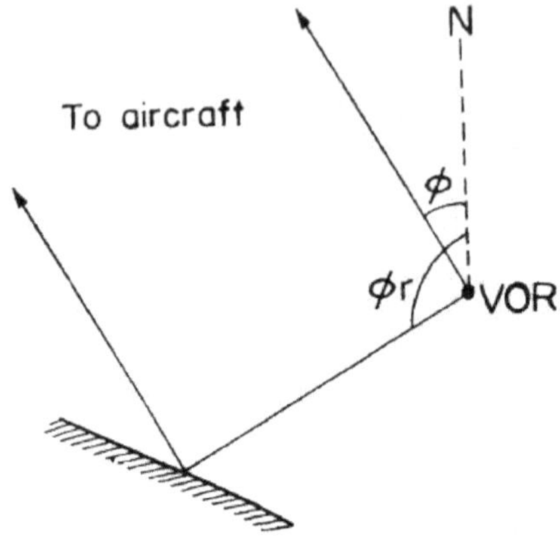

FIGURE 13.8 (*a*) Magnetic dipole (*b*) Pair of magnetic dipoles with opposite polarity (*c*) Practical form of a magnetic dipole (*d*) Alford aerial.

The leads can be connected in an alternative way with the source at the center as shown in Figure 13.8 (*c*). Here also the radiation from the closely spaced leads will be negligible. A modification of the magnetic dipole is called an Alford antenna, as shown in Figure 13.8 (*d*). The azimuthal pattern for the antennas shown in Figures 13.8 (*c*) and (*d*) are almost similar to that of Figure 13.8 (*a*), although there will be a minor difference due to the current not being continuous in Figures 13.8 (*c*) and (*d*) as it is in Figure 13.8 (*a*). Alford antennas are simpler in design and have been used in some VOR systems, such as that developed by the Civil Aeronautics Administration (CAA).

Mechanical Rotation of an Antenna. In some VOR systems such as the FTL equipment the rotating limacon pattern is produced by the rotation of vertical dipoles mechanically about a vertical axis with a motor. However, the length of the dipole antennas at VHF is not convenient for arranging the rotation. Such mechanical rotation will be easier at UHF, and an improved radio range called TACAN (Tactical Air Navigation) employs the mechanical rotation.

Performance of VOR. *Range*. The range of a VOR is about 400–500 km as determined mainly by the line of sight range between the ground antenna system and the aircraft antenna. Refraction of the VHF signals in the air allows extension of the range by about 10–15%. The range will be about 335 km for an aircraft flying at an altitude of 20,000 ft (6 km).

Error in Bearing. Bearing errors in a VOR may originate from sites containing stray reflectors, cross-polarization, and equipment error.

Error Due to Site and Terrain Features. If the VOR transmission is reflected from objects such as bridges, buildings, towers, or trees and arrives at the aircraft, it may interfere with the signal reviewed by the line of sight path. The plan view in a typical case as shown in Figure 13.9.

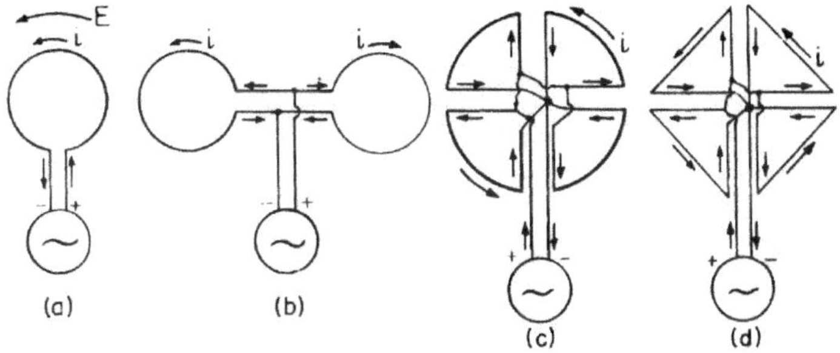

FIGURE 13.9 Error in VOR due to stray reflections.

The phase difference between the transmission from the rotating pattern antenna system and that from the omnidirectional antenna is equal to Φ, the bearing of the aircraft. Likewise, the reflected component will have a difference of phase, ϕr. The resultant of the direct and reflected signals will have a phase different from ϕ, which is required for a correct indication of the bearing by the VOR.

A similar situation may arise due to reflection from hills, and thus terrain conditions play a major role in introducing errors in bearing.

Error Due to Polarization. Due to the fact that the radiation patterns of the transmitting antenna are generally different for horizontal and vertical polarizations, errors in bearing may occur due to the cross-polarized component.

The VOR is usually designed for horizontally polarized antenna systems. The origin of the vertically polarized component from a horizontally polarized dipole is illustrated in Figure 13.10. Although the polarization will indeed be horizontal for any point in the yz plane perpendicular to the dipole, a vertical component arises at a points in the xz plane along the dipoles. In fact, a vertical component will be present at a point in any plane other than the yz plane. The vertical component is considered here as the cross-polarized component. The polarization error may be minimized by a suitable antenna design to reduce the vertical component and ensure that the aircraft antenna is insensitive to the vertical component.

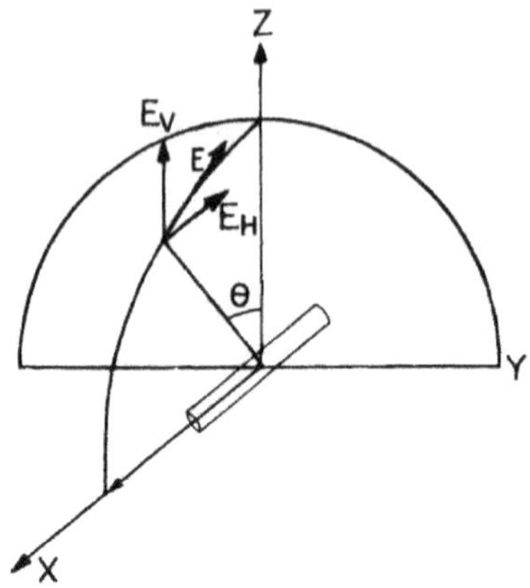

FIGURE 13.10 Error in VOR due to a cross-polar component.

Equipment Error. Besides the errors in bearing arising from the considerations of ratio wave generation and propagation mentioned previously, there may be errors due to the equipment. The ground station equipment errors may be due to octantal errors, while the aircraft equipment errors arise from the imperfections in the receiver and indicator leading to errors in phase measurement and hence in the bearing. Statistics of equipment error indicate that the error is within 2% with a 95% probability.

13.4.2 Doppler VOR

The accuracy of bearing determinated by VOR is limited mainly by site and polarization errors. Such errors are drastically reduced in recent developments of omnirange, which make the use of the Doppler shift of frequency resulting from a physical or electronic rotation of the antenna system.

The antenna system in a Doppler VOR system consists of 17 separate antennas arranged along the circumference of a circle of diameter about 5 λ, and spaced equally from each other, together with a central antenna. In Figure 13.11, 16 circumferential antennas and one central antenna are shown. The transmitter power is switched on to successive antennas, the frequency of the transmitter being made 9960 Hz higher than that of a continuous wave transmission from the central antenna. At the aircraft the transmissions from the central and outer antennas will produce a beat of 9960 Hz at the receiver, modulated additionally in frequency, due to the Doppler shift of the transmission frequency arising from the effective rotation of the antenna element. The shift will be largest for antenna pairs at the ends of a diameter, perpendicular to the direction of the aircraft, and the shift will be upward for the elements to the right half plane shown by the shaded area, and it will be downward for those at the left half plane (unshaded). The shift will in fact, vary sinusoidally at the effective rotation frequency of the antennas, while the amount of Doppler shift is proportional to the line of sight velocity of an antenna element, which in turn depends on the speed of effective antenna rotation, diameter of the circle, and direction of the aircraft. If may be noted that for the direction of the aircraft shown, the shift will be largest and upward for the elements near 2, and it will be lowest and downward for the elements near 7. This means that the phase of the effective frequency modulation arising from the Doppler effect depends on the direction of the aircraft. The 9960 beat note will thus be frequency modulated by a 30 Hz modulation due to the effective antenna rotation, the phase of this 30 Hz signal carrying the bearing information, ϕ. The central antenna is modulated by a 30 Hz reference phase signal as in a conventional VOR. Thus, the received signal is similar in nature to that of the simple VOR, although the bearing information is transferred to the transmission from the switched antenna system by making use of the Doppler effect. Site errors in Doppler VOR are reduced by a factor of four to seven. However, to make the system compatible with conventional VOR, the diameter of the antenna circle required may be as large as 50 m for producing a 480 Hz maximum frequency deviation arising from the Doppler effect, with a rotation frequency of 30 Hz.

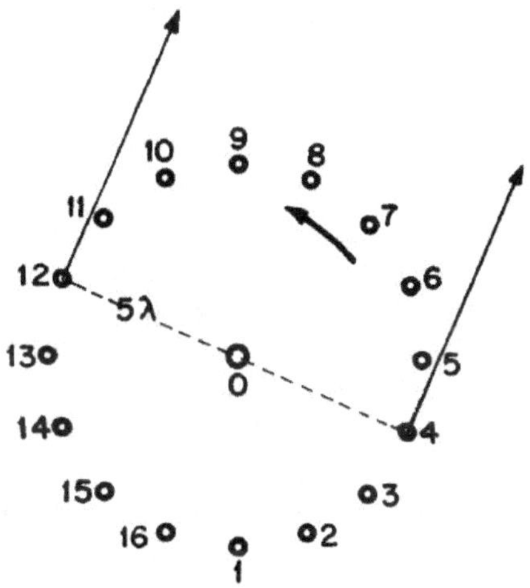

FIGURE 13.11 Doppler VOR antenna configuration.

The concept of Doppler VOR has been extended to develop a Doppler direction finder which serves as an efficient aid for radio monitoring of aircraft from the airport for air traffic control. This is discussed in detail in the last section of this chapter.

13.5 TACTICAL AIR NAVIGATION (TACAN)

An improved radio range TACAN (Tactical Air Navigation) was developed to facilitate for both range and precision bearing determination from an aircraft even from long ranges. The principle of the bearing determination is similar to that of VOR with refinements for a higher accuracy of bearing. The range measurement is based on the principle of radar, and this facility is separately called DME (Distance Measuring Equipment).

Bearing Facility. A rotating limacon pattern modified to generate a toothed pattern as shown in Figure 13.12 (*a*) is generated by a rotating antenna system. This consists of a central vertical antenna radiating with an omnidirectional pattern. The radiation from this and that from a single parasitic element p_1 will produce a limacon pattern, the element behaving as a reflector. Nine more parasitic elements arranged along a circle with the central antenna as the center, modify the limacon pattern, imparting to it the toothed appearance. Rotation of the entire antenna system about the center produces a signal at the aircraft having an amplitude varying at the rotation frequency with a superimposed finer amplitude variation at 9 times the rotation frequency, as shown in Figure 13.12 (*b*). The field pattern of the rotating antenna system may represented by

$$F(\phi, \phi_s, t) = a + b \cos(pt - \phi) + C \cos(p_s t - \phi)$$
$$= a + b \cos(pt - \phi) + C \cos(9\,pt - \phi)$$

where $p = 2\pi f_p$, f_p is the rotation frequency of the antenna and the factor 9 inside the third cosine term is due to tooths of the pattern. Usually f_p = 15 revolutions/sec. = 900 RPM.

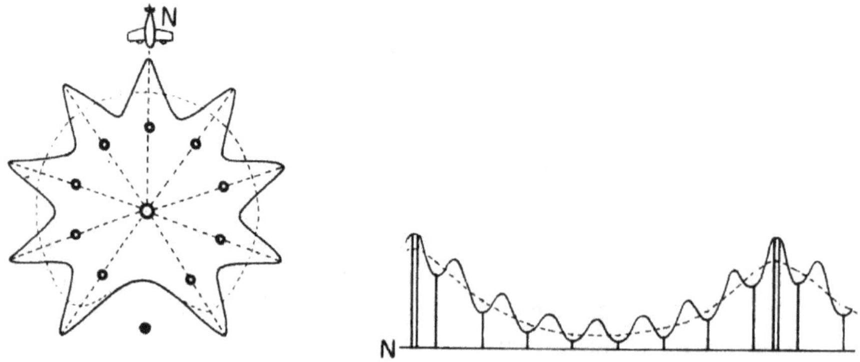

(*a*) Radiation pattern for TACAN. (*b*) Waveform of detector output in the onboard TACAN receiver.

FIGURE 13.12

The received signal at the aircraft will then be amplitude modulated simultaneously at 15 Hz and 9 × 15 = 135 Hz. The respective demodulated signals will be represented by

$$E_1 = a + b \cos (pt - \phi)$$

and $$E_2 = a + c \cos (p_s t - \phi) = a + c \cos (9\, pt - \phi).$$

The reference phase signals at p and p_s, that is, at f_p and $9f_p$, may be provided either by interruption of a continuous wave transmission at the maxima of the limacon pattern and at the maxima of each tooth for the distinctive periods or by sending distinct groups of coded pulses; at such times, E_2 allows bearing determination with nine times the precision obtainable without the toothed pattern. The ambiguity of the tooths are resolved by first determining the course bearing using the limacon pattern as in a VOR through the 15 Hz component and then obtaining the finer bearing information from the 135 Hz components arising from the toothed pattern.

To facilitate distance measurement the transmission is pulse modulated. The amplitude of the received pulse train, however, undergoes cyclic variations at 15 Hz and 135 Hz due to pattern rotation. In the aircraft receiver these two components are separated at the detector output by narrow band filters tuned to the respective frequencies, to facilitate phase comparison with the reference phase signals needed for bearing determination. The distance measurement facility is detailed in Chapter 10.

13.6 INSTRUMENT LANDING AIDS

Radio aids for landing of an aircraft at the airport are called Instrument Landing Aids. While the homing systems help in guiding the aircraft toward the destination airport, the final phase of the flight during landing starts when aircraft is near the airport within a certain range to enable the pilot to see the runway visually. Under good visibility conditions, such visual landing is possible in daytime as well as in night hours when lights marking the boundaries of the runway are visible from the aircraft. However, during foggy weather, the visibility may be too poor for landing, and the radio aids will be useful. If the horizontal visibility is 5 km or more and vertical visibility greater than 300 m, visual flight rules (VFR) are followed. At other times such as that during poor visibility conditions, the landing may be made under instrument flight rules (IFR) using the radio aids.

The radio aids may be incorporated within the aircraft operated and controlled by the pilot. These are called instrument landing systems (ILS). Alternatively, it may be systems controlled from the ground, when the pilot is to land following instructions from the ground control operator obtained from the Ground Control Approach (GCA) systems.

13.6.1 Instrument Landing Systems

The instrument landing systems (ILS) enable blind landing of an aircraft under poor visibility conditions. The ILS effectively guides the aircraft both in elevation and azimuth, usually supported by a measurement of the altitude above the ground by an aid called a radio altimeter.

Elevation and Azimuth Guidance. The radiation pattern of antennas at the airport may be utilized for both elevation and azimuth guidance. A very simple way of achieving this will be to make use of the fact that a radiation pattern is basically a contour of constant field strength as shown in the following. Figure 13.13 shows the radiation pattern of a half-wave dipole antenna in a plane passing through the antenna. At a distance R from the center of the antenna in a direction θ, the field strength E will be proportional to the intercept r on the horizontal line cutoff by the pattern. Thus, $E \propto r$, when R is constant. Again, $E \propto 1/R$, when θ is constant.

$\therefore$ $E \propto r/R$ when both R and θ vary

or $E = k\dfrac{r}{R}$

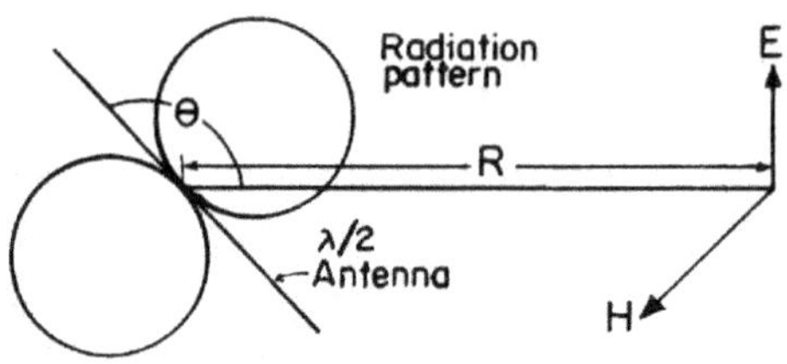

FIGURE 13.13 Radiation pattern of a half-wave dipole.

Now, if an aircraft flies along the line describing the pattern, $R = r$, the field strength will be given by

$$E(r, \theta) = k\frac{r}{r} = k$$

which is a constant. Thus, the aircraft may be guided along the line of the pattern if it follows a path of constant field strength. An elevation guidance system based on this principle is illustrated in Figure 13.14. The radiation patterns of the antenna *A* at the airport drawn on different scales are represented by 1, 2, and 3, each one of which is a locus of constant field strength. However, the field is progressively lower for patterns 1, 2, and 3, the field being inversely proportional to the distance from *A*.

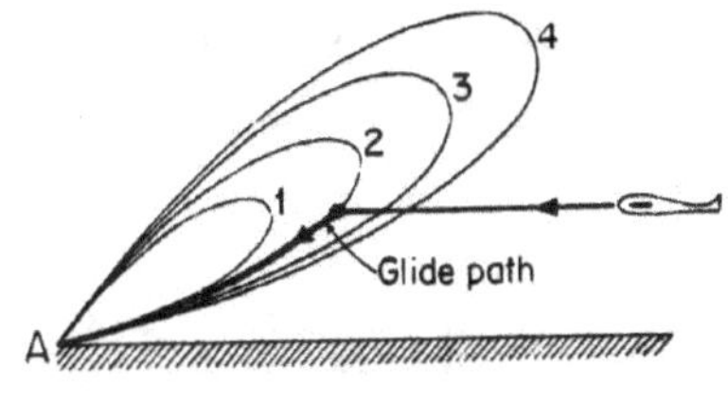

FIGURE 13.14 Elevation guidance using radiation patterns.

An aircraft approaching the airport horizontally picks up the field due to the transmission from *A* after it approaches the airport within a certain range, touching the pattern 2. At greater ranges, at pattern 3, for instance, the field is too weak to be detectable by the aircraft. If the aircraft descends now following pattern 2, the signal received will be of constant amplitude. The aircraft is thus guided in elevation, although the path of the aircraft will be curved as indicated by pattern 2. The aircraft may, in fact, follow any of the patterns lying within pattern 2.

For elevation guidance in a straight path, which is desirable for an easier maneuver of the aircraft, lobe switching may be employed. This is illustrated in Figure 13.15. If the radiation patterns 1 and 2 correspond to the two switched positions of the antenna pattern, an aircraft flying along the line of intersection of the patterns will receive signals of equal strength in the two switched positions of the pattern. The equisignal course thus enables the pilot to guide the aircraft in elevation, in a straight path.

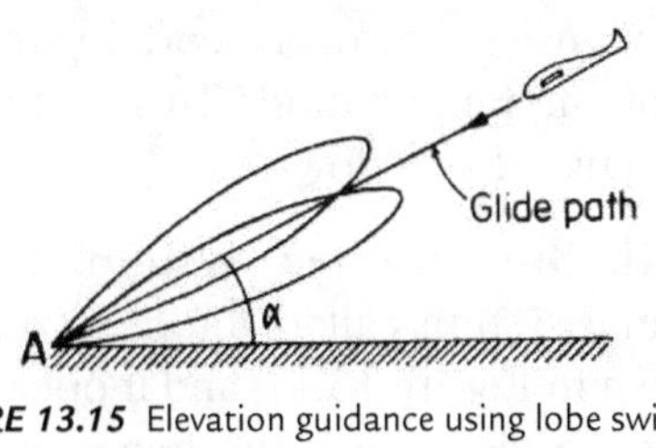

FIGURE 13.15 Elevation guidance using lobe switching.

In ILS systems, the lobe switching is achieved by switching the transmitter between the antennas 1 and 2 as shown in Figure 13.16. Horizontal dipoles 1 and 2, at right angles to the runway, are placed at heights h_1 and h_2, from the ground. Reflection of the transmission at the ground produces image antennas 1′ and 2′ at depths h_1 and h_2 respectively. Interference between the direct and reflected components produces interference patterns 1 and 2 above the ground for antennas 1 and 2 respectively, similar to that in a Lloyd's mirror in optics. Intersection of the resulting patterns 1 and 2 defines equisignal courses I and II. Course I is around an elevation of 3°, which may be utilized for elevation guidance in a straight path.

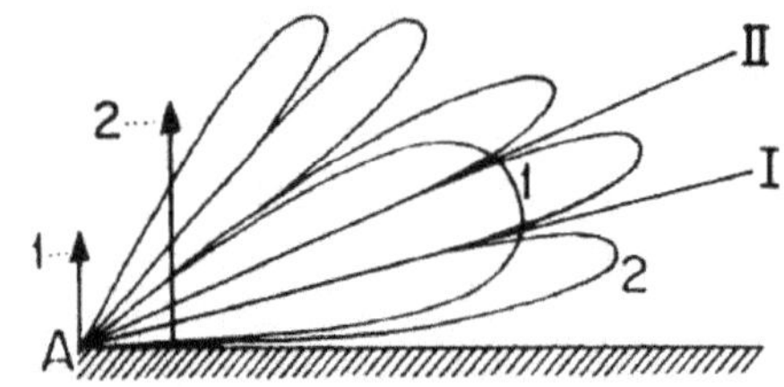

FIGURE 13.16 Elevation guidance using interference patterns due to ground reflection.

Azimuth Guidance. The lobe switching technique is also employed for azimuth guidance. For this purpose the same antenna pairs employed for elevation guidance may be utilized, as shown in Figure 13.17 (*a*). The antenna elements 1 and 2 may be displaced horizontally, at right angles to the runway and oriented at a small angle, α, to each other. The pair is next backed by a parabolic cylinder reflecting surface, producing azimuthal patterns 1 and 2 for the antennas 1 and 2 respectively, which are at the respective heights h_1 and h_2, The cylindrical reflectors do not cause significant modification of the patterns in the vertical plane, allowing the elevation guidance to be obtained as discussed previously. The beams 1 and 2 in the horizontal plane generated by inclined antenna pairs backed by the reflector are employed for obtaining azimuth guidance by following the equisignal course. Practical forms of ILS are based on the principles discussed previously; however, certain modifications are needed for convenience of implementation. Two systems, localizer and glide slope systems are discussed as follows:

Localizer with the Antenna System. In this system, the transmitter with the antenna system called the localizer is located at the end of the runway as shown in Figure 13.17, and it operates in the VHF band between 108–110 MHz. To distinguish the two switched positions of the antenna pattern at the aircraft, the transmission during one of the switched positions is modulated by a 90 Hz tone, while during the other it is modulated by a 150 Hz tone.

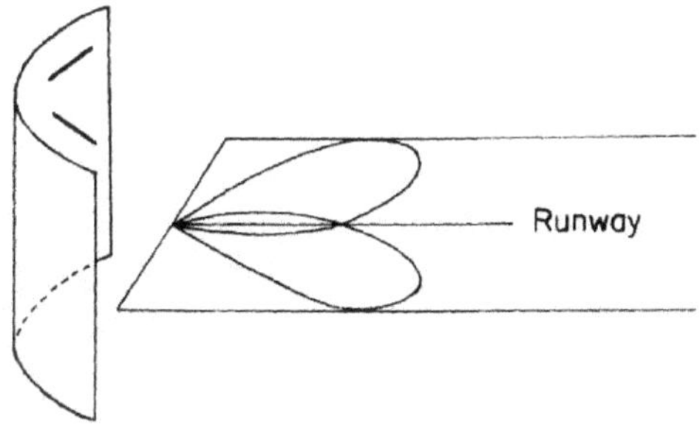

(*a*) Azimuth guidance in ILS.

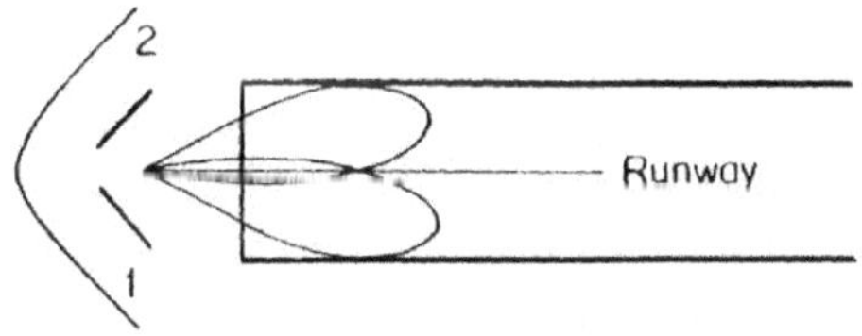

(*b*) Plane view of ILS azimuth guidance.

FIGURE 13.17

For azimuthal guidance along an equisignal course, the antenna system consists of 7–8 Alford loops, arranged in a line at right angles to the runway and located at a distance of about 300 m from the end of the runway. The loops are divided to form three separate antenna arrays, with one at the center and the remaining two on either side of it. The azimuthal patterns of the three antenna arrays are shown in Figure 13.18.

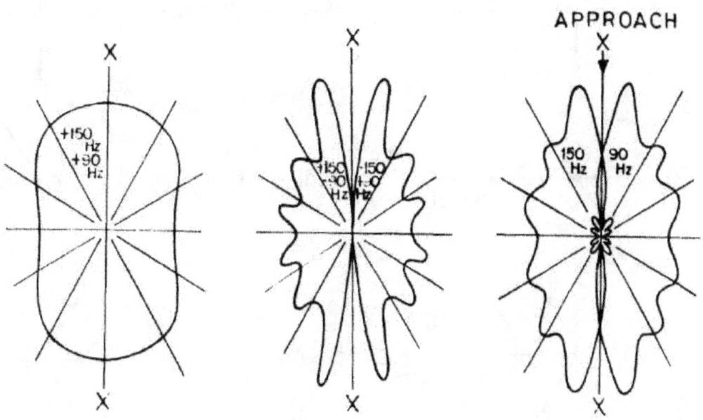

(*a*) Carrier antenna pattern. (*b*) Sideband antenna pattern. (*c*) Combined pattern.

FIGURE 13.18 Radiation patterns of a localizer of ILS.

The central antenna is fed with a carrier modulated by both 90 Hz and 150 Hz signals of equal amplitudes to obtain the same amplitude of sidebands at each frequency as that employed for the other arrays, which are fed only with sideband signals at the frequencies. However, the phase of modulation of each of the sideband antenna arrays is in quadrature with that of the central array, while the sidebands in the two sideband antennas are mutually in anti-phase to each other.

Further, the phase of each of the 90 and 150 Hz modulations applied to one of the sideband antennas is reversed when applying to the other sideband antenna. The phasor diagrams of the resulting modulation at the three antenna arrays are shown in Figure 13.19. When the aircraft is on course, sidebands from both the left and right antennas arrive at the same instant, traveling through an equal phase path and, therefore, cancel each other, as they are mutually in anti-phase for each of the 90 and 150 Hz modulations. At such times, the transmission from the central array is only received by aircraft containing both the 90 and 150 Hz modulations in equal amplitudes. If the aircraft is toward the left of the correct course, the 90 Hz sideband from the left antenna is advanced in phase as the antenna now becomes nearer, while that from the right antenna lags by the same amount. The resultant of the 90 Hz sidebands subtracts from the 90 Hz modulation of the central array and the 90 Hz modulation becomes weaker. The situations are reversed for the sidebands at 150 Hz. In this case the resultant of the 150 Hz sidebands add up to that due to the central array and becomes stronger.

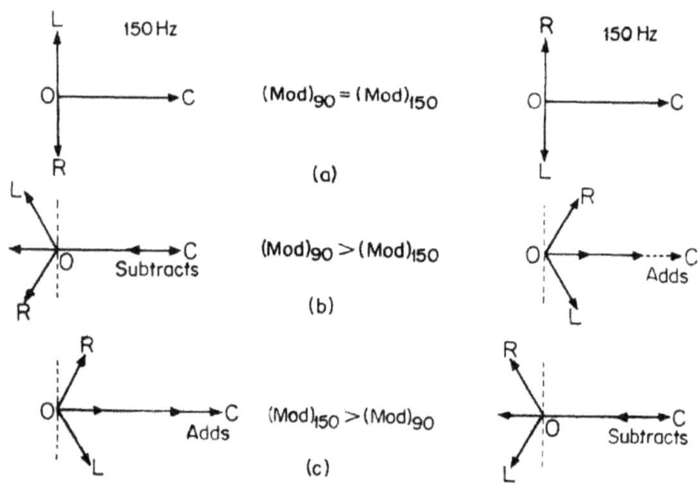

(*a*) On course (*b*) Left of the course (*c*) Right of the course

FIGURE 13.19 Vector diagram of the field due to localizer antennas.

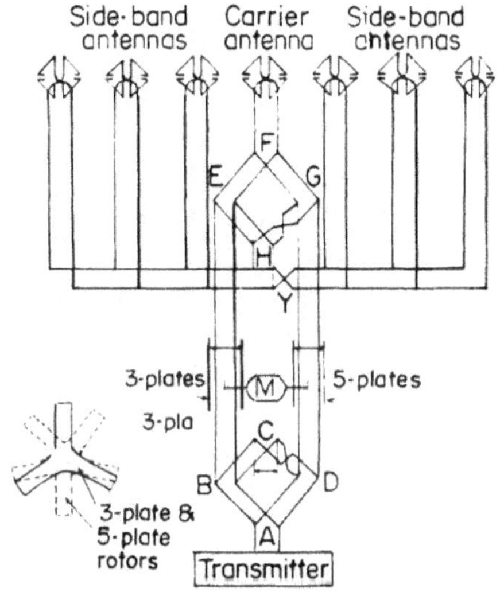

FIGURE 13.20 Localizer antenna configuration.

If the aircraft is to the right of the course, a similar argument shows that the 90 Hz tone will be stronger than the 150 Hz tone. The correct course is thus detected by the equality of the amplitude of the tones at 90 and 150 Hz. In the aircraft receiver, 90 and 150 Hz modulations of the VHF signal from the localizer are then separated by narrow band filters. The filter outputs at 90 and 150 Hz are rectified separately and the resultant DC voltage is substracted as illustrated in the circuit diagram, Figure 13.20. This ensures a null of indicator in the correct course.

Glide-Slope System. The ILS called the Glide-slope system is now in use in airports. It operates in the UHF band 339.3–335 MHz using only two antenna with distinctive elevation and azimuthal patterns as shown in Figure 13.21. Otherwise, the operation is similar to that of a localizer. The patterns are produced by horizontal half-wave dipole antennas placed at different heights from the ground to generate distinctive fringe patterns above the ground using the reflection from the ground, and the lower antenna produces the board beam 1 while the upper antenna produces a number of sharper beams marked 2. Besides these, to concentrate the transmission along the runway, each dipole is backed by a vertical reflecting screen. The variations of the signal strength at the aircraft with elevation angle for the two patterns are shown in Figure 13.21 (*a*). From this figure it is evident that an aircraft will receive a signal with equal depths of the modulation at 90 and 150 Hz if it flies at an elevation along the null between the two finer patterns, at which the broader pattern will have maximum. For higher elevation angles the 90 Hz sideband will be reinforced, while for a lower elevation the sideband for 150 Hz will be strengthened, as is the case during azimuth guidance in a localizer. Because of the use of the null of the finer vertical pattern, the ILS is also called a "null type glide slope system." If the pilot chooses to fly along a course for which the difference in the depth of modulation at 90 and 150 Hz is a constant other than zero (null), the course defined will not be straight, as shown in Figure 13.21 (*a*) but will be a hyperbola around the equisignal course, and this may not touch the runway. To correct this a modifier antenna array has often been recommended, but its utility is not well recognized for which it has been obsolete. The pilot has, therefore, to follows only the equisignal course for elevation as well as azimuth guidance on which the depths of modulation at 90 and 150 Hz are equal and, therefore, produce a null in the indicator. The transmitter and antenna systems are located about 450 m away from the end of the runway.

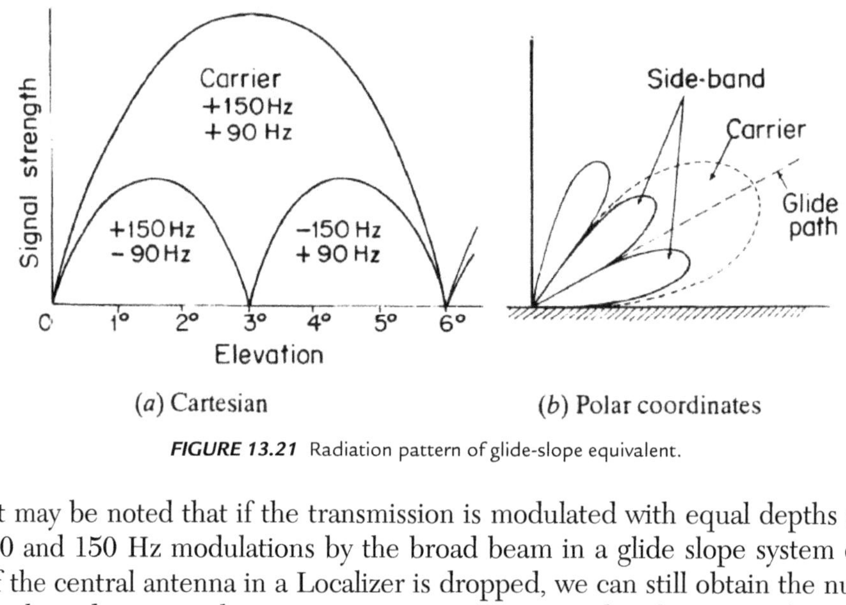

(*a*) Cartesian (*b*) Polar coordinates

FIGURE 13.21 Radiation pattern of glide-slope equivalent.

It may be noted that if the transmission is modulated with equal depths of 90 and 150 Hz modulations by the broad beam in a glide slope system or if the central antenna in a Localizer is dropped, we can still obtain the null in the indicator on the correct course. However, as the phase cancellations of the sidebands occur at the transmission frequency, the signal strength around null will be very weak and may be drowned out in the noise level picked up by the receiver. Introduction of the carrier transmission modulated by 90 and 150 Hz thus ensures a good signal strength well above the noise level even in the position of null, the null being created only after detection of a large signal, modulated with both signals, by employing post detection filters, rectifiers, and subtractors.

13.7 GROUND CONTROLLED APPROACH

13.7.1 Radar Control

The usual technique of the Ground Controlled Approach (GCA) system is to employ a high-precision radar system which may be a Surveillance Radar Element (SRE) or a Precision Approach Radar (PAR). On the display system of these radar systems, the location of an aircraft in the air can be viewed directly with high precision, and this positional information of the aircraft is sent from the ground control operator to the pilot by a radio link for this guidance. The details of these radar systems are discussed in a separate chapter.

Recently, the principles of doppler VOR have been adopted with certain modifications to develop a Doppler direction finder, which is very suitable for a busy modern airport.

13.7.2 Doppler Direction Finder

A Doppler Direction Finder (DDF) is based on the Doppler shift of frequency of the signal from an aircraft received on the ground by a single antenna moving along a horizontal circle. The motion is obtained by a electronic switching of successive antenna elements uniformly spaced along the circle as shown in Figure 13.22. Element 1 is effectively approaching the aircraft when switching from antennas 1 to 2, producing an increase of frequency due to the Doppler effect. Element 5 is receding at the same speed when it is switched to antenna 6, causing a decrease of frequency by f_d due to the Doppler effect. If the electronic switching which may be called electronic commutation occurs at 170 Hz revolutions per second, the frequency of the received signal will be deviated sinusoidally at 170 Hz by an amount f_d due to the Doppler shift of frequency having a phase depending on the direction of arrival of the signal. The deviation will be largest at instants when the vertical plane of travel of the signal is perpendicular to the line joining an antenna element to the center of the antenna circle, while the deviation becomes zero when the line coincides with the vertical plane of travel. The motion of the aircraft toward the direction finder as a whole causes an additional Doppler shift of frequency in each of the antenna elements related to the velocity of approach of the aircraft. To eliminate this part of the Doppler shift, an antenna element is placed at the center, which will pick up only that part of the Doppler shift arising from the aircraft motion. By combining the outputs of the commentated antenna and the central antenna, the Doppler shift due to aircraft motion is eliminated.

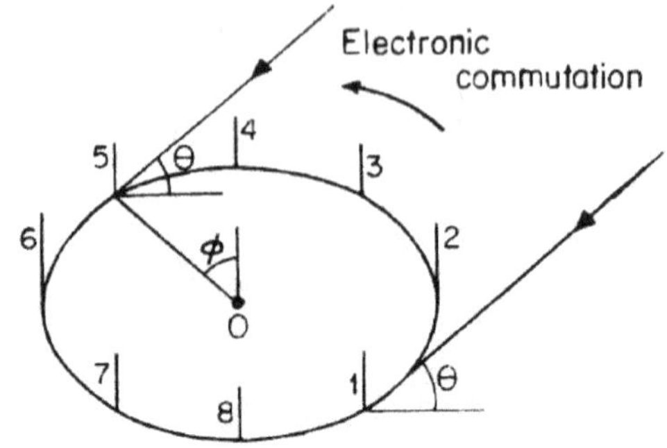

FIGURE 13.22 Electronic commutation in a Doppler direction finder.

A block diagram of the Doppler direction finder is shown in Figure 13.23. The rotating or commutation antenna picks up the signal from the aircraft that is Doppler shifted both due to the commutation and the aircraft motion, and is received by a direction finder double superheterodyne receiver having a first IF of 10.7 Hz and a second IF of 555 kHz. The signal in a reference stationary antenna picks up the same signal Doppler shifted due to aircraft motion only and is brought down in frequency to produce an IF of 455 kHz by a communication receiver meant to receive the communication signal

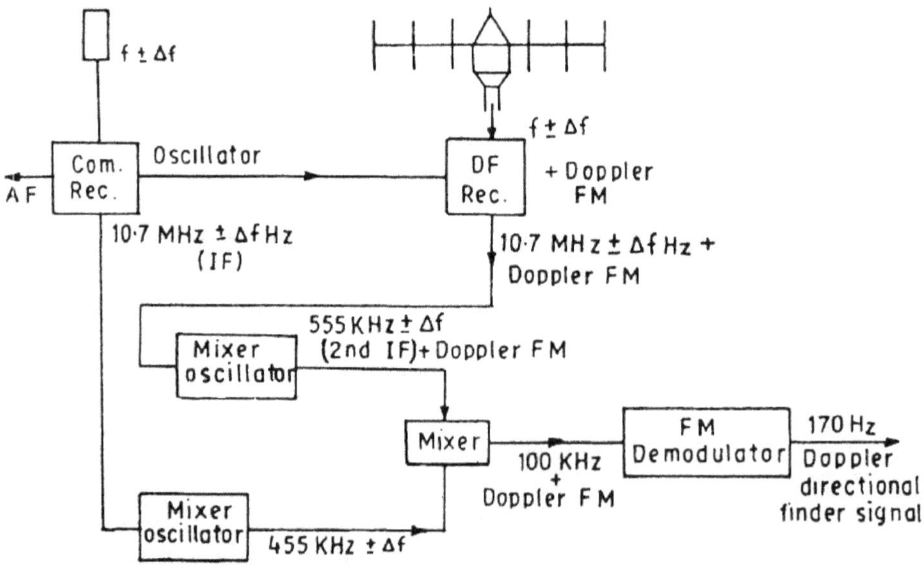

FIGURE 13.23 Doppler direction finder equipment.

from the aircraft. The two IF signals at 555 kHz and 455 kHz from the two receivers are mixed to produce a 100 kHz difference IF signal in which the Doppler shift due to aircraft motion is cancelled, and contains only the 170 Hz frequency modulation due to commutation of the antenna. The phase of this signal bears the bearing information and can be obtained by comparing it with that of a reference phase signal received from the generator employed for electronic commutation. For the purpose a direct reading, an electronic phase meter may be employed as in the case of a VOR.

A Doppler direction finder is designed for operation in the VHF and UHF bands for use at airports as an aid to landing as well as for air traffic control. As in the case of radar control, the operator sends the bearing information obtained from the DDF system to the pilot for his guidance in landing.

13.8 RADIO ALTIMETER

13.8.1 Triangular Modulation

A radio altimeter is designed to obtain an indication of the altitude of an aircraft during the landing phase as well as during higher altitude flights, and is based on a frequency modulated CW radar (FM-CW). A block diagram of the altimeter is shown in Figure 13.24. The modulating waveform may be steadily rising frequency [Figure 13.25 (*a*)] or triangular as shown in Figure 13.25 (*b*). Due to the frequency modulation, the transmitted frequency will follow the modulation as shown in Figure 13.25 (*c*). The transmitted frequency at the instant *O*, when the modulation waveform passes through a zero, will be the carrier frequency f_c. This transmission will be received back after reflection at the ground after a delay

$$\Delta t = \frac{2h}{c} \tag{13.6}$$

where h is the altitude of the aircraft and c is the velocity of propagation of the radiowave, 3×10^8 m/s. During this period Δt the transmitted frequency is changed to a new value f'_c differing from f_c by $f_b = f'_c - f_c$. From the similar triangles *OPQ* and *ONM*, we have

$$\frac{PQ}{OP} = \frac{NM}{ON}$$

or
$$\frac{f_b}{\Delta t} = \frac{\Delta f / 2}{1/4\, f_m}$$

where Δf is the peak-to-peak frequency deviation and f_m is the frequency of the triangular modular modulation signal.

or
$$f_b = \frac{\Delta f}{2} . 4 f_m \times \Delta t$$

$$= \frac{4h\, f_m\, \Delta f}{c} \qquad (13.7)$$

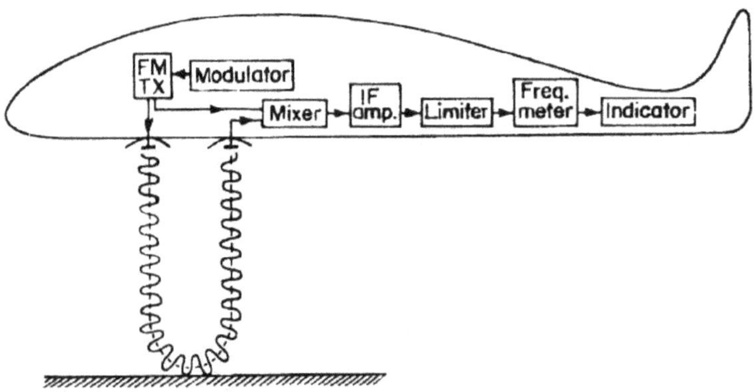

FIGURE 13.24 Radio altimeter equipment.

It may be noted that the pattern of the received frequency is displaced as a whole from the transmitted one along the time axis by an amount t due to the round trip delay between the aircraft and the ground. The beat frequency f_b is measured by the counter and the altitude obtained from

$$h = \frac{Cf_b}{4 f_m\, \Delta f} \qquad (13.8)$$

Effect of the Doppler Shift. If the aircraft starts descending rapidly due to any failure of the flight equipment of the aircraft, then in addition to the delay Δt due to the altitude, the pattern of frequency variation of the received signal will be shifted up along the frequency axis due to the sizable Doppler shift for the rapid motion, as shown if Figure 13.25 (d), and this can be utilized to detect the velocity of descent. Due to the Doppler effect the beat frequency during rising portions of the received frequency pattern will be

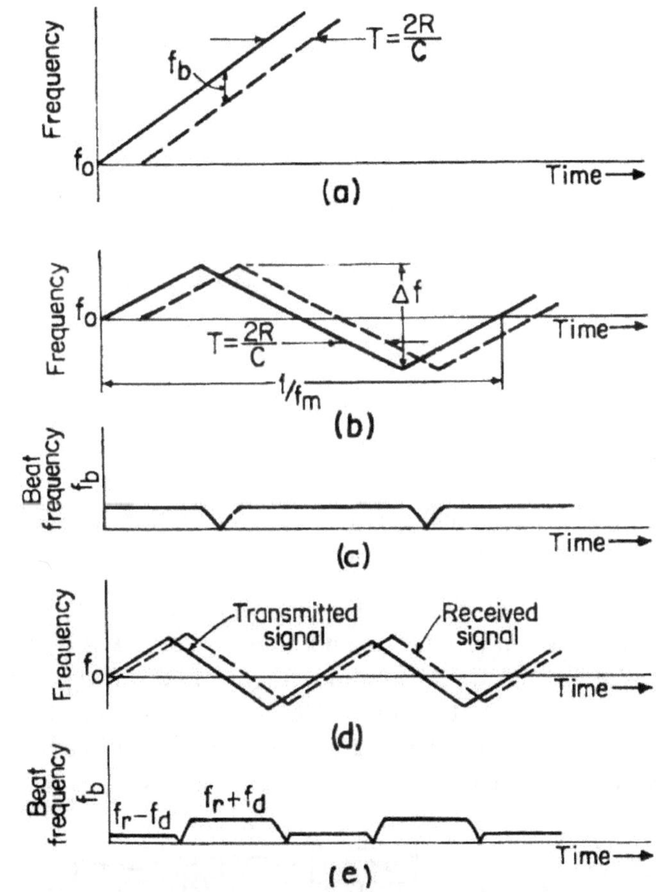

FIGURE 13.25 Radio altimeter waveforms. (*a*) Linearly rising transmitter frequency (*b*) Triangular frequency modulation (*c*) Beat frequency waveform with triangular modulation (*d*) Doppler shifted waveform of the modulation in the receiving signal (*e*) Beat frequency waveform with Doppler shift.

lowered by an amount f_d, while during the falling portions it will be increased by the same amount f_d, as indicated in Figure 13.25 (*e*). Accordingly, the received beat frequency will be $f_b(np) = f_b + f_d$ and f_b (down) $= f_b + f_d$ during the rising and falling portions respectively. By a separate counting facility for the rising and falling parts, both f_b (up) and f_b (down) can be measured from which the Doppler shift as well as the altitude can be obtained electronically utilizing the relations:

$$f_b = \frac{1}{2}[f_b(\text{up}) + f_b(\text{down})] \text{ and } f_d = \frac{1}{2}[f_b(\text{down}) - f_b(\text{up})] \quad (13.9)$$

To deduce the velocity of descent from the measured Doppler shift f_d, we note that the round trip phase path is given by

$$\psi = \frac{2\pi}{\lambda} \times 2h$$

and
$$2\pi f_d = \frac{d\psi}{dt} = \frac{2\pi}{\lambda} \cdot 2\frac{dh}{dt} = \frac{2\pi}{\lambda} \cdot 2v_r \tag{13.10}$$

where v_r is the velocity of descent of the aircraft.

Therefore,
$$f_d = \frac{2v_r}{\lambda}$$

or
$$v_r = \frac{\lambda f_d}{2} \tag{13.11}$$

Thus v_r is proportional to f_d and an indicator for f_d may be directly calibrated to indicate the velocity of descent.

Sensitivity. Leakage of transmitter powers to the receiver limits the sensitivity and hence the altitude range of an FM-CW radio altimeter. The paths of leakage are illustrated in Figure 13.26. The paths include (1) direct leakage from the transmitter to the receiver, (2) leakage between the transmitting and receiving antennas, and (3) multiple hop leakage, which is, however, very weak.

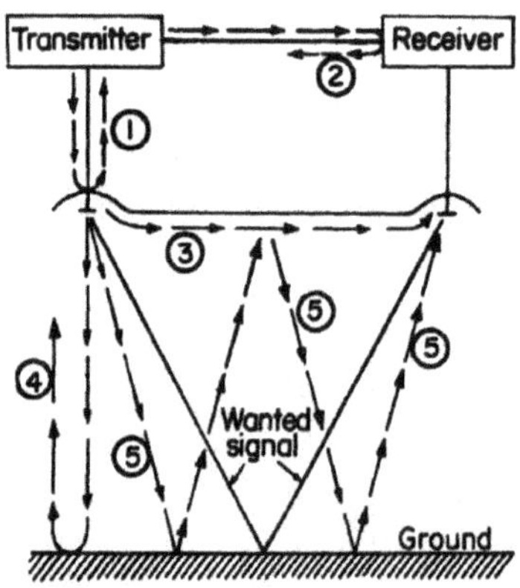

FIGURE 13.26 Leakage paths in a radio altimeter.

Range error. In a radar altimeter the error in range may occur due to the limitation of the beat frequency counter which can count the total number of cycles in discrete numbers but is unable to measure fractions of a cycle, leading to what is called quantization error in range. Besides this, range error may also occur due to the finite stability of the transmission frequency.

The quantization error in range can be estimated as follows:

The average number of cycles N of the beat frequency f_b in one period of the modulation cycle f_m is given by $n\bar{f}_b / f_m$ where the bar over f_b denotes time average. Equation (13.7) may be rewritten as

$$h = \frac{c\bar{f}_b}{4 fm\, \Delta f} = \frac{cN}{4\Delta f} \tag{13.11a}$$

Since the output of the frequency counter N is an integer, the altitude error or range error will be an integral multiple of $\frac{c}{4\Delta f}$ giving rise to a quantization error equal to

$$\Delta h = \frac{c}{4\Delta f} = \frac{75}{\Delta f(\text{MHz})}\, m \tag{13.11b}$$

This equation indicates that the range error Δh due to quantization is independent of the range of altitude and, therefore, it is also called a fixed error or step error.

13.8.2 FM-CW Altimeter with Sinusoidal Modulation

The problem of transmitter leakage can be minimized by employing a sinusoidal frequency modulation in place of the triangular one and processing a certain higher order sideband of the resulting frequency modulated signal. With triangular modulation the number of such higher order FM sidebands with appreciable amplitude will be too many, and restricting to a certain sideband will cause a drastic reduction of signal-to-noise ratio. With sinusoidal modulation, on the other hand, such a reduction in signal-to-noise ratio for a higher order sideband is not significant, but enables a reduction of transmitter leakage.

With sinusoidal modulation the waveform of the transmitted and received frequencies in the absence of Doppler shift are shown in Figure 13.27. The waveform of the frequency, f_r, of the received signal is effectively delayed with respect to that of the transmitted one by an amount $\Delta t = 2h/c$ due to the round trip travel of the radio waveform on the ground.

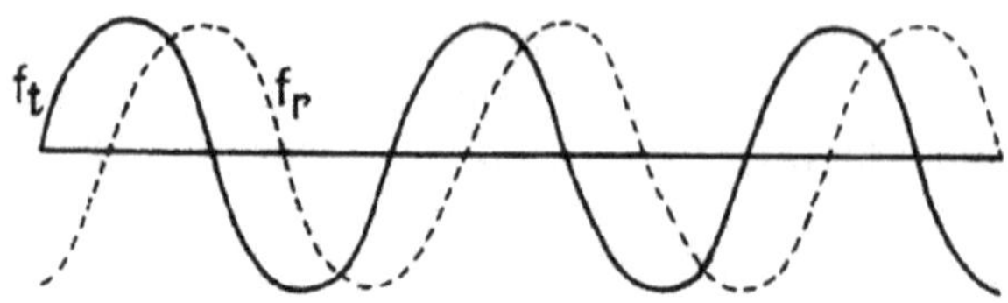

FIGURE 13.27 Sinusoidal modulation waveforms in a radio altimeter.

If the transmitted voltage waveform is represented by

$$e_t = E_0 \sin\left(2\pi\, f_0 t + \frac{\Delta f}{2}\frac{1}{f_m}\sin 2\pi\, f_m t\right)$$

$$= E_0 \sin A \qquad (13.12)$$

where $\Delta f/2$ is the frequency deviation and f_m is the modulation frequency. The voltage waveform of the received signal may be represented by

$$e_r = k\, E_0 \sin\left[2\pi f_0\,(t-\Delta t) + \frac{\Delta f}{2}\frac{1}{f_m}\sin 2\pi\, f_m (t-\Delta t)\right]$$

$$= k\, E_0 \sin B \qquad (13.13)$$

At the aircraft, the transmitted and received signals are heterodyned by a mixer, which effectively multiplies the two signals and selects the difference frequency signals by a filter at its output.

The multiplied output is given by

$$e_m = e_t\, e_r = k\, E_0^2 \sin A \sin B$$

$$= \frac{k\, E_0^2}{2}\, 2 \sin A \sin B$$

$$= \frac{k\, E_0^2}{2}\left[\cos(A-B) - \cos(A+B)\right]$$

The first term within the bracket corresponds to the difference frequency signal, and the filter output at the frequency is represented by

$$e_b = k_f k \frac{E_0^2}{2} \cos(A-B)$$

where k_f is the insertion loss of the filter.

Putting the values of A and B, the difference frequency signal is given by

$$e_b = k_f k\, E_0^{\,2} \cos\left[2\pi f_0 t + \frac{\Delta f}{2f_m}\sin 2\pi f_m t - 2\pi f_0 (t-\Delta t) - \frac{\Delta f}{2f_m}\sin 2\pi f_m (t-\Delta t)\right]$$

$$= \frac{k_f k\, E_0^{\,2}}{2}\cos\left[2\pi f_0\,\Delta t + \frac{\Delta f}{2f_m}.\,2\cos 2\pi f_m\left(t-\frac{\Delta t}{2}\right)\right]\times \sin\frac{2\pi f_m\,\Delta t}{2}$$

$$= \frac{k_f k\, E_0^{\,2}}{2}\cos\left[2\pi f_0\,\Delta t + \frac{\Delta f}{f_m}\sin \pi f_m\,\Delta t\cos 2\pi f_m\,(t-\Delta t)\right]$$

If $\Delta t << \dfrac{1}{f_m}$, $\sin \pi f_m t \cong \pi f_m\,\Delta t$ and $\dfrac{\Delta t}{f_m}\sin \pi f_m t = \pi \Delta f \Delta t$.

Therefore, we get

$$e_b = \frac{k_f k\, E_0^{\,2}}{2}\cos\left[2\pi f_0\,\Delta t + \pi\Delta f\,\Delta t\cos(2\pi f_m t - \pi f_m\,\Delta t)\right]$$

The instantaneous frequency of this different frequency signal is the beat frequency, f_b, which may be obtained by differentiating the argument, $A - B$ with respect to time. Accordingly, the beat frequency f_b is given by

$$2\pi f_b = \frac{d}{dt}(A-B) = \frac{d}{dt}\left[2\pi f_0\,\Delta t + \pi\Delta f\,\Delta t\cos(2\pi f_m t - \pi f_m\,\Delta t)\right]$$

$$= -\pi\Delta f\,\Delta t.\,2\pi f_m\sin(2\pi f_m t - \pi f_m\,\Delta t)$$

$$\therefore\quad f_b = -\frac{\pi\Delta f\,\Delta t}{2\pi}2\pi f_m\sin(2\pi f_m t - \pi f_m\,\Delta t)$$

$$= \pi\Delta f\,\Delta t f_m\sin(2\pi f_m t - \pi f_m\,\Delta t + \pi) \qquad (13.14)$$

The minus sign being considered equivalent to a phase shift of π radians. The beat frequency is thus varying sinusoidally at the modulating frequency f_m, becoming positive and negative on alternate half cycles. The negative signs, again, may be considered to be as due to a reversal of phase of the beat frequency waveform given by equation (13.14). If the subsequent electronic system measures the positive and negative half cycles separately ignoring the sign and taking the average, then we get an average beat frequency given by

$$f_b(av) = \frac{1}{T/2}\int_0^{T/2} f_b\,dt$$

Nothing that $T = \dfrac{1}{f_m}$, we have

$$f_b(av) = \frac{1}{1/2f_m}\int_0^{1/2f_m} f_b\, dt = 2f_m \int_0^{1/2f_m} f_b\, dt$$

$$= 2f_m \int_0^{1/2f_m} \pi\Delta f\ \Delta t\ f_m \sin(2\pi f_m t - \pi f_m\ \Delta t)\, dt \qquad \text{(from Equation 4.14)}$$

$$= 2f_m \int_0^{1/2f_m} \pi\Delta f\ \Delta t f_m (\sin 2\pi f_m t \cos \pi f_m\ \Delta t - \cos 2\pi f_m t \sin \pi f_m\ \Delta t)\, dt$$

and $\Delta t << \dfrac{1}{f_m}$ or $f_m\ \Delta t << 1$

$$\cos \pi f_m\ \Delta t \cong 1 \quad \text{and} \quad \sin \pi f_m\ \Delta t \cong \pi f_m\ \Delta t$$

$$\therefore\ f_b(av) = 2\pi\Delta f\ \Delta t\ f_m \left[-\frac{\cos 2\pi f_m t}{2\pi f_m} - \pi f_m\ \Delta t \frac{\sin 2\pi f_m t}{2\pi f_m}\right]_0^{1/2f_m}$$

$$= 2f_m\ \Delta f\ \Delta t = 2f_m\ \Delta f . \frac{2h}{c}$$

$$= \frac{4hf_m\Delta f}{c} \tag{13.15}$$

Equation (13.15) for f_b (av) is identical with that for f_b, in equation (13.7), for triangular modulation. Thus, a sinusoidal modulation produces the same average beat frequency as that for a triangular modulation, while at the same time reducing complexity of the hardware needed in a sinusoidal modulation, and also reducing the total bandwidths of the significant FM sidebands. The Doppler shift of the received frequency due to a rapid rate of descent of an aircraft can be determined by taking half of the difference of the beat frequencies during the positive and negative half cycles of the beat frequency waveform, as in the case of triangular modulation. However, in this case, a more sensitive indication of the Doppler shift, much less affected by a carrier leakage, can be obtained by processing of the sidebands of a higher harmonic of the modulation frequency. This can be shown by the following analysis:

If the sample of the transmitted waveform is fed to the mixer represented by

$$e_t = E_0 \sin\left(2\pi f_0 t + \frac{\Delta t}{2f_m} \sin 2\pi f_m t\right)$$

and the received waveform will be delayed by

$$\Delta t = \frac{2h}{c} \text{ and is given by}$$

$$e_r = E_r \sin\left[2\pi f_0 (t - \Delta t) + \frac{\Delta f}{2 f_m} \sin(2\pi f_m t - \Delta t)\right]$$

then the Fourier expansion of the different frequency signal with E_0 adjusted equal to E_r may be written as

$$\begin{aligned} v_D &= J_0(D)\cos(2\pi f_d t - \phi_0) + 2J_1(D)\sin(2\pi f_d t - \phi_0) \times \cos(2\pi f_m t - \phi_m) \\ &- 2J_2(D)\cos(2\pi f_d t - \phi_0)\cos 2(2\pi f_m t - \phi_m) \\ &- 2J_3(D)\sin(2\pi f_d t - \phi_0)\cos 3(2\pi f_m t - \phi_m) \\ &+ 2J_4(D)\cos(2\pi f_d t - \phi_0)\cos 4(2\pi f_m t - \phi_m) \\ &+ 2J_5(D)\ldots\ldots.. \end{aligned}$$

where J_0, J_1, J_2 etc., = Bessel function of first and order 0, 1, 2, and so forth, respectively.

$D = (\Delta f/f_m)$

$f_d = 2v_r f_o/c$ = Doppler shift of frequency

v_r =velocity of the target relative to the radar

ϕ_o = phase shift $\cong 2\pi f_o h/c$

ϕ_m = phase shift $\cong 2\pi f_m h/c$.

The spectrum of the difference in a frequency signal represented by the Fourier expansion is shown in Figure 13.28 [*a* (*i*)]. The difference frequency signal contains a Doppler-frequency component of amplitude $J_o(D)$ and a series of carriers at f_m, $2f_m$, $3f_m$, and so on, each one suppressed carrier double-sideband modulated by the Doppler frequency with amplitudes proportional to the respective Bessel functions $J_1(D)$, $J_2(D)$, $J_3(D)$, and so forth, shown in Figure 13.28 [*a* (*ii*)].

For a stationary target $v_r = 0$ resulting in $f_D = 0$ when the Doppler sidebands degenerate into the frequency modulation sidebands due to the modulation at f_m, having amplitude proportional to $J_n(D)$. The magnitude of a Bessel coefficient increases to a maximum at zero range, when all other

Bessel coefficients fall to zero. Therefore, the carrier leakage which can be looked upon as equivalent to reflection from a target at zero range dominates if we consider the DC term with the Bessel coefficient $J_0(D)$. This DC term is in fact also modulated by noise and vibration components producing what is called microphonics, and limits the detectibility of target echo in radio altimeters.

At greater range corresponding to normal altitudes above the ground the $J_0(D)$ term decreases causing the echo signal to decrease in addition to increased attenuation expected at the greater range involved. The other terms containing $J_1(D)$, $J_2(D)$, $J_3(D)$, and so forth, on the other hand, approach zero at zero range and show maxima at certain optimum ranges. The microphonics due to carrier leakage can thus be avoided by selecting one or more of these other harmonic terms. However, each of these terms also becomes zero at certain ranges which are spaced by one RF wavelength, the amplitude of the modulation frequency harmonics for a stationary target with zero Doppler shift being proportional to $J_n(D) \sin \phi_0$ or $J_n(D) \cos \phi_0$. Such closely spaced zeros of the terms, therefore, make it difficult to use these higher order terms in radio altimetry. In the presence of Doppler shift, on the other hand, the terms no longer exhibit such zeros at closely spaced ranges but contain the Doppler frequency sidebands with ψ_0 entering as a phase term of the Doppler frequency component. The higher order terms at the harmonics of f_m are, therefore, useful in detecting the Doppler component free from the effect of microphonics prevalent in the term containing $J_0(D)$. Further, the parameter D which depends on the modulation parameters Δf and f_m can be adjusted to have a maximum of $J_n(D)$ at the desired target range. The loss of echo signal energy for selection of a harmonic term containing $J_n(D)$ is compensated largely by the reduction of microphonics contained in the DC term containing $J_0(D)$. The loss for 3rd harmonic operation is reported to be 4 to 12 dB. The amplitude of the Doppler frequency output, which is proportional *to* $J_3(D)$, varies with distance as shown in Figure 13.28 [*a* (*ii*)]. The nulls in the echo signal indicated by a zero of $J_3(D)$ may be adjusted by an adjustment of D to suppress the echoes from certain ranges. A typical setup of aircraft equipment for the third harmonic operation of an FM-CW altimeter for detecting the velocity of descent is shown in Figure 13.28 (*b*). The frequency modulator provides the sinusoidal modulation of the transmitter, which is fed to the antenna through the circular. The received echo after mixing with output of the filter at the third harmonic is heterodyned with the fourth harmonic of modulation frequency and low pass filters. A plot of $J_3(D)$ as a function of distance is shown in Figure 13.29.

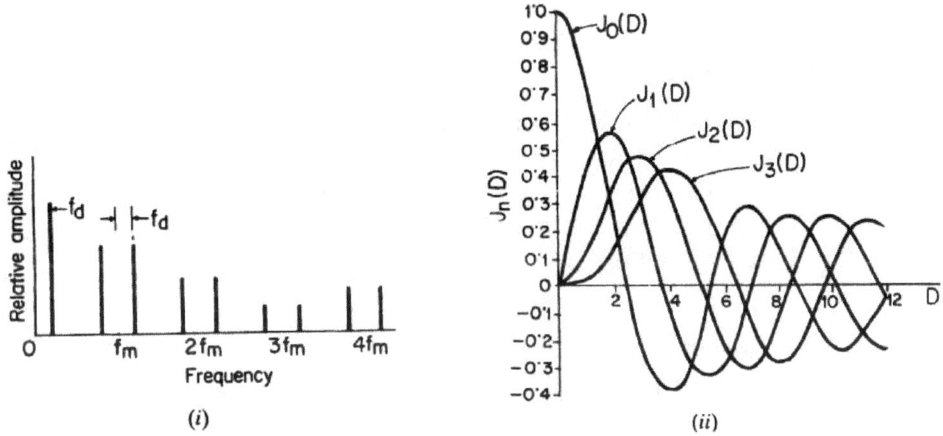

FIGURE 13.28 *(a)* Spectrum of beat frequency signal with sinusoidal modulation in the radio altimeter *(i)* Graphical presentation of Bessel functions of order 0, 1, 2, and 3.

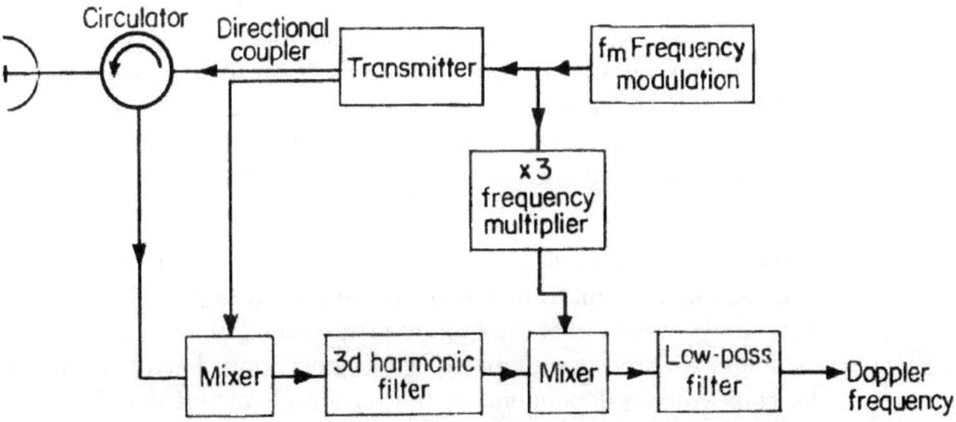

FIGURE 13.28 *(b)* Radio altimeter using sinusoidal modulation and third harmonic extraction.

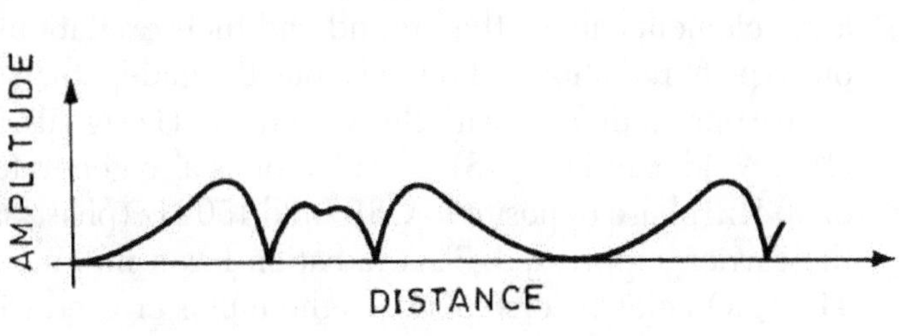

FIGURE 13.29 $J_3(D)$ Plotted as a function of distance.

13.9 MICROWAVE LANDING SYSTEM

13.9.1 Introduction

The Instrument Landing System (ILS) is for providing the pilot of an aircraft with steering information to make a controlled and accurate approach to a runway even under adverse weather conditions. This is accomplished by the provision of azimuth and elevation guidance and also distance from threshold information.

13.9.2 Limitation of ILS

The conventional ILS has been in operation for more than four decades, yet it has some technical and operational limitations.

1. **Terrain Requirements.** Localizer problems arise in ILS due to vertically oriented obstructions which are mostly manmade. Reduction in localizer course bends is achieved by restricting the radiations in a particular direction (*s*). Of course, such a scheme may not be practicable always, as preventing the radiation along possible reflections could leave certain directions without significant localizer coverage. Further, at most of the airports the extended center line of the runway is not available for the localizer sighting. In such cases it may become necessary to go in for offset installations, which may result in an increase in the landing minimums.

   In the case of a glideslope array, the reflections for the ground must be taken into account as they enter the approach region. The design of these arrays assume that the ground plane is flat, infinite, and perfectly conducting. Three basic types of glideslope arrays have been in use, the choice being dependent on the terrain characteristics. These are (*a*) Null Reference, (*b*) Sideband Reference, and (*c*) Capture Effect Arrays. The main difference in these three types lies in the height of the array elements above the ground and their excitations. In all the cases, one type of radiation is directed along the glidepath, which consists of an RF carrier modulated with the 90 and 150 Hz signals. This is referred as CSB. A sideband only (SBO) radiation is also generated, which consists of 90 Hz (phase opposite to CSB) and 150 Hz (phase same as CSB) with the carrier suppressed. This radiation has a null along the glide path. The SBO fields reverse phase as the null is crossed with the sum of the CSB and SBO fields producing the carrier variation.

Ground Plane Requirement. All the previous arrays have a minimum ground plane requirement. This is because the sideslopes that are directed downward must be controlled so that no spurious image radiation enters the flight path region.

It can be shown that the reflecting ground required must satisfy the relations

$$x \geq \lambda/2(1 - \cos \theta) \qquad (i)$$

and

$$x > h/\tan \theta \qquad (ii)$$

Here x is the length of the reflecting ground required, θ is the glide angle, and λ is the wavelength. Since the Glidepath is required to furnish elevation guidance from 0.18 to 6 degrees, applications of the above formulae give the figures of 4663 m and 732 m respectively. On many occasions leveling a large area involves huge expenditure and also may not be practicable. Considerable work has been done to design glidepath antennas which provide a satisfactory course even with non-ideal terrain. The End fire arrays and the Flush mounted types are examples of such designs.

2. **Traffic Handling Capability.** The present ILS is not capable of providing high sustained traffic rates when aircraft have varying approach speeds. At present aircraft are required to follow a common path just before landing. This common path includes (*a*) the final approach during which alignment with extended center line of the runway is established, and (*b*) rollout along the runway during which speed is decreased until a turnoff is safely made. During these operations, neither vertical nor lateral separation of aircraft can be provided and hence require sufficient longitudinal separation. But in an environment of several aircraft approaching at different speeds, their separation will have to be determined in each case to avoid overtaking and/or close encounters. The larger the velocity spread among the types of aircraft using a runway, the larger must be the average spacing between successive aircraft.

   When the traffic into the terminal airspace is high, then the total traffic rate imposed by ILS would necessiate in-flight holding of aircraft until they can be accommodated. This process is very expensive, particularly for jet aircraft which consume large amount of fuel per unit time in low level flight. The chief limitation of ILS is that the aircraft are required to follow a single flight path for a significant length before touchdown. Flexibility in landing is attained by permitting the aircraft to approach within an extended sector without restricting the flight paths to be straight lines. Further the size of the antenna (about 20 m wide and 3.5 m high for the localizer and 14 m high for the glidepath) makes it unsuitable for small landing sites as installation of such big structures poses problems. Thus, the search for a new system that will overcome the shortcomings of ILS started in the early sixties, and that led to the development of the MLS (microwave landing system).

13.9.3 MLS Development

In the early 1960s, a number of systems were proposed by different countries/agencies as replacement to current ILS. In 1967, the Radio Technical Commission for Aeronautics (RTCA) formed a special committee for evaluating the various systems, and came up with a single proposal for endorsement by International Civil Aviation Authority (ICAO).

The SC 117 working group evaluated 23 different systems and recommended the Time Reference Scanning Beam developed by the United States for worldwide use. MLS has all the features to accommodate the explosive growth of aviation in the next century. These include:

1. Availability of 200 channels
2. Continuous range and angle information
3. Improved signal quality
4. Reduced requirement of siting and environment
5. Wider guidance coverage
6. Multilevel system design using advanced technology

13.9.4 Types of Antennas in MLS

Two basic types of antennas are used in the MLS system, namely, (*a*) the Azimuth type, and (*b*) the Elevation Type. The basic line array design is used for both; the only difference is in the radiating apertures. The radiating aperture for the elevation scanning antenna is a vertical array of dipoles, while that for the azimuth scanner is an array of vertical slotted waveguide radiating elements. Such an approach will result in a high degree of antenna design and component communality, which in turn will result in reduced system costs.

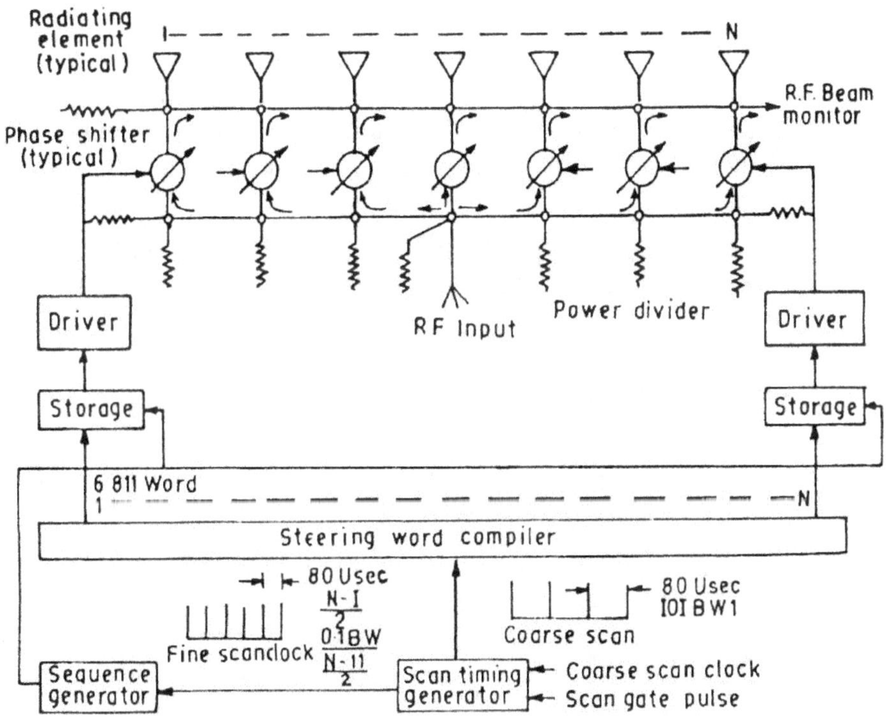

FIGURE 13.30 Antenna for a microwave landing system.

This array is well classified into four major sections: (*a*) Radiating Elements, (*b*) Phase Shifters, (*c*) R.F. Power Dividers, and (*d*) Beam Steering Logic. Phase shifters are electrically controlled, ferrite latching types with an integral solid state driver. All transmissions are FM/CW and are radiated only when the antenna is scanning. The beam steering logic consists of a scan timing generator, a sequence generator, and steering word compiler with storage units. Upon receipt of a scan gate pulse from the central timing generator, the scan timing generator supplies a series of coarse scan clock pulses to a steering word compiler which computes a new set of six bit steering words for each phase shifter in the array. It also produces a series of fine scan clock pulses for each coarse scan clock pulse, which are then sent to the sequence generator. The storage units are switched on two at a time at the fine scan clock rate by the sequence generator. As only two phase shifters are reset at any one time, the switching transient effect on the antenna pattern can be ignored. The MLS antenna structure is small and will not pose any installation problem even at small landing strips.

13.9.5 Principle of MLS

The MLS uses two narrow beams which are scanned to and fro in the azimuth and elevation sectors. The various azimuth and elevation scanning beam signals are time multiplexed into an allotted time frame. Each ground element transmits the beam sequentially at an appropriate time in the scan format.

For a given scanning speed and pause time, the angular position of the aircraft, θ, in degrees, can be calculated from the equation (shown in Figure 13.31),

$$\theta = V(t_0 - t)/2$$

where, V = angular speed of the scanning beam (0.02 degree/microsec)

t = actual time interval between to and fro scans

t_0 = value of t in microsec for θ^0 (4800 microsec for AZ and 3350 microsec for EL).

This type of MLS is often referred to as Time Reference Scanning Beam (TRSB) MLS.

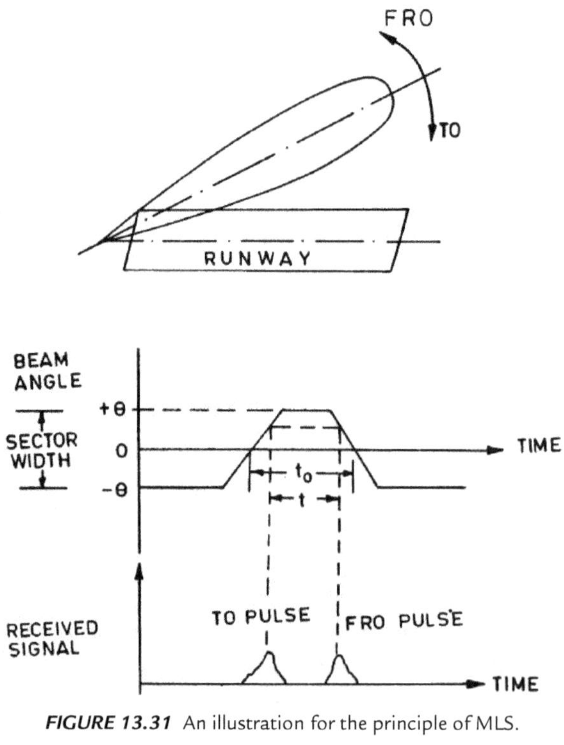

FIGURE 13.31 An illustration for the principle of MLS.

13.9.6 System Composition

A block schematic of the system composition is given in Figure 13.32. The MLS consists of (*a*) Azimuth Equipment providing azimuth information, (*b*) Elevation Equipment providing elevation guidance, (*c*) Back Azimuth providing azimuth angle information to aircraft that have missed the approach, and (*d*) Flare Equipment. The approach taken is to generate a flare trajectory on line as a function of the glide slope angle, the desired touchdown flight path angle, and touchdown point so that when the glide slope is steeper, the flare initiation altitude is higher (*e*) DME (*P*), which provides distance information in the approach, landing, and missed approach operations.

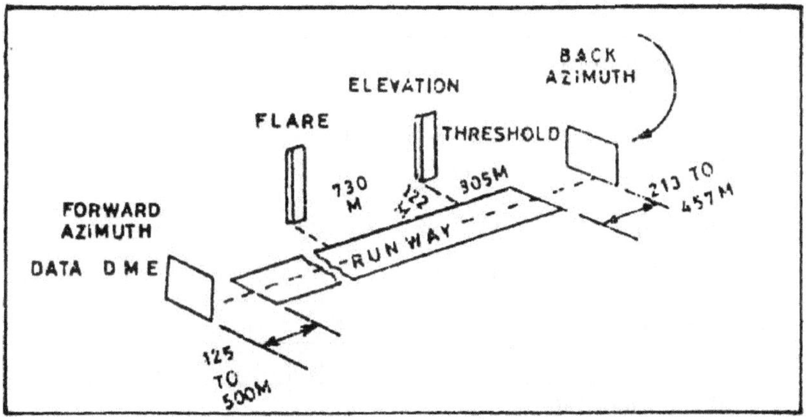

FIGURE 13.32 A block diagram of MLS composition.

MLS Coverage. The MLS coverage requirements are shown in Figure 13.33. This coverage provides three-dimensional data in azimuth, elevation, and range anywhere within the approach and landing coverage area. The wide angle proportional guidance allows aircraft to acquire the MLS signals and verify that they are correct before they turn on to final approach. The wide angle approach is also useful during VER operations to preclude erroneous approaches to wrong runways. Moreover, the wide angle guidance allows aircraft to better anticipate the turn on to final approach and thus reduce overshoots on final approach. Unlike the 1LS, the landing process consists of curved paths, vertical, curved, or segmented guidance, and transition to final center line approach.

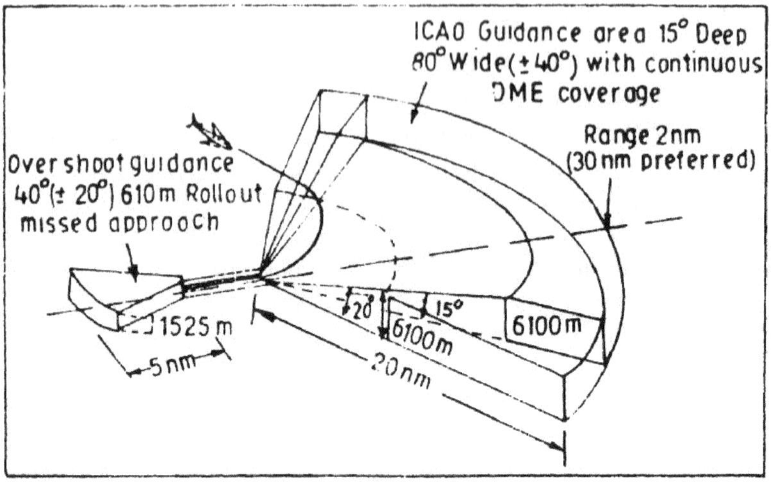

FIGURE 13.33 MLS coverage.

MLS Characteristics. The MLS characteristics are summarized in Table 13.1. The prototype specification is given in Table 13.2.

TABLE 13.1
MLS Overall Characteristics

Frequency:	5031 to 5095.7 MHz.
	200 Channels spaced 300 kHz apart.
Information Rate:	
(*i*) Azimuth	13 Hz for wide coverage (± 40 deg.)
(*ii*) Elevation	39 Hz for narrow coverage (± 10 deg.)
Information Coding:	
(*i*) Angular information	Pulse space coding with a time reference scanning beam.
(*ii*) Permeable and Data	Biphase coding in DPSK.
Polarization	Vertical
Coverage deg.	
(*i*) Azimuth	± 40
	± 60 (for wide aperture systems)
(*ii*) Elevation	0.9 to 15 deg
Distance	20 NM
	16

TABLE 13.2
Levels of MLS Prototype Specification

	Prototype			Production	
	SC	BNA	BWA		
Antenna Beamwidth (deg.)					
(*i*) AZ	3	2	1	2	1
(*ii*) EL	2	1.5	1	1.5	1
Coverage (deg.)					
(*i*) AZ	± 10 (P) ± 40 (CL)	± 40 (P)	± 60 (P)	± 10 (P) ± 60	± 40 (P)
(*ii*) EL	1 to 15	1 to 15	1 to 15	1 to 15	
Transmitter	Solid state	TWT	TWT	Solid state	
Antenna Tech.	PA/MO	MO	PA	PA	
DME/P	Optional	Yes	Yes	Yes	

Notes:
SC = small community,
BNA = basic (narrow aperture),
BWA = basic (wide aperture),
P = proportional guidance,
CL = clearance signal,
PA = phased Array,
MO = Microwave optics.

13.10 ADVANTAGES OF MLS

The MLS has the potential to provide significant fuel and time savings. It has all the essential features to support the goals of safety, capacity, and economy in the approach and landing functions. The TRSB MLS appears to be adequate to cater to the needs of aviation well into the twenty-first century.

WORKED OUT PROBLEMS

1. (*i*) For a FM-CW radio altimeter operating at 4 GHz, calculate the beat frequency f_b between the transmitted and received signals when $\Delta f = 30$ kHz, $R = 2.5$ km and $f_m = 10$ kHz.

$$f_b = \frac{4Rf_m \, \Delta f}{c} \tag{1}$$

$$= \frac{4 \times 2.5 \times 10^3 \times 10^4 \times 3 \times 10^4}{3 \times 10^8}$$

$$= \frac{10 \times 10^{11}}{10^8} = \frac{10^{12}}{10^8} = 10 \text{ kHz.}$$

2. (*ii*) Calculate the frequency range of f_b for R varying between 10 km to 250 m.

For $R = 10$ km $= 4 \times 2.5$ km

$f_v = 4 \times 10$ kHz $= 40$ kHz, using the previous result in (1)

While for $R = 250 \text{ m} = \frac{1}{10} \times 2.5 \text{ km}$

$f_v = \frac{1}{10} \times 10 \text{ kHz} = 1 \text{ kHz}$, also using the previous result in (1).

Thus, the frequency range of f_b will be 1 – 40 kHz for ranges varying from 250 m to 10 km.

(*iii*) If the frequency stability of the transmitter is 1 part in 10^6 calculate the error in range for the 250 m and 10 km.

For $R = 250\ m\ f_b = 1$ kHz $= 1000$ Hz

Frequency of the carrier being 4 GHz $= 4 \times 10^6$ Hz,

the fluctuation in carrier frequency for the stability of 1 part in 10^6 will be $4 \times 10^6 \times 10^{-6}$ Hz $= 4$ Hz

error *in* $f_b = 4$ Hz in 1000 Hz

$= 1$ part in 250

as the range $R \;\alpha\; f_b$, error in range ΔR will also be 1 part in 250

Likewise, for the range of 10 km.

for which $f_b = 40$ kHz $= 4 \times 10^4$ Hz, the error in range is given by

$\Delta R = 4$ Hz in 4×10^4 Hz $= 1$ part in 10^4 (4 Hz is the fluctuation in carrier as deduced previously)

(*iv*) Calculate the percentage fixed error in range due to quantization, not dependent on range but dependent on the frequency swing Δf of the transmitter

The fixed error $\Delta R_f = \dfrac{c}{4\Delta f} = \dfrac{75}{\Delta f(\text{MHz})}$

For $\Delta f = 30$ kHz $\Delta R_f = \dfrac{75}{.03} = 2.5$ km

For a range $R = 10$ km

the percentage fixed error $\Delta R_f = \dfrac{2.5}{10} \times 100 = 25\%$.

(*v*) Estimate the frequency swing Δf required to bring down the percentage fixed error in range to 2.5%.

As ΔR_f *is* $\alpha \dfrac{1}{\Delta f}$, to reduce R_f from 25% to 2.5%, Δf has to be increased by a factor of 10 for which $\Delta f = 30$ kHz $\times 10 = 300$ kHz.

CHAPTER 14

HYPERBOLIC NAVIGATION

14.1 INTRODUCTION

The position of a ship or an aircraft over the seas can be known by measurement of the difference of the times of arrival of pulsed transmissions from two fixed locations on the shore or islands, separated by a distance. The courses defined for a fixed difference in the time of arrival are hyperbola as can be shown in the following.

From Figure 14.1 in which *A* and *B* are the two fixed stations, and *P* is the location of a ship:

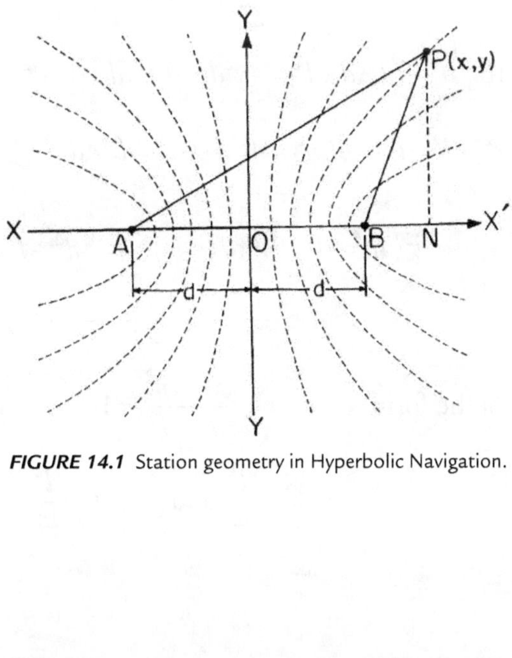

FIGURE 14.1 Station geometry in Hyperbolic Navigation.

$$t_1 = \frac{AP}{C} \text{ and } t_2 = \frac{BP}{C}$$

But $$AP = \sqrt{(x+d)^2 + y^2}$$

and $$BP = \sqrt{(x-d)^2 + y^2}.$$

The time difference is given by $t_D = t_1 - t_2$

$$= \sqrt{(x+d)^2 + y^2} - \sqrt{(x-d)^2 + y^2} \tag{14.1}$$

for a constant delay between the arrival of pulses transmitted at the same instant, we have

$$t_d = \sqrt{(x+d)^2 + y^2} - \sqrt{(x-d)^2 + y^2} = l$$

where l is a constant.

Therefore, $$\sqrt{(x+d)^2 + y^2} = l + \sqrt{(x-d)^2 + y^2}$$

or $$(x+d)^2 + y^2 = l^2 + 2l\sqrt{(x-d)^2 + y^2} + (x-d)^2 + y^2$$

or $$4xd = l^2 + 2l\sqrt{(x-d)^2 + y^2}$$

or $$(4xd - l^2)^2 = \left[2l\sqrt{(x-d)^2 + y^2}\right]^2$$

or $$16x^2d^2 - 8l^2xd + l^4 = 4x^2d^2 - 8l^2xd + 4l^2d^2 + 4l^2y^2$$

or $$4x^2[4d^2 - l^2] - 4l^2y^2 = 4l^2d^2 - l^4 = l^2[4d^2 - l]$$

or $$\frac{x^2}{l^2(4d^2 - l^2)/4(4d^2 - l^2)} - \frac{y^2}{l^2(4d^2 - l^2)/4l^2} = 1$$

or $$\frac{x^2}{l^2/4} - \frac{y^2}{d^2 - l^2/4} = 1$$

which is of the form $$\frac{x^2}{a^2} - \frac{y^2}{b^2} = 1$$

where $$a^2 = \frac{l^2}{4} \text{ and } b^2 = d^2 - \frac{l^2}{4}.$$

The equation represents hyperbola with foci at A and B. For each possible delay t_d between the arrival of pulses, we obtain a new value of l, thus giving a family of confocal hyperbolae, with the same foci A and B. It may be noted that a measurement of the difference in the time of propagation of the radio pulses along AP and BP is equivalent to the measurement of the difference of the ranges AP and BP, for the difference $(AP - BP) = C \times$ (time of difference). The system is, therefore, called a *range difference system.* If the pulses are sharp, having a short duration, then the measurement of time delay between these will be precise, for the smallest measurable delay is equal to the width of a pulse envelope. However, modulation with a short pulse width calls for a large spectral bandwidth of the RF pulse. While a pulse width of a few μ sec is useful in the MF band, and is used in LORAN (long range navigation) at IF/VIF, a much longer pulse is needed to keep the sidebands within the allowable limits, as determined by the limitation of the equipment as well as that of radio propagation characteristics. In fact, special long duration modulation waveforms have been involved for the purpose to keep the bandwidth small as in LORANC, while a CW transmission in place of the pulsed ones will require negligible bandwidth as is employed in Decca and *Omega* systems. For CW transmissions the measurement of the delay of the pulse envelope is replaced by a measurement of the phase difference between the two transmissions. Such systems are, therefore, also called the phase difference systems.

14.2 LORAN-A

LORAN-A, also called Standard LORAN, is a hyperbolic navigational aid operating on the range difference principle in the MF band around 2 MHz, in four equally spaced frequency bands. The ground stations for such a system transmit a train of pulse modulated ratio waves each of duration 40 m sec and having controllable basic pulse repetition frequencies (PRF) 20, 25 or $33\frac{1}{3}$ Hz, with provision for a finer control in smaller steps of 1/25, 1/16, and 1/9 Hz, stabilized by crystal clocks. This enables eight subdivisions of each basic frequency, resulting in a total number of selectable PRFs of $3 \times 8 = 24.3 \times 3 = 9$ different pulse repetition frequencies. Figure 14.2 shows the ground stations A and B separated by a distance 2, which is about 400–700 km. Ambiguity of course: If the pulses are transmitted simultaneously

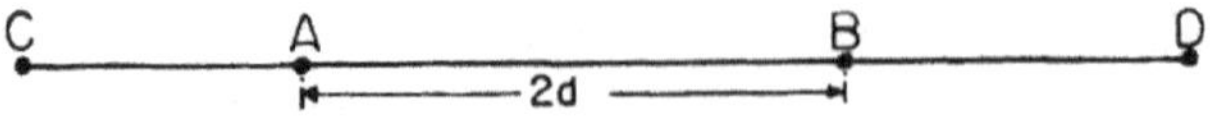

FIGURE 14.2 Station geometry in LORAN-A.

from A and B, an aircraft or ship midway between the stations along the perpendicular bisector of the line joining A and B will receive the signals at the same instant, and the course defined will be unique. However, for any other course there will, in fact, be a pair of courses, each pair being symmetrically disposed about the perpendicular bisector for a fixed time delay. Removal of ambiguity of courses: To avoid such an ambiguity of courses, a fixed absolute delay is introduced between the pulses from A and B, with the pulse from B delayed by a second after that from A, the master station. The station B serves as a slave station and is triggered by the pulses from A delayed by the absoluting delay. In this case, for a ship along the line AB, the apparent delay of the transmission from B after that from A will be maximum and is given by

$$\tau_C + \tau_{max} = \tau + \frac{2d}{c}$$

if the ship is at C, while for a ship at D, the delay will be minimum as given by

$$\tau_D = \tau_{min} = \tau - \frac{2d}{c} = \tau - \beta$$

where $2d/c = \beta$ the time of travel of the pulsed radio wave between A and B. If the absolute delay τ is not introduced, then $\tau_c = -\tau_{AD} = 2d/c = \beta$, and τ_{AC} and τ_{AB} are then not distinguishable as positive, and negative time delays are not resolved by the delay measurement systems. The absolute delay τ is, therefore, made greater than β so that even the minimum apparent delay $\tau - \beta$ may be positive, making both τ_{AC} and τ_{AB} positive and distinguishable. For any point midway between A and B, the apparent delay will then be equal to the absolute delay, τ.

14.3 MEASUREMENT OF DELAY

For measurement of the delay, an oscilloscopic display of the demodu pulses are employed. The horizontal trace during the interpulse period, T is divided into two halves alternately displaying the pulses due to transmitters at A and B (Figure 14.2) in separate traces, one below the other, as shown in Figure 14.3

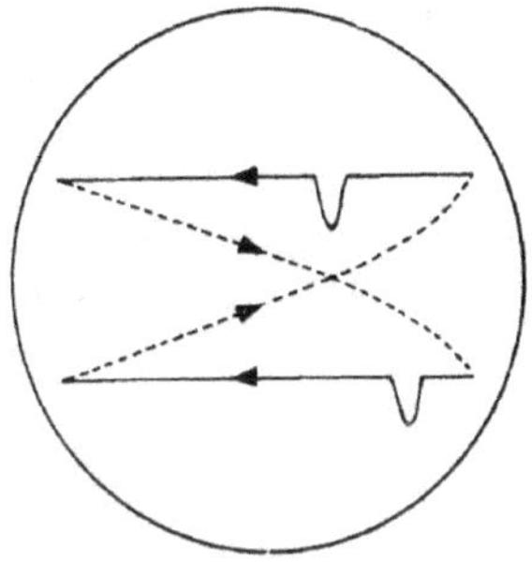

FIGURE 14.3 LORAN-A display.

The duration of the sweep for each sweep is, therefore, $T/2$, if the flyback or retrace times are neglected. In order that B may be displayed always on the lower trace, the minimum delay between A and B must be $T/2$. Accordingly, a small additional delay δ is introduced in the pulse from B, so that

$$\tau - \beta = \frac{T}{2} + \delta$$

or

$$\tau = \frac{T}{2}\beta + \delta \tag{14.2}$$

Substituting τ from the equation (5.4) in equations (5.2) and (5.3), the maximum and minimum delays are respectively given by

$$t_{max} = \left(\frac{T}{2} + \beta + \delta\right) + \beta$$

$$= \frac{T}{2} + 2\beta + \delta \tag{14.3}$$

and

$$t_{min} = \left(\frac{T}{2} + \beta + \delta\right) - \beta$$

$$= \frac{T}{2} + \delta \tag{14.4}$$

The sweep is synchronized with the pulse repetition frequency (PRF), f_R which is the reciprocal of the interpulse period, T, to obtain a stationary waveform pattern on the oscilloscope. After the synchronization with one particular PRF, presence of any other PRF will result in a moving pattern and can thus be eliminated. In that way, any one of the 24 controllable PRFs can be synchronized, leaving the others to produce a moving pattern.

Number of Channels. As each LORAN-A setup can work in four different frequency bands covering 750 to 1950 Hz, the total number of independent channels effectively becomes $4 \times 24 = 96$. This means that identical LORAN-A equipment may operate simultaneously, each one working on a different channel chosen from the 96 possible channels, without any mutual interference.

14.4 TRANSMITTER POWER

The transmitter LORAN-A has a power typically 100 kW peak feeding into a vertical $\lambda/4$ antenna.

14.4.1 Accuracy

The accuracy of delay measurement is about 1 μ sec, although pulses are each of duration 40 μ sec. The accuracy is, in fact, ensured by controlling waveshapes of the modulating pulses from *A* and *B* that allows matching of the pulse shapes during the oscilloscopic display. Also, the crystal clocks used in controlling the pulse parameters as well as the transmission frequency help greatly in obtaining accuracy of delay measurements. The accuracy is reduced at night when the sky wave is received with a variable delay and amplitude due to reflection from the ionosphere, in fact, the synchronization between the master station *A* and the slave station *B* is determined both by the stabilities of the ground wave propagation characteristics, between *A* and *B*. The range LORAN-A for ground wave coverage is about 600 km for sea in temperates latitudes and is about 500 km in equatorial regions. The system was developed during World War II, and still continues to serve, principally in the Pacific and Atlantic sea boards of North America as well as the region around Japan.

14.5 LORAN-C

Lowering the transmission frequency of a LORAN from the MF to IF band between 90–110 kHz increases the ground wave range greatly due to the efficient propagation conditions at such low frequencies. This apparently led to the development of LORAN-C, which can operate for ranges up to a few thousand kilometers, typically 3500 km over sea and 2700 km over the land, with base line lengths between 1000–1800 km. However, for the limited bandwidth available in the IF range, the duration of the pulses must be made rather long. In the extreme case if the duration tends to infinity, the bandwidth tends to zero corresponding to CW transmission. To allow time delay measurements, the duration must be finite. The accuracy of measurement is ensured by a special wave shape of the pulse modulation as shown in Figure 14.4, allowing matching between the RF cycles along with its envelope due to the two stations A and B. The waveshapes chosen allow 90% of the energy to be contained within the 90–110 kHz band. By adjusting the amplitudes of the received pulses due to A and B to equality, they may be superimposed on the oscilloscope by adjustable delay in the pulse due to A at the receiving end to facilitate cycle matching, which ensures accuracy of time delay measurement down to the duration of half an RF cycle. Usually the first three cycles are matched and nine successive pulses are stored and processed to improve the signal-to-noise ratio. In practice, the cycle matching is achieved by fully automatic servo-controlled equipment. However, the larger baseline length calls for corrections introduced by the changes in the velocity of propagation of the ground wave. For the distance $2d$ between the stations A and B the time of travel of the ground wave is given by

$$t = \frac{2d}{v} + C$$

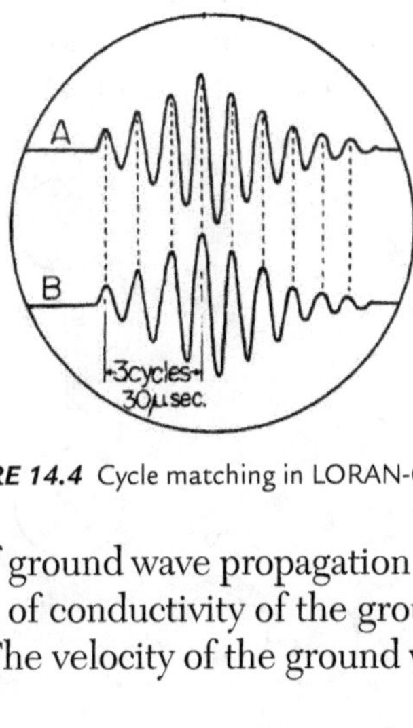

FIGURE 14.4 Cycle matching in LORAN-C pulse.

where v is the velocity of ground wave propagation and c is a correction term allowing for the changes of conductivity of the ground, including the associated pulse retardation. The velocity of the ground wave is related to the free space velocity as

$$v = c/n$$

where n in the refractive index of the ground, which in turn depends on the conductivity and dielectric constant of the ground. An average velocity v is assumed while the changes of v due to predictable changes of the ground constants are included in the correction term C. The unpredictable or uncertain part of the ground constants limits the accuracy of LORAN-C. Nevertheless, the precision of time delay measurement with all the corrections and waveform matching is the order of 0.1 μ sec for the ground wave range. The peak power of a LORAN-C transmitter is 1 MW.

14.6 DECCA

The DECCA system is based on the measurement of phase difference of continuous wave transmission from a pair of stations operating in the IF band between 70 and 120 kHz. The frequencies for the two stations of a pair are made different to distinguish the signals during reception at the ship or aircraft. After reception these frequencies are separately multiplied by a suitable number for conversion to a common frequency to facilitate a phase comparison.

14.6.1 DECCA Chain

In practice three DECCA stations together with a central master station constitute a *DECCA Chain* as shown in Figure 14.5. Each pair of stations with the master station as common and a slave system provides a family of hyperbolic courses to serve effectively a wide area all around.

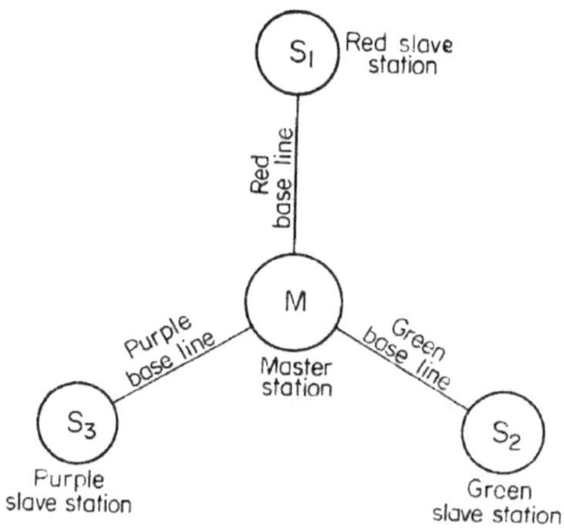

FIGURE 14.5 Station geometry in a DECCA chain.

The master station transmits at a frequency about 14 kHz, while the stations 1, 2, and 3 transmit at $8f$, $5f$, and $9f$ respectively. The slave stations are distinguished by the color of the charts for the hyperbolic lines, the color being red, purple, and green for the stations 1, 2, and 3 respectively, each one generated with the master station. After the reception of the master and slave transmissions in a ship separately at the frequencies, the difference of phase between the master transmission and that of a slave is measured by multiplying each by a suitable number to obtain a common frequency which is equal to their L.C.M. Thus $6f$ and $8f$ pair have a common frequency of $24f$. Likewise, $6f$ and $5f$ have a common frequency of $30f$ while that for $6f$ and $9f$ is $18f$.

In general, if the master frequency is $n_1 f$ and that of a slave station, S, is $n_2 f$, then the radiated field at the two stations may be represented as

$$E_M = E_{OM} \cos 2\pi n_1 ft \quad \text{For the master station}$$

and

$$E_S = E_{OS} \cos 2\pi n_2 ft \quad \text{For the slave station}$$

The fields at a point P at a distance r_1 and r_2 from the master and slave stations will be given by

$$E_{r1} = kE_{OM} \cos\left(2\pi n_1 ft - \frac{2\pi}{\lambda_1} r_1\right) \tag{14.5}$$

and

$$E_{r2} = kE_{OS} \cos\left(2\pi n_2 ft - \frac{2\pi}{\lambda_1} r_2\right) \tag{14.6}$$

where k is a constant containing the distance factor.

Let the m_1th harmonic of the first and the m_2th harmonic of the second be equal so that $m_1 n_1 = m_2 n_2$. These harmonics are generated after reception of the signals E_{r1} and E_{r2} by a suitable multiplier, producing outputs given by

$$E_{m1} = k_1 kE_{OM} \cos m_1 \left(2\pi n_1 ft - \frac{2\pi}{\lambda_1} r_1\right) \tag{14.7}$$

and

$$E_{m2} = k_2 kE_{OS} \cos m_1 \left(2\pi n_2 ft - \frac{2\pi}{\lambda_2} r_2\right) \tag{14.8}$$

where k_1 and k_2 are the multiplier constants.

The phase difference between E_{m1} and E_{m2} is given by

$$\Delta\phi = \frac{2\pi m_1 r_1}{\lambda_1} - \frac{2\pi m_2 r_2}{\lambda_2}$$

$$= 2\pi m_1 \frac{n_1 f}{v} r_1 - 2\pi m_2 \frac{n_2 f}{v} r_2$$

Since $$\lambda_1 = \frac{v}{n_1 f}$$

and $$\lambda_2 = \frac{v}{n_2 f},$$

v being the velocity of propagations of the radio wave

or $$\Delta\phi = \frac{2\pi}{\lambda mn}(r_1 - r_2)$$

where $$\lambda mn = \frac{v}{m_1 n_1 f} = \frac{v}{m_2 n_2 f}$$

as $$m_1 n_1 = m_2 n_2.$$

The phase difference measured is identical with that due to transmission at the same common harmonic frequency $m_1 n_1 = m_2 n_2$ by both the stations. However, actual transmission at the common harmonic frequency would make the signals indistinguishable at the receiver, besides producing interference fringes showing nulls for a phase difference of 180°.

14.6.2 Ambiguity of Courses and Lanes

Unlike the time difference system, the phase difference system suffers from the ambiguity of courses for phase differences exceeding 360°, corresponding to a path difference of λ_{mn}. Movement along a baseline formed by joining a master station with a slave station, by $\lambda_{mn}/2$ courses a phase advance of 180° of the signal from the station which is approached and a phase log of 180° for the other signal resulting in a net phase difference of 180° – (– 180°) = 360°. Within the baseline lengths of the order of 120 – 200 km, there may be such 360° phase changes occurring several hundred times for movement between the stations. The region between two consecutive hyperbolic corresponding to a phase change of 360° or a path difference of λ_{mn}, is called a line, the width of a lane along the baseline being $\lambda_{mn}/2$, which is typically about 500 m.

14.6.3 Lane Identification

The ambiguity of course due to the presence of a large number of indistinguishable lanes may be partly removed by arranging transmissions differing by f = 14 kHz from each station, in succession, at certain intervals, for lane identification. This produces hyperbolic courses corresponding to the rather low difference frequency, the lane width for which will be much larger. For the frequency of 14 kHz, the lane width $\lambda/2 = v/2f = 10.5$ km. This lane width for the difference frequency signal is called a zone. If the ship or aircraft is certain about its location within a zone width 10.5 km, it is possible to determine the particular lane within the zone by committing the limited number of lanes it crosses within the zones as indicated by the number of 360° phase charges, and thus the particular lane is identified without ambiguity. Land identification is called 'coarse fixing' while determination of the particular hyperbolic lane on which it is located by phase measurement is called fine fixing. For lane indentification the zones are created by transmitting signals differing by f for short periods three times every minute, each time the master transmitting at $6f$ and $5f$ together and one of the slaves making simultaneous transmissions at $8f$ and $9f$ as indicated in Figure 14.6.

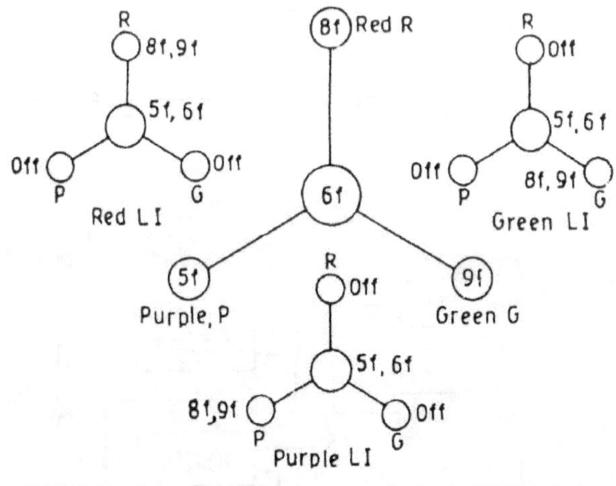

FIGURE 14.6 Lane identification transmission in DECCA chain.

At the beginning of each lane identification cycle starting at each full minute, the master station transmits $6f$—60 Hz for 1/12 sec, initiating the red lane identification cycle. The master station then transmits $6f$ and $5f$ and the red slave transmits $8f$ and $9f$ for half a second, after which normal transmission is resumed. At the beginning of the 16th sec, the green lane identification is initiated following a similar sequence, out with the master at $6f + 50$ Hz, while the purple lane identification cycle starts at the 30th second and continues for half a second after the master starting at $6f + 60$ Hz for 1/12 sec followed by $6f - 60$ Hz for 1/25 sec. The rest of the period of 1 minute is devoted to normal transmission.

14.7 DECCA RECEIVER

DECCA receiver structures are shown in Figures 14.7 (*a*) and (*b*). These are described as follows.

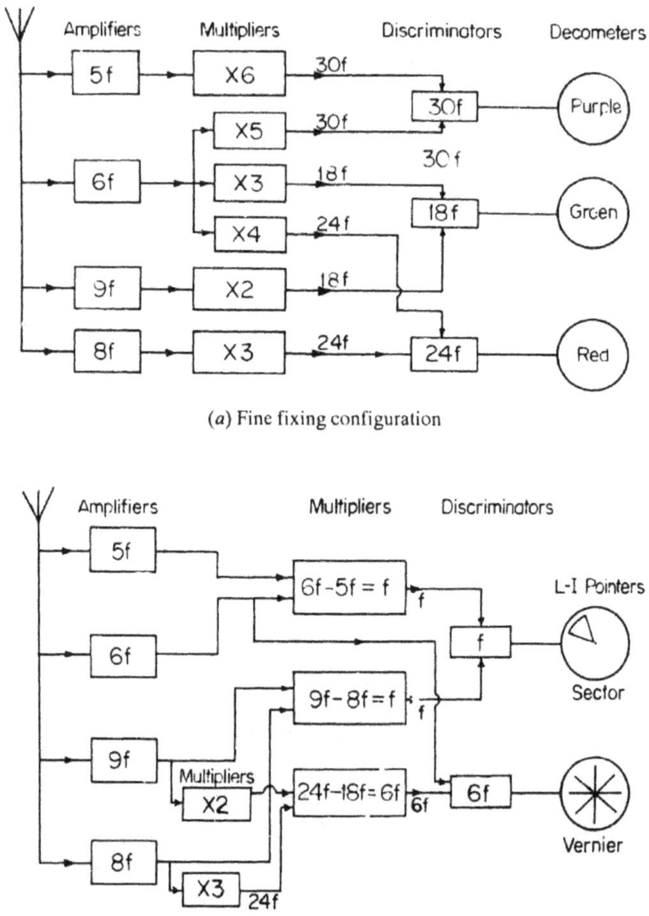

(*a*) Fine fixing configuration

(*b*) Lane identification configuration

FIGURE 14.7 Block diagram of DECCA receiver.

14.7.1 Fine Fixing Configuration

The signal from the DECCA chain received by an antenna at the ship is divided into four and led to amplifiers tuned to $5f$, $6f$, $9f$, and $8f$. The output of the amplifier at $6f$ due to the master transmission is divided into three multipliers producing outputs at $30f$, $18f$, and $24f$, to facilitate phase comparison for the purple, green, and red channels separately. For the purpose a discriminator followed a decometer, which indicates the phase difference between the two inputs to the discriminator, is employed in each channel.

14.7.2 Lane Identification Configuration

In this configuration the amplifier outputs are processed differently. The master transmissions at $5f$ and $6f$ during the lane identification cycle are received and mixed to produce a difference frequency f, which is compared in phase with a difference frequency output of a second mixer due to signals received at $9f$ and $8f$ from a slave station at such time. Besides this, for a finer vernier indication of lane, the signals at $8f$ and $9f$ are multiplied to $24f$ and $18f$ respectively and mixed to produce a $6f$ output, which is compared in phase directly with the output amplifier at $6f$. In this way 6 fold is better resolution with, however, a 6 times higher probability of ambiguity. To avoid the ambiguity with the vernier, it is geared down in the ratio 1 : 6.

The receiver configuration described is employed in a Mark *V* receiver suitable mainly for marine navigation, where a high accuracy is needed. The receiver is, however, somewhat sluggish to respond quickly, as will be needed for use in aircraft.

Special receivers for use in aircraft employ phase comparison directly at $5f$, $8f$, and $9f$, which effectively makes the lanes wider by factors of 6, 3, and 2 respectively. To achieve this, the master signal at $6f$ is divided by 6 to obtain the signal at f, which is next multiplied to the slave frequencies $5f$, $8f$, and $9f$ by multipliers; for phase comparison, other improved receivers for use in aircraft have been developed.

Range and Accuracy. The range is limited by the tolerable accuracy, which decreases with range. For a radial error of 100 m, for instance, the range is about 300 km when the master slave distance is 200 km. It may be noted that for any pair in hyperbolic navigation, the lateral motion causes greater change in phase difference, and hence the precision is greater than that for motion toward or away from the station. In particular, there will be no change in phase difference for motion along a hyperbolic line. The use of the three pairs of stations ensures that no motion can be along the hyperbolae of two pairs, simultaneously, and the precision is of the same order for motion in any direction.

In recent years, the precision of DECCA chains has increased greatly by the advent of atomic standards, which are so stable in frequency that they gain or lose by 1 sec in 3000 years. Separate Rubidium vapor frequency standards, are, in fact, used to control the frequency of each station of a DECCA chain. The hyperfine transition of Rubidium vapor at a microwave frequency is employed to control the frequency of an X-tal oscillator in a Rubidium vapor standard.

14.8 OMEGA

OMEGA is a phase difference system similar to DECCA, but operating at lower frequencies in the VLF band around 10 kHz, at which the reliable

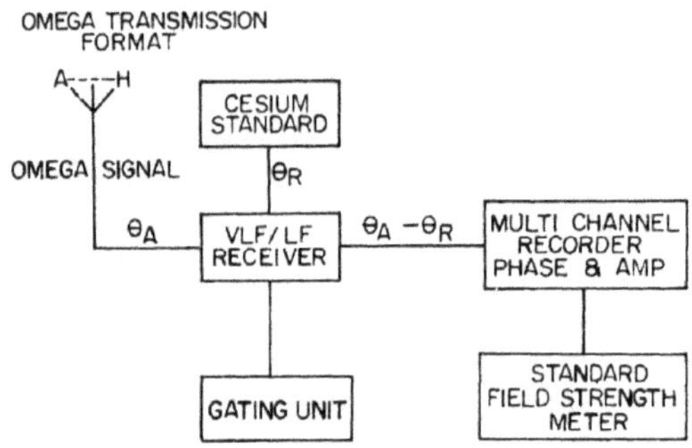

(*a*) Block Diagram of OMEGA receiving equipment

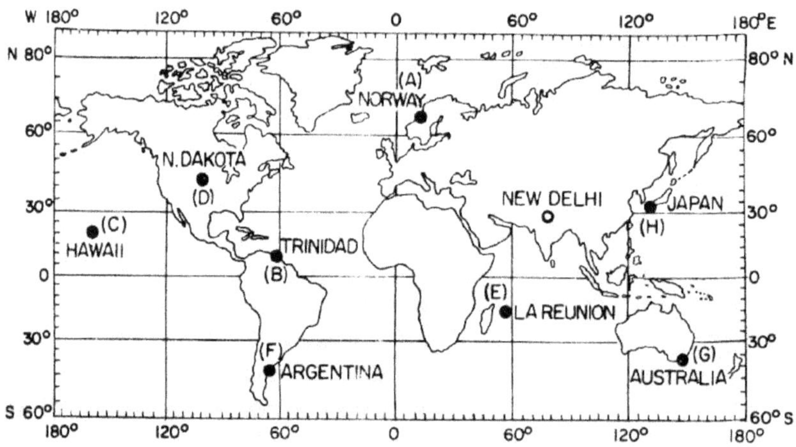

(*b*) Global station geometry of OMEGA transmitters

FIGURE 14.8

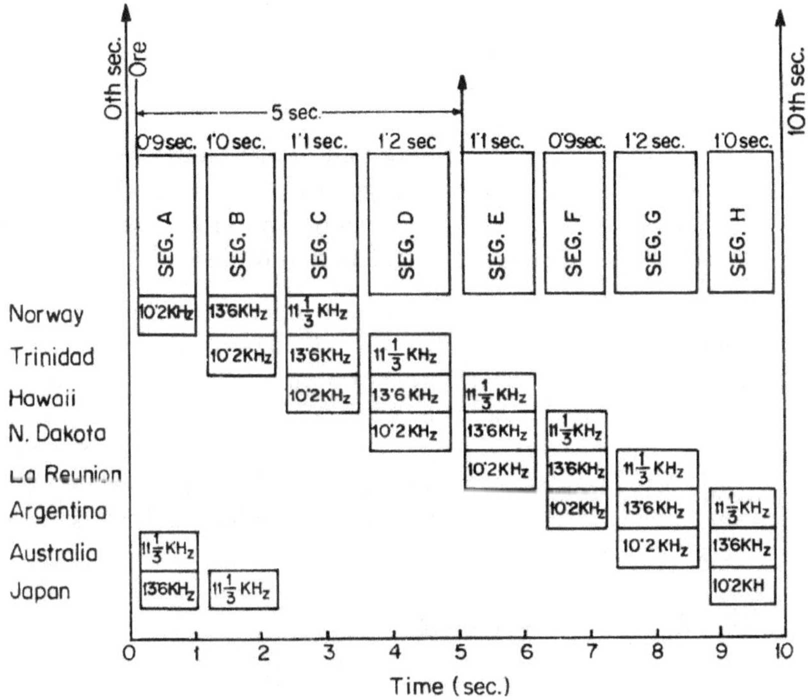

FIGURE 14.9 OMEGA transmission format.

ground wave coverage may be increased to a few thousand kilometers, with base line lengths of several thousand kilometers. Even longer ranges are possible by making use of the sky wave reflected in the space between the lower ceiling of the ionosphere and the ground. The ionospheric characteristics at VLF are relatively stable and predictable, while the ground conductivity is rather high. By using atomic frequency standards to control the transmission frequencies at each of a worldwide network of stations, a truly global coverage of hyperbolic navigational aid has become a reality for use by ships and aircraft. A typical OMEGA transmitter is shown in Figure 14.8 (*a*). The world network with eight OMEGA stations marked by letters A-H is shown in the map in Figure 14.8 (*b*). The table shows the latitude and longitude of the stations, the transmitter frequency, and power. The format of the transmission cycle for the eight stations is shown in Figure 14.9.

14.8.1 Transmission Cycle

At a station the transmission is made successively at 10.2, 13.6, and 11.33 Hz for periods of the order of 1 sec, the exact period being 0.9, 1.0, and 1.1 sec respectively, followed by a transmission for 6.2 sec at 10.462 kHz, which is the characteristic carrier frequency for each station. Each of these periods are separated by a 0.2 sec interval of no transmission, thus making a total period $T = 0.9 + 1 + 1.1 + 4 \times 0.2 + 6.2 = 10$ sec, for a complete cycle. For lane identification, phase comparison at the difference frequencies of 13.6 – 10.2 = 3.4 kHz and 11.33 – 10.2 = 1.13 kHz are made. The cycles for different stations A to H are progressively delayed by a 1 sec transmission period of the preceding one as indicated in Figure 14.9.

At each 30 sec interval of standard time, all carrier and modulation frequency currents pass through zero with a positive slope, and this eliminates the need for master slave synchronization. The stability of such a zero is ensured by employing atomic beam controlled transmission from each station. Comparison of phase for the intermittent transmission is possible by the use of phase lock loops, which effectively store the phase information received at a frequency for a brief period, until the same frequency is received during the next transmission cycle.

14.8.2 Accuracy and Range

The errors in the OMEGA system have predicable and random parts. With corrections for predicable errors, the positional accuracy is about 2 km. A world wide coverage with such an accuracy is, indeed, an achievement for navigation over the seas by aircraft or ships. As the destination is approached, however, much higher precision of course is available through the use of a DECCA chain.

14.9 DECTRA

DECTRA is an acronym for Decca tracking and ranging, operating at 70 kHz for long-range coverage over the seas, such as in the Atlantic. A master-slave pair at each coast with a short baseline generates hyperbolic lines which are almost straight lines, and helps in determining the lateral positions. The position along a hyperbolic line is determined by comparing the phases of the transmission from the two pairs, transmitting at frequencies simply related to each other by phase lock loops, to generate orthogonal hyperbolic lines. The master and slave of a DECTRA transmit alternatively at the same frequency, while the master transmission is utilized for the phase comparison between the two pairs across the sea.

14.10 DERLAC

This is an acronym for DECCA long-range area coverages. DERLAC operates in the VIF band between 10 to 14 kHz, and like the OMEGA system is meant for worldwide coverage. The decrease in accuracy in the determination of lateral position with increasing range is minimized in DERLAC by combining two sets of hyperbolae, the intersection of which gives accurate positional information as shown in Figure 14.10.

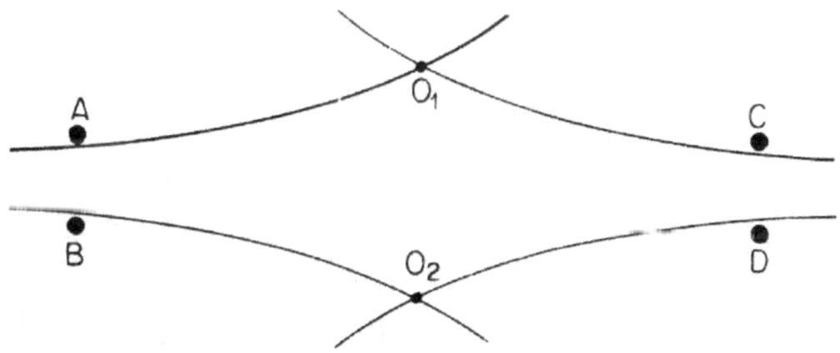

FIGURE 14.10 Station geometry in DERLAC.

14.11 INLAND SHIPPING AIDS

For shipping along any length of a river well inland, the problem is usually to guide the ship along a route where the depth of the river is greatest. Otherwise, the great depths to which the base of a ship usually extends may cause obstruction at the base with the river bed. To ensure such guidance along a river, the accuracy of position fixing should be fairly high. For this purpose, the distance or range of the ship along the desired route from the shore of the river may be measured by short-range navigational aids; such aids are more economical than a DECCA navigator and are lightweight. The aids may be based on the range difference system as in hyperbolic navigation or on the range-range system generating circular in place of hyperbolic patterns.

14.11.1 HIFIX

HIFIX is an accurate electronic navigation aid for inland shipping operating in the MF band between 1700–2000 kHz. Unlike the DECCA chain only one transmission frequency is shared by the master and two slave stations employing samples of transmission sequentially as shown in Figure 14.11 (*a*).

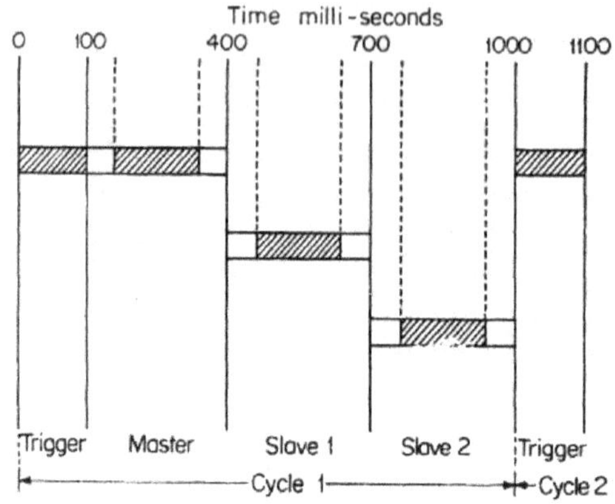

(*a*) HIFIX time sharing scheme

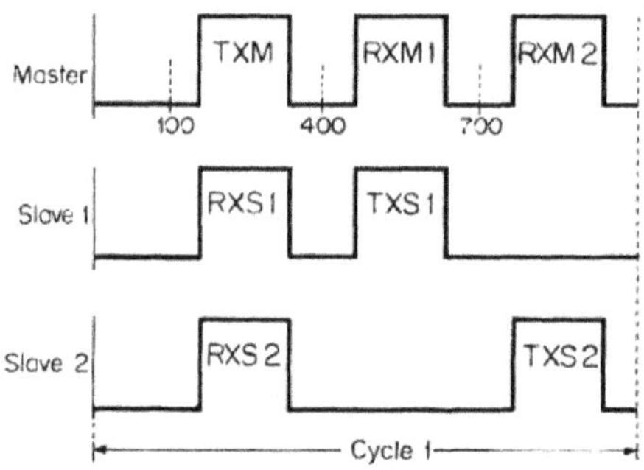

(*b*) HIFIX transmission and reception cycles

FIGURE 14.11

The receiver at each of the stations picks up a desired slot of transmission from another station and is used to lock a lock oscillator by a phase lock loop on a continuous basis. The transmission and reception cycles of the HIFIX are shown in Figure 14.11 (*b*).

The master starts a transmission for 100 m sec, which is received by each slave station and triggers the electronic timers controlling the transmission and reception cycles at the respective stations. The next period of master transmission locks the slave receivers at each station to the phase and frequency of the master signal. Each slave receiver then in turn injects a pulse into its associated transmitter. The transmissions from the master station as well as the two slave stations may be picked up by a ship where the phase difference between the master and each of the slave stations is measured separately to generate two sets of hyperbolic lines. The intersection of these lines fixes the position of the ship. Alternatively, the master station equipment may be installed in the ship, where the ranges from the two slave stations are determined by receiving the two slave transmissions and measuring the phase difference of each with that of the master transmission, when the contours of equal range will be circles centerd around the two slave stations. The intersection of the two sets of circles fixes the position of the ship.

14.11.2 Accuracy and Range

Accuracy of the fix in such a range-range system is equal at all distances from the slave stations, while in the hyperbolic system the accuracy decreases with increasing distance from the three station chain including the master. When the master is aboard the ship, as the range difference is measured along a base lines between the master and a slave station, accuracy is not dependent on the distance. Typical accuracies on the baseline are $2\frac{1}{2}$ feet for hyperbolic working and about $3\frac{3}{4}$ feet for range-range working. The range is about 160–320 km.

14.11.3 Lane Identification

A lane defined by a region over which the phase difference equals 360° becomes ambiguous for a greater difference of phase as discussed in connection with the DECCA chain. For an identification of lane number within a zone, a second transmission at the master station operating at frequency 10% lower is employed. Beating between the two transmissions of the master will produce a zone covering about ten lanes. Due to the difference between the two operating frequencies, the number of lanes on the lower frequency over a given distance will be exactly ten percent less than that on the higher frequency. If the higher frequency is 1900 kHz, for instance, then within a distance covered by 10 lanes there will be 9 lanes on a lower frequency of the second transmission at 1710 kHz. The lane identification with such a two frequency transmission is illustrated in Figure 14.12, which shows when the higher frequency (HF) lane reading is 1.0, the lower frequency (IF) lane reading is 0.9, since the HF is 10% less (1.0 – 0.1 = 0.9). Similarly, where the HF reads 2.0, the LF reads 1.8 and so on up to 10 lanes on HF, which corresponds to 9 lanes on IF Subtracting the IF from the HF reading results in

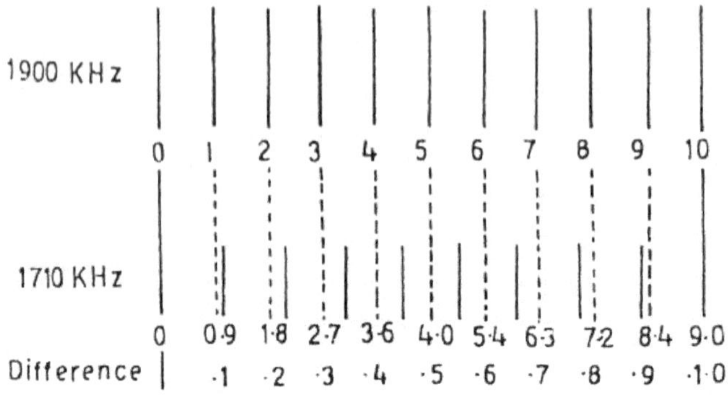

FIGURE 14.12 Lane identification in HIFIX with two-frequency transmission.

a difference which is always one-tenth of the HF reading. A lane identification computer computes this difference, which when multiplied by 10 gives the lane numbers.

14.11.4 Transmitter Power

The radiated power is about 40 watts from a vertical antenna of height 31 ft. and having a capacity of 300 pf.

14.12 SYLEDIS

The SYLEDIS is a radio navigational aid for inland shipping operating in the UHF band between 420 and 450 MHz as a circular or a hyperbolic position finding system. Operation in the UHF band makes it less affected by the changes of the conditions of propagation and noise, while the use of the digital delay lock loop makes the accuracy of position finding to be superior to that of the HIFIX system.

14.12.1 Principle of SYLEDIS

In the SYLEDIS system the distances of a ship from two stations on the shore are measured by measuring the respective radio propagation times. The interrogater onboard a ship transmits the UHF radio signal modulated by a train of coded pulses. The signal receivers at each location demodulate to recover the pulse train, which is used to modulate the beacon reply transmission from the respective Responder. At the Interrogator the two reply signals are received separately, demodulated, and compared with the original modulating pulse train to measure the propagation delays.

14.12.2 The Principle of Distance Measurement

If the modulation pulse of the interrogator is a single one of duration and the pulse to the reply signals from the responders 1 and 2 is delayed by T_1 and T_2 respectively, which are the respective round trip delays as shown in Figure 14.13, then the distances r_1 and r_2 will be given by

$$T_1 = \frac{2r_1}{v} \tag{14.9}$$

or

$$T_2 = \frac{2r_2}{v} \tag{14.10}$$

where v is the velocity of the wave in the atmosphere and r_1 and r_2 are the distances of the responders. The distances are displayed on the SYLEDIS after introducing corrections required to compensate for the additional small delay due to the air medium, assuming an average value of refractive index of 1.000325, the average value found at sea levels. The phase path in the presence of air is given by

$$\phi = \frac{2\pi}{\lambda}\mu r$$

where μ is the refractive index of air and r is the distance.

FIGURE 14.13 Station geometry in SYLEDIS.

Since $v = f\lambda, \frac{1}{\lambda} = \frac{f}{v}$. putting this in the previous equation

equation, we get
$$\phi = \frac{2\pi f}{c} - \mu r = \frac{\omega}{v}\mu r$$

$$\tau \frac{db}{dw} = \frac{1}{v}\mu r$$

Hence, T is given by
$$T = 2\tau = \frac{2\mu r}{v} \tag{14.11}$$

Hence, we get one way propagation delay and round trip delay. Comparing equation (14.11) with equations (14.9) and (14.10), we note that the effect of atmospheric refraction is to multiply the apparent delay by the refractive index μ, and this multiplication is achieved automatically in the SYLEDIS display equipment.

The accuracy of measurement of delay between an interogation pulse and the pulse received from the responder is limited by the finite duration of the pulse. For the change in delay less than the duration of the pulse is not resolved and, therefore, the shorter the pulse, the greater is the accuracy obtained. However, a shorter pulse calls for a higher peak power to maintain the range capability. In the presence of noise the range is, in fact, dependent on the energy of the pulse, which in turn is equal to the product of the duration and the peak power of the pulse. In SYLEDIS, the necessity of a high peak power is avoided by involving a special digital signal processing technique which enables a rather long coded sequence of pulse train of total duration 2.66 to 10.66 milliseconds, each pulse having a duration of 0.42 m sec. The signal processing allows an accuracy equivalent to that of the signal pulse, while at the same time ensuring the signal energy corresponding to the total duration of the pulse train. Thus, by means of a fully transistorized transmitter, the SYLEDIS emits a signal with a peak antenna power of only 10 W, but with an effective peak power due to the processing equipment to

$$\frac{10 \times 2.66 \times 10^{-3}}{0.52 \times 10^{-6}} = \text{W} = 50 \text{ kW}.$$

A complete sequence of SYLEDIS code comprises 127 sample pulses each of duration 52 μ sec, making a total duration of 66.66 m sec for the code. This code sequence is repeated 40 times so as to make up a long coded pulse train with a total length of 66.66 × 40 μ sec. = 2.66 m sec as shown in Figure 14.14 (*a*). A gap or a dead time of 666 μ sec. is then introduced, making a total time slot of 3.33 m sec allotted for emitting the interrogation from the master or for obtaining responses due to replies from the slaves.

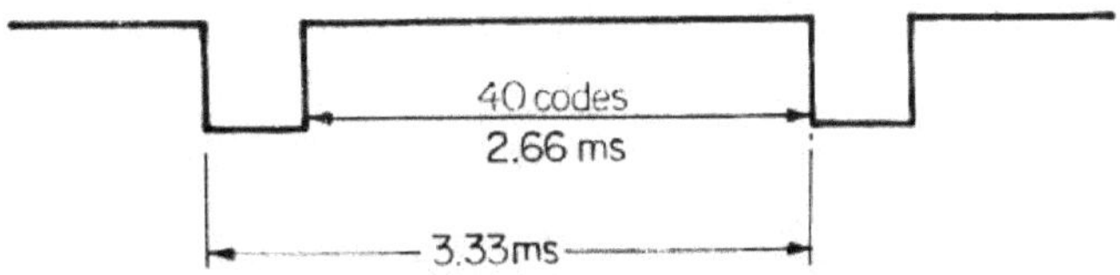

(*a*) Code length in a SYLEDIS transmission

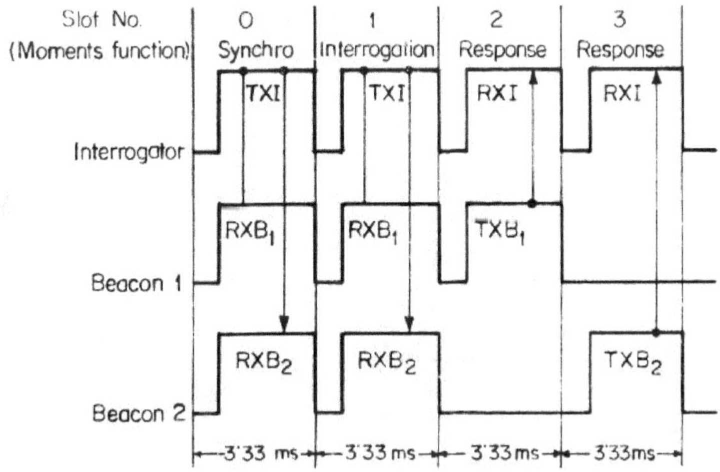

(*b*) SYLEDIS transmission and reception cycle

FIGURE 14.14

The sequence of operation of the master and slave stations indicating the respective slots is shown in Figure 14.14 (*b*).

In the time slot No. 0, the Interrogator emits a synchronizing pulse train, which is received by both beacons 1 and 2. The received synchronizing pulse train is processed to synchronize the local code generators of the beacon with that of the interrogator. The interrogator emits the coded interrogating pulse train in slot 2. This is received by both the beacons triggering reply pulses in the slots 2 and 3 for the beacon 1 and 3 respectively, while the receiver of the interrogation responds to these replies in the respective slots. The delay between the interrogation signals and the responses in the beacons are used to deduce the round trip delays of the radio wave, with a precision of 52 μ sec, and the ranges r_1 and r_2 from the Beacons are determined from equations (14.1) and (14.2) respectively.

Modes of SYLEDIS Operation.

Range-Range Mode. This is the basic use of the SYLEDIS, when the turn ranges of Beacons 1 and 2 from the interrogator are measured, and hence it is called the range-range mode of operation. This range-range mode makes it possible to fix the position of a mobile station, carrying the interrogator by the intersection of two circles with radio corresponding to the ranges r_1 and r_2 as shown in Figure 14.15. The range-range mode can work with only one mobile interrogator at a time. For time slots of the system are all occupied by the emissions from one or more of the stations. The system is thus saturated by a single mobile station.

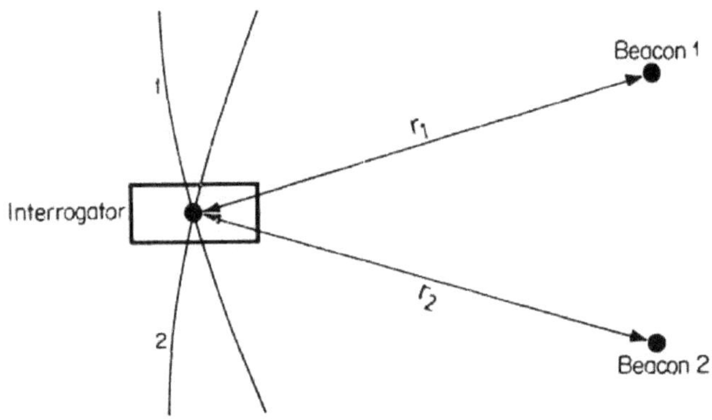

FIGURE 14.15 Station geometry for a range-range mode of operation.

Hyperbolic Mode. The saturation can be avoided if there are three beacons instead of two and the interrogator acts only as a receiver, the emissions occurring from each of the three beacons. The configuration of such a system is shown in Figure 14.16. Synchronizing emissions from the master station are received in slot 0 at the slave stations *B* and *C*, while emissions from *A*, *B*, and *C* are received in the mobile station, in the three remaining time slots. The delay between the responses due to *B* and *A* gives the range difference between r_1 and r_3. Likewise, the delay between *C* and *A* gives the range difference $r_2 - r_3$. The loci of a constant range difference being hyperbolic, the two range difference data enables us to locate the mobile station of two particular intersecting hyperbolics. The point of intersection fixes the position of the mobile station. Due to the involvement of hyperbolic courses, the range-range mode is also called the hyperbolic mode. In a similar way, due to the involvement of a circular course the range-range mode is also called circular mode.

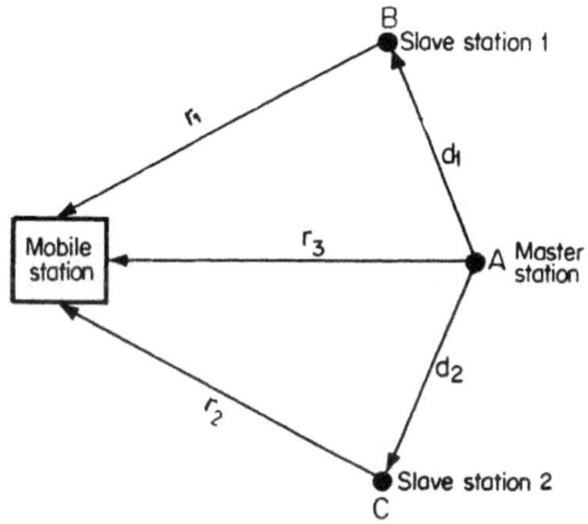

FIGURE 14.16 Station geometry for hyperbolic mode of operation.

WORKED OUT PROBLEMS

1. (*i*) In a DECCA receiver for fine fixing, calculate the output frequency of the multipliers for Purple, Green, and Red Decameters, assuming the Master frequency to be 14 kHz

For the Purple Decameter the output frequencies of the multipliers are

$$5f \times 6 = 30 \times f = 30 \times 14 \text{ kHz } 420 \text{ kHz and } 6f \times 5 = 30 \times f$$
$$= 30 \times 14 \text{ kHz} = 420 \text{ kHz}$$

For the Green Decameter the output frequencies of the multipliers are

$$6f \times 3 = 18f = 18 \times 14 \text{ kHz} = 252 \text{ kHz}$$

and $$9f \times 2 = 18f = 18 \times 14 \text{ kHz} = 252 \text{ kHz}$$

For the Red Decameter

the output frequencies of the multipliers are

$$6f \times 4 = 24f = 24 \times 14 \text{ kHz} = 336 \text{ kHz}$$

and $$8f \times 3 = 24f = 24 \times 14 \text{ kHz} = 336 \text{ kHz}$$

2. (*ii*) In the DECCA receiver configured for lane identification, calculate output frequencies of the multipliers for the sector and vernier LI Pointers

For the sector LI Pointer

output frequencies of the multiplier are

$$6f - 5f = f = 14 \text{ kHz} \quad \text{and} \quad 9f - 8f = f = 14 \text{ kHz}$$

while for the vernier LI Pointer

output frequencies of the multiplier is

$$8f \times 3 - 9f \times 2 = 24f - 18f = 6f = 6 \times 14 \text{ kHz} = 84 \text{ kHz}$$

(iii) Calculate zone width indicated in the sector pointer for different frequencies of transmissions equal to Master frequency $f = 14$ kHz.

$$\text{Zone width} = \lambda / 2 = \frac{c}{2f} = \frac{3\times10^8}{2\times14\times10^3}\text{m} = \frac{3\times10^5}{28}\text{km}$$

$$= \frac{3}{28}\times10^2\text{km} = \frac{300}{28}\text{km} = 10.7 \text{ km}$$

(iv) Calculate the Lane width within a zone, indicated by a vernier pointer.

$$\text{Lane width} = \frac{c}{2\times6f} = \frac{3\times10^8}{12\times14\times10^3}\text{m} = 1.7\,\text{km}.$$

(v) Calculate the Lane widths for fine fixing configuration of a Decca receiver for Purple, Green, and Red Decameter, and find the average lane width.

For purple Decameter

$$\text{Lane width} = \frac{c}{2\times30f} = \frac{3\times10^8}{2\times30\times14\times10^3}\text{m} = 357\,\text{m}$$

For Green Decameter

$$\text{Lane width} = \frac{c}{2\times18f} = 594\,\text{m}$$

Fox Red Decameter

$$\text{Lane width} = \frac{c}{2\times24f} = 445\,\text{m}$$

Average Lane width for Purple, Green, and Red Decameters

$$= \frac{357+594+445}{3} = \frac{1396}{3} = 465\,\text{m} \simeq 500\,\text{m}.$$

CHAPTER 15

SATELLITE NAVIGATION

15.1 INTRODUCTION

Radio navigational aids like VOR and Instrument Landing Systems (ILS) are now in use in airports around the world as an aid to aerial navigation, while LORAN and OMEGA are in use for marine navigation. Such navigational aids have several limitations. These include propagation error and other errors. Also, the coverage offered by those is neither universal for aircraft and ships nor global, except the OMEGA system with which, however, the positioning accuracy attainable is insufficient for use in aerial navigation. The advent of satellites led to the development of a new generation of universal navigational aids called (*a*) Doppler Navigation, followed by a more recent development called (*b*) Global positioning system (GPS). In GPS, radio signals from a number of orbiting satellites are received by an aircraft or ship, which determines its own position in space relative to the satellites whose coordinates are transmitted by the respective satellites to the aircraft or ship, as the case may be. The cost of the terminal equipment involved is also small, making the Global Positioning System to be the most cost effective and versatile universal navigational aid for the coming years. The GPS was, in fact, evolved to realize a cost effective Universal Satellite Navigational aid, free from the limitations of Doppler Navigation.

15.1.1 Doppler Navigation

The radio signal transmitted from a satellite exhibits a Doppler shift of frequency when received on the ground because of the rapid change of range of the satellite from the ground as it passes from one horizon to the other through the zenith, which corresponds to the position of minimum range. On either side of this position, the range is larger.

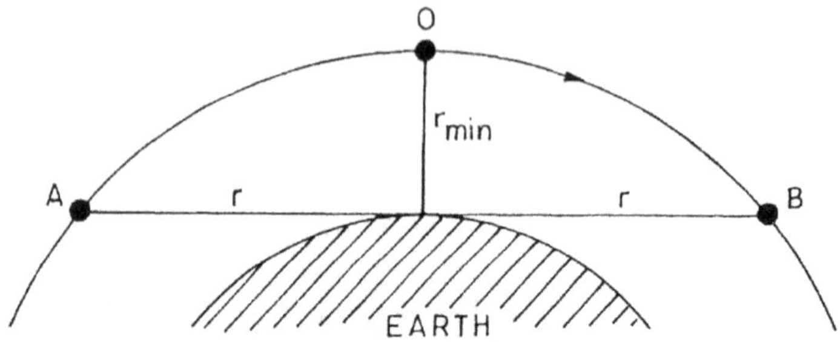

FIGURE 15.1 Satellite orbit: satellite rises at A and sets in at B passing through the overhead position O. Range r_1 is minimum at O, where it is r_{min}.

Initially after the satellite comes within the field of view near one horizon at A, the range decreasing from r attains a minimum value r_{min} when it is overhead at O, and then starts increasing again until it sets in below the other horizon at B, as shown in the Figure 15.1.

The line of sight range r of the satellite corresponds to a phase path $\phi = \frac{2\pi}{\lambda} r$. Due to changes of the range with time, we have a Doppler shift of frequency given by $\omega_d = \frac{d\omega}{dt}$.

The line of sight velocity of the satellite changes sign around the overhead position, the rate of change of range being zero to O when the satellite is overhead at O. Accordingly, the sign of the Doppler shift also changes around the overhead position O, with the frequency shift initially decreasing with the satellite approaching the zenith reaching a stage of zero Doppler shift. The Doppler shift of frequency with time may be used to determine the angular position of the satellite on the sky. The situation is analogous

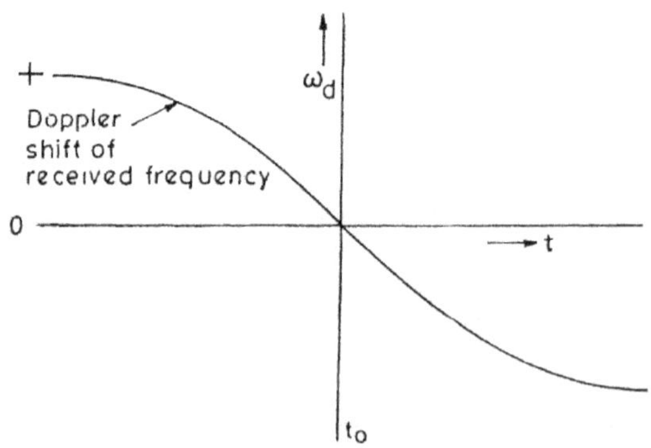

FIGURE 15.2 Doppler shift of received frequency with time, passing through ⟶ zero in overhead position of the satellite, at t_o.

to the Doppler shift of the whistling tone emitted by a train passing by the platform, as can be heard by anybody standing on the platform. The whistle initially appears high pitched, when the train is approaching, comes down to a pitch corresponding to the transmitted one when the whistle is passing nearest to the observer, beyond which it decreases further due to the receding train.

In the case of a satellite pass, the satellite position can be tracked by monitoring the Doppler shift. Conversely, if the position of the satellite is known, the position of the observer can be determined in relation to the satellite. If a number of orbiting satellites are within view at different positions, each with its position in the sky known accurately as a function of time from their known orbital parameters, then the position of an observer can be determined accurately with reference to the set of satellites within view, if the Doppler shift of the radio signals from each satellite is monitored and the results obtained are processed by a computer to derive the position of the observer. This is the principle of the Doppler satellite navigating system. This was first suggested by F. T. Clure as early as 1958, and a practical system based on the satellite Transit IB were operational for three months during the 1960s to prove the feasibility of single pass Doppler Navigation method.

The principle of operation of the transit system is shown in Figure 15.3. A typical operating sequence for the system is as follows: One of the system's receiving stations receives the Doppler signal and corrects it for ionospheric refraction effect, for which satellite transmissions at two phase locked frequencies are received simultaneously.

The ionospheric effect will be different at the two frequencies, the effect decreasing rapidly with increasing frequency which, in fact, allows the two frequency data to be utilized to extract the ionospheric correction. The corrected data are forwarded to the computing center where future orbital parameters are predicted. These predictions are transmitted over land lines in digital form to an injection station, meant for injecting the predicted data to the future positions of the satellite, after conversion into command format, together with the latest time corrections and any housekeeping commands for satellite operation. This information is in fact transmitted to the satellite when it becomes visible to the injection station, and stored in nondestructive headout for updating, after receiving the previous data. The success of such an injection is checked through the satellite's telemetry link, which continues to broadcast these data along with the corrected time in every minute. Any ground station or even a mobile station in an aircraft or ship can receive these broadcasts from the satellite with a simple receiver and then compute its own latitude and longitude, in addition to making any desired use of the accurate time information. Besides these, messages like weather news, traffic information, and similar other information could be added to the broadcasts.

In the Transit I system, the four onboard transmitters were organized in two systems, *A* and *B*. The *A* system is equipped with a 54 MHz unmodulated transmitter with 200 mW output power. In the *B* system, a 162 MHz class AB common emitter power amplifier was driven by a frequency trippler for the 54 MHz input producing 54 × 3 = 162 MHz, followed by a buffer amplifier at 162 MHz of power output 100 mW. The 54 MHz transmitter of the A system drives its input from a separate multiplier chain, excited by the output of a highly stabilized 3 MHz crystal controlled Colpitts transistor oscillator. AT-cut type crystals were used having a frequency stability of 1.10^{-8} for 30 min period and 2.5×10^{-7} for a three-month period.

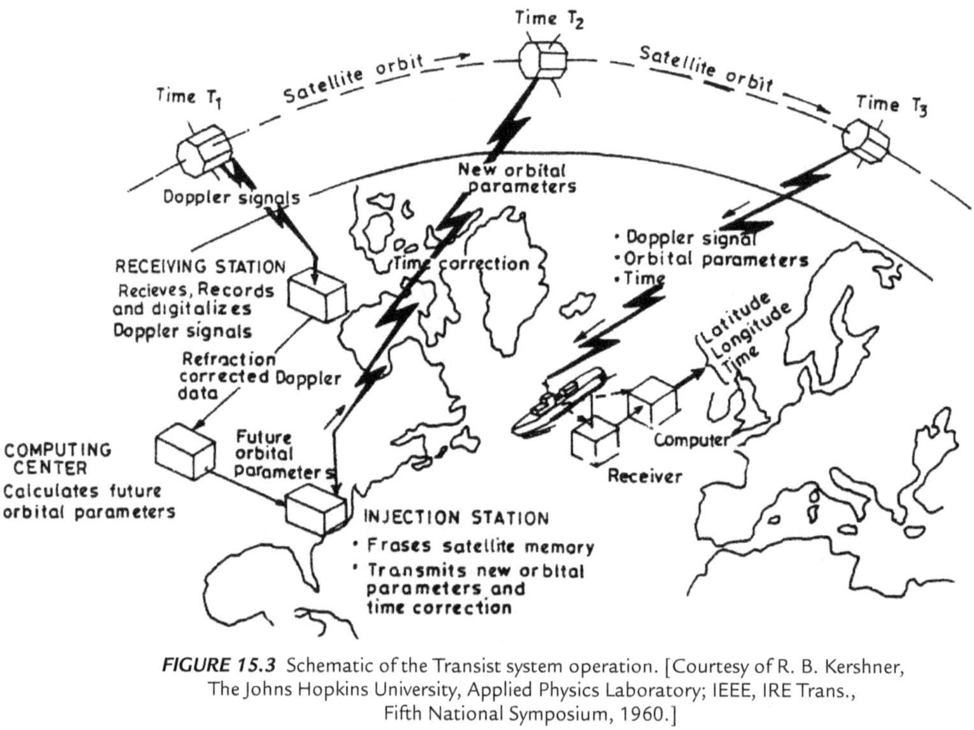

FIGURE 15.3 Schematic of the Transist system operation. [Courtesy of R. B. Kershner, The Johns Hopkins University, Applied Physics Laboratory; IEEE, IRE Trans., Fifth National Symposium, 1960.]

A second transmitter in the A system operating at 324 MHz derived its frequency from the same crystal oscillator multiplier chain as that of the 54 MHz transmitter. Likewise, the B system had a second transmitter at 216 MHz with an output power of 100 mW, also deriving its frequency from the same source as the 162 MHz transmitter with a single broadband antenna system. Transit satellite radiates all four, transmitting 54, 162, 216, and 324.

Besides Transit I A and I B, Transit II A was launched in June 22, 1960 using a separate FM/PM telemetry with all four navigation frequencies kept unmodulated. It worked for two years.

After this, transit III B, launched in February 21, 1961, carried the Transit Memory System (TMS) to receive, store, and periodically read out orbital parameters and other digital information, in addition to the previously described equipment. Transit IV A, launched in June 29, 1961, proved that Earth's equator is elliptical. Transit IV AB, similar to Transit IV A, was launched in November 15, 1961 carrying into orbit a piggyback satellite TRAAC (Transit Research and Altitude Control). TRAAC had a mission to test the feasibility of a spacecraft stabilization system using Earth's gravitational field.

It may be mentioned here that in all these cases of Doppler Satellite Navigation systems, an aircraft may obtain its coordinates by receiving the satellite Doppler signals. However, the motion of the aircraft itself relative to the satellite will lead to some additional line of sight Doppler effect which must be taken into consideration while estimating the positional coordinates of the aircraft.

15.1.2 Global Positioning System

The Global Positioning System (GPS) is a second-generation Satellite Navigation based on the measurement of the times of arrival of the time signals received from 3 or more orbiting satellites, whose positional coordinates in space are also transmitted. There are now 21 GPS satellites including 3 spares, in 12-hour circular orbits inclined at 55° to the equatorial plane, orbiting at an altitude of 20,000 Km, about halfway lower than the geostationary satellite as shown in Figure 15.4 (*a*). This will ensure that at least four satellites will be visible at a time from any location around the world for position determination from aircraft and ships. Signals from up to ten GPS Satellites may be received in a GPS receiver, and typically up to seven satellites are involved at a time for position determination of the GPS station.

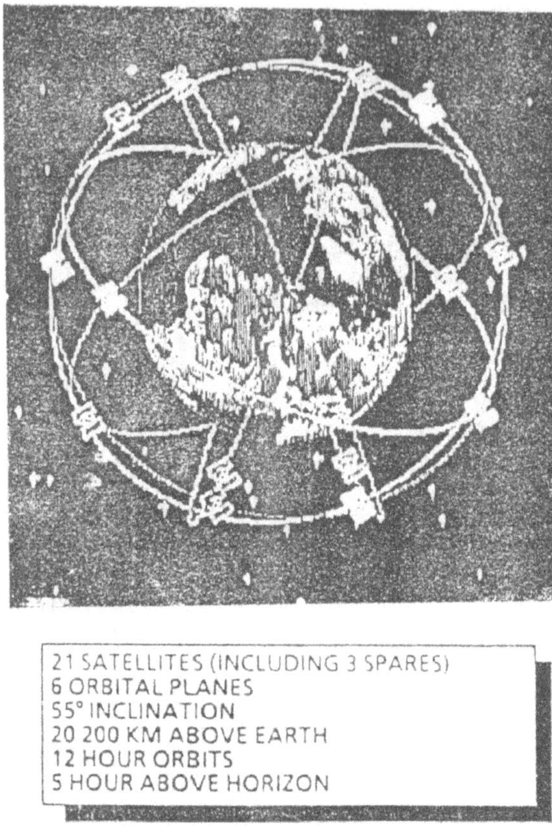

FIGURE 15.4 (*a*) Navstar Satellites Constellation.

The timing pulses are sent out by each satellite in the L-band using spread spectrum modulation and are received by the GPS receivers in aircraft or ships, as the case may be, processed by matched filters which are required to increase the precision of the arrival-time measurement of the pulse. The onboard GPS receivers are, however, much simpler than those required for Doppler navigation, and are expected to determine the position of an aircraft or ship with a precision of 16 m or better in three dimensions. It is planned by the ICAO (International Civil Aviation Organisation) to install GPS systems around the world and this may eventually render the conventional navigational aids for airports and ships redundant.

It may be mentioned here that satellite-based navigation and radio-positioning systems were first implemented in the United States in January 1964 with the launch of a Navy Navigation Satellite called the *Transit* system. The accuracy

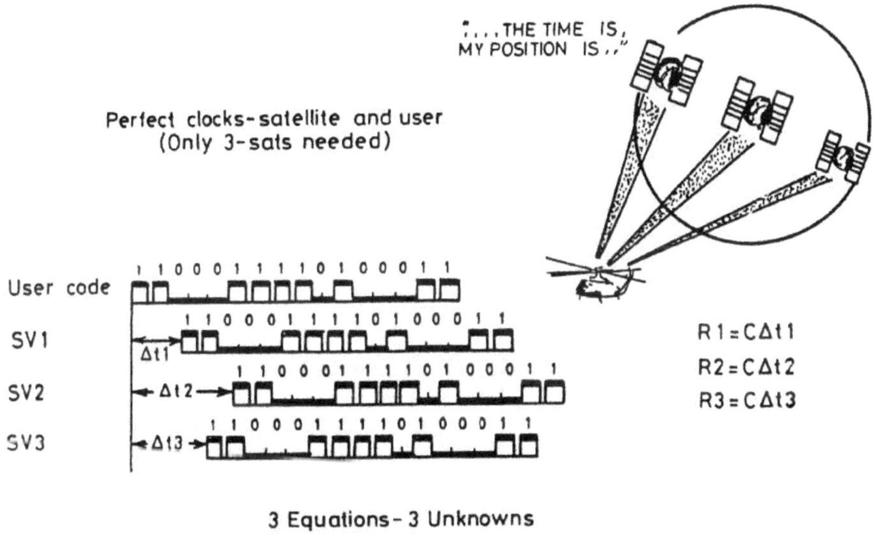

FIGURE 15.4 (*b*) Space Borne 1-way ranging.

of position fixing with Transit system, which was based on Doppler Navigation was, however, limited. With the introduction of the Global Positioning System, this limitation was overcome. GPS was, in fact, conceived in 1974 in the United States as a better and faster navigation system capable of providing three-dimensional position and velocity information for an aircraft or ship along with precise information to any user anywhere, on a twenty-four-hour basis. Besides GPS, a similar system called *Glonass* was developed in the USSR. There are also other satellite-based radio positioning systems being considered by other agencies, including GEOSTAR of the European Space Agency. Likewise, INMARSAT (International Maritime Satellite) has been in use for a decade, but later on its role has been primarily focused on communication. Similarly, the COSPAS/SARSAT System has been evolved for search and rescue operations, based on data communication. However, the GPS remains the only satellite-based universal positioning system that is available now for both military and civilian users.

15.1.3 Principles of Operation of GPS Navigation

The Global Positioning System (GPS) for Navigation is used to determine the positional coordinates of the user vehicle, which may be an aircraft or ship or a land mobile vehicle, by measuring its distance or range from 3 or more Satellites, whose positional coordinates are telemetered to the user by radio links using pseudorandom coded radio transmissions, as shown in Figure 15.4 (*b*). If the delay of the received codes relative to a locally generated identical reference code at the user location is Δt_1, Δt_2, and Δt_3 for the satellite transmissions from SV1, SV2, and SV3 respectively, then we have the respective ranges R_1, R_2, and R_3 of the Satellite given by

$$R_1 = c\Delta t_1 \tag{1}$$

$$R = c\Delta t_2 \tag{2}$$

$$R = c\Delta t_3 \tag{3}$$

The point of intersection of these three lines representing the ranges, R_1, R_2, and R_3 defines the special coordinates of the user vehicle, and that can be converted to latitude, longitude, and altitude by transformation of coordinates using a computer.

In a GPS Navigation System, the information of time is also required for which the signal from a fourth GPS satellite at a range R_4 also has to be utilized to obtain 4 equations relating R_1, R_2, R_3 and R_4 to Δt_1, Δt_2, Δt_3 and Δt_4 respectively. Three of these equations will allow us to determine the positional coordinates, while the fourth is used to derive the time information. Another way of looking at the problem would be to consider that we must have at least four equations to solve for the four unknowns X, Y, Z (positional coordinates) and T (time reference).

To measure the values of time delays for wave propagation between the satellite transmitter and the user receiver, the phase modulated radio signals in the L-band are first received by the GPS receiver from four GPS satellites. The received signals are then demodulated by phase demodulator and the resulting pseudonoise code is compared with the locally generated code to measure the delay between the envelopes of transmitted and the local pseudonoise codes. From these envelope delays, the range of each satellite is determined. For a higher precision of the range estimates, the phases of the carriers from the satellite transmissions are compared with the phase of the local reference oscillator serving as the clock. The PN sequence demodulation is, in fact, made by a coherent, clocked detection process. The initial estimate of range based on envelope delay measurement is called pseudorange measurement, while the finer estimate based on carrier phase measurement, is called pseudophase measurement, from which the phase path is, in fact, determined to estimate the range with a higher precision. In the equation for range the C/A code wavelength is divided by 1540 for the following reason: The C/A code operates at a chipping rate of 1.023 MHz, therefore, in the second the 1575.42 MHz wave will contain 1575.42/1.023 cycles = 1540 Hz.

15.1.4 GPS Segments

The Global Positioning System (GPS) is comprised of three distinct segments: (*a*) Space Segment, (*b*) Control Segment, and (*c*) User segment. The different GPS segments are as shown in Figure 15.5.

(*a*) GPS Space Segment. The GPS space segment comprises 18 satellites orbiting in circular orbits at an altitude of 20,200 Km at an inclination of 55° in a period of 12 hours. The relative positions have been arranged so that a minimum of 4 satellites will be visible to any user at a given instant of time, thus ensuring a global coverage. With the type of broadbeam antennas used in GPS receiving systems, up to 10 GPS satellite signals can be received, although typically 7–8 satellites are involved at a time.

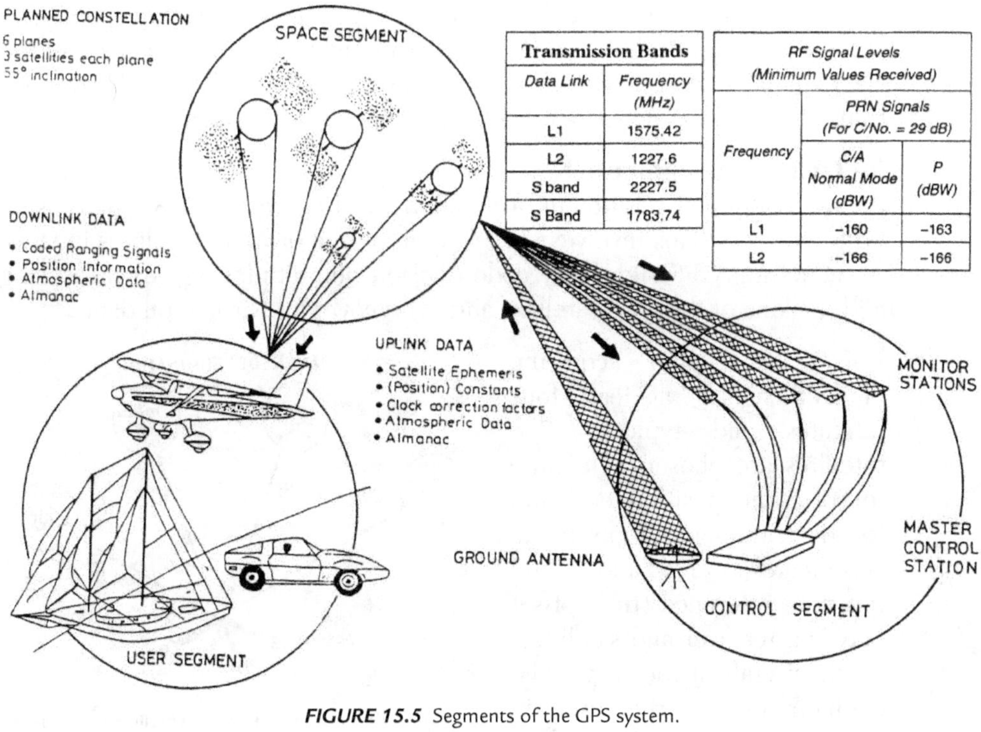

FIGURE 15.5 Segments of the GPS system.

Each satellite is designed to transmit radio signals at two L-band frequencies, one at 1575.42 MHz for commercial services and the other at 1277.6 MHz for military services only. The radiation pattern of the L-band transmitting antennas at the GPS satellites are specially designed to produce a uniform signal strength at the earth's surface independent of the position of the user. Besides these, for allowing control and telemetry function by ground stations, the satellite has an S-band antenna operating on a downlink frequency of 2227.5 MHz together with an uplink frequency of 1783.74 MHz. A typical GPS satellite weighs about 816 kg, consuming an average electric power of 700 watts derived from solar cell arrays, which, in turn, are backed by Ni-Cd (105 AH) batteries; the life span of a GPS satellite is typically 7–10 years.

***(b)* GPS Control Segment.** The control segment of GPS, required for monitoring and maintaining the performance of GPS satellites, involves four Monitoring Ground Stations distributed throughout the globe at these locations: (1) Hawaii, (2) Kwaje-lein (Pacific Ocean), (3) Diego Garcia (Indian Ocean), and (4) Accension Island (Atlantic Ocean). Each station, in fact, uses a GPS receiver to passively track all satellites in view to accumulate ranging data from the satellite signals, and the information obtained is further processed at the Master Control Station (MCS), in Colorado, to derive the orbits of the GPS satellite and update the navigation message of each satellite for onward transmission to the respective satellites through the ground stations.

***(c)* GPS User Segment.** The user segment of GPS is meant for providing the user with their position information by decreasing the GPS signals using GPS receivers. This involves (1) selection of one or more satellites in view, (2) acquiring GPS signals by code division demultiplexing, (3) measuring and tracking of the GPS satellite, and (4) recovering Navigation data.

For tracking and acquiring of GPS signals, at least four satellites and typically 7–8 satellites are chosen at a time to solve the position equation for determining the positional coordinates, *X*, *Y*, *Z* and *T*, and the time reference (time offset between receiver and satellites) of the aircraft or the ships. The configuration of the user in relation to the four GPS satellites is shown in Figure 15.6.

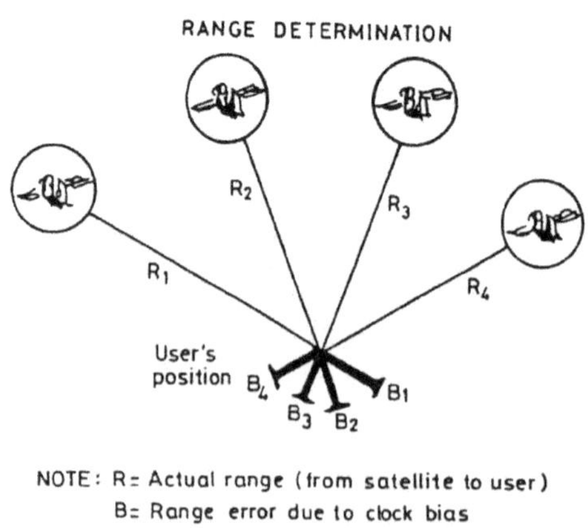

FIGURE 15.6 GPS navigation concept.

15.1.5 Format of GPS Navigation Message

The radio navigation signals transmitted by satellites consist of accurately timed L-band transmissions, at L(1575.42 MHz) and L(1277.6 MHz), using a code Division Multiple Access (CDMA) format on two orthogonal channels in phase quadrature. Each of the channels utilize two pseudorandom codes, known as (1) Coarse Acquisition (C/A) code and (2) Precision (P) code. The resulting (C/A) code equation of the GPS signal is shown in Figure 15.7.

It may be mentioned here that the data stream of ranging and timing information is superimposed on both the (C/A) code and (P) code. The resulting waveforms of the Data stream superimposed on the C/A code, as shown in Figure 15.8, indicate that the GPS data frame is divided into five 300 bit subframes each with a 6 sec duration. Each subframe consists of 10 words of 30 bits each, as shown in Figure 15.9. These words are encoded in the form of patterns of binary 1s and 0s that stand for various parameter values. The first two words in each sub-frame are Telemetry Word (TLM) and Handover Word (HOW) in which bits 1 through 17 are the most significant Bits (MSB) of the Time of Week (TOW) count. The remaining bits of the HOW words are used to present the registers generating the P code.

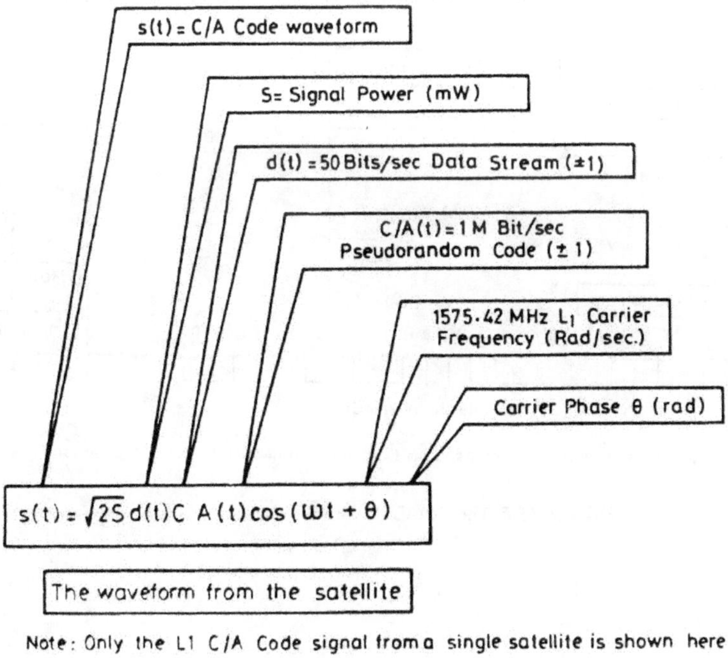

FIGURE 15.7 The GPS Signal waveform.

The third word contains the MSBs of the Z count. It stands for the GPS week number, and is incremented every 1.5 second. The zero time point of GPS was defined as midnight on Saturday night January 5, 1980/Sunday morning, that is, January 6, 1980. Thus, for instance, the first week of July 1990 corresponds to GPS week No. 547.

15.1.6 GPS Data Subframe

The contents of the five GPS data subframes include subframe 1, for clock correction parameters, subframes 2 and 3 containing satellite ephemeris, and subframe 4 reserved for alphanumeric messages for future operational applications, while subframe 5 contains almanac data for satellites.

The contents of subframes 1, 2, and 3 are fixed, while those of subframes 4 and 5 change on a page roll basis (1 to 25).

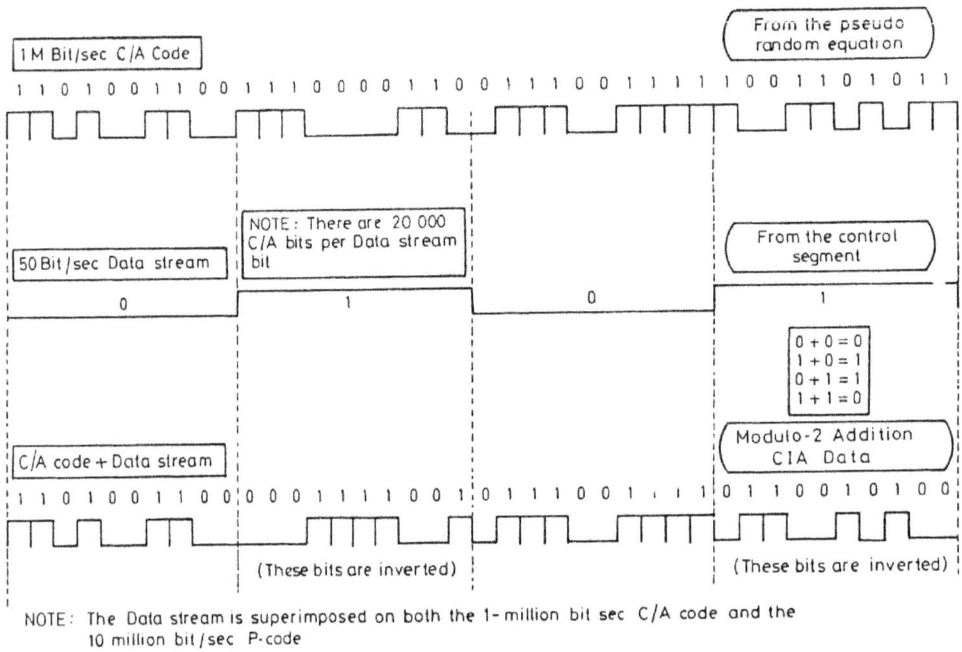

FIGURE 15.8 The C/A-Code with the Data stream superimposed*.

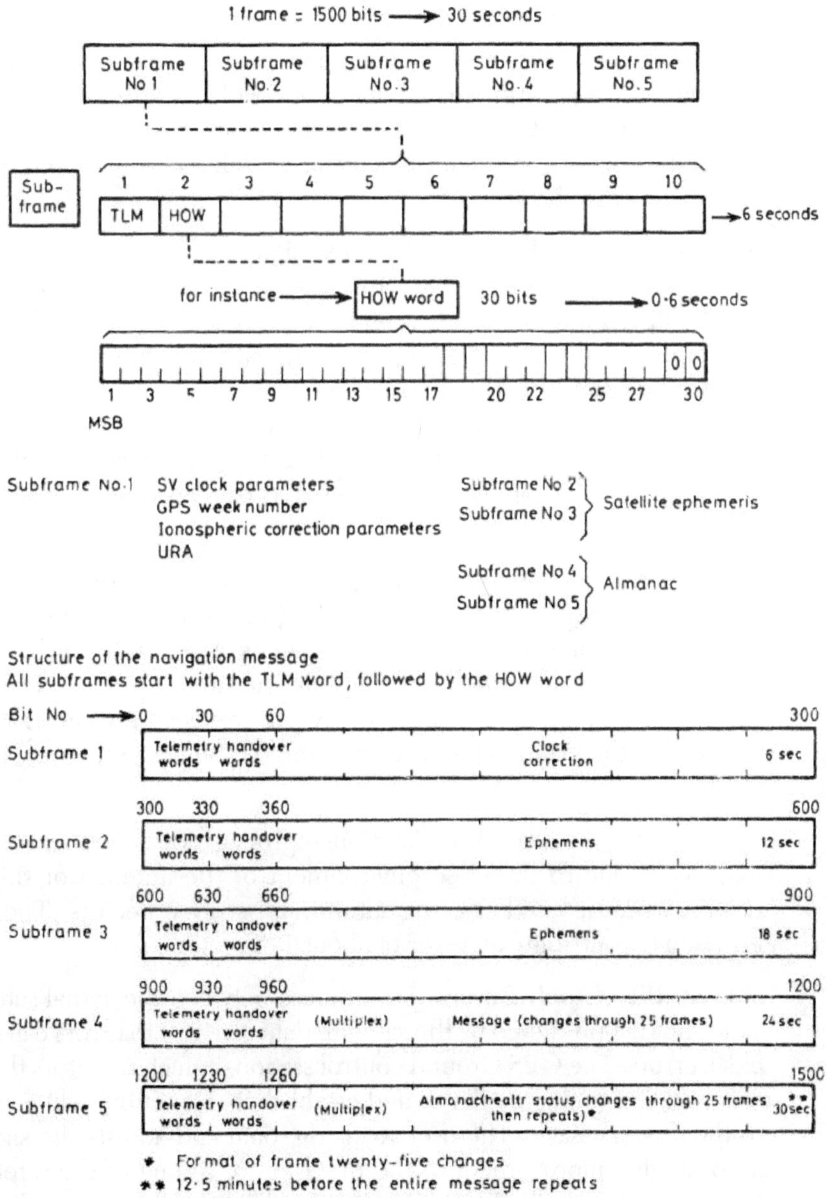

FIGURE 15.9 The content of the GPS data stream.

15.1.7 Sources of Errors in GPS

Three major sources of error may be introduced in GPS operation. These include: (*a*) propagation errors due to ionospheric and tropospheric propagation and multipath effects, (*b*) satellite clock errors, and (*c*) deliberately injected errors in codes and selective availability of unpolluted code. These are detailed as follows:

(*a*) Propagation Errors. The most significant propagation error is the ionospheric error, which may normally be as much as 20–30 meters during the day and 3–6 meters at night. However, over the midlatitudes such as that over India, where a strong latitudinal gradient of ionization prevails due to the proximity of equational anomaly, the ionospheric error may be as high as 50–150 meters. In the two frequency modes of GPS operation, using L (1227.6 MHz) and L (1575.42 MHz), the ionospheric effect may be largely removed by applying the inverse square law dependence of delay on frequency.

Besides the ionospheric errors, propagation errors due to propagation delay in the troposphere corresponding to 30 m may occur at low satellite elevation angles, but the errors are consistent and can be adequately modeled to eliminate the errors largely. Variation of index of refraction of the troposphere can also cause delay differences between the reference station and the user station by an amount of about 1–3 meters for low lying satellites.

In addition to the previously mentioned propagation errors, multipath errors may occur due to the close environment of the antenna or the presence of obstructions, buildings, vegetation, and stray reflectors. The multipath errors may contribute an error of about 1–3 meters.

(*b*) Satellite Clock Errors. Differences between the actual satellite clock time and that predicted by the satellite data give rise to errors called Satellite clock errors. The GPS Ground control station, in fact, monitors the free running satellite timing oscillator and establishes corrections which are sent up to the data message. The user reads the data and adjusts the signal timing accordingly. Improvement of the precision of timing of the satellite clocks by introducing the hydrogen MASER atomic clock in place of the existing cesium clock may improve the accuracy by at least two orders of magnitude, reaching a stability about 10^{-14}.

15.2 DIFFERENTIAL GLOBAL POSITIONING SYSTEM (DGPS)

The accuracy of position determination in Global Positioning System (GPS) can be improved greatly by using a Differential GPS (DGPS) technique, in which a reference station of known coordinates is used to receive the GPS station codes to estimate the error in positioning and then the error information transmitted by radio link to the GPS at the user location for incorporating the corrections in positioning information obtained by the user GPS. The DGPS concept is shown in Figure 15.10 (*a*), the basic concept being similar to that employed in differential OMEGA, differential Loran-C, and the translocation mode using TRANSIT. In DGPS, a GPS receiver installed at the Reference Station as shown in Figure 15.10 (*b*), located on a well surveyed site for its positional coordinates, serves as means of calibration for the GPS navigation system. Any errors common to both the receivers, at the Reference Station and user location, is reduced significantly by using the Differential technique; the degree of reduction of errors decreases with increase

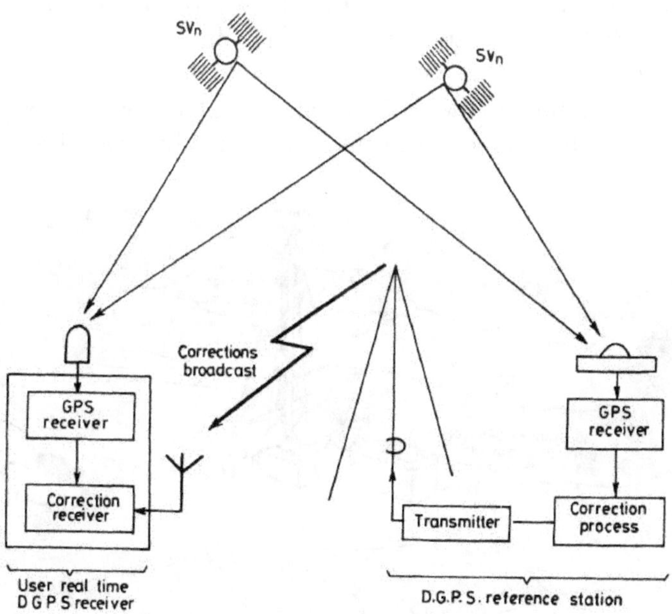

FIGURE 15.10 (*a*) Concept of differential GPS.

of spacing between the Reference and user locations, due to the ionospheric errors becoming uncorrelated with the increased separation of the stations. The radio communication of the GPS errors measured at the Reference Station to the user location is made using the existing VHF/UHF ratio links or the MF/HF ionospheric radio links, used as data link systems. Practical DGPS systems have, in fact, been developed successfully by several companies using MF Radio Beacon band in the range 250 kHz to 400 kHz, HF band in the range 1.5 MHz to 3.5 MHz, and UHF band.

It may be mentioned here that accuracies of GPS systems using only the C/A code is 100 m, while with DGPS, the accuracy is improved to 5–20 meters for dynamic navigation and to 3 meters for stationary applications. Using the P-Code the accuracies may be improved further. However, the access to P-Code is often restricted for military use with C/A code only being accessible to civilian users.

It may be mentioned here that the accuracy of DGPS is degraded by the ionospheric effects like scintillation, particularly over tropical regions, where the ionospheric electron density reaches a high value due to equatorial anomaly. Recently, attempts are being made to provide correction data to the users of GPS from a geostationary satellite which collects the data of ionospheric corrections over various user regions, so that an user of GPS may introduce the corrections to the position data instantaneously, allowing the GPS system to be useful even for Instrument Landing Systems (ILS).

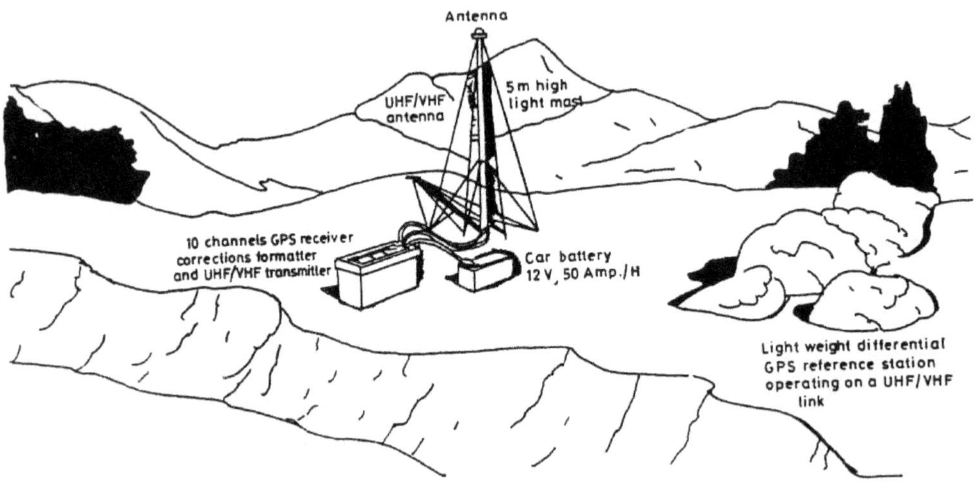

FIGURE 15.10 (*b*) Short range DGPS reference station configuration (SERCEL).

15.2.1 DGPS Configurations

Three different configurations of DGPS are possible, which include the following:

1. The Differential corrections are sent to the user in the form of simple additions of positional coordinates such as ΔX, ΔY, ΔZ using the same satellites for both stations.
2. The differential GPS Reference Station generates its own pseudorandom (PRN) signal, along with correction data with the 3 GPS satellites together with the Reference Station forming the group of GPS transmitters. The correction data, in the form of range errors to all the visible satellites, is superimposed on the satellite navigation message. This approach is useful where only 3 satellites are visible due to terrain topology or due to GPS satellite constellation configuration.
3. The errors in the pseudorange to all the visible satellites are calculated and transmitted to the users by the Differential GPS Reference Station. In this approach a user is getting the correction data in the form of errors in pseudoranges for all satellites, and he is free to use any constellation.

From the previous list it appears that there are two forms of differential corrections: (*i*) The user is applying the differential corrections prior to calculating his navigation fix, or (*ii*) he is adding in the form of simple additions (ΔX, ΔY, ΔZ) as in case 1 indicated previously.

15.2.2 Applications of GPS and DGPS

For marine applications an accuracy of 100 m, obtainable in GPS with C/A code, is adequate. However, for applications such as that in (*a*) precise landing of aircraft, (*b*) geophysical surveys of remote locations, (*c*) conducting search and rescue operations, (*d*) off shore oil exploration, (*e*) Seismic surveys, (*f*) Marine gravity, (g) Magnetic measurements, (*h*) Hydrographic surveys, (*i*) Physical oceanography operations, (*j*) Fixing, (*k*) Collision avoidance, (*l*) Harbor approach, and (*m*) Entry, the accuracy of simple GPS is inadequate and one may have to use DGPS to obtain accuracies better than 30 m, down to perhaps 3 meters. Some of the GPS applications for Military and Civil Services are illustrated in Figures 15.11 and 15.12.

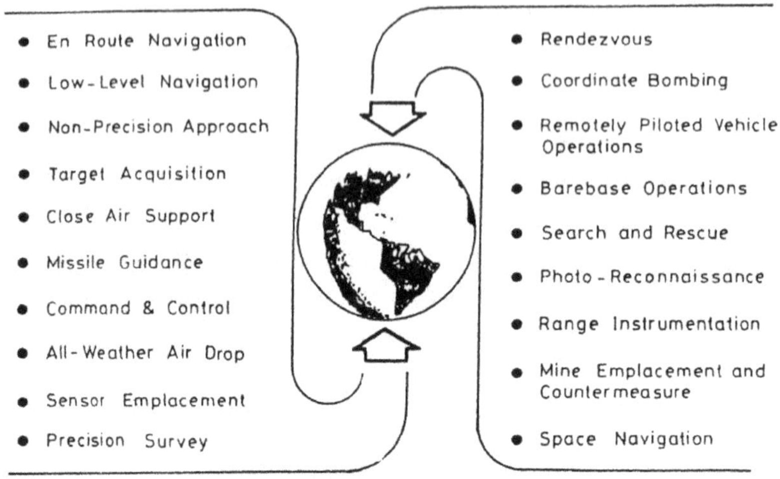

FIGURE 15.11 GPS Military applications.

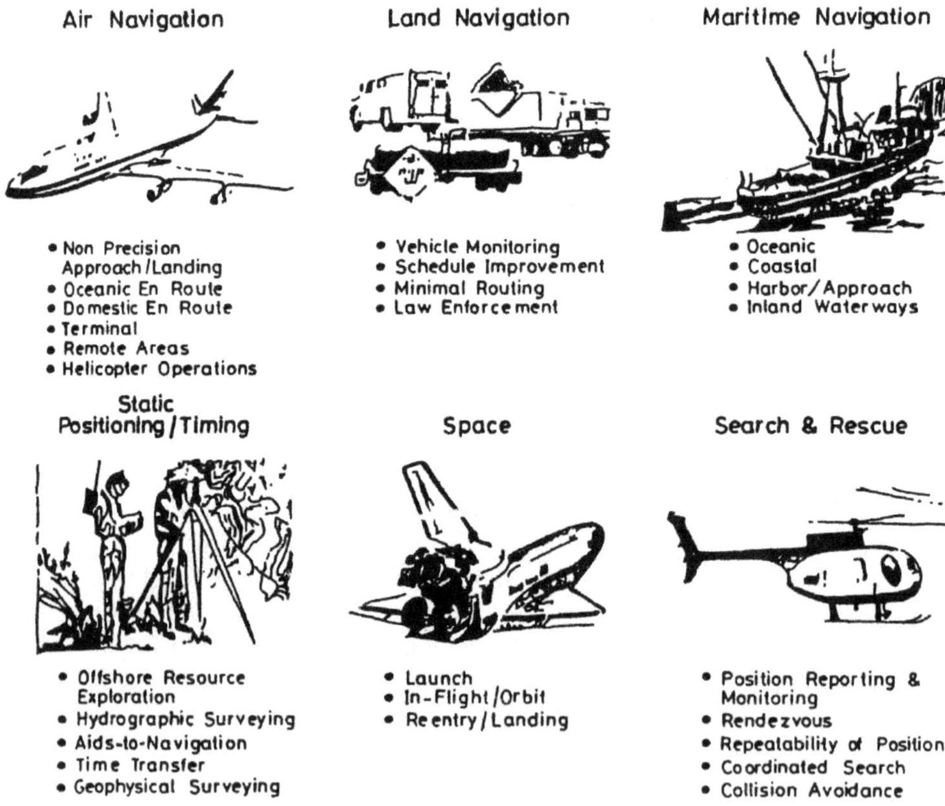

FIGURE 15.12 GPS Civil applications.

15.2.3 GPS Receivers

The configuration of a typical GPS/DGPS receiver is shown in Figure 15.13 (*a*) while a block diagram of the RF front end is shown in Figure 15.13 (*b*).

Referring to the figure, the RF front end of the Global Positioning System Receiver converts the frequency of the signal received by the helical antenna to an intermediate frequency by a double superheterodyne technique to process the data digitally for determining the user's geographic position. Low noise preamplification is obtained by using either low noise HEMT (High Electron Mobility Transistor), PHEMT (Pseudomorphic HEMT), Ga As MESFETs, low noise bipolar transistors, or silicon MMICs. Mixer functions can be accomplished by using silicon MMICs or Schottky diodes. Silicon MMICs or PIN attenuator diodes are used for the AGC. Bipolar transistors provide a low phase noise for the local oscillator, which is an LC oscillator

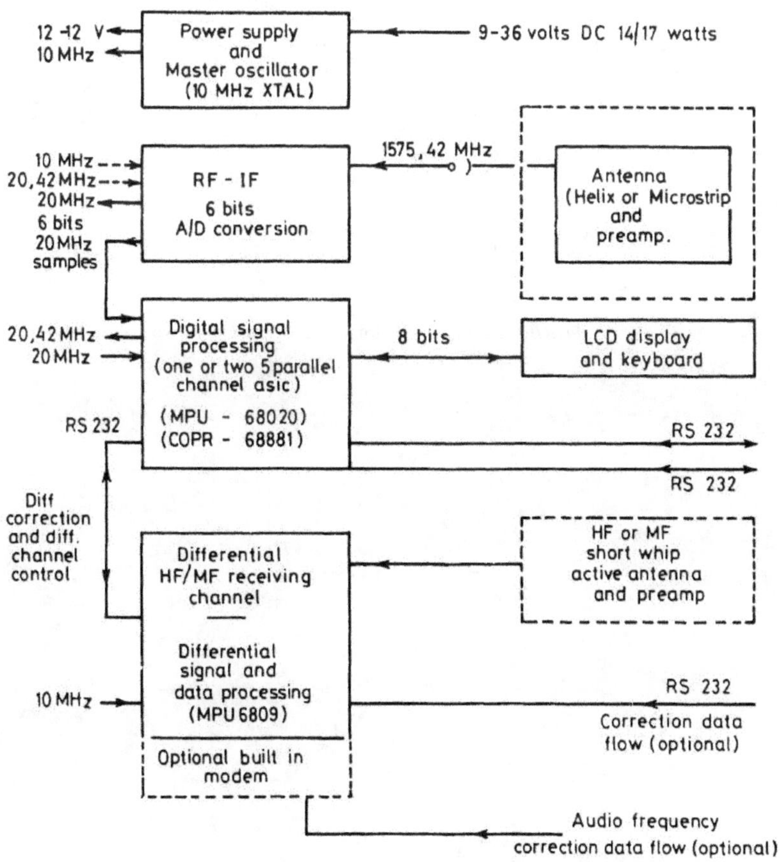

FIGURE 15.13 (*a*) Block diagram of a typical GPS-DGPS receiver.

phase locked to an X-tal reference oscillator. The signal processing circuit comprises a coherent clocked detector for the C/A code and/or P-Code, estimator for latitude, longitude, and altitude of the GPS station together with their errors, ΔX, ΔY, ΔZ, and the time error, ΔT.

It may be mentioned here that a comprehensive review of GPS and DGPS has been made by the Department of Electronics, Government of India in December 1991, in the form of a Technical Report entitled "Differential Global Positioning System (DGPS)," to initiate R & D activities in India in this emerging area of strategic importance. Much of the information about GPS and DGPS highlighted previously are based on this Technical Report.

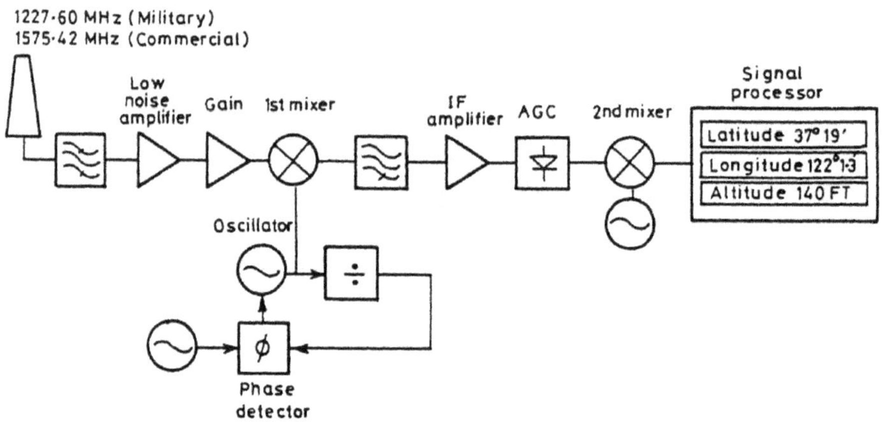

FIGURE 15.13 (*b*) RF Solutions for the Global Positioning System.

CHAPTER 16

VESSEL TRAFFIC MANAGEMENT SYSTEM

16.1 INTRODUCTION

Like Radar Control of air traffic, the movement of ships in and around ports are recently being controlled by Radar control networks, called Vessel Traffic Management System (VTMS). The basic requirements to be fullfilled by VTMS include:

1. Obtaining and recording of pre-arrival and pre-departure information on all vessels calling at the port.
2. Acquisition and tracking of vessels using the port area and its approaches.
3. Monitoring of vessel movement within the surveillance area, the allocation of water and berth space, and the coordination of traffic movement.
4. Provision of information to the vessel underway to enable navigation to be safe and expeditious.
5. The recording and transmission of vessel data to the VTMS Control Center.

To carry out the previously mentioned basic functions of VTMS, the hardware and software requirements include the following:

a. Surveillance radar to suit the specific needs of the port or the coastal area to be surveyed.

b. Communication links for transmitting and receiving radar surveillance information and other informations associated with VTMS.

c. Display consoles and workstations with all necessary software packages to operate VTMS.

d. Database to handle and analyze all the information associated with VTMS.

16.1.1 Configuration of VTMS

A typical configuration of the Vessel Traffic Management System (VTMS) is shown in Figure 16.1.

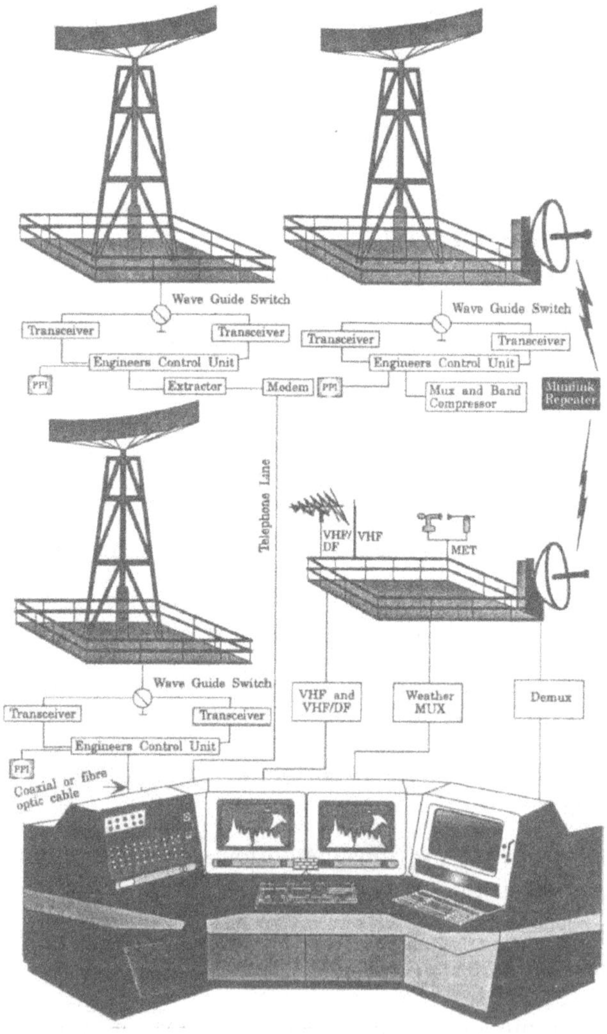

FIGURE 16.1 Possible vessel traffic system configuration.

The three Radars, often located on opposite sides of the route, are distributed along the route of the ships, in and around the port. The frequency of operation of the radars may be *S*-band or *X*-band. The *X*-band units have higher resolution of display than the *S*-band unit. However, the *S*-band unit will be less affected by rain environment. An *X*-band unit with circular polarization will be useful to provide better accuracy in adverse rainy weather and an improved clutter performance.

In a typical VTMS, the parameters of the three types of surveillance Radars, used simultaneously, are given as follows:

	Radar 1	Radar 2	Radar 3
Antenna band and polarization:	*X*-band	*S*-band	*X*-band
	horizontal polarization	horizontal polarization	H/V/circular polarization
Antenna diameters	18′	14′	22′
Antenna beamwidth (H)	0.44°	1.8°	0.40°
Antenna beamwidth (V)	± 6°	± 12.5°	± 6
Antenna Gain	35 dB	30 dB	35 dB
Antenna rotation rate	20 RPM	20 RPM	20 RPM
Transmitter Power	0.75 kW	1 kW	1.1 kW

The antenna beamwidth in the vertical (V) plane, as indicated in the previous list, is more than an order of magnitude bigger than the beamwidth in the horizontal (H) plane, thus making a vertical fan beam of the antenna, which helps in high resolution azimuthal scanning of targets. To produce the fan beam, a parabolic cylindrical reflector antenna is used as shown in Figure 16.1.

One unit of each type of radar may be used in a port, installed at different locations along the route of ships, to take the specific advantages of each type of unit indicated previously. In one of these locations, the master control unit of the VTMS may be installed. The three units may then be interconnected to form a network by using a microwave communication link, fiber optic or cable link, or even by using a telephone line to send the digital radar data from different locations to the master station for display in the master control operators workstation, shown at the bottom of Figure 16.1. Besides the VTMS, a VHF Doppler Direction finder and weather monitor are also used as a supporting facility to VTMS by connecting those to the operators workstations.

16.1.2 Operator's Workstation

The operators workstation may be of Single Display Version, Dual Diplay Version, or Triple Display Version, as shown in Figure 16.2, depending on the number of surviellance radars used for the VTMS. The latest display and electronics technology are utilized in the operator's workstation to ensure high reliability and mean time between failure. In case of any system malfunction of VTMS, the built-in test and performance monitoring system will immediately altert the traffic operator by means of audible alarms and flashing text on the data display.

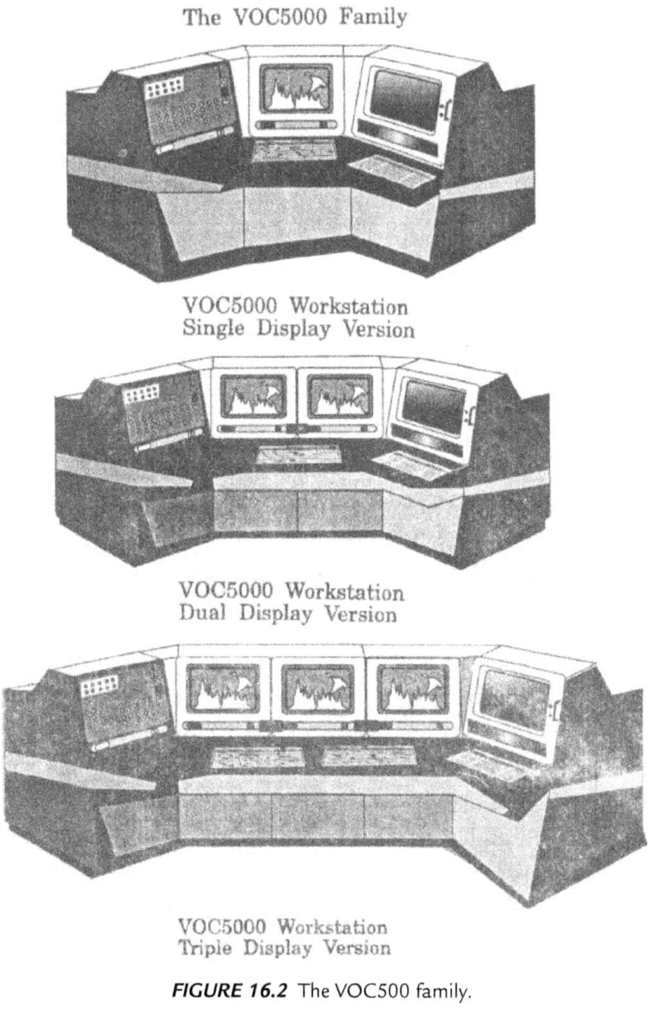

FIGURE 16.2 The VOC500 family.

The main basic components of the operator's workstation are shown in Figure 16.3.

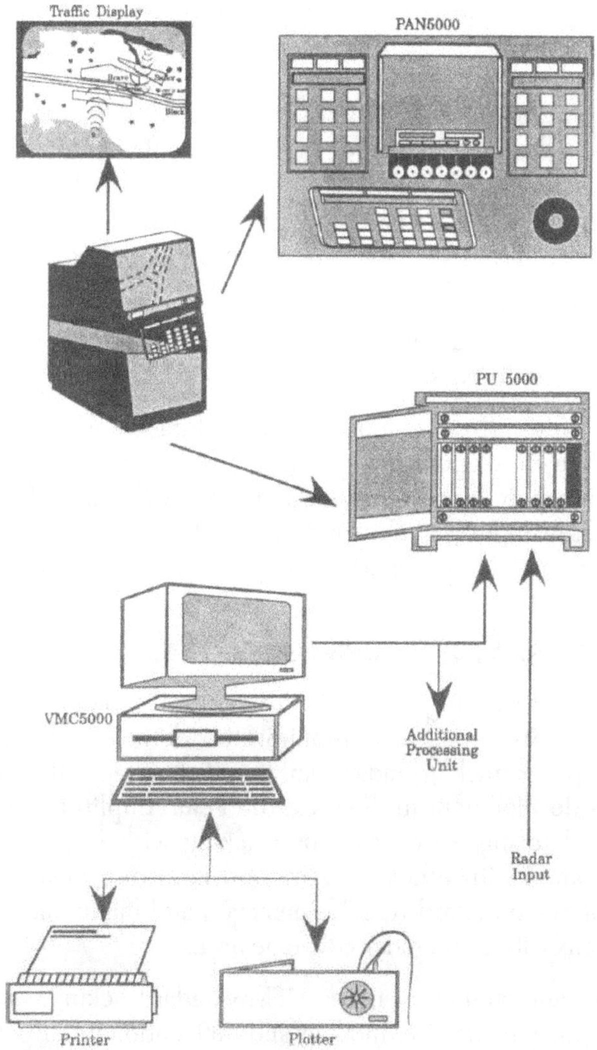

FIGURE 16.3

A single display version of an operations workstation, as shown at the middle of the figure on the lefthand side, consists of traffic display, operator panel, and a high-capacity processing unit as shown by arrows directed radially from the workstation. The processing unit also processes the input from the multipurpose computer shown at the bottom of Figure 16.3 on the left side. The processing unit comprises all necessary softwares and hardwares, such as printed circuit boards, fan tray, and power supply. The operator panel comprises dedicated buttons for the most frequently used functions, an alphanumeric display for target data information, a standard size keyboard for operator interaction, and a tracker ball for selective display functions. The multipurpose computer is used for editing and modification of the synthetic maps such as that for the coast line, buoys, and sea marks. In addition to this, the computer is used for logging and replay and for implementation of a complete vessel database for port management. For a hard copy of the vessel database, the computer is capable of handling printers and plotters.

The traffic displays used for presentation of the scan converted raw radar video, tracked targets, and synthetic maps are obtained as high resolution multicolor displays.

The multipurpose computer is a multitask desktop computer system run under a unix operating system in an 80486 hardware environment. The multipurpose computer communicates with the traffic display workstation shown in Figure 16.1 through a Local Area Network (LAN) and may also contain one or more of a series of services programs.

16.1.3 Radar Tracking System for VTMS

The radar tracking system is designed to process radar signals in order to automatically track and monitor the detected vessels. For this purpose monopulse tracking radar may be used, as described in section 3.16. The latest development in the area, the Peak Amplitude Detection mechanism, is used to ensure continuous tracking with high accuracy at any range. Combined with other software and hardware facilities, continuous tracking of any observed vessel is achieved and maintained as long as the target remains within the radar coverage area.

Important parameters in the VTMS Radar tracking system include: (*a*) high data capacity; (*b*) 500 moving and 500 stationary targets; (*c*) 32 echoes per radar sweep; (*d*) high speed real-time processing of target data with a sample rate typically 25 MHz, providing a 6.25 m range resolution and 8 bit signal amplitude resolution; (*e*) high position accuracy, being about + / – 2 meter for range and 0.1 deg for bearing; and (*f*) monitoring of every detected target, as long as required, and calculating position, speed, and other related information, using high speed digital circuits for target detection and automatic clutter suppression.

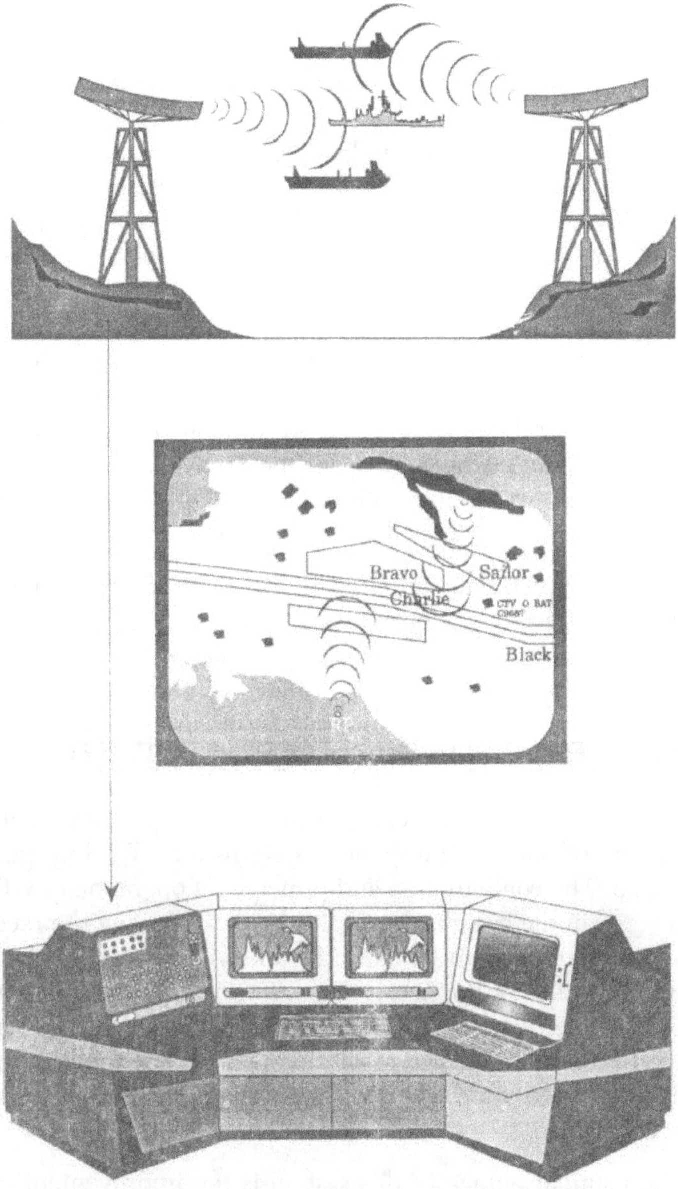

FIGURE 16.4 Multiradar tracking.

In practice, Multiradar Tracking is employed by a setup shown in Figure 16.4. The purpose of multiradar tracking is to combine and integrate navigational target data, such as speed and course of vessels, from several radars into one unique integrated target table, to have a more accurate and wider vision than that possible with a Single Tracking Radar. The reliability of Multiradar Tracking is higher than that of single radar tracking, as the failure of a radar will continue collecting the data by the remaining radars, to continue the display of the tracking data. The operator, in fact, also has the freedom to display the data from any individual Tracking Radar in a multiradar Tracking system. This is often very important, particularly in adverse weather when the *S*-band radar, even with its lower resolution compared to the *X*-band one, may work better. Also a particular target may be tracked more precisely by a radar near it compared to one located at a distance. The descision about which radar is to be selected as the prime sensor for a specific target in a particular environment is taken by the operator through critical quality tests such as "handover" and match. The extensive number of target parameters evaluated for the purpose in the software algorithm used includes: (1) target speed, (2) target course, (3) target position, (4) target size, (5) target hit-rate, (6) target signal-to-clutter ratio, (7) target distance to closest radar, (8) target distance to closest "neighbor" target, and (9) target maneuvering state.

It may be mentioned here that VTMS has recently been installed in India at the Calcutta and Bombay ports.

16.2 GPS TRANSPONDER SYSTEM FOR VTMS

Substantial effort has been put in over the past few years into the development of satellite transponder systems for VTMS purposes using GPS or DGPS. The configuration and principle of operation of GPS and DGPS are covered in the last chapter of this book. Following the recommendations of international authorities like IMO and IALA, it is possible to develop VTMS systems utilizing the accurate position information obtained from GPS or DGPS. Such a system enables a VTMS Control Center to improve its coverage beyond the limitations of radar. Automatic identification and information exchange of preselected ship-related data is another important feature of the system.

The main advantage of this system is the improvement of safety and efficiency in the area of interest of the VTMS Control Center. It is expected that, within a few years, the number of satellite communication systems would be sufficient for global coverage, when the GPS transponder systems may easily be converted into a feet Management system with worldwide coverage.

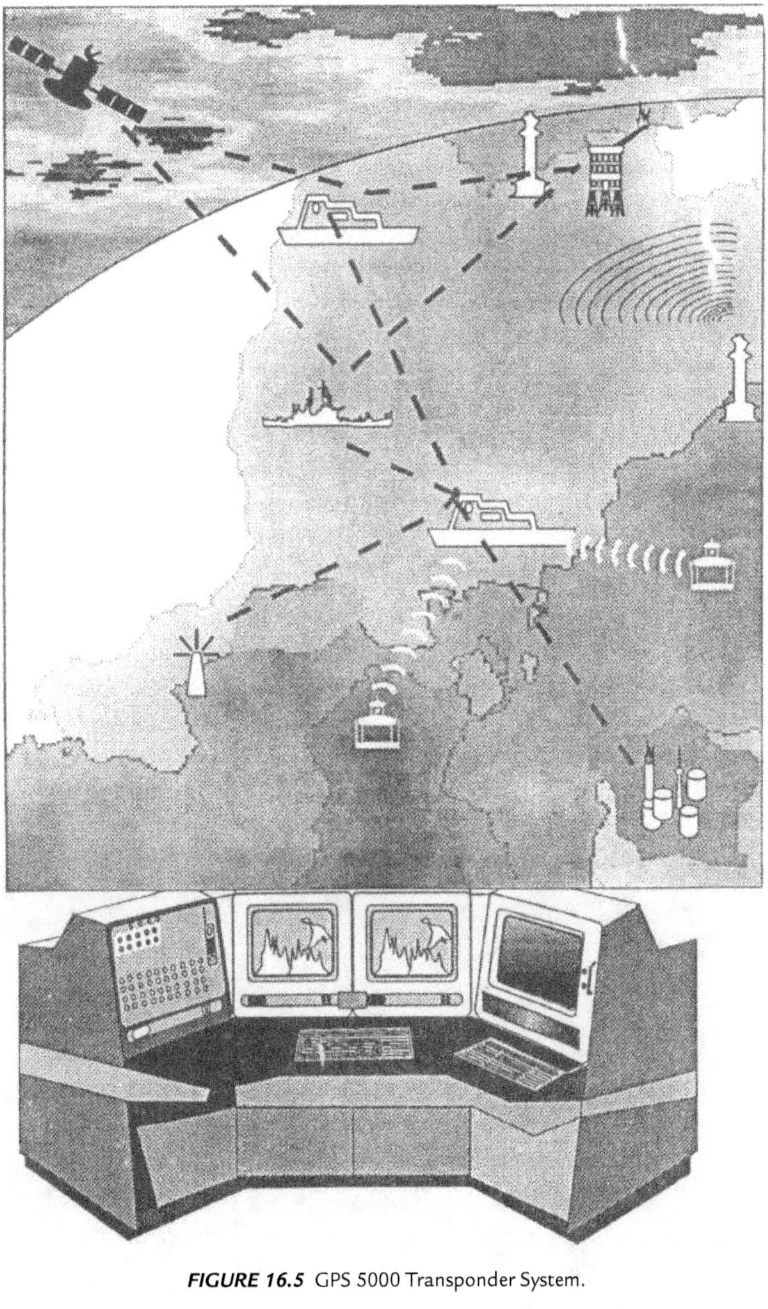

FIGURE 16.5 GPS 5000 Transponder System.

CHAPTER 17

NAVIGATION AND MODERN MEASURING TECHNIQUES

17.1 INTRODUCTION

Navigation is the process of planning, reading, and controlling the movement of a vehicle or craft from one place to another. The word *navigate* is derived from the Latin word *navis* meaning "ship" and *agere* meaning "to move" or "to direct."

All the navigational techniques involve locating the navigator's position compared to known patterns or locations.

17.2 LATITUDE AND LONGITUDE OF A PLACE

The latitude of a place on the earth's surface is the angular distance north or south of the equator. Latitude is expressed in **degrees** ranging from 0° at the **Equator** to 90° at the North and South poles. The latitude of the **North Pole** is 90°N, while the latitude of the **South Pole** is 90°S. Historically, mariners calculated latitude in the Northern Hemisphere by observing the North Star **Polaris** with a **Sextant** and sight reduction tables to take out error for height of eye and atmospheric refraction. In general, the height of **Polaris** in degrees of arc above the horizon is the latitude of the observer.

Similarly, the longitude of a place on the earth's surface is the angular distance east or west of the **Greenwich meridian**. Longitude is also expressed in **degrees** ranging from 0° at the Greenwich meridian to 180° east and west. For example, **Sydney, Australia** has a longitude of about 151° east while

New York City has a longitude of about 74° west. Mariners struggled for a long time to determine precise longitude. The problem was solved with the invention of the **marine chronometer**. Longitude can be determined correctly if the precise time of a sextant sighting is known.

17.3 MODERN NAVIGATIONAL METHODS

Most modern navigation relies mainly on positions obtained electronically by receivers gathering information from satellites. Other techniques rely on crossing **lines of position** (LOP), which refer to two different things: a line on a chart and a line between the observer and an object in real life. A bearing is a measure of the direction to an object. If the navigator measures the direction in real life, the angle can then be drawn on a **nautical chart** and the navigator will be on that line on the chart. In addition, navigators often measure distances to objects. On the chart, a distance produces a circle or arc of position. Circles, arcs, and hyperbolae of positions are referred to as lines of position.

If the navigator draws two lines of position and they intersect each other, then he must be at that position. A **fix** is the intersection of two or more LOPs. When only one line of position is available, this may be evaluated against the dead reckoning position to establish an estimated position.

Lines or circles of position can be obtained from many sources, some of which are as follows:

1. celestial observation (a short segment of the circle of same altitude, but represented as a line, in general)
2. compass bearing to a charted object
3. terrestrial range (manmade or natural) when two charted points are found to be in line with each other
4. on certain coastlines, a depth sounding from **echo sounder** or land **lead line**
5. radar range to a charted object, etc.

There are other methods hardly used today like "dipping a light" to find the geographic range from observer to lighthouse.

Navigation methods have changed from time to time. Each new method has enhanced the mariner's ability to complete his voyage. Some modern types of navigation are depicted in Table 17.1.

TABLE 17.1
Navigation Methods and Applications

Modern navigation methods	
Description	**Application**
Dead reckoning (DR) in which one advances a prior position by using the ship's course and speed. The new position is called a DR position; only course and speed determine the DR position. Correcting the DR position for **leeway**, current effects, and steering error result in an estimated position (EP). An **inertial navigator** develops highly accurate EP.	Used at all the times.
Pilotage involves navigating in **restricted waters** with quick determination of position relative to geographic and hydrographic features.	When within sight of land.
Celestial navigation involves reducing celestial measurements to lines of position using suitable tables, **spherical trigonometry**, and **almanacs**.	Used mainly as a backup to **satellite** and other **electronic systems** in the open ocean.
Radio navigation uses radio waves for determining position by either **radio direction finding systems** or hyperbolic systems, like **DECCA**, **OMEGA**, and **LORAN-C**.	Losing ground to GPS.
Radar navigation uses radar for determining the distance from or bearing of objects whose position is known. This process is different from radar's use as a collision avoidance system.	Primarily when within radar range of land.
Satellite navigation uses artificial earth satellite systems, like GPS, to determine position.	Used in all situations.

Usually the practice of navigation involves a combination of all these different methods.

17.4 DEAD RECKONING

Dead reckoning is a process of estimating present position by projecting course and speed from a known past position. It is further used to predict a future position by projecting course and speed from a known present position. The DR position is, however, only an approximate position as it does not allow for the effect of leeway, current, helmsman error, compass error, or any other external influences. The navigator uses dead reckoning by various means, such as

1. to find sunrise and sunset,
2. to evaluate the accuracy of electronic positioning information,
3. to predict landfall, sighting lights, and arrival times, and
4. to predict which celestial bodies will be available for future observation.

The most important use of DR is to project the position of the ship into the immediate future and avoid hazards to navigation. The navigator tends the DR plot, updating it, and uses it to evaluate external forces acting on the ship. The navigator always considers the DR plot to avoid navigation hazards. A fix taken at each DR position exhibits the effects of current, wind, and steering error, and allow the navigator for staying on track by correcting for them.

Before the development of the **lunar distance method** or the **marine chronometer**, the dead reckoning was the primary method to determine **longitude** available to mariners.

17.5 PILOTING

Piloting also known as pilotage involves navigating a vessel in restricted waters and to fix its position frequently. Procedures differ from vessel to vessel, between military, commercial, and private vessels.

A military navigation team usually consists of several people. A military navigator might have bearing takers stationed at the gyro repeaters on the bridge wings to take simultaneous bearings. The civilian navigator on the other hand must often take and plot them himself. The military navigator has a bearing book and someone to record entries for each fix, but the civilian navigator simply plots the bearings on the chart as they are taken and does not record them at all.

17.6 CELESTIAL NAVIGATION

Celestial navigation systems are based on the findings of the positions of the **Sun, Moon, Planets,** and **navigational stars**. Such systems are also used for terrestrial navigating as for interstellar navigating. By knowing which point on the rotating earth a celestial object is above and determining its height above the observer's horizon, the navigator can find the distance from that subpoint. A **Nautical almanac** and a **chronometer** are used generally to compute the subpoint on earth a celestial body is over, and a **sextant** is used for measuring the body's angular height above the horizon. That height can then be used to compute distance from the subpoint for creating a circular line of position. A navigator shoots a number of stars in succession for giving a series of overlapping lines of position. The position where they intersect is the celestial fix. The moon and the sun may also be used. In fact the sun can be used by itself to shoot a succession of lines of position.

17.7 MARINE CHRONOMETER

To accurately measure longitude, precise time of a sextant sighting must be recorded. The spring-driven marine chronometer is a timepiece used aboard ship for providing accurate time for celestial observations. A chronometer differs from a spring-driven watch principally in that it contains a variable lever device for the purpose of maintaining even purpose of pressure on the mainspring, and a special balance designed to compensate for temperature variations.

A spring-driven chronometer is set to Greenwich means time (GMT) approximately and is not reset until the instrument is overhauled and cleaned. The difference between GMT and chronometer time is determined and then applied as a correction to all chronometer readings. Spring-driven chronometers must be wound at about the same time each day.

Quartz crystal marine chronometers have replaced spring-driven chronometers aboard many ships due to their higher accuracy. They are maintained on GMT directly from radio time signals. This eliminates both chronometer error and watch error corrections.

The basic element used for time generation is a quartz crystal oscillator. The crystal is temperature compensated. It is hermetically sealed in an evacuated envelope. A calibrated adjustment capability is provided to adjust the aging of the crystal.

The chronometer can be operated for a minimum of 1 year on a single set of batteries. Observations may be timed and ship's clocks set with a comparing watch. This is then set to chronometer time and taken to the bridge wing for recording sight times. For practical purposes, a wrist watch coordinated to the nearest second with the chronometer is adequate. A stop watch, either spring wound or digital, may be used for celestial observations. In this case, the watch is started at a known GMT by chronometer, and the elapsed time of each sight added to this for getting GMT of the sight. All chronometers and watches should be checked with a radio time signal regularly. Times and frequencies of radio time signals are listed in publications such as **Radio Navigational Aids**.

17.8 MARINE SEXTANT

A critical component of celestial navigation is to find the angle formed at the observer's eye between the celestial body and the sensible horizon. The sextant, an optical instrument, can be serve this purpose. It consists of two primary assemblies. The frame is a rigid triangular structure with a pivot at the top and a graduated segment of a circle (referred to as the "arc") at the bottom. The next component is the index arm, attached to the pivot at the top of the frame. At the bottom there is an endless vernier which clamps into teeth on the bottom of the "arc." The optical system consists of two mirrors and a low proper telescope. One mirror, called the "index mirror," is fixed to the top of the index arm over the pivot. With the movement of the index arm, this mirror rotates, and the graduated scale on the arc indicates the measured angle, that is, altitude. The other mirror, known as the "horizon glass," is fixed to the front of the frame. One half of this horizon glass is silvered while the other half is clear. Light from the celestial body strikes the index mirror and is reflected to the silvered portion of the horizon glass, then back to the observer's eye through the telescope. The index arm is manipulated by the observer so the reflected image of the body in the horizon glass is just resting on the visual horizon and then seen through the clear side of the horizon glass.

Adjustment of the sextant consists of aligning the optical elements to eliminate "index correction." The practice of taking celestial observations from the deck of a rolling ship through cloud cover and with a hazy horizon is yet the most challenging part of celestial navigation.

17.9 INERTIAL NAVIGATION

It is a DR type of navigation system. It computes the position based on motion sensors. Once the initial latitude and longitude is established, the system receives impulses from the motion detector. Thus, it measures the acceleration along three or more axes, enabling it continually and accurately to calculate the current latitude and longitude. It has some advantages over other navigation systems, namely, (*i*) once the starting position is set, it does not require outside information, (*ii*) it is not affected by adverse weather conditions, and (*iii*) it cannot be detected or jammed by the enemy. The U.S. Navy has successfully developed a Ships Inertial Navigation System (SINS) during the **Polaris missile** program to insure a safe, accurate, and reliable navigation system for its missile submarines. Inertial navigation systems were in use until **satellite navigation** systems (GPS) became available.

17.10 RADIO DIRECTIONS FINDER

A **radio direction finder** (RDF) is a device to find the direction to a **radio** source. Due to the ability of radio to travel very long distances "over the horizon," it makes a particularly good navigation system for ships and aircraft that might be flying at a distance from land.

The working of RDFs can be done by rotating a directional **antenna** and listening for the direction where the signal from a known station comes. This system was widely used in the 1930s and 1940s. In navigational applications, RDF signals are provided in the form of *radio beacons*, a radio version of a **lighthouse**. The signal is a simple **AM** broadcast of a **morse code** series of letters, which the RDF can tune in to find if the beacon is "on the air." Most modern detectors can also be tuned in any commercial radio stations, useful due to their high power and location near major cities.

DECCA, OMEGA, and **LORAN-C** are three hyperbolic navigation systems of similar type. DECCA is a **hyperbolic low frequency radio navigation** system that was first deployed during **World War II** which could be used to achieve accurate landings. The primary use of LORAN C was for ship navigation in coastal waters. Fishing vessels were mostly postwar users, but it was also used on aircraft, including an early (1949) application of moving-map displays. The system was deployed in the North Sea and was used by helicopters operating to **oil platforms**. After being shut down in 2000, it has been superseded by systems like the American **GPS** and the planned European **Galileo positioning system**.

The OMEGA Navigation System was the first global **radio navigation** system for aircraft, operated by the **United States**. It was developed by the United States Navy for military aviation users. Initially, the system was used for navigating nuclear bombers across the North Pole to Russia and it was found to be useful for submarines too. Owing to the success of the **Global Positioning System,** the use of OMEGA declined during the 1990s, was terminated in 1997, and all stations ceased operation.

LORAN is a terrestrial **navigation** system using **low frequency** radio transmitters that use the time interval between radio signals received from three or more stations to determine the position of a ship or aircraft. The current version of LORAN in common use is LORAN-C, which operates in the **low frequency** portion of the EM spectrum from 90 to 110 **kHz**. Many nations are users of the system, including the **United States, Japan**, and several European countries. Russia uses a nearly exact system in the same frequency range, called **CHAYKA**. LORAN use is in steep decline, with **GPS** being the primary replacement. However, there are attempts to enhance and re-popularize LORAN.

17.11 RADAR NAVIGATION

If a vessel is within radar range of land or special radar aids to navigation, the navigator can find distances and angular bearings to charted objects and use these to establish arcs of position and lines of position on a chart. A fix consisting of only radar information is known as a radar fix.

Types of radar fixes include "range and bearing to a single object," "two or more bearings," "tangent bearings," and "two or more ranges." Parallel indexing is a technique defined by William Burger in 1957. This technique involves creating a line on the screen parallel to the ship's course, but offset to the left or right by some distance. This parallel line permits the navigator to maintain a given distance away from hazards.

Some techniques have been developed for special purposes; for example, the "contour method" involves marking a transparent plastic template on the radar screen and moving it to the chart to fix a position. Another technique, known as the Franklin Continuous Radar Plot Technique, involves drawing the path a radar object should follow on the radar display when the ship stays on its planned course. At the time of the transit, the navigator can check that the ship is on track by checking that the pip lies on the drawn line.

17.12 SATELLITE NAVIGATION

Global Navigation Satellite System (GNSS) is the term for satellite navigation systems to provide positioning with global coverage. A GNSS permits small **electronic** receivers for determining their location (viz. **longitude, latitude,** and **altitude**) to within a few **meters** using **time signals** transmitted along a **line of sight** by **radio** from **satellites**. Receivers on the ground with a fixed position can further be used to calculate the precise time as a reference for the purpose of scientific experiments. The **United States** NAVSTAR **Global Positioning System** (GPS) is a fully operational GNSS.

Since the first experimental satellite was launched in 1978, GPS has become an indispensable aid to navigation around the world. GPS is also an important tool for **map-making** and **land surveying** and provides a precise **time reference** utilized in many applications including scientific study of **earthquakes**, and **synchronization** of telecommunications networks. Developed by the **Department of Defense** in the United States, GPS is named NAVSTAR GPS (Navigation Satellite Timing And Ranging Global Positioning System).

17.13 NAVIGATION APPLICATION WITH MOBILE TELEPHONY

The navigation application is divided into two modules: (*i*) the "Shortest Path Calculation" module, which is the core part of the study, and (*ii*) the "Display" module that shows mainly the user interface design according to the functionality of the application on WAP phones.

The "Shortest Path Calculation" module is executed in the GIS server and the "Display" module is executed on the Web server. The process of the "Shortest Path Calculation" module includes the inputs and validation of the roads and the intersections, and calculation of the shortest path between the two locations. The final product from this module is a map in a wireless bitmap format displaying the highlighted shortest path. The map is next sent to the "Display" module for display purposes.

CHAPTER 18

MILLIMETER WAVE RADARS AND RADIOMETERS

18.1 INTRODUCTION

The millimeter wavelength portion of the electromagnetic spectrum has many advantages over the microwaves such as broad bandwidths, higher spatial resolution, low probability of interception and interference, and small antenna and equipment size. The present decade is in fact characterized by efforts to extend communication and radar systems and navigational aids to frequencies of mm-wave range. The advent of millimeter wave solid state devices and passive components in recent years apparently triggered a revolution in millimeter waves. The excellent performance, ruggedness reliability, precision, and maneuverability of millimeter wave systems are very attractive features for defense applications like radar and navigational aids and missile trackers. Also in civil applications like intercity broadband communication of voice, video, facsimile, and data, video conferencing systems and countrywide satellite computer networking, millimeter waves offer a unique advantage of its broad bandwidth capability. Besides these, the millimeter wave radiometers with broad bandwidths offer excellent opportunity of detecting enemy targets by measuring the temperature differentials with respect to the surroundings.

In all these applications the crucial design parameters are often determined by the condition of propagation of millimeter waves through the atmosphere. During World War II, an attempt to build a radar in the 1 cm wavelength range was in fact a disappointment, apparently due to the choice of the exact frequency near the wave vapor molecular absorption resonance line, at which range it was seriously curtailed by the absorption. This fact inhibited the development of radar technology at millimeter waves for a long period. Also, J. H. Van Vleck's theoretical prediction about a strong atmospheric absorption line at 60 GHz was classified during World War II for its millitary significance in developing severe communications. After World War II development of millimeter wave telecommunications began at NRL and Bell Telephone Laboratories (BTL) in the United States.

For satellite links where the propagation path is slant and is quite often nearly vertical, the total atmospheric attenuation could be estimated by integrating the attenuation in various horizontal slabs of atmosphere of elementary thickness, existing at various altitudes along the propagation path. The total attenuation for vertical propagation is of the order of a few decibels at the atmospheric window frequencies.

The millimeter waves are relatively less affected by fog, smoke, and dust compounds than those at infrared and optical wavelengths. However, the rain attenuation is a major problem in millimeter waves, and studies of rain attenuation in a tropical region would be very informative in obtaining preliminary data for application in the design of millimeter wave systems like Radar, Radiometer LOS, and satellite links.

18.2 MM-WARE RADARS AND RADIOMETERS FOR DEFENSE APPLICATIONS

Some of the interesting applications of millimeter wave radars are mentioned as follows:

Battlefield Surveillance radars at 70 GHz have been designed to cover a range of up to about 50 km with a beamwidth of 0.2–0.3° and a scan sector of 30°.

In a millimeter wave beamrider guidance system missile, the missile electronics senses and proceeds along the maximum of the antenna beam of the millimeter wave radar. A luncher for the missile close to the radar initially rides or locks on to a broad first beam. As the missile advances, the beam is progressively made sharper in the second and third beams, each of which point toward the enemy target, like a tank. A very sharp beam not only ensures precision of the attack but also ensures a comfortable signal strength at the missile received right up to the final phase for the beamrider electronics in the missile to be actuated.

18.3 DEVELOPMENT OF A COHERENT MM-WAVE RECEIVER FOR 94 GHZ

The coherent millimeter wave receiver developed for the radio link at 94 GHz is configured as shown in Figure 18.1. It differs from the detailed configuration included in RPE-4 in that the subsystems for polarization, phase shift, and online calibration are excluded during the development of the receiver, to be used over the short path length of 100 m.

Also, the receiving antenna used for the short path is a 24 dB pyramidal horn instead of the 1′ Cassegrainian dish antenna.

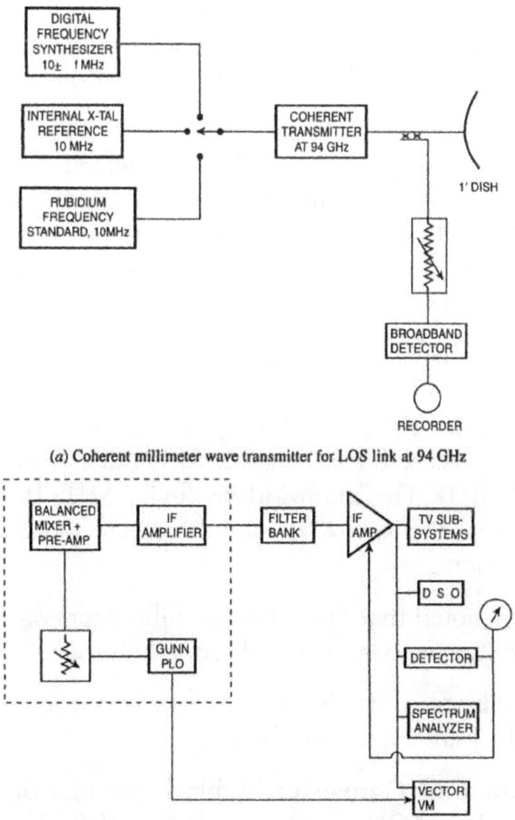

(*a*) Coherent millimeter wave transmitter for LOS link at 94 GHz

(*b*) Coherent millimeter wave receiver for 94 GHz radio link.

FIGURE 18.1

The low noise front end converter is developed from millimeter wave components and subsystems. The main IF amplifier is AGC controlled to have a wide dynamic range of the receiver. The IF output is at 200 MHz, and it is divided into 4 parts by power dividers and fed simultaneously to

1. an HP digital storage oscilloscope of range dc – 300 MHz,
2. an Takeda Riken Spectrum analyzer of range 10 KHz – 4.5 GHz,
3. an IF detector with an Omnispectra Schottky barrier diode to monitor the power of IF output,
4. a Vector Voltmeter to measure the amplitude and phase of the 200 MHz IF output, and
5. a TV subsystem for FM demodulation and display.

The digital storage oscilloscope (DSO) displays the IF waveforms which can also be recorded with an 8-pen graphics plotter. The DSO is also useful in recording the spectral patterns of the IF output on the graphics plotter.

The power level of the If output is monitored and recorded on a strip chart recorder by using the output of the IF detector, operated in the square law region, while the amplitude and phase of the IF output is measured with an HP vector voltmeter fed with a 200 MHz reference signal derived from the PLO reference oscillator. The IF detector is also used to control the gain of the IF amplifier using an Automatic Gain Control (AGC) circuit.

The bandwidth of the IF amplifier can be reduced from 500 MHz down to 20 MHz by using passive IF filters in the IF chain. It can be reduced further by feeding the 200 MHz IF output to a TV receiver subsystem, which is a double superheterodyne receiver with a first IF of 36.5 MHz and a second IF of 5.5 MHz. The bandwidth of the 5.5 MHz IF stage is 100 kHz and this IF output is fed to an FM demodulator to recover any frequency modulation of the carrier.

It may be noted that the coherent millimeter wave receiver along with the TV subsystem constitute a triple superheterodyne FM receiver for 94 GHz.

The development of different blocks of the coherent receiver shown in Figure 18.1 are detailed as follows:

(*a*) **Front end Converter.** A block diagram of the coherent front end converter for 94 GHz is shown in Figure 18.2.

The Front end converter consists of:

1. a waveguide balanced mixer using Silicon Schottky barrier diodes followed by a low noise IF preamplifier in an integrated unit, and

2. a 22 mW Gunn PLO as the LO fed to the mixer LO input through a level set attenuator. The LO power input required is + 5 dBm for the silicon mixer employed. The noise figure of the front end is 6 dB and the conversion gain is 20 dB.

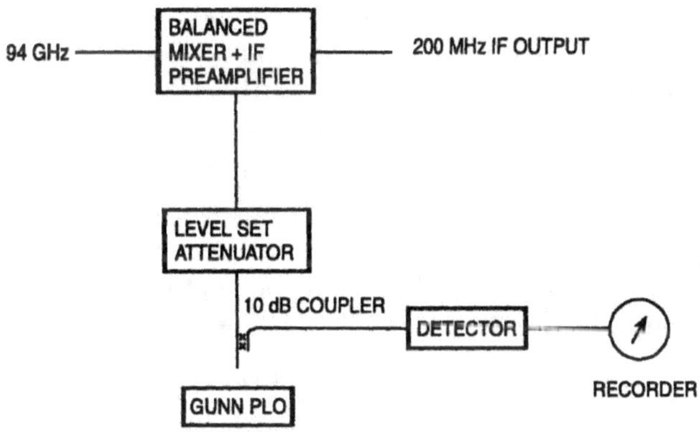

FIGURE 18.2 Coherent front end converter for 94 GHz.

(*b*) **Main IF Amplifier.** The main IF amplifier of the coherent receiver consists of two IF modules:

1. an AGC controlled 1st IF module based on Avantek AGC-553 followed by
2. an IF output amplifier, Avantek UTC-223 feeding
3. an omnispectra IF detector 2087-6001-00 with silicon point contact diode as shown in Figure 18.3.

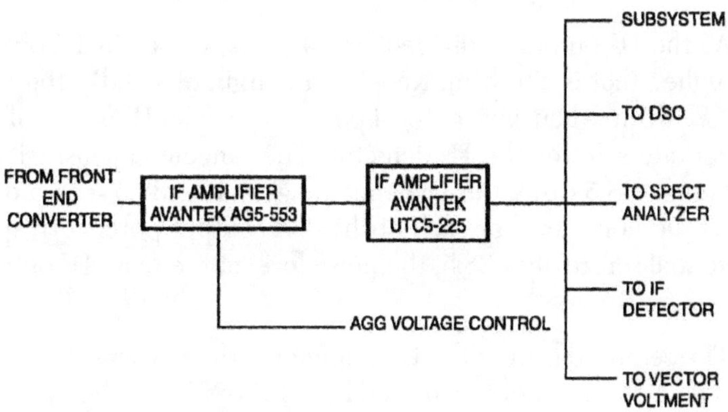

FIGURE 18.3 IF amplifier and detector.

(*c*) **If Detector, AGC, and Dynamic range.** The tangential sensitivity of the detector at a bandwidth of 1 MHz is –45 dBm and can be operated with an input up to –20 dB, in square law region. For a 20 MHz IF bandwidth the tangential sensitivity is $(-45 + 5 \log_{10} 20)$ dBm = –38.5 dBm. The departure from square law behavior at input levels exceeding –20 dBm can be corrected by post detector digital processing with a microprocessor. The detector can handle power levels up to 100 mW or + 20 to (–38.5) dBm = 58.5 dBm. Application of AGC to the IF by AGC-553 will increase the dynamic range by about 30 dB, making it 88.5 dB. Application of AGC, however, renders the output input characteristic of the IF Amplifier roughly logarithmic. Deviation from the true logarithmic law can also be corrected by post detector digital processing with a microprocessor. In practice, a margin of safety of about + 3 dB may be kept to minimize the risk of detector burnout. The dynamic range then becomes 85.5 dB.

(*d*) **IF Amplifiers.** The total gain of IF amplifiers, other than the IF preamplifier which is built-in with the balanced mixer, may be estimated as follows:

Receiver Noise level referred to its input

$$N_i = FkTB = 4 \times 4 \times 10^{-21} \times 20 \times 10^6 = 3.2 \times 10^{-13} = -95 \text{ dBm}$$

assuming $F = 6$ dB = 4, the front end noise figure

and $\quad B = 20$ MHz $= 20 \times 10^6$ Hz: the IF bandwidth

$[kT = 1.38 \times 10^{-23} \times 300 = 4 \times 10^{-21}]$

Noise level at the input of the IF Amplifier = –95 + 20 = –75 dBm, assuming a conversion gain of front end, 20 dB.

The largest IF output that the detector can handle is + 20 dBm.

As the IF output is divided into 4 parts, the main IF output can be 6 dB higher, that is, 26 dBm. Keeping a margin of + 3 dB, the IF output may be + 23 dBm, when each of the 4 ports receives an IF power of + 17 dBm, which is quite safe for the IF detector. The tangential sensitivity of the detector being –38.5 dBm, the Dynamic range = 17 + 38.5 = 55.5 dB, without AGC. As the noise level at each of the 4 IF output ports corresponds to the tangential sensitivity –38.5, the noise level at the main IF output is –38.5 + 6 = –32.5 dBm.

The required gain of the IF amplifier is then $G_{IF} = -32.5 - (-75) = 42.5$ dB. A gain of this order is achieved by UTC5-222, which also produces an output of + 23 dB, as desired. The dynamic range with such a linear IF amplifier without AGC is thus 55.5 dB as indicated previously.

To introduce AGC, an AGC controlled IF amplifier, AGC-553, is inserted between the Front end converter and the IF amplifier. The AGC 553 amplifier has a gain of 44 dB, an AGC range of 45 dB, and an output power –4 dBm at 1 dB compression. The overall gain of the AG5-553 and a UTC5-221 (35dB) can be controlled between the limits 34 to 35 + 44 = 79 dB. By proper AGC control the Dynamic range is increased by, at least, 30 dB to attain a value of 85.5 dB as the dynamic range of the IF amplifier and hence that of the coherent millimeter wave receiver for 94 GHz.

18.4 TUNABLE MM-WAVE TRANSMITTER FOR 53 GHZ

A block diagram of the millimeter wave transmitter developed for 53 GHz, tunable over a 1 GHz band from 52.2 to 53.2, is shown in Figure 18.4.

The transmitter is based on a 50 mw varactor tuned Gunn oscillator with voltage controlled frequency tuning and frequency modulation facilities. The modulation may be frequency division multiplexed subcarriers modulated by multichannel time division multiplexed data or it may be due to a TV camera/TV pattern generator or a multichannel telephone terminal. The transmitter is tunable over 50 channels each of width 20 MHz, which may accommodate a Frequency modulated video channel or a number of PSK/FM data transmission channels or FM/FM telephone channels, or a suitable combination of these, as desired.

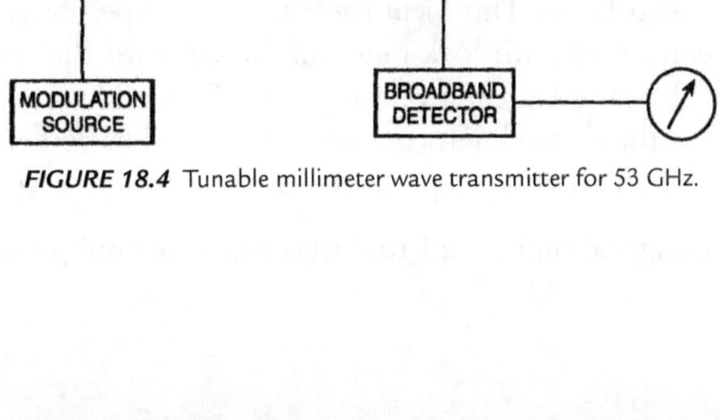

FIGURE 18.4 Tunable millimeter wave transmitter for 53 GHz.

The transmitter output is monitored both in frequency and power, using a tunable frequency meter and an HP power meter with Hughes power sensor, respectively. The broadband detector, which serves as an indicator for detecting the null during frequency monitoring, can also be used to monitor the amplitude of the millimeter signal with the tuning meter set to off resonance. The amplitude data may also be converted to power level data by using the known input impedance of the detector. This indirect estimate of transmitter power may, in fact, be calibrated periodically by direct measurement with the power meter.

The Gunn source is temperature stabilized, electronically, to reduce the thermal drift of the frequency of the transmission to within acceptable limits.

18.5 TUNABLE MM-WAVE RECEIVER

A block diagram of the tunable millimeter wave receiver developed for 53 GHz is shown in Figure 18.5.

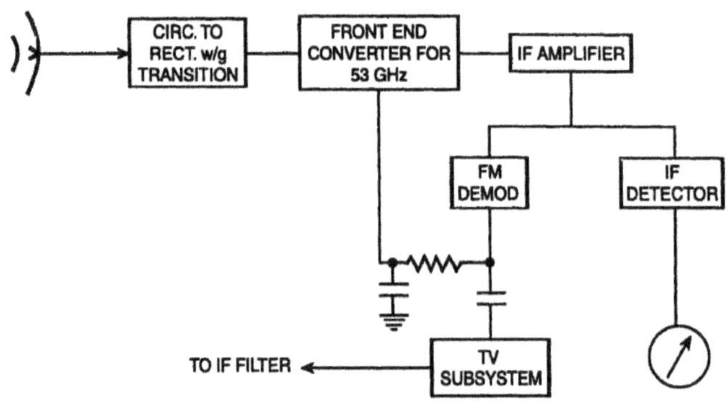

FIGURE 18.5 Millimeter wave tunable receiver for 53 GHz.

The front end converter is based on a Silicon diode balanced mixer-IF preamplifier fed with a varacter tuned Gunn VCO as LO tunable over a 1 GHz band from 52.3 to 53.3 GHz to receive millimeter wave radio signals in the range 52.5 – 53.5 GHz. The local oscillator is temperature stabilized and is also AFC controlled with lock range of 20 MHz for the receiver tuning to track any drift of the transmitter frequency. The FM demodulator for AFC also serves as the demodulator for the video modulation. The multichannel PSK data/telephonic signal is processed further for the respective destinations.

The configuration in the Front end converter unit is shown in Figure 18.6.

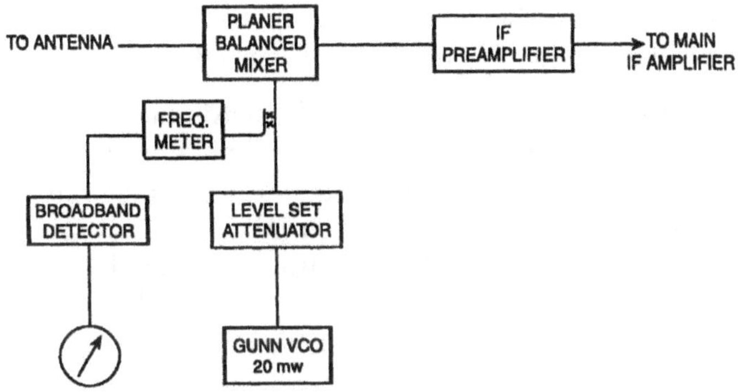

FIGURE 18.6 Front end converter.

18.5.1 IF Main Amplifier and IF Demodulator

A block diagram of the IF Main amplifier and IF demodulators for FM and AM signals for the 53 GHZ receiver are shown in Figure 18.7.

Dynamic range and bandwidth of IF amplifier and sensitivity:

Sensitivity $= F.kT.B$

$$= 3.2 \times 4 \times 10^{-21} \times 20 \times 10^{6} = -96 \text{ dBm}$$

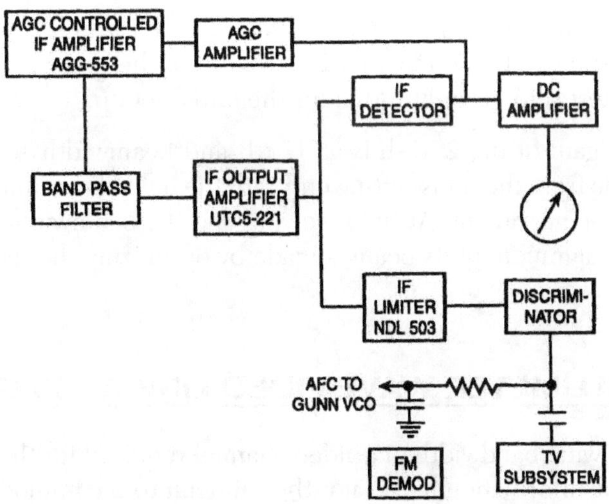

FIGURE 18.7 IF main amplifier and IF demodulator for a 63 GHz receiver.

assuming a noise-figure of 5 dB (= 3.2)

and $kT = 4 \times 10^{-21}$ for $T = 30$ K, and B = 20 MHz

Noise level at the input of the IF amplifier = –96 + 20 = –76 dBm assuming a conversion gain of 20 dB for the balanced mixer IF preamplifier.

Gain of UTC-5 221 is 35 dB and the maximum gain of AGC-553 is 44 dB. The maximum total gain of the main amplifier is thus 35 + 44 = 79 dB while the minimum gain is 34 dB since the AGC range of AGC-553 is 45 dB. Largest detector output = 20 dBm. Allowing a margin of 3 dB, the largest IF detector input should be + 17 dBm. Tangential sensitivity of the detector for 20 MHz bandwidth is –38.5 dBm, as calculated earlier for the 94 GHz receiver. Dynamic range of the detector is 55.5 dB [17 – (–38.5)]. The IF gain required to raise the noise level at its input to –38.5 – (–76) = 37.5 dB. Assuming a loss of 3 dB in the IF filter and an additional loss of 3 dB due to division of 1 Fontput in two parts, the required gain of the IF main amplifier is 37.5 + 6 = 43.5 dB.

Application of AGC control increases the dynamic range by about 30 dB making it 55.5 + 30 = 85.5 dB. It may be mentioned here that correction for deviations from the square law relation of the detector as well as that due to the deviation from the logarithmic response with AGC, are made by digital processing with a microprocessor.

18.5.2 Antenna and Beam Alignment

The antenna for the test LOS link for 53 GHz in a 2′ Cassegrainian dish at the transmitting end and a 21 dB horn at the receiving end as the range of the test setup is only 100 m for the initial set up.

The gain of the 2′ dish is at 47 dB and beamwidth 1°. Beam alignment is made from the transmitting end with the help of the built-in telescope meant for the alignment. At the receiving end the beamwidth is broad, about 10°, and alignment of its beam is made by optimizing the receiver output level.

18.6 MILLIMETER WAVE VIDEO LINK AT 53 GHZ

The wide bandwidth of a video channel required for the live transmission of visual information makes it rather difficult to accommodate such wide bandwidths in most of the conventional communication links.

In the telecommunication networks a single video channel would consume the available bands for about 1000 telephone channels and hence become rather expensive. Further, to maintain the conditions of distortionless transmission through a video channel, elaborate equalizing or video compensation networks are to be incorporated in a line or in a radio link as the case may be, to ensure (1) a frequency-independent propagation delay, (2) flat frequency response over the video band, and (3) minimal transient distortion on video transient excitation. At millimeter waves the bandwidths available in each of the atmospheric windows or absorption bands are wide enough to accommodate a large numbers of video channels with a millimeter wave band without the need of any video compensation circuit. The choice of the particular millimeter wave band depends to a large extent on the (1) rain attenuation characteristics in the band, and the (2) cost of the terminal equipment which, in turn, depends on range to be covered by the link.

All these considerations lead to a frequency around 40 to 55 GHz to be optimum for use on millimeter wave video links or video conferencing systems.

If the modulating signal $f(t)$ is considered to be a sinusoid, then it can be shown that

$$\phi_{FM}{}^{(t)} = A \sum_{n=-\infty}^{\infty} J_n(\beta)\cos(W_C + nW_m)t$$

It is evident from the previous equation that the modulating signal $f(t)$ of frequency W_m gives rise to sideband frequencies $(W_C \pm W_m)$, $(W_C \pm 2W_m)$, $(W_C \pm 3W_m)$.... and so forth. It therefore appears that an FM carrier contains components of infinite frequencies and has an infinite band width. In practice, the amplitudes of the spectral components of higher frequencies become negligible, and hence almost all of the energy of the FM carrier is contained in the spectral component lying within a finite bandwidth.

18.7 GENERATION OF FM SIGNALS

FM signals may be generated directly by frequency modulating the carrier. A simpler scheme of FM generation using a mm wave (53 GHz) carrier may be as shown in Figure 18.8. A CW varactor tuned Gunn oscillator with built-in temperature controller is used as the basic millimeter wave source. A typical 1 volt peak to peak signal from one of three video sources (coaxial switch selectable) is fed directly to the varactor terminal through a video amplitude control potentiometer (or it may be called an FM Deviation control).

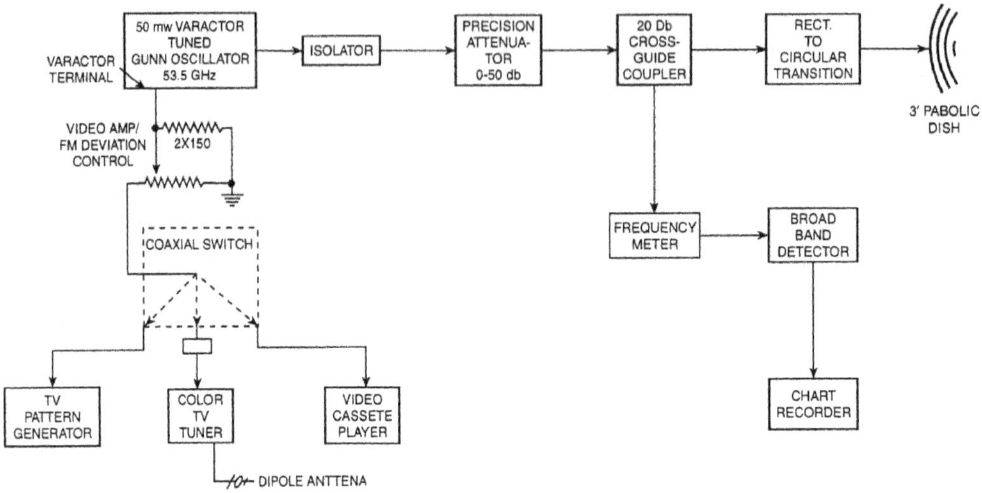

FIGURE 18.8 FM generation at millimeter waves (53 GHz).

To ensure 75 ohms, two 150 ohm resistors are connected in parallel at the varactor terminal. Also frequency stability of the transmitter is ensured by elevating and maintaining temperature of the Gunn at 60 degrees centigrade by connecting a 15 V, 1A power supply at its temperature controller terminals. The FM modulating signal at 53 GHz is then transmitted through a 3 foot parabolic dish antenna.

18.8 FM MM-WAVE RECEIVER

A block diagram of a 53 GHz receiver is also shown in Figure 18.9. As shown, in the conventional superheterodyne receiving system, special emphasis is given on the stabilization of the varactor tuned Gunn oscillator. As a primary source of stabilization, the buit-in temperature controller is powered using a 15 V, 1A power supply which in turn raises the junction temperature of the Gunn at an elevated temperature of 60 degrees centigrade. For a faithful reliable communication link, the stabilization of the Gunn may be more stringent. So, we also have to think of an secondary means of stabilization. We can easily implement this using a close loop

AFT discriminator and a commercial color TV tuner with built-in AFT, and the response curve is as shown in Figure 18.10. Normally, the Gunn varactor terminal accepts negative voltage from 0 to – 15 V. This is ensured using an op amp LM 308 where both the correction voltage from the AFT discriminator and the manual tuning voltage are fed together. A diode clamper is also used to clamp the positive voltage at the varactor terminal to 0.6V. IF at 200 MHz is branched to two parts using a matched tee. One part goes directly to a broadband detector preceded by a chart recorder to record the DC voltage and its variation due to atmosphere. The second part goes to a 3D directional coupler where the IF signal is again split into two to have inputs for FM and AFT discriminators. The recovered video signal is fed directly to the video input of a color TV receiver to have a picture on its screen. To ensure the linearity in operation of both the transmitter and receiver, we have connected a spectrum analyzer at one port of the matched tee just by opening the broadband detector for a while and connecting the IF output of the spectrum analyzer to a digitizing oscilloscope.

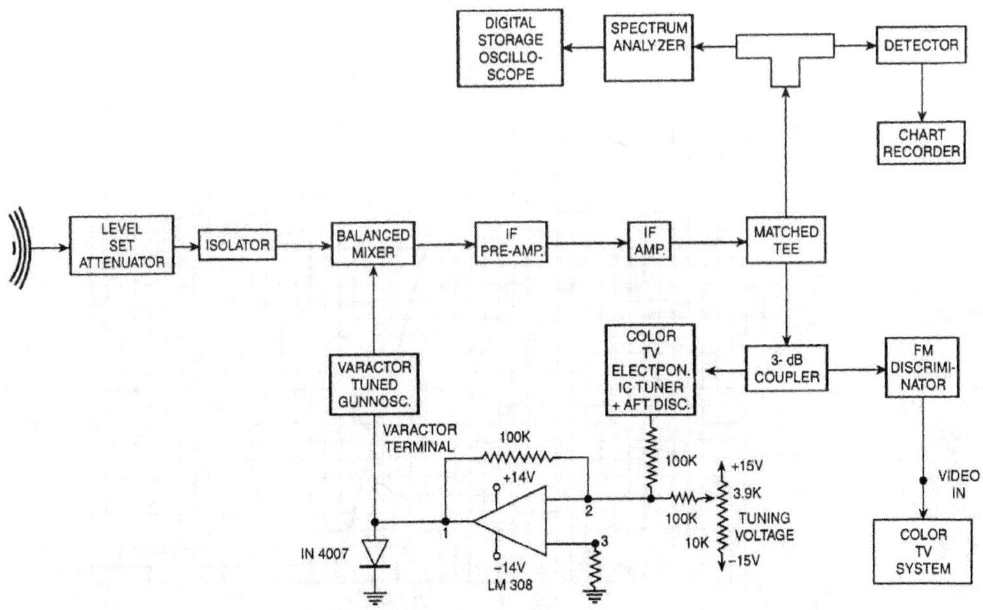

FIGURE 18.9 FM millimeter wave receiver.

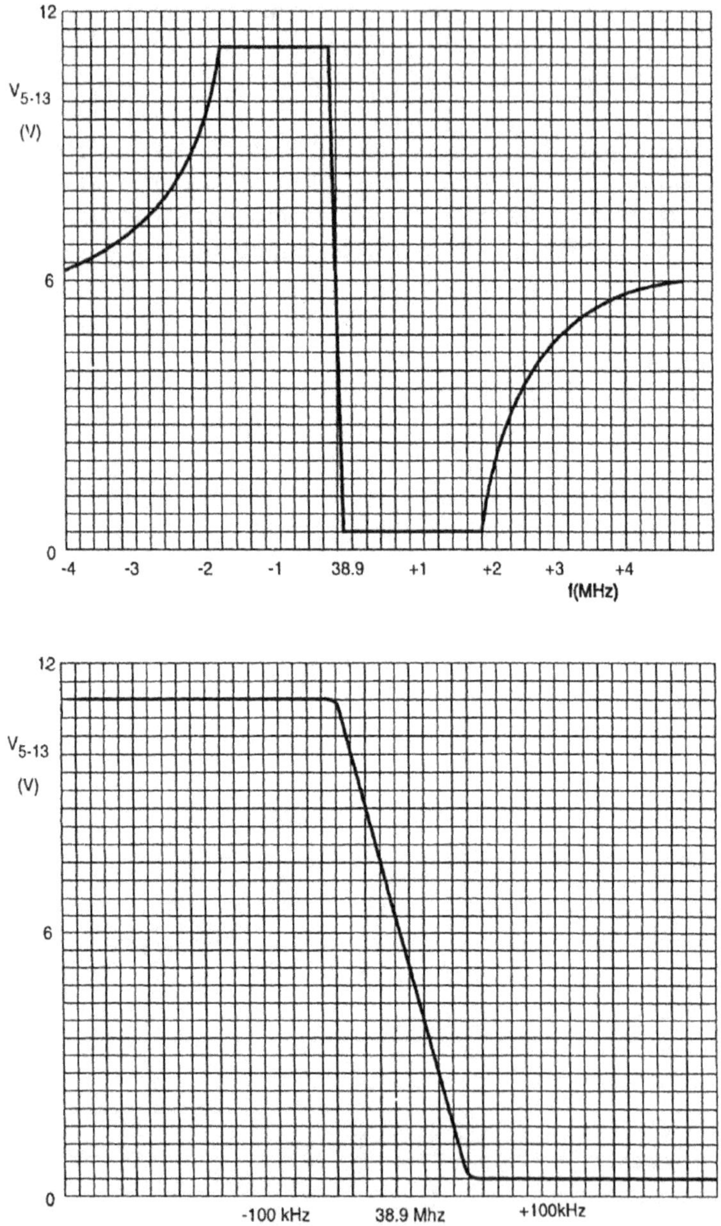

FIGURE 18.10 A.F.C. output voltage ($V_{5.13}$) as a function of the frequency.

This results in a perfect sinewave of frequency 10 MHz, which is as shown in Figure 18.11.

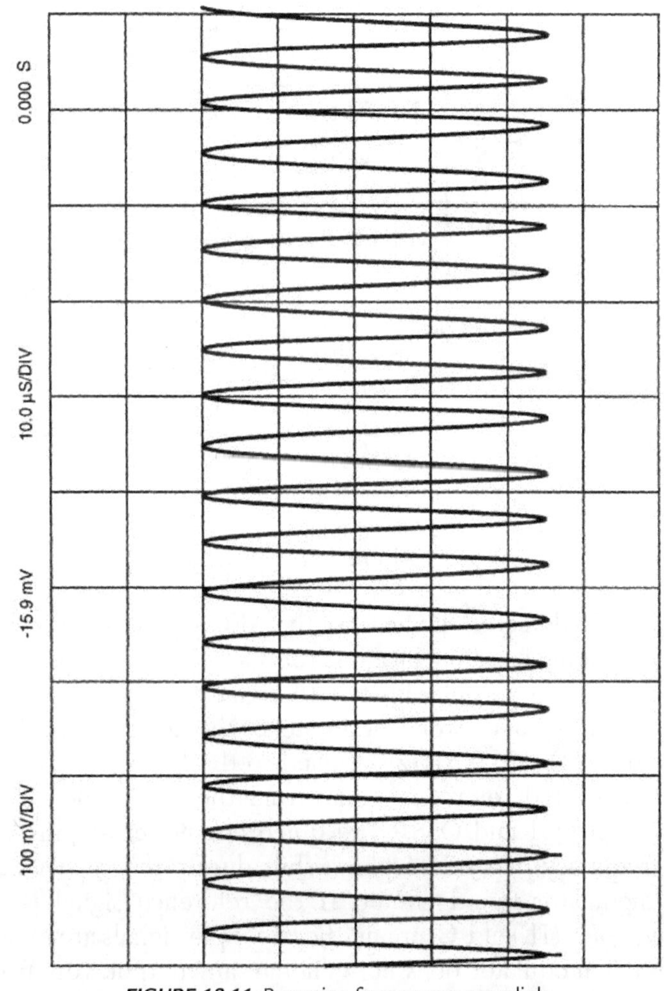

FIGURE 18.11 Reception from an mm-wave link.

18.9 AFT DISCRIMINATOR (OR DETECTOR, BUILT-IN INSIDE THE CTV TUNER)

The discriminator is shown in Figure 18.12. The details are:

Pins 5 and 6: Balanced AFT output

Pins 7 and 10: Tuned circuit to obtain 90 degree phase shift of the reference carrier.

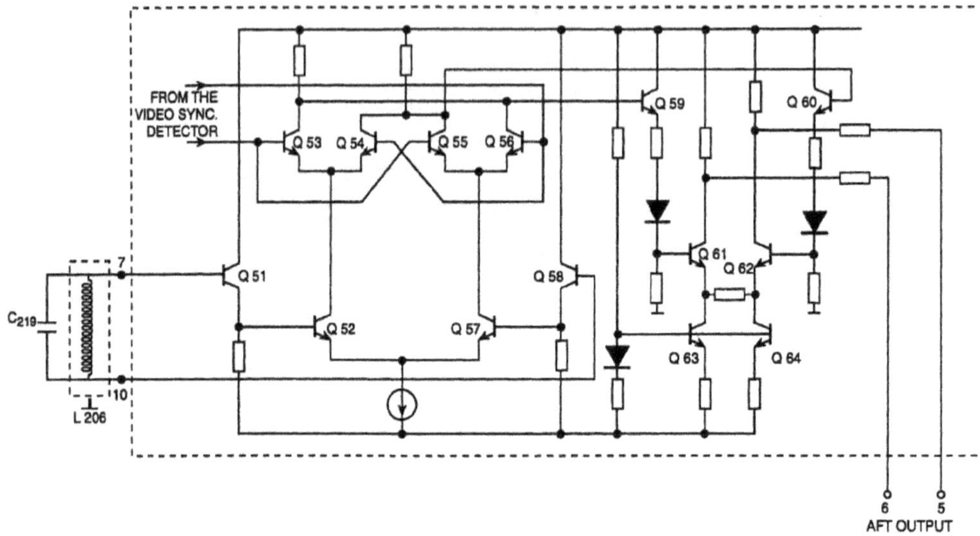

FIGURE 18.12 AFT detector.

The AFT detector, the heart of the AFC circuit, is a double balanced phase comparator consisting of transistors Q51 through Q58. The circuit compares the reference signal with the IF signal and gives an output proportional to the frequency error. The inputs of the detector are the reference signal from the IF (38.9 MHz) (applied to the bases of O.53 and O56) and the phase shifted reference signal from the AFT tuned circuit (applied the bases of O51 and O58). The tuned circuit across pins 7 and 10 provides a frequency dependent phase shift due to the capacitive coupling to the demodulator tuned circuit. If the reference signal is at the resonance frequency (IF) of LC circuit, the two input signals are in quadrature and the AFT demodulator does not generate any output. Any frequency deviation of the IF carrier causes a corresponding phase difference between the two signals of the AFT demodulator and a proportional variation of outputs is obtained. The differential output of the detector is further processed in the AFT output stage consisting of transistors O59 through O64, and the AFT voltage is available at pins 5 and 6. The sensitivity of the AFT detector, the slope of the detector output, is proportional to the value of the load resistors and loaded 9 of the tuned circuit. The damping due to the IC on the tank circuit is negligible. The capacitive coupling required for the AFT tuned circuit is achieved by two capacitors. The overall AFT sensitivity is typically 12 KHz/v. In case the receiver has to be tuned manually, the AFT facility has to be inhibited. This is achieved either by open circuiting the AFT outputs or short circuiting pins 5 and 6.

18.10 DEMODULATOR OF FM SIGNALS

To recover the modulating signal $f(t)$ from the FM carrier, we must provide a circuit whose output varies linearly with frequency of input signal. In general, the frequency discriminator consists of a circuit whose gain varies linearly with frequency. The AM converted signal is then detected by an envelope detector using a diode and an R-C circuit.

(*i*) Slope Detection. An ideal frequency discriminator will be a pair of slope circuits with their complex transfer functions related by

$$\tilde{H}_2(f) = \tilde{H}_1(f)$$

followed by envelope detector and a summer. This scheme is called the balanced frequency discriminator.

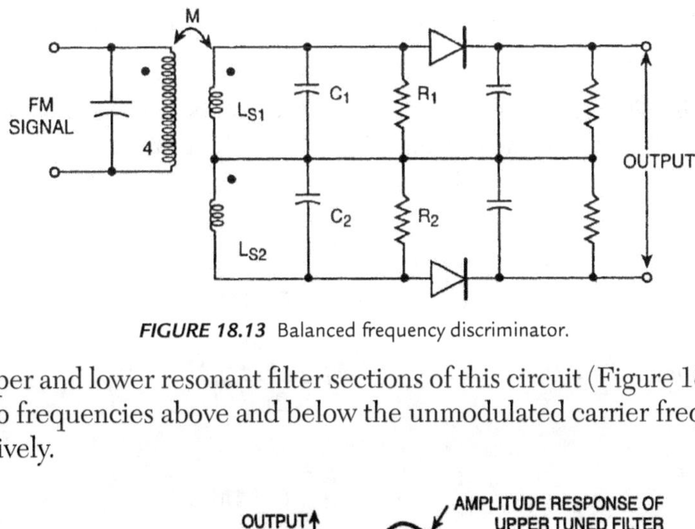

FIGURE 18.13 Balanced frequency discriminator.

The upper and lower resonant filter sections of this circuit (Figure 18.13) are tuned to frequencies above and below the unmodulated carrier frequency f_c respectively.

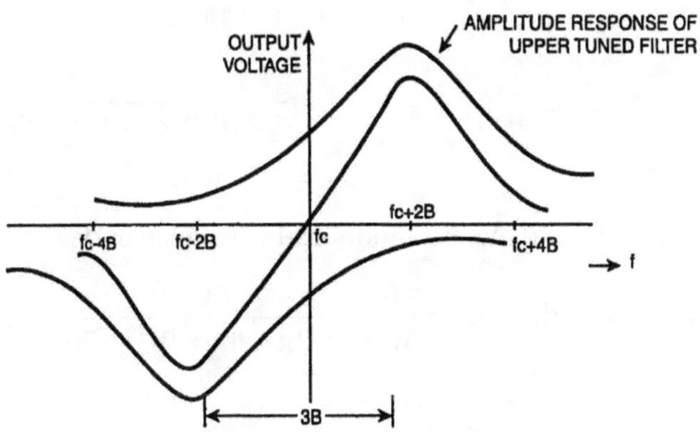

FIGURE 18.14 Amplitude response.

In the previous figure the amplitude response of these two tuned filters are plotted, together with their total response, assuming that both filters have high Q-factor.

The linearity of the useful portion of this total response, centered at f_c, is determined by the separation of the two resonant frequencies. However, these will be distorted in the output of this frequency discriminator due to the following factors.

1. The spectrum of the input FM wave is not exactly zero for frequencies outside the range $f_c - \frac{B_T}{2} \leq f \leq f_c \pm \frac{B_T}{2}$ where B_T = transmission bandwidth.
2. The tuned filter outputs are not strictly band limited and some distortion is introduced by the low pass RC filters following the diodes in the envelope detectors.
3. The tuned filter characteristics are not linear over the whole frequency band of the FM wave.

***(ii)* The Design.** In calculating the component values for the circuit, we shall take the mean frequency of 200 MH and assume the circuits to be identical except for the inductance values L_{s1} and L_{s2}

$$f_1 = 150 \text{ MC/s}$$

$$C = 5 \text{ PF (say)}$$

then
$$f = \frac{1}{2\pi\sqrt{LC}}$$

or
$$L = \frac{1}{C \times 4\pi^2 \times f^2}$$

$$= \frac{1}{5\times 10^{-12} \times 4 \times 10 \times 225 \times 10^{14}_{10}2}$$

$$= 0.22 \ \mu\text{H}.$$

$R_D = 0.1$ M the dynamic anode load resistance and

$$Q = 150 \text{ (say) then } L_{s1} = \frac{R_D}{W_m Q} = \frac{0.1\times 10^6}{2\pi \times 200 \times 10^6 \times 150}$$

$$= \frac{5\times 10^{-4}}{4\pi \times 15}$$

$$= 5.30 \times 10^{-7} \text{ H} = 0.53 \ \mu\text{H}.$$

$$L_{s2} = \frac{1}{C \times 4\pi^2 \times f_2^2}$$

$$= \frac{1}{5 \times 10^{-12} \times 4 \times 10 \times 250 \times 250 \times 10^{12}}$$

$$= \frac{10^{-3}}{5 \times 4 \times 625} = 0.08\,\text{H}.$$

The reflected resonant impedance across the primary is $R_D\left(\frac{M}{L}\right)^2$ < resonant impedance of tuned circuit (500 say) $\frac{M}{L}$ taken to be 0.9.

or, $M = 0.477\ \mu\text{II}.$

The coupling coefficient $K = \frac{M}{\sqrt{L_p L}}. = 0.8$ (air core)

$\therefore \quad L_P = \frac{M^2}{L \times 0.64} = 0.67\ \mu\text{H}.$

Following the design specifications one may modify circuit a little bit by adding two series resistors in the tuned circuit (in series with L) and damping the response so as to obtain the best Linearity. A computer package Spice was employed in realizing the response where the FM signal was replaced by a constant current sweep in frequency from 50 HHz to 500 MHz. The modified circuit is as shown in Figure 18.15.

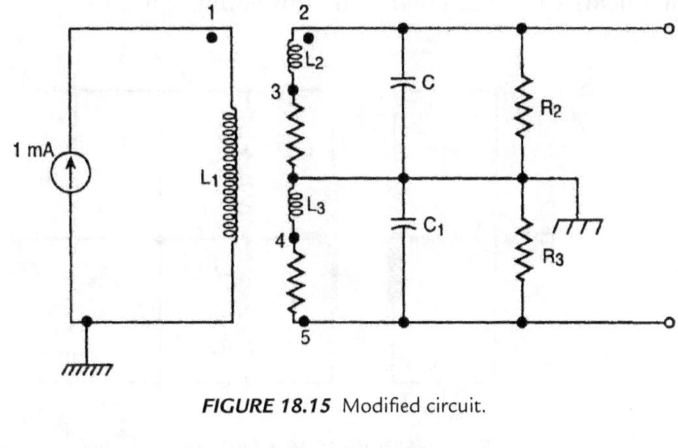

FIGURE 18.15 Modified circuit.

The nodes were specified and the circuit specification to SPICE were made as follows:

Double tuned	*Node*	*Node*	
I_1	1	0 AC1 mA	
L_1	1	0 0.6 H.	
L_2	2	3 0.5 H.	
RS	3	0 10	
C	2	0 5 PF	
R_2	2	0 500 ohms	
K_{12}	L_1	L_2 0.8	
L_3	0	4 0.08 H.	
R_{s1}	4	5 70	
C_1	0	5 4 PF	
R_3	0	5 500	
K_{13}	L_1	L_3 0.8	
AC DEC	100	50 MEGhz	500 MEGhz
PROBE			
PRINT AC V (2, 5)			
END.			

The problem that may be faced with the circuit is that the two coils differ widely in their inductance values which would pose a problem in the coupling from the input side.

Attempts were made to modify the circuit which involved a 1 : 1 transformer with doubly tuned circuit (Figure 18.16). The rest of the circuit being same the linearity was obtained using this configuration.

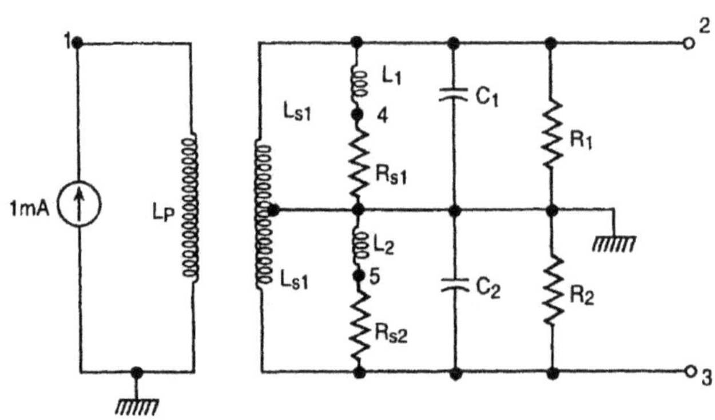

FIGURE 18.16 A further modification.

Balanced Discriminator

I_{in}	1	0	AC 1 mA	
L_p	1	0	0.5 µH	
L_{s1}	2	0	0.25 µH	
L_{s2}	0	3	0.25 µH	
L_1	2	4	0.6 µH	
R_{s1}	4	0	5	
C_1	2	0	7 PF	
R_1	2	0	500	
L_2	0	5	0.08 µH	
R_{s2}	5	3	5	
C_2	0	3	7 PF	
R_2	0	3	550	
K_{ps1}	L_p	L_{s1}	0.8	
K_{ps2}	L_p	L_{s2}	0.8	
AC DEC	100	59 MEGhz		500 MEGhz

PRINT AC V(2), V(3)
PRCBE
END.

The response obtained employing the package is as shown in Figure 18.17. The linearity obtained in this configuration is highly satisfactory.

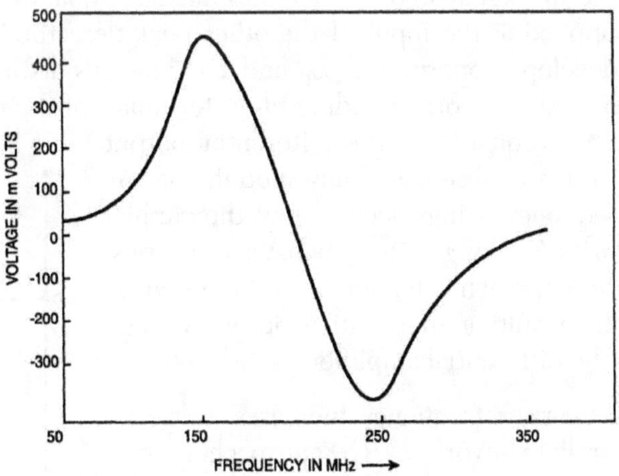

FIGURE 18.17 Response of the frequency discriminator.

***(iii)* Differential Peak PM Detector.** In this system a new technique called differential peak detection is used to develop the signal voltage from the FM signal.

The two peak detectors employ differential amplifier configuration with emitters follower at their inputs (Figure 18.18). The output circuit of the two detectors have identical RC networks of suitable time constants to provide peak detectors of the two signals fed to them.

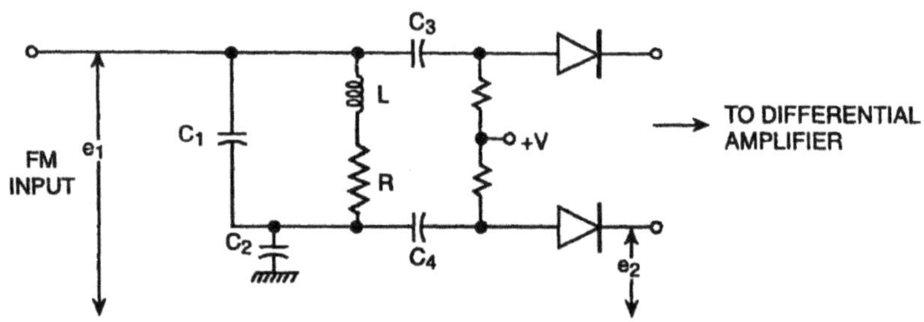

FIGURE 18.18 Peak detector.

The external frequency sensitive network comprises a parallel L-C circuit and a series capacitor C_2. The FM input is fed to this frequency selective network and one of the peak detectors.

At the center frequency (175 MHz) the parallel circuit behaves like an inductor in series with the capacitor C_2. This combination provides a slight boost to the signal across C_2 so as to make e_2 equal to e_1. The signal voltage e_2 is applied at the input of the other peak detector. The two peak detectors develop proportionate L_C and DC voltages at their output terminals, which feed into corresponding input terminals of the differential amplifier. Since e_1 is equal to e_2, the differential output is zero in the absence of any modulation. As the frequency swings become low, the parallel network LC_1 along with C_2 behave as a series. Thus e_2 approaches higher magnitude than e_1 thereby resulting in negative output voltage from the differential amplifier.

As the carrier frequency increases above f_C, the parallel network (L_1, C_1) approaches a parallel resonant circuit and its impedance rises to make $e_2 < e_1$ and thereby the output of the differential amplifier becomes positive.

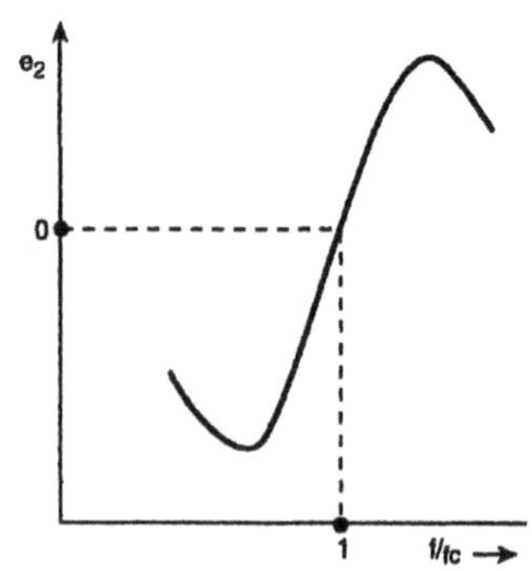

FIGURE 18.19 The characteristic variation.

The linearity of this detector is good, and total harmonic distortion in the output is less than one percent.

$$f_C = 175 \text{ MHz}$$

$$f_1 = 150 \text{ MHz. If } C_1 = 4.7 \text{ pF (taken)}$$

Then for $f_2 = 200$ MHz.

$$f_2 = \frac{1}{2\pi\sqrt{LC_1}} = \frac{1}{2\pi\sqrt{L \times 4.7 \times 10^{-12}}}$$

$$\therefore \qquad L = 0.132 \ \mu\text{H}.$$

$\therefore \quad f_1 = \dfrac{1}{2\pi\sqrt{L(C_1 + C_2)}}$ substituting values of L and C_1, we get $C_2 = 4$ pF.

18.11 SOLAR TRACKING MM-WAVE RADIOMETERS

At millimeter waves the sun appears as a blackbody radiator at 6000 K with occassional bursts and continuum radiation originating from a nonthermal process. Using the sun as a black body radiator, it is possible to measure the absorption of the solar radiation at microwaves and millimeter waves by the atmosphere. The technique is called the "absorption mode" measurement. The antenna beamwidth for such a measurement should be comparable to the angular width, 32, of the sun and the beam should be arranged to track the sun, either by a preprogrammed tracker or by an automatic optical sun follower. The solar emission is in fact deflected by a plane millimeter wave reflector along the axis of the antenna beam of the radiometer, kept stationary inside an air-conditioned room, with a view to protecting the radiometer from the outside weather. The reflector kept outside the room is, therefore, arranged to track the sun. An automatic optical sun follower, developed for the purpose, was reported in an earlier Technical Report No. RPE-11. Three solar tracking millimeter wave radiometers at 53, 94, and 140 GHz were being developed for the (*a*) study of millimeter wave attenuation in the atmosphere for "slant" (non-horizontal) paths under clear, cloudy, and rainy conditions; (*b*) study of solar millimeter wave bursts and continuum radiation; and (*c*) estimation of solar millimeter wave interference in radio receiving systems for radars, radiometers, and satellite links.

18.11.1 Solar Tracking Radiometer for 53 GHz

For an antenna beamwidth comparable to that of the sun, the antenna temperature is also comparable to the blackbody temperature of the sun, 6000 K. A low noise millimeter wave receiver at 53 GHz will have a noise temperature of about 600 K, which is one order of magnitude lower than the solar brightness temperature and as such Dicke switching is not essential. However, in the study of clear air attenuation and noise temperature after sunset, during night time, the antenna temperature is about an order of magnitude lower than the receiver noise temperatures and Dicke switching will be helpful under these conditions. As a first step in the development we left out Dicke switching as well as any facility for polarization switching. These will be introduced at a later stage after preliminary observations of the solar radio emissions are continued without Dicke switching for some period of statistical significance. A block diagram of the 53 GHz solar tracking radiometer is shown in Figure 18.20.

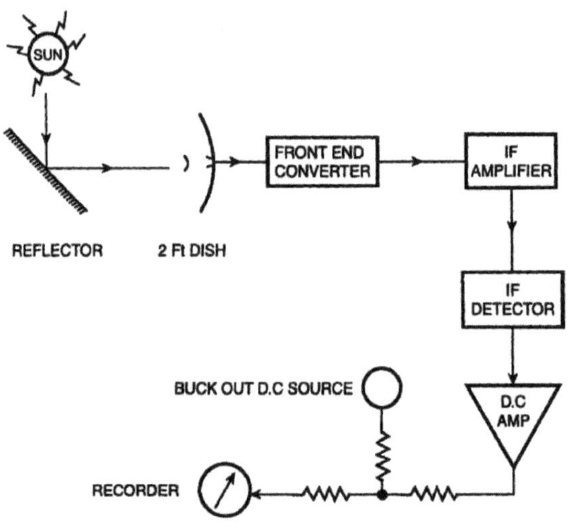

FIGURE 18.20 53 GHz solar tracking radiometer.

18.11.2 Dynamic Range of Antenna Temperature and IF Amplifier

The IF operates in the square law region. The DC level at the IF detector output is, therefore, proportional to the system noise temperature, which is the sum of antenna noise temperature T_S, with the antenna having beamwidth of 1° pointed toward the sun and the Front-end noise temperature, T_R. The radiometer system noise temperature is thus, $T_S + T_R = 6600$ K, while, with the antenna pointed away from the sun toward the cold sky, the system noise temperature is only $T_R = 600°K$, since the galactic noise at 53 GHz is negligible.

The dynamic range of noise level is, therefore, $= \frac{6600}{600} = 11$

Taking account of atmospheric noise temperature, 100 K, the dynamic range $= \frac{6700}{700} = 9.6$.

However, for amplification of the noise level by the IF amplifier, the ratio of peak to average for the noise waveform must be taken into consideration. Ideally, for a random noise the ratio of peak to average is taken as 3 without leading to appreciable error in measurement. Accordingly the dynamic range of the IF amplifier should be $11 \times 3 = 33 = 15$ dB. During solar millimeter wave bursts the peak noise level may increase by about 50%, when the dynamic range is increased to about 50 = 17 dB.

18.11.3 Dynamic Range of the Detector

The tangential sensitivity of the point contact silicon IF detector is –45 dBm at a 1 MHz, bandwidth. For a DSB noise output of the balanced mixer of bandwidth 2 × 500 MHz, the tangential sensitivity would be $(-45 + 5 \log_{10}1000) = (-45 + 15) = -30$ dBm. The point contact detector operates in the square law region up to an input level of –20 dBm. The dynamic range of the detector is therefore, –20 – (30) = 10 dB. The dynamic range of the IF amplifier should also be made equal to 10 dB for a linear power law of detection.

In practice, by adopting microprocessor based digital processors we can extend the useful operating region of the detector up to an input of +17 dBm, keeping a margin of 3 dB to avoid the risk of detector burnout. In that case, the dynamic range is extended from 10 dB to 17 – (–30) = 47 dB, offering a margin of 47 – 17 = 30 dB in dynamic range to accommodate the most outstanding solar millimeter wave radio noise outbursts.

18.11.4 IF Gain

The noise figure of the Front end convertor is 6 dB and the conversion gain is 20 dB. Noise temperature for 6 dB noise figure is 600 K.

Noise level at input with antenna pointed toward the cold sky would be kT_RB

$$= 1.38 \times 10^{-23} \times 600 \times 1000 \times 10^{6}\ \text{W}$$

$$= (1.38 \times 10^{23} \times 300) \times 2 \times 1000 \times 10^{6}\ \text{W}$$

$$= 4 \times 10^{-21} \times 2 \times 10^{9}\ \text{W}$$

$$= 8 \times 10^{-21}\ \text{W} = 8 \times 10^{-9}\text{mW}$$

$$= 9 - 90\ \text{dBm} = -81\ \text{dBm}$$

At the output of the Front-end converter the noise level is –81 + 20 = –61 dBm.

To bring this noise level up to –30 dBm we need IF gain of –30 – (–61) = 31 dB.

With a dynamic range of 17 dB for noise level, the IF output would be (–30 + 17 dB) dBm = –13 dBm.

With microprocessor based digital processors the dynamic range of the IF amplifier would be higher by 30 dB, that is, (17 + 30) = 47 dB making the IF output (–13 + 30) dBm = + 17 dBm, requiring an IF gain of 31 + 30 = 61 dB. To attain these two options of the dynamic range, 17 dB and 47 dB, we use a combination of an AGC controlled IF amplifier followed by an IF output amplifier as shown in Figure 18.21.

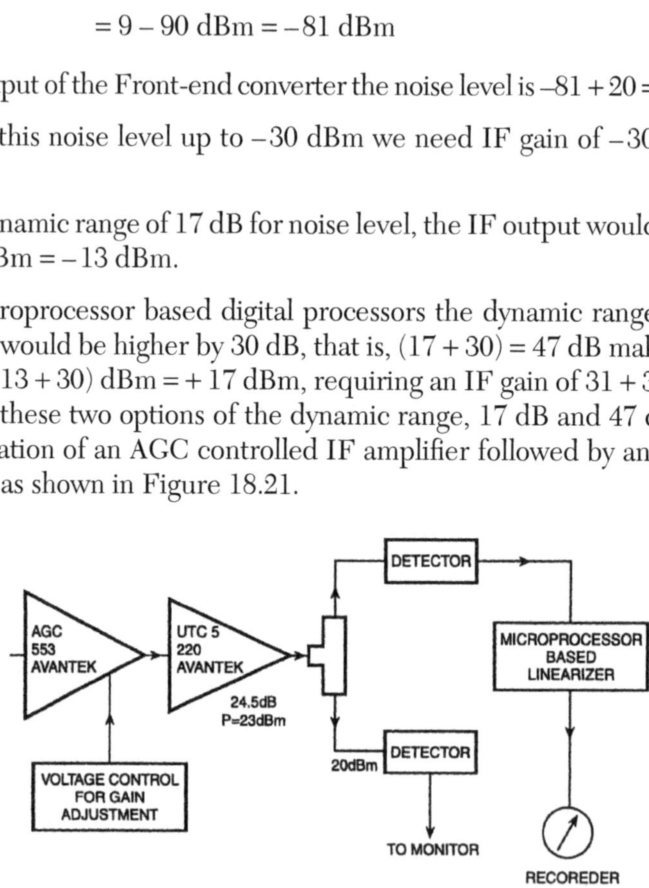

FIGURE 18.21 IF amplifier of 53 GHz radiometer.

18.11.5 Front-End Converter for a 53 GHz Radiometer

It is based on a balanced planer Crossbar Mixer with GaAs Schottky barrier diode pair. The mixer is fed from a varactor tuned Gunn VCO. The level set attenuator is used to adjust the L.O. input to an optimum level of

+ 13 dBm for the mixer. The balanced mixer IF preamplifier combination offers a noise figure of 4.5 dB. Adding the losses in the isolator and other transmission systems at the input would result in an effective noise figure of the front-end to be 6 dB. The isolator reduces the leakage of L.O. power to the mixer input by an additional 20 dB to reduce reflection of the leakage power from the antenna mismatch.

The frequency of the radiometer is tunable over 1 GHz from 52.5 to 53.5 GHz by a voltage control of the Gunn VCO. Both the upper and lower sideband noise is accepted by a 10–510 MHz IF amplifier, resulting in an effective RF bandwidth of the radiometer to be $2 \times (500 - 10) = 2 \times 490 - 1000$ MHz.

18.11.6 Calibration and Drifts in the Radiometer

Calibration of the 53 GHz solar radiometer is made periodically by an IMPATT noise source acting as a hot load of noise temperature 10,000°K.

The noise source may be connected through a calibrated attenuator to the input of the radiometer by disconnecting the antenna. Controlling the attenuator provides the calibration marks in the radiometer output recorder. Alternatively, the radiation with its antenna connected may be left undisturbed and the noise radiometer from a gain standard horn connected to the noise source may be used for calibration. The noise power fed to the horn antenna may be controlled by an attenuator, or the horn antenna with noise source may be steered to produce a controlled misalignment between the axes of the horn and the dish of the radiometer. A detailed report of the techniques of radiometer calibration was included in an earlier Report No. RPE-8.

If the operating voltages and temperature of the radiometer are held fixed within close limit, the frequency of calibration can be reduced to once a day or even once a week. For the purpose, the following measures may be adopted:

1. The DC supplies to the radiometer are obtained from a storage battery which is under trickle charging from a float charger. In this system the float charger effectively supplies the power to the load, the battery acting as a low impedance shunt to absorb any drifts or transients in the supply voltage.
2. The reference load with radiometer electronics is enclosed in a temperature controlled enclosure.

18.12 SOLAR TRACKING RADIOMETER FOR 94 GHZ

The front-end converter for a 94 GHz radiometer is shown in Figure 18.22.

The post detector integration time of the 94 GHz is selectable by a switch, as in the 53 GHz radiometer, to 2, 5, 10, and 100 m sec, 1 sec, and 10 sec.

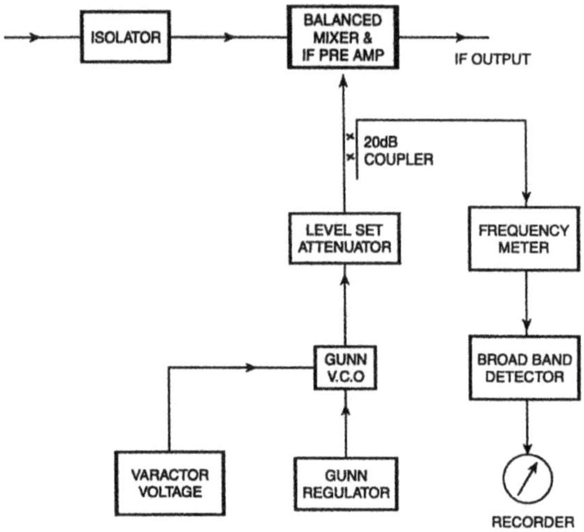

FIGURE 18.22 Front-end convertor for a 94 GHz solar tracking radiometer.

18.12.1 Calibration and Drifts

Calibration of the 94 GHz solar tracking radiometer is made periodically by an IMPATT diode noise source fed to a standard gain horn, with the beam directed along the main beam of the radiometer antenna. This calibration is compared initially with that made using a direct feed from the hot noise source to the radiometer input terminal through a calibrated attenuator, and the use of any antenna.

As in the 53 GHz radiometer, the DC power supplies are stabilized by the floating battery technique, to minimize the need for frequency calibration.

18.13 SOLAR TRACKING RADIOMETER FOR 140 GHZ

A block diagram of the 140 GHz solar tracking millimeter wave radiometer is shown in Figure 18.23. In this figure, except for the front-end converter, the remaining blocks are similar to that of the 53 and 94 GHz Radiometer.

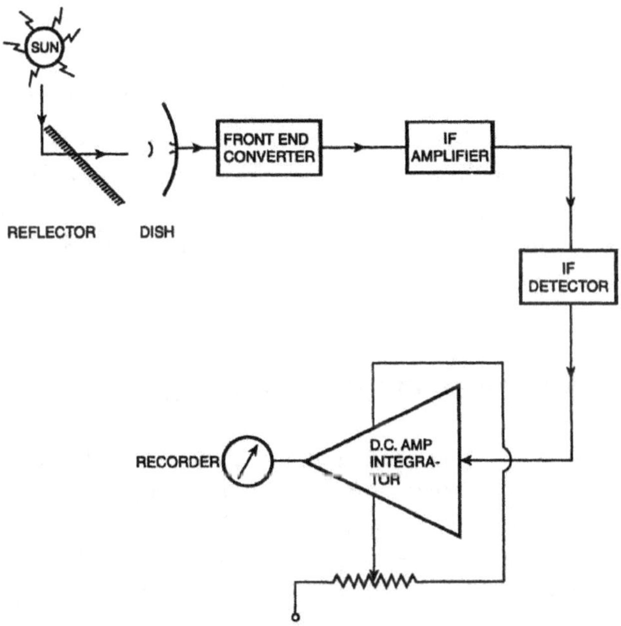

FIGURE 18.23 Block diagram of a 140 GHz solar tracking radiometer.

The antenna is a 1′ dish with Cassegrain feed. The gain of the IF amplifier is adjusted to optimize the dynamic range of the radiometer.

Front-End Converter. The front-end converter of the 140 GHz solar tracking radiometer is shown in Figure 18.24. The local oscillator is based on a 46.67 GHz Gunn oscillator followed by a varactor trippler to generate the desired 140 GHz L.O. power. The L.O. power is monitored by a tunable broadband detector.

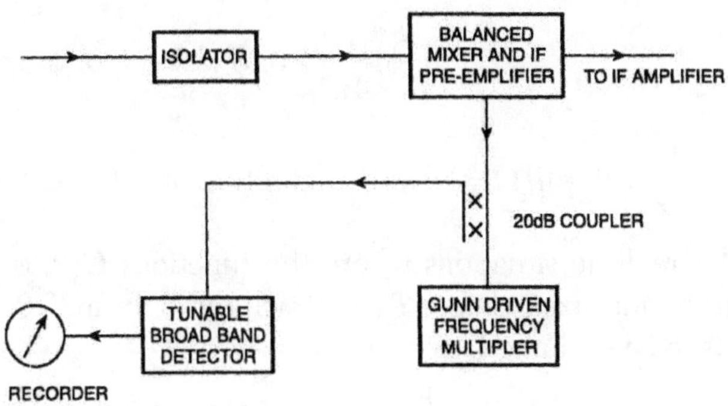

FIGURE 18.24 Front-end converter for a 140 GHz solar tracking radiometer.

The noise figure of the balanced mixer IF preamplifier is 6 dB. Adding up the contribution due to the losses in isolator and transmission systems to the antenna, the effective noise figure of the front-end is about 7 dB.

Calibration. A 140 GHz IMPATT noise source is used for calibration in the same way as with the 53 and 94 GHz radiometers. The DC power supplies for the radiometer are also derived from the same floating battery supply used for the 53 and 94 GHz radiometers to reduce the drifts and the need for frequent calibration.

18.14 ESTIMATION OF SOLAR BRIGHTNESS TEMPERATURE

The antenna beam of the radiometers at each of the frequencies 53, 94, and 140 GHz are measured by pointing the beam-axis to the center of the solar disc by using three solar tracking plane reflectors placed outside the air-conditioned radiometer hutment, at an angle such that solar radio emission is directed along the main antenna beam of the three radiometers.

If the antenna Gain function is $G(\theta, \phi)$ and the brightness temperature of the sun is $T(\theta, \phi)$ then the measured antenna temperature $T_a(\theta, \phi)$ can be expressed as

$$T_a(\theta, \phi) = \frac{1}{4\pi} \int T(\theta, \phi) G(\theta, \phi) d\Omega$$

This relation was proved by Nyquist from thermodynamic consideration (RPE-6). If we assume idealized rectangular shapes of $G(\theta, \phi)$ and $T(\theta, \phi)$, and the antenna beamwidth is equal to the angular width of the sun, then we have $T_a(\theta, \phi)$

$$T(\theta, \phi) \frac{1}{4\pi} G(\theta, \phi) d\Omega = T(\theta, \phi).$$

Since $\frac{1}{4\pi} G(\theta, \phi) d\Omega = 1$ by the definition of antenna gain.

Under realistic situations where the functions $G(\theta, \phi)$ and $T(\theta, \phi)$ depart from the idealized shape, $T_a(\theta, \phi)$ will differ from $T(\theta, \phi)$ and equation (1) needs to be evaluated.

In the present case the shape of the antenna beam is nearly Gaussian with half power beamwidths of 1°, 0.75°, and 0.5° at 53, 94, and 140 GHzs respectively, although $T(\theta, \phi)$ is almost rectangular in shape at millimeter waves, except for minor limb "brightning" or "darkening" effects. If the antenna beam is swept across the solar disc, then the antenna gain function will smooth out the true brightness temperature distribution, in just the same way in the spatial domain as the shape of a rectangular pulsed carrier waveform is smoothed out in the time domain by a bandpass filter. The smoothing out of the temperature distribution by the antenna can be looked upon as a case of two dimensional "aerial smoothing" as indicated by Prof. R. N. Bracewell, who first dealt with problem in connection with radioastronomical measurements of the brightness temperature distribution of "radio stars." In case of filters we know that the source function $s(t)$ can be derived from the response of a smoothing filter of impulse response $h(t)$ by using a convolution intergral. The same technique was applied in the space domain by Prof. Bracewell to restore the original temperature distribution of the source from the spatial response with a broadbeam antenna acting as a spatial smoothing filter. A computer program for the restoration of true distribution of the solar millimeter wave disk in the presence of sizable two-dimensional aerial smoothing will be made to find the true distribution. It may be mentioned here that for the solar tracking radiometers, the axis of the antenna beam tracks the center of the solar disk throughout the day between sunrise and sunset, and the response of the antenna always corresponds to that for perfect beam alignment to the solar disk. Once the true solar temperature distribution is evaluated by the technique of Bracewell indicated previously, it is possible to estimate the blackbody temperature of the sun from the measured antenna temperature.

18.14.1 Programmed Solar Tracking

A plane reflector for millimeter waves, kept outside the radiometer hutment, is used to deflect the solar emission toward the antenna beam of the radiometer by a preprogrammed tracking system driven by stepper motor. Independent tracking reflectors are required for each of the radiometers at 53, 94, and 140 GHz. Such a tracking reflector in fact directs the solar millimeter wave emission toward a fixed direction along the pole star, and the antenna beam of the radiometer is arranged to look downward along the same direction towards the south. The platform of the radiometer is arranged to shift the axis of the antenna beam by 3° away from the direction of the sun periodically to sample the noise background around the sun. This motion of the antenna beam is called "noding."

18.14.2 Automatic Tracking of the Sun

An automatic solar tracker was also developed using optical sensors to orient a plane mirror such that the reflected solar rays always point toward a fixed direction along the pole star. On cloudless days, the optical tracker may be used to drive the radiometer antenna drive motor to point the beam toward the sun. Usually the preprogrammed tracker is employed for the purpose, as it can work also in cloudy conditions, and the accuracy of tracking is checked against the optical autotracker occasionally when the sky is clear.

18.15 CALIBRATION OF MILLIMETER WAVE RADIOMETERS

Dicke type radiometers can detect thermal emission noise from sources placed within its antenna beam, with a sensitivity typically about l°K. To calibrate the radiometer we should ideally have to place it inside a black body enclosure whose temperature is controllable in steps of l°K, starting perhaps from a lowest temperature of the order of 20°K, which corresponds to the lowest observable emission noise temperature of the atmosphere. However, the radiometers are usually designed to indicate directly the antenna noise temperature with a linear scale. In fact, the square detectors at the IF output produce an AC component at the switching frequency proportional to the received noise power which, in turn, is proportional to the antenna noise temperature. This linear response of a radiometer to the noise temperature allows one to calibrate it at only two reference temperatures from which to deduce the entire temperature scale. One of these reference temperatures may be that due to a liquid nitrogen source for calibration at 77°K, while the other may be room temperature, about 290°K. For calibration points at lower temperatures, one can choose gaseous helium cooled to 20°K or even liquid helium at 4.7°K. in either case, however, we need a liquid nitrogen environment to sustain the liquid or cooled gaseous helium source and the arrangements are rather expensive. For propagation studies of the atmosphere, only a liquid nitrogen calibration together with one at the room temperature is usually considered sufficient. The liquid nitrogen load may be connected directly to the radiometer receiver excluding the antenna, or one may use the thermal emission from a liquid nitrogen source picked up by the radiometer antenna directly or through reflection in a plane or parabolic focuser. The various calibration methods are highlighted in this report and their performances critically assessed.

18.15.1 Brightness Temperature and Antenna Noise Temperature

If the radiometer antenna is placed inside a blackbody enclosure maintained at a temperature T, the thermal emission noise picked up by the antenna matched to the radiometer input will try to raise the temperature of the equivalent input resistance. Thermal radiation due to this input resistance will be balanced by what is picked up by the antenna under a thermodynamic equilibrium (Sen et al., 1987). The brightness temperature of the blackbody source will, here, be T. If the enclosure is not at a uniform temperature over its surface but exhibits a brightness temperature distribution $T(\theta, \phi)$, the antenna temperature T_a is given by

$$T_a = \frac{1}{4\pi}\int_0^{4\pi} T(\theta, \phi)G(\theta, \phi)d\Omega$$

In the case of atmosphere $T(\theta, \phi)$ is a slowly varying function of θ and ϕ, when we have

$$T_a = T(\theta, \varphi)\frac{1}{4\pi}\int G(\theta, \phi)\, d\Omega = T(\theta, \phi)\times 1$$

Since $\frac{1}{4\pi}\int_0^{4\pi} G(\theta, \phi)\, d\Omega = 1$, by the definition of antenna gain.

Thus, if we can measure the antenna temperature of the radiometer, it will be equal to the brightness temperature of the atmosphere viewed by the antenna beam. The brightness temperature may also be called the emission noise temperature.

18.15.2 Measurement of Antenna Noise Temperature and Liquid Nitrogen Calibration

The fact that antenna noise temperature is equal to the brightness temperature of the thermal emitter, under thermodynamic equilibrium, may be utilized to calibrate a radiometer. With an experimental set up as shown in Figure 18.25, the antenna main beam with all its side lobes looks into liquid nitrogen contained in a flask, acting as a thermal emitter at 77°K, which is the liquid nitrogen temperature. The system is equivalent to that of a radiometer placed inside a blackbody enclosure maintained at 77°K. There is no part of the antenna beam falling outside the liquid nitrogen source to view other sources at room temperature. The ecosorb absorber placed at the bottom of the flask ensures that there may be no reflection from the bottom for a radio wave incident on the source; which means that the source behaves as a perfect absorber like a blackbody and hence can be treated as a blackbody source. In practice the radiometer is usually mounted with the axis of the antenna horizontal, due to other constraints in atmospheric emission noise measurements. Under the circumstances, a reflector has to be employed to direct the liquid nitrogen emission to a horizontal direction as shown in Figure 18.26 (a). In doing so, if the total path length from the antenna mouth to the liquid nitrogen surface is large, as shown in Figure 18.26 (b), a part of the antenna beam may viewed outside the liquid nitrogen surface, leading to errors in calibration. A concave reflector instead of a plane one may be helpful in this case, as illustrated in Figure 18.26 (c). Its convergent property will reduce the divergence of the antenna beam, confining it to within the boundaries of the flask.

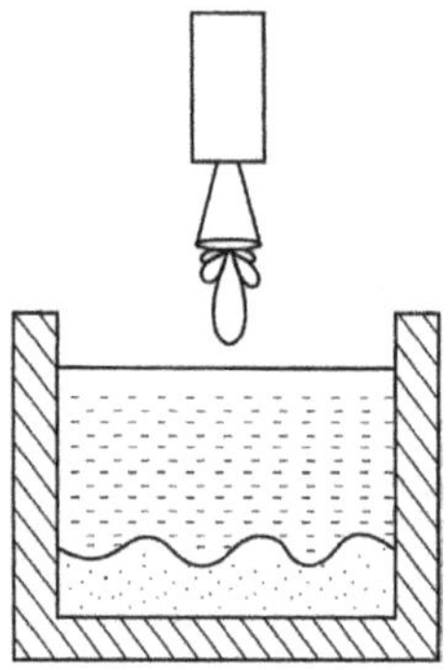

FIGURE 18.25 Experimental setup.

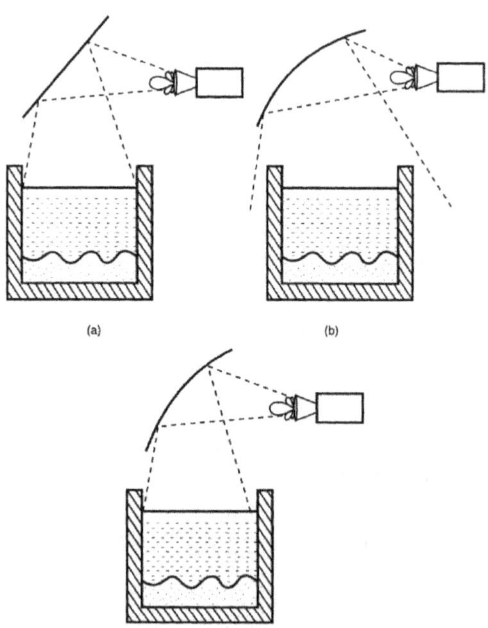

FIGURE 18.26 Use of plane and concave reflectors under different condition.

Calibration at Room Temperature. The liquid nitrogen load is replaced by an ecosorb absorber at room temperature to obtain a calibration mark for room temperature.

18.15.3 Continuous Calibration Points between Liquid Nitrogen and Room Temperature

In an alternative method of calibration, the antenna is disconnected and a matched termination immersed in a liquid nitrogen bath is connected directly to the receiver input. If the antenna efficiency were unity, the deflection of the radiometer would correspond to the brightness temperature of liquid nitrogen. In the presence of antenna losses, one has to incorporate some correction in the result.

The matched load cooled to 77°K may be connected through a direct reading precision attenuator instead of connecting it directly to the receiver input as shown in Figure 18.27.

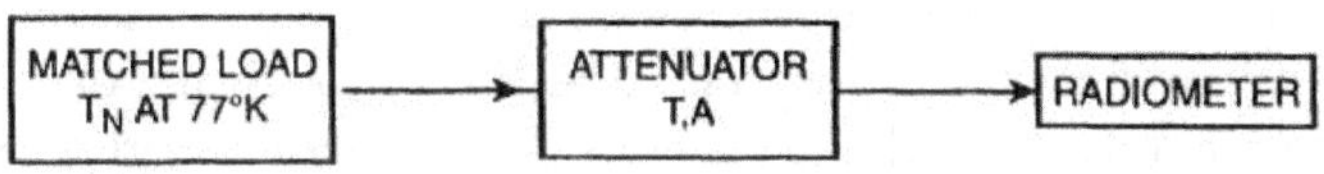

FIGURE 18.27 The matched load connected through an attenuator.

With a setting A (ratio) of the attenuator, the noise power input to the radiometer is given by

$$K\,T_a B = K\,T_N B e^{-A} + k\,TB\,(1 - e^{-A})$$

From this equation, we note that

if $\qquad A = 0\ T_a = T_N$, while with $A = \infty\ T_a = T$

Between these two limits T_N and T one can obtain any intermediate calibration mark.

Thus, there can be calibration points at intervals of 1 K at 77 K, 78 K, 79 K, 80 K, 81 K, 90 K, and 300 K (room temperature), depending on the setting of the attenuator. Calibration marks at a closer spacing of 0.1 K or even 0.01 K may be obtained if the attenuator setting can be reliable to that degree of resolution, and if the sensitivity of the radiometer is of that order.

18.15.4 Liquid and Gaseous Helium Calibration

For studies of ultramicrowaves from the cosmos, calibration at temperatures lower than that of liquid nitrogen are important. For instance, to measure the isotopic background emission at 3 K from the expanding universe from a space platform, we need calibration mark around 3 K. The lowest temperature of a source that can be easily realized is a liquid helium source at 4.7 K, while gaseous helium can provide us with a source at about 20 K. In either case, however, we need a liquid nitrogen environment. A liquid helium flask is in fact enclosed inside another flask containing liquid nitrogen. Likewise, the tubes of a helium cooling system are covered with another tube, the intervening space between the tubes filled up with liquid nitrogen.

18.15.5 Calibration of Solar Tracking Radiometers

In solar tracking radiometers a sharp antenna beam, of width comparable to the angular width of the sun, is steered to track the sun, acting as a hot blackbody source, with a brightness temperature of 6000°K of the photosphere. During active phases of sun, around the sunspot maximum years, enhanced solar radio emission may occur in the form of microwave and millimeter wave bands when the apparent brightness temperature of the sun will be higher. For calibration of a solar tracking radiometer we must, therefore, have a calibration source having a noise temperature significantly higher than the room temperature. Assuming a room temperature of 300°K, the noise temperature of the source should be about 6000/300 = 20 times that of the room temperature. The calibration source generating this amount of noise will, accordingly, have an Excess Noise Temperature Ratio (ENR) = 20 (ratio) = 13 dB. Solid state millimeter wave noise sources based on IMPATT diodes act as hot loads, and they are available with ENR = 25 dB. Such a noise source can be controlled by a variable attenuator to calibrate the solar tracking radiometer, by connecting the attenuator output to the radiometer input after disconnecting the antenna. Equation 1, given earlier, will enable one to calibrate the solar tracking radiometer, with T_N being replaced by T_H, the noise temperature of the hot load.

A solar tracking radiometer usually operates at night as a zenith looking fixed radiometer, when the antenna temperature will be lower than the room temperature. At such times, therefore, a liquid nitrogen load, acting as a cold load, will have to be used again replacing the hot load for calibration of the radiometer. Often, a waveguide switch is incorporated to switch between the cold and hot loads for calibration and to switch the antenna to the third position of the switch. The losses in the waveguide switch have to be taken into account during the calibration.

For calibration of emission noise records during night, we plan to employ calibration with liquid nitrogen emission instead of using a liquid nitrogen cooled matched load. The antenna need not be disconnected in that case. Even the hot load calibration can be arranged with the antenna connected to the radiometer, if a second antenna connected to the hot load is employed to illuminate the radiometer antenna. In that case we can calibrate the radiometer over the entire range from 77°K to 6000°K without disconnecting the antenna. A plane reflector may be used to direct either the hot or cold source emissions by adjusting the plane reflector with the two sources being kept on either side of the line passing through the axis of the antenna, as shown in Figure 18.28. The reflector positions 1 and 2 correspond to calibration with the liquid nitrogen source and noise sources respectively, while position 3 is for reception of solar or sky emissions.

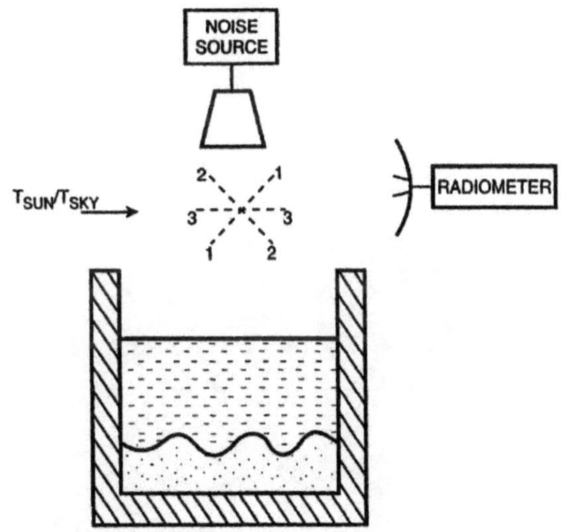

FIGURE 18.28 Use of plane reflector to direct either the hot or cold source emissions.

CHAPTER 19

DOPPLER RADAR DEVELOPMENT AND REMOTE SENSING RESEARCH

19.1 INTRODUCTION

Significant accomplishments have been made in the upgrade of the research and development of the WSR-88D radar. Conceptual design of the hardware and associated software have been completed very successfully. Detailed design of the synchronizer board for timing and an interface board have been completed including fabrication. A modern development environment and prototype signal processors and host computers have been acquired. Development of base data estimation algorithms on the new digital signal processor has begun. Final design of the building to house the radar and to provide a workshop and office space has been completed. Modifications of the microwave circuits include installation of a dual polarization feed, associated waveguides, and an elevation rotary joint. A dual azimuthal rotary joint has been procured as well as all other microwave components. Pattern measurements were made on the antenna before these changes and after the changes. Both the initial pattern and patterns for vertically polarized and horizontally polarized waves confirm to specifications. Good matches exists down to 20 dB below the main lobe, and the cross polar pattern is more than 35 dB below the co-polar pattern.

Software to simulate various phase encoding methods has been developed and a novel phase coding has been proposed. The coding and decoding procedure allows recovery of overlaid echoes that are 30 dB below the strong echo.

19.2 POLARIMETRIC MEASUREMENTS OF PRECIPITATION

It was determined that polarimetry offers valuable diagnostic capability concerning the type of weather radar echoes. Besides providing inferences on the type a hydrometeors, polarimetry can be used to discern between meteorological and biological scatterers. The fields of differential reflectivity and differential phase for bird echoes have distinctly different signatures than those of insect echoes. Although this finding has significance to entomologists and ornithologists, its effects are important to operational meteorologists, as radar measured winds can have vastly different bias in the presence of birds versus insects. One obstacle in precipitation measurements is the presence of ground clutter caused by anomalous propagation. Polarimetry not only can be applied to identify regions of anomalous propagation, but can also be used to measure rain in the contaminated region.

19.2.1 Electrical Effects of Storms and Lightning

After the completed study on long-term changes of lightning deaths during the twentieth century and their connection with demographic and climatological changes during the same period, the nationwide distribution of lightning casualties and damage was investigated and documented. The physical and quantitative relation between cloud charges, potentials, and electric fields, and the induced charges, currents, and electric field changes in the lightning channel have been determined by various scientists.

19.3 AUTOMOTIVE RADAR APPLICATIONS

The demand for radar applications in the automotive sector is growing rapidly. Such applications are, for example, distance measurement for automotive cruise control, parking air, side crash detectors, or blind spot detection. Combining all the intended applications, the car will be included in a radar bubble. The applications will support the driver and increase road safety. First radar systems are already on the market. By now they are only implemented in upper class cars. However, in order to extend the applications and to open up the mass market for these systems, additional research work is still required. For the mass market costs, reliability and easy fabrication are the most important factors. These needs can be satisfied best with hybrid or monolithically intended circuits. By now no single chip implementation for the radar front-end exists or makes economic sense, and therefore the hybrid assembly offers some advantages. It is easy to combine different types of semiconductors like GaAs and silicon. Furthermore, these active semiconductor devices can be mounted onto a cheaper substrate with the printed structures of the passive elements. Even more there is flexibility for the antenna design. The objective of the design presented herein was to reduce costs and losses by reducing the elements and the required substrate area. Reducing the size is especially an advantage for MMIC circuits, but it is also a design goal also for the less costly hybrid assembly.

In a traditional transceiver the oscillator signal is divided into the transmitted and the LO signal by a directional coupler. In addition, a second directional coupler is required to merge the received and the LO signal at the non-linear devices of the mixer. In such a concept losses are considerable due to large structures. In this design the purpose of splitting and merging the signals is realized with only one directional coupler. A planar monostatic radar front-end is particularly suited for low-cost application, as it uses only one antenna for transmission and reception and hence it minimizes the needed chip area. For small relative speed in a CW-Doppler-radar or short distance in a FMCW-radar, the IF is very small. Therefore l/*f*-noise is a significant value. In order to achieve good 1/*f*-noise characteristics, silicon Schottky barrier diodes are used. Compared to GaAs-based diodes they provide a lower l/*f*-noise cut-off frequency. The diodes were flip-chip bonded onto a microstrip circuit on a Al 2 O 3 substrate with microstrip waveguide structures. The flip-chip assembly using thermal compression bonding is suited for fully automated fabrication. The largest noise contribution in a transceiving mixer is the LO AM noise. With a view to gain good AM noise suppression at the IF port, a singly balanced mixer is used as the down-converting device. Thereby, the signal is applied in phase to the diodes of the balanced mixer, whereas the AM LO noise is applied out of phase to the diodes and therefore is rejected. The use of the singly balanced mixer principle is a good tradeoff between low losses and sufficiently high AM noise suppression. The designs basically differ in RF-LO isolation, but not in noise suppression. This work deals with the transceiving mixer. The purposes of transmitting a part of the signal power to the antenna, driving the diodes, and the mixing are all included in this element. The goal of the presented design is to realize this functioning with a minimum of components. Within this reduction the components usually allocated to a single function are merged into a single component. Thus, in the resulting element, there is no more clear allocation between a special function and a discrete element. In general, in a standard homodyne radar a ratio-frequency signal generated by an oscillator is used for both the transmitted signal and the local oscillator. Well controlled distribution of the input power between the antenna and mixer is crucial. There is a tradeoff between the output power, that is, the fraction of power, which is transmitted, and the LO power level at the diodes determining the conversion loss of the mixer. Both parameters influence the achievable range of the radar. The microstrip circuit was designed based on a full-wave analysis. The whole front-end was simulated using a harmonic-balance method. The diodes are grounded by a virtual short realized with a radial stub. As the properties of the diodes were not completely known, due to the absence of an appropriate model, the position of the virtual short was determined experimentally. The IF is applied at the rat-race ring. The detachment of the IF from the RF matched part should not influence the symmetrical power distribution and the AM noise suppression. A very low symmetry distortion was achieved by applying a standard RF block with a low-pass filter consisting of two stubs in a certain angle to the rat-race ring. The singly balanced mixer principle permits to serially bias the diodes. The DC-bias part is isolated from the RF by use of standard bandstop structures using radial stubs. For the measurement with a waver-prober, coplanar-to-microstrip transitions were included at the LO and the RF port.

19.3.1 Measurement

To verify the simulation the passive circuit was first measured separately. In the following the complete mixer was measured with variations of the diode match and the virtual short. The measurement frequency was 76.5 GHz with an LO-power level of 7.1 dBm. The IF was measured with a True RMS Voltmeter. The optimum load impedance was determined to be 39. The frequency shift of the applied RF signal compared to the LO frequency was 100 kHz with a RF power of –20 dBm. The bias voltage of the diodes was adjusted separately for each mixer under test. It depended on the match of the diodes and the resulting drive of the diodes by the LO. For both diodes in series, the applied voltage was about 400 mV. The conversion loss was calculated from the measured IF voltage and the impedance of the load. The noise figure of the mixer was characterized also. In order to measure the $1/f$-noise, the IF signal was amplified by use of a low-noise amplifier and measured using an spectrum analyzer. Only LO signal and DC bias voltage were applied to the mixer. The RF port was terminated using a 50 match.

The presented transceiving mixer at 76.5 GHz takes advantage of the elimination of a second hybrid. Thus, losses and the required substrate area are reduced significantly compared to existing realizations. The LO-RF port isolation of the hybrid is bypassed via the reflections at the diodes of the singly balanced mixer. A matching of the diodes was found to decrease the performance because of the additional losses. The measurements show low conversion loss of 8.9 dB at 3 dBm input power and excellent AM noise suppression.

19.4 WEATHER RADAR CORRECTION PROCEDURES

Mean areal rainfall intensity estimates are widely used in hydrology as input to lumped water and energy balance models and as "ground truth" for radar and satellite rainfall measurement systems. In general, mean areal rainfall intensity must be estimated by averaging available rain gauge readings. These estimates are subject to various sources of error: (*a*) inadequate spatial coverage or configuration, (*b*) inadequate gauge density, and (*c*) instrument errors. The issue of the accuracy of mean areal rainfall intensity estimates is particularly important when these estimates are computed in real-time as input to flood forecasting models, as in this case the high cost of equipping and maintaining the required telemetric facilities often results in low-density observation networks. A promising development in this field is related to the gradual change of weather radar from a tool for qualitative rainfall estimation to a tool for more quantitative rainfall measurement. The advantages of radar derived rainfall data with respect to rain gauge networks are manifold. Radar covers extended areas, notwithstanding the fact that it is easily measuring from a single point; it allows rapid access for real-time hydrological applications, and its spatial and temporal resolution can be high. However, many sources of error can affect radar rainfall estimation. Their magnitude depends on numerous parameters: radar features, distance from the radar, rainfall characteristics, and meteorological context. Adjustment with rain gauges remains the most widely used method to correct radar data. This approach, however, suffers from limitations which are mainly related to the fundamental differences in the sampling characteristiscs of the two sensors.

An intercomparison between different techniques for areal rainfall evaluation can be done on rain gauges, on radar data blended with rain gauge measurements through a case study on a catchment area equipped with a dense rain gauge network. Finally, a lumped hydrological model is used to assess the effects of errors in estimates of areal rainfall on predictions of river flow.

Kriging, a statistical technique for performing spatial interpolation, can be used to interpolate gauge data on a catchment area. Kriging has an unbiased minimum variance estimator, and it is preferable to other rainfall interpolation methods.

19.4.1 Radar Rainfall Data Preprocessing

Radar data used for the study are corrected for wavelength attenuation effects, beam blocking, and variability of the Vertical Profile of Reflectivity (VPR). Ground clutter rejection is accomplished through the use of Doppler measurements. A short account of the procedures employed for beam blocking and the VPR is given as follows:

Correction of the beam blocking was performed, taking into account the propagation of the beam into the atmosphere and its interaction with the terrain, modeled through a Digital Terrain Model characterized by a 200 m size dimension.

The procedure is able to account for the variability of the blocking effects within the single radar bin. As a result the percentage of not-intercepted energy emitted by the radar is expressed as a function of the antenna azimuth, antenna elevation, and range based on the assumptions that the beam propagates following the effective earth radius model. In this way, for each bin a correction factor computed once for all computed corrections can be added to the measured reflectivities in the preprocessing stage, with the assumption that the reflectivity is constant with height on the beam.

The concept of VPR was used to perform range-related corrections (brightband, partial beam filling), taking into account conjointly the beam geometrical properties and the vertical structure of the atmosphere. The identification of the VPR and compensation for its effect in real-time is still an open problem, and it is receiving considerable attention. This procedure requires the use of radar reflectivity data from at least two different elevation angles, and it is based on the computation of the ratio of reflectivity (reflectivity at high elevation divided by reflectivity at the lower elevation angle) versus distance from the radar. According to the method the ratio function depends on both the radar beam geometry and the VPR. The procedure is based on the use of an inverse method to deduce the VPR from the observed ratios of reflectivity.

19.5 COMPARISON OF MEAN AREAL RAINFALL ESTIMATES

Six different mean areal hourly rainfall estimates are compared for any event under investigation:

1. estimates obtained through kriging applied to a lower density network made of 9 rain gauges (Kriging 9, hereafter);
2. radar based estimates corrected for beam occlusion and the VPR (P Radar);
3. radar based estimates corrected for beam occlusion and the VPR and adjusted to agree with the rain gauge measurements from the lower density network (CP Radar 9);
4. radar based estimates from the 0.50 scan not corrected for beam occlusion and the VPR (NP Radar);
5. radar based estimates from the 0.50 scan not corrected for beam occlusion and the VPR and adjusted to agree with the rain gauge measurements from the lower density network (CNP Radar 9);
6. mean areal intensities obtained on the basis of a single rain gauge located centrally over the basin (Raingage SG-S. Giovanni Ilarione).

These estimates can be compared with that obtained by applying kriging to the complete rain gauge network (Kriging 19).

It should be pointed out that just eight out of the nineteen rain gauges (for the complete network) and four out of nine (for the lower density network) have non-negligible weights for mean areal rainfall estimation over the Alpone basin. The accuracy es can be investigated on the basis of variogram analysis. Then, while the complete network can lead to mean areal rainfall estimates of acceptable accuracy, the low density network appears to be more appropriate for estimation on larger areas or longer duration time steps. The sampling provided by one rain gauge is used here as representative of operational telemetric network densities used for real-time forecasting.

The following is a comparison of cumulative mean areal estimates obtained through the application of kriging on the complete network and from raw and processed radar data, both unadjusted and adjusted to agree with rain gauges.

19.6 THE RAINFALL-RUNOFF MODEL

The probability Distributed Model may be used to partition rainfall into direct runoff, groundwater recharge, and soil moisture storage. This model is complemented with a linear transfer function derived on a geomorphologie basic to simulate the runoff routing process. This section provides a brief description of these two components. Following the probability-distributed principle, only the frequency of hydrological variables of certain magnitudes over the basin is considered, without regard to the localization of a particular occurrence within the basin. The theory is applied to retention of water in upper soil layers, that is described by means of a statistical population of stores instead of a single storage component. Direct runoff production at a point in the catchment is controlled by the capacity of the soil to take up water. This is conceptualized as a single store with a given storage capacity. The river basin is considered to be made of a population of stores, each one characterized by its store capacity. This is viewed as a random variable with probability density function (pdf). The Rayleigh function, which is completely specified by the parameters, was adopted as the pdf for the depth. This is used to derive an algebraic expression that integrates the point direct runoff to yield the response from the basin. Moreover, it is assumed here that the relation between different storage elements allows the redistribution of water into the storages in such a way as to cause all storage elements to have an equal depth of water. The direct runoff rate generated from the spilling of a full storage element is assumed to travel independently of runoff from neighboring elements. It is added to the baseflow rate and routed to the basin outlet by means of a linear transfer function (Instantaneous Unit Hydrograph, IUH). In practice, since data are available at discrete time intervals (Dt = 1 hour) both are convoluted through a discrete convolution integral (Unit Hydrograph, UH).

A geomorphologic approach is followed for the derivation of the UH. In the investigation described here, the catchment response function is broken down into a hillslope response function and a network response function. As to the channel network, its IUH is seen as the holding time probability distribution function of a drop of water once it enters the channel network. This pdf is computed by using the network geometry and a single routing function governing water transport from any point in the network to the outlet $g(t, x)$. The river network geometry is synthesized by the width function $N(x)$, which represents the frequency distribution of channels with respect to flow distance x from the outlet. In this study the width function was derived from the river network automatically extracted from a 50 m × 50 m digital elevation model of the basin. The Geomorphologic Instantaneous Unit Hydrograph (GIUH) is then computed by summing, over the network. The assumption made here is that the drainage density is constant everywhere in the basin, and this average length is approximately half the inverse of the drainage density.

Analysis of results obtained by comparing different mean areal rainfall estimates indicates that the underestimation affecting raw radar rainfall data is more efficiently removed through correction for beam blocking and the VPR than by adjusting raw radar data with rain gauges. However, in terms of correlation, a negligible gain is found through the application of the correction procedures, while adjusting with rain gauges allows considerable improvement. This is mainly due to the rainfall overestimation during the second part of the event. Radar estimates are in that case affected by bright-band, and the capability of the VPR correction technique to remove this error is limited by shielding effects over the region used for the VPR Identification and correction.

A substantial improvement in hydrograph simulation is found by using unadjusted, both raw and corrected, radar rainfall data in place of a single rain gauge located centrally on the basin. Results obtained by using nine rain gauges to adjust radar rainfall data indicate that the use of either the same rain gauges or unadjusted radar (both raw and corrected) provides worse results than adjusted radar which combines the two. The results presented here suggest that improving the knowledge of the vertical structure of radar rainfall observations and adjusting with rain gauges can significantly enhance the accuracy of both mean areal rainfall estimation and resulting flood simulations. However, it is required to confirm these results with more rainfall and flood events. Research is underway to provide additional testing and evaluation on other catchments characterized by different conditions as to shielding and range.

19.7 RADAR AND STORM MODEL

The United Nations International Decade for Natural Disaster Reduction (IDNDR) is focusing upon the need to understand the occurrence and behavior of major national hazards like earthquakes, hurricanes, tidal and fluvial floods, and drought. However, little attention has been paid to dam bursts due to overtopping, which may, in the extreme, bring death and destruction to communities with little or no warning. This type of event effects equally both the developed and developing world. In the United Kingdom many large dams were built during the last century. These dams, and those built recently, are inspected regularly as a result of legislation passed during the first half of this century in response to several dam failures. However, the design parameters to which they are built, including the Probable Maximum Precipitation (PMP), were sometimes based upon limited, or poorly understood, data. Overtopping is prevented by building a spillway, the crest of which is below the height of the top of the dam wall. Should overtopping occur then earth core dams, which account for around 78% of all dams in the world, would be washed away. The spillway is designed to take the maximum flow occurring during the Probable Maximum Flood (PMF) arising from the PMP over the catchment for the critical storm duration. Therefore, estimation of the PMP and the PMF are a vital part of dam design.

The Storm Model seeks to interpret dynamically processes of the type mentioned in the previous section in terms of modifications to the thermodynamics of the ascent of a single parcel of air. This approach has the advantage of using maximum measured values of atmospheric parameters with clearly defined geometrical characteristics of a catchment. However, there is no requirement to measure the maximum likely surface dew-point temperature, which is required to evaluate precipitable water and hence, if a storm efficiency is assumed, PMP. The maximum dew-point temperature, Tdm, is calculated objectively from the model for a particular river or reservoir catchment. This is important as surface dew-point temperatures are often not available for many areas of interest. Even over lowland areas, hourly dew-point temperatures may not represent local effects sufficiently well, as the measurements are made at sites which are chosen specifically to be unaffected by, for example, areas of water such as reservoirs. By running the model with input wind fields and efficiency values at minute intervals, and obtaining the catchment PMP for each time interval, it is possible to construct a hyetograph. The distribution of wind fields are assumed to be Gaussian across the storm, that is the last half of the event is a reflection of the wind fields for the first half. This assumption greatly reduces the calculation times without compromising these calculations to any large degree.

19.8 DOPPLER WEATHER RADAR

The wind field is important in assessing the spatial distribution of PMP, although the exact form of the wind field is less important for the aerial estimation of PMP, particularly over catchment of 50 km^2 or greater. Dual Doppler radar estimates can provide detailed wind fields. However, even a single Doppler radar can provide useful wind field estimates, albeit using techniques which are subject to approximations. For example, the Uniform Wind Technique assumes that the horizontal wind is constant as a function of azimuth near an analysis point, and the vertical velocity is zero. More sophisticated techniques are possible, but they may not be necessary for PMP studies, which only need wind fields having the appropriate spatial and temporal kinematics features.

A simple Storm Model has been used, with detailed wind fields derived from a complex Large Eddy Simulation model representing the wind fields that might be derived from Doppler radar for estimating PMP for a catchment in Peninsular Malaysia. These results are preliminary, but they suggest that recent estimates of PMP using transposition and maximization techniques may be overcautions. However, further work using more appropriate wind fields is essential.

Many countries are now installing Doppler weather radars, and the data from these systems should provide useful information to assess the viability of the Storm Model approach. More complex numerical models do provide wind fields which have the required spatial and temporal scales, but should only be used after detailed verification using Doppler radar data.

19.9 SPACE WEATHER EFFECTS ON GPS AND RATIO PROPAGATION

Modern navigational systems which use radio-wave signals reflecting from or propagating through the ionosphere as a means of determining range or distance are vulnerable to a variety of effects that can degrade performance. In particular, systems like the Global Positioning Systems (GPS), that use constellations of earth-orbiting satellites, are affected by space weather phenomena. In principle, the GPS uses known positions of satellites and their distances from a receiver to determine the location of the receiver (Figure 19.1). When charged particles ejected from the sun arrive at the earth, they cause perturbations in the geomagnetic field. Another effect is the in the ionosphere variation of electron density considerably, both in time and space.

A GPS receiver uses radio signals from several orbiting satellites for determining the range or distance from each satellite and determines from these ranges the actual position of the receiver. The radio signals must pass through the ionosphere, and in doing so they are subjected to variations in the electron density structure of the ionosphere. Changes in the electron density due to space weather activity can change the speed at which the radio waves travel, introducing a "propagation delay" in the GPS signal. The propagation delay can vary from minute to minute, and such intervals of rapid change can last for several hours, particularly in the polar and auroral regions. Changing propagation delays cause errors in the determination of the range, or "range errors." The performance of single-frequency GPS receivers using Code Phase Tracking techniques can be significantly degraded by the ionospheric propagation delays. Use of dual-frequency GPS receivers can largely compensate for most of the ionospheric propagation delays by measuring the different delays at the two frequencies. Ionospheric delay corrections for a region can be determined from a network of precisely-positioned dualfrequency receivers and then be transmitted in near-real-time to users of single frequency GPS receivers in the region. Such a system is operated by the Canadian Active Control System of Natural Resources Canada.

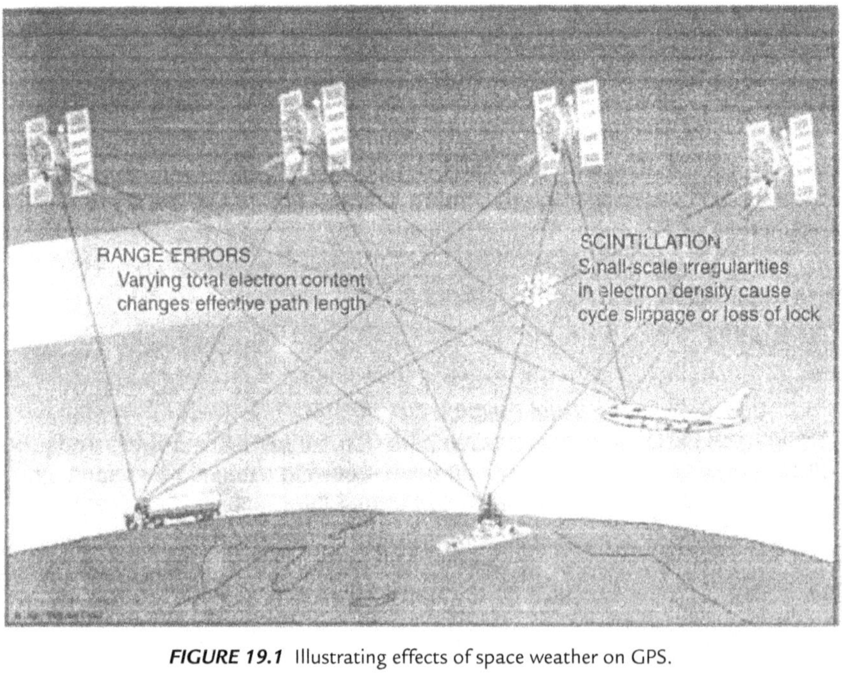

FIGURE 19.1 Illustrating effects of space weather on GPS.

Another GPS technique uses Carrier Phase Tracking. In this technique, the phases of individual cycles of the carrier waves are compared. However, if the electron density along a signal path from a satellite to a receiver changes very rapidly, as a result of space weather disturbances, the resulting rapid change in the phase of the radio wave may cause difficulties for the GPS receiver, in the form of "loss of lock." Temporary loss of lock results in "cycle slip," a discontinuity in the phase of the signal. Very rapid variations (less than about 15 seconds) in the signal's strength and phase are known as "ionospheric scintillations." Scintillations can be particularly troublesome for receivers that are making carrier-phase measurements and may result in inaccurate or no-position information. Code-only receivers are less susceptible to these effects. The GPS system provides continuous routine measurements of the Total Electron Content, that is, the aggregate of electrons along each radio wave propagation path from satellite to receiver along the multitude of varying signal paths to each receiving station in a regional or global network. These measurements permit the mapping of variations in the ionospheric electron density over a region. Such information can be of use for studying space weather phenomena themselves.

The electromagnetic radiation emitted from the sun spans a continuum of wavelengths from radio, through microwave, infrared, visible, ultraviolet x-rays and beyond. The ultraviolet radiation interacts with the upper atmosphere to form an ionized layer known as the ionosphere. Radio waves interact with the ionosphere in a variety of ways, depending on their frequencies. For frequencies below 30 MHz, the ionosphere acts as a reflector and it permits very long distance radio communications around the globe. At frequencies greater than 30 MHz, radio signals pass through the ionosphere, in general. The ionosphere sometimes becomes disturbed due to reaction by some types of solar activity. Consequently, radio wave propagation may be degraded or disrupted. Solar flares emit electromagnetic radiation like x-ray emissions and increases ionization in the lower ionosphere, with consequent phase shifts in low frequency radio signals and increased absorption called fading in HF and VHF radio signals. The wide spectrum of radio noise as emitted from a flare may be responsible for interfering with a wanted radio signal. These effects have been experienced at all latitudes. If the frequencies are above 30 MHz, unexpected reflections of the radio waves by the ionosphere may cause radio interference. Ionospheric irregularities may produce fluctuating signals, a phenomenon known as scintillation, and may distort paths of radio waves. During the period of geomagnetic storms and the associated ionospheric disturbances, scintillation activity may affect certain applications of navigational aids such as the Global Positioning System (GPS).

Solar flares may be accompanied by streams of very energetic particles, mainly protons and electrons which enter the upper atmosphere in the regions near the magnetic poles. Due to this the lower levels of the polar ionosphere become very ionized, with severe absorption of HF and VHF radio signals. Such an event known as a polar cap absorption (PCA) event may last from days to weeks, depending on the strength of the stream of solar particles and the location of the source region on the sun. In polar regions HF radio communication is severely disturbed during PCA events. Large clouds of plasma (ionized gases), known as Coronal Mass Ejections (CME), can be emitted from the sun. When they reach the earth they cause disturbances both in the geomagnetic field and in the ionosphere. Coronal holes, regions of the solar corona with diminished x-ray emissions, also emit streams of charged particles resulting in disturbances of the ionosphere. Ionospheric disturbances are significant at auroral latitudes, and during magnetic storms and substorms at these latitudes, HF radio communication may be largely unreliable.

CHAPTER 20

MATHEMATICAL PRELIMINARIES RELATED TO RADAR SYSTEMS AND ASSOCIATED PROBLEMS

20.1 FOURIER ANALYSIS

20.2 PERIODIC FUNCTION

20.2.1 Fourier Series

A periodic function $f(t)$ of period T can be represented by a Fourier series as follows:

$$f(t) = \frac{a_0}{2} + \sum_{n=1}^{\infty} a_n \cos n\omega t + \sum_{n=1}^{\infty} b_n \sin n\omega t \tag{20.1}$$

where $\omega = \dfrac{2\pi}{t}$ is the circular frequency corresponding to the period T. The function $f(t)$ has been decomposed into a DC component $(a_0/2)$ and into harmonic components of the fundamental frequency ω and its multiples.

The trigonometric functions are orthogonal. This is to say that these functions satisfy the equations:

$$\int_0^T \cos n\omega t \sin m\omega t \, dt = 0 \tag{20.2}$$

$$\frac{2}{T}\int_0^T \cos n\omega t \cos m\omega t \, dt = \delta_{nm} \tag{20.3}$$

$$\frac{2}{T}\int_0^T \sin n\omega t \sin m\omega t \, dt = \delta_{nm} \tag{20.4}$$

where δ_{nm} is unity for $m = n$, and zero otherwise.

Using (20.2) through (20.4), the following expressions may be derived for the coefficients of the Fourier series (20.1):

$$a_0 = \frac{2}{T}\int_{-T/2}^{T/2} f(t) \, dt \tag{20.5}$$

$$a_n = \frac{2}{T}\int_{-T/2}^{T/2} f(t) \cos n\omega t \, dt \tag{20.6}$$

$$b_n = \frac{2}{T}\int_{-T/2}^{T/2} f(t) \sin n\omega t \, dt \tag{20.7}$$

20.2.2 Complex Representation

Equations (20.1), (20.5), (20.6), and (20.7) can be written more compactly by using complex notation. Recall that

$$\cos\theta = \left[\exp(j\theta) + \exp(-j\theta)\right]/2 \tag{20.8}$$

and

$$\sin\theta = \left[\exp(j\theta) - \exp(-j\theta)\right]/2j \tag{20.9}$$

Using the previous equations, and admitting negative frequencies, we have

$$f(t) = \sum_{n=-1}^{n=\infty} C_n \exp(jn\omega t) \tag{20.10}$$

and

$$C_n = \frac{1}{T}\int_{-T/2}^{T/2} f(t) \exp(-jn\omega t) \, dt \tag{20.11}$$

A good geometrical representation of expressions of the form exp $(j\omega t)$ is in terms of rotating vectors on an Argand diagram. In this representation positive frequencies correspond to vectors rotating in the positive (i.e., anticlockwise) sense, while negative frequencies correspond to vectors rotating in the negative (i.e., clockwise) sense.

The coefficients a, b, and C are related by

$$C_n = \frac{1}{2}(a_n - jb_n) \tag{20.12}$$

where we formally admit negative values of n on the right-hand side of this equation. Using (20.12) the term C_n exp $(jn\omega t)$ may be written A exp $(jn\omega t + \phi_n)$

where
$$\left.\begin{aligned} A &= \frac{1}{2}(a_n^2 + b_n^2)^{1/2} \\ \phi_n &= \tan^{-1}(-b_n / a_n) \end{aligned}\right\} \tag{20.13}$$
and

In our work, we shall always be concerned with real valued function $f(t)$.

In this case the a's and b's must be real. However, the C's must be complex, since these occur as the coefficients of the complex quantities exp $(jn\omega t)$.

From equation (20.11) it is seen that

$$C_n{*} = C_{-n} \tag{20.14}$$

20.2.3 Symmetry Properties

If $f(t)$ possesses certain symmetry properties, then the coefficients of the Fourier series may be related. For example, (*a*) If $f(t) = f(-t)$, even symmetry, then all the sine terms in the trigonometric Fourier series vanish. (*b*) If $f(t) = -f(-t)$, odd symmetry, then all the cosine terms in the trigonometric series vanish. (*c*) If $f(t) = -f(t \pm T/2)$, rotation symmetry, then all even harmonics vanish. (*d*) If $f(t) = f(t \pm T/2)$, then all odd harmonics vanish, including the Fundamental Example of case (*c*) is given in Figure 20.1 (*b*). Note that any arbitrary function $f(t)$ may be decomposed into (*i*) even and odd parts

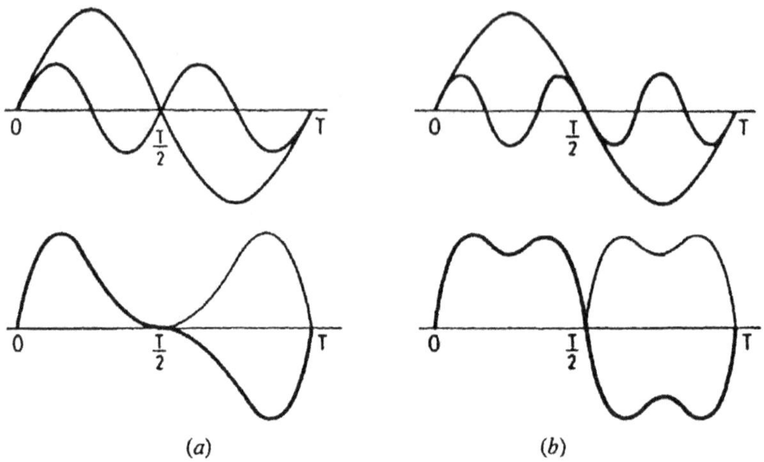

FIGURE 20.1 The effect of even harmonics (*a*), and odd harmonics (*b*).

$$f(t) = \underbrace{\frac{f(t)+f(-t)}{2}}_{\substack{\text{even} \\ \text{cosine terms}}} + \underbrace{\frac{f(t)-f(-t)}{2}}_{\substack{\text{odd} \\ \text{sine terms}}}$$

or (ii) parts having only even or odd harmonics

$$f(t) = \underbrace{\frac{f(t)+f(t+T/2)}{2}}_{\text{even harmonics}} + \underbrace{\frac{f(t)-f(t+T/2)}{2}}_{\text{odd harmonics}}$$

20.2.4 Further Developments

(*a*) Expansion in terms of orthogonal functions. The previous representation of a function in terms of trigonometric or complex exponential functions is a special case of the general problem of approximation of function by means of a set of orthogonal functions. The functions $\phi_i(t)$, $i = 1, 2, \ldots\ldots$ are said to be orthogonal if

$$\int \phi_i(t)\phi_j{}^*(t)dt = \begin{cases} 0, & i \neq j \\ K_{i,} & i = j \end{cases} \tag{20.15}$$

where the integration is over an appropriate range of t and * denotes a complex conjugate. If, in addition, $K_i = 1$, then the functions are called orthonormal (orthogonal and normalized).

Suppose it is required to approximate a function $f(t)$ by a linear combination of n of the function $\phi_i(t)$:

$$f(t) \cong \lambda_1\phi_1(t) + \lambda_2\phi_2(t) + ... + \lambda_n\phi_n(t), \tag{20.16}$$

and the coefficients λ_i are to be determined so that the approximation is optimum in some sense. The error of the approximation is

$$\varepsilon = f(t) - \Sigma\lambda_i\,\phi_i\,(t) \tag{20.17}$$

Let the coefficients be determined such that the integral of the square of the error as given by

$$E = \int \varepsilon^2\,dt = \int | f - \Sigma\lambda_i\,\phi_i\,|^2\,dt \tag{20.18}$$

is a minimum. By differentiating with respect to λ_i show that E is a minimum when

$$\lambda_i = \int f\phi_1 *\,dt / \int \phi_i\,\phi_i * dt \tag{20.19}$$

Verify that (*i*) if ϕ_i are the trigonometric functions, then (20.15) and (20.19) reduce to (20.16) through (20.17); (*ii*) If ϕ_i are the complex exponentials, then (20.15) above is satisfied, and (20.19) reduces to (20.4). In this way the expansion in terms of complex exponentials is obtained without going through the intermediary of the trigonometric expansion.

(*b*) **Explicit expression for the error.** By expanding (20.18), and making use of (20.15), with $K_i = 1$, and (20.19), show that the error E is given by

$$E = \int ff * dt - \sum_{i}^{n} \lambda_i\lambda_i * \tag{20.20}$$

If the set ϕ_i, is complete and f satisfies certain regularity conditions, which is almost always the case in physical situations, then $E \to 0$ as $n \to \infty$. Then we have

$$f(t) = \sum_{i=1}^{\infty} \lambda_i\phi_i(t) \tag{20.21}$$

and

$$\int | f |^2\,dt = \sum_{i=1}^{\infty} | \lambda_i |^2 \tag{20.22}$$

Equation (20.22) is an expression of Parseval's theorem. For trigonometric functions, we have

$$\frac{1}{T}\int_{-T/2}^{+T/2} f^2(t)\,dt = \frac{a_0^{\,2}}{4} + \Sigma\left(\frac{a_i^{\,2}}{2} + \frac{b_i^{\,2}}{2}\right) \tag{20.23}$$

For the interpretation of this relation, see Section (1.10) of Stuart.

20.3 NON-PERIODIC FUNCTIONS

20.3.1 Transition to the Fourier Integral

Suppose the function $f(t)$ is not periodic. We may view this case as the limit of the periodic case as the period $T \to \infty$. In this limit we can write,

$$f(t) = \frac{1}{2\pi}\int_{-\infty}^{+\infty} F(\omega)\exp(j\omega t)\,d\omega \tag{20.24}$$

and

$$F(\omega) = \int_{-\infty}^{+\infty} f(t)\exp(-j\omega t)\,dt \tag{20.25}$$

This transition may be made plausible in the following way. The spectrum of the periodic function of period T consists of lines at frequencies 0, ω, 2ω, ... where $\omega = 2\pi/T$. As $T \to \infty$, C_n and $\omega \to 0$, and the lines of the spectrum become closer and "dense." In the limit we have a "continuous" spectrum instead of the discrete line spectrum, the summation of the Fourier series being replaced by an integration. We find that $(C_n\, T)$ remains finite as $T \to \infty$. Let us call this limit $F(\omega)$, and identify $n\omega$ with the continuous frequency variable ω. Then

$$F(\omega) = \int_{-\infty}^{+\infty} f(t)\exp(-j\omega t)\,dt \tag{20.26}$$

Now we can write

$$f(t) = \frac{1}{2\pi}\sum (C_n\, T)\exp(jn\omega t)\frac{2\pi}{T} \tag{20.27}$$

As $T \to \infty$, $\dfrac{2\pi}{T} \to 0$ and may be identified with the infinitesimal frequency increment $d\omega$, $n\omega$ may be identified with the continuous frequency variable ω, and $(C_n T)$ has already been called $F(\omega)$; the summation becomes an integration and in the limit we have

$$f(t) = \frac{1}{2\pi} \int_{-\infty}^{+\infty} F(\omega) \exp(j\omega t)\, d\omega \tag{20.28}$$

The functions $f(t)$ and $F(\omega)$ are referred to as a Fourier integral transform pair. The Fourier transform $F(\omega)$ is also denoted by $(f(t))$, and is termed the spectral density function. Sometimes the transformation is also indicated by an arrow: $f(t) \leftrightarrow F(\omega)$.

If $f(t)$ is real valued then it follows that

$$F(\omega) = F*(-\omega) \tag{20.29}$$

Since $F(\omega)$ is, in general, complex it may be written as

$$F(\omega) = |F(\omega)| \exp|j\theta(\omega)| \tag{20.30}$$

$|F(\omega)|$, and $\theta(\omega)$ are referred to as the magnitude and the phase spectrum, respectively. $|F(\omega)|^2$ is referred to as the energy density spectrum (but does not necessarily have the dimensions of energy: its dimensions are $\{[f]\ \mathrm{Hz}^{-1}\}^2$, where $[f]$ denotes the dimension of $[f]$.

20.3.2 Symmetry Properties

The Fourier transform of $f(t)$ may also be expressed as

$$F(\omega) = \int_{-\infty}^{+\infty} f(t) \cos \omega t\, dt - j \int_{-\infty}^{+\infty} f(t) \sin \omega t\, dt \tag{20.31}$$

It is seen that if $f(t)$ is an even function of t, then

$$F(\omega) = 2\int_{-\infty}^{+\infty} f(t) \cos \omega t\, dt \tag{20.32}$$

and if $f(t)$ is an odd function of t, then

$$F(\omega) = -2j\int_{0}^{\infty} f(t) \sin \omega t\, dt \tag{20.33}$$

With the help of the previous it is easily seen that:

If $f(t)$ is a	*Then $F(\omega)$ is a*
Real and even function of t	Real and even function of ω
Real and odd	Imaginary and odd
Imaginary and even	Imaginary and even
Complex and even	Complex and even
Complex and odd	Complex and odd

20.3.3 Examples

(*a*) Gate function. In pulsed radar the scattered signal is received between pulses. In order to examine the signal from a given range interval, the incoming signal is range gated, which amounts to multiplying it by a so-called gate function.

The gate function, $G_T(t)$, is defined by

$$G_T(t) = \begin{cases} 1 & |t| < T/2 \\ 0 & |t| > T/2 \end{cases} \tag{20.34}$$

The Fourier transform of this function is easily calculated

$$\begin{aligned} F(\omega) &= \int_{-T/2}^{+T/2} (-j\,\omega\, t)\, dt = T\frac{\sin(\omega T/2)}{(\omega T/2)} \\ &= T\,Sa(\omega T/2) \end{aligned} \tag{20.35}$$

The function ($\sin x/x$) is called the sampling function and is denoted by $Sa(x)$. The gate function is also called the rectangular pulse. The gate function and its Fourier transform are shown in Figure 20.2. Note that the width of the frequency interval to the first zeros of $F(\omega)$ is ($4\pi/T$).

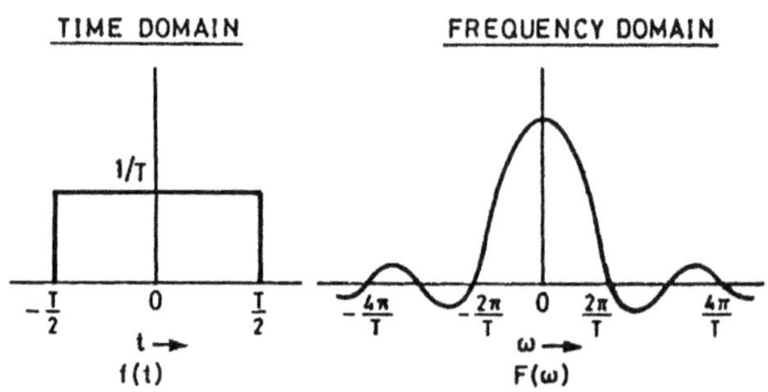

FIGURE 20.2 The gate function and its Fourier transform.

Note the following definite integral

$$\int_{-\infty}^{+\infty} Sa(x)\,dx = \int_{-\infty}^{+\infty} (\sin x / x)\,dx = \pi \tag{20.36}$$

(*b*) Cosine wave of finite duration. In pulsed radar harmonic wave trains of finite duration are transmitted periodically. Let us examine the Fourier transform of one such wave train.

The cosine wave of frequency ω_0 and duration T may be written

$$f(t) = \begin{cases} \cos \omega_0 t & -\dfrac{T}{2} < t < +\dfrac{T}{2} \\ 0 & \text{elsewhere} \end{cases} \tag{20.37}$$

Its Fourier transform is

$$F(\omega) = \frac{T}{2}\left\{ \frac{\sin[(\omega-\omega_0)T/2]}{[(\omega-\omega_0)T/2]} + \frac{\sin[(\omega+\omega_0)T/2]}{[(\omega+\omega_0)T/2]} \right\} \tag{20.38}$$

$$F(\omega) = \frac{T}{2}\left[Sa\{(\omega-\omega_0)T/2\} + Sa\{(\omega+\omega_0)T/2\} \right] \tag{20.39}$$

Figure 20.3 shows the cosine wave of finite duration and its Fourier transform. Note that (*a*) the cosine wave of finite duration is the product of a cosine wave of infinite duration and a gate function, and (*b*) the Fourier transform is the sum of the Fourier transform of the gate function shifted by ω_0 on either side of the origin. The significance of this will become clear when we consider the frequency shift or modulation theorem.

Here we may note some orders of magnitude. If the pulse duration τ is 1μ sec, then the spread of the frequency about the center frequency f_0 is about $1/\tau$ or 1 megahertz. This spread is small compared to the radar frequency f_0 of some thousands of megahertz, but must be carefully considered when dealing with problems of pulsed radar; that is, the receiver bandwidth must accept a frequency band $\Delta f \simeq 1/\tau = 1$ MHz when $\tau = 1$ μ sec.

Before considering further examples, we shall extend the Fourier integral to include certain cases in which the integral for the Fourier transform fails to exist; this extension is made through the introduction of the Dirac delta function.

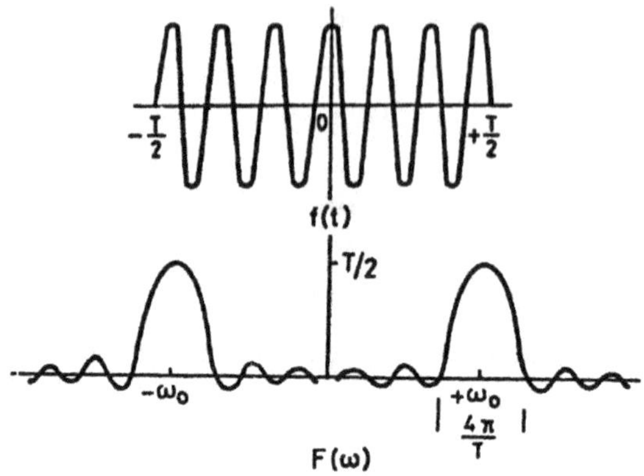

FIGURE 20.3 Cosine wave of finite duration and its Fourier transform.

20.4 SINGULARITY FUNCTIONS

Consider the function $f(t)$ to be a constant of magnitude unity. In the case of the Fourier series, the computation of the spectrum is trivial; it is a line of magnitude unity at $\omega = 0$. The Fourier integral of the constant $f(t) = 1$ is given by

$$F(\omega) = \int_{-\infty}^{+\infty} \exp(-j\,\omega\,t)\,dt \tag{20.40}$$

This integral does not exist. However, let us consider the Fourier transform of the constant as the limit of that of the gate function as its duration $T \to \infty$. Thus

$$F(\omega) = \lim_{T\to\infty} T\frac{\sin(\omega T/2)}{(\omega T/2)} \tag{20.41}$$

It can be shown that as $T \to \infty$ $F(\omega) \to 0$ for $\omega \neq 0$ and $F(\omega) \to \infty$. For $\omega = 0$. Further, the area under the curve of $F(\omega)$ is 2π irrespective of the value of T. Thus as $T \to \infty$ the gate function tends to a constant, and the Fourier transform of the gate function tends to a spike at $\omega = 0$, the area under it being 2π. Hence, we may write

$$F(\omega) = \int_{-\infty}^{+\infty} \exp(-j\,\omega\,t)\,dt = 2\,\pi\,\delta(\omega) \tag{20.42}$$

where the Dirac function $\delta(\omega)$ is such that it is zero for $\omega \neq 0$, and infinite at $\omega = 0$, in such a way that

$$\int_{-\varepsilon}^{+\varepsilon} \delta(\omega)\, d\omega = 1 \tag{20.43}$$

for any $\varepsilon > 0$. A useful property of the delta function is that

$$\int_{-\infty}^{+\infty} f(x)\, \delta(x - x_0)\, dx = f(x_0) \tag{20.44}$$

In view of (20.43) and (20.44), the Dirac delta function is also referred to as the unit impulse function or the impulse sampling function. With the help of the Dirac delta function it may be shown that:

$$\Im(e^{j\omega_0 t}) = 2\,\pi\, \delta(\omega - \omega_0) \tag{20.45}$$

$$\Im(\cos \omega_0\, t) = \pi\left[\delta(\omega - \omega_0) + \delta(\omega + \omega_0)\right] \tag{20.46}$$

$$\Im(\sin \omega_0\, t) = j\,\pi\left[\delta(\omega + \omega_0) - \delta(\omega - \omega_0)\right] \tag{20.47}$$

$$\Im(\delta(t)) = 1 \tag{20.48}$$

$$\Im(u(t)) = \pi(\omega) + 1/(j\omega) \tag{20.49}$$

$$\Im\left[\operatorname{sgn}(t)\right] = 2/(j\omega) \tag{20.50}$$

The function $u(t)$ is the so-called step function which is zero for $t < 0$, and unity for $t > 0$. The function sign (t) is the signum function, which is –1 for $t < 0$ and + 1 for $t > 0$. The above transforms are illustrated in Figure 20.4.

With the help of the Dirac delta function, the Fourier transform of a periodic function $f(t)$ of period T may be easily written. Let the Fourier series decomposition in complex notation be given by

$$f(t) = \sum_{n=-\infty}^{+\infty} C_n \exp(jn\, \omega_0\, t),\ \omega_0 = \frac{2\pi}{T}, \tag{20.51}$$

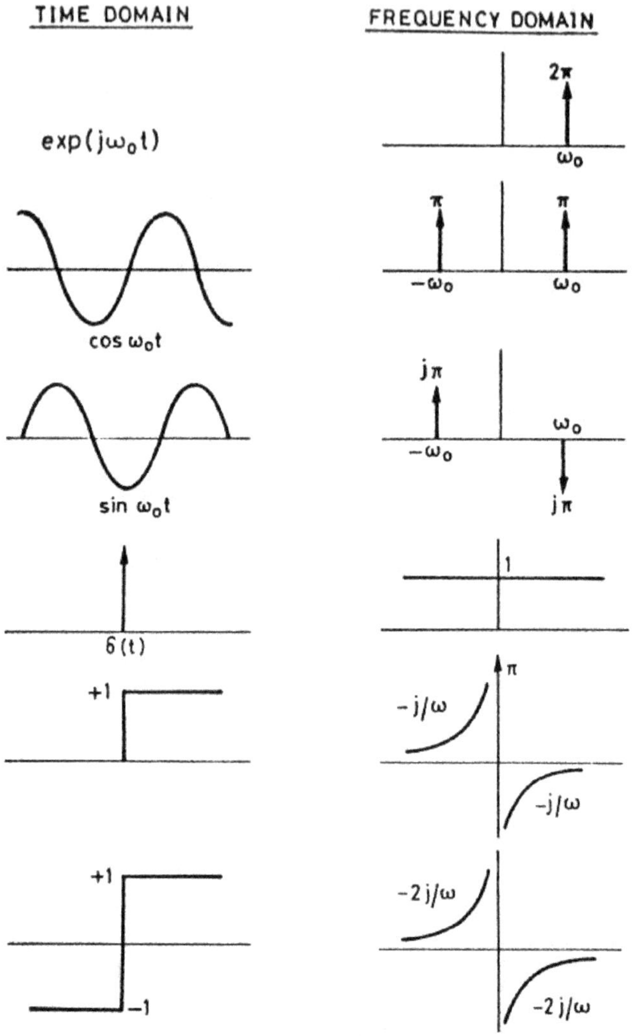

FIGURE 20.4 Some Fourier transform pairs involving the Dirac delta function.

Take the Fourier transform of both sides (here we are anticipating the linearity of the Fourier transform discussed in the following section):

$$\begin{aligned}\Im(f(t)) &= \sum_{n=-\infty}^{+\infty} C_n \Im(\exp(jn\,\omega_0\,t)) \\ &= 2\,\pi \sum_{n=-1}^{+\infty} C_n\; \delta(\omega - n\omega_0)\end{aligned} \tag{20.52}$$

Thus, the Fourier transform consists of a series of impulses at $\omega = 0, \pm\omega_0, \pm 2\omega_0, \ldots$, the magnitudes of the impulses being equal to 2π times the coefficients in the Fourier series representation.

A very useful transform is that of a periodic train of delta function repeating every T seconds. This function, denoted $\delta_T(t)$, may be written

$$\delta_T(t) = \sum_{n=-\infty}^{n=\infty} \delta(t - nT) \tag{20.53}$$

With the help of (5) and (13) show that:

$$\begin{aligned}\Im(\delta_T(t)) &= \omega_0 \sum_{n=-\infty}^{\infty} \delta(\omega_{-n}\,\omega_0),\ \omega_0 \equiv \frac{2\pi}{T} \\ &= \omega_0\, \delta_{\omega_0}(\omega)\end{aligned} \tag{20.54}$$

The function $\delta_T(t)$ is sometimes referred to as a Dirac comb. The Fourier transform of a Dirac comb in the time domain is a Dirac comb in the frequency domain. The transform pair (20.53), (20.54) is shown in Figure 20.5.

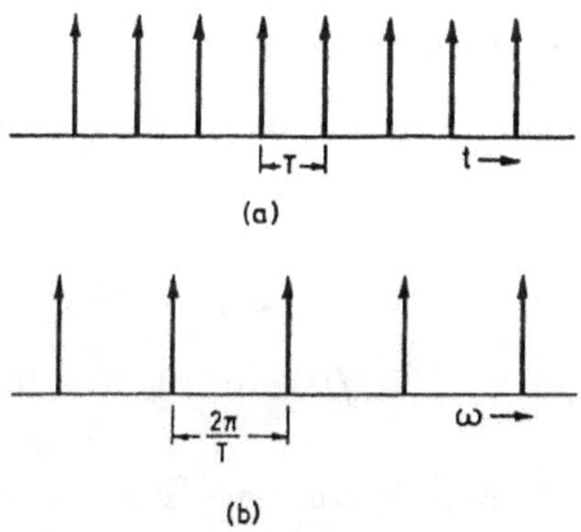

FIGURE 20.5 The transform of a Dirac comb in the time domain (*a*) is a Dirac comb in the frequency domain (*b*).

20.5 SOME PROPERTIES OF THE FOURIER TRANSFORM

(*a*) Duality

If

$$\Im(f(t)) = F(\omega) \tag{20.55}$$

then

$$\Im(f(t)) = 2\pi f(-\omega) \tag{20.56}$$

If f is even $$\Im(F(t)) = 2\pi f(\omega) \tag{20.57}$$

(*b*) Linearity

If $$\Im(f_i(t)) = F_i(\omega) \tag{20.58}$$

then $$\Im(\Sigma a_i f_i(t)) = \Sigma a_i F_i(\omega) \tag{20.59}$$

where the a_i's are constants.

(*c*) Scaling

If $$\Im(f(t)) = F(\omega) \tag{20.60}$$

then for a real constant a

$$\Im(f(at)) = \frac{1}{|a|} F\left(\frac{\omega}{a}\right) \tag{20.61}$$

Thus, if a function is compressed in the time scale by a factor a, it is expanded in the frequency scale by the same factor.

(*d*) Frequency-shifting property

This is also sometimes referred to as the modulation theorem:

It states that if

$$\Im(f(t)) = F(\omega) \tag{20.62}$$

then $$\Im[f(t)\exp(j\omega_0 t)] = F(\omega - \omega_0) \tag{20.63}$$

$$\Im[f(t)\cos\omega_0 t] = \frac{1}{2}[F(\omega+\omega_0) + F(\omega-\omega_0)] \tag{20.64}$$

$$\Im[f(t)\sin\omega_0 t] = \frac{j}{2}[F(\omega+\omega_0) - F(\omega-\omega_0)] \tag{20.65}$$

This theorem shows how the spectrum of a function is shifted when it is modulated (or multiplied) by a carrier of frequency ω_0. The spectrum of the truncated cosine was considered previously. The truncated cosine may be thought of as the product of a gate function and an infinite cosine wave. The transform of the truncated cosine may be seen to be that of the gate function shifted by $\pm\,\omega_0$ in accordance with equation (20.65).

(*e*) Time shifting property

Corresponding to *d*, we have

If $$\Im(f(t)) = F(\omega)$$

$$\Im(f(t-t_0)) = F(\omega)e^{-j\omega t_0} \tag{20.66}$$

A time shift is therefore equivalent to a phase shift; that is, the spectral component initially at ω is shifted in phase by ωt_0.

(*f*) Time differentiation and integration (Stuart: Section 3.8)

If $$\Im(f(t)) = F(\omega)$$

then $$\Im(df/dt) = (j\omega)\,F(\omega), \tag{20.67}$$

$$\Im(d^n f/dt^n) = (j\omega)^n\,F(\omega),$$

and $$\Im\left[\int_{-\infty}^{+\infty} f(t)\,dt\right] = \frac{F(\omega)}{j\omega} \tag{20.68}$$

Equation (16.68) holds provided $F(\omega)/\omega$ is bounded at $\omega = 0$.

(g) Frequency differentiation

If $$\Im(f(t)) = F(\omega)$$

then $$\Im(-jtf(t)) = \frac{dF}{d\omega} \tag{20.69}$$

20.6 CONVOLUTION THEOREM

The convolution of two functions $f_1(x)$ and $f_2(x)$ is defined as follows:

$$f(x) = \int_{-\infty}^{+\infty} f_1(X)\,f_2(x-X)\,dX \tag{20.70}$$

The graphical interpretation of the convolution is discussed in the following section. A shorthand for the convolution of two functions is $f_1(x)*f_2(x)$. The convolution operation is commutative [*i.e.*, $f_1(x)*f_2(x) = f_2(x)*f_1(x)$], distributive and associative.

If $$\Im(f_1(t)) = F_1(\omega)$$

and $$\Im(f_2(t)) = F_2(\omega)$$

then $$\Im[f_1(t) * f_2(t)] = f_1(\omega)F_2(\omega) \tag{20.71}$$

and $$\Im[f_1(t)f_2(t)] = \frac{1}{2\pi}[F_1(\omega) * F_2(\omega)] \tag{20.72}$$

Equations (20.71) and (20.72) are referred to as the time convolution and frequency convolution theorems respectively.

An important theorem, the so-called multiplication theorem, may be deduced as a corollary of (20.72). Writing (20.72) in full, we have

$$\int_{-\infty}^{+\infty} f_1(t)f_2(t)e^{-j\omega t}\,dt = \frac{1}{2\pi}\int_{-\infty}^{+\infty} F_1(\Omega)F_2(\omega-\Omega)\,d\Omega \tag{20.73}$$

Putting $\omega = 0$, we have

$$\int_{-\infty}^{+\infty} f_1(t)f_2(t)dt = \frac{1}{2\pi}\int_{-\infty}^{+\infty} F_1(\omega(\Omega))\,F_2(-\Omega)\,d\Omega$$

$$= \frac{1}{2\pi}\int_{-\infty}^{+\infty} F_1(\Omega)F_2{}^*(\Omega)\,d\Omega \tag{20.74}$$

If the two functions are identical then

$$\int_{-\infty}^{+\infty} f(t)^2\,dt = \frac{1}{2\pi}\int_{-\infty}^{+\infty} |F(\omega)|^2\,d\omega \tag{20.75}$$

This is the analog of Parseval's theorem and can be interpreted in terms of energy dissipation by various harmonic components.

The convolution theorem has important applications and facilitates the computation of Fourier transforms. We consider two examples: (a) an infinite train of gate functions (or rectangular pulses) of width τ repeated every T seconds ($T > \tau$), and (b) an infinite train of cosine waves of frequency ω_c, duration τ, repeated every T seconds. Such a repetitive wave train is used in pulsed radar. Case a can be represented as the convolution of a gate function and an infinite Dirac comb, and case b can be represented as the product of case a and an infinite cosine wave of frequency ω_c. Then we can find the Fourier transforms by using the convolution theorem. Show that the Fourier transforms are:

Case (a): $$F_a(\omega) = \tau\,\omega_0 \sum_{-\infty}^{+\infty} Sa(n\omega_0\,\tau/2)\,\delta(\omega - n\omega_0) \tag{20.76}$$

and **Case(b):** $$F_b(\omega) = \frac{1}{2}[F_a(\omega + \omega_c) + F_a(\omega - \omega_c)] \tag{20.77}$$

where $\omega_0 \equiv 2\pi/T$ and F_a is given by (20.76). It is seen that the spectrum of the pulsed radar transmission consists of lines at frequencies $\omega_c \pm n\omega_0$, the magnitudes being determined by the sampling function. The previous two Fourier transform pairs are shown in Figure 20.6.

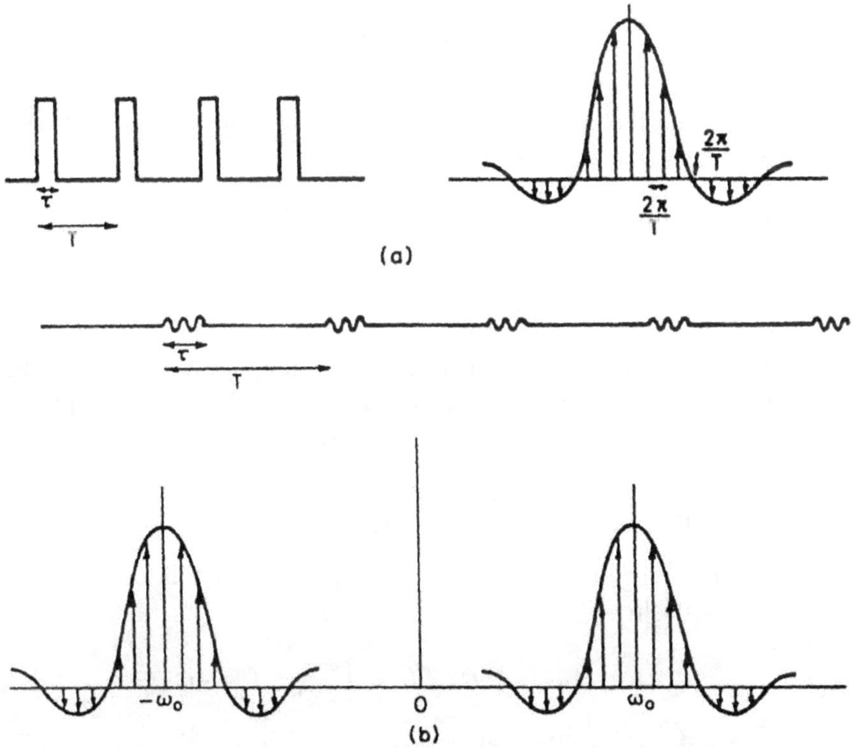

FIGURE 20.6 (*a*) An infinite train of gate functions of width τ repeated every *T* seconds and its Fourier transform. (*b*) An infinite train of cosine waves of frequency ω_c and duration τ repeated every *T* seconds and its Fourier transform.

20.7 AUTO-CORRELATION FUNCTION AND ENERGY DENSITY SPECTRUM

The auto-correlation, $\phi_{11}(\tau)$, of a function, $f_1(t)$, with itself is defined by

$$\phi_{11}(\tau) = \int_{-\infty}^{+\infty} f_1(t) f_1(t+\tau)\, dt \tag{20.78}$$

By changing from t to $t - \tau$, it can be shown that ϕ_{11} (τ) is an even function of τ:

$$\phi_{11}(\tau) = \phi_{11}(-\tau) \tag{20.79}$$

From (20.78) we also have the result

$$\phi_{11}(0) = \int_{-\infty}^{+\infty} [f_1(t)]^2 \, dt \tag{20.80}$$

It can also be proved that the maximum value of $\phi_{11}(\tau)$ occurs at $\tau = 0$.

The auto-correlation theorem, also called the Wiener-Khintchine theorem, states that the auto-correlation function and the spectral energy density are a Fourier transform pair. In symbols:

If
$$\Im(f_1(t)) = F_1(\omega), \tag{20.81}$$

then
$$\Im(\phi_{11}(\tau)) = |F_1(\omega)|^2 \tag{20.82}$$

Since ϕ_{11} is an even function of τ (Equation 20.79), the Fourier transform may be written as a cosine transform. By putting $\tau = 0$, in the explicit expression of (20.82), Parseval's theorem can be deduced.

We give in the following a proof to show the connection between the auto-correlation and convolution theorems.

For comparison with (20.78), we repeat the definition of the convolution f_1 with itself:

$$f_1(t) * f_1(t) = \int_{-\infty}^{+\infty} f_1(t) f_1(\tau - t) \, dt \tag{20.83}$$

The following steps are involved in the calculation of the auto-correlation function according to equation (20.78) (*a*) displace $f_1(t)$ by $-\tau$ to from $f_1(t + \tau)$, (*b*) form the product $f_1(t)$ $(t + \tau)$, (*c*) integrate to find $\phi_{11}(\tau)$, and (*d*) repeat for each τ. In the case of the convolution, we first fold $f_1(t)$ to form $f_1(-t)$, then proceed by displacing $f_1(-t)$ by $-\tau$, and so on, exactly as in the case of the autocorrelation. Thus, the auto-correlation and (self) convolution of an even function coincide. Alternatively, we may view the auto-correlation of $f_1(t)$ as the convolution of $f_1(t)$ with $f_1(-t)$. This follows from the definition of the convolution:

$$f_1(t) * f_1(-t) = \int_{-\infty}^{+\infty} f_1(t) f_1(t + \tau) \, dt = \phi_{11}(\tau) = \phi_{11}(-\tau) \tag{20.84}$$

Applying the convolution theorem to (20.84), we get

$$\begin{aligned}\Im(\phi_{11}(\tau)) &= \Im[f_1(\tau) * f_1(-\tau)] \\ &= \Im[f_1(\tau)].\Im[f_1(-\tau)] \\ &= F_1(\omega).F_1(-\omega) = F_1(\omega).F_1{}^*(\omega) \\ &= |F_1(\omega)|^2\end{aligned} \tag{20.85}$$

It may be remarked that the Fourier transform of $\phi_{11}(\tau)$ is pure real in keeping with its even character. It may also be noted that the computation of the spectrum *via* the auto-correlation function gives only the energy spectral density or the square of the magnitude of the spectral density; all information about the phase spectrum is lost in this procedure.

The Wiener-Khintchine theorem is the basis for the spectral decomposition of random function.

In the last two sections, we have considered only real valued functions. If the functions are complex valued, then the definitions of convolution and auto-correlation are generalized as follows:

$$f_1(t) * f_2(t) = \int_{-\infty}^{+\infty} f_1(\tau) f_2{}^*(t-\tau)\,d\tau \tag{20.86}$$

$$\phi_{11}(\tau) = \int_{-\infty}^{+\infty} f_1(t) f_1{}^*(t+\tau)\,dt \tag{20.87}$$

Note that these definitions reduce to the previous ones if f_1 and f_2 are real valued functions.

20.8 APPLICATIONS

Fourier analysis is indispensable in communication theory and almost all phenomena involving transmission, reception, or processing of information. We have already seen one important application to amplitude modulation. The multiplication of a signal $f(t)$ by cos $(\omega_c t)$ simply shifts the spectrum of $f(t)$ by $\pm\,\omega_c$ in the frequency domain. Such a process occurs in mixers, detectors, and modulators. We have seen how the frequency content of pulses transmitted by radar may be calculated. Generally speaking, the frequency spread around the central radar frequency is of the order of $1/\tau$ where τ is the pulse duration. The effect of range-gating has also been discussed. Fourier analysis can also be applied to find the frequency content of FM-CW transmission.

20.9 SAMPLING THEOREM

The sampling theorem is very important in communication theory. In the time domain it may be stated as follows (uniform sampling theorem):

A band-limited signal which has no spectral components above a frequency f_m Hz is uniquely determined by its values at uniform intervals* less than $(1/2 f_m)$ seconds apart.

For completeness, we also state the frequency domain version of the uniform sampling theorem:

A time limited signal which is zero for $|t| > T$ is uniquely determined by the samples of its frequency spectrum at uniform intervals less than $(1/2T)$ Hz apart.

According to the time domain version of the theorem, the maximum sampling interval is $(1/2 f_m)$ in order that the signal may be completely reconstructed from its samples. This interval is called the Nyquist interval. Thus, human speech which is limited to about 10 kHz may be completely reconstructed from samples every 1/20 millisecond. This is made use of in telephone communication where several conversations may be transmitted over the same channel; the method used is referred to as division multiplexing.

In pulsed radar, the scattered signal from a give-range element is received every T seconds where T is the duration between pulses. According to the theorem, we can completely recover the spectral structure of the signal if it is band-limited to $|f| < (1/2T)\ Hz$ or $\frac{p.r.f}{2}$ (p.r.f. = pulse repetition frequency).

The theorem (time domain version) may be proved as follows. Supposed the signal $f(t)$, band-limited to ω_m, is sampled every T seconds (see Figure 20.7).

The sampled signal, $f_s(t)$ say, may be written as the product of $f(t)$ and the infinite Dirac comb, $\delta_T(t)$. Hence the Fourier transform $F_s(\omega)$ of the sampled signal $f_s(t)$ is the convolution of $F(\omega)$ and $\omega_0\delta_{\omega_0}(\omega)$;

$$F_s(\omega) = \frac{1}{2\pi}[F(\omega) * \omega_0\, \delta_{\omega_0}(\omega)]$$

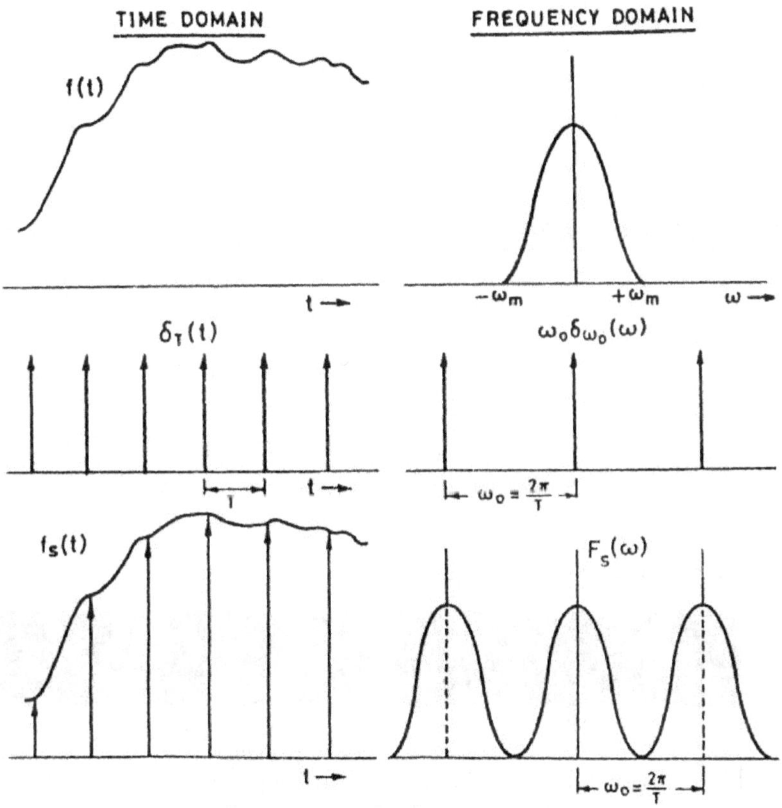

FIGURE 20.7 The sampling theorem.

From Figure 20.6, it is clear that $F_s(\omega)$ will consist of non-overlapping repetitions of $F(\omega)$ provided $\omega_0 > 2\omega_m$ or $T \leq (1/2f_m)$. If this condition is satisfied, then $F(\omega)$ and therefore $f(t)$ may be recovered by passing $F_s(\omega)$ through a low pass filter. If $T > \dfrac{1}{2f}$, then it is clear that the spectrum $F_s(\omega)$ will consist of displacements of $F(\omega)$ which overlap, and therefore $F(\omega)$ cannot be recovered.

In the Previously, we have considered sampling by a train of Dirac functions; this is called impulse sampling. However, any other repetitive pulse form is adequate for sampling. In radar practice, usually the signal is sampled and held.

20.10 SUMMARY

The following table gives the important properties of the Fourier transform. A table of some Fourier transforms is also included for reference.

20.11 ENERGY AND POWER SIGNALS

Signals for which the integral

$$\int_{-\infty}^{+\infty} f(t)\, f(t+\tau)\, dt \tag{20.88}$$

converges, as has been implicitly assumed in the foregoing, are called energy signals. If the previous integral diverges, we form the limit of the average:

TABLE 20.1
Fourier Transform Theorems

$f(t) = \frac{1}{2\pi}\int_{-\infty}^{\infty} F(\omega)e^{j\omega t}d\omega$	$F(\omega) = \int_{-\infty}^{\infty} f(t)e^{-j\omega t}dt$
$f(at)$	$\frac{1}{\lvert a\rvert}F\left(\frac{\omega}{a}\right)$
$f^*(t)$	$F^*(-\omega)$
$F(t)$	$2\pi f(-\omega)$
$f(t-t_0)$	$F(\omega)e^{-jt_0\omega}$
$f(t)e^{j\omega_0 t}$	$F(\omega-\omega_0)$
$f(t)\cos\omega_0 t$	$\frac{1}{2}[F(\omega+\omega_0)+F(\omega-\omega_0)$
$f(t)\sin\omega_0 t$	$\frac{j}{2}[F(\omega+\omega_0)-F(\omega-\omega_0)]$

(continued)

$f(t) = \frac{1}{2\pi}\int_{-\infty}^{\infty} F(\omega)e^{j\omega t}d\omega$	$F(\omega) = \int_{-\infty}^{\infty} f(t)e^{-j\omega t}dt$
$\frac{d^n f(t)}{dt^n}$	$(j\omega)^n F(\omega)$
$(-jt)^n f(t)$	$\frac{d^n F(\omega)}{d\omega^n}$
$m_n = \int_{-\infty}^{\infty} t^n f(t)\,dt$	$F(\omega) = \sum_{n=0}^{\infty} \frac{m_n}{nl}(-j\omega)^n)$
$\int_{-\infty}^{\infty} f_1(r) f_2(t-r)dr$	$F_1(\omega)F_2(\omega)$
$\int_{-\infty}^{\infty} f(t+r) f^*(r)dr$	$\lvert F(\omega)\rvert^2$
$\int_{-\infty}^{\infty} \lvert f(t)\rvert^2\, dt = \frac{1}{2\pi}\int_{-\infty}^{\infty} \lvert F(\omega)\rvert^2\, d\omega$	

TABLE 20.2
Examples of Fourier Transforms

$f(t) = \frac{1}{2\pi}\int_{-\infty}^{\infty} F(\omega)\, e^{j\omega t}d\omega$	$F(\omega) = \int_{-\infty}^{\infty} f(t)\, e^{-j\omega t}dt$
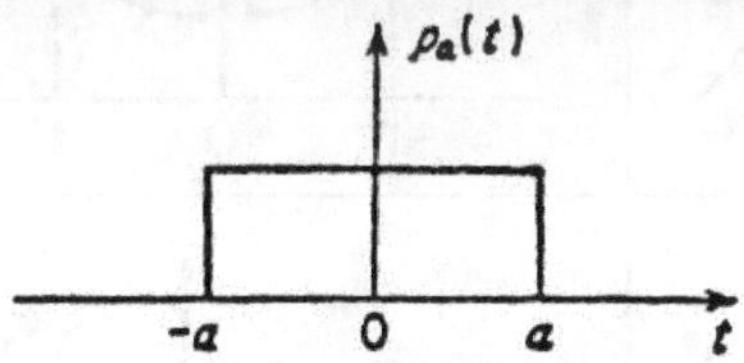	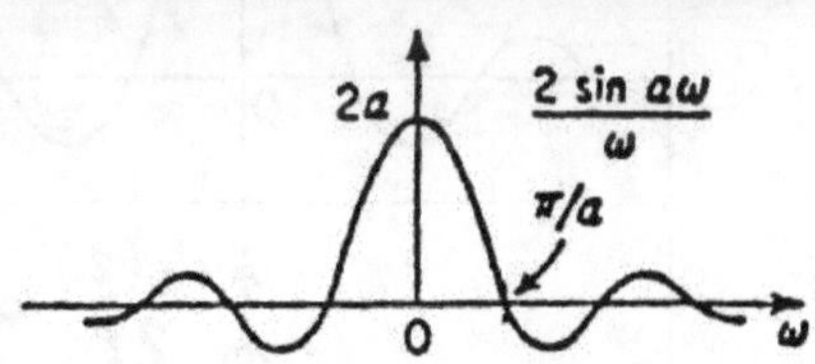
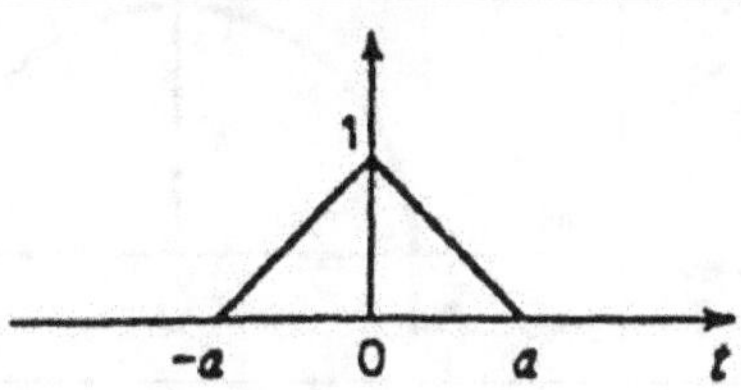	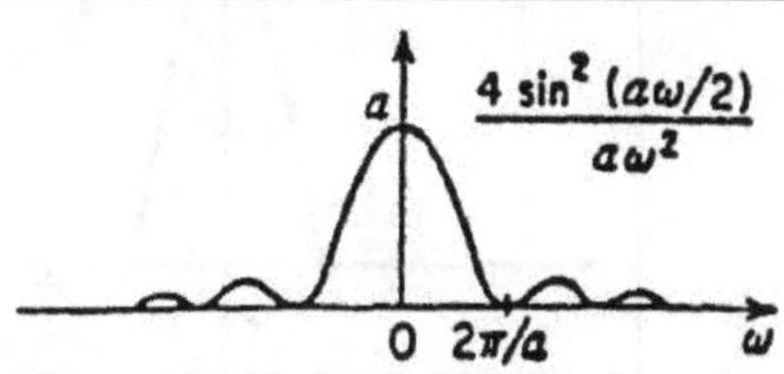

(continued)

$f(t) = \frac{1}{2\pi}\int_{-\infty}^{\infty} F(\omega)\, e^{j\omega t}\, d\omega$	$F(\omega) = \int_{-\infty}^{\infty} f(t)\, e^{-j\omega t}\, dt$
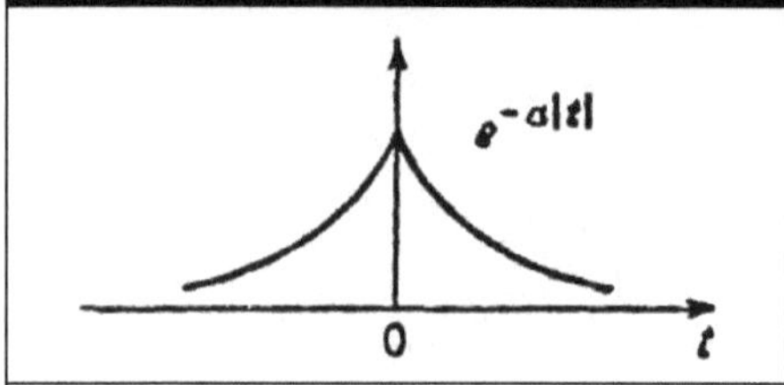	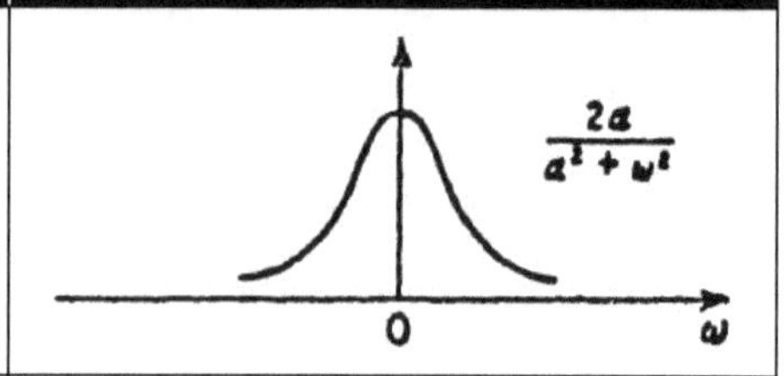
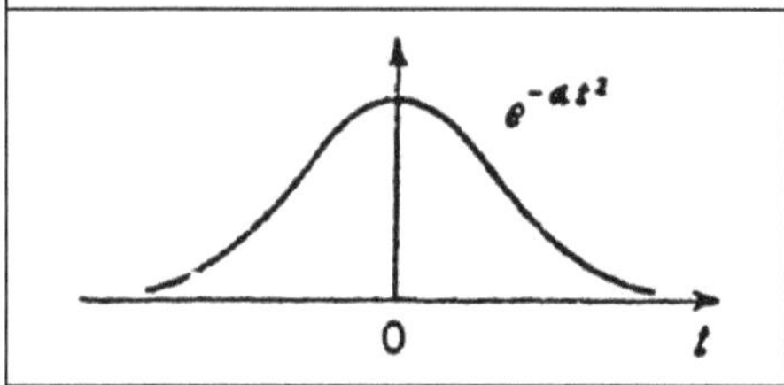	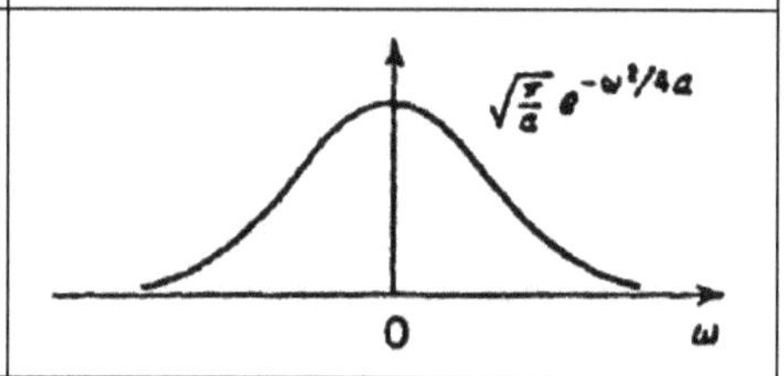
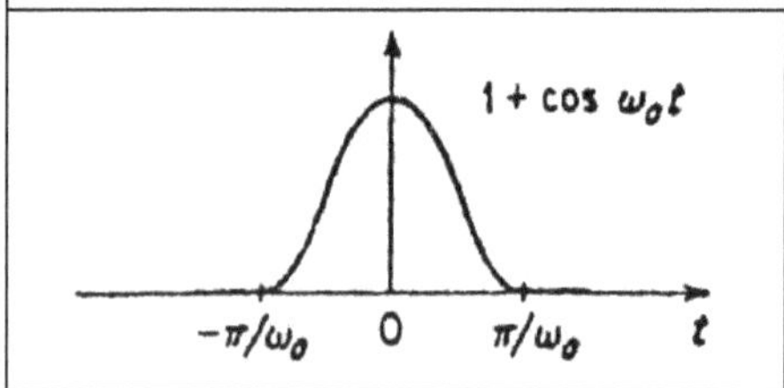	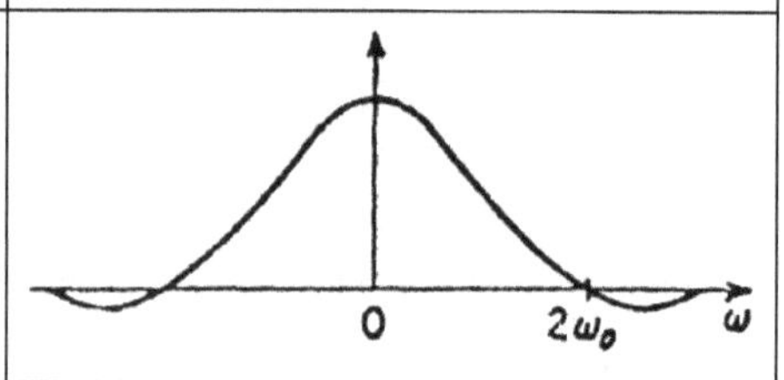
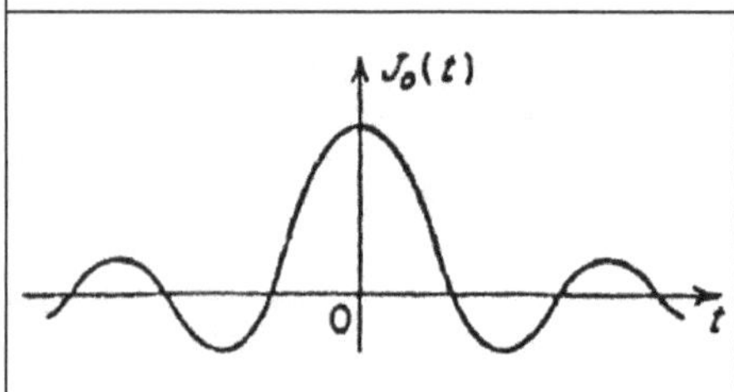	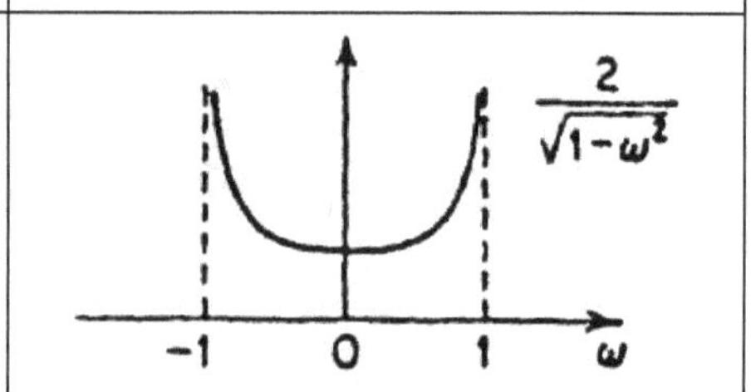
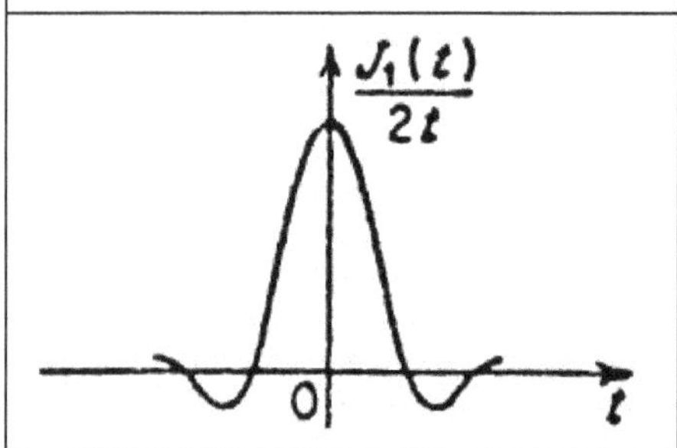	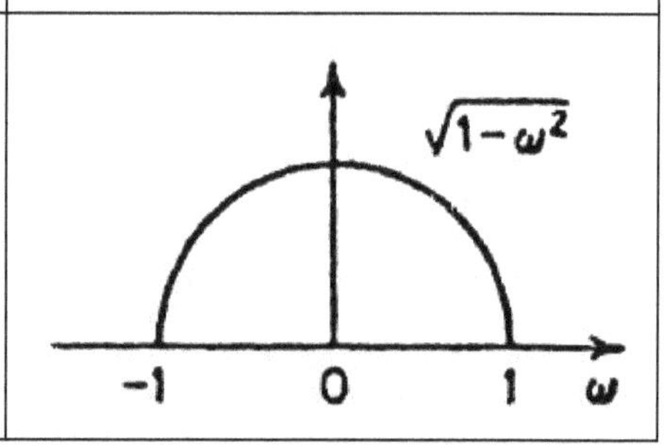

TABLE 20.3
Examples of Fourier Transforms (continued)

$f(t) = \frac{1}{2\pi}\int_{-\infty}^{\infty} F(\omega)\, e^{j\omega t}\, d\omega$	$F(\omega) = \int_{-\infty}^{\infty} f(t)\, e^{-j\omega t}\, dt$
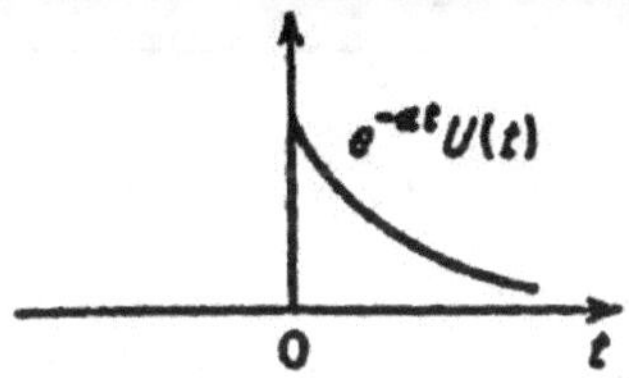	$\frac{1}{\alpha + j\omega}$
$\frac{j}{\neq t}$	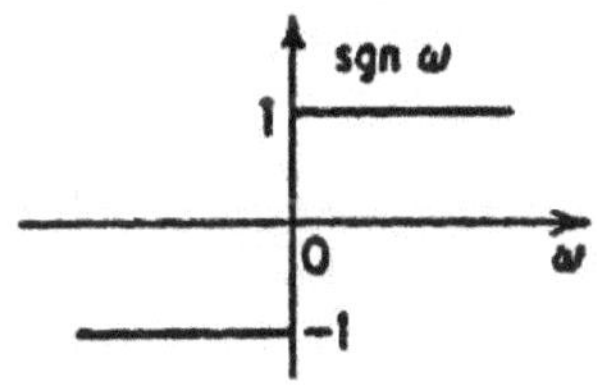
$t\alpha U(t)\ \alpha > -1$	$\frac{\Gamma(\alpha+1)}{\lvert\omega\rvert^{\alpha+1}}\, e^{\pm\frac{j\pi(\alpha+1)}{2}}$ −if $\omega > 0$, +if $\omega < 0$
$t^n e^{-\alpha t}\ U(t) > 0$	$\frac{n!}{(\alpha + j\omega)^{n+1}}$
$J_n(t)$	$\begin{cases} \frac{2\cos(n \arcsin \omega)}{\sqrt{1-\omega^2}} & n \text{ even} \lvert\omega\rvert < 1 \\ \frac{-2j\sin(n \arcsin \omega)}{\sqrt{1-\omega^2}} & n \text{ odd} \lvert\omega\rvert 1 \\ 0 & \lvert\omega\rvert > 1 \end{cases}$
$\frac{J_n(t)}{t^n}$	$\frac{2(1-\omega^2)^{n-1/2}}{1.3.5\ldots(2n-1)}$ $\lvert\omega\rvert < 1$; 0 $\lvert\omega\rvert > 1$
$e^{j\alpha t^2}$	$\sqrt{\frac{\pi}{\alpha}}\, e^{j\pi/4}\, e^{-j\omega^2/4\alpha}$

(continued)

$f(t) = \frac{1}{2\pi}\int_{-\infty}^{\infty} F(\omega)\, e^{j\omega t} d\omega$	$F(\omega) = \int_{-\infty}^{\infty} f(t)\, e^{-j\omega t} dt$
$\cos \alpha t^2$	$\sqrt{\frac{\pi}{\alpha}} \cos\left(\frac{\omega^2}{4\alpha} - \frac{\pi}{4}\right)$
$\sin \alpha t^2$	$-\sqrt{\frac{\pi}{\alpha}} \sin\left(\frac{\omega^2}{4\alpha} - \frac{\pi}{4}\right)$
$e^{j\alpha t^2} \quad 0 < t < T$	$\sqrt{\frac{\pi}{2\alpha}}\, e^{-j\omega^2/4\alpha}\left[F\left(\sqrt{\alpha}\, T - \frac{\omega}{2\sqrt{\alpha}}\right) + F\left(\frac{\omega}{2\sqrt{\alpha}}\right)\right]$
0 otherwise	$F(x) = \sqrt{\frac{2}{\pi}} \int_0^x e^{iy^2}\, dy$

$$R(\tau) = \lim_{T\to\infty} \frac{1}{T} \int_{-T/2}^{+T/2} f(t) f(t+\tau)\, dt \tag{20.89}$$

If (20.88) diverges but (20.89) converges, the signal is called a power signal. We can form the Fourier transform, $S(\omega)$, of $R(\tau)$. By the auto-correlation theorem:

$$S(\omega) = \lim_{T\to\infty} \frac{|F(\omega)|^2}{T} \tag{20.90}$$

$S(\omega)$ is called the power density spectrum, or power spectral density, as opposed to $|F(\omega)|^2$ which was called the energy spectral density.

20.12 RANDOM FUNCTIONS

To fix ideas, let us suppose that we have a resistor under specific conditions. The voltage across the resistor is a random function of time, say $f(t)$. The voltage is due to the thermal agitation of the electrons and is not predictable in detail. We may make similar measurements on ostensibly similar resistors under ostensibly similar conditions. The measurements yield the voltages $f_n(t)$ where n refers to the resistor. Each $f_n(t)$ is one realization of an infinite ensemble of possible realizations. Suppose we take all the functions $f_n(t)$ and from probability distributions of all order. The probability distribution function of the nth order gives the probability that the value of $f(t)$ is less than A_1, A_2, … A_n at the instants, t_1, t_2, … t_n. The probability distributions provide a complete description of the random process.

If the probability distributions of all orders are independent of that time, then the process is called stationary in the strict sense, or simply stationary. If the mean and the auto-correlation are independent of time, then the process is referred to as stationary in the wide sense. A stationary process is necessarily stationary in the wide sense. If all the probability distributions formed from the ensemble of the functions $f_n(t)$ are the same as the corresponding distributions formed from *a* realization of the process, then the process is called ergodic, that is, one sample is enough to represent the entire ensemble. An ergodic process is necessarily stationary. These relationships are illustrated in Figure 20.8.

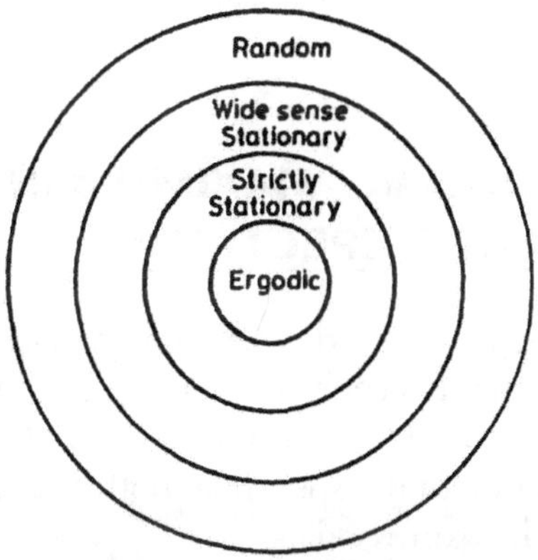

Classification of random processes

FIGURE 20.8

In the theory of the spectral analysis of random functions, the auto-correlation theorem is basic. For a random function, the auto-correlation is defined as

$$R(\tau) = < f(t)\, f(t+\tau) >, \tag{20.91}$$

where $< >$ denotes an ensemble average. For an ergodic process, this will be the same as the auto-correlation defined as

$$R(\tau) = \lim_{T\to\infty} \frac{2}{T} \int_{-T}^{+T} f(t)\, f(t+\tau)\, dt \tag{20.92}$$

For a realization of the random process, form $R(\tau)$ according to (2) and take its Fourier transform. This yields the so-called power (as opposed to energy) spectral density function $S(\omega)$. For the ergodic process, the same result is obtained if $R(\tau)$ is formed according to (20.91) and then its Fourier transform is taken:

$$S(\omega) = \Im(< f(t)\, f(t+\tau) >) \tag{20.93}$$

We use (20.93) to define the spectral power density for the case of the wide sense stationary process also. The relation between the spectral power density of a realization of the process, and that formed according to (20.93), from the ensemble average, is not as straightforward for the wide-sense stationary process as for the ergodic process. Note that for any stationary process, $R(\tau)$ is an even function of τ, and therefore, the Fourier transform can be written as a cosine transform.

20.13 THE SECONDARY MAXIMUM IN THE GENERALIZED FLUCTUATION SPECTRUM

In the symmetrical case the secondary maximum in the fluctuation spectrum is a reasonably accurate measure of the difference in mean Doppler velocities in the two beams when $2(v_1 - v_2)/\sigma \geq 4$, where σ is the (velocity) standard deviation of the spectra in both beams. The measurable fluctuation spectrum can be expressed as:

$$\begin{aligned} S_p(F) = \frac{1}{2\sqrt{\pi}\,\sigma}&[\exp(-F^2/4\sigma^2) + \frac{1}{2}\exp\{-(F-\Delta)^2/4\sigma^2\} \\ &+ \frac{1}{2}\exp\{-(F+\Delta)^2/4\sigma^2\}]F > 0 \end{aligned} \tag{20.94}$$

Clearly a secondary maximum will occur at $|F| \simeq |\Delta|$ when the second term in (20.94) is at its maximum and the first term is negligible with respect to the second. Let us then take

$$\left.\frac{\exp\left[-\dfrac{|F-\Delta|^2}{4\sigma^2}\right]}{2\exp[-F^2/4\sigma^2]}\right|_{F \simeq \Delta} \simeq \frac{e^0}{2\exp\left[-\Delta^2/4\sigma^2\right]} \geq 10$$

or $\exp + (\Delta^2/4\sigma^2) \geq 20$. This requires that $\Delta^2/4\sigma^2 \geq 3.0$ or $k = \Delta/\sigma > 3.46$ which is sufficiently close to the value $k \simeq 4$. In actual fact, the secondary maximum never occurs precisely at $F = \Delta$ because the derivative of the first term is never zero at this point.

Problem: For the symmetrical case of equal Doppler spectra, find the condition of $\Delta / \sigma_f = |\overline{f}_1 - \overline{f}_2| \sigma_f$ such that the secondary maximum in the fluctuation spectrum occurs at $|\hat{F}_2| = |\Delta|(1-\varepsilon)$ where ε is any arbitrarily small fraction. Prove that the solution is given by $2(1-\varepsilon)/\varepsilon = \exp|+\Delta^2|4\sigma^2|$ and plot ε vs. Δ/σ_f. [For this solution, neglect the last term in Equation (2.23)].

In the case of the unequal Doppler spectra, the occurrence of a secondary maximum in S_p ($|F|$) at $|\hat{F}_2| - |\Delta|$ must depend upon ψ, σ_1, σ_2, and the ratio of $\Delta/(\sigma_1^2 + \sigma_2^2)^{1/2}$. The conditions under which the secondary peak $|\hat{F}_2| = |\Delta|(1-\varepsilon)$, where ε is a small fraction. Furthermore, since the fluctuation spectra will be "noisy," the identification of the secondary peak requires that its power density exceed that at the minimum by a factor of about 2.

20.14 DUAL BEAM

Dual Beam Wind Measurements from the Variances of the Fluctuation spectra. In the case of equal Doppler spectra in both beams of the dual beam system, the variance of the dual beam *relative velocity* spectrum is

$$s^2 = 2(\sigma_v^{\ 2} + \delta^2\ W^2) \tag{20.95}$$

where $\delta W = \overline{v}_1 = \overline{v}_2 = v_0$. In terms of the corresponding variance of the fluctuation spectrum, where $F = 2u/\lambda$ and $\overline{F^2} = \Sigma^2 = (4/\lambda^2)s^2$

$$\Sigma_{1,2}^{\ 2} = (8/\lambda^2)(\sigma_v^{\ 2} + \delta^2\ W^2) \tag{20.96}$$

where the subscript 1, 2 denotes the variance of the dual beam spectra, $\sigma_v^{\ 2}$ is the variance of Doppler velocity spectrum on either beam (assumed equal), $\sigma_f = 2\sigma_v/\lambda$, and the variance of the single beam fluctuation spectrum is $\Sigma_1^{\ 2} = \Sigma_2^{\ 2} = 2\sigma_f^{\ 2} = 8\sigma_v^{\ 2}/\lambda^2$. In actuality $\delta W = (\overline{v}_1 - \overline{v}_2)/2$ in the non-symmetrical case, so that

$$\Sigma_{1,2}^{\ 2} = \Sigma_1^{\ 2} + 2(\overline{v}_1 - \overline{v}_2)^2/\lambda^2 \tag{20.97}$$

$$\Sigma_{1,2}^{\;2} = \Sigma_1^{\;2} + \frac{1}{2}(\bar{f}_1 - \bar{f}_2)^2 \qquad (20.98)$$

Let us now consider the case of the fluctuation spectrum variance corresponding to the sum signals of two unequal Doppler spectra. This was derived for the incoherent turbulence detector in which the signals (and doppler spectra) from two pulse volumes were added in the RF or IF stages prior to detection. The result is given by

$$\Sigma_{1,2}^{\;2} = \frac{P_1\,\Sigma_1^{\;2} + P_2\,\Sigma_2^{\;2}}{(P_1 + P_2)} + \frac{2P_1P_2}{(P_1 + P_2)^2}[\bar{f}_1 - \bar{f}_2]^2 \qquad (20.99)$$

where P_1 and P_2 are the powers contained in spectra (*i.e.*, beams) 1 and 2, respectively, and $\Sigma_1^{\;2}$ and $\Sigma_2^{\;2}$ are the variances of the fluctuation spectra in each of the beam spectra taken individually. Noting that $P_1/(P_1 + P_2) = \psi \equiv$ fractional total power in spectrum 1, and $P_2/(P_1 + P_2) = (1 - \psi) \equiv$ fractional total power in spectrum 2, then (20.99) becomes

$$\Sigma_{1,2}^{\;2} = \psi\,\Sigma_1^{\;2} + (1 - \psi)\,\Sigma_2^{\;2} + 2\psi(1 - \psi)\,[\bar{f}_1 - \bar{f}_2]^2 \qquad (20.100)$$

In the equal spectra case $\psi = (1 - \psi) = 1/2$ and $\Sigma_1^{\;2} = \Sigma_2^{\;2}$ thus demonstrating that (20.100) reduces to (20.98).

Equations (20.99) and (20.100) tell us that we can determine the difference in the mean Doppler frequencies (and thus the mean Doppler velocities) between the two beams simply be measuring the echo power and fluctuation spectrum variance in each of the two beams separately, and then of the simultaneously combined beams. With sufficient instrumentation all of these measurements can be made simultaneously. Such measurements obviate the need to measure the entire fluctuation spectrum of the dual beam signals and to search for the secondary maximum in that spectrum. Indeed, in principle at least, it is no longer necessary that a secondary maximum exist in the combined fluctuation spectrum since the term involving $(\bar{f}_1 - \bar{f}_2)^2$ must always increase the variance over the sum of the individual beam variances. It is only necessary that the amount of this increase be large with respect to the "noise" in the other measurements. This will undoubtedly turn out to require that the term $(\bar{f}_1 - \bar{f}_2)$ be relatively large, which is equivalent to stating that the beam spacing must also be sufficiently large. Also, it can be seen that for a given difference $(\bar{f}_1 - \bar{f}_2)$, the last term in (20.100) is a maximum when the product $\psi(1 - \psi) = \psi - \psi^2$ is a maximum. This occurs when $\psi = 1/2$, corresponding to equal powers in both beams.

However, it should be noted that the derivation of (20.99) and (20.100) makes no assumptions whatsoever about either the form or relative power of the individual beam Doppler spectra, or of the orientation of the beams relative to the wind. But, in order to determine the difference in mean Doppler velocities between the two beams with ease, it is desirable to assume that the mean Doppler velocities correspond to the radial velocities on the respective beam axes. This is equivalent to the assumption that the individual Doppler spectra on each of the beams is symmetrical.

APPENDIX A

ACHIEVING INDEPENDENCE OF SUCCESSIVE SAMPLES BY FREQUENCY SHIFTING FROM PULSE TO PULSE

Goldstein (1952) has shown that the signal intensities received from the identical assembly of random scatterers illuminated at frequencies f and $f + \Delta f$ would be decorrelated without any change in the positions of the scatterers within the pulse volume provided that Δf were sufficiently large. Wallace (1953) arrived at the same result using another approach. The correlation coefficient $\rho(\Delta f)$ between the intensities received at f and $f + \Delta f$ was shown to be

$$\rho(\Delta f) = \left[\frac{\sin(\pi \tau \Delta f)}{\pi \tau \Delta f}\right]^2 \qquad (A.1a)$$

under the assumption that the pulse is rectangular and the distance to the pulse volume is large so that all scatterers are illuminated uniformly. While the approaches of both Goldstein and Wallace are elegant, they are unnecessarily complex. In the following we arrive at the same solution in a simpler but indirect fashion.

Consider a uniformly illuminated pulse volume of range depth $h/2 = c\tau/2$ containing a large number of randomly distributed scatterers, and let the range distance $x = 0$ at the leading edge of the pulse volume and $h/2$ at its far edge as shown in Figure A.1a. Counting the phase as 0 to $x = 0$, we then have

$$\phi(f) = 4\pi x/\lambda = 4\pi x f/c \qquad (A.2a)$$

at frequency f and $$\phi(f+\Delta f) = 4\pi x(f+\Delta f)/c \tag{A.2b}$$

Accordingly, a shift in frequency from f to $f+\Delta f$ produces an effective phase shift of the scatterers at x by

$$\Delta\phi = 4\pi\, x\, \Delta f / c \tag{A.3}$$

It will be noted that when $x = h/2$ at the trailing edge of the pulse volume

$$\Delta\phi = 2\pi\, h/c\, \Delta f \; (x = h/2) \tag{A.4a}$$

and since $$h = c\tau$$

$$\Delta\phi = 2\pi\, \tau\, \Delta f \; (x = h/2) \tag{A.4b}$$

and $$\Delta\phi = 2\,\pi \left(\frac{x}{h/2}\right) \tau\, \Delta f \; (\text{at } x) \tag{A.5}$$

Thus a frequency shift of $\Delta f = +\,1/\tau$ produces an effective phase shift of $+\,2\pi$ for the scatterers at the trailing edge of the pulse volume. This is equivalent to increasing the number of wavelengths in the **two way** pulse depth h by exactly 1, as can be seen readily by noting that the number of wavelengths, n, in this distance is

$$n = h/\lambda = c\tau\, f/c = \tau f \tag{A.6}$$

and $$\Delta n = \tau\, \Delta f \tag{A.7}$$

At distances $x < h/2$, the phase shift is given by (A.5) so that the average phase shift over the pulse depth under the condition $\Delta f = \tau^{-1}$ corresponds to that at $x = h/4$ and amounts to π. Clearly, since the initial complex amplitude of the signal is

$$A = \Sigma\, a_i\, e^{j\,\phi_i}$$

and the average phase shift is π, then

$$A' = \Sigma\, a_i\, e^{j\,(\phi_i + \pi)}$$

In other words the frequency shift of $\Delta f = \tau^{-1}$ rotates the net complex amplitude by a phase angle π without a change in amplitude. This is the maximum possible change in the net phase angle and so we may expect that this is the condition for complete independence. But we are getting slightly ahead of ourselves.

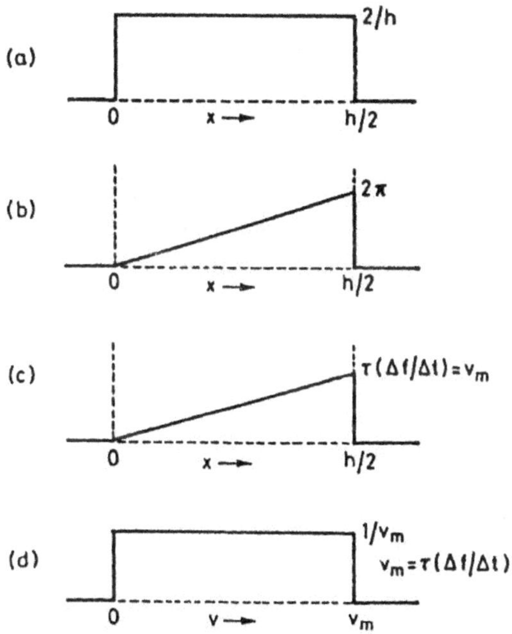

FIGURE A.1 (*a*) Pulse shape or average power density function $S_1(x)$.
(*b*) Normalized relative phase shift $\Delta\phi/(\tau\Delta f)$ *vs* *x*.
(*c*) Equivalent Doppler frequency shift *v* *vs* *x*.
(*d*) Equivalent normalized Doppler frequency spectrum, $S_1(v)$.

Returning to equation (A.5) we have plotted the relative phase change $\Delta\phi/\tau\Delta f$ vs x in Figure A.l*b*. Now let us imagine that the frequency shift Δf occurs in time Δt. From (A.5) we then get

$$\frac{\Delta\phi}{\Delta t} = 2\pi\left(\frac{x}{h/2}\right)\tau\frac{\Delta f}{\Delta t} \tag{A.8a}$$

Since $d\phi/dt = \omega = 2\pi\,v$, we see that $d\phi/dt$ is equivalent to a Doppler shift of radian frequency ω and cyclic frequency v, so that

$$v = \left(\frac{x}{h/2}\right)\tau\frac{\Delta f}{\Delta t} \tag{A.8b}$$

Accordingly the **equivalent** Doppler shift varies linearly with x from 0 to $v_{max} = \tau(\Delta f/\Delta t)$ from the front to the back of the pulse volume, respectively. This is plotted in Figure A.1*c*.

If we now divide the pulse depth into slabs of thickness Δx we may be confident that upon averaging over a large enough number of realizations of all possible random configurations of the scatters, that the intensity (power) returned by each slab Δx_i will be same and equal to

$$S(x)\,\Delta x = \frac{\Delta x}{h/2}\,\Sigma\, a_i^2 \tag{A.9}$$

where the summation on the right is over the entire pulse volume, Since $\Sigma\, a_i^2$ is the total average intensity from the pulse volume, the normalized distribution of power with x is simply

$$S_1(x)\,dx = \frac{2\,dx}{h} \tag{A.10}$$

Accordingly the normalized power **density** distribution is $S_1(x) = 2/h$ and is simply represented by the illuminating pulse with intensity $2/h$ as in Figure A.1*a*.

Using the relation

$$S_1(x)\,dx = S_1(v)\,dv \tag{A.11a}$$

we have

$$S_1(v) = \frac{S_1(x)}{dv/dx} \tag{A.11b}$$

From (A.8*b*), we get

$$\frac{dv}{dx} = \frac{\tau\,(\Delta f/\Delta t)}{h/2} = \frac{v_{max}}{h/2} \tag{A.12}$$

Substituting (A.12) and $S_i(x) = 2/h$ (from A.10), we then find the equivalent normalized Doppler spectrum

$$S_1(v) = \frac{1}{\tau(\Delta f/\Delta t)} = \frac{1}{v_{max}} \tag{A.13}$$

This is shown in Figure A.1*d*. Clearly it is an image of $s_1(x)$ with the x axis replaced by v. Since the maximum Doppler shift $v_{max} = \tau\,(\Delta f/\Delta t)$, the area under the $S_1(v)$ spectrum is 1 so that the spectrum is indeed normalized. Of course any single spectrum will show fluctuations in $S(v)\,dv$ corresponding to the fluctuations in number concentration or in $\Sigma\, a_i^2$ from slab to slab. Thus, $S(x)\,dx$ and $S(v)\,dv$ given by (A.9) and (A.13), respectively, are average spectral powers.

Corresponding to the equivalent Doppler spectrum $S_1(v)$ we also find the equivalent normalized fluctuation spectrum

$$Sp(F) = \int S_1(v)\, S_1(v+F)\, dv \tag{A.14}$$

The FT of $S_p(F)$ is the equivalent auto-correlation function $\rho(\Delta t)$. (Δt is used rather than τ because τ represents the pulse duration.) Since S_1 (v) is a rectangle, $S_p(F)$ is obviously an isosceles triangle. Both normalized spectra are shown in Figure A.2.

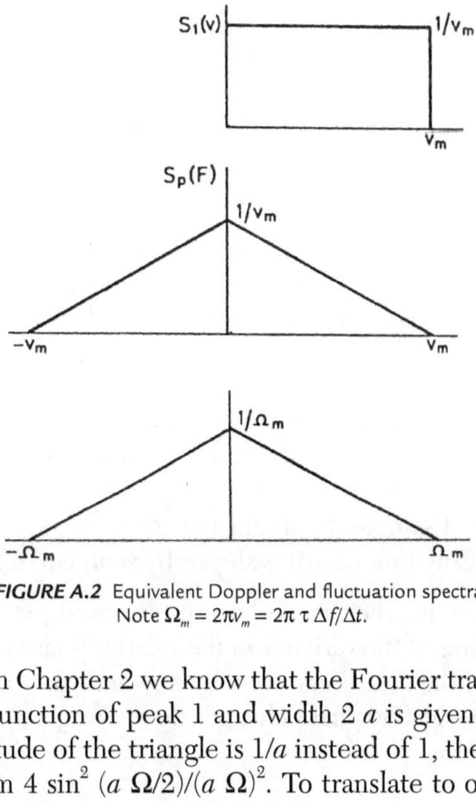

FIGURE A.2 Equivalent Doppler and fluctuation spectra. Note $\Omega_m = 2\pi v_m = 2\pi\, \tau\, \Delta f/\Delta t$.

From Table 2.1 in Chapter 2 we know that the Fourier transform of an isosceles triangular function of peak 1 and width 2 a is given by 4 $\sin^4$ $(a\ \Omega/2)/(a\ \Omega^2)$. If the altitude of the triangle is $1/a$ instead of 1, then we have for the Fourier transform 4 $\sin^2$ $(a\ \Omega/2)/(a\ \Omega)^2$. To translate to our case, we interchange Ω and Δt, and put $a = \Omega_m$, so that

$$\rho(\Delta t) = \left[\frac{\sin(\Omega_m\ \Delta t/2)}{(\Omega_m\ \Delta t/2)}\right]^2 = \left[\frac{\sin y}{y}\right]^2 \tag{A.15}$$

where $\omega = \Omega_m\ \Delta t/2 = \pi\ \tau\ \Delta f$

Since y is independent of Δt, we may write

$$\rho(\Delta f) = \left[\frac{\sin(\pi\,\tau\,\Delta f)}{(\pi\,\tau\Delta f)}\right]^2 \tag{A.16}$$

The result in equation (A.16) could also have been obtained directly from (A.14) by noting that

$$\begin{aligned}\rho(\Delta t) &= \Im\,[S_p\,(F)] = \Im\,[S_1(v)]\,\Im\,[S_1(-v)]\\ &= G_A\,(\Delta t)\,G_A\,(-\Delta t)\\ &= [G_A\,(\Delta t)]^2\end{aligned} \tag{A.17}$$

Since $S_1(v)$ is symmetrical. Since $G_A(\Delta t)$ equal to ($\sin y/y$), equation (A.16) follows immediately.

$\rho\,(\Delta f)$ has its first zero at $\tau\,\Delta f = 1$ for $\Delta f = \pm\,1/\tau$, indicating what we anticipated earlier; that complete decorrelation occurs when the frequency is shifted by just $1/\tau$, the average phase shift across the pulse volume is $\pm\,\pi$ and the number of wavelengths in the pulse length is changed by $\pm\,1$. Accordingly, when the signal intensities on successive pulses are correlated, due either to a narrow Doppler spectrum or to a long wavelength, effectively independent samples may be achieved on successive pulses by frequency shifting by $\Delta f = 1/\tau$ or more from pulse to pulse. With incoherent radars, this is a most useful way of reducing the time required to average a sufficient number of independent samples to obtain a measure of the average intensity with small variance. Equation (A.16) is shown by solid curve in Figure A.3.

In the same manner as we have interpreted $\rho(\tau)$ of Equation 2.3 *a* (main text) in terms of the variance of the relative phase changes due to the relative motions of the scatterers, we may now interpret $\rho(\Delta f)$ in terms of the variance of the relative phase changes due to frequency shifting. From equation (A.5) we have

$$\Delta\phi = \left(\frac{4\pi\,\tau\,\Delta f}{h}\right)x$$

Thus $$\sigma^2_{\Delta\phi} = \overline{\Delta\phi}^2 - (\overline{\Delta\phi})^2 = \left(\frac{4\pi\,\tau\,\Delta t}{h}\right)^2 [\overline{x^2} - \overline{x}^2]$$

Now $$\overline{x^2} = \frac{2}{h}\int_0^{h/2} x^2\,dx = \frac{h^2}{12} \quad \text{and} \quad \overline{x} = \frac{h}{4}$$

Therefore $$\sigma^2_{\Delta\phi} = \left(\frac{4\pi\,\tau\,\Delta f}{h}\right)^2 \cdot \frac{h^2}{48}$$

or $$\sigma_{\Delta\phi} = \frac{\pi\,\tau\,\Delta f}{\sqrt{3}} \tag{A.18}$$

Hence $$\rho(\Delta f) = \rho(\sigma_{\Delta\phi}) = \left[\frac{\sin\sqrt{3}\,\sigma_{\Delta\phi}}{\sqrt{3}\,\sigma_{\Delta\phi}}\right]^2 \tag{A.19}$$

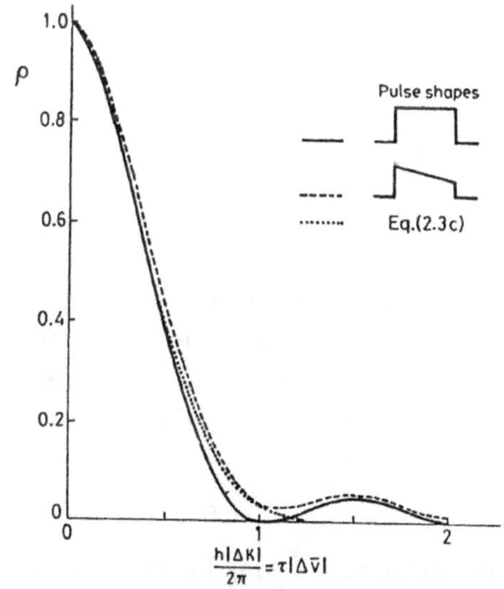

FIGURE A.3 Variation of ρ, the autocorrelation coefficient between signal intensities at frequencies *f* and $f + \Delta f$ respectively. The solid curve is for a rectangular pulse (Equation A.16); the dashed curve is for the trapezoidal pulse shown (as calculated by Wallace, 1963).

In other words the $\tau\,\Delta f$ axis of Figure A.3 may be transformed to a $\sigma\Delta\phi$ axis using (A.18). Such a scale is also shown in Figure A.3. For purposes of comparison between the decorrelation associated with the Gaussian Doppler spectrum and that due to a rectangular spectrum with the same $\sigma\Delta\phi$, we have replotted equation (2.3 *c*) of the main text in Figure A.3.

Scattering of Radio Waves by Turbulent Fluctuations of Refractive Index.

(Continuum viewpoint)

Given that (*a*) plane monochromatic waves from a transmitter *T* are incident on a volume *V*, which is remote from *T* and the receiver *R*, and (*b*) inside *V* the refractive index has small random variations due to the existence of turbulent motion, it is required to determine the characteristics of the scattered wave received at R. The characteristic in which we shall be primarily interested is the mean intensity of the scattered signal.

FIGURE A.4

First we find a wave equation. It is assumed that the conductivity of the medium is zero, and that there are no free charges or currents in the medium. Also assume that the permeability is constant and equal to that of free space (this assumption is justifiable in the atmospheric case). Let the dielectric permittivity, ε, be a random function of position:

$$\varepsilon = \varepsilon_0 + \delta\varepsilon$$

where ε_0 is the mean value. The fluctuation $\delta\varepsilon$ is assumed to be random and weak (*i.e.*, $\delta\varepsilon/\varepsilon_0 < < 1$).

Maxwell's equations may be written:

$$\text{curl}\,\vec{H} = \partial\,(\varepsilon\vec{E})\,\partial\,/\,t \tag{1}$$

$$\text{curl}\,\vec{E} = -\,\partial\,(\mu_0\,H)\,/\,\partial t \tag{2}$$

$$\text{div}\,(\varepsilon\vec{E}) = 0 \tag{3}$$

$$\text{div}\,(\mu\,\vec{H}) = 0 \tag{4}$$

Assume a harmonic time dependence [~exp $(-i\,\omega\,t)$] for all field vectors, and that ε is independent of time. The latter assumption is clearly not applicable to the atmosphere. The justification for the assumption is as follows: Let the distribution of ε at any given instant be frozen, then the signal scattered by this distribution of ε (at the particular instant) is the same as that scattered by the time varying distribution of ε.

Assuming the harmonic time dependence [~exp (– *i* ω*t*)] in Maxwell's equations, we get:

$$\text{curl}\,\vec{H} = -i\,\omega\varepsilon\vec{E} \tag{5}$$

$$\text{curl}\,\vec{E} = i\omega\mu_0\vec{H} \tag{6}$$

$$\text{div}\,(\varepsilon\vec{E}) = 0 \tag{7}$$

$$\text{div}\,\vec{H} = 0 \tag{8}$$

(Note that we are using the same symbol for the field quantities with and without the harmonic time dependence). A wave equation can be deduced from (5) through (8). Perform curl on (6) and substitute from (5):

$$\text{curl curl}\,\vec{E} = \text{grad div}\,\vec{E} - \text{div grad}\,\vec{E}$$

$$= i\omega\mu_0\,\text{curl}\,\vec{H} = \omega^2\mu_0\varepsilon\vec{E} \tag{9}$$

or,

$$\text{grad div}\,\vec{E} - \text{div grad}\,\vec{E} = \omega^2\mu_0\varepsilon_0(\varepsilon/\varepsilon_0)\,.\,\vec{E} \tag{10}$$

Expanding (7), we can get div $\vec{E}$:

$$\varepsilon\,\text{div}\,\vec{E} + \vec{E}.\,\text{grad}\,\varepsilon = 0$$

or

$$\text{div}\,\vec{E} = -\vec{E}.\frac{1}{\varepsilon}\,\text{grad}\,\varepsilon = -\vec{E}.\,\text{grad}\,(\text{In}\,\varepsilon) \tag{11}$$

Substituting this value of div $\vec{E}$ into (10), we get

$$\text{div grad}\,\vec{E} + \omega^2\mu_0\varepsilon_0\,(\varepsilon/\varepsilon_0)\,\vec{E} + \text{grad}\,[\vec{E}.\,\text{grad}\,(\text{In}\,\varepsilon)] = 0 \tag{12}$$

Using

$$\varepsilon = \varepsilon_0\,(1 + \delta\varepsilon/\varepsilon_0) \tag{13}$$

and

$$\mu_0\varepsilon_0 = 1/c^2$$

$$k = \omega/c \tag{14}$$

where c is the velocity of electromagnetic waves in the homogeneous medium of the mean properties and k the wave number, Equation (12) may be rewritten:

$$\text{div grad}\,\vec{E} + k^2\,\vec{E} + k^2(\delta\varepsilon/\varepsilon_0).\vec{E} + \text{grad}\,[\vec{E}.\text{grad In}\,(1+\delta\varepsilon/\varepsilon_0)] = 0 \quad (15)$$

or $$\text{div grad}\,\vec{E} + k^2\vec{E} = -k^2(\delta\varepsilon/\varepsilon_0)\,\vec{E} - \text{grad}\,[\vec{E}.\text{grad In}\,(1+\delta\varepsilon/\varepsilon_0)] \quad (16)$$

This is the basic wave equation (with the time dependence removed) for the kind of inhomogeneous medium we are considering. For a homogeneous medium, we would have $\varepsilon = \varepsilon_0 =$ constant, $\Delta\varepsilon = 0$, and the right hand side of (16) would be zero, giving in place of (16), the standard wave equation:

$$\text{div grad}\,\vec{E} + k^2\vec{E} = 0 \quad (17)$$

The presence of inhomogeneities, which gives rise to the right hand side of (16), may be viewed as equivalent to sources of radiation. This is to say that the incident field polarizes the medium which then acts as a secondary source of radiation.

To solve the inhomogeneous wave equations, we make use of $\delta\varepsilon/\varepsilon_0 << 1$, and adopt a perturbation approach. Assume that the solution may be written:

$$\vec{E} = \vec{E}_0 + \vec{E}_1 + \vec{E}_2 + \ldots, \quad (18)$$

where the term $\vec{E}_1$ is of order $(\delta\varepsilon/\varepsilon)$. $\vec{E}_0$, the term $\vec{E}_2$ is of order $(\delta\varepsilon/\varepsilon_0)^2$. $\vec{E}_0$, and so on. Substituting (18) into (16), expanding the logarithm and equating terms of equal order we get:

$$\text{div grad}\,\vec{E}_0 + k^2\,\vec{E}_0 = 0 \quad (19)$$

$$\text{div grad}\,\vec{E}_1 + k^2\vec{E}_1 = -k^2(\delta\varepsilon/\varepsilon_0)\vec{E}_0 - \text{grad}\,[\vec{E}_0.\text{grad}\,(\delta\varepsilon/\varepsilon_0)] \quad (20)$$

etc. etc.

Thus, in the zeroth approximation, the equation is the same as for wave propagation in the homogeneous medium with the parameters ε_0, μ_0. Assume that this is a plane wave which may be written (time dependence understood):

$$\vec{E}_0 = \vec{A}_0\,\exp\,(i\,\vec{k}.\vec{r}),$$

where $\vec{r}$ is the position vector, and $\vec{k}$ the wave vector. Of course $|\vec{k}| = k = \dfrac{2\pi}{\lambda}$, where λ is the wave length of the radiation. Because the wave is transverse, we also have $\vec{A}_0.\vec{k} = 0$.

In the first approximation, we have an inhomogeneous wave equation for $\vec{E}_1$. It is seen that on the right hand side [see Equation (20)] we have terms involving $\vec{E}_0$ and $\delta\varepsilon$. The incident wave $\vec{E}_0$ may be considered to polarize the medium, the polarized elements then acting as secondary sources of radiation which produce the field $\vec{E}_1$. In general, the equation for $\vec{E}_n$ will be an inhomogeneous wave equation for which the sources may be considered to arise as a result of excitation by $\vec{E}_{n-1}$. We shall consider only as far as $\vec{E}_1$. This approximation is sometimes referred to as the Born approximation. This means that we consider radiation by the sources excited by the primary field $\vec{E}_0$, but not that by sources excited by $\vec{E}_1$. This is also expressed by saying that multiple scattering is neglected. Obviously, in order for this to yield a good approximation, we must have $\delta\varepsilon/\varepsilon_0 << 1$. Further analysis shows that neglect of multiple scattering places a restriction not only on the magnitude of the fluctuations, but also on their length scale.

Equation (20) is an inhomogeneous **vector** wave equation. From the theory of differential equations (see, for example, Sneddon: *Elements of Partial Differential Equations*), the formal solution of the scalar wave equation

$$\text{div grad } \phi + k^2\phi = f(\vec{r})$$

is known to be $$\phi(\vec{r}) = -\frac{1}{4\pi}\int_V f(\vec{r}').\frac{\exp[ik|\vec{r}-\vec{r}'|]}{|\vec{r}-\vec{r}'|}dV' \qquad (21)$$

The integration is over the scattering volume V. The various vectors are identified in Figure A.5: O is the origin of coordinates taken inside the scattering volume. The point $P(\vec{r})$ is the "observation" point at which the field is required. dV' is the element of integration, and $\vec{r}'$ is the vector to this volume element.

Before applying (21), we simplify it. Assume that $|\vec{r}'|$ is small compared to $|\vec{r}|$. This will certainly be the case if the linear dimensions of the scattering volume are small compared to the distance from the scattering

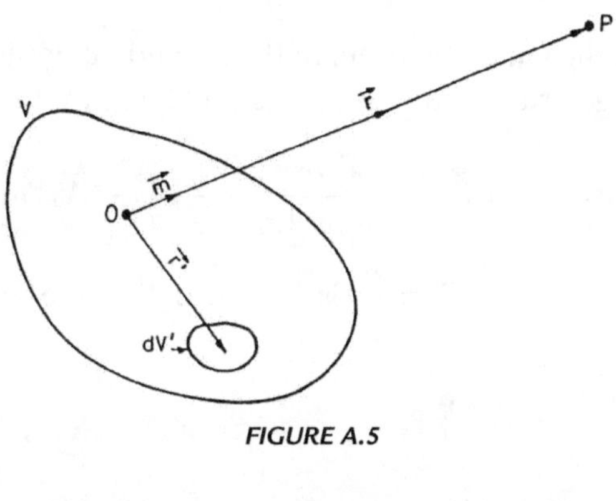

FIGURE A.5

volume to the receiver. In this case we can put $|\vec{r}-\vec{r}'| \cong |\vec{r}| = \vec{r}$ in the denominator of the integrand in (21) and take it out of the integral sign. For the argument of the exponential in (21), we use a binomial expansion:

$$|\vec{r}-\vec{r}'| = [(\vec{r}-\vec{r}').(\vec{r}-\vec{r}')]^{1/2}$$

$$= [r^2 - 2rr'\cos\psi + r'^2]^{1/2}, \ \psi = \vec{r}, \vec{r}'$$

$$\cong r - r'\cos\psi + [r'^2 - r'^2\cos^2\psi]/2r, \text{ up to terms of order } (r'/r)^2$$

$$= r - \vec{m}.\vec{r}' + \frac{1}{2r}[r'^2 - (\vec{m}.\vec{r}')^2], \qquad (22)$$

where $\vec{m}$ is a unit vector in the direction $\vec{r}$. In (21), $|\vec{r}-\vec{r}'|$ occurs multiplied by k in the argument of the exponential.

Assume $$k[r'^2 - (\vec{m}.\vec{r}')^2]/2r << 1 \qquad (23)$$

If L is a linear dimension of the scattering volume, (23) will certainly be satisfied if $\lambda r >> L^2$. Thus, neglect of the last term in (22) is permissible in the Fraunhoffer zone. With these simplifications (21) may be written:

$$\phi(\vec{r}) = \frac{\exp(ikr)}{4\pi r}\int_V f(\vec{r}')\exp.[-ik\vec{m}.\vec{r}']\,dV' \qquad (24)$$

To apply (24) to (20), replace ϕ by $\vec{E}_1$, and $f(\vec{r}')$ by the right hand side of (20). In addition, put $\vec{E}_0 = \vec{A}_0\exp.(i\vec{k}.\vec{r})$. Making these substitutions, we have:

$$\vec{E}_1(\vec{r}) = \frac{k^2}{4\pi}.\frac{\exp(ikr)}{r}\int_V (\delta\varepsilon/\varepsilon_0).\exp(i\vec{k}.\vec{r} - ik\vec{m}.\vec{r}')\,dV'$$

$$+\frac{\exp(ikr)}{4\pi r}\int_V \text{grad}\,[\exp(i\vec{k}.\vec{r})\vec{A}_0 . \text{grad}\,(\delta\varepsilon/\varepsilon_0)]\exp[-ik\vec{m}.\vec{r}']\,dV' \qquad (25)$$

Applying the Gauss theorem to the second integral in the previous equation, neglecting surface integrals, and simplifying we have

$$\vec{E}_1(\vec{r}) = \frac{k^2\exp(ikr)}{4\pi r}[\vec{A}_0 - (\vec{A}_0.\vec{m})\vec{m}]C_1 \qquad (26)$$

where $$C_1 \equiv \int_V (\delta\varepsilon/\varepsilon_0).\ \exp[i(\vec{k} - k\vec{m}).\vec{r}']dV' \qquad (27)$$

From (26), the scattered wave is seen to be spherical, and transverse (since $\vec{m}\,.\,[\vec{A}_0 - (\vec{A}_0\,.\,\vec{m})\,\vec{m}] = 0)$.

We are interested in the scattered energy. The time mean value of the Poynting vector is given by (for harmonic time variation):

$$\vec{S} = \frac{1}{2}\mathrm{Re}\,[\vec{E}_1 \times H_1{}^*] \tag{28}$$

where *Re* and * denote real part and complex conjugate respectively. Now we need to find $\vec{H}_1$. From (6), we obtain, after some simplification using vector identities,

$$\vec{H}_1 = \frac{k^3}{4\pi r}\frac{C_1}{\omega\mu_0}.\exp[ikr]\,\vec{m} \times \vec{A}_0 \tag{29}$$

Substituting from (26) and (29) into (28) gives:

$$\vec{S} = \frac{1}{2}\vec{E}_1\; x\,\vec{H}_1{}^* = \frac{k^5}{32\pi^2\,\omega\mu_0}\frac{1}{r^2}\{C_1C_1{}^*\}\,[A_0{}^2 - \vec{A}_0\,.\,\vec{m}_0)^2]\vec{m} \tag{30}$$

the energy flow is in the direction $\vec{m}$. Since $\vec{m}$ is a unit vector in the direction $\vec{OP}$, and $\vec{A}_0$ is in the direction of the electric vector of the incident wave, $\vec{A}_0$ $\vec{m} = A_0 \cos X$, where $A_0 = |\vec{A}_0|$, and X is the angle between the direction of the incident electric field and the direction of scattering. Hence:

$$A_0{}^2 - (\vec{A}_0\,.\,\vec{m})^2 = A_0{}^2 \sin^2 X \tag{31}$$

Equation (30) may now be written:

$$S = |\vec{S}| = \frac{k^5}{32\pi^2\,\omega\mu_0}.\frac{A_0{}^2}{r^2}\sin^2 X[C_1C_1{}^*] \tag{32}$$

The random part of S is contained in $C_1C_1{}^*$, since this involves the fluctuations of the dielectric permeability which are supposedly random. The mean (this is now a statistical average, as opposed to the time mean of the Poynting vector considered previously) density of energy flow is:

$$\vec{S} = \frac{k^5}{32\pi^2\omega\mu_0}.\frac{A_0{}^2}{r^2}\sin^2 X\,\overline{C_1C_1}\,{}^* \tag{33}$$

where the bar denotes the statistical average. From (27)

$$C_1C_1{}^* = \int_V [\delta\varepsilon(r'_1)/\varepsilon_0] \exp[i(\vec{k} - k\vec{m}) . \vec{r'}_1] dV_1$$

$$\int_V [\delta\varepsilon(\vec{r'}_2)/\varepsilon_0] . \exp[i(\vec{k} - k\vec{m}) . \vec{r'}_2] dV_2,$$

or $$\overline{C_1C_1{}^*} = \int_V \int_V \overline{[\delta\varepsilon(\vec{r'}_1) . \delta\varepsilon(\vec{r'}_2)/\varepsilon_0{}^2]} . \exp[i(\vec{k} - k\vec{m}) . \vec{r'}_1 - \vec{r'}_2)] dV_1 dV_2 \quad (34)$$

The dielectric permittivity is related to the refractive index through the relation:

$$n^2 = \varepsilon/\varepsilon_0,$$

Therefore, $$\delta\varepsilon/\varepsilon_0 = 2n\delta n \cong 2\,\delta n. \quad (35)$$

Substituting in (34)

$$\overline{C_1C_1{}^*} = 4\int_V \int_V \overline{\delta n(\vec{r_1'}) . \delta n(\vec{r_2'})} \exp[i(\vec{k} - k\vec{m}) . (\vec{r'}_1 - \vec{r'}_2)] dV_1 dV_2 \quad (36)$$

Assume that the field of refractive index fluctuations is spatially homogenous. Then $\overline{\delta n(\vec{r_1'}) . \delta n(\vec{r_2'})}$ depends only on $(\vec{r'}_1 - \vec{r'}_2)$:

$$\overline{\delta n(\vec{r_1'}) . \delta n(\vec{r_2'})} = B_n(\vec{r'}_1 - \vec{r'}_2) \quad (37)$$

The argument of the exponential in (36) also depends only on $\vec{r'}_1 - \vec{r'}_2$.

Therefore, change the variables of integration in (36) from $\vec{r'}_1, \vec{r'}_2$ to $\vec{\rho} \equiv \vec{r'}_1 - \vec{r'}_2$ and $\vec{R} = (\vec{r'}_1 + \vec{r'}_2)/2$. Since the integrand does not depend on $\vec{R}$, the integration with respect to $\vec{R}$ will give V. Hence,

$$\overline{C_1C_1{}^*} = 4V \int_V B_n(\vec{\rho}) \exp[i(\vec{k} - k\vec{m}) . \vec{\rho}] dV_\rho \quad (38)$$

If the volume V were infinite, then the integral on the right hand side would by definition equal $(2\pi)^3\,\phi_n\,(\vec{k} - k\vec{m})$, where $\phi_n\,(\vec{k} - k\vec{m})$ is the spectral power density of the refractive index fluctuations at the wave number $(\vec{k} - k\vec{m})$. When the volume V is finite, we get in place of $\phi_n\,(\vec{k} - k\vec{m})$, an average of ϕ_n, say $\tilde{\phi}_n$ over a range of wave numbers centered at $\vec{k} - k\vec{m}$. For practical radar and bistatic radar cases, this range is very small, so that $\tilde{\phi}_n \cong \phi_n$. In place of (38), we now have:

$$\overline{C_1C_1{}^*}\ 4V . 8\pi^3 . \phi_n(\vec{k} - k\vec{m}) \quad (39)$$

Substituting in (33), we get

$$\bar{S} = \frac{k^5 \pi V}{\omega \mu_0} \cdot \frac{A_0^{\,2}}{r^2} \sin^2 \times \phi_n(\vec{k} - k\vec{m}) \tag{40}$$

This mean density of energy flow is best expressed in terms of a differential scattering cross section. The energy flow density in the incident wave is

$$\left| \frac{1}{2} Re\,(E_0 \times H_0{}^*) \right| = \frac{A_0^{\,2} k}{2\omega\mu_0}$$

If $d\sigma$ is the differential scattering cross section for scattering in the direction $\vec{m}$ in a small solid angle $d\Omega$, then we must have:

$$d\sigma \frac{A_0^{\,2} k}{2\omega\mu_0} = \bar{S}\, r^2\, d\Omega \tag{41}$$

Using (40):

$$\frac{d\sigma}{d\Omega} = 2\pi k^4 V \,.\, \sin^2 \times \,.\, \phi_n(\vec{k} - k\vec{m}), \tag{42}$$

Figure A.6 shows the direction and magnitude of the vector $(\vec{k} - k\vec{m})$. Only refractive index fluctuations of this scale are effective in producing the scattering in the direction $\vec{m}$.

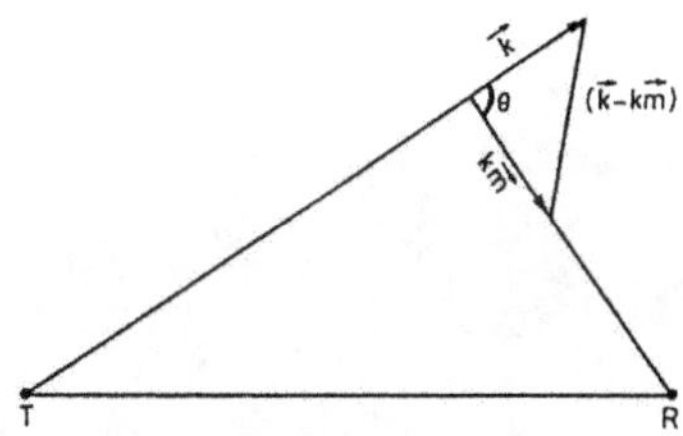

FIGURE A.6

The magnitude of $(\vec{k} - k\vec{m})$ is given by

$$|(\vec{k} - k\vec{m})| = 2k \sin\frac{\theta}{2} = \frac{4\pi}{(\lambda / \sin \theta/2)}$$

The corresponding length scale $l(\theta)$ is given by:

$$l(\theta) = \frac{2\pi}{|\vec{k} - k\vec{m}|} = \frac{\lambda}{(2 \sin \theta/2)} \tag{43}$$

This may be interpreted as a Bragg condition.

APPENDIX B

REVIEW QUESTIONS

Chapter 1. Navigation

1.1. What is meant by radio location technique?

1.2. What are the significant navigational parameters for a moving vehicle?

1.3. Mention the different types of navigational aids. What are instrument landing systems?

1.4. What are the uses of marine radar?

Chapter 2. Antenna for Radar and Radio Navigational Aids

2.1. What is the purpose of a Radar Antenna?

2.2. Illustrate the antenna parameters. What is the Hertzian dipole?

2.3. Show mathematically that the power pattern can be obtained from the field pattern.

2.4. Discuss the (*i*) Vertical pattern (*ii*) Horizontal pattern and (*iii*) Three dimensional pattern of a short wire.

2.5. Explain the following terms of an antenna:

(*a*) Beamwidth (*b*) Gain

(*c*) Effective Aperture.

2.6. Calculate the radiated power of an antenna. Show that the radiation resistance of an antenna is

$$R_r = 80\pi^2 \left(\frac{1}{\pi}\right)^2.$$

2.7. Show that the radiating efficiency is high at higher frequencies?

2.8. Explain clearly a half-wave dipole.

2.9. Describe a yagi antenna with single wire reflector and two directors.

2.10. Give the principle of operation of a folded dipole antenna. Calculate the impedance of a folded dipole antenna.

2.11. How can the directive gain of an antenna be increased?

2.12. Outline the principle of a transmitting and receiving antenna. Find a relation between gain and capture area.

2.13. Explain the polarization of the radiated electric field from a straight wire antenna. How it is modified when the electric field is considered due to two crossed dipoles?

2.14. Describe a widely used microwave pyramidal antenna. Explain how the gain of a horn antenna can be estimated.

2.15. Show how a Fourier transform allows the aperture distribution $A(Z)$ to be found for a given field intensity.

2.16. Explain the principle of one-dimensional aperture distribution.

2.17. Consider the antenna pattern due to two-dimensional distribution across a circular aperture.

2.18. Describe a parabolic reflector-type microwave "antenna."

2.19. What are the different type of lens antennas? Describe those.

2.20. What is meant by pattern synthesis in antennas?

2.21. From suitable mathematical consideration, discuss Fourier Integral Synthesis.

2.22. Describe the Woodward-Levinson method related to the antenna-synthesis technique.

2.23. Mention the different errors in a radiation pattern. Clearly explain the effect of random errors in reflectors.

2.24. Draw the typical structures of parabolic reflector antennas for radar and discuss.

2.25. Explain what you mean by the aperture field method and induced current method.

2.26. Give an approximate design method for a parabolic reflector antenna. Discuss about the tolerance of reflectors.

2.27. Write down an expression for the phase error due to tolerance and hence consider its effect on directivity.

2.28. What are the essential requirements in order to design feeds of parabolic reflector antennas?

2.29. Give names of different types of feeds and discuss.

2.30. Clearly explain the waveguide feeding method for a dipole feed.

2.31. Briefly discuss the waveguide horn feeds.

Chapter 3. Principles of Radar

3.1. What is meant by the range equation of a radar? Establish the fundamental form of the radar equation. What is the range of radar frequencies?

3.2. Give the block diagram of a typical pulse radar and hence describe its different parts.

3.3. Mention the different applications of Radar.

3.4. Discuss how the radar range equation is affected by receiver noise.

3.5. What is duty cycle of a Radar? Obtain an expression of the radar equation in terms of the average power.

What do you mean by pulse repetition frequency and pulse duration?

3.6. Discuss briefly: (*a*) Air Surveillance Radar (*b*) Tracking Radar.

3.7. What is meant by sequential lobing? Describe a Conical-scan, Tracking Radar.

3.8. Give a schematic diagram of a monopulse tracking radar.

3.9. Give the principle of radio interferometers.

3.10. Discuss the accuracy of radar measurements in the presence of noise.

3.11. Explain the (*a*) effect of pulse shape on signal-to-noise ratio (*b*) effect of gating waveform on range accuracy.

3.12. Discuss about the accuracy angular position measurement.

3.13. Explain how an idea of the form of the matched filter response is obtained.

Chapter 4. Radar Targets

4.1. Define radar cross section. Show that the RCS is equal to its geometrical projected area.

4.2. Explain clearly that for the radar cross section there are three distinct regions of behavior for a sphere of radius r.

4.3. Discuss the features in Rayleigh, Mie, and optical regions.

4.4. Write a note on "Back Scatter cross section."

4.5. Consider the geometry of a two scatter complex target mathematically.

4.6. Explain the different models of cross section fluctuations.

4.7. Show how the radar cross section is dependent on frequency.

4.8. Mention the different measurement systems of radar cross section.

4.9. What is meant by sensitivity of RCS measurement? Find an expression for maximum permissible gain in terms of RCS.

4.10. What are the different type of instrumentation radar?

4.11. Draw a block diagram of CW radar and discuss the different sections.

4.12. Draw a block diagram of a gated CW instrumentation radar for RCS measurement and discuss.

4.13. Describe a pulse radar. What do you mean by COHO and STALO?

Chapter 5. Radar Transmitters and Receivers

5.1. Describe the magnetron power oscillator used in Radar.

5.2. What is a magnetron resonator? Draw the cross section of a coaxial cavity magnetron and discuss.

5.3. What are the different parameters that determine the operation of the magnetron?

5.4. Show the performance characteristics of the coaxial magnetron and explain.

5.5. Explain the principle of operation of a magnetron. Give a detailed configuration of magnetron oscillator.

5.6. Give the principle of operation of Rlystron amplifier.

5.7. What is a reflex Klystron? Explain how it works.

5.8. With a neat diagram show the principal parts of a drawling wave tube and explain their operations.

5.9. What are the different types of crossed field amplifiers? What is an amplitron?

5.10. Describe the basic elements of a radar pulse modulator. Draw the circuit diagram of a line type modulator and explain its operation.

5.11. What are the different methods for employing solid state devices as radar transmitters?

5.12. What is meant by the noise figure of a receiver? If N number of networks are connected in Cascade, then calculate the resulting noise figure. Calculate the effective noise temperature.

5.13. Define the noise temperature ratio of a crystal mixture. Explain the principle of a balanced mixer.

5.14. What is the use of a duplexer? Describe a branch type duplexer and a balanced duplexer.

5.15. Describe (*a*) Matched Filter Receiver (*b*) CFAR Receiver.

Chapter 6. Propagation of Radio Waves

6.1. Give the principle of propagation of radar waves considering the earth as a flat reflecting surface. Explain the effect of Earth's curvature.

6.2. Discuss the effects of atmospheric refraction on propagation of radar waves.

6.3. Plot the electric field strength at the target as a function of the distance from the transmitting antenna and discuss.

6.4. What is super-refraction? Explain how the radar ranges are extended with ducted propagation. What are the effects of elevated duct and evaporation duct?

6.5. Discuss how the attenuation of radar energy is produced by the oxygen and water vapor molecules.

6.6. Discuss briefly the characteristics of the sources of ambient noise.

6.7. Find an expression of the noise power radiated by the atmosphere itself. What do you mean by atmospheric and urban noise?

6.8. Explain the atmospheric and galactic contributions to the effective noise temperature of receiving system.

Chapter 7. Radar Clutter

7.1. What is radar clutter? Define surface clutter and volume clutter.

7.2. Establish the surface clutter radar equation. Hence show that the clutter power varies inversely as the square of the range.

7.3. Write down the Bragg back scattering resonance condition. Express the probability density function of sea clutter at the output of the envelope detector.

7.4. Why is land clutter more complicated than the sea clutter?

7.5. Information about radar back scatter due to land is required for various applications. Mention some important applications.

7.6. What do you mean by inter clutter visibility? Discuss the effects of weather on Radar.

7.7. Establish the radar equation for meteorological echoes. Show that the choice of Z-R relationship should be made on the basis of the particular type of rain.

7.8. Why is radar less affected by snow and ice than by rain?

7.9. Classify the different types of angle echoes. How is the PPI affected by birds?

Chapter 8. MTI Radar

8.1. What is MTI radar? Explain the operation of MTI radar.

8.2. Give two block diagrams of MTI radar with (*a*) power amplifier (*b*) oscillator transmitter and discuss.

8.3. Briefly describe the method for extracting Doppler information for display on the PPI scope with a delay line canceler.

8.4. Find an expression of the output of the delay line canceler. What do you mean by blind speeds?

8.5. In the design of MTI Doppler filters, what is the need of using multiple pulse repetition frequencies?

8.6. Give the block diagram of MTI radar using range gates and filters. Hence, discuss the different blocks. Explain the frequency response characteristics of an MTI using range gates and filters.

8.7. Describe a digital MTI signal processor. Give the principle of a moving target detector radar processor.

8.8. Explain how MTI radar works from a moving platform.

8.9. What are the limitations to MTI performance?

Chapter 9. Modem Radars

9.1. Classify the four different types of modern radar.

9.2. Give the block diagram of pulse Doppler radar and explain it.

9.3. Describe how detection of multiple targets moving with different velocities can be done in practice.

9.4. What are the applications of pulse Doppler radars?

9.5. What are the advantages of pulse Doppler radars over CW and MTI radar?

9.6. Give the block diagram of a linear FM pulse compression radar.

9.7. Explain the envelope waveform for discrete frequency coding. Draw the waveform of the envelope of a matched filter output.

9.8. Explain with a diagram the purpose of a tapped delay line autocorrelator.

9.9. What are the applications of frequency coded radars?

9.10. What do you mean by phase coding and decoding?

9.11. Give the block diagram of phase coded CW radar.

9.12. What do you mean by a code demodulator for a phase coded radar?

9.13. Explain the use of a cross correlator and tracker. What is the use of range trackers?

9.14. Make a Comparison of phase-code and linear FM pulse compressions.

9.15. Discuss the propagation of millimeter waves through the atmosphere and consider the attenuation.

9.16. Write notes on: (*i*) Military radars (*ii*) Bean rider (*iii*) Missile seeker (*iv*) FMCW Sensor.

9.17. What is an anti-aircraft weapon system? Explain how millimeter wave radars are employed for precision guidance of missiles.

9.18. What do you mean by jamming and anti-jamming techniques? Give the theory of electronic counter-measures.

Chapter 10. Navigational and Remote Sensing Radars

10.1. What are the two different types of air surveillance radars? Give their principles.

10.2. Explain precision approach radar (PAR) considering (*i*) elevation and azimuth display (*ii*) coverage of PAR antenna and (*iii*) position of PAR with respect to the runway.

10.3. Why are meteorological radars installed at airports? Give the outline of its operation.

10.4. Give names of different airborne radars.

10.5. Give the principle of the Doppler radar.

10.6. Clearly explain the Doppler radar beam configurations.

10.7. Obtain the equations for Doppler shift. What is the effect of beam angle on sensitivity?

10.8. Discuss the effect of beamwidth on Doppler shift. Calculate an expression for the spectral width.

10.9. Illustrate the incoherent and coherent pulsed Doppler systems.

10.10. Obtain the Doppler radar range equation. Hence, show that the range equation indicates an inverse square law dependence of (S/N) on range.

10.11. Give the principle of operation of distance measuring equipment (DME). With a block diagram explain the Airborne DME interrogator.

10.12. Write notes on (*i*) Navy radar (*ii*) Remote Sensing radar.

10.13. Give the principle of pattern synthesis. What is meant by discrete array synthesis?

10.14. Give the principle of directional pattern.

10.15. Explain: (*i*) Beam steering by phasing (*ii*) Controllable phase shifts (*iii*) Conformal and non-uniform arrays (*iv*) Adaptive arrays.

10.16. What is remote sensing? Explain atmospheric windows for remote sensing.

10.17. Describe briefly the following satellite instruments for microwave remote sensing: (*a*) Radar altimeter (*b*) Scatterometer (*c*) Synthetic aperture radar (*d*) Radiometer (*e*) Pressure sounder (*f*) Microwave limb sounder.

10.18. Give the geometry corresponding to a side looking imaging radar. Also consider the geometry perpendicular to the satellite track. Find expressions for range resolution and azimuth resolution.

10.19. Give the principles of (*i*) synthetic aperture radar and (*ii*) CW radar.

10.20. What is imaging radar? What are the sources of back scatter in imaging radar?

10.21. What do you mean by monopulse radar imaging? Discuss the three-dimensional imaging by monopulse radar.

10.22. Describe multifunction array radar. Define multifunction capability of radar.

10.23. Design fixed antenna arrays for search and track.

Chapter 11. Radars for Atmospheric Probing

11.1. Give the antenna configuration of the MST radar system. What are the different echoing processes?

11.2. What are the different uses of MST radar? Describe its forecasting applications.

11.3. What is a dual polarization radar? Describe some typical results of atmospheric wind parameters as obtained from dual polarization radar.

11.4. Explain cloud and precipitation radar reflectivity at millimeter waves.

11.5. Describe in detail the configuration of dual polarization Doppler radars for millimeter waves.

11.6. Estimate updraft and particle size spectra from Doppler data.

11.7. Explain how meteorological radar is used for rainfall measurement. Give the relevant theory.

11.8. Give the principle of wind measurement by dual beam incoherent radar.

11.9. Give the theory for mean and variance of the Doppler velocity and the spectrum of turbulent velocity.

11.10. Give a simple CW radar diagram and explain its operation. What is the purpose of the heterodyning mixing process?

11.11. Describe a pulse Doppler radar and give its theory.

11.12. Give some examples of normalized Doppler power spectra.

11.13. Show that the rate of level crossings should increase with the variance of the Doppler spectrum. What is zero-crossing rate?

11.14. Calculate the mean Doppler velocity and variance of the Doppler spectrum. Find also the median velocity of the Doppler spectrum.

Chapter 12. Direction Findings

12.1. What is a Loop Antenna? Give the theory of a rectangular loop antenna.

12.2. Draw a circuit diagram of a sense finder antenna system and explain its theory.

12.3. Explain how the sensitivity of direction finder can be increased.

12.4. Describe different methods of increasing loop voltage.

12.5. Find an expression for radiation resistance of the loop.

12.6. Discuss how the Q of the loop can be increased. Explain how the loop voltage can be increased by large fixed cross loops.

12.7. What are the different errors in direction finding? Discuss in detail the errors due to anomalous propagation effect.

12.8. Describe how direction finding can be made fully automatic without any human operator.

Chapter 13. Aircraft Homing System and Instrument Landing System

13.1. Describe briefly switched cardioid homing systems.

13.2. What do you mean by omnidirectional changes?

13.3. Describe an antenna pattern with horizontal polarization.

13.4. Describe a Doppler VOR antenna configuration.

13.5. Explain the radiation pattern of TACAN. Give the waveform of detector output in the onboard TACAN receiver and explain.

13.6. Explain the elevation guidance using a radiation pattern. What do you mean by azimuth guidance?

13.7. Give the localizer antenna configuration. Clearly explain the vector diagram of field due to localizer antennas.

13.8. Describe one usual technique of the ground controlled approach system.

13.9. Explain the principle of Doppler direction finder.

13.10. What is a radio altimeter? Give the relevant theory of it. What is the effect of Doppler shift. Show how the leakage of transmitter power to the receiver limits the sensitivity.

13.11. Draw the sinusoidal modulation waveforms in radio altimeter and explain it mathematically.

13.12. What do you mean by an Instrument Landing System (ILS)? What are its technical and operational limitations?

13.13. Describe different types of antennas in microwave landing systems.

13.14. Give a block diagram of MLS and describe. Explain the principle of MLS. What are the advantages of MLS?

Chapter 14. Hyperbolic Navigation

14.1. Explain the station geometry in hyperbolic navigation.

14.2. Give a brief description of LORAN-A. How is the delay measured? What is the accuracy of delay measurement?

14.3. Explain how the cycle is matched in LORAN-C pulse.

14.4. Give the station geometry in a DECCA chain. Explain its principle.

14.5. Give the block diagram of a DECCA receiver and explain the different blocks.

14.6. Give the block diagram of OMEGA receiving equipment and describe it. What are the errors in the system?

14.7. Write notes on: (*a*) DECTRA (*b*) DERLAC (*c*) HIFIX and (*d*) Syiedis.

Chapter 15. Satellite Navigation

15.1. What is Doppler navigation? Give the principle of its operation.

15.2. What is GPS? Give the principle of operation of GPS navigation.

15.3. What are the three different segments of GPS? Briefly describe those.

15.4. Explain the GPS signal waveform. What are the sources of errors in GPS?

15.5. What do you mean by Differential Global Positioning Systems (DGPS)? Give its configurations.

15.6. Point out some applications of GPS and DGPS.

15.7. Give the block diagram of a typical GPS receiver and describe the blocks.

Chapter 16. Vessel Traffic Management System

16.1. What are the basic requirement to be fulfilled by a Vessel Traffic Management System (VTMS)?

16.2. Give the configurations of VTMS.

16.3. What do you mean by an operator's workstation?

16.4. Describe the Radar tracking system for VTMS.

16.5. Write a note on a GPS Transponder System for VTMS.

Chapter 17. Navigation and Modern Measuring Techniques

17.1. What do you mean by the process of navigation? Define latitude and longitude of a place with examples.

17.2. Outline the different modern navigational methods and mention applications in each case.

17.3. What is dead reckowing? Mention some of its important uses. What is piloting?

17.4. Discuss briefly about celestial navigation.

17.5. What is a marine chronometer. How it is used in practice?

17.6. What is marine sextant? What is its role in celestial navigation?

17.7. What is inertial navigation? Explain briefly the working of a radio direction finder.

17.8. Discuss briefly about (*i*) Radar Navigation and (*ii*) Satellite Navigation.

17.9. Give an outline of navigation applications with mobile telephony.

Chapter 18. Multimeter Wave Radars and Radiometers

18.1. What are the advantages of mm wavelengths portions of the e.m. wave over the microwave?

18.2. Briefly discuss mm-wave radars and radiometers for defense applications.

18.3. Outline the development of a coherent mm-wave receiver for 94 GHz explaining (*a*) Front end converter, (*b*) Main IF amplifier, (*c*) IF detector, AGC, and dynamic range (*d*) IF amplifier.

18.4. Discuss about the development of tunable mm-wave transmitters and receivers for 53 GHz with suitable diagrams.

18.5. Give the working principle of the mm-wave video link at 53 GHz.

18.6. Explain with a block diagram the FM generation at mm-waves.

18.7. Write a technical note on AFT discriminators.

18.8. With suitable diagrams explain the design of the demodulator of FM signals.

18.9. Give the outline of a solar tracking radiometer with all technical details. How is the calibration done?

18.10. Explain how the solar brightness temperature can be estimated.

18.11. What is meant by antenna noise temperature? How it is measured? Discuss about the calibration.

Chapter 19. Mathematical Preliminaries Related to Radar Systems and Associated Problems

19.1. Calculate the coefficient of the Fourier series.

19.2. What are orthogonal functions? What do you mean by gate function?

19.3. Explain the following properties of Fourier Transforms:
(*i*) Duality (*ii*) Linearity (*iii*) Scaling (*iv*) Frequency-shifting Property (*v*) Time-shifting Property.

19.4. State and explain the convolution theorem. What are its applications?

19.5. State and explain the sampling theorem.

19.6. What do you mean by random functions? Classify random processes?

19.7. Explain dual beam wind measurement from the variances of the fluctuation spectra.

APPENDIX C

MULTIPIE CHOICE QUESTIONS AND ANSWERS

1. There are radar systems that can be used
   (*a*) to determine both range as well as velocity of a moving target
   (*b*) to identify a target
   (*c*) to do imaging of a target
   (*d*) both (*a*) and (*c*)
   (*e*) (*a*), (*b*) and (*c*)
2. If (Ω) is the solid angle in steradians of the beam radiated by a certain lossless antenna, then the power gain of the antenna can be expressed by
   (*a*) $G = \dfrac{4\pi A}{\Omega}$
   (*b*) $G = \dfrac{4\pi}{\Omega}$
   (*c*) $G = 4\pi\Omega$
   (*d*) $G = \dfrac{\Omega}{4\pi}$
3. A given lossless antenna produces a beam that is 10° in azimuth and 20° in elevation. The power gain of this antenna would be
   (*a*) $648/\pi$
   (*b*) $324/\pi$
   (*c*) $\pi/50$
   (*d*) 200
4. A radar operating at 3 GHz and using a 1 m diameter parabolic reflector antenna with an efficiency factor of ($1/\pi$) would radiate with a power gain of
   (*a*) 3140
   (*b*) 314
   (*c*) 3.14
   (*d*) 10

5. If in Question 4, the efficiency factor was 1, the power gain in decibels would be
   (*a*) 10 dB (*b*) 20 dB
   (*c*) 30 dB (*d*) 40 dB
6. A pulse radar with a transmitted pulse width of 10 μs would have a range resolution of
   (*a*) 150 m (*b*) 1.5 km
   (*c*) 15 m (*d*) none of these
7. The radar bandwidth for the radar in Question 6 would be
   (*a*) 100 kHz (*b*) 10 MHz
   (*c*) 50 kHz (*d*) none of these
8. A beamwidth of 1° in the azimuth direction and 1° in the elevation direction implies a radiated beam solid angle (in steradians) of about
   (*a*) 1 (*b*) 0.1
   (*c*) 1/3240 (*d*) π/180
9. With reference to accuracy of range measurement, if the received signal to noise ratio is increased by a factor of 4, the root mean square value of the range error would
   (*a*) increase by a factor of 2 (*b*) reduce by a factor of 4
   (*c*) increase by a factor of 4 (*d*) reduce by a factor of 2
10. If the received signal to noise ratio is unity, the error in making range measurement approaches
    (*a*) zero (*b*) infinity
    (*c*) unity too (*d*) range resolution
11. With reference to the integration time required by the radar to resolve two Doppler shifts separated by (Δfd)
    (*a*) the integration time is independent of (Δfd)
    (*b*) the integration time is directly proportional to (Δfd)
    (*c*) the integration time is inversely proportional to (Δfd)
    (*d*) the integration time is proportional to square root of (Δfd)
12. The Doppler shift produced by a radar operating at 1 GHz from a target, moving with a velocity that has a radial component only the radar axis of 100 m/s, would be
    (*a*) 667 Hz (*b*) 667 kHz
    (*c*) 6.67 kHz (*d*) 333.3 Hz

13. When the target is moving away from the CW Doppler radar, the received frequency

(*a*) is higher than the transmitted frequency

(*b*) is lower than the transmitted frequency

(*c*) has a lot of harmonic content

(*d*) none of these

14. A ship sailing at a radial speed of 10 m/s toward an HF radar operating at 3 MHz would produce a Doppler shift of

(*a*) 0.1 Hz (*b*) 1 Hz

(*c*) 0.2 Hz (*d*) 2 kHz

15. If the Doppler shift in Question 14 is to be distinguished from stationary ground clutter, the observation must be made for a period over

(*a*) 500 μs (*b*) 5 s

(*c*) 5 ms (*d*) 10 s

16. The error ($\delta f d$) in the measurement of Doppler frequency as a function of integration time (t) and signal to noise ratio (S/N) is given by

(*a*) $\delta fd \simeq \dfrac{1}{t\sqrt{2\,(S/N)}}$ (*b*) $\delta fd \simeq \dfrac{1}{2\sqrt{(S/N)}}$

(*c*) $\delta fd \simeq \dfrac{1}{2\,t\sqrt{(S/N)}}$ (*d*) none of these

17. For a Doppler radar operating at 300 MHz, the ratio of error ($\delta \upsilon_r$) in velocity measurement to the error (δfd) in Doppler frequency measurement is approximately

(*a*) 50 cm (*b*) 100 cm

(*c*) 200 cm (*d*) 25 cm

18. Radar range is given by

(*a*) $(R_{max})^4 = \dfrac{P_t\, G_t\, A_r \sigma}{(4\pi)^2 S_{min}}$ (*b*) $R_{max} = \dfrac{P_t\, G_t\, A_r \sigma}{(4\pi)^2 S_{min}}$

(*c*) $R_{max} = \left(\dfrac{P_t\, P_r\, G_t\, G_r \sigma}{(4\pi)^2 KTB}\right)^{1/4}$ (*d*) none of these

19. A radar antenna with a power gain of 4000 needs to probe

(*a*) 4000 directions to survey one hemisphere

(*b*) 2000 directions to survey an entire sphere

(*c*) 2000 directions to survey one hemisphere

(*d*) 1000 directions to survey an entire sphere

20. An antenna beamwidth is 0.314 steradians. The number of beam positions required to fill the hemisphere would be

(*a*) 10 (*b*) 20

(*c*) 5 (*d*) 150

21. The beamwidth of a radar antenna is 0.1 radians in the azimuth direction and 0.05 radians in the elevation direction. The number of beam positions needed to fill the complete sphere would be

(*a*) 4000π (*b*) 100π

(*c*) 300π (*d*) 800π

22. The maximum PRF (f_r) that can be used for an unambiguous range (R_{max}) is given by

(*a*) $f_r = \frac{C}{2\,R_{max}}$ (*b*) $f_r = \frac{C}{R_{max}}$

(*c*) $f_r = \frac{2\,C}{R_{max}}$ (*d*) $f_r = \frac{R_{max}}{2\,C}$

23. For a maximum unambiguous range of 150 km, the maximum PRF that can be used is

(*a*) 500 Hz (*b*) 2 kHz

(*c*) 1 kHz (*d*) 10 kHz

24. In Question 23, if the transmitted pulse width is 10 μs, the duty cycle of the transmitted waveform would be

(*a*) 0.01 (*b*) 0.1

(*c*) 0.001 (*d*) 1

25. With reference to tracking radars, the problems of pulse to pulse variations in echo amplitudes

(*a*) occurs in monopulse tracking technique

(*b*) is overcome in simultaneous lobing technique

(*c*) is overcome in lobe switching technique

(*d*) is overcome in conical scan technique

26. One of the following is an HF radar system

(*a*) Moving Target Indicator (MTI)

(*b*) Pulse Doppler radar

(*c*) FM-CW radar

(*d*) Over-the-horizon radar

27. One of the following radar systems is mainly used in remote sensing applications.

(*a*) Pulse compression radar

(*b*) FM-CW radar

(*c*) Synthetic aperture radar

(*d*) Phased array radar

28. In one of the following radar systems, the radar beam is steered electronically

(*a*) Phased array radar

(*b*) Synthetic aperture radar

(*c*) Over-the-horizon radar

(*d*) MTI

29. In a Synthetic aperture radar, if the real length of the antenna is 1 m, the synthetic length at a range of 1000 m for the radar operating at 10 cm wavelength would be

(*a*) 10 m (*b*) 100 m

(*c*) 1000 m (*d*) none of these

30. The radar cross section of a target with a scattering area (*A*) is given by

(*a*) $4\pi A/\lambda^2$ (*b*) $4\pi A^2/\lambda^2$

(*c*) $4\pi A^2/\lambda$ (*d*) $4\pi A/\lambda$

31. The MTI radar system

(*a*) is used to determine the velocity of a moving target without any velocity ambiguity

(*b*) can be used to determine both range as well as velocity of moving target

(*c*) is a very high PRF radar system

(*d*) is basically used to track moving targets

32. When same antenna is used for transmission as well as reception, the radar range is

(*a*) directly, proportional to the antenna gain

(*b*) directly, proportional to one fourth power of antenna gain

(*c*) directly, proportional to square root of antenna gain

(*d*) inversely proportional to square of antenna gain

33. In a pulse compression radar, the lowest and the highest frequency components in the transmitted pulse are 50 and 70 MHz respectively. The width of the compressed pulse after matched filter detection in this case would be

(*a*) 50 ns (*b*) 20 ns

(*c*) 14.3 ns (*d*) none of these

34. The radar system that has the range determining capability of a long pulse radar and range resolution capability of a short pulse system is the

(*a*) MTI (*b*) Chirp radar

(*c*) Pulse Doppler radar (*d*) FM-CW radar

35. The radar system in which the velocity of a moving target is determined by the frequency with which the amplitude of received echo pulses changes is the

(*a*) MTI (*b*) Chirp radar

(*c*) FM-CW radar (*d*) Pulse Doppler radar

36. If in a linear array of antenna elements, the distance between the successive elements is (d) and the phase difference between the successive element is (ψ), then the radiated wavefront will make an angle (θ) with the broadside axis given by

(*a*) $\psi\left(\frac{2\pi}{\lambda}\right)d\sin\theta$ (*b*) $\psi\frac{d}{\lambda}\sin\theta$

(*c*) $\psi\left(\frac{\lambda}{d}\right)\sin\theta$ (*d*) $\psi\left(\frac{d}{\lambda}\right)^2\theta$

37. With reference to radars, better range resolution requires

(*a*) larger bandwidth

(*b*) larger transmitted power

(*c*) higher pulse repetition rate

(*d*) larger antenna

38. With reference to radar, better angular resolution electrically means

(*a*) smaller antennas (*b*) larger antennas

(*c*) larger bandwidth (*d*) larger transmitted power

39. The concept of Doppler resolution to get a better resolution in the cross range dimension is put to use in

(*a*) Pulse Doppler radar

(*b*) FM-CW radar

(*c*) Synthetic Aperture Radar (SAR)

(*d*) Chirp radar

40. The higher the PRF of a pulse radar, the

(*a*) larger is the radar's unambiguous range

(*b*) smaller is the radar's unambaguous range

(*c*) larger is the range resolution

(*d*) poorer is the angle resolution

41. A certain pulse compression radar transmits a coded pulse of 10 μs duration. The received echo pulse after passing through the matched filter in the receiver is only 100 ns. The pulse compression ratio is

(*a*) 0.01 (*b*) 100

(*c*) 200 (*d*) indeterminate from given data

42. The transmitted coded pulse in Question 41 has a spectral bandwidth of

(*a*) 100 KHz (*b*) 10 MHz

(*c*) 20 MHz (*d*) none of these

43. In a pulse compression radar, the transmitted coded pulse is 5 μs long and is frequency modulated with (f_{min}) and (f_{max}) being 50 and 70 MHz respectively. The pulse compression ratio is

(*a*) 350 (*b*) 250

(*c*) 100 (*d*) indeterminate from given data

44. With reference to a high PRF pulse Doppler radar and a low PRFMTI,

(*a*) the former is more prone to velocity ambiguities

(*b*) the former is more prone to range ambiguities

(*c*) the former has a far higher maximum unambiguous range

(*d*) none of these

45. With reference to the statement of Question 44,

(*a*) the latter is more prone to Doppler ambiguities

(*b*) the latter is more prone to range ambiguities

(*c*) the latter has a lower maximum unambiguous range

(*d*) none of these

46. The lowest blind speed is given by

(*a*) $\frac{\text{PRF} \times \lambda}{4}$ (*b*) $\frac{\text{PRF} \times \lambda}{2}$

(*c*) $\frac{\lambda}{2 \times \text{PRF}}$ (*d*) $\frac{\lambda}{4 \times \text{PRF}}$

47. When the target is moving with a velocity equal to the lowest blind speed,

(*a*) the Doppler shift equals twice the PRF

(*b*) the Doppler shift equals one half of the PRF

(*c*) the Doppler shift equals PRF

(*d*) the radar fails to determine its range

48. A pulse compression radar transmits an encoded pulse with a frequency chirp from 990 MHz to 1010 MHz centered around 1 GHz. The range resolution capability of this radar would be

(*a*) 15 m (*b*) 30 m

(*c*) 3.75 m (*d*) 7.5 m

49. The maximum unambiguous range of a pulse radar with a PRF of 5 kHz would be

(*a*) 30 km (*b*) 150 km

(*c*) 300 km (*d*) 60 km

50. One of the following radar systems does not make use of an echo signal

(*a*) Bistatic radar

(*b*) Synthetic Aperture Radar (SAR)

(*c*) Secondary Surveillance Radar (SSR)

(*d*) Tracking radar

51. One of the undermentioned radar systems uses separate antennas for transmission and reception which may sometimes even be miles apart.

(*a*) Monostatic radar

(*b*) Secondary Surveillance Radar (SSR)

(*c*) Chirp radar

(*d*) Bistatic radar

52. One of the following is surely a tracking radar

(*a*) Monostatic radar

(*b*) Bistatic radar

(*c*) Secondary Surveillance Radar (SSR)

(*d*) Monopulse radar

53. A radar receives an echo pulse 2 ms after transmission of signal pulse. The target range is

(*a*) 300 km (*b*) 150 km

(*c*) 600 km (*d*) 100 km

54. A radar transmits 100 μs wide pulses at a repetition rate of 5 kHz. The range resolution is

(*a*) 3 km (*b*) 1.5 km

(*c*) 0.75 km (*d*) 150 m

55. A radar system uses a rotating antenna to

(*a*) determine the target range

(*b*) determine the target bearing

(*c*) identify the target

(*d*) determine target velocity

56. A-scan type radar display can

(*a*) indicate only the target range

(*b*) indicate only the target bearing

(*c*) indicate both range as well as direction of the target

(*d*) none of these

57. PPI-scan type radar display can

(*a*) indicate only the target bearing

(*b*) indicate only the target range

(*c*) indicate both the target range as well as direction

(*d*) be used to produce target image

58. In a pulse radar system, the peak transmitted power is increased by a factor 81, keeping all other parameters unchanged. The maximum range determining capability of the radar increases by a factor of

(*a*) 3 (*b*) 9

(*c*) 27 (*d*) 81

59. In a Doppler radar, the transmitted frequency was 100 MHz whereas that of received echo was 110 MHz. This indicates that

(*a*) the target is moving away from the radar

(*b*) the target is moving towards the radar

(*c*) the target is at an angle with the horizontal determined by change in frequency

(*d*) the target is moving toward the radar with a speed of 10 m/s

60. A radar transmits a pulse to detect a target thought to be located at a distance of 100 km to 150 km. The echo shall be received after a time period of

(*a*) 0.66 ms to 1 ms (*b*) 0.33 ms to 0.5 ms

(*c*) 1.33 ms to 2 ms (*d*) none of these

61. A CW Doppler radar located on a stationary platform is emitting at 10 GHz. The Doppler shift observed in case of a moving vehicle traveling along the axis of the radar is 1 kHz. The speed of the vehicle would be

(*a*) 54 km/hr (*b*) 27 km/hr

(*c*) 108 km/hr (*d*) 150 m/s

62. In a Question 61, if the moving vehicle instead of traveling along the radar axis is moving in a direction making angle of 60° with the radar axis, the vehicle speed for the same observed Doppler shift would be

(*a*) 300 m/s (*b*) 30 m/s

(*c*) 30 km/hr (*d*) 54 km/hr

63. The third blind speed starting from the lowest in case of a 3 cm MTI emitting at a pulse repetition rate of 1 kHz would be

(*a*) 54 km/hr (*b*) 108 km/hr

(*c*) 162 km/hr (*d*) 216 km/hr

64. The capability of a tracking radar to resolve to targets moving perpendicular to the radar axis for 1° antenna beam width at a range of 100 km would be

(*a*) 1.5 km (*b*) 150 m

(*c*) 1.744 km (*d*) 3.5 km

65. In a monopulse tracking radar system, three signals needed to generate angular error infromation are the azimuth difference signal, the elevation difference signal, and the

(*a*) sum singal

(*b*) difference of azimuth and elevation difference signals

(*c*) average signal

(*d*) none of these

66. If (ΔA), (ΔE), and (Σ) are the azimuth difference, the elevation difference signal, and the sum signal in a monopulse tracking radar system, for an angle (θ) between the azimuth difference signal, the azimuth angular error detector output is given by

(*a*) $\dfrac{\Delta A}{\Sigma}\cos\theta$ (*b*) $\dfrac{\Delta E}{\Delta A}\cos\theta$

(*c*) $\dfrac{\Delta E}{\Sigma}\sin\theta$ (*d*) $\dfrac{\Delta A}{\Sigma}\sin\theta$

67. If in Question 66, (0) is the angle between the elevation difference signal and the sum signal, the elevation angular error detector output is given by

(*a*) $\dfrac{\Delta A}{\Delta E}\cos\theta$ (*b*) $\dfrac{\Delta E}{\Sigma}\sin\theta$

(*c*) $\dfrac{\Delta E}{\Sigma}\cos\theta$ (*d*) $\dfrac{\Delta A\,\Delta E}{\cos\theta}$

68. If the antenna diameter in a radar system is increased by a factor of 16, the maximum range will be increased by a factor of

(*a*) 2 (*b*) 4
(*c*) 8 (*d*) 16

69. The biggest disadvantage of CW Doppler radar is that

(*a*) it does not give the target position
(*b*) it does not give the target range
(*c*) it does not given the target velocity
(*d*) none of these

70. If the target cross section is changing, the best choice for accurate tracking is

(*a*) monopulse (*b*) lobe switching
(*c*) conical scanning (*d*) sequential lobing

71. One of the possible solutions to the blind speed problem is to

(*a*) use an MTI radar system
(*b*) use a low pulse repetition rate
(*c*) vary the pulse repetition rate
(*d*) change Doppler frequency

72. The ratio of antenna diameter to the transmitted wavelength is high in a given radar system. One of the following statements is false:

(*a*) The radar has a large maximum range
(*b*) The radar has a good target discrimination capability
(*c*) The radar has an increased capture area
(*d*) The target acquisition is difficult

73. A missile guidance technique using a radar in which a tracking radar located outside the missile continuously tracks the target and it is imperative that the target remains illuminated by the radar beam right from the time of launch until the time of engagement is the

(*a*) command guidance
(*b*) homing guidance
(*c*) active homing guidance
(*d*) beam rider guidance

74. In the radar guidance of missiles where two tracking radars are used, one tracking the missile and the other tracking the target, and the data from the two is fed to a computer which generates the missle trajectory, the technique is called

(*a*) homing guidance (*b*) command guidance
(*c*) semi-active homing (*d*) beam rider guidance

75. The homing guidance technique where the radar is located elsewhere (ground based or airborne) and the missle homes onto the target reflected energy is

(*a*) active homing (*b*) passive homing
(*c*) semi-active homing (*d*) none of these

76. A radar receives an echo signal 2 ms after the signal pulse was transmitted. The target is at a distance of

(*a*) 300 km (*b*) 600 km
(*c*) 150 km (*d*) none of these

77. A radar transmits a signal to detect a target thought to be located at a distance of 100 km. to 150 km. The echo shall be received after a time period of

(*a*) 0.65 ms to 1 ms (*b*) 1 ms to 3 ms
(*c*) 1 ms to 1.5 ms (*d*) none of these

78. In a radar system, a rotating antenna is used to

(*a*) determine the target range
(*b*) determine the target cross section
(*c*) determine the target bearing
(*d*) none of these

79. In a Doppler radar, the transmitted signal frequency was 100 MHz whereas that of received echo was 110 MHz. This indicates

(*a*) that the target is moving toward the radar
(*b*) that the target is moving away from the radar
(*c*) that the target is at an angle with the horizontal determined by change in frequency
(*d*) none of these

80. A device that is used to detect objects under water is known as

(*a*) Radar (*b*) Sonar

(*c*) MTI (*d*) Pulse compression radar

81. A-Scan type display can

(*a*) indicate the range of the target only

(*b*) indicate the direction of the target only

(*c*) indicate both the range as well as the direction of the target

(*d*) none of these

82. PPI-scan type indicators can

(*a*) indicate the range of the target only

(*b*) indicate the direction of the target only.

(*c*) indicate both range as well as direction

(*d*) none of these

83. A radar system that can determine the range of the target along with the target velocity is

(*a*) CW radar (*b*) MTI

(*c*) Pulse radar (*d*) Doppler radar

84. The radar system that has both a higher as well as a higher range resolution capability and which makes use of a matched filter design is

(*a*) Pulse compression radar

(*b*) MTI

(*c*) Pulse Doppler radar

(*d*) Monopulse radar

85. A radar system that does not make use of the echo signal is

(*a*) Pulse Doppler radar (*b*) COHO-STALO MTI

(*c*) Pulse compression radar (*d*) Secondary surveillance radar

86. One of the following is tracking radar:

(*a*) Secondary surveillance radar

(*b*) Bistatic radar

(*c*) Monostatic radar

(*d*) FM-CW radar

(*e*) Monopulse radar

87. In one of the following radars, the transmitted beam is steered in angle electronically

(*a*) Monopulse radar (*b*) Pulse compression radar

(*c*) Phased array radar (*d*) Moving target indicator

88. In a pulse-radar, the peak transmitted power is increased by a factor of 81, keeping the other parameters unchanged. The maximum range capability of the radar increases by a factor of

(*a*) 3 (*b*) 9

(*c*) 81 (*d*) 27

89. Mark the correct expression

(*a*) Doppler shift $= \frac{2v_r f}{c}$ (*b*) Doppler shift $= \frac{2cf}{v_r}$

(*c*) Doppler shift $= \frac{2f}{cv_r}$ (*d*) Doppler shift $= \frac{v_r f}{c}$

90. The maximum unambiguous range of a pulsed radar depends upon

(*a*) width of transmitted pulses

(*b*) peak power of the transmitted pulses

(*c*) pulse repetition rate of the transmitted signal

(*d*) signal to noise ratio of the radar receiver

91. The navigational aid in which the user can only know the quadrant in which he is located and not the exact position and also in which the information is in the form of an audio tone as well as a visual indication is

(*a*) Visual Omnirange (VOR)

(*b*) Visual Aural Range (VAR)

(*c*) LORAN

(*d*) none of these

92. The navigation system that makes use of accelerometers mounted on gyro stabilized platforms is known as

(*a*) Dead reckoning navigation system

(*b*) Instrument landing system

(*c*) Inertial navigation system

(*d*) LORAN

93. The navigation system that allows the pilot of an aircraft make a proper approach to the runway while landing is

(*a*) Aircraft Landing System (ALS)

(*b*) Visual Aural Range (VAR)

(*c*) Instrument Landing System (ILS)

(*d*) Visual Omni Range (VOR)

94. One of the following navigation systems cannot be used for finding exact location of the plane or ship.

(*a*) Visual Aural Range (VAR)

(*b*) Visual Omni Range (VOR)

(*c*) LORAN

(*d*) Radio Direction Finding (RDF)

95. The terms LOCALIZER and GLIDE PATH are associated with

(*a*) LORAN (Long Rang Navigation)

(*b*) VOR (Visual Omni Range)

(*c*) Instrument Landing System (ILS)

(*d*) Inertial Navigation System (INS)

ANSWERS

1.	(*e*)	**2.**	(*b*)	**3.**	(*a*)	**4.**	(*b*)
5.	(*c*)	**6.**	(*b*)	**7.**	(*a*)	**8.**	(*c*)
9.	(*d*)	**10.**	(*d*)	**11.**	(*c*)	**12.**	(*a*)
13.	(*b*)	**14.**	(*c*)	**15.**	(*b*)	**16.**	(*a*)
17.	(*a*)	**18.**	(*a*)	**19.**	(*c*)	**20.**	(*b*)
21.	(*d*)	**22.**	(*a*)	**23.**	(*c*)	**24.**	(*a*)
25.	(*b*)	**26.**	(*d*)	**27.**	(*c*)	**28.**	(*a*)
29.	(*b*)	**30.**	(*b*)	**31.**	(*b*)	**32.**	(*c*)
33.	(*a*)	**34.**	(*b*)	**35.**	(*a*)	**36.**	(*a*)
37.	(*a*)	**38.**	(*b*)	**39.**	(*d*)	**40.**	(*b*)
41.	(*b*)	**42.**	(*b*)	**43.**	(*c*)	**44.**	(*b*)
45.	(*a*)	**46.**	(*b*)	**47.**	(*c*)	**48.**	(*d*)

49.	(*a*)	**50.**	(*c*)	**51.**	(*d*)	**52.**	(*d*)
53.	(*a*)	**54.**	(*b*)	**55.**	(*b*)	**56.**	(*a*)
57.	(*c*)	**58.**	(*a*)	**59.**	(*b*)	**60.**	(*a*)
61.	(*a*)	**62.**	(*b*)	**63.**	(*c*)	**64.**	(*c*)
65.	(*a*)	**66.**	(*a*)	**67.**	(*c*)	**68.**	(*d*)
69.	(*b*)	**70.**	(*a*)	**71.**	(*c*)	**72.**	(*c*)
73.	(*d*)	**74.**	(*b*)	**75.**	(*c*)	**76.**	(*a*)
77.	(*a*)	**78.**	(*c*)	**79.**	(*a*)	**80.**	(*b*)
81.	(*a*)	**82.**	(*c*)	**83.**	(*b*)	**84.**	(*a*)
85.	(*d*)	**86.**	(*e*)	**87.**	(*c*)	**88.**	(*a*)
89.	(*a*)	**90.**	(*c*)	**91.**	(*b*)	**92.**	(*c*)
93.	(*c*)	**94.**	(*a*)	**95.**	(*c*)		

REFERENCES

Andrieu, H., and J. D. Creutin. 1995a. "Identification of Vertical Profiles of Radar Reflectivities for Hydrological Applications Using an Inverse Method. Part I: Formulation." *J. Appl. Meteor.* 34: 225–239.

Andrieu, H., Delrieu, G., and J. D. Creutin. 1995b. "Identification of Vertical Profiles of Radar Reflectivities for Hydrological Applications Using an Inverse Method. Part 2: Sensitivity Analysis and Case Study." *J. Appl. Meteor.* 34: 240–259.

Aoyagi, J. 1969. "Mean Doppler Velocities of Precipitation Near the Ground." *Proc. Thirteenth Radar Meteor. Conf.* 22–25. Boston, MA: Amer. Meteor. Soc.

Armijo, L. 1969. "A Theory for the Determination of Wind and Precipitation Velocities with Doppler Radars." *J. Atoms. Sci.* 26: 570–573.

Armstrong, G. M., and R. J. Donaldson Jr. 1969. "Plan Shear Indicator for Real-Time Doppler Radar Identification of Hazardous Storm Winds." *J. Appl. Meteor.* 8: 376–383.

Aston, P. S. 1964. "Ecological Studies in and Mixed Depterocarp Forecast of Brunei State." *Oxford Forest Memoir*, 26. Oxford, United Kingdom (from Lockwood, 1974).

Atlas, D. 1958a. "Radar Lightning Echoes and Atmospherics in Vertical Cross Section." In *Recent Advances in Atmospheric Electricity*, 441–459. New York, NY: Pergamon Press.

Atlas, D. 1958b. "Radar as a Sferics Detector." *Proc. Seventh Wea. Radar Conf.*, 1–8. Boston, MA: Amer. Meteor. Soc.

Atlas, D. 1959. "Radar Studies of Meteorological 'Angel' Echoes." *J. Atmos. Terrest. Phys.* 15: 262–287.

Atlas, D., and S. C. Mossop. 1960. "Calibration of a Weather Radar by Using a Standard Target." *Bull. Amer. Meteor. Soc.* 41: 377–382.

Atlas, D., Harris, F. I., and J. H. Richter. 1970. "Measurement of Point Target Speeds with Incoherent Non-Tracking Radar: Insect Speeds in an Atmospheric Wave." *J. Geophys. Res.* 75: 7588–7596.

Atlas, D., Matcalf, J. I., Richter, J. H., and E. E. Gossard. 1970. "The Birth of 'CAT' and Microscale Turbulence." *J. Atmos. Sci.* 27: 903–913.

Atlas, D., Srivastava, R. C., and P. W. Sloss. 1969. "Wind Shear and Reflectivity Gradient Effects of Doppler Spectra." *II. J. Appl. Meteor.* 8: 384–388.

Atlas, D., and R. S. Srivastava. 1971. "A Method of Radar Turbulence Detection." *IEEE Trans. Aero. and Elect. Systems* AES-7: 179–187.

Austin, B. N., Cluckie, I. D., Collier, C. G., and P. J. Hardaker. 1995. *Radar-Based Estimation of Probable Maximum Precipitation and Flood.* Bracknell, United Kingdom.

Bacchi, B., and M. Borga. 1994. "Spatial Correlation Patterns and Rainfall Field Analysis." *Excerpta* 7: 7–40.

Barancourt, C., Creutin, J. D., and J. Rivoirad. 1992. "A Method for Delineating and Estimating Rainfall Fields." *Wat. Resour. Res.* 28, no. 4: 1133–1144.

Battan, L. J. 1971. "Radar Attenuation by Wet Ice Spheres." *J. Appl. Meteor.* 10: 247–252.

Battan, L. J., and J. B. Theiss. 1970. "Measurements of Vertical Velocities in Convective Clouds by Means of Pulsed-Doppler Radar." *J. Atmos. Sci.* 27: 293–298.

Bean, B. R., and E. J. Dutton. 1968. *Radio Meteorology.* N: Dover.

Berson, F. A., and K. G. Simpson. 1971. "Radar Line Echoes and Bird Movements." *Weather* 26: 23–32.

Black, P. G., Senn, H. V., and C. L. Courtright. 1972. "Airborne Radar Observations of Eye Configuration Changes, Bright Band Distribution and Precipitation Tilt During the 1969 Multiple Seeding Experiments in Hurricane Debbie." *Mon. Wea. Rev.* 100: 208–217.

Borga, M., and P. Giaretta. 1991. "Errors in Beam Modelling and Correction of Partial Blocking."

Borga, M., Capovilla, A., Cazorzi, F., and S. Fattorelli. 1991. "Development and Application of a Real-Time Flood Forecasting System in the Veneto Region of Italy." *Wat. Resour. Manage.* 5: 209–216.

Boston, R. C. 1970. "Radar Attenuation and Reflectivity Due to Size Distributed Hydrometeors." *J. Appl. Meteor.* 9: 188–191.

Boston, R. J., and R. R. Rogers. 1969. "Hail Detection by Doppler Radar." *J. Appl. Meteor.* 8: 837–840.

Boucher, R. J., and H. Ottersten. 1971. "Doppler Radar Observation of Wind Structure in Snow." *J. Appl. Meteor.* 10: 228–233.

Browning, K. A., and C. D. Watkins. 1970. "Observations of Clear Air Turbulence by High Power Radar." *Nature* 227: 260–263.

Carlson, P. E., and J. S. Marchall. 1972. "Measurement of Snowfall by Radar." *J. Appl. Meteor.* 11: 494–500.

Cheang, B. K. 1978. "Structure of a Cyclone Disturbance Over the South China Sea and the Malaysian Region During the Winter Monsoon." *Indian J. Met., Hydrol. Geophys.* 29, no. 1 & 2: 16–25.

Cheang, B. K. 1987. "Short-and Long-Range Monsoon Prediction in Southeast Asia." Pap*er 18 in Monsoons*, editor J.S. Fein and P.L. Stephens, 580–606. Wiley.

Collier, C. G., and J. M. Knowels. 1986. "Accuracy of Rainfall Estimates by Radar, Part III: Application to Short Term Forecasting." *J. Hydrol.* 83: 225–249.

Creutin, J. D., and C. Obled. 1982. "Objective Analyses and Mapping Techniques for Rainfall Fields: An Objective Comparison." *Water Resour. Res.* 18, no. 2: 413–431.

Currie, N. C., and C. Brown (Eds.). 1987. *Principles and Applications of Millimeter-Wave Radar*. Artech House.

Currie, N. C. (Ed.). 1984. *Techniques of Radar Reflectivity Measurement*. Artech House.

Da Ros, D., and M. Borga. 1995. "On the Use of Digital Elevation Model Data for the Derivation of the Geomorphologic Instantaneous Unit Hydrograph." *Hydrological Processes.*

Dawson, G. A. 1972. "Radar as a Diagnostic Tool for Lightning." *J. Geophys. Res.* 77: 4518–4528.

Donaldson, R. J. Jr. 1970. "Mapping a Thunderstorm Anvil Flow by Doppler Radar." *J. Appl. Meteor.* 9: 911–915.

Donaldson, R. J. Jr., and R. Wexler. 1969. "Flight Hazards in Thunderstorms Determined by Doppler Velocity Variance." *J. Appl. Meteor.* 8: 128–133.

Ekpenyong, B. E., and R.C. Srivastava. 1970. *Radar Characteristics of the Melting Layer: A Theoretical Study.* Tech. Rept. no. 16. Chicago, IL: Lab. for Atmos. Probing, Univ. of Chicago.

FSR. 1975. *Flood Studies Report.* London, United Kingdom: Natural Environment Research Council. Five Volumes.

Fujita, T., and P. G. Black. 1970. "In and Outflow Fields of Hurricane Debbie as Revealed by Echo and Cloud Velocities from Airborne Radar and ATS-III Pictures." *Preprint vol. Fourteenth Radar Meteor. Conf.*, 353–358. B: Amer. Meteor. Soc.

Geotis, S. G. 1971. "Thunderstorm Water Contents and Rain Fluxes Deduced from Radar." *J. Appl. Meteor.* 1: 1233–1237.

Glover, K. G., Hardy, K. R., Konrad, T. G., Sullivan, W. N., and A. S. Michaels. 1966. "Radar Observations of Insects in Free Flight." *Science* 154: 967–972.

Gold, E. 1933. *Maximum Day Temperatures and The Tephigram (T-0 Diagram)*. Met Office Professional Notes No. 63.

Goldstein, H. 1952. Sects. 6.13–6.21. In *Propagation of Short Radio Waves. MIT Radiation Laboratory Series, Vol. 13*, edited by D. E. Kerr. New York, NY: McGraw-Hill.

Gossard, E. E., Richter, J. H., and D. Atlas. 1970. "Internal Waves in Atmosphere from High-Resolution Radar Measurements." *J. Geophys. Res.* 75: 3523–3536.

Gupta, V. K., Waymire, E., and I. Rodriguez-Itrube. 1986. "On Scales, Gravity and Network Structure in Basin Runoff." In *Scale Problems in Hydrology*, edited by V. K. Gupta, I. Rodriguez-Itrube, and E. F. Wood(eds.), 159–184. Dordrecht, Netherlands: Reidel.

Hardaker, P. J., and A. H. Auer. 1994. "The Separation Of Rain and Hail Using Single Polarisation Radar Echoes and IR Cloud Top Temperatures." *Met. Apps* 1: 201–204.

Hooper, J. E. N., and A. A. Kippax. 1950a. "The Bright Band: A Phenomenon Associated with Radar Echoes from Falling Rain." *Quart. J. Roy. Meteor. Soc.* 76: 125–131.

Hooper, J. E. N., and A. A. Kippax. 1950b. "Radar Echoes from Meteorological Precipitation." *Proc. Inst. Elect. Engineers* Pt. 1, 97: 89–97.

Hoose, J. M., and J. A. Colon. 1970. "Some Aspects of the Radar Structure of Hurricane Beulah on September 9, 1967." *Mon. Wea. Rev.* 98: 529–533.

Hunter, I. M. 1954. "Polarization of Radar Echoes from Meteorological Precipitation." *Nature* 173: 165–166.

IDNR. 1991. *International Decade For National Disaster Reduction Scientific and Technical Committee, Report Of The First Session*, Bonn, 4–8 March IDNOR Secretariat, World Met. Org. Geneva.

Imai, I. 1960. "Raindrop Size-Distribution and Z-R Relationships." *Proc. Eighth Wea. Radar Conf.*, 211–218. Boston, MA: Amer. Meteor. Soc.

Joss, J., and A. Waldvogel. 1990. "Precipitation Measurement and Hydrology." In *Radar in Meteorology: Battan Memorial and 40th Anniversary of Radar Meteorology Conf.*, edited by David Atlas, 577–606.

Joss, J., and A. Waldvogel. 1970. "Raindrop Size Distributions and Doppler Velocities." *Preprint vol. Fourteenth Radar Meteor. Conf.*, 153–156. Boston, MA: Amer. Meteor. Soc.

Kerr, D. E. 1951. "Propagation of Short Radio Waves." New York, NY: McGraw-Hill.

Konard, T. G. 1970. "The Dynamics of the Convective Process in the Clear Air as Seen by Radar." *J. Atmos. Sci.* 27: 1138–1147.

Laroche, S., and I. Zawadzki. 1994. "A Variational Analysis Method for Retrieval of Three Dimensional Wind Field from Single Doppler Radar Data." *J. Atm. Sci.* 51, no 18: 2664–2682.

Lebel, T., Bastin, G., Obled, C., and J. D. Creutin. 1987. "On the Accuracy of Areal Rainfall Estimation: A Case Study.", *Water Resour. Res.* 23, no 11: 2123–2134.

Lhermitte, R. M. 1970. "Dual-Doppler Radar Observations of Convective Storm Circulation." *Preprint Vol. Fourteenth Radar Meteor. Conf.*, 139–144. Boston, MA: Amer. Meteor. Soc.

Lhermitte, R. M. 1971. "Probing of Atmospheric Motions by Airborne Pulse Doppler Radar Techniques." *J. Appl. Meteor.* 10: 234–246.

Lin, Y-J., T-C, C., Wang, R. W. Pasken, H. Shen, and Z-S Deng. 1990. "Characteristics of A Subtropical Squall Line Determined From TAMEX Dual-Doppler Data. Part II: Dynamic and Thermodynamic Structures and Momentum Budgets." *J. Atm. Sci.* 47, no. 20, 2382–2399.

Lockwood, J. G. 1974. *World Climatology: An Environmental Approach.* Edward Arnold.

Mayer, J. H. 1971. "Radar Observations of Land Breeze Fronts." *J. Appl. Meteor.* 10: 1224–1232.

Mesa, O. J., and E. Mifflin. 1986. "On the Relative Role of Hillslope and Network Geometry in Hydrologic Response." In *Scale Problems in Hydrology*, edited by V. K. Gupta, I. Rodriguez-Iturbe, and E. F. Wood, 159–184. Dordrecht, Netherlands: Reidel.

Monai, M., Pesci, A. D., and E. Trolese. 1994. "Application of Advanced Techniques at Monte Grande Weather Radar." *CSIM Report*, Teolo, Padova, Italy.

Moore, R. J., Mar, B. C., Jones, D. A., and K. B. Black. 1991. "Local Calibration of Weather Radar over London." In *Advances in Radar Hydrology,* edited by M. E. Almeida-Telxeria, R. Fantechi , R. J. Moore, and V. M. Silva, 186–195.

Moore, R. J., and R. T. Clarke. 1981. "A Distribution Function Approach to Rainfall Runoff Modeling.", *Water Resour. Res.* 17, no. 5: 1367–1382.

Moore, R. J. 1985. "The Probability-Distributed Principle and Runoff Production at Point and Basin Scales." *Hydrological Sciences Journal* 30, no. 2: 273–297.

National Center for Atmospheric Research. 1970. "Coherent Cloud Physics Radar." In *Facilities for Atmospheric Research, No. 13*, 26–28. Boulder, CO: Nat. Ctr. Atmos. Res.

Oblcd, C., Wendling, J., and K. Beven. 1994. "The Sensitivity of Hydrological Models to Spatial Rainfall Patterns: An Evaluation Using Observed Data." *J. Hydrol.* 159: 305–333.

Pasquallucci, F. 1981. "Millimeter-Wave Radar Applications in Meteorology." *Recent Progress in Radar Meteorology, Nat. Centre for Atoms. Res.*: 46.

Peace, R. L. Jr., and R. A. Brown. 1968. "Comparison of Single and Double Doppler Radar Velocity Measurements in Convective Storms." *Proc. Thirteenth Radar Meteor. Conf.*, 464–470. Boston, MA: Amer. Meteor. Soc.

Pell, J. 1971. "The Use of Broad-Beam Radar for Quantitative Analysis of Severe Storms." *J. Appl. Meteor.* 10: 1238–1251.

Persson, P. O. G., and T. Andersson. 1987. "Automatic Wind Field Interpretation of Doppler Radial Wind Components." *Swedish Met. Hydo. Inst., PROMIS Report* No. 6, 72 .

Prosser, N. E. 1970. "Mesoscale Surface Analysis Techniques Utilizing Radar Data." *Preprint Vol. Fourteenth Radar Meteor. Conf.*, 343–346. Boston, MA: Amer. Meteor. Soc.

Ramage, C. S. 1971. *Monsoon Meteorology*. New York, NY: Academic Press.

Ready, E. K., and G. W. Ewell. 1981. "Millimeter-Wave Radar." In Chapt. 2 of *Infrared and Millimeter Waves, Vol. 4*, edited by Kenneth J. Button and James C. Wittse, 23–94. Academic Press.

Richter, J. H. 1969. "High Resolution Tropospheric Radar Sounder." *Radio Science* 4: 1261–1268.

Seed, A. W., and G. L. Austin. 1990. "Sampling Errors for Raingauge-Derived Mean Areal Daily and Monthly Rainfall." *J. Hydrol.* 118, 163–173.

Sekhon, R. S., and R. C. Srivastava. 1970. "Snow Size Spectra and Radar Reflectivity." *J. Atmos. Sci.* 27: 299–307.

Sekhon, R. S., and R. C. Srivastava. 1971. "Doppler Radar Observations of Drop-Size Distribution in a Thunderstorm." *J. Atmos. Sci.* 28: 983–994.

Shedd, R., Smith, J. A., and M. L. Walton. 1991: "Sectorized Hybrid Scan Strategy of the NEXRAD Precipitation-Processing System." In *Hydrological Applications of Weather Radar*, edited by I. D. Cluckie, C. G. Collier, and Ellis Horwood, 151–159.

Sirmans, D., Watts, W. L., and J. H. Horwedel. 1970. "Weather Radar Signal Processing and Recording at the National Severe Storms Laboratory." *IEEE Transactions on Geoscience Electronics*, GE-8: 88–94.

Skolnik, M. I. 1962. *Introduction to Radar Systems*. New York, NY: McGraw-Hill.

Smith, D. I. 1989. "A Dam Disaster Waiting to Break." *New Scientist*, 11 November, 42–46.

Starr, J. R. and K. A. Browning. 1972. "Observation of Lee Waves by High-Power Radar." *Quart. J. Roy. Meteor. Soc.* 98: 73–85.

Stratmann, E., Atlas, D., Richer, J. H., and D. R. Jensen. 1971. "Sensitivity Calibration of a Dual-Beam Vertically Pointing EM-CW Radar." *J. Appl. Meteor.* 10: 1260–1265.

Tabios G. Q., and J. D. Salas. 1985. "A Comparative Analysis of Techniques for Spatial Interpolation of Precipitation." *Water Resour. Bull.*, 365–380.

Tsuboki, K., Fujiyoshi, Y., and G. Wakahama. 1989. "Structure of a Land Breeze and Snowfall Effects." In *Advances in Radar Hydrology,* edited by M. E. Almeida-Teixeria, R. Fantechi, R. J. Moore, and V. M. Silva, 15–114.

Wallace, R. R. 1953. "Interpretation of the Fluctuating Echo from Randomly Distributed Scatterers." *II. Canad. J. Phys.* 31: 995–1009.

Wiley, R. L., Browning, K. A., Joss, J., and A. Waldvogel. 1970. "Measurement of Drop Size Distribution and Vertical Air Motion in Widespread Rain Using Pulsed Doppler Radar and Disdromcter." In *Preprint Vol. Fourteenth Radar Meteor. Conf.,* 167–170. Boston, MA: Amer. Meteor. Soc.

Wilk, K. E., and E. Kessler. 1970. "Quantitative Radar Measurements of Precipitation." *Meteor. Monographs* 11: 315–329.

Wilson, J. W. 1970. "Integration of Radar and Rainfall Data for Improved Rainfall Measurement." *J. Appl. Meteor.* 9: 489–497.

Zawadzki, I. 1984. "Factors Affecting the Precision of Radar Measurements of Rain." *Preprints of Conf. of Radar Meteorology*, 10–13 Sept. 1984, Zurich, AMS, 251–256.

Zawadzki, I.I., and Ballantyne, E. 1970. "HARPI: A New Weather Radar Display." *Quart. J. Roy. Meteor. Soc.* 96: 144.

INDEX

A

"absorption mode" measurement, 625
active electronic counter measures, 269–270
adaptive arrays, 304–305
Adcock antenna, 483–484
AFT detector, 615, 617–618
AFT discriminator, 615, 617–618
AGC. *See* automatic gain control
agility, 121
airborne automatic direction finder, 487–490
airborne interrogator equipment, 290–291
airborne moving target indication (AMTI), 227
airborne radars, 273, 278
aircraft homing system, 491
aircraft navigation, radar application, 68
airport radars, 274–277
airport-surveillance radar (ASR), 274
air-route surveillance radar (ARSR), 274
air surveillance radar (ASR), 273, 274
  beam shaping, 275
  sensitivity time control, 275–276
air traffic control (ATC), radar application, 68
altimeters, 310
ambiguity diagram, 101–102
ambiguity function, 102
amplitron, 156–157
amplitude weighing, 243
AMTI. *See* airborne moving target indication
analogue all range decoders, 250–252
angel echoes, 212–213
angular beamwidth, 346
angular position measurement, 99–101
anomalous propagation effects, 481–482
antenna, 66
  as an aperture, 26–28
  and beam alignment, 612
  configuration, MST radars, 353–355
  feeding of, 14–16
  gain, 346, 375
  length greater than half-wave length, 14–16
  mechanical rotation of, 500
  parabolic reflector. *See* parabolic reflector antenna
  parameters
    aperture efficiency, 8
    beamwidth, 6–7
    definition of, 9
    effective aperture, 8
    field pattern, 6
    gain, 7–8
    radiated power, 9
    radiating efficiency, 10–11
    radiation resistance, 9–10
    side lobe radiation, 9
  radiation resistance of, 22
  stabilization of, 42
antenna effect, 479–481
antenna noise temperature, 634–635
  measurement of, 635–637
antenna scanning modulation, 230
anti-aircraft weapons systems, 260–261
anti-jamming techniques, radars, 269
aperture blocking, 31
aperture efficiency, 8
aperture field method, 45, 46
aperture illumination, errors in, 39–41
Argand diagram, 657

array factor, 301
array synthesis, 295
ARSR. *See* air-route surveillance radar
A-scope, 67, 77
ASR. *See* air surveillance radar
atmospheric gas absorption spectrum, 358
atmospheric radars, 294, 337–338
  interpretation of product, 342–343
  terminology, 338–341
  weather surveillance radar. *See* weather surveillance radar
atmospheric windows, for remote sensing, 306–307
ATR tubes, 167, 168
attenuation, 211, 257, 340
auto-correlation, 681
auto-correlation function, 402–404, 671–673
automatic direction finders, 486–490
automatic gain control (AGC), 606, 608
automatic solar tracker, 634
automatic tracking of Sun, 634
automotive radar applications, 642–645
average power return, 345
azimuth angle, 338
azimuth guidance, 508
azimuth resolution, 313

B

back scatter cross section, 110–113
backscatter in imaging radar, 321–322
balanced discriminator, 623
balanced duplexers, 167–168
balanced frequency discriminator, 619
balanced mixers, 163–164
BAM. *See* bird activity modulation
bandwidth, 22
Barker code, 246–247
Battlefield surveillance radars, 604
beam alignment, 612
beam rider, 262
beam shaping, air surveillance radar, 275
beam steering, 301–302
beamwidth, 6–7
bearing
  errors, 500
  measurement of, 498
Bellini-Tosi direction finder, 479, 487
Bell Telephone Laboratory (BTL), 257, 604
binomial expansion, 698
bipolar, 394
bird activity modulation (BAM), 212
birds, echoes, 212–213
blind speeds, 220–221
boxcar generator, 225
branch-type duplexer, 167
brightness temperature, 634–635
broad band, 161
broadband detector, 610
bunching parameter, 151

C

calibration, 630, 632
  and drifts, 629
  emission noise, 639
  gaseous helium, 638
  liquid helium, 638
  of millimeter wave radars, 634
    antenna noise temperature, 634–635
    brightness temperature, 634–635
    liquid and gaseous helium, 638
    liquid nitrogen calibration, 635–637
    room temperature, 637
    solar tracking radiometers, 638–639
  of millimeter wave radiometers, 634–639
  in radiometer, 629
  solar tracking radiometer, 638–639
capillary waves, 199
capture area, 20–22
carrier phase tracking, 653
Cassegrain antenna, 44
Cassegrainian feeds, 53
CAT. *See* clear-air turbulence
cathode ray tube (CRT) display, 165
ceilometer, 239
celestial navigation systems, 595, 597
CFAR. *See* constant false alarm rate
CHAYKA, 600
Chebyshev array, 298
"cheese" antenna, 43
Chilbolton radar, 356
CHILL radar, 356
chi-square distribution, 120

chronometer, 597
circuit impedance, 24
circular aperture, 31–32
circularly polarized radio waves, 23
circular polarization, 23
clear-air turbulence (CAT), 213
cloud droplets, 212
clutter, 195. *See also* land clutter; sea clutter; surface clutter
    internal fluctuation, 229
clutter Doppler shift, 227–228
clutter Doppler spread, 228
coaxial magnetron, 143
Code Phase Tracking techniques, 652
coherent instrumentation radar, characteristics of, 138
coherent integration, 75, 238
coherent millimeter wave receiver for 94 GHz, 605–606
    automatic gain control, 608
    dynamic range, 608
    front end converter, 606–607
    IF amplifiers, 608–609
    IF detector, 608
    main IF amplifier, 607
coherent oscillator (COHO), 136, 234
coherent reference signal, 393
COHO. *See* coherent oscillator
compact range radar cross section (RCS) measurements, 129–130
complex representation, 656–657
complex targets, 115–118
composite surface model, sea clutter, 199
computer controlled phased array radar, 305
conformal arrays, 304–305
conical horn, 26
conical scan, 82–85
constant false alarm rate (CFAR), 230
    receiver, 173
continuous trackers, 81
continuous wave (CW) radar, 131–133, 215, 249–250, 317–320, 389–392
    frequency modulated, 278
    gated, 133–135
controllable phase shifts, 303
conversion efficiency, crossed-field amplifier, 156
conversion loss, 161
convolution theorem, 669–671
Coronal Mass Ejections (CME), 654
correlation detection, 171–172
correlation function, 171
cosecant-squared antenna pattern, 41
cosine aperture distribution, radiation pattern for, 29
cosine wave of finite duration, 663–664
COSPAS/SARSAT system, 569
course finding aids, 2
cross-correlators, 253–254
crossed-field amplifier (CFA), 156–157
cross section fluctuations, 118–121
current distribution, 11–13
cutler dual aperture feed, 55–56
CW Doppler radars, 380–392

D

DDF. *See* Doppler direction finder
dead reckoning, 595, 596
DECCA, 599
DECCA receiver, 548
    fine fixing configuration, 549
    lane identification, 549
    range and accuracy, 549–550
DECCA system, 544
    ambiguity of courses and lanes, 546
    DECCA chain, 544–546
    lane identification, 547–548
Decca tracking and ranging (DECTRA), 552
decoders, 250
    analogue all range, 250–252
    digital, 252–253
decoding, 246–249
delay line canceler, 219
    blind speeds, 220–221
    double cancellation, 222
    response of, 220
    transversal filter, 222–223
demodulator of FM signals, 619
    design, 620–623
    differential peak detection, 624–625
    slope detection, 619–620
Department of Defense in United States, 601
depression angle, 198

DERLAC, 553
detection radar frequency agility, 122
detector, 67, 608
  in CTV tuner, 617–618
  dynamic range of, 627
DGPS. *See* differential global positioning system
Dicke switching, 626
Dicke type radiometers, 634
dielectric lenses, 33–34
  index of refraction of, 33
dielectric permittivity, 700
Differential Global Positioning System (DGPS), 577–578
  applications, 579–580
  configurations, 579
differential peak detection, 624–625
differential peak PM detector, 624–625
digital decoders, 252–253
digital signal processing, MTI radar, 225–226
digital storage oscilloscope (DSO), 606
diode detector, 67
dipole feed, 54–55
  electric, 45
  magnetic, 45
  waveguide feeding method for, 55
Dirac comb, 667
Dirac delta function, 664
Dirac functions, 675
direction finders, 465
  automatic, 486–490
  increasing sensitivity of, 471–472
direction finding errors
  anomalous propagation effects, 481–482
  antenna effect, 479–481
  night effect, 481, 482–483
directivity, of parabolic reflector, 50, 52
discrete array synthesis, 298
discrete frequency waveform coding, 241–242
display, 67
  cathode ray tube, 165
  module, 601
  plan position indicator, 67
distance measuring equipment (DME), 2, 278, 288, 504
  airborne interrogator equipment, 290–291
  noise pulse modulation, 288–289
  principle of operation, 289–290
  search and track modes, 291
  transponder beacon, 292
distributed angels, 212
DME. *See* distance measuring equipment
Doppler direction finder (DDF), 513–515
Doppler filter, 238
Doppler FM-CW radars, 351
Doppler frequency, 394
Doppler navigation, 563–567
  equipment, 285–288
Doppler navigation system, 278
  beam angle on sensitivity, 282–283
  beamwidth on Doppler shift, 283–284
  Doppler radar, 279–280
  equations for Doppler shift, 281–282
  errors, 280–281
  heading and track stabilizations, 283
  multibeam configurations, 280
  spectral width, 284
Doppler radar, 279–280
  application to meteorology, 388–389
    continuous wave radar, 389–392
    detection of sign of Doppler shift, 391–392
    heterodyning process, 389–392
    pulse-Doppler radar, 392–399
  range equation, 286–288
  velocities, 342–343
Doppler radar probing with MST radars, 350
Doppler shift
  detection of sign, 391–392
  effect of beamwidth, 283–284
  equations for, 281–282
Doppler spectra, 683–685
  characteristics of precipitation, 412–423
  examples, 400–404
  median velocity of, 422
  peak of, 421–422
  reflectivity density, 421
  variance of, 408, 418–421
    rate of crossing, 409–411
    zero-crossing rate, 411–412
  velocity parameters of, 423
  Z′(v) spectrum, for exponential size distributions, 414–418
Doppler spectral velocity parameters, 424–429

Doppler velocity
  measurement, 99
  variance of, 385
Doppler VOR system, 502–503
Doppler weather radar, 651
dot angels, 212
double canceler, 222
drifts, 630
  in radiometer, 629
dry snow particles, 210
DSO. *See* digital storage oscilloscope
dual beam, 683–685
  relative velocity spectrum, 683
duality, 667
dual-polarization Doppler radars, 357–358
  characteristics of, 358
  millimeter waves
    cloud and precipitation radar reflectivity, 359–363
    configuration for, 363–368
  updraft and particle size spectra, 368–371
dual polarization radar, 278, 356
dual-polarized radar
  signals, 430–431
    simulation of, 434–437
    simulation of log receiver method, 437–439
    Z_{DR} using *logarithmic receivers*, 431–434
duplexer, 66, 166–168
duty cycle of radar, 76
dynamic range, 608
  antenna temperature, 626–627
  of detector, 627
  IF amplifier, 626–627

E

early-late gate range tracker, 255
earthquakes, 601
ECCM. *See* electronic counter counter measure
echo, 77–78, 342
echoing processes, 353–354
echo sounder, 595
ECM. *See* electronic counter measures
effective aperture, 8
effective radiated power (ERP), 7
electrical pattern rotation, 496–497
electric dipole feed, 45
electric surface current, 46
electro magnetic interference (EMI), 332
electron beam velocity, 150
electronic counter counter measure (ECCM), 271–272
electronic counter measures (ECM), 269–271
electronic receivers, 601
electronic support measures (ESM), 269, 271
electronic warfare (EW), 269
element factor, 301
elevation angle, 338
EMI. *See* electro magnetic interference
emission noise, calibration of, 639
emission noise temperature, 635
energy, 676–680
energy density spectrum, 671–673
energy flow density, 701
envelope processing, in MTI radar, 231
equator, 593
equipment error, 502
equivalence theorem, 27
equivalent Doppler shift, 689
equivalent normalized fluctuation spectrum, 691
error function, 73
ESM. *See* electronic support measures
excess noise temperature ratio (ENR), 638
exciter, 24
explicit expression for error, 659–660

F

false alarm probability, 71–72
false alarm time, 71
feeding of antennas, 14–16
feeds
  cutler dual aperture, 55–56
  defined, 32
  dipole, 54–55
  monopulse, 58–59
  types, 53–54
  waveguide horn, 56–58
field pattern, 6
fixed antenna array design, for search and track, 333–335

fluctuation spectra, examples, 400–404
FM-CW altimeter, 519–525
FM-CW radar, 133, 134
FMCW sensor. *See* frequency modulated continuous-wave sensor
FMCW technique, 350
FM millimeter wave receiver, 614–617
FM signals
  demodulator of, 619–625
  generation of, 613–614
folded dipole, 18–19
  impedance of, 18
forecasting applications, 355
four course radio range, 493–495
Fourier analysis, applications, 673
Fourier integral
  synthesis, 36–39
  transition to, 660–661
Fourier integral transform pair, 661
Fourier series, 655–656
Fourier transform, 98, 664–667, 670–671, 682
  examples of, 677–679
  of gate function, 663, 664
  of isosceles triangular function, 691
  properties of, 667–669
  theorems, 676–677
  theory, 28
frame effect, 473
frequency agility effects, on target detection and tracking, 121–123
frequency coded radars, 239
  applications of, 245
  block diagram of, 240–241
  discrete frequency waveform coding, 241–242
  linear-stepped frequency pulse, 244–245
  matched filter realization, 243
  side lobe reduction, 243
frequency-coded waveforms, 243
frequency dependence, of land clutter, 205
frequency differentiation, 669
frequency modulated continuous-wave (FMCW) sensor, 266
frequency modulated CW radar, 278
frequency-response function, 171
frequency shifting from pulse to pulse, 687–701
frequency-shifting property, 668
frequency, variation with, 202
Fresnel region, 27
front-end converter, 606–607, 611, 630–632
  for 53 GHz radiometer, 628–629
Frounhofer region, 27

G

GaAs FET (field-effect transistor), 66
gain, 7–8
  antenna, 375
  and capture area, 20–22
Galileo positioning system, 599
gaseous helium calibration, 638
gated CW radar, 133–135
gate function, 662–663
  Fourier transform of, 663
Gaussian beam shape equation, 207
Gaussian noise, 71
Gauss theorem, 698
GCA systems. *See* ground control approach systems
generalized dual beam fluctuation spectrum, 379–384
Geomorphologic Instantaneous Unit Hydrograph (GIUH), 649
Glide-slope system, 511–512
Global Navigation Satellite System (GNSS), 601
Global Positioning System (GPS), 567–569, 651–654
  applications, 579–580
  data subframes, 574–575
  navigation message, 573–574
  principles of operation, 569–571
  receivers, 581–582
  segments
    control segment, 572
    space segment, 571–572
    user segment, 572
  sources of errors, 576
    propagation error, 576
    satellite clock errors, 576
Glonass, 569
GMT. *See* Greenwich means time
GNSS. *See* Global Navigation Satellite System
GPS. *See* Global Positioning System
GPS transponder systems, 590–591

gravity waves, 199
grazing angle, 198
  variation with, 202
Greenwich means time (GMT), 597
Greenwich meridian, 593
Gregorian feed, 53
ground-based radars, 273
ground control approach (GCA) systems, 274
ground controlled approach (GCA) systems
  Doppler direction finder, 513–515
  radar control, 513
Gunn-Kinzer velocity data, 417
Gunn source, 610

H

half-wave dipole, 13–14
heading-stabilized antenna, 283
heterodyning process, 389–392
HF/UHF parabolic antennas, 53
HIFIX, 553–555
high resolution radar (HRR), 327
horizontal polarization, 498–500
horn antenna, 24–26
HRR. *See* high resolution radar
Huygens principle, 27
hyperbolic low frequency radio navigation system, 599
hyperbolic mode, SYLEDIS, 560–561
hyperbolic navigation, 537–539
  DECCA receiver, 548
    fine fixing configuration, 549
    lane identification, 549
    range and accuracy, 549–550
  DECCA system, 544
    ambiguity of courses and lanes, 546
    DECCA chain, 544–546
    lane identification, 547–548
  Decca tracking and ranging, 552
  DERLAC, 553
  inland shipping aids. *See* inland shipping aids
  LORAN-A. *See* LORAN-A
  LORAN-C, 542–544
  OMEGA, 550–551
    accuracy and range, 552
    transmission cycle, 551–552
  SYLEDIS. *See* SYLEDIS

I

iceberg detection, navy radar, 293–294
IF amplifiers, 66, 70, 608–609, 626–627
IF demodulator, 611–612
IF detector, 606
IF gain, 627–628
IF main amplifier, 611–612
ILS. *See* instrument landing systems
image-enhanced mixer, 164
imaging radar, 320–321
  inverse synthetic aperture radar, 322–326
  sources of backscatter in, 321–322
impulse sampling, 675
incidence angle, 198
induced current method, 46
inertial navigation, 599
inhomogeneous vector wave equation, 697
inhomogeneous wave equations, 696
inland shipping aids, 553
  accuracy and range, 555
  HIFIX, 553–555
  lane identification, 555–556
  transmitter power, 556
input impedance, 15
insect echoes, 213
Instantaneous Unit Hydrograph (IUH), 648–649
instrumentation radars, for RCS measurements, 131–138
instrument landing aids, 505
instrument landing systems (ILS), 2, 491, 506
  elevation and azimuth guidance, 506–508
  glide-slope system, 511–512
  limitations
    ground plane requirement, 527
    terrain requirements, 526
    traffic handling capability, 527–528
  localizer, 508–511
integration, 669
integration process, 75
intensity integrator, 430
inter clutter visibility, 206
internal fluctuation of clutter, 229
inverse synthetic aperture radar (ISAR), 322–326
ionospheric effect, 565
ISAR. *See* inverse synthetic aperture radar

J

jamming techniques, radars, 269
Janus system, 280
Jet Propulsion Laboratory, 309

K

Klystron amplifiers, 149–151
  reflex, 152–155
  three-cavity, 152
knots, 339
Kriging, 646

L

land clutter, 198, 203
  applications, 204
  examples, 204
  frequency dependence of, 205
  models, 206
  properties of, 206
land surveying, 601
latitude, 593
layer structure measurements, 354
lead line, 595
lens antennas, 33–36
light detection and ranging (LIDAR), 78
lightning, electrical effects of, 642
limiter, MTI radar, 230
linear FM radar, 239
  vs. phase-coded radar, 239
linearity, 668
linearly stepped frequency waveform, 244–245
linear receiver method, 440
line of sight, 601
lines of position (LOP), 594
line type modulator, 158
liquid helium calibration, 638
liquid nitrogen, 637
  calibration, 635–637
lobe switching, 81–82
Local Area Network (LAN), 588
log receiver method, simulation of, 437–439
long-focal-length type parabolic reflector antenna, 44
longitude, 593–594, 596
long-range radars, 294
loop antenna, 465–469
loop voltage, methods of increasing
  increase of *A*, 472–473
  increase of *N*, 473
  large fixed crossed loops, 477–479
  Q of loop, 475
  radiation resistance, 474–475
  tuning of loop antenna, 474
  using ferrites, 475–477
LOP. *See* lines of position
LORAN-A, 539–540
  measurement of delay, 540–542
  transmitter power, 542
LORAN-C, 542–544, 599, 600
low pass filter output, 97
LSA diodes, 159
lunar distance method, 596
Luneburg lens, 35–36

M

magnetic dipole feed, 45
magnetron oscillator, 142–143
  performance characteristics of, 145
  performance chart and Rieke diagram, 144–145
  principle of operation of, 146–148
main IF amplifiers, 607
marine chronometer, 594, 596, 597–598
marine navigation, radar application, 68
marine sextant, 598
matched filter, 101–102, 171
  characteristic, 170–171
  and correlation function, 171–172
  efficiency of, 96
  for pulse compression, 243
  signal-to-mean-noise power ratio of, 94
matched filter receiver, 169–172
maximum transmitted power, 345
Maxwell's equations, 694, 695
mean areal rainfall estimates, comparison of, 647–648
mean Doppler velocity, 418–421
mean square noise voltage, 68
median velocity, of Doppler spectrum, 422
Mesosphere, Stratosphere and Troposphere (MST) radars, 349
  advanced meteor wind observations, 441–442
  antenna configuration

echoing processes, 353–354
forecasting applications, 355
layer structure measurements, 354
turbulence, 354
wind measurement, 355
Doppler radar probing with, 350
evolution of, 350–351
systems, 351–353
metal-plate lens, 34–35
index of refraction, 35
meteorological echoes, radar equation for, 207–208
meteorological radar (MET) radar, 273, 277–278
meteorological radar, rainfall measurement
adjustment, 375–376
antenna gain, 375
determination of radar parameters, 372–374
factors affecting, 376–377
reflectivity factor, 374
signal averaging, 374–375
theory, 371–372
MET radar. *See* meteorological radar; meteorological radar
microwave antenna, 23, 24
microwave landing system (MLS)
advantages, 533
characteristics, 532–533
coverage requirements, 531–532
development, 528
instrument landing systems. *See* instrument landing systems
principle of, 530
system composition, 531
types of antennas in, 528–529
microwave limb sounder (MLS), 310
microwave or millimeter wave signal, 157
Mie scattering, 359–360, 366–367
military radars, 67, 258–259
millimeter wave absorbtion spectrum, 358
millimeter wave beamrider guidance system missile, 604
millimeter wave pressure sounder (MPS), 309
millimeter wave radars, 257, 603–604
AFT discriminator, 617–618
anti-aircraft weapons systems, 260–261
beam rider, 262
calibration of, 634
antenna noise temperature, 634–635
brightness temperature, 634–635
liquid and gaseous helium, 638
liquid nitrogen calibration, 635–637
room temperature, 637
solar tracking radiometers, 638–639
coherent millimeter wave receiver, 605–609
for defense applications, 604
demodulator of FM signals, 619–625
FM millimeter wave receiver, 614–617
FM signals generation, 613–614
military radars, 258–259
power sources for, 267–269
propagation of, 257–258
tunable millimeter wave receiver, 610–612
tunable millimeter wave transmitter, 609–610
video link, 612–613
millimeter wave radiometers, calibration of, 634–639
millimeter wave tracking radars, 90–91
millimeter wave video link at 53 GHZ, 612–613
missile guidance and seeker systems, 261–262
missile seekers, 262–265, 271
missile seeker sensors, configurations of, 265–266
mixers, 162–165
MLS. *See* microwave landing system
mobile tracking radar, 81
modern navigational methods, 594–595
celestial navigation systems, 595, 597
dead reckoning, 595, 596
inertial navigation, 599
marine chronometer, 597–598
marine sextant, 598
navigation application, 601
piloting, 595, 596
radar navigation, 595, 600
radio directions finder, 599–600
satellite navigation, 595, 601
modern radars, 233
frequency coded radars, 239
applications of, 245
block diagram of, 240–241
discrete frequency waveform coding, 241–242
matched filter realization, 243
side lobe reduction, 243
waveform analysis, 244–245
millimeter wave radars, 257
propagation of, 257–258

phase-coded radars, 246
block diagram of, 249–250
cross-correlators and trackers, 253–254
decoders, 250–253
phase coding and decoding, 246–249
range trackers, 254–256
pulse Doppler radar, 233–234
advantages of, 239
applications, 238–239
block diagram of, 234–236
coherent integration, 238
multiple targets moving with different velocities, 236–238
modulators, 65, 158–159
moment of power distribution, 341
monopulse feeds, 58–59
monopulse radar imaging, 326–327
stepped-frequency 3D, 328–331
three-dimensional imaging, 327
monopulse tracking, 85–90
millimeter wave tracking radars, 90–91
Moon, 597
morse code, 599
moving target detector (MTD), 226
moving target indication (MTI) radar, 215, 396
block diagram of, 218
delay line canceler. *See* delay line canceler
digital signal processing, 225–226
example of, 226
limitations to, 228–231
antenna scanning modulation, 230
envelope processing, 231
equipment instabilities, 228–229
internal fluctuation of clutter, 229
limiting in, 230
on moving platform, 227–228
multiple or staggered pulse repetition frequencies, 223–224
operation of, 215–219
range-gated Doppler filters, 224–225
signal processing, 438, 440–441
MTD. *See* moving target detector
MTI radar. *See* moving target indication radar
multifunction array radar, 331–332
fixed antenna array design, 333–335
multifunction capability of radar, 333
multiple time around echoes, 77
MU radar, 442

N

National Oceanic and Atmospheric Administration (NOAA), 351, 357
National Research Laboratory (NRL), 257
National Severe Storms Laboratory (NSSL), 429
National Weather Service, 337
nautical almanac, 597
nautical chart, 595
navigation, 593
navigational aids, 2–3
navigational and remote sensing radars, 273
airborne radars, 278
airport radars. *See* airport radars
continuous wave radar, 317–320
distance measuring equipment, 288–292
Doppler navigation equipment, 285–288
Doppler navigation system. *See* Doppler navigation system
imaging radar, 320–326
meteorological radar, 277–278
monopulse radar imaging, 326–331
multifunction array radar, 331–335
navy radar, 292–294
pattern synthesis, 295–298
phased array. *See* phased array
remote sensing. *See* remote sensing
remote sensing radars, 294
navigational parameters, 1–2
target coordinates for, 2
navigational stars, 597
navy radar, 292–293
detection of icebergs, 293–294
scan-to-scan integration, 293
NEXRAD WSR-88D, 347
night effect, 481
NOAA. *See* National Oceanic and Atmospheric Administration
noding, 633
noise figure
of receiver, 68–70, 160
due to RF losses, 164
measurement, 162
of networks in cascade, 160–161
noise pulse modulation, for radar beacons, 288–289
noise temperature, 161
noise temperature ratio, crystal mixer, 162

noise voltage, 71
noncoherent integration, 75
nonmatched filters, efficiency of, 96
non-periodic functions, 660–664
non-recursive filter, 222
non-uniform arrays, 304–305
normalized power density distribution, 690
normalized radiation pattern, 30
nose-on radar cross section, 113
Nyquist interval, 674

O

off-set feed, 53
oil platforms, 599
OMEGA, 550–551, 599, 600
  accuracy and range, 552
  transmission cycle, 551–552
omnidirectional ranges, 495
  Doppler VOR system, 502–503
  VHF omnirange. *See* VHF omnirange
one-dimensional aperture distribution, 28–30
open-circuited transmission line, 18
operator's workstation, VTMS, 586–588
orthogonal functions, 658–659
over-the-horizon (OTH) radars, 245

P

PAR. *See* precision approach radar
parabolic reflector, 19
parabolic reflector antenna, 32–33
  curvilinear surface current, 45
  design of, 46–48
  design of the feeds of, 52–53
  geometric properties of, 43
  radiation pattern, 31
  structures of, 43
parabolic reflector radar antennas, design of, 42–44
parasitic elements to increase directivity, 16–17
Parseval's theorem, 670, 672
passive electronic counter measures, 269
passive radiometric sensor, 262
pattern synthesis, 36, 295–298
peak detector, 624–625
periodic function, 655–660
phase-coded radars, 246
  block diagram of, 249–250
  cross-correlators and trackers, 253–254
  decoders, 250–253
  phase coding and decoding, 246–249
  range trackers, 254–256
  vs. linear FM pulse compressions, 253–254
phase-comparison monopulse radar, 88
phased array, 298–299
  adaptive arrays, 304–305
  beam steering, 301–302
  computer controlled phased array radar, 305
  conformal and non-uniform arrays, 303–304
  controllable phase shifts, 303
  directional pattern, 299–301
  implementation of, 302
  two-dimensional array, 301
phased array antenna, 294
phased array radar, 294
phase error, 49–52
phase shift, 688
"pillbox" antenna, 43
piloting, 595, 596
PIN diode protector, 174
plan position indicator (PPI), 340
  display, 79
PN code. *See* pseudorandom noise code
polar cap absorption (PCA), 654
polar coordinate, 1–2
polarimetric measurements of precipitation, 642
Polaris, 593
Polaris missle, 599
polarization, 22–23, 501–502
  errors, 501–502
  variation with, 202
polarization scattering matrix, 114–115
pole, 593
post detection integration, 75
power density distribution, 690
power signals, 676–680
power sources, for millimeter wave radars, 267–269
Poynting vector, 699
PPI. *See* plan position indicator
PPI display, 67
precipitation, polarimetric measurements of, 642
precipitation rate, 425
precision approach radar (PAR), 273, 276–277

pressure sounder, 308–309
PRF. *See* pulse repetition frequency
probability density function, 72, 118–120, 206
  of sea clutter, 199–201
probability-density function, 70
probability distributed model, 648
probability distribution function, 680–681
Probable Maximum Flood (PMF), 650
Probable Maximum Precipitation (PMP), 650
programmed solar tracking, 633
propagation effects, 79
pseudorandom noise (PN) code, 246, 247
pulse compression technique, 239
pulsed FM, 239
pulse Doppler radar, 233–234, 392–399
  advantages of, 239
  applications, 238–239
  block diagram of, 234–236
  coherent integration, 238
  multiple targets moving with different velocities, 236–238
pulse duration, 78–79
pulse length, 346–347
pulse radar, 136–137
pulse repetition frequency (PRF), 77–78, 223–224, 337
pulse-to-pulse fluctuation, 119
pulse width, 203

Q

quadrantal error, 486
quantization error, 519

R

radar
  aids, 2
  applications, 67–68
  characteristics of precipitation echo, 440
  defined, 61
  early developments, 349–350
  equation, 62–64, 76
    for meteorological echoes, 207–208
  frequencies, 64–65, 340
  measurements, 92–93
    accuracy of angular position measurement, 99–101
    accuracy of Doppler-velocity measurement, 99
    ambiguity diagram, 101–102
    effect of gating waveform on range accuracy, 96–98
    effect of pulse shape on signal-to-noise ratio, 93–96
    matched filter, 101–102
  model, 650
  navigation, 595, 600
  pulse, scattering of, 340
  resolution, 206
  sensitivity, 162
  set, 65–67
radar altimeter, 307
radar antenna, 43
radar ceilometers, 357
radar cross section (RCS), 107–108, 207, 208
  measurements, 124
    compact range, 129–130
    instrumentation radars for, 131–138
    problems in, 125–126
    sensitivity of, 126–129
  measurement systems, 124
  at microwave frequencies, 118
  Mie or resonance region, 108–110
  optical region, 108–110
  Rayleigh region, 108–110
  reduction in, 116
radar CRT phosphor characteristics, 166
radar parameters, determination of, 372–374
radar pulse modulator, basic elements of, 158
radar rainfall data preprocessing, 646–647
radar reflectivity factor, 208
radar tracking system, for VTMS, 588–590
radial velocity, 342
radiated power, 6, 9, 18
radiating efficiency, 10–11
radiation field pattern, 14
radiation patterns, 29
  characteristics of, 30
  for cosine aperture distribution, 29
  errors in, 39–41
radiation resistance, 9–10, 22, 474–475
radio altimeter, 2
  effect of Doppler shift, 516–518
  FM-CW altimeter with sinusoidal modulation, 519–525

range error, 518–519
sensitivity, 518
triangular modulation, 515–516
radio compass, 487, 491
radio detection and ranging (RADAR), 337
radio directions finder (RDF), 599–600
radio interferometric tracking, 88
radiometers, 262, 308
calibration, 629
of millimeter wave, 634–639
for defense applications, 604
Dicke type, 634
drifts in, 629
millimeter wave radars and, 604
solar tracking millimeter wave. *See* solar tracking millimeter wave radiometer
radiometric sensor, 265, 266
radio monitoring equipment (RME), 3
radio range system, 491
radio transmission, 2
rain attenuation, 211
rainfall-runoff model, 648–649
random errors in reflectors, 39–41
random functions, 680–682
random phase error, 49, 50, 52
range difference system, 539
range equation, 62–64
range-gated Doppler filters, 224–225
range height display, 67
range-range mode, SYLEDIS, 560
range resolution of radar, 312–313
range trackers, 254–256
rate of crossing, 409–411
Rayleigh clutter, 200
Rayleigh distribution, 120, 441
Rayleigh function, 648
Rayleigh scattering, 211, 212
RCS. *See* radar cross section
RDF. *See* radio directions finder
real-time Doppler radar data processing and display, 429–430
receiver, 66
receiver noise, 68–70
receiver protector, 174
receiving antenna, 20–23
reference phase, 498
reflection bunching, 152
reflectivity, 342
reflectivity Doppler spectral density, 416
reflectivity factor, 374, 425
reflectivity measurement, 377
reflectors, tolerance of, 49
reflector type antenna, radiation pattern of, 45–46
reflector wire, 17
reflex Klystron, 152–155
refractive index, 700
remote sensing, 306
atmospheric windows for, 306–307
radars. *See* remote sensing radars
satellite instruments, 307
remote sensing, radar application, 68
remote sensing radars, 294, 310
azimuth resolution, 313
range resolution, 312–313
side looking airborne radar, 311
synthetic aperture radar geometry, 313–317
repeater jammer, 272
re-radiated fields, 485
return signal, 340
RF hybrid junction, 164
Rice distribution, 121
Rice probability distribution function, 73
Rieke diagram, 144–145
room temperature, 637

S

sample mean, variance of, 404–408
sampling function, 662
sampling theorem, 674–675
SAR. *See* synthetic aperture radar
satellite instruments, for microwave remote sensing, 307
microwave limb sounder, 310
pressure sounder, 308–309
radar altimeter, 307
radiometer, 308
scatterometer, 307–308
synthetic aperture radar, 308
satellite navigation, 563, 595, 601
Differential Global Positioning System, 577–578
applications, 579–580
configurations, 579

Doppler navigation, 563–567
Global Positioning System. *See* Global Positioning System
satellites, 601
scaling, 668
scan angle, 341
scanning modulation, 230
scanning radars, 79–80
scan-to-scan integration, navy radar, 293
scattering
from clouds, 212
of radar pulse, 340
from rain, 208–210
from snow, 210
scatterometer, 307–308, 310
Schwartz inequality, 170
"scintillating echo", 115
scintillations, 653
sea clutter, 198
amplitude of, 199–201
composite surface model, 199
probability density function, 199–201
properties, 201–203
theory, 199
search-and-lock system, 83
SEA TRACKS system, 91
secondary maximum, in generalized fluctuation spectrum, 682–683
secondary radar, 290
second-time-around clutter, 224
sector scan, 340
sense finder, 469–471
sensitivity
of direction finder, 471–472
of radar cross section (RCS) measurements, 126–129
sensitivity time control (STC), 174, 275–276
sextant, 593, 597
ships inertial navigation system (SINS), 599
shore effect, 485
"Shortest Path Calculation" module, 601
side lobe radiation, 9
side lobe reduction, frequency coded radars, 243
side looking airborne radar (SLAR), 311, 323
signal averaging, 374–375
signal-to-noise ratios (SNRs), 70–76, 442
effect of pulse shape on, 93–96
single-delay line canceler, 219
singularity functions, 664–667
SINS. *See* ships inertial navigation system
sinusoidal modulation, 519–525
site error, 485–486
SLAR. *See* side looking airborne radar
SNRs. *See* signal-to-noise ratios
solar brightness temperature, 632–633
automatic tracking of Sun, 634
estimation of, 632–634
programmed solar tracking, 633
solar flares, 653–654
solar tracking millimeter wave radiometer, 625
antenna temperature and IF amplifier, 626–627
calibration and drifts, 629
dynamic range of detector, 627
front-end converter, 628–629
IF Gain, 627–628
solar tracking radiometer
calibration of, 638–639
for 53 GHz, 626
for 94 GHZ, 630
for 94 GHz, 630
for 140 GHZ, 630–632
solid state transmitters, 159
South African Council of Scientific Research, 357
space factor, 301
space weather effects, on GPS and ratio propagation, 651–654
spectrum width, 342
spit-gate range tracker, 255
squint angle, 83
SRE. *See* surveillance radar element
stabilization of antennas, 42
stable local oscillator (STALO), 234, 364
standard LORAN. *See* LORAN-A
STARTLE radar, 91
stationary process, 681, 682
stepped-frequency 3D monopulse radar, 328–331
storm model, 650, 651
storms, electrical effects of, 642
storm warning radars, 3

stripmap, 325
Sun, 597
  automatic tracking of, 634
surface clutter, 195, 275
  radar equations, 195–198
surface fitting, 51
Surveillance Radar Element (SRE), 274
surveillance radar element (SRE), 274
surveillance scan, 340
swath width, 312
switched cardioid homing system, 491–493
SYLEDIS, 556
  distance measurement, 557–559
  modes
    hyperbolic mode, 560–561
    range-range mode, 560
  principle of, 556–557
symmetry properties, 657–658, 661–662
synchronization, 601
synthetic aperture radar (SAR), 294, 308
  geometry, 313–317
systematic errors, 39

T

tactical air navigation (TACAN), 504–505
terminal impedance, 14–15
three-cavity klystron, 152
three-dimensional imaging, monopulse radar, 327
time differentiation, 669
time reference, 601
time reference scanning beam (TRSB), 530
time shifting property, 669
time signals, 601
TMS. *See* transit memory system
tolerance of reflectors, 49
Total Electron Content, 653
tracking radar, 80–81
  frequency agility, 122
track-stabilized antenna, 283
track-while-scan (TWS) radar, 81
TRAKX dual frequency range instrumentation radar, 91, 92
transit memory system (TMS), 567
transit system, 568
transmitter, 65
transmitter power, 76–77
transmitting antenna, 20
transponder beacon equipment, 292
transversal filter, 222–223
traveling wave tube (TWT) amplifiers, 155–156
TR tubes, 167
tunable millimeter wave receiver for 53 GHz, 610–611
  antenna and beam alignment, 612
  IF demodulator, 611–612
  IF main amplifier, 611–612
tunable MM-wave transmitter for 53 GHZ, 609–610
turbulence, 354
  measurement, 384–388
turbulent velocity, 385
two-dimensional array, 301
two scatterer complex target, 115

U

unequal element-to-element spacing, 298
United Nations International Decade for Natural Disaster Reduction (IDNDR), 650

V

VCO. *See* voltage controlled oscillator
vector processing MTI, 219
velocity processor, 430
Vertical Profile of Reflectivity (VPR), 646–647, 649
Vessel Traffic Management System (VTMS), 583–584
  configuration of, 584–585
  GPS transponder systems for, 590–591
  operators workstation, 586–588
  radar tracking system, 588–590
VHF omnirange (VOR), 495–496
  bearing errors, 500
  electrical pattern rotation, 496–497
  equipment error, 502
  horizontal polarization, 498–500
  measurement of bearing, 498
  mechanical rotation of antenna, 500

performance, 500
polarization errors, 501–502
reference phase, 498
site and terrain errors, 501
video amplifier, 67
voltage controlled oscillator (VCO), 266
voltage-standing wave ratio (VSWR), 145
volume clutter, 195
VOR. *See* VHF omnirange
VPR. *See* Vertical Profile of Reflectivity
VTMS. *See* Vessel Traffic Management System

W

wave equation, 694
waveguide feeding method, for dipole feed, 55
waveguide horn feeds, 56–58
weather, effects of, 207
attenuation by precipitation, 211
radar equation for meteorological echoes, 207–208
scattering
from clouds, 212
from rain, 208–210
from snow, 210
weather radars, 3
correction procedures, 645–647
weather surveillance radar
angular beamwidth, 346
antenna gain, 346
average power return, 345
maximum range, 344–345
maximum transmitted power, 345
NEXRAD WSR-88D, 347
pulse length, 346–347
WSR-88D radar, 343–344
Weibull clutter model, 201
Weibull clutter parameters, 206
Wiener-Khintchine theorem, 672
wind barbs, 338–339
wind measurement, 355
by dual beam incoherent radar, 379–384
wind, variation with, 202
Woodward-Levinson method, 37–39
World War II, 599, 603
WSR-88D radar, 343–344

X

X-band GaAs FET amplifier, 159

Y

Yagi antennas, 17, 18

Z

zero-crossing rate, 411–412
zero-velocity clutter, 226
Z′(v) spectrum, for exponential size distributions, 414–418

www.ingramcontent.com/pod-product-compliance
Lightning Source LLC
LaVergne TN
LVHW082000200926
848481LV00010B/2776
9781683921189